Innovations in Landscape Research

Series Editor

Lothar Mueller, Leibniz Centre for Agricultural Landscape Research (ZALF),
Müncheberg, Brandenburg, Germany

Aims & Scope

The Springer series "Innovations in Landscape Research" presents novel methodologies and technologies to understand, monitor and manage landscapes of the Anthropocene. The aim is to achieve landscape sustainability at high productivity. This includes halting degradation of landscapes and their compartments, developing cultural landscapes, and preserving semi-natural landscapes. Clean water and air, fertile and healthy soils for food and other ecosystem services, and a green and bio-diverse environment are attributes of landscapes for the survival and well-being of humans who inhabit them.

How do landscapes function? How do future landscapes look like? How can we sustainably develop intensively used and stressed kinds of landscapes? Scientific innovations and decision tools are key to answer and solve those challenging questions. The series will inform about advanced methods and results of disciplinary, interdisciplinary and transdisciplinary work in landscape research. It presents a broad array of methods to measure, assess, forecast, utilize and control landscapes and their compartments. These include field and laboratory measurement methods, methods of resource evaluation, functional mapping and risk assessment, and sensing methods for landscape monitoring, advanced methods for data analysis and ecosystem modeling, methods and technologies for optimizing the use of multi-functional landscapes, for the bioremediation of soil and water, and basics and procedures of landscape planning. The series provides a new view on landscapes with some focus on scientific and technological innovations, on soils and problems of optimizing agricultural landscapes under conditions of progressive urbanization. Landscape research in a globalized world of the Anthropocene is based on gathering big data and scenario modeling. International long-term experiments and agri-environmental monitoring systems will deliver data for ecosystem models and decision support systems.

Edited volumes of this series will address the following topics at high priority: Status and Trends of Landscape Research; Understanding Key Landscape Processes; Landscape Services, Functions and Biodiversity; Assessing Soil Resources and Quality; Water Resource and Quality Monitoring; Landscape Monitoring Concepts and Studies; Landscape Sensor and Monitoring Technologies; Landscape Modeling and Decision Support; Agricultural Soil and Plant Management; Basics and Tools for Landscape Planning; Tools for Water and Wetland Management; Forest Management and Agroforestry; Rehabilitation of Degraded Landscapes.

The books of this series are a source of information for researchers, teachers, students, and stakeholders interested in the topics of landscape science and related disciplines. They present status analyses, methodical chapters and case studies showing the practical relevance and feasibility of novel decision tools and technologies. Thus, the books of this series will be a particular valuable information basis for managers and decision makers at various levels, from local up to international decision bodies.

An author/editor questionnaire, instructions for authors and a book proposal form can be obtained by contacting the Publisher.

More information about this series at http://www.springer.com/series/16118

Wilfried Mirschel · Vitaly V. Terleev ·
Karl-Otto Wenkel
Editors

Landscape Modelling and Decision Support

Springer

Editors
Wilfried Mirschel
Research Platform Models & Simulation
Leibniz Centre for Agricultural
Landscape Research (ZALF)
Müncheberg, Brandenburg, Germany

Vitaly V. Terleev
Faculty of Land and Water
Resources Management
Peter the Great St. Petersburg
Polytechnic University
Saint Petersburg, Russia

Karl-Otto Wenkel
Leibniz Centre for Agricultural
Landscape Research (ZALF)
Müncheberg, Brandenburg, Germany

ISSN 2524-5155 ISSN 2524-5163 (electronic)
Innovations in Landscape Research
ISBN 978-3-030-37420-4 ISBN 978-3-030-37421-1 (eBook)
https://doi.org/10.1007/978-3-030-37421-1

This Springer imprint is published by the registered company Springer Nature Switzerland AG
The registered company address is: Gewerbestrasse 11, 6330 Cham, Switzerland

Preface

We are living in a rapidly changing world. Land use changes caused by specific human activities such as new forms and intensities of land use, especially in agricultural landscapes, changes in infrastructure and an increased touristic use of attractive landscape regions as well as the current climate change influence the entire landscape with all its processes. To achieve sustainable solutions in landscape development and land use management we therefore need better methods and tools to assess the long-term consequences of land use and climate changes. These assessments should be consistent and linked to a probability statement in order to prevent negative consequences for the abiotic and biotic environment as well as for the landscape structure.

Regional models and algorithms for landscape indicators, including decision support for landscape-related measures, must be developed on a fundamental knowledge basis. This basis comprises namely the natural conditions and functional relationships as they exist in and between ecosystems, and therefore landscapes. These form the fundaments for developing, scientifically accompanying and furthering concepts for landscape use and designs, well as their sustainable rehabilitation. Such an approach requires the consideration not only of nature and landscape values such as various balances and biodiversity, but also of socio-economic and political factors. This shows that landscape research oriented towards sustainability is only possible across disciplines.

In the international research landscape, the accepted view of the necessity of a holistic approach to the landscape is still contrasted by many purely disciplinary and sectoral approaches and a general methodology that corresponds to them. The answer to the question "Which variables or indicators sufficiently characterise a landscape, and how can conditions and processes in the landscape as well as landscapes in their entirety be described, represented, interpreted and predicted?" is much more complicated than the answer to isolated sectoral questions.

A major deficit in the estimation and evaluation of complex processes on large spatial and temporal scales is that experiments on a landscape scale are in practice excluded for spatial, temporal and financial reasons. In order to see landscapes holistically, to combine sectoral and reductionist approaches and to draw conclusions from the many disciplinary information regarding the behaviour of a whole landscape, one has to take a new course. Based on the current technological state and practice standards seen in landscape modelling, the landscape can be represented virtually at the computer level. Advanced information acquisition and processing, a variety of complex model approaches, including algorithms for individual landscape indicators, as well as highly effective simulation frameworks all contribute notably to the field potential. Thus the prerequisites for the realisation of virtual landscape experiments within very small timeframes are satisfied. On this basis, strategies can be derived for the use of landscapes and for the provision of anthropogenic intervention contributing to future landscapes which are sustainable with regard to both the landscape as a whole and their ecosystem interactions.

This book is intended to contribute to a deeper understanding of landscape and regional modelling in the broad spectrum of its facets with regard to various landscape indicators. Additionally, it seeks to underline the necessity of using these models within the framework of decision support systems in finding sustainable decision and solution strategies for the protection of a comprehensively functioning landscape. The focus of this book is mainly on research results concerning landscape and regional modelling as well as landscape-related decision support systems and simulation tools from Europe and Russia.

In accordance with this requirement, the 29 chapters of this volume are subdivided into three parts.

Part I "Landscape Modelling—Requirements, Understanding and General Methods" addresses the necessity of landscape modelling, the model-based approaches in landscape research, possible solutions, methodological aspects and problems in the model choice for the landscape scale.

Part II "Landscape Indicators—Model Approaches for Specific Ecological and Socio-economic Indicators" deals in separate chapters with regional modelling approaches for various ecological and socio-economic landscape indicators. These modelling approaches are applied to landscapes of different scales and regions in Europe and Russia.

In Part III "Landscape Decision Support—Systems, Tools and Frameworks", some decision support systems and simulation tools are presented, which on the one hand, consider the landscape as a complex system and on the other hand, address specific aspects of the landscape or land use and offer decision support.

Many scientists from different countries and institutions have actively contributed to this publication. We have also received support from various ministries. We would like to thank the German Federal Office for Agriculture and Food (BLE) for funding the projects 45, 128 and 3/09-11 of the German–Russian programme for cooperation in the field of agricultural research, the Ministry of Science, Research and Culture of the Federal State of Brandenburg and the Federal Ministry of Food, Agriculture and Consumer Protection.

The Springer publishing house ensured that the editing and printing process was handled and completed smoothly. The editorial board thanks all sponsors and other supporters for their help and their commitment. Special thanks goes to Mrs. K. Luzi from the Leibniz Centre for Agricultural Landscape Research Müncheberg for the intensive processing of the individual chapters of this volume.

It was our pleasure to serve as editors of this book by coordinating and reviewing the contributions of very motivated scientists. We hope that this book can contribute to a better understanding in the field of landscape modelling and decision support with the aim to find sustainable solutions in the conflictual field between mankind, environment and the use of landscapes.

Müncheberg, Germany — Wilfried Mirschel
Saint Petersburg, Russia — Vitaly V. Terleev
Neuenhagen, Germany — Karl-Otto Wenkel
July 2019

Contents

Editors and Contributors

About the Editors

Wilfried Mirschel is a Senior Scientist at the Leibniz Centre for Agricultural Landscape Research (ZALF) in Müncheberg, Germany. He studied Mathematics at the State University Donezk, Ukraine (graduate mathematician, 1977) and graduated in 1982 with a doctorate in agriculture (Dr. agr.) at the Academy of Agricultural Sciences of the GDR with a dissertation on problems of model-based irrigation water distribution. From 1977 to 1991, he worked at the Research Centre for Soil Fertility Müncheberg and from 1992 to 2019 at ZALF (Institute of Ecosystem and Process Modelling, Institute of Landscape Systems Analysis, Research Platform "Models and Simulation"). His main fields of work and experience are applied landscape systems analysis, regional modelling, regional biomass and yield estimation, model-based climate impact research, agroecosystem modelling, complex decision support systems for sustainable agricultural landscape development, methods and models for operational irrigation management.

During more than 40 years of research activity, Wilfried Mirschel was involved in many national and international research projects and was leading some of them. He has published more than 300 scientific papers, including more than 100 peer-reviewed publications and book chapters as well as four books/monographs. He is actively involved in the scientific community and with his knowledge and experience enriches the work of various organisations and working groups.

In the last three decades, he has established a fruitful international cooperation in the fields of process and regional modelling and has been integrated into education with lectures and seminars on the topics "Modelling of biological systems" and "Ecological modelling" at the University of Potsdam, the University for Sustainable Development Eberswalde, both Germany and the Peter the Great St. Petersburg Polytechnic University, Russia.

Vitaly V. Terleev is a Professor at Peter the Great St. Petersburg Polytechnic University in St. Petersburg, Russia. He studied physics at Omsk State University, USSR (graduate of physics, 1981), in 1989 received a candidate (technical) degree in Agricultural Meteorology at Agrophysical Research Institute (Laboratory of Mathematical Simulation of Agroecosystem), Leningrad, USSR. He has a degree of Doctor in Agricultural Science (Agrophysics) since 2002 (Agrophysical Research Institute, St. Petersburg, Russia).

From 1991 to 2001, he worked at St. Petersburg State University (Faculty of Agricultural Chemistry). From 2004 to the present, he has been working at Peter the Great St. Petersburg Polytechnic University (Department of Water- and Hydroengineering Construction).

Vitaly V. Terleev is responsible for the mathematical modelling of the dynamics of water and nutrients in the soil. His research aimed at quantitative description of the hydrophysical properties of the soil, movement and absorption of the soil moisture and agricultural chemicals in the root zone to control the use of water and fertilisers in irrigation farming.

For more than 30 years of scientific activity, Vitaly V. Terleev has participated in many national and international research projects and led some of them. He has published over 230 scientific papers, including more than 30 peer-reviewed publications and book chapters, as well as five books and monographs.

He takes an active part in international scientific cooperation and enriches the work of various organisations and working groups with his knowledge and experience. From 2006 to 2017, he completed four scientific traineeships at the Leibniz Centre for Agricultural Landscape Research in Müncheberg (Germany) and established fruitful international cooperation in the field of modelling processes in agroecosystems.

Karl-Otto Wenkel is a Guest Scientist at the Leibniz Centre for Agricultural Landscape Research (ZALF) in Müncheberg, Germany. He studied Melioration Science and Crop Science at the University of Rostock and completed a postgraduate study in Agricultural Informatics at the Martin-Luther-University Halle/Saale, Germany. In 1975, he graduated with a doctorate in agriculture (Dr. agr.) at the Academy of Agricultural Sciences of the GDR with a dissertation on problems of computer-aided irrigation planning and management. In 2004, he was appointed as an Honorary Professor of Integrated Land Use and Water Management at Peter the Great St. Petersburg Polytechnic University in St. Petersburg, Russia.

From 1971 to 1991, he was the head of several research departments at the Research Centre for Soil Fertility Münchenberg and led several research projects. After the reunification of Germany and the reorganisation of the research landscape in East Germany, he was appointed as the Head of the new founded Institute of Landscape Systems Analysis of the Leibniz Centre for Agricultural Landscape Research Müncheberg in 1992. He was in this position until his retirement from active employment in 2013.

Karl-Otto Wenkel has a great experience in the development and use of both dynamic agroecosystem models and integrated landscape models for impact assessment of land use and climate change on landscape functions and services. He also has experience in the fields of irrigation, soil–plant interactions and agro-management as well as in the development and use of model-based decision support systems to adapt agriculture to climate change. He has been involved in several national and international collaborative research projects, including irrigation scheduling system, precision agriculture, river basin management, computer-aided soil and crop management, regional guidelines for sustainable land use and impact assessment of climate change. He has published more than 200 scientific papers, including more than 80 peer-reviewed publications and book chapters.

Contributors

Blal Adem Esmail Department of Civil, Environmental and Mechanical Engineering, University of Trento, Trento, Italy

José Alves Júnior Research Group on Climate and Water Resources of the Cerrado Biome, College of Agronomy, Federal University of Goiás, Goiânia, GO, Brazil

Maria Andreoli Department of Agricultural, Food and Agro-Environmental Sciences (DAFE), University of Pisa, Pisa, Italy

Giacomo Assandri Section of Vertebrate Zoology, Museo delle Scienze, Trento, Italy

Anne-Gaelle Ausseil Manaaki Whenua - Landcare Research, Wellington, New Zealand

Vladimir Badenko Peter the Great St.Petersburg Polytechnic University, Saint Petersburg, Russia

Rafael Battisti Research Group on Climate and Water Resources of the Cerrado Biome, College of Agronomy, Federal University of Goiás, Goiânia, GO, Brazil

Michael Berg-Mohnicke Leibniz-Centre for Agricultural Landscape Research (ZALF) e.V., Müncheberg, Germany

Vladimir L. Bogdanov St. Petersburg State University, St. Petersburg, Russia

Mattia Brambilla Section of Vertebrate Zoology, Museo delle Scienze, Trento, Italy

Hamish Brown New Zealand Institute for Plant & Food Research Limited, Lincoln, New Zealand

Eric Burgueño New Zealand Institute for Plant & Food Research Limited, Lincoln, New Zealand

Derblai Casaroli Research Group on Climate and Water Resources of the Cerrado Biome, College of Agronomy, Federal University of Goiás, Goiânia, GO, Brazil

Rogerio Cichota New Zealand Institute for Plant & Food Research Limited, Lincoln, New Zealand

Marcus Davy New Zealand Institute for Plant & Food Research Limited, Lincoln, New Zealand

Natalia N. Dergunova A.N. Severtsov Institute of Ecology and Evolution, Russian Academy of Sciences, Moscow, Russia

Mark Deubert Institute for AgroEcology (IfA), RLP AgroScience, Neustadt, Germany

Vasiliy V. Dmitriev St. Petersburg State University, St. Petersburg, Russia

Ielizaveta Dunaeva Federal State Budget Scientific Institution "Research Institute of Agriculture of Crimea", Simferopol, Russia

Josef Eitzinger Institute of Meteorology, University of Natural Resources and Life Sciences Vienna, Vienna, Austria

Alexandra A. Ershova Russian State Hydrometeorological University (RSHU) Russian Federation, St. Petersbourg, Russia

Adão Wagner Pêgo Evangelista Research Group on Climate and Water Resources of the Cerrado Biome, College of Agronomy, Federal University of Goiás, Goiânia, GO, Brazil

Frank Ewert Leibniz Centre for Agricultural Landscape Research (ZALF), Müncheberg, Germany

Irina V. Fedorova St. Petersburg State University, St. Petersburg, Russia

Uwe Franko Department of Soil System Science, UFZ—Helmholtz Centre for Environmental Research, Halle, Germany

Vitaly V. Garmanov St. Petersburg State University, St. Petersburg, Russia

Davide Geneletti Department of Civil, Environmental and Mechanical Engineering, University of Trento, Trento, Italy

Roman Ginevsky Peter the Great St. Petersburg Polytechnic University, St. Petersburg, Russia

Jan Göpel University Koblenz-Landau, Landau (Pfalz), Germany

Vasily Grabovsky Center for Forest Ecology and Productivity, Russian Academy of Sciences, Moscow, Russia

Nadezhda V. Grits Federal State Budgetary Educational Institution of Higher Education "Tver State Agricultural Academy" (FSBEI HE Tver State Agricultural Academy), Sakharovo, Russia

Karin Groth Leibniz Centre for Agricultural Landscape Research (ZALF), Research Platform "Models & Simulation", Müncheberg, Germany

François Guerrin SELMET Cirad, Inra, Montpellier SupAgro, University of Montpellier, Montpellier, France

Jing Guo Manaaki Whenua - Landcare Research, Wellington, New Zealand

Allister Holmes Foundation for Arable Research (FAR), Christchurch, New Zealand

Dean Holzworth Commonwealth Scientific and Industrial Research Organisation (CSIRO), Toowoomba, Australia

Nikolai V. Hovanov St. Petersburg State University, St. Petersburg, Russia

Wei Hu New Zealand Institute for Plant & Food Research Limited, Lincoln, New Zealand

Ellen Hume New Zealand Institute for Plant & Food Research Limited, Lincoln, New Zealand

Natalia V. Ignatyeva Institute of Limnology, Russian Academy of Sciences (IL RAS), Russian Federation, St. Petersburg, Russia

Dmitry A. Ivanov Federal State Budgetary Scientific Institution "All-Russian Research Institute of Reclaimed Lands" (FNBI VNIIMZ), Emmaus, Russia

Linley Jesson New Zealand Institute for Plant & Food Research Limited, Lincoln, New Zealand

Bosko Jezerkic Forecasting and Reporting Service for Plant Protection of AP Vojvodina (PIS), Novi Sad, Serbia

Paul Johnstone New Zealand Institute for Plant & Food Research Limited, Lincoln, New Zealand

Anna A. Karachenkova National Research Institute on Agricultural Meteorology (NRIAM), Obninsk, Russia

Karin Kauer Estonian University of Life Sciences, Institute of Agricultural and Environmental Sciences, Tartu, Estonia

Kurt Christian Kersebaum Leibniz Centre for Agricultural Landscape Research (ZALF), Müncheberg, Germany

Lyudmila A. Khlyap A.N. Severtsov Institute of Ecology and Evolution, Russian Academy of Sciences, Moscow, Russia

Yuri Kirsta Institute for Water and Environmental Problems of Siberian Branch, Russian Academy of Sciences, Barnaul, Altai Krai, Russia;
Altai State Technical University, Barnaul, Altai Krai, Russia

Ivan Koci Forecasting and Reporting Service for Plant Protection of AP Vojvodina (PIS), Novi Sad, Serbia

Raimo Kõlli Estonian University of Life Sciences, Institute of Agricultural and Environmental Sciences, Tartu, Estonia

Sergey A. Kondratyev Institute of Limnology, Russian Academy of Sciences (IL RAS), Russian Federation, St. Petersburg, Russia

Hymmi Kong New Zealand Institute for Plant & Food Research Limited, Lincoln, New Zealand

Michael Korzukhin Institute of Global Climate and Ecology, Moscow, Russia

Barbara Köstner Department of Meteorology, Technical University Dresden, Tharandt, Germany

Valeriy P. Kulesh St. Petersburg State University, St. Petersburg, Russia

Branislava Lalic Faculty of Agriculture, University of Novi Sad, Novi Sad, Serbia

Mait Lang University of Tartu, Tartu Observatory, Tartumaa, Estonia;
Estonian University of Life Sciences, Institute of Forestry and Rural Engineering, Tartu, Estonia

Viktor Lazarev Peter the Great St. Petersburg Polytechnic University, St. Petersburg, Russia

Linda Lilburne Manaaki Whenua - Landcare Research, Wellington, New Zealand

Jian Liu New Zealand Institute for Plant & Food Research Limited, Lincoln, New Zealand

Mirjana Ljubojevic Faculty of Agriculture, University of Novi Sad, Novi Sad, Serbia

Athanasios Loukas Department of Rural and Surveying Engineering, Aristotle University of Thessaloniki, Thessaloniki, Greece

Reimo Lutter Estonian University of Life Sciences, Institute of Forestry and Rural Engineering, Tartu, Estonia

Karin Luzi Leibniz Centre for Agricultural Landscape Research (ZALF), Research Platform "Models & Simulation", Müncheberg, Germany

Milena Marcic Forecasting and Reporting Service for Plant Protection of AP Vojvodina (PIS), Novi Sad, Serbia

Sergey Medvedev Agrophysical Research Institute, Grazhdansky, Saint Petersburg, Russia

Sathiyamoorthy Meiyalaghan New Zealand Institute for Plant & Food Research Limited, Lincoln, New Zealand

Marcio Mesquita Research Group on Climate and Water Resources of the Cerrado Biome, College of Agronomy, Federal University of Goiás, Goiânia, GO, Brazil

Elena A. Minakova Kazan Federal University, (KFU), Russian Federation, Kazan, Russia

Wilfried Mirschel Leibniz Centre for Agricultural Landscape Research (ZALF), Research Platform "Models & Simulation", Müncheberg, Germany

Kirill Moiseev Agrophysical Research Institute, St. Petersburg, Russia

Francesco Monacci University of Pisa, Pisa, Italy

Lothar Mueller Leibniz-Centre for Agricultural Landscape Research (ZALF) e.V., Müncheberg, Germany

Aleksandr O. Nikonorov Peter the Great St. Petersburg Polytechnic University, St. Petersburg, Russia

Aleksandr N. Ogurtsov St. Petersburg State University, St. Petersburg, Russia

Tatiana I. Oliger Natural State Reserve "Nizhnesvirsky", Kovkenitsy, Russia

Aleksey G. Osipov A.F. Mozhaysky's Military-Space Academy, St. Petersburg, Russia

Fedor A. Osipov A.N. Severtsov Institute of Ecology and Evolution, Russian Academy of Sciences, Moscow, Russia

George K. Osipov Military Spaces Academy Named After A. F. Mozhaisky, St. Petersburg, Russia

Vitaly V. Osipov State Nature Reserve "Privolshky Forest-Steppe", Penza, Russia

Jéssica Sousa Paixão Research Group on Climate and Water Resources of the Cerrado Biome, College of Agronomy, Federal University of Goiás, Goiânia, GO, Brazil

Giorgos Papadavid Agricultural Research Institute of Cyprus, Nicosia, Cyprus

Vladimir S. Pashtetsky Federal State Budget Scientific Institution "Research Institute of Agriculture of Crimea", Simferopol, Crimea

Vera N. Pavlova National Research Institute on Agricultural Meteorology (NRIAM), Obninsk, Russia

Viktoria A. Pavlova St. Petersburg State Agrarian University, St. Petersburg-Pushkin, Russia

Paolo Pedrini Section of Vertebrate Zoology, Museo delle Scienze, Trento, Italy

Tara Petric Faculty of Agriculture, University of Novi Sad, Novi Sad, Serbia

Varos G. Petrosyan A.N. Severtsov Institute of Ecology and Evolution, Russian Academy of Sciences, Moscow, Russia

Valentina V. Popovich Federal State Budget Scientific Institution "Research Institute of Agriculture of Crimea", Simferopol, Crimea

John Powell National Institute of Water and Atmospheric Research (NIWA), Wellington, New Zealand

Alexander Puzanov Institute for Water and Environmental Problems of Siberian Branch, Russian Academy of Sciences, Barnaul, Altai Krai, Russia

Kaire Rannik Estonian University of Life Sciences, Institute of Agricultural and Environmental Sciences, Tartu, Estonia

Kate Richards New Zealand Institute for Plant & Food Research Limited, Lincoln, New Zealand

Martina Roß-Nickoll Institute for Environmental Research (Biology V), RWTH Aachen University, Aachen, Germany

Pablo Rosso Leibniz Centre for Agricultural Landscape Research (ZALF), Müncheberg, Germany

Massimo Rovai Department of Civil and Industrial Engineering (DICI), University of Pisa, Pisa, Italy

Tamara Rozhdestvenskaya Institute for Water and Environmental Problems of Siberian Branch, Russian Academy of Sciences, Barnaul, Altai Krai, Russia

John de Ruiter New Zealand Institute for Plant & Food Research Limited, Lincoln, New Zealand

Olga V. Rukhovich All-Russian Scientific Research Institute of Agrochemistry, Russian Academy of Sciences, Moscow, Russia

Rüdiger Schaldach Center for Environmental Systems Research (CESR), University of Kassel, Kassel, Germany

Björn Scholz-Starke Darwin statistics, Aachen, Germany

Jan Schüngel Center for Environmental Systems Research (CESR), University of Kassel, Kassel, Germany

Rocco Scolozzi Department of Sociology and Social Research, University of Trento, Trento, Italy

Yuriy N. Sergeyev St. Petersburg State University, St. Petersburg, Russia

Larisa S. Sharaya Institute of Volga Basin Ecology, Russian Academy of Sciences, Toglyatti, Samarskaya Oblast, Russia

Peter A. Shary Institute of Physicochemical and Biological Problems in Soil Science, Russian Academy of Sciences, Pushchino, Moscow, Russia

Marina V. Shmakova Institute of Limnology, Russian Academy of Sciences (IL RAS), Russian Federation, St. Petersburg, Russia

Nikita M. Sinitsyn Russian Ministry of Foreign Affairs, Moscow, Russia

Marios Spiliotopoulos Department of Civil Engineering, University of Thessaly, Volos, Greece

Ana Firanj Sremac Faculty of Agriculture, University of Novi Sad, Novi Sad, Serbia

Roy Storey New Zealand Institute for Plant & Food Research Limited, Lincoln, New Zealand

Lucas Streib Institute for AgroEcology (IfA), RLP AgroScience, Neustadt, Germany

Andrew Tait National Institute of Water and Atmospheric Research (NIWA), Wellington, New Zealand

Edmar Teixeira New Zealand Institute for Plant & Food Research Limited, Lincoln, New Zealand

Anton V. Terekhov Institute of Limnology, Russian Academy of Sciences (IL RAS), Russian Federation, St. Petersburg, Russia

Vitaly V. Terleev Peter the Great St. Petersburg Polytechnic University, St. Petersburg, Russia

Issa Togo Peter the Great St. Petersburg Polytechnic University, St. Petersburg, Russia

Tõnu Tõnutare Estonian University of Life Sciences, Institute of Agricultural and Environmental Sciences, Tartu, Estonia

Alex Topaj Agrophysical Research Institute, St. Petersburg, Russia;
Bureau Hyperborea Ltd, St. Petersburg, Russia

Andreas Toschki Gaiac - Research Institute for Ecosystem Analysis and Assessment, Aachen, Germany

Leonidas Toulios Institute of Industrial and Forage Crops, Greek National Agricultural Organization-Demeter, Larisa, Greece

Matthias Trapp Institute for AgroEcology (IfA), RLP AgroScience, Neustadt, Germany

Svetlana E. Varcheva National Research Institute on Agricultural Meteorology (NRIAM), Obninsk, Russia

Yulia Volkova Peter the Great St. Petersburg Polytechnic University, St. Petersburg, Russia

Tony van der Weerden AgResearch, Hamilton, New Zealand

Karl-Otto Wenkel Leibniz-Centre for Agricultural Landscape Research (ZALF) e.V., Müncheberg, Germany

Ralf Wieland Leibniz Centre for Agricultural Landscape Research (ZALF), Research Platform “Models & Simulation”, Müncheberg, Germany

Felix Witing Department of Computational Landscape Ecology, UFZ—Helmholtz Centre for Environmental Research, Leipzig, Germany

Elena Zakharova Agrophysical Research Institute, Grazhdansky, Saint Petersburg, Russia

Nikolay A. Zavyalov Natural State Reserve “Rdeysky”, Kholm, Russia

Part I
Landscape Modelling—Requirement, Understanding and General Methods

Chapter 1
Modelling and Simulation of Agricultural Landscapes

Wilfried Mirschel, Michael Berg-Mohnicke, Ralf Wieland, Karl-Otto Wenkel, Vitaly V. Terleev, Alex Topaj and Lothar Mueller

Abstract An agricultural landscape is a section of a region shaped by its natural landscape features primarily involving agricultural land use and land management. Intensive anthropogenic activities have left a permanent mark on agricultural landscapes, which have been developed over hundreds of years. Agricultural landscapes constitute a spatiotemporal structure. Hence they represent a complex system in which a large number of processes occur continuously that, due to their temporal dynamics, lead to constant changes in the state of the system. Traditional experiments are inappropriate for the impact assessment of anthropogenic and naturally occurring changes in agricultural landscapes. The only option here is to conduct virtual landscape experiments at the computer level. To this end, a relevant set of spatial, quantifiable landscape indicators is defined that can be used to map the landscape on the computer. Building on the extensive expertise in agricultural landscape research, indicators can be mapped using validated and robust models and their

---

W. Mirschel (✉) · M. Berg-Mohnicke · R. Wieland · K.-O. Wenkel · L. Mueller
Leibniz-Centre for Agricultural Landscape Research (ZALF) e.V., Eberswalder Str. 84, 15374 Müncheberg, Germany
e-mail: wmirschel@zalf.de

M. Berg-Mohnicke
e-mail: Michael.Berg@zalf.de

R. Wieland
e-mail: rwieland@zalf.de

K.-O. Wenkel
e-mail: kowenkel@web.de

L. Mueller
e-mail: agrarlandschaft@gmail.com

V. V. Terleev
Peter the Great St. Petersburg Polytechnik University, Politekhnicheskaja ul. 29, 195251 St. Petersburg, Russia
e-mail: vitaly_terleev@mail.ru

A. Topaj
Agrophysical Research Institute, Grazhdansky Prospekt 14, 195220 St. Petersburg, Russia
e-mail: topaj@hotmail.ru

W. Mirschel et al. (eds.), *Landscape Modelling and Decision Support*, Innovations in Landscape Research,
https://doi.org/10.1007/978-3-030-37421-1_1

dynamics described involving temporal aspects. Various model types can be used in this process. Special simulation environments involving the use of spatial data and accounting for possible land use and global changes enable forward-looking scenario simulations to provide answers to the question of the sustainability of agricultural land use systems. Decision support systems (DSS) that exploit the latest possibilities offered by information technology, statistics and artificial intelligence provide the framework for integrating models, spatial data concerning the state of the landscape and scenario data, simulation techniques as well as tools for interpreting and visualising results. Such DSS are also the basis for quantifying the complex impact of site conditions, changes in land use or management, and of potential climate change on individual landscape parameters or landscape indicators. A number of examples show how indicator-based models of different types can be used to assess the impact and sustainability of land use systems on a landscape scale. As a prerequisite for the development and validation of integrated dynamic landscape models, more long-term ecological studies and monitoring systems are required. This also means that more resources are necessary to support these activities. The use of models and virtual simulation experiments within a DSS framework at the computer is a very promising way of finding suitable site-specific complex measures for the adaptation of agriculture to climate change.

Keywords Agricultural landscape · Modelling · Scenario simulation · Landscape indicator · Decision support system · Climate change · Growing season · Ontogenesis · Yield · Additional water demand · Water erosion risk · Cultivation strategy

1.1 Introduction

Land surfaces (mainland and islands) account for only 29.3% of the Earth's surface, equating to an area of approximately 149.4 × 10^6 km^2. With reference to the land surface, 72.1 × 10^6 km^2 were under continued human use at the start of the twenty-first century (SEDAC 2016). In 2015, a total of approximately 48.6 × 10^6 km^2 was utilised globally for intensive agricultural, corresponding to 37.3% of land surfaces (WBG 2018). Utilised agricultural areas are usually—or in the case of Europe, always—part of an agricultural landscape. Human activities have left a permanent mark on agricultural landscapes, which have been developed over hundreds of years (Fig. 1.1). An agricultural landscape is therefore always a cultural landscape too.

Agricultural landscapes like all landscapes are multifunctional and holistic systems (van Huylenbroeck et al. 2007; Antrop and van Eetvelde 2017). They are characterised by agriculture, a human activity interacting with nature and with soil and local climate in particular (Mueller et al. 2010, 2014). Primary, agricultural landscapes are the basis for satisfaction of a basic human right: having enough food (United Nations 2015). However, their function to provide further ecosystem services and biodiversity must be also maintained (Buckwell et al. 2014).

Fig. 1.1 A diverse agricultural landscape from the *Barnimer Höhenzug* in north-eastern Germany, formed in the Pleistocene (photo: ZALF Müncheberg)

Profound knowledge of the structure of an agricultural landscape and a sound understanding of the underlying processes and how they interact on all relevant spatial and temporal scales are the prerequisites for understanding such complex systems and for being able to influence and change them sustainably in a targeted manner. Generally sustainable developments must be found for agricultural landscapes, particularly with regard to multifaceted global change, which is also influenced by the existing effects of climate change. A new type of experiments with agricultural landscapes is needed first because long time periods are required to initiate and implement such developments, but the decisions underlying such developments must be made now with the help of impact assessment, and second because conventional agricultural landscape experiments are infeasible as a basis for decisions, being too costly and protracted. This new type of experiments is only possible at the computer level, i.e. using agricultural landscapes mapped semi-realistically onto the computer by modelling. In this way, it is possible to conduct simulation experiments concerning scenarios of different probability with any desired temporal and spatial resolution, and practically without any temporal or spatial limitations. The results of such simulation experiments create the basis for assessing the potential consequences of anthropogenically induced and naturally occurring changes in agricultural landscapes. This, in turn, represents the basis for decisions that must now be taken to pave the way for a generally sustainable development of agricultural landscapes.

1.2 Agricultural Landscape

An agricultural landscape is a more or less intensively anthropogenically influenced part of a region shaped by its natural landscape features, primarily involving agricultural land use and land management. An agricultural landscape constitutes a spatiotemporal structure in which a large number of complex processes occur continuously and in parallel. The temporal dynamics of these complex processes result in constant changes in the system state and in a constant occurrence of complex interactions between nature and human society. Consequently, the most stable of an agricultural landscape is its permanent change.

According to Haase et al. (1991), agricultural landscapes are, in geographic terms, a mosaic of ecotopes with a choric dimension and, in functional terms, an ensemble of more or less anthropogenically influenced ecosystems. The "choric dimension" is defined as associations or mosaics of units in a humans usually still manageable size of the area that corresponds roughly to what is commonly referred to as "landscape". In this respect, lateral processes and interactions play a central role. In an agricultural landscape, different ecosystems (agroecosystems, forest ecosystems, grassland ecosystems, limnic ecosystems and urban ecosystems) usually coexist and influence each other (see Fig. 1.2).

One example of interactions between ecosystems could be altered light and soil water conditions on arable land adjacent to the forest. These conditions are triggered by shading from trees and the water uptake by the extensive root system of the trees, which in turn change the temperature and water supply regimes of the crops grown on the arable land. Productivity of the arable crop on this land is negatively affected as a result (Schmitt et al. 2019a, b). Another example is the emission of nutrients from agricultural areas into kettle holes, lakes or ditches, which influences the composition of the water and may, as a result, lead to the emergence of altered plant and animal communities in these waters (Lischeid et al. 2018). These are only two easily comprehensible examples that reflect and emphasise the complexity of agricultural landscapes.

Agricultural landscapes have a broad spectrum of abiotic and biotic elements and at the same time provide habitats for a large number of plants and animals. Consequently, they serve as a basis for biodiversity, valued by society, at the levels of ecosystem diversity, species diversity and genetic diversity within species. The variety of sustainable land use forms in an agricultural landscape intended in the future will increase biodiversity as a whole and genetic diversity within many groups of species.

Humans use natural resources and processes in agricultural landscapes, e.g. for producing human food and animal feed, raw materials and bioenergy. The interaction of targeted human activities with natural resources and processes enables agricultural landscapes to provide a variety of benefits for society, referred to as ecosystem services (Helming et al. 2013). In addition, agricultural landscapes provide space for human recreation and for tourism.

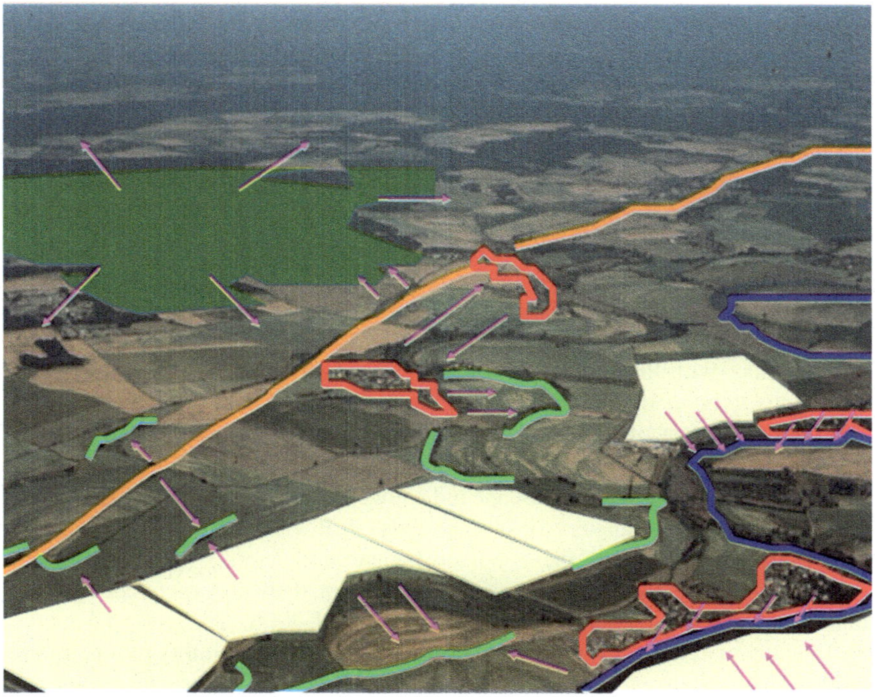

Fig. 1.2 Schema of an agricultural landscape in north-eastern Germany featuring different ecosystems (yellow—agroecosystem; dark green—forest ecosystem; red—urban ecosystem; blue—limnic ecosystem), diverse landscape features (orange—roads; light green—hedges and field rains) and various interactions between the landscape elements (arrows)

1.3 Agricultural Landscape and Landscape Modelling

The subject matter of agricultural landscape research is the agricultural landscape itself as a complex system. Agricultural landscape research attempts to investigate, analyse and describe this complex system in all its components and to characterise and map it using relevant landscape indicators. To achieve this, knowledge must be brought together about the different natural, technical, political/economic, social and cultural aspects and components of the system and their behaviour at various spatial and temporal scales of the agricultural landscape. This necessitates collaboration between the natural sciences, engineering, life sciences and the social sciences. This is also the essential requirement for creating a fundamental knowledge base for the sustainable use of agricultural landscapes.

Landscape indicators must be used to map and characterise the complexity, the current status and the dynamics of agricultural landscapes. After all, it is impossible to capture and describe such complex systems as agricultural landscapes using only one single indicator. Due to system complexity, there are a large number of landscape indicators. These include, among others, natural scientific, socio-economic and landscape structural indicators. Concrete examples of such indicators include plant productivity, transpiration, humus content in the soil, percolation, soil fertility, nitrate leaching into groundwater or surface waters, surface runoff, erosion or erosion risk, trace gas emissions, degree of biodiversity, as well as water and land quality (Bouma 2002). Additional indicators include the farm structure, labour demand and labour potential, degree of landscape fragmentation due to linear elements such as hedges, field edges, roads/paths, ditches as well as land use portions and the degree of the structuring of the agricultural landscape. However, there are also a number of other important landscape indicators.

The spatial reference of these indicators is typical for landscapes and therefore also for agricultural landscapes, enabling them to provide geographic information and, together with the use of geographic information systems (GIS), area-related statements to be made on the individual landscape indicators in the form of maps. For the description of landscapes, spatial information on weather and climate is required in addition to spatial status information such as soil (soil map) or elevation (elevation map). Individual landscape indicators are influenced directly or indirectly by meteorological parameters for short periods (weather) or longer periods (climate). In agricultural landscapes, these meteorological parameters always have a spatial reference, because, for temperature and precipitation, there are pronounced gradients that can have different orientations (e.g. Mirschel et al. 2006). The produced maps can be edited, processed or overlaid within GIS and offer the possibility to link different spatial datasets with each other. If these spatial landscape indicators, landscape status and landscape structure information for different time slices from the past over the present up to the future are interlinked, it is possible to represent and evaluate the agricultural landscape in space and time and thus also with regard to spatial and temporal changes.

Based on the analysis of measured and observed values, landscape indicators from the past and the present can be spatially mapped, and hence quantified. Remote sensing data also play an important role in this regard. In the case of forward-looking statements on the behaviour of individual landscape indicators, well-validated and robust models relating to the individual landscape indicators are an essential prerequisite, as well as forward-looking driving force data, such as data on different climate scenarios. In addition, extensive knowledge of the individual indicator-related processes in an agricultural landscape is a precondition for the model development. And yet models can only be as precise as the knowledge incorporated into them. The more knowledge scientists accumulate on processes in the agricultural landscape, the more precise and reliable the models will be. With indicator models, and hence landscape models, it is therefore also possible to describe the temporal dynamics of complex agricultural landscapes and the landscape indicators that characterise them. What, then, is a dynamic landscape model? “A dynamic landscape model is the complex mathematical-cybernetic representation of a landscape on the basis of a natural landscape description based on temporally invariant properties and a dynamic landscape description taking into account temporally variable landscape indicators” (modified slightly according to Lutze et al. 1993). Being part of a large landscape, this is of course also the case for agricultural landscapes.

With future-oriented scenarios and model-based simulation experiments at the computer level, the effects of possible changes, such as land use and climate changes, on the agricultural landscape or parts thereof can be assessed using special simulation environments such as SAMT (Wieland et al. 2006, 2015), LandCaRe-DSS (Wenkel et al. 2013) or SEAMLESS (van Ittersum et al. 2008). This, in turn, is the basis for the evaluation of forward-oriented options for action in agricultural landscapes. A prerequisite for this is, of course, that appropriate modelling approaches and the underlying spatial data exist and are available for the individual landscape indicators that map the landscape and its development. For this, the complete arsenal of methods known from mathematics can be used, i.e. from statistics to differential equations. Moreover, neural network models, expert-based models, multi-agent models, machine learning approaches or models that combine various approaches can also be used (Wieland and Mirschel 2017; Murgue et al. 2016). Great care should be taken to select model approaches and model types dependent on the landscape indicators (Lutze et al. 1993; Mirschel et al. 1997, 2004). For broad and, above all, robust application at the landscape scale, it makes sense to select models of intermediate complexity (*RE*gional *M*odels of *I*ntermediate *C*omplexity (REMICs), Wenkel et al. 2008).

At present, a wide range of reliably applicable regional models of intermediate complexity exist that can be used to describe individual landscape indicators in time and space. Examples for a spatial use include the growing season, the phenology of indicator plants, stages of development of agricultural crops, the productivity of agricultural crops, the irrigation water demand for a full exploration of the site-specific yield potential, and the erosion risk on arable land, to name just a few (Wenkel et al. 2010).

Due to the performance of the currently available computing technology, the use of parallel processing, cluster and cloud computing, the use of enormous data storage capacities and the huge amount of digitally available regional data, experiments with agricultural landscapes at the computer level, which would otherwise take decades, can be carried out in a matter of seconds. In addition, a very large number of simulation scenarios can be carried out for long time periods, even for time periods in the future.

This requires appropriate assumptions, e.g. for a possible land use development or possible spatial climate developments in the future, which can be calculated for different climate scenarios using global climate models. All this is already possible with sufficient accuracy and reliability.

The research questions relevant to agricultural landscape research and landscape modelling thus cover a broad field, starting with the processes in agricultural landscapes and the impact of different land uses and management systems on the individual landscape indicators, through to the resulting land use conflicts and their regulation (Topaj et al. 2018). Long-term ecological studies and monitoring systems are required not only to understand ecosystem complexity but also to develop and evaluate integrated dynamic ecosystem and landscape models.

1.4 The Agricultural Landscape in Simulation Experiments

In order to derive decisions with regard to the sustainable development of agricultural landscapes, it is absolutely necessary for the society to know what effects planned or expected changes (e.g. land use changes, changes in agricultural production systems or climate change) can have in and on agricultural landscapes in the future, as well as the risks associated with those impacts. In science, specific experiments are usually carried out to clarify scientific questions. However, such experiments with large regions and thus also agricultural landscapes in the conventional sense are not possible in order to clarify a wide variety of landscape-related issues, neither from a temporal nor from a financial point of view. Imagine experimentally investigating in an agricultural landscape what the long-term effects of converting arable land to grassland or of converting a forest from a pure pine forest into a mixed forest or deciduous forest would be on the individual landscape indicators, possibly also superimposed by the effects of expected climate change. It would take decades to answer these questions, racking up costs of many millions. And not only these two issues are relevant for changes in agricultural landscapes—there are considerably more issues of relevance. As described above, only one viable alternative exists—model-based simulation experiments involving different scenarios at the computer level. A number of examples will be shown below to demonstrate this.

1.4.1 Climate Change and the Growing Season

Continuous measurements taken since 1893 at the long-term meteorological secular station Potsdam, Germany, show that until 2015 the annual mean temperature has increased by 1.15 °C, accompanied on the one hand by 4 and 16.5 fewer ice days and frost days, respectively, and, on the other hand, 6 and 15 more hot days and summer days, respectively. For Central Europe, global climate models indicate that the annual mean temperature could increase by a further 0.8–2 °C until 2050, and by an estimated further 2.5–4.5 °C until 2100 (IPCC 2013). In agricultural landscapes, rising temperatures have a significant impact on plant development, biomass accumulation and therefore also on yield. Rising temperatures mean an earlier start of the growing season and an end of the growing season that goes well into autumn, equating to an extension of the growing season as a whole. Based on the model by Chmielewski (2003) for calculating the thermal growing season, using the climate scenario STARS2 (Orlowsky et al. 2008), an earlier start of the growing season by 37 days, an extension of the end of the growing season by 27 days, and hence an extension of the growing season by 64 days, is computed for the year 2060 compared to 1951 for the Central German Lowland, represented by the Potsdam site. Figure 1.3 shows the start and end of the growing season for Potsdam for the period 1951–2060, in addition to annual fluctuations and the corresponding linear trends. For the climate data, the STARS (STatistical Analogue Resampling Scheme) method

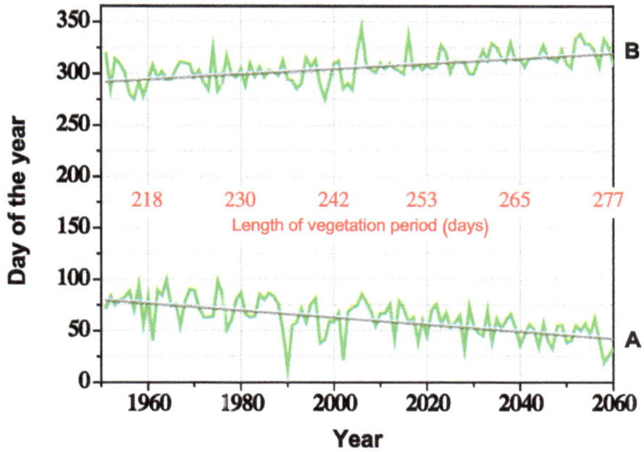

Fig. 1.3 Annual fluctuations and linear trends for the start of the growing season (A), the end of the growing season (B) and the length of the growing season at the Potsdam site for the period 1951–2060 on the basis of STARS2 climate data (Orlowsky et al. 2008)

for regional climate simulation developed at the Potsdam Institute for Climate Impact Research (PIK) is used, assuming a possible temperature increase of 2 K.

A longer growing season also has its advantages for agriculture. It would enable German farmers to grow two crop species each year rather than just one (Graß et al. 2015), provided that there is sufficient rainfall in the future. A longer growing season with higher temperatures would also be relevant for the cultivation of heat-loving crop plants, such as the very protein-rich soybean as fodder for livestock or for other crop species with high biomass production as a basis for generating bioenergy. Rising temperatures could improve growing conditions in regions where previous temperature conditions have restricted the cultivation of high-yield crops and varieties. Examples of such regions include higher areas of the Harz, the Thuringian forest and the Ore Mountains (Kersebaum and Nendel 2014). In addition, as temperatures rise, the cultivation boundaries for some agricultural crops will shift further north (Endlicher and Gerstengarbe 2007).

1.4.2 Climate Change and Plant Development

Due to the earlier start of the growing season and the spreading of the vegetation period, there is a change in the temporal development of a landscape's phenological indicator plants. Rising temperatures lead to a shift in the temporal development of the indicator plants such as snowdrop, gooseberry, apple, the common oak, elder and summer linden, which are the basis for the phenological calendar. The spatially applicable model PHENO for the phenology of indicator plants developed by Chmielewski (Chmielewski and Hennings 2007) computes what is known as chilling and forcing rates, and takes into account dormancy, the resting period in winter

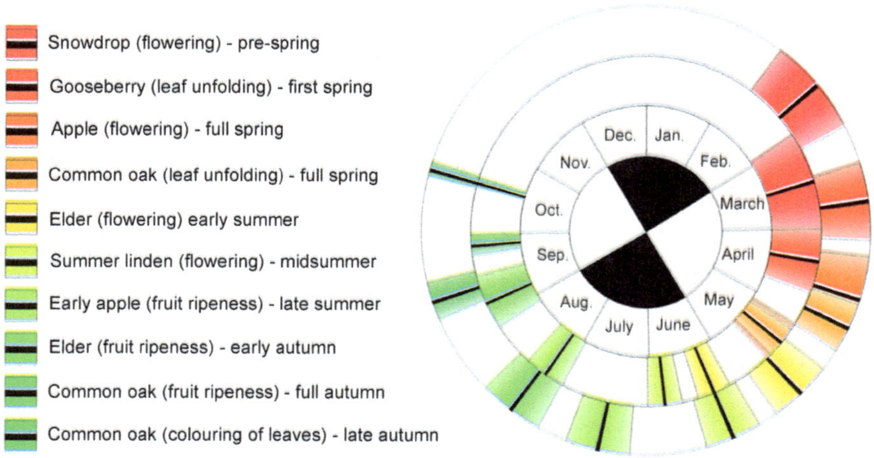

Fig. 1.4 The phenological calendar comparing climatic periods 1951–1980 (inner circle) and 2031–2060 (outer circle), valid for the Uckermark region, Germany, on the basis of STARS2 climate data

when woody plants exhibit no active growth. Dormancy is broken once the chilling requirement or cold stimulus has been satisfied, i.e. the plant must be exposed to chilling temperatures for a certain length of time. Only then can temperatures that are conducive to plant growth (forcing temperatures) force the development and bursting of buds in spring, the subsequent unfolding of leaves and the blossoming of trees and shrubs. Figure 1.4 shows, taking the example of the Uckermark region, how the phenological calendar could change, comparing climatic periods 1951–1980 and 2031–2060 using STARS2 climate data (Orlowsky et al. 2008). The coloured segments characterise the fluctuation range around the 30-year average in each case.

The earlier start of the vegetation period results in the earlier development of agricultural crops. Winter crops start developing earlier in spring, and the cultivation of fields with summer crops also starts earlier. This results in a change in the regime for agro-technical measures, which in turn results in an adjustment of the whole farm management process. Here, too, models can help researchers to identify and consider tendencies. Using the ONTO model (Wenkel et al. 2010), based on temperature sums, for example, it is possible to compute the starting dates of the individual developmental stages of main agricultural crops between sowing/planting and harvest. Using STARS2 climate data (Orlowsky et al. 2008), Fig. 1.5 shows the ontogenesis for winter wheat as a winter crop and sugar beet as a spring crop for the agricultural region of the Uckermark in north-eastern Germany, comparing the two periods 1951–1980 and 2031–2060. The numbers in the individual colour segments indicate the length of the developmental stage in days. It is clearly visible for both agricultural crops as to how higher temperatures will shorten the length of the individual development stages in the future, considerably shortening the period between sowing and harvest.

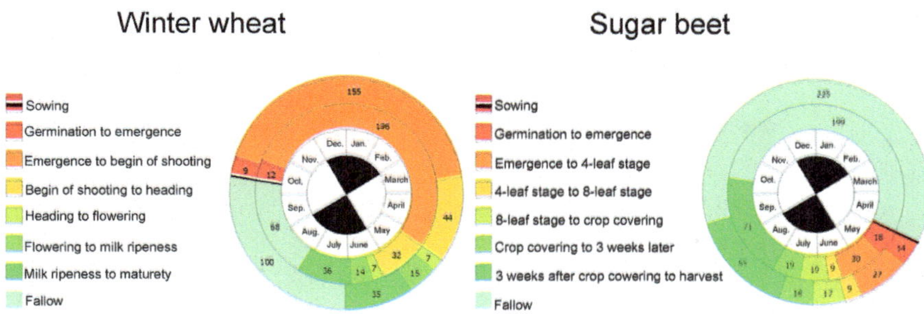

Fig. 1.5 Ontogenesis for winter wheat (sowing date: 10 October) and sugar beet (sowing date: 27 April), comparing climatic periods 1951–1980 (inner circle) and 2031–2060 (outer circle) for the agricultural region of the Uckermark, Germany, using STARS2 climate data

1.4.3 Climate Change and Yield

Plant development is the "biological clock" according to which all processes are controlled in the plant, i.e. they are switched on or switched off, accelerated or slowed down. Biomass accumulation and yield formation are also controlled according to plant development. In particular, climatic parameters such as temperature, radiation and precipitation have an essential impact on this control mechanism. Biomass and yield formation are also determined to a significant extent by the level of plant supply with nutrients and water. The site and therefore the soil, with its subsequent delivery capacity of nutrients and water, play an important role in this case. Since soil is usually very heterogeneous in landscapes, due to its genesis history, this is also reflected, for example, in the yield of agricultural crops which then, seen from a regional perspective, can also be very heterogeneous. Yield heterogeneity is not only due to regional differences in climates but is mainly dependent on soil characteristics. In other words, the best soils produce the highest yields, and the poorest, sandiest soils produce the lowest yields. This is presented as an example for the whole Free State of Thuringia using winter wheat as a winter crop and silage maize as a spring crop. The yields in Thuringia are as heterogeneous as the soil, site and climate conditions, as well as the differences in altitude. The yields of both agricultural crops, winter wheat and silage maize, are calculated for all arable locations within the Free State of Thuringia using the YIELDSTAT model (Mirschel et al. 2014). In the model-based simulation experiment, the WETTREG 2010 climate data (Spekat et al. 2010) with the emission scenario A1B, which assumes an expected temperature increase of 2 K, is used. Figure 1.6 shows a summary of the regional heterogeneity of the expected climate-induced changes in yields for winter wheat and silage maize for the Free State of Thuringia, comparing 1981–2010 and 2021–2050.

The simulation calculations for winter wheat in Thuringia show that the yields in the projection period 2021–2050 are lower in some regions and can also increase in other regions, especially in regions with higher altitudes. In the case of silage maize, on the other hand, yield losses can be expected throughout Thuringia's arable

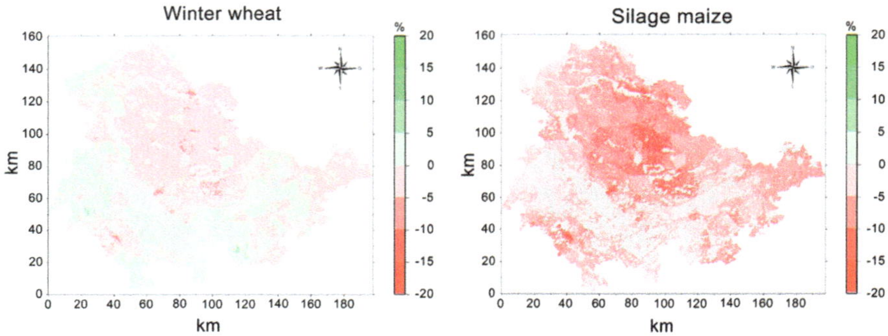

Fig. 1.6 Relative climate-induced changes in yield for winter wheat and silage maize, comparing 2021–2050 and 1981–2010, regionalised for the Free State of Thuringia (climate data: WETTREG 2010/A1B)

land in 2021–2050. The main reasons for these losses are rising temperatures, and consequently also increasing evaporation, as well as less favourable precipitation distributions throughout the year, accompanied by more frequently occurring dry periods.

Simulation calculations using the YIELDSTAT model have shown that, averaged overall arable land in Thuringia, there will only be a slight yield reduction for the agricultural crops under consideration, with the exception of silage maize with a higher yield reduction, in the projection period 2021–2050 compared to the reference period 1981–2010 (winter wheat: −2.40%; winter barley: −0.25%; winter oilseed rape: −3.35%; spring barley: −2.90%; and silage maize: −10.95%). For all five agricultural crops, however, there will be a widening of both the yield variance (winter wheat: +19.6%; winter barley: +17.6%; winter oilseed rape: +47.8%; spring barley: +34.2%; and silage maize: +33.1%) and the yield range between the lowest yield year and the highest yield year (winter wheat: +18.2%; winter barley: +20.6%; winter oilseed rape: +52.6%; spring barley: +22.2%; and silage maize: +21.1%).

1.4.4 Climate Change and Irrigation Water Demand

In response to climate change, leading to rising temperatures, increasing evaporation and decreasing precipitation during the main growing season, irrigation is one of the potential measures that farmers can take to adapt to climate change. If sufficient regional water resources are available, irrigation ensures yield stability over the years. For economic reasons, however, only important agricultural crops will be irrigated. For the example of an irrigation scenario for 2015 in the following, the required irrigation water demand and the additional yield caused by irrigation is shown county-wise for the Federal State of Brandenburg. It is assumed, in the process, that current trends, defined by breeding, the climate and the agricultural market, will be continued up to 2025. It is also assumed that the cropping areas required for forage production remain constant; that the maximum percentage of cultivated crops restricted by phytosanitary requirements are not exceeded; and that the typical crop rotation systems are retained. In the irrigation scenario, all winter wheat, winter oilseed rape, silage maize and sugar beet fields in Brandenburg are extensively irrigated (Mirschel et al. 2016). All other crops are grown without irrigation. The climate data used are based on STARS2 (Orlowsky et al. 2008). The ZUWABE model (Mirschel et al. 2019) is used to compute additional water demand for the four agricultural crops throughout Brandenburg.

Figure 1.7 shows the spatial distribution of irrigation water demand in the Federal State of Brandenburg at the county level for the four irrigated agricultural crops in the irrigation scenario. As can be seen, the counties with the best soil and site conditions, such as the counties of Prignitz and Elbe-Elster, require less irrigation water. Accordingly, for the four irrigated agricultural crops the yield increases caused by irrigation in 2025 are also distributed spatially for Brandenburg. In order to better compare the effects of irrigation on crop yields and for a better spatial presentation of these effects in Fig. 1.7, the yields were converted into cereal units (CU).

Comparing districts, the quantities of irrigation water demand per irrigation season for silage maize at the climate level of 2025 vary from 109 to 134 mm, and the increase of yield caused by irrigation range from 13.0 to 16.1 t ha^{-1}; the figures for winter wheat range from 72 to 89 mm with increases of yield between 1.1 and 1.3 t ha^{-1}; in the case of winter oilseed rape, the corresponding figures range from 18 to 28 mm with increases of yield between 0.2 and 0.3 t ha^{-1}; for sugar beet, the quantities range from 98 to 119 mm, and increases of yield are between 9.3 and 11.3 t ha^{-1}.

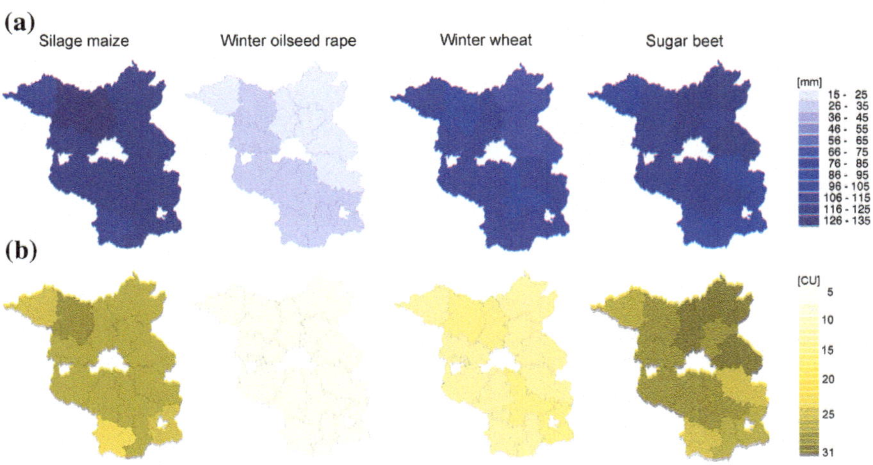

Fig. 1.7 Irrigation water demand (a, in mm) and yield increase caused by irrigation (b, in CU) in 2025 for silage maize, winter oilseed rape, winter wheat and sugar beet at the county level in the Federal State of Brandenburg for the irrigation scenario (modified according to Mirschel et al. 2016)

1.4.5 Climate Change and Agricultural Management Strategies

As the climate changes, farmers will need to adapt their management strategies to these changes in the interests of profitable management of their farms. Altered crop rotations, adapted soil cultivation methods, the use of irrigation, and the cultivation of energy crops play an important role in this connection. In order to choose the right and best management strategies for the future, taking into account these adaptation measures, a model-based decision support system and resilient climate scenarios are required, so that the best management strategy can be derived from the results of regional scenario simulations. The LandCaRe-DSS (Wenkel et al. 2013) is such a decision support system. Using LandCaRe-DSS, scenario simulations were run for the 2,600 km^2 agricultural Uckermark region in north-eastern Germany. The crop yield was used to assess how the productivity in the agricultural Uckermark region was affected by irrigating economically important agricultural crops within the crop rotation; by changing from conventional to conservation tillage in winter oilseed rape and winter cereals; and by increasing the proportion of energy maize to a maximum of 25% in the crop rotation. WETTREG 2010 climate data with the emissions scenario A1B (Spekat et al. 2010) were used for this purpose. The simulation results are shown in Fig. 1.8.

The simulation results show that the greater integration of energy maize in the crop rotation increases productivity in the agricultural Uckermark region compared to the conventional crop rotation. In contrast, the increased application of conservation or no-tillage soil cultivation methods in winter crops results in a slight decrease in productivity of the agricultural Uckermark region. These slight decreases in yields can already be observed in farms that systematically use minimal or no-tillage systems.

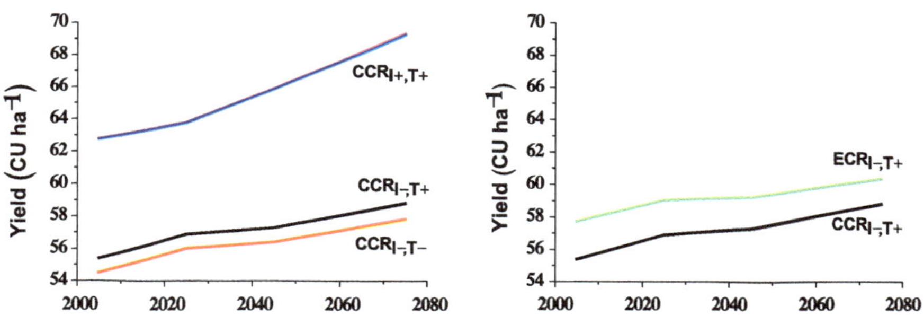

Fig. 1.8 Crop yield (in cereal units per hectare (CU ha^{-1}) (GE (2019)) produced in the agricultural Uckermark region for the current crop rotation (CCR) for farming without irrigation and with conventional tillage ($CCR_{I-,T+}$), with irrigation and conventional tillage ($CCR_{I+,T+}$) and for conservation tillage (no-tillage) farming without irrigation ($CCR_{I-,T-}$), and for a crop rotation enriched with energy maize without irrigation but with conventional tillage ($ECR_{I-,T+}$)

The results of simulations run using the EROSION model (Wieland 2010; DIN19708 2005) for assessing the potential water erosion risk show for the agricultural Uckermark region that, compared to conventional tillage, the only application of no-tillage systems in winter crop cultivation year by year also leads to a reduction in water erosion risk in the relatively flat agricultural landscape of the Uckermark. Taking into account the WETTREG 2010/A1B climate scenario, the decrease of erosion is 0.22 t soil $ha^{-1}a^{-1}$ (2005), 0.18 t soil ha^{-1} a^{-1} (2025), 0.17 t soil ha^{-1} a^{-1} (2045) and 0.30 t soil ha^{-1} a^{-1} (2075). In a considerably hillier agricultural region, such as the catchment area of the Weißeritz river in Saxony, Germany, where absolute soil erosion due to water erosion is 4–5 times higher, the reduction in soil erosion that can be achieved by applying no-tillage systems is significantly larger. In this case, it is 0.73 t soil ha^{-1} a^{-1} (2005), 0.69 t soil ha^{-1} a^{-1} (2025), 0.86 t soil ha^{-1} a^{-1} (2045) and 1.15 t soil ha^{-1} a^{-1} (2075) (Wenkel et al. 2013).

1.5 Conclusions and Perspectives

Profound and reliable knowledge of the structure of the agricultural landscape and of landscape-related processes and their interactions at all relevant temporal and spatial scales is required to be able to specifically and sustainably change such a complex system as the agricultural landscape for the benefit of nature and mankind. For the characterisation and description of the landscape status, a representative set of spatially related landscape indicators is necessary, which covers the different areas of natural science, sociology, economy and landscape structure in a balanced ratio. With increasing knowledge acquisition in the fields of landscape structure and landscape processes, the number of quantifiable landscape indicators grows and the informative quality of existing landscape indicators improves.

Based on the current state and the current human use of landscapes or agricultural landscapes, if we are to continue to use these landscapes sustainably, there is a need to assess how planned uses and future changes in use will affect the agricultural landscape, which landscape indicators will change as a result and how they will do so, whether they could thus exceed certain permissible areas and lead to irreversible changes in the agricultural landscape, which will disrupt the ecological balance. A model- and simulation-based approach is required to be able to answer such forward-related questions and assessments in and with agricultural landscapes. This means that

I. The individual landscape indicators must be described by models, without having to define a particular model type, but instead involving the most appropriate model types in each case,
II. These models must be embedded in simulation tools that are optimised in terms of computing capacity and computing time, enabling a very large number of simulation runs to be realised swiftly using parallel processing, cluster and cloud

computing, and facilitating the rapid statistically based and spatially visualised interpretation and presentation of the results, and

III. A wide range of models and scenarios for future probable land uses and climate developments and their spatial generated data must be available.

Decision support systems (DSS) that exploit the latest possibilities offered by information technology, statistics and artificial intelligence provide the framework for integrating models, spatial data concerning the state of the landscape and scenario data, simulation techniques as well as tools for interpreting and visualising results. Such DSS are also the basis for quantifying the complex impact of site conditions, changes in land use or management, and of potential climate change on individual landscape parameters or landscape indicators. They are also the prerequisite for making statements on the degree of uncertainty of the results gained. This is required, in turn, in order to be able to carry out appropriate risk assessments, which is of great importance for the adaptation of agriculture to climate change, and hence for the sustainable use of agricultural landscapes under a changing climate.

Due to the many uncertainties both in the processes necessitated by climate change and in the adaptation of agriculture to climate change, adaptation measures must be chosen that are robust against uncertain developments, flexible and easy to redirect. These are certainly not measures aimed at profit and maximum yield, but measures aimed at sustainability and the long-term existence of the agricultural enterprises. Examples of such measures include

- Increasing the diversity of agricultural crops and varieties in the context of extended crop rotations,
- Stabilising and improving the humus content of soils,
- Ensuring a high supply of nutrients to soils on arable land and grassland,
- Reducing unproductive water losses on the fields,
- Reducing soil erosion by ensuring all-year-round ground cover, where possible,
- Developing rooting zones that are as deep as possible,
- Regulating the water and stabilising landscape hydrology and
- Establishing new farming and production systems.

Intact soil is much better able to buffer events that are unfavourable to plant growth. Therefore, all the measures which serve to sustainably maintain and improve the soil as the main production resource in agricultural landscapes, and hence its fertility, are the essential measures for a promising adaptation of agriculture to climate change. But breeding can also contribute to the adaptation to climate change through new, stress-tolerant and physiologically flexible varieties.

Due to the complexity of processes in an agricultural landscape, it becomes apparent that the consequences of climate change cannot be counteracted by taking individual measures—a balanced combination of different measures is required. The use of models and simulation experiments within the framework of decision support systems at the computer level is a very promising way of finding suitable site-specific complex measures for the adaptation of agriculture to climate change.

As prerequisites for the development and validation of integrated dynamic landscape models more long-term ecological studies and monitoring systems are required. However, this also means that more resources need to be made available to support these projects.

Acknowledgements This study was funded by the Ministry of Science, Research and Culture of the Federal State of Brandenburg and the Federal Ministry of Food, Agriculture and Consumer Protection.

References

Antrop M, Van Eetvelde V (2017) Landscape perspectives: the holistic nature of landscape. Landscape series, vol 23. Springer Science+Business Media, p 436

Bouma J (2002) Land quality indicators os sustainable land management across scales. Agr Ecosyst Environ 88(2):129–136

Buckwell A, Heissenhuber A, Blum WEH (2014) The sustainable intensification of european agriculture: a review. Sponsored by the rise foundation, 96 pp. http://www.risefoundation.eu/images/files/2014/2014_%20SI_RISE_FULL_EN.pdf. Accessed 19 Dec 2018

Chmielewski FM (2003) Phenology and agriculture. In: Schwartz MD (ed) Phenology: an integrative environmental science. Kluwer Academic Publishers, Boston/Dordrecht/London, pp 505–522

Chmielewski FM, Hennings Y (2007) Phänologische Modelle als Grundlage zur Abschätzung des Klimaimpact. Berichte Meteorologisches Institut Freiberg (6. Fachtagung BIOMET) 16:229–235

DIN19708 (2005) Bodenbeschaffenheit – Ermittlung der Erosionsgefährdung von Böden durch Wasser mit Hilfe der ABAG (Soil quality – Predicting soil erosion by water by means of ABAG). DIN 19708:2005–02, Normenausschuss Wasserwesen (NAW) im DIN, p 25

Endlicher W, Gerstengarbe F-W (eds) (2007) Der Klimawandel – Einblicke. Rückblicke und Ausblicke, Potsdam-Institut für Klimafolgenforschung, p 134

GE (2019) Getreideeinheit (GE) – Ausführliche Definition. https://wirtschaftslexikon.gabler.de/definition/getreideeinheit-ge-35840/version-259314. Last accessed 13 March 2019

Graß R, Thies B, Kersebaum KC, Wachendorf M (2015) Simulating dry matter yield of two cropping systems with the simulation model HERMES to evaluate impact of future climate change. Eur J Agonomy 70:1–10

Haase G, Barsch H, Schmidt R (1991) Zur Einleitung: Landschaft, Naturraum und Landnutzung. Beiträge zur Geographie 34:19–25

Helming K, Diehl K, Geneletti D, Wiggering H (2013) Mainstreaming ecosystem services in European policy impact assessment. Environ Impact Assess Rev 40:82–87

van Ittersum M, Ewert F, Hechelei T, Wery J, Olsson JA, Andersen E, Bezlepkina I, Brouwer F, Donatelli M, Flichmann G, Olsson L, Rizzoli AE, van der Wal T, Wien JE, Wolf J (2008) Integrated assessment of agricultural systems—a component-based framework fort the European Union (SEAMLESS). Agric Syst 96:150–165

IPCC (2013) Climate Change 2013: the physical science basis. In: Stocker TF, Qin D, Plattner G-K, Tignor M, Allen SK, Boschung J, Nauels A, Xia Y, Bex V, Midgley PM (eds) Contribution of working group I to the fifth assessment report of the intergovernmental panel on climate change. Cambridge University Press, Cambridge, United Kingdom and New York, NY, USA, 1535 S

Kersebaum KC, Nendel C (2014) Site-specific impacts of climate change on wheat production across regions of Germany using different CO_2 response functions. Eur J Agronomy 52:22–32

Lischeid G, Kalettka T, Holländer M, Steidl J, Merz C, Dannowski R, Hohenbrink T, Lehr C, Onandia G, Reverey F, Pätzig M (2018) Natural ponds in an agricultural landscape: external drivers, internal processes, and the role of the terrestrial-aquatic interface. Limnologica 68:5–16

Lutze G, Schultz A, Wenkel K-O (1993) Vom Populationsmodell zum Landschaftsmodell – Neue Herausforderungen und Wege zur Nutzung von Modellen in der Agrarlandschaftsforschung. Zeitschrift für Agrarinformatik 1:19–25

Mirschel W, Lutze G, Schultz A (2006) Luzi K (2006) Klima und Wetter in der Agrarlandschaft Chorin—gestern, heute, morgen. In: Lutze G, Schultz A, Wenkel K-O (eds) Landschaften beobachten, nutzen und schützen – Landschaftsökologische Langzeit-Studie in der Agrarlandschaft Chorin 1992–2006. G.B.TeubnerVerlag, Wiesbaden, pp 49–59

Mirschel W, Schultz A, Wenkel K-O (1997) Agroökosystemmodelle als Bestandteile von Landschaftsmodellen. Arch Nat Conserv Landsc Res 35:209–225

Mirschel W, Schultz A, Wenkel K-O, Wieland R, Poluektov RA (2004) Crop growth modelling on different spatial scales—a wide spectrum of approaches. Arch Agronomy Soil Sci 50(3):329–343

Mirschel W, Wieland R, Gutzler C, Helming K (2016) Luzi K (2016) Auswirkungen landwirtschaftlicher Anbauszenarien auf Ertrag und Zusatzwasserbedarf im Land Brandenburg im Jahr 2025. In: Nguyen XT (ed) Modelling and simulation of ecosystems: workshop Kölpinsee 2015. Rhombos-Verlag, Berlin, pp 1–19

Mirschel W, Wieland R, Luzi K, Groth K (2019) Model-based estimation of irrigation water demand for different agricultural crops under climate change, presented for the Federal State of Brandenburg, Germany (in this book)

Mirschel W, Wieland R, Wenkel K-O, Nendel C, Guddat C (2014) YIELDSTAT—a spatial yield model for agricultural crops. Eur J Agron 52(2014):33–46

Mueller L, Schindler U, Ball BC, Smolentseva E, Sychev VG, Shepherd TG, Qadir M, Helming K, Behrendt A, Eulenstein F (2014) Productivity potentials of the global land resource for cropping and grazing. In: Mueller L, Saparov A, Lischeid G (eds) Novel measurement and assessment tools for monitoring and management of land and water resources in agricultural landscapes of Central Asia. Environmental science and engineering. Springer International Publishing, Cham, pp 115–142. https://doi.org/10.1007/978-3-319-01017-5_6

Mueller L, Schindler U, Mirschel W, Shepherd TG, Ball B, Helming K, Rogasik J, Eulenstein F, Wiggering H (2010) Assessing the productivity function of soils: a review. Agron Sustain Dev 30(3):601–614. https://doi.org/10.1051/agro/2009057

Murgue C, Therond O, Leenhardt D (2016) Hybridizing local and generic information to model cropping system spatial distribution in an agricultural landscape. Land Use Policy 54(2016):339–354

Orlowsky B, Gerstengarbe F-W, Werner PC (2008) A resampling scheme for regional climate simulations and its performance compared to a dynamical RCM. Theoret Appl Climatol 92:209–223

Schmitt M, Liescheid G, Nendel C (2019a) Microclimate and matter dynamics in transition zones for forest to arable land. Agriculture and Forest Meteorology 268:1–10. https://doi.org/10.1016/j.agrformet.2019.01.001

Schmitt M, Nendel C, Funk R, Mitchel MGE (2019b) Lischeid G (2019b) Modeling yields response to shading in the field-to-forest transition zones in heterogeneous landscapes. Agriculture 9(1):6. https://doi.org/10.3390/agriculture9010006

SEDAC (2016) Last of the Wild (Version Two). Socioeconomic date and application center (SEDAC) in NASA's earth observing system data an information system (EOSDIS), hosted by CIESIN at the Columbia University. http://sedac.ciesin.columbia.edu/data/collection/wildareas-v2. Last accessed 15 Jan 2019

Spekat A, Kreienkamp F, Enke W (2010) An impact-oriented classification method for atmospheric patterns. Phys Chem Earth 35:352–359

Topaj A, Badenko V, Medvedev S, Terleev V (2018) Dynamically adjusted forecasting of agro-landscape productivity using massive computations of crop model in GIS environment. In: Sychev VG, Mueller L (eds) Novel methods and results of landscape research in Europe, Central Asia and Siberia. Monograph in 5 Volumes. Vol III, Landscape monitoring and modelling © «FSBI VNII Agrochemistry» 2018, pp 253–257. https://doi.org/10.25680/3309.2018.28.99.246, http://vniia-pr.ru/monografii/pdf/tom3-53.pdf. Last accessed 15 Jan 2019

United Nations (2015) Agenda 2030: sustainable development goals. 17 goals to transform our world. https://www.un.org/sustainabledevelopment/. Last accessed 19 Dec 2018

Van Huylenbroeck G, Vandermeulen V, Mettepenningen E, Verspech A (2007) Multifunctionality of agriculture. A review of definitions, evidence and instruments. Liv Rev Landsc Res 1:3. http://www.livingreviews.org/lrlr-2007–3. Last accessed 19 Dec 2018

WBG (2018) Agriculture and rural development—agricultural land (% of land area). World Bank open data—free and open access to global development data. https://data.worldbank.org/topic/agriculture-and-rural-development. Last accessed 15 Jan 2019

Wenkel K-O, Berg M, Mirschel W, Wieland R, Nendel C, Köstner B (2013) LandCaRe DSS—an interactive decision support system for climate change impact assessment and the analysis of potential agricultural land use adaptation strategies. J Environ Manage 127(Supplement):S168–S183

Wenkel K-O, Berg M, Wieland R, Mirschel W (2010) Modelle und Entscheidungsunterstützungssystem zur Klimafolgenabschätzung und Ableitung von Adaptionsstrategien der Landwirtschaft an veränderte Klimabedingungen (AGROKLIM-ADAPT)—Decision Support System (DSS). Forschungs-Abschlußbericht: BMBF 01 LS 05104, Leibniz-Zentrum für Agrarlandschaftsforschung Müncheberg, 51 pp, 14 Annexes (133 pp.)

Wenkel K-O, Wieland R, Mirschel W, Schultz A, Kampichler C, Kirilenko A, Voinov A (2008) Regional models of intermediate complexity (REMICs)—a new direction in integrated landscape modelling. In: Jackeman A, Voinov AA, Rizzoli AE, Chen SH (eds) Environmental modelling, software and decision support-state of the art and new perspectives–developments in integrated environmental assessment, vol 3. Elsevier, Amsterdam, pp 285–295

Wieland R (2010) EROSION—Modell zur Berechnung der potentiellen Erosionsgefährdung. In: Wenkel K-O, Berg M, Wieland R, Mirschel W Modelle und Entscheidungsunterstützungssystem zur Klimafolgenabschätzung und Ableitung von Adaptionsstrategien der Landwirtschaft an veränderte Klimabedingungen (AGROKLIM-ADAPT)—Decision Support System (DSS). Forschungs–Abschlußbericht: BMBF 01 LS 05104, Leibniz-Zentrum für Agrarlandschaftsforschung (ZALF), Müncheberg, pp A6/1–A6/7

Wieland R, Groth K, Linde F, Mirschel W (2015) Spatial analysis and modeling tool version 2 (SAMT2), a spatial modeling tool kit written in Python. Ecol Inform 30(2015):1–5

Wieland R, Mirschel W (2017) Combining expert knowledge with machine learning on the basis of fuzzy training. Ecol Inform 38:26–30

Wieland R, Voss M, Holtmann X, Mirschel W, Ajibefun IA (2006) Spatial analysis and modeling tool (SAMT): 1. Struct Possibilities Ecol Inform 1(2006):67–76

Chapter 2
Challenges and Perspectives for Integrated Landscape Modelling to Support Sustainable Land Use Management in Agricultural Landscapes

Karl-Otto Wenkel, Ralf Wieland and Wilfried Mirschel

Abstract This contribution discusses scientific challenges and new possibilities for better modelling of the consequences of changes in land use and land management in whole landscapes on ecosystem functions in space and time. The regional dimension under discussion is an area of up to several thousand square kilometres. Main problems on this scale are high complexity, structural diversity, ecological heterogeneity, inadequate representation of the governing processes in the model with respect to a given application and uncertainty in data and in understanding of the process dynamics.

Keywords Sustainable land use · Climate change · Reduced resources availability · Ecosystem models of intermediate complexity · Fuzziness · Dynamic processes · Software tool boxes · Landscape indicators · Neural network · Genetic algorithms

2.1 Introduction

Sustainable development and sustainable land use management together with increased awareness of climate change and reduced resources availability have been received increasing attention around the world (Hurni 2000; Wenkel et al. 1997). This is also the case for sustainable land use management of agro-landscapes. Today we can observe an emerging consensus that a sustainable and more productive agriculture is needed that can meet the local, regional and global food security challenges of the twenty-first century. This consensus implies there would be value in new and

K.-O. Wenkel · R. Wieland · W. Mirschel (✉)
Leibniz-Centre for Agricultural Landscape Research (ZALF) e.V., Eberswalder Str. 84, 15374 Müncheberg, Germany
e-mail: wmirschel@zalf.de

K.-O. Wenkel
e-mail: kowenkel@web.de

R. Wieland
e-mail: rwieland@zalf.de

W. Mirschel et al. (eds.), *Landscape Modelling and Decision Support*, Innovations in Landscape Research,
https://doi.org/10.1007/978-3-030-37421-1_2

improved tools that can be used to assess the sustainability of current and prospective systems, design more sustainable systems, and manage systems sustainably (Antle et al. 2017). Profound knowledge of the structure of an agricultural landscape and a sound understanding of the underlying processes of an agricultural landscape and how they interact on all relevant spatial and temporal scales are the prerequisite for understanding such complex systems and for being able to influence and change them sustainably in a targeted manner. Our environment with its dynamic and spatial processes is recognised as complex, highly interacting and spatially distributed. These properties make analysing, describing, modelling and even simulating our environment a challenging task (Seppelt 2003).

A framework that enables us to study, for example, the consequences of human influences and climate change on ecological systems without even disturbing these is a valuable and important tool for environmental management and sustainable development. Recent development of environmental models has shown that a multitude of possible approaches and theories have been developed. This multitude of different approaches refers to the considered temporal scale, the considered mathematical languages and the chosen concept of regionalization and spatial extent. The incorporation of spatial attributes into the modelling process causes a mismatch between the scale at which attributes are obtained, and the scale at which the processes occur.

Despite the progress in mathematical modelling of ecosystems, practical tools and integrated models which can be used for a better sustainable landscape management at local and regional levels have emerged, however, only recently. The development of such models, which can be used for the quantitative or qualitative characterization the sustainability of land use systems still remains a challenge for ecological and environmental modelling worldwide.

In order to the different dimensions of agriculture and food systems at different scales, ideally we would have a virtual laboratory containing models, data, analytical tools and IT tools to conduct studies that evaluate outcomes and trade-offs among alternative technologies, policies, or scenarios. The virtual laboratory would allow users to define scenarios, specify analyses covering different social, political and resource situations and different spatial and temporal scales, and produce outputs suitable for interpretation and use by decision-makers. However, that virtual laboratory does not exist (Jones et al. 2017).

2.2 Modelling at Landscape Scale

2.2.1 Next Generation Models for the Landscape Scale–An Outlook

Significant progress has been made in ecosystem modelling in recent decades. Process and agro-ecosystem models are developed as point model (one-dimensional case), i.e. for individual points within an agricultural used region. Examples for such agro-ecosystem models (soil-plant-atmosphere models) are AGROSIM (Mirschel et al. 2001; Wenkel and Mirschel 1995), AGROTOOL (Poluektov et al. 2002), APSIM (Keating et al. 2003), CERES (Ritchie and Godwin 1993), DAISY (Hansen et al. 1990), HERMES (Kersebaum 1995), MONICA (Nendel et al. 2011, 2014), STICS (Brisson et al. 2003), THESEUS (Wegehenkel et al. 2004) or WOFOST (Supit et al. 1994), each model with a different focus.

An overview of agro-ecosystem and ecological models is given in the register of Ecological models (REM) which is found in the WWW Server for Ecological Modelling (ECOBAS 2011; Benz et al. 2001). Models on the landscape scale are generally based on four-dimensional data sets. Three dimensions are used for the spatial components, one dimension for the temporal component. The spatial components in landscape research are mainly processed using Geographic Information Systems (GIS). Typical digital maps in GIS consist of raster data or vector data. In GIS, attribute maps for the study area are managed via layers. Models read the digital maps from the GIS and so they add the temporal component. Models allow to look into the future and to calculate developments which are not visible today.

The question how much information is necessary to model a process and which information is available to develop a model is central. Local mechanistic point models describe processes in great detail and demand a lot of information as driving forces or as parameters. At a regional scale, it is difficult to provide all this information for all the spatial nodes. Here the available information is usually characterised by significant fuzziness, and heterogeneity. Monitoring is usually conducted only at several points in the landscape, and vast areas are not supported by any data at all. At the same time, remote sensing information that can cover considerable areas and could provide good spatial information, are usually available only for a few points in time, and rarely provide any understanding of the dynamics of processes. It is almost a given that the GIS-based models will be under-parameterised with much uncertainty about the spatiotemporal dynamics. Another limitation of such comprehensive ecosystem models is their high computational requirements, making it very difficult to calibrate and analyse them (Specka et al. 2015).

An agricultural used region is characterised by many processes that are related to different temporal and spatial scales. For example, biotic processes described by crop growth or habitat models usually are very local. Abiotic processes such as material flow driven by wind or water (erosion), on the other hand, are strongly influenced by spatial factors within water catchments such as the heterogeneity of land use types or altitude (Siegmund et al. 2018). On the political side, it must be noted that most people' concerns are quite local and focus on the local impacts of different policies, but require effective plans and regulations so that regional processes and trends need to be taken into account (Wechsung 2005). An example is energy production by wind or the use of crop biomass for biogas reactors. New land use options can bring new benefits to a whole agricultural used area. However, the associated ecological problems can be very local (increased nutrient pollution, increased water extraction from the groundwater resources, changes in habitats for birds). In any case, it is necessary to investigate the effects of different locally realised land use activities on the environment and the whole region in space and time. Landscapes are complex, spatially and temporally multilayered systems, which change and develop naturally but are also subject to anthropogenic changes (Lausch 2003). In order to analyse, evaluate and manage them, tools and models are needed that can represent and interpret the diversity and complexity of the relationships between biotic and abiotic landscape structures and functions. Recently increasing problems related to climate change and the exploitation of limited fossil resources will intensify problems at regional level and require even more urgent and reliable decision-making and political support (Wenkel et al. 2008). However, in spite of significant progress in the development of mechanistic, integrated agro-ecosystem and ecosystem models in the last decade, integrated spatial and temporal dynamic landscape models have been the exception so far. The development of integrated regional models is a very ambitious task and requires its 'own inter-methodical approach' (Lausch 2003). However, this is still a challenge as the variety and diversity of knowledge from various scientific disciplines is increasing, requiring ever more creativity, versatility and openness in research and communication (Wenkel et al. 2008).

2.2.2 The Future Direction of Development

In the coming years, the development of integrated dynamic models and their use for sustainable landscape management is one of the most important challenges for landscape ecology and landscape science. Notwithstanding the progress achieved in agro-ecosystem and ecosystem modelling, there are only a few models that can be used for sustainable resource management at regional or landscape level. New frameworks and methods are necessary in landscape and regional environmental modelling. From the scientific point of view we are confronted with the following problems (Wenkel et al. 2008):

- How do we reduce the complexity of landscape processes within models? How much detail in process dynamics can be integrated in regional landscape models for capturing the most important feedback loops?
- How can we adequately capture the functional consequences of structural diversity and spatial heterogeneity in landscape models?
- What are the suitable compromises between short-term and long-term development processes in landscape models?
- Which model types and tools are best suited to landscape modelling, considering the fuzziness and uncertainties in ecological functions and data?
- How it is possible to bring together conceptually different ecological and economic models?
- Which spatial resolution is required to appropriately capture processes with regional significance?

For finding answers to these questions, we need new ideas, new thinking and a paradigm shift in integrated regional landscape modelling. We propose the following steps as crucial ones (order not essential) (Wenkel et al. 2008):

- Establishment of a set of landscape indicators that can be used to assess the sustainability of land use systems and the dynamics of landscape development for whole regions.
- Development of a new generation of integrated regional models of intermediate complexity.
- Creation of hybrid models by combining simplified or aggregated process-oriented dynamic models with statistical, fuzzy or other model types.
- Development and use of new software toolboxes for the model development and for model-based scenario simulations at landscape scale, such as SAMT (Wieland et al. 2006; Mirschel et al. 2006; Ajibefun et al. 2004) or LandCaRe-DSS (Wenkel et al. 2013; Mirschel et al. 2019).
- Development of a new generation of tools for multi-criteria evaluation and visualisation at landscape scale.
- Improvement of regional model validation using long-term monitoring data from landscape studies around the world, such as ILTER (International Long-Term Ecological Research Network) data sets (ILTER 2006) or TERENO (TERrestrial ENvironmental Observatories) data sets (TERENO 2019).
- Promotion of open source in program development and data as basis for a free flow of ideas, models, landscape-related data sets and communication.

2.3 Landscape–Indicators, Models and Data

2.3.1 Landscape Indicators

Since landscapes are highly complex systems, one possible approach is to deviate from process-based modelling and instead focus on indicator-based modelling. Because of the system complexity, there are a large number of landscape indicators which describe the landscape in an aggregated way. They come from different disciplines such natural science, ecology, geology, socio-economics, landscape structure or soil science. Examples of such indicators are plant productivity, Normalised Difference Vegetation Index (NDVI 2018), humus content in soil, evapotranspiration, percolation, soil fertility, nitrate leaching, erosion, trace gas emissions, degree of biodiversity, farm structure, labour potential, degree of landscape fragmentation. Many other indices can be developed, but the real challenge is to select the most representative ones and learn to quantify and measure them. We must keep in mind that these indices represent processes that take place in space and time and can therefore be modelled in time and space. It is plausible that not a single set of landscape indicators can be universal to address all problems. As with other modelling efforts also it is a problem-specific process that is driven by the goals. The selection of a problem- and scale-oriented indicator system cannot be free of subjective preferences. To describe a regional landscape by indicators derived from the main regional processes and functions, the indicator selection should be undertaken in participative discussions with stakeholders and scientists. In this process, the chosen indicators and their limitations should be critically evaluated and determined. The following challenges are, on the one hand, the interpretation of landscape indicators and model-based simulation results and, on the other hand, the implementation of these interpretation results in practical measures to influence political decisions (Wenkel et al. 2008).

2.3.2 Regional Models

As in modelling in general, the question is how detailed the process dynamics must be. Another question to be answered is what spatial resolution is required to obtain sufficiently precise answers concerning the social and ecological impacts of land use changes at the regional level, as a basis for decision-making in landscape management. Neither simple models nor three-dimensional comprehensive models seem to be the right solution for all problems. Regional models of intermediate complexity (REMICs) could be an alternative. REMICs would appear to be the type of models that would address most management problems at the regional landscape scale. REMICs describe most processes in a simplified, reduced form taking into account the most important driving forces (Wenkel et al. 2008). They simulate the interactions between different components of the landscape system, including the necessary feedback loops (Claussen et al. 2002). However, they are simple enough to allow long-term simulations at the landscape scale. REMICs models are characterised by the fact that they describe the processes with a lower level of detail and have a smaller number of input variables, but may have more variables to parameterize some of these processes. For a broad and robust application at the landscape scale, it makes sense to select and to use such Regional Models of Intermediate Complexity.

It is very unlikely that a single model will be sufficient to address all the requirements of landscape management because, on the one hand, many processes and feedbacks need to be taken into account and, on the other hand, the individual processes are very different not only in their functional focus but also in their temporal and spatial resolution. Otherwise, experimental data and information at landscape level are only sparsely or incompletely available and are also connected with great uncertainties. Depending on the specific spatial question on the one hand and the spatial data availability on the other hand, it is necessary to have an extensive model base, a large module spectrum and different simulation tools from which can be selected to describe different landscape indicators at different complexity levels and for different types of input and output data requirements.

Extensive landscape-scale simulation studies will require hydride models that are an intelligent combination of simplified or aggregated process-oriented, robust, dynamic models with static and/or stochastic simulation models (empirical or statistical models, neural network models, fuzzy models and others). These hybrid models have to organise and manage the data flows between the sub-models via data interfaces for both spatial (maps) and temporal (time series) data types as well as combinations of those (dynamic maps, animations) (Wenkel et al. 2008).

Models, including hybrid models, are also classified according to their complexity. In this paper models of intermediate complexity are discussed. But how do we measure complexity? How can models be evaluated? How can the data be evaluated?

In the next paragraph, this question is discussed using very general but also widely applicable measures. These measures are evaluated according to their degree of abstraction and their applicability to the concrete question. No distinction is made between the complexity of models and the complexity of data. The measures are applicable to the complexity of models as well as data.

2.3.3 *Spatial Data–Entropy, Mixing and Configuration*

In landscape modelling one has to deal with spatial data, often organised as a grid. These grids change over time, so they are dynamic spatial data. This chapter examines which information entropy, mixing and configuration carry, how they are calculated and which information is lost. The entropy goes back to Boltzman:

$$S = k * \ln(W)$$

$$\text{with}\, k = 1.38064852 \times 10^{-23}\text{J/K} \,\text{and}\, W = \frac{N!}{\prod_i N_i!}$$

To approach these terms, a simple example of a stylized landscape with different land use types is introduced (see Fig. 2.1). It consists of a grid with arable land (yellow), grassland (green), forest (dark green), urban areas (red) and water bodies (blue).

In this first example we have N = 16, N1 = 5, N2 = 4, N3 = 3, N4 = 2, N5 = 2

$$\text{W} = 16!/(5! * 4! * 3! * 2! * 2!) = 302,702,400$$

Fig. 2.1 Stylized landscape with five different land use types (example 1) highlighted by different coloured grids

$$S = 2.696 * 10^{-22}\,J/K$$

To make this more handy we will use the S/k = ln(W) as entropy in this paper:

$$S/k = 19.528$$

The Boltzmann Entropy measures the disorder of a system. If the system consists of only one component, then S/k = 0 and the system is in a state of complete order. Another land use example will be introduced as a comparison (Fig. 2.2).

In example 2 there is: N = 16 but N1 = 8, N2 = 4, N3 = 2, N4 = 1, N5 = 1 which leads to the Entropy S/k = ln(16!/(8! * 4! * 2! * 1! * 1!) = 16.196. Example 2 has less disorder than example one. The entropy gives us information about the disorder of a system but it tell us nothing about the number of objects N and the number of the parts N1, N2, N3, …

The mixing theory (Seitz and Kirwan 2014) addresses this by providing a vector containing N1, N2, N3, … ordered by the values. To underline this let´s start with a picture for example 1 (Fig. 2.3).

This picture can be coded in a mixing-vector $x1 = [5, 4, 3, 2, 2]$. The second example looks more ordered compared to example 1 (Fig. 2.4).

The mixing-vector for example 2 is $x2 = [8, 4, 2, 1, 1]$. The mixing theory provides more information in form of a vector but the possibility for comparability is lost.

The mixing theory gives us more information in the form of a vector, but at the same time we lose the ability to compare two grids by a number, as was possible with entropy. The majorization can be used to compare two mixing-vectors:

$$x2 > x1 \leftrightarrow \sum_{i}^{m} x2_i \geq \sum_{i}^{m} x1_i \forall_m \in [1, N']$$

The majority will not always build a strict order, because some mix-vectors are not comparable. There are so-called incomparable objects which lead to a partial order

Fig. 2.2 Stylized landscape with five different land use types (example 2) highlighted by different coloured grids

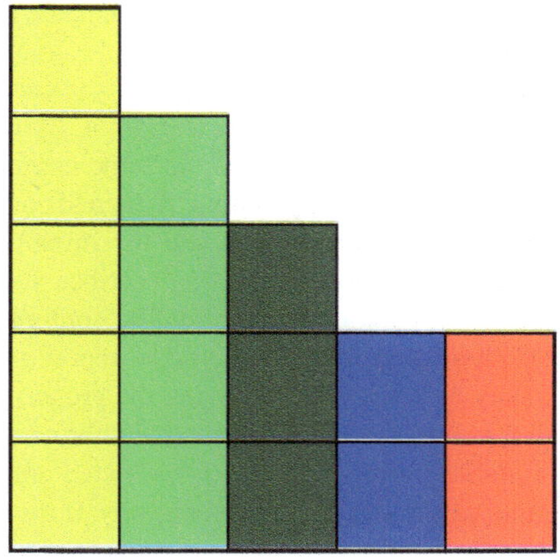

Fig. 2.3 Land use type ordering from land use example 1 (mixing example 1)

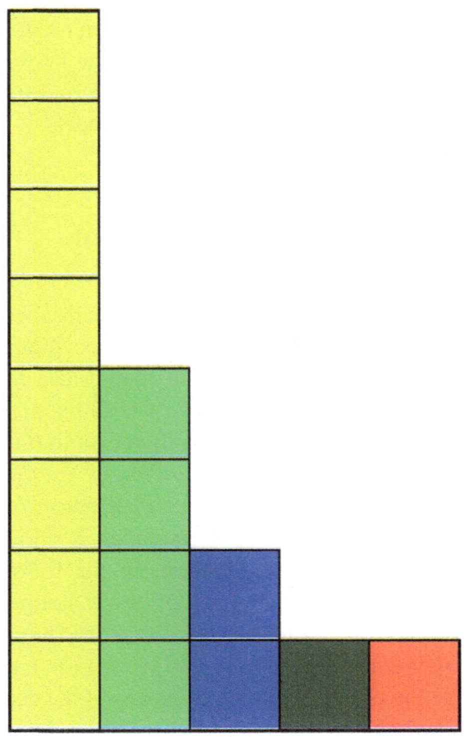

Fig. 2.4 Land use type ordering from land use example 2 (mixing example 2)

set. For the example 1 and example 2 the inequality $x1 < x2$ is fulfilled. Therefore, a strict order between example 1 and example 2 is given. However, it can also happen that an example ($x3$) consists of only two components with [5, 5] units and an example ($x4$) of three components with [4, 4, 2] units. Although both examples (example 1 and example 2) consist of the same number of units, they cannot be compared due to the different components. A typical case of incomparability is given.

But both, the entropy and the mixing are abstractions of the configuration of the grids. The configuration describes not only the number of objects in a grid but also their position in the grid. The configurations are therefore the grids themselves. Modern deep learning methods import grids as images and break them down into many partial images, deriving the properties of a grid as a whole. This can be used, for example, to extract suitable subimages from a remote sensing image that are suitable, such as habitats for selected animals. A change of the configuration over time can lead to a changed entropy. If the entropy becomes smaller, the order in the system is increased, which is connected with an expenditure of energy. If the entropy increases, the disorder increases too. Systems tend to increase disorder when no external energy is used to maintain order. A change in the mixing-vectors indicates a change in the components of the system. Mixing-vectors describe the state of a system by its components. In modern deep learning the configuration that means the pictures itself are the inputs of models.

2.4 Spatial Modelling Tool

Simulation systems, such as ModelMaker (https://modelmaker.co.za) or MATLAB (https://de.mathworks.com/products/matlab.html) can handle simulations quite well, but have insufficiencies in spatial data handling. GIS platforms like ArcGIS, can handle with spatial data very well but are not well suited for dynamic simulations. What is needed is an open-source toolbox that can combine both, perform simulations and handle geodata sets.

The Python ecosystem (PES, https://www.anaconda.com/distribution) for simulation and modelling includes a wide range of modern and powerful modules for spatial simulation. In the following only some key modules are mentioned, such as *numpy* (http://www.numpy.org) a basic module for multidimensional numerical calculations, *scikit learn* (https://scikit-learn.org) for machine learning, *matplotlib* (https://matplotlib.org/) for visualisation. Python enables the implementation of new modules by means of the compiler *Cython* (https://cython.org), i.e. the modules run almost at the speed of C programs. Using the PES, the Spatial Analysis and Modelling Tool SAMT2 (Wieland et al. 2015) was developed as a Python module. SAMT2 includes the import of raster data from GIS and their very efficient storage using the HDF5 format (HDF5 2019). SAMT2 allows the development of fuzzy models and their effective application to spatial data. A variety of important spatial methods, such as statistics (including entropy and majorization), normalisation, scaling of data, mathematical operations on grids, etc., can be used. It is important that all these operations respect the ‘nodata’ values common in GIS. If certain operations are not included, the data can be exported to *numpy* arrays and all the operations of the PES are applicable. The user easily can switch between the SAMT2 data and the Python data.

In order to be able to use SAMT2 with a graphical user interface, the implementation in Ipython (https://ipython.org/) notebooks is recommended. Here models can be implemented together with their documentation. Ipython notebook allows the step-by-step processing of the model, the visualisation of the results and the outsourcing of the calculation to powerful computers such as computer clusters. SAMT2 combines interactive modelling with model testing, model optimization and control of parallel processes on a computer cluster. It should not go unmentioned that modern deep learning algorithms are available in the PES, one must even say that Python is the language of choice for machine learning. In this context, SAMT2 can serve as a data provider for machine learning and can apply the resulting models to spatial data.

2.5 Conclusions and Outlook

The development of dynamic landscape simulation models to support sustainable resource management and the political decision-making for the regional scale represents a major challenge for landscape research and modelling in the coming decades. We recommend the development of new models of intermediate complexity that describe landscape indicators and landscape variables and thus enable modularity and flexibility. This provides the basis for recombining these modules in different ways, depending on the type and objective of the simulation studies to be carried out.

SAMT2 offers very promising possibilities as a new developed effective and variable toolbox that can support the new type of modelling very well. There is little hope that the major problems leading to changes in ecosystems, to losses of habitats or to biodiversity erosions can only be addressed through a mosaic of subject-specific models. Addressing these questions on the large scale will require the development of ecological models aimed at achieving generality at the expense of detail. This does not reduce the usefulness of site- and subject-specific models when data are available for them. The search for the optimal compromise both in complexity and in the empirical vs. mechanistic components of models must be based on a few criteria and is an important milestone for the successful development of models on the landscape scale. Indeed complex models should be rigorously evaluated to be made as ‘minimalistically complex’ as possible. This criterion would make it possible to focus more on a more specific or relevant mechanistic understanding of a particular process and would provide politicians and managers with a useful tool for justifying appropriately validated changes.

In the field of integrated landscape modelling there are still many substantial tasks to be solved (Wenkel et al. 2008, adapted):

- **Scales**. At the landscape level, the temporal scales of ecological systems and models are very variable. Because most environmental management decisions are considered over years to decades, landscape models for assessing the impacts of land use management for sustainability should run over that time period in first line. In any case, the time scale of the models needs to relate to the time scale of the management questions and their implications.
- **Validation**. A network of long-term experimental study areas or, if feasible, specially designed landscape experiments are the prerequisite for scientifically rigorous validation of complex integrated landscape simulation models. Back-casting to historical conditions and comparing model results offer a useful way to validate a model. The core research areas within the International Long-Term Ecological Research Network (ILTER) (ILTER 2006) could be the basis for the needed data sets.
- **Stakeholder**. The development of landscape simulation models is an integrative, interactive and iterative process. By involving stakeholders in the process of landscape modelling, it is possible to include their knowledge and thus improve the understanding of landscape dynamics as well as ensure that the model results are better accepted by decision-makers.
- **Software tools**. The integration of models into decision-making requires user-friendly and transparent tools for landscape management. These tools have to include analysis tools, have to include graphical user interfaces and have to guarantee a user-friendly visualisation of the simulation results. In order to use the possibilities of the modelling community, both the integrated landscape models and the simulation tools have to be developed as open-source software.
- **Uncertainties**. Model results are always uncertain because they are based on an incomplete understanding of processes and interactions in the landscape, on field and laboratory studies that are always approximate (Dale 2003), and on an incomplete availability of all necessary spatial input data. Model results are rather estimates of future possibilities than predictions, i.e. what is explained in advance (Dale and van Winkle 1998). Models are approximations to real situations and are only as good as the assumptions on which they are based. The results of the model simulations should therefore be considered with caution. Decisions on landscape management issues should always be made on the basis of the best available information. However, the absence of complete information does not imply that the development of models has no scientific value.

References

Ajibefun IA, Wenkel K-O, Wieland R, Mirschel W (2004) Modelling of landscape and land use changes: with application of spatial analysis modelling tool to Quillow region of North East Germany. J Food Agric Environ 2(2):364–368

Antle JM, Basso B, Conant RT, Charles H, Godfray J, Jones JW, Herrero M, Howitt RE, Keating BA, Munos-Carpena R, Rosenzweig C (2017) Towards a new generation of agricultural system data, models and knowledge products: design and improvement. Agric Syst 155(2017):255–268

Benz J, Hoch R, Legovic T (2001) ECOBAS—modelling and documentation. Ecol Model 138(1–3):3–15

Brisson N, Gary C, Justes A, Roche R, Mary B, Ripoche D, Zimmer D, Sierra J, Bertuzzi P, Burger P, Bussiére F, Cabidoche YM, Cellier P, Debaeke P, Gaudillér JP, Hénault C, Maraux F, Seguin B, Sinoquet H (2003) An overview of the crop model STICS. Eur J Agron 18:309–332

Claussen M, Mysak LA, Weaver AJ, Crucifix M, Fichefet T, Loutre M-F, Weber SL, Alcamo J, Alexeev VA, Berger A, Calov R, Ganopolski A, Goosse H, Lohmann G, Lunkeit F, Mokhov II, Petoukhov V, Stone P, Wang Z (2002) Earth system models of intermediate complexity: closing the gap in the spectrum of climate system models. Clim Dyn 18:579–586

Dale VH (2003) Opportunities for using ecological models for resource management. In: Dale VH (ed) Ecological modeling for resource management. Springer, New York, pp 3–19

Dale VH, van Winkle W (1998) Models provide understanding, not belief. Bull Ecol Soc Am 79:169–170

ECOBAS (2011) WWW server for ecological modelling. http://ecobas.org/www-server. Accessed 07 Mar 2019

Hansen S, Jensen HE, Nielsen NE, Svendsen H (1990) DAISY: soil plant atmosphere system model. The national agency for environmental protection, Copenhagen, NPO Report No. A10, 272 pp

HDF5 (2019) Introduction to HDF5. https://support.hdfgroup.org/HDF5/doc/H5.intro.html. Accessed 11 Mar 2019

Hurni H (2000) Assessing sustainable land management. Agric Ecosyst Environ 81(2):83–92

ILTER (2006) International long term ecological research network (ILTER). https://www.ilter.network. Accessed 07 Mar 2019

Jones JW, Antle JM, Basso B et al (2017) Toward a new generation of agricultural system data, models, and knowledge products: state of agricultural systems science. Agric Syst 155:269–288

Keating BA, Carberry PS, Hammer GL, Probert ME, Robertson MJ, Holzworth D, Huth NI, Hargreaves JNG, Meinke H, Hochmann Z, McLean G, Vergurg K, Snow V, Dimes JP, Silburn M, Wang E, Brown S, Bristow KL, Asseng S, Chapman S, McCown RL, Freebairn DM, Smith CJ (2003) An overview of APSIM, a model designed for farming systems simulations. Eur J Agron 18:267–288

Kersebaum KC (1995) Application of a simple management model to simulate water and nitrogen dynamics. Ecol Model 81:145–156

Lausch A (2003) Integration of spatio-temporal landscape analysis in model approaches. In: Helming K, Wiggering H (eds) Sustainable development of multifunctional landscapes. SpringerVerlag, Berlin, Heidelberg, pp 111–123

Mirschel W, Berg-Mohnicke M, Wenkel K-O, Wieland R, Köstner B (2019) LandCaRe-DSS—an interactive model-based decision support system for assessing the impacts of climate change on agriculture and agricultural landscapes (in this book)

Mirschel W, Schultz A, Wenkel K-O (2001) Assessing the impact of land use intensity and climate change on ontogenesis, biomass production, and yield of northeast German agro-landscapes. In: Tenhunen JD, Lenz R, Hantschel R (eds) Ecosystem approaches to landscape management in central Europe. Ecological studies, vol 147. Springer-Verlag, Berlin, Heidelberg, New York, pp 299–313

Mirschel W, Wieland R, Voss M, Ajibefun IA, Deumlich D (2006) Spatial analysis and modeling tool (SAMT), 2. Applications. Ecol Inform 1:77–85

NDVI (2018) What is NDVI (normalized difference vegetation index). https://gisgeography.com/ndvi-normalized-difference-vegetation-index. Accessed 4 Mar 2019

Nendel C, Berg M, Kersebaum KC, Mirschel W, Specka X, Wegehenkel M, Wenkel KO, Wieland R (2011) The MONICA model: testing predictability for crop growth, soil moisture and nitrogen dynamics. Ecol Model 222(9):1614–1625

Nendel C, Kersebaum KC, Mirschel W, Wenkel K-O (2014) Testing farm management options as a climate change adaptation strategy using the MONICA model. Eur J Agron 52(2014):47–56

Poluektov RA, Fintushal SM, Oparina IV, Shatskikh DV, Terleev VV, Zakharova ET (2002) AGROTOOL—a system for crop simulation. Arch Agron Soil Sci 48:609–635

Ritchie JT, Godwin DC (1993) Simulation of nitrogen dynamics in the soil plant system with the CERES models. Agrarinformatik 24:215–230

Seitz W, Kirwan AD (2014) Entropy vs. majorization: what determines complexity? Entropy 16:3793–3807

Seppelt R (2003) Computer-based environmental management. Wiley-VCH Verlag, Weinheim, p 281

Siegmund N, Funk R, Koszinski S, Buschiazzo D, Sommer M (2018) Effects of low-scale landscape structures on aeolian transport processes on arable land. Aeol Res 32:181–191

Specka X, Nendel C, Wieland R (2015) Analysing the parameter sensitivity of the agro-ecosystem model MONICA for different crops. Eur J Agron 71:73–87

Supit I, Hooijer AA, van Diepen CA (eds) (1994) EUR 15956—system description of the WOFOST 6.0 crop simulation model implemented in CGMS European Commission. Office for Official Publications of the European Communities, Luxembourg. Agricultural Series, Catalogue Number: CL-NA-15956-EN-C, Theory and Algorithms, 1, 146 pp

TERENO (2019) TERENO—terestrial environmental observatories. http://www.tereno.net/overview-de. Accessed 05 Mar 2019

Wechsung F (2005) Herausforderungen des globalen Wandels für die Elbe-Region. In: Wechsung F, Becker A, Gräfe P (eds) Auswirkungen des globalen Wandels auf Wasser, Umwelt und Gesellschaft im Elbegebiet. Weißensee Verlag Berlin, Kap I-1

Wegehenkel M, Mirschel W, Wenkel K-O (2004) Predictions of soil water and crop growth dynamics using the agro-ecosystem models THESEUS and OPUS. J Plant Nutr Soil Sci 167(6):736–744

Wenkel K-O, Berg M, Mirschel W, Wieland R, Nendel C, Köstner B (2013) LandCaRe DSS—an interactive decision support system for climate change impact assessment and the analysis of potential agricultural land use adaptation strategies. J Environ Manage 127(Supplement):S168–S183

Wenkel K-O, Mirschel W (1995) Agroökosystemmodellierung—Grundlage für die Abschätzung von Auswirkungen möglicher Landnutzungs und Klimaänderungen. ZALF-Bericht 24, Müncheberg, 187 pp

Wenkel K-O, Schultz A, Lutze G (1997) Landschaftsmodellierung—Anspruch und Realität. Arch für Nat-Lands 36:61–85

Wenkel K-O, Wieland R, Mirschel W, Schultz A, Kampichler C, Kirilenko A, Voinov A (2008) Regional models of intermediate complexity (REMICs)—a new direction in integrated landscape modelling. In: Jakeman AJ, Voinov AA, Rizzolia AE, Chen SH (eds) Environmental modelling, software and decision support—state of the art and new perspectives—, developments in integrated environmental assessment, vol 3. Elsevier, Amsterdam, pp 285–295

Wieland R, Groth K, Linde F, Mirschel W (2015) Spatial analysis and modeling tool version 2 (SAMT2), a spatial modeling tool kit written in Python. Ecol Inform 30(2015):1–5

Wieland R, Voss M, Holtmann X, Mirschel W, Ajibefun IA (2006) Spatial analysis and modeling tool (SAMT): 1. Structure and possibilities. Ecol Inform 1:67–76

Chapter 3
Application of Ecosystem Modelling Methodology on Rural Areas of Crimea—Systematic Approach

Ielizaveta Dunaeva, Valentina V. Popovich, Vladimir S. Pashtetsky, Vitaly V. Terleev and Aleksandr O. Nikonorov

Abstract The possibility of adapting agricultural ecosystems and modelling their condition under changing economic conditions is determined by agricultural status indicators under current conditions compared to retrospective indicators characterizing the socio-economic and ecological status of rural areas and the agricultural ecosystem as a whole. The analysis of their development involves a systematic approach, the first phase of which is the development of a verbal model taking into account statistical data on the economic, social and environmental situation of agricultural areas. For this purpose, three districts of the Republic of Crimea (Krasnogvardeysky, Dzhankoysky and Saksky) were selected as pilot regions. The indicators were calculated to identify strong and weak development trends and the ability to adapt the regions to the conditions of changes in economic activity. It can be seen that at this stage the Krasnogvardeysky region occupies the best position (integral coefficient: 0.3), where the economic and social coefficients are highest; the position of the Dzhankoysky and Saksky regions is approximately the same (integral coefficients: 0.19 and 0.18, respectively). The development of a methodology for ecosystem modelling initially involves the assessment of individual elements. First simulations of winter crops were carried out, thereafter an agro-ecological evaluation of soils based on soil parameter modelling. Further development steps of the methodology are (1) the analysis of the status of all regions of the Republic of Crimea (RC) and (2) the identification of optimization and sustainability regions using logical-mathematical models and fuzzy cognitive maps (FCM). In particular, they will address the questions of how rational solutions can be found for their development and how emerging problems can be eliminated.

I. Dunaeva · V. V. Popovich · V. S. Pashtetsky
Federal State Budget Scientific Institution "Research Institute of Agriculture of Crimea", Kievskaya, 150, Simferopol 295453, Crimea
e-mail: dunaeva_e@niishk.ru

V. V. Terleev · A. O. Nikonorov (✉)
Peter the Great St. Petersburg Polytechnic University, Polytechnicheskaya, 29, St. Petersburg 195251, Russia

W. Mirschel et al. (eds.), *Landscape Modelling and Decision Support*, Innovations in Landscape Research,
https://doi.org/10.1007/978-3-030-37421-1_3

Keywords Agricultural territories · Ecosystem · Development indicators · Integral coefficient · Modelling

3.1 Introduction

Sustainable development of rural areas, which comprehends as the stable elaboration of the rural community, ensuring an adequate level of agricultural production, growth, and improvement of the quality of life of agricultural producers, maintaining the ecological balance in the biosphere, is an essential component of the development of the entire agro-industrial complex of Crimea. Analysis of indicators characterizing the level of economic, social and ecological condition of the territories makes it possible to assess the level of production and agro-ecological potential, as well as to model development options, ways to adapt them to the new development conditions (Vasantha Kandasamy and Smarandache 2003).

Agricultural areas are the complex socio-economic and ecological system. The method of system analysis is used in building models of their development (Bertalanffy 1969). Such scientist as Averyanov (Averyanov 1985) was engaged on the issues of system analysis, as well as the other scientists (Spitsnadel 2000; Surmin 2003; Chernikov et al. 2004; Sovetov and Yakovlev 2007) and nowadays further development of these issues is also continued by the others. A systematic approach to solving the problems of agricultural production assumes a comprehensive study of the processes occurring in agroecosystems. Logical and mathematical modelling is one of the main tools of system analysis, the end result of which is a dynamic model of possible development scenarios. It includes four stages: the creation of a verbal model; the mathematical implementation of the model logical structure; the model verification; the study of possibilities for optimizing the structure and controlling the behaviour of the modelled system and creation of forecasts (Sovetov and Yakovlev 2007; Nikonorov et al. 2016a, b). Based on the results of creating models of the dynamics of agricultural systems, issues of optimizing the structure of land use, rational use of fertilizers and plant protection products, and land reclamation work were solved; developments are performed for landscape optimization of agricultural land use depending on local conditions (topography, climate, soil conditions, distribution of economic objects) (Terleev et al. 2016). In addition, the modelling of development options for territories can significantly improve their ability to adapt to external and internal risks associated with climate change, droughts, etc. (Dunaeva et al. 2017).

The development of the verbal model, consisting in the description and analysis of agricultural lands state in the form of text, tables, illustrations, and diagrams, was made for three municipal districts of the Republic of Crimea as an initial step for further research on the ecosystem state of agricultural lands in all regions of RC (see Fig. 3.1).

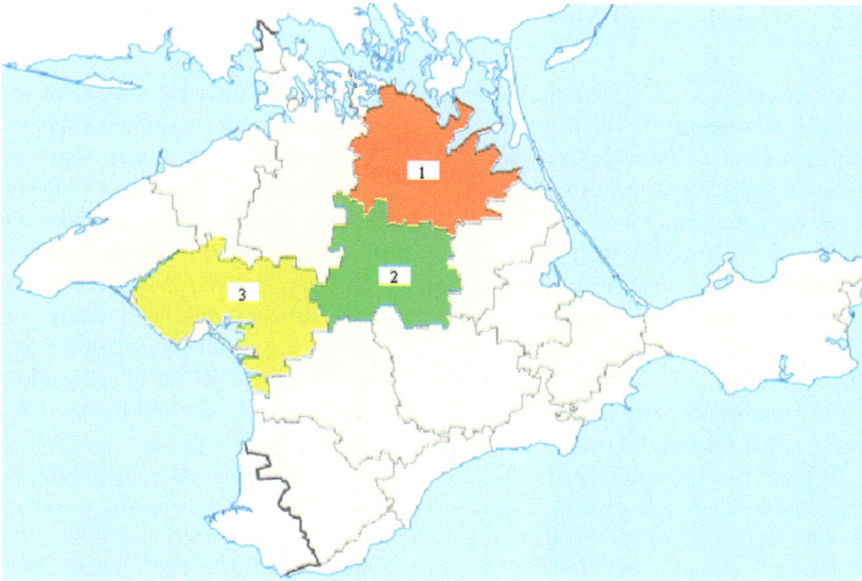

Fig. 3.1 The map of the Republic of Crimea (1—Dzankoysky region, 2—Krasnogvardeysky region, 3—Saksky region)

One of the most important areas of sustainable agricultural development is land use, which has an impact on all fields of activity: economic—production; social—the standard of living for agricultural workers; environmental—the state of topsoil, fertility. The study of the current state of the fields and cultivated crops, as well as modelling their future state are carried out using modern GIS technologies and Earth remote sensing (Dunaeva et al. 2018a, b). With their application, the approbation of individual elements of ecosystem modelling methodology focused on sustainable development was carried out. Using regional agro hydrological characteristics, the simulation of winter crops and estimation of possible variation of the results, caused by the absence or range of the initial values of the parameters (taking into account the characteristics of the data source—field measurements, remote sensing data, etc.) were made. One of the elements of the methodology is agro-ecological land assessment, which includes: landscape-ecological analysis of the territory, agro-ecological soil assessment, agro-ecological typification and land classification, agro-geographic information systems on agro-ecological land assessment (Kiryushin 2005).

One of the approaches to modelling the behaviour and performance of complex systems, such as agricultural lands ecosystem, is to make Fuzzy Cognitive Maps (FCMs) (Leon et al. 2010; Carvalho 2012). FCMs are fuzzy weighted oriented feedback graphs that create models to simulate the behaviour of complex decision-making processes using fuzzy cause-effect relationships.

3.2 Materials and Methods

The current state of agricultural land is characterized by integral indices of sustainable development. The method of its calculation takes into account comparable indicators and coefficients, expressing the links between the state of ecosystems and the anthropogenic impact on them. Comparable indicators in natural or monetary forms are calculated for three regions and take into account the load on one person or per hectare of area. For their calculation, statistical data for 2017 was used (Official statistics of the Department of Federal State Statistics Service for the Republic of Crimea and the city of Sevastopol for 2017). Conversion of indices into relative values can be done by several methods. So, when comparing a small number of objects of observation, for example, the three municipal regions of RC, it is advisable to use the method of pairwise comparisons (Saaty algorithm) (Saaty and Kearns 1991; Saaty 1993; Andreychikov and Andrejchikova 2004).

If there is a large number of data for comparison, a rating assessment with the calculation of the aggregated (integral) index (Bobylev 2001) using the worst and the best indicators to convert them into relative values can be used.

Using the Saaty method (Tikhomirova and Sidorenko 2012; Gorkovenko 2015) suggests a scale of pairwise comparisons from 1 to 9, which makes it possible to identify not only the best-worst-average indicator, but also to take into account what the difference between them is, how much more or less are indicators from each other (how much this or that direction of the indicator differs from those included in the sample). Using the scale of comparisons, the matrix of pairwise comparisons is filled with dimension (m*n), where m is the number of rows, n is the number of columns (Akhmedkhanova et al. 2015). In our case the number of rows and columns is equal (A1, A2, A3); the matrix looks like this:

	A1	A2	A3
A1	a11	a12	a13
A2	a21	a22	a23
A3	a31	a32	a33

After compositing the matrix for each indicator and making calculations on them, the aggregate indicators for each group of indicators and the total integral index for each pilot region using a rating assessment (Gazizov 2014) was determined. The rating estimation is the most effective for a large number of compared indicators since it includes in the calculation the coefficients 0 and 1, which significantly affect the final result with a small number of indicators.

For each indicator, a coefficient is calculated. It allows converting dimensional values to dimensionless ones: the difference between the actual and worst values of the indicator is divided by the difference between the best and worst values of the indicator.

$$Yi = \frac{Bi - Bk}{Ba - Bk} \tag{3.1}$$

where Yi—rating assessment of the indicator; Bi—actual value of the indicator; Bk—the worst value of the indicator; Ba—the best value of the indicator.

Aggregated indicators (of a specific sector of the economy, social sphere, or ecology) are calculated as the geometric mean of rating scores of characterizing indicators

$$I = \sqrt[n]{V1 \times V2 \times V3 \times V4 \times \ldots \times Vn} \tag{3.2}$$

where I—aggregated indicator of a specific direction; V—rating assessment of the indicators; n—number of indicators.

The integral index of sustainable development is calculated as a geometric average of indicators of social, economic and environmental development according to the following dependency:

$$D = \sqrt[3]{Econ \times Soc \times Ecol} \tag{3.3}$$

where D—aggregated indicator of sustainable development; Econ—indicator of economic development; Soc—indicator of social development; Ecol—indicator of environmental development.

The calculated aggregated indicators for each region show the degree of their development and the possibility of comparison within each direction, and the integral index gives a general assessment of the level of development for comparing the regions among themselves.

An important tool for predicting ways to solve complex social, economic and environmental problems are Fuzzy cognitive maps. FCMs in the form of an acyclic oriented graph, including causal relationships between concepts (indicators) related to this problem (concepts has a positive and negative impact and expressed quantitatively or qualitatively) are the model, that reflects all areas of the problem, as well as their internal relations. After building, FCM provides insights into how the problem changes due to interactions of indicators.

3.3 Results and Discussion

Socio-economic conditions of agricultural production are the most important factor having a significant impact on agro-ecological systems. Therefore, the analysis and development of indicators of this factor for the territory of the Crimea, where there is an imbalance in land use, deterioration of ecological landscapes, reduction of their biodiversity, are a necessary condition for solving social and agro-ecological problems.

Assessment of the territories state is carried out using an integral indicator, which is calculated from 3 types of indicators (economic, social, environmental), including a number of basic indicators characterizing their condition, as a geometric average of indicators of areas of analysis.

The verbal model of socio-economic and environmental development of pilot territories (comparable indices for each area of analysis) is presented in the form of Table 3.1.

The cognitive map of the main factors of sustainable development of the territory is presented in Fig. 3.2, compiled in the software product Vensim PLE 7.2a (https://appdb.winehq.org/objectMa-nager.php?sClass=version&iId=37298). Further processing of the cognitive map was carried out in the software FCMapper vs. 1.0 (http://www.FCMappers.net). This approach allows modelling the development scenarios of territory as the situation changes.

The mathematical formalization of the analysed system, its subsystems and elements, including three areas of analysis are presented in the form of Tables 3.2 and 3.3.

In the Table 3.2, the indicator values are calculated using a rating method. The data show that this method is the best for comparing a large number of indicators, a small number does not give a correct assessment. In the Table 3.3 the calculation of indicators within each direction of development is carried out by the method of paired comparisons; the integral index is calculated as a geometric average.

The calculations show the level of development of each region as a whole and the level of development of individual spheres of society, industries, and areas (social, economic, and environmental). Calculations, carried out by comparing pair matrices, showed that the best situation among three areas was in the Krasnogvardeysky district, and in the other two—approximately the same, which corresponds to real data. Analysis and calculation of indices for all regions of the Republic of Crimea can be made using both these methods. It allows conducting a deeper analysis of the sustainability of areas, to identify less developed areas.

3.4 Conclusions

Sustainable development of rural areas, which comprehends as the stable development of the rural community, ensuring an adequate level of agricultural production, growth, and improvement of the quality of life of agricultural producers, maintaining the ecological balance in the biosphere, is an essential component of the development of the entire agro-industrial complex of Crimea. Analysis of indicators characterizing the level of economic, social and ecological condition of the territories makes it possible to assess the level of production and agro-ecological potential, as well as to model development options, ways to adapt them to external and internal challenges, including, for example, problems associated with climate change or significant reduced water availability.

The calculations of indicators for identifying strong and weak development trends and the ability to adapt to the conditions of changes in economic activity showed that

Table 3.1 Sustainable development Indicators for pilot areas (2017 data)

Indicators	Variable	District		
		Dzankoysky	Krasnogvardeysky	Saksky
1	2	3	4	5
Economic indicators				
Shipped goods of own production, thousand rubles/person	A1	9645	95881	11231
Total amount of all food products for 1 person, thousand rubles	A2	14.5	16.65	10.9
Profit (loss) before tax for the reporting year, total, rubles/person	A3	658.2	2838.5	1197.5
Share of profitable organizations, (%)	A4	71	80	100
Arable lands, ha per 1000 inhabitants	A5	1124	1244	1095
Total grain yield, thousand t/ha	A6	25.7	29.7	25.9
Milk production, t/1 cow	A7	3.4	3.9	3.8
Local budget revenues per 1 inhabitant, thousand rubles	A8	3.88	5.25	10.05
The average monthly salary in agriculture, $	A9	252.5	350	364.1
The volume of investment in fixed assets (excluding budget funds) for 1 person, $	A10	11.5	159.3	64
Social indicators				
Fertility rate, persons per 1000 inhabitants	A11	8.9	10	9.1
Working age population, % of total	A12	53.5	54.5	58.5
The proportion of pensioners in the total number of population, %	A13	32.9	30	28.4
Population density, persons /sq. km	A14	25.1	41	33.7
The average area of dwelling (average per 1 inhabitant), m^2	A15	19.5	21.5	23.1
Lead-in housing per year, per 1 inhabitant, m^2	A16	0.042	0.1	0.5

(continued)

Table 3.1 (continued)

Indicators	Variable	District		
		Dzankoysky	Krasnogvardeysky	Saksky
Coverage of children by preschool educational institutions (% of the number of children of the appropriate age)	A17	52.1	48.7	41.5
Provision with doctors per 10 thousand inhabitants	A18	18.6	20.5	26.2
Provision with medium–medicines per 10 thousand inhabitants, units	A19	69	62.9	55.9
Consumer services, % per 10 thousand inhabitants	A20	1.3	11.1	2
Number of sports facilities, units/1000 inhabitants	A21	2.6	2.3	1.9
Ecologic indicators				
Current expenses for environmental protection, rub./ha	A22	1.6	236	164.3
The number of objects with stationary sources of pollution, units	A23	4	17	14
Exported for the year of solid and liquid household waste, m^3/person	A24	0.24	1.38	1.48
Environmental load on the territory (due to contaminant emissions into the atmosphere), kg/ha	A25	0.8	7.7	2.1
The environmental load on the population (due to contaminant emissions to the atmosphere), kg/person	A26	3.2	16.1	6.2

at this stage the Krasnogvardeysky district occupies the best position (the integral coefficient is 0.3). Its coefficients of economic and social areas are the highest. Positions of Dzankoysky and Saksky regions are approximately the same (the integral coefficients are 0.19 and 0.18, respectively).

Further analysis of the state of all regions of the Republic of Crimea and identification of areas of optimization and sustainable development using GIS technologies will allow comparing the state of agricultural territories of all regions of the Crimea, modelling the development of areas that more than others need to be adapted and

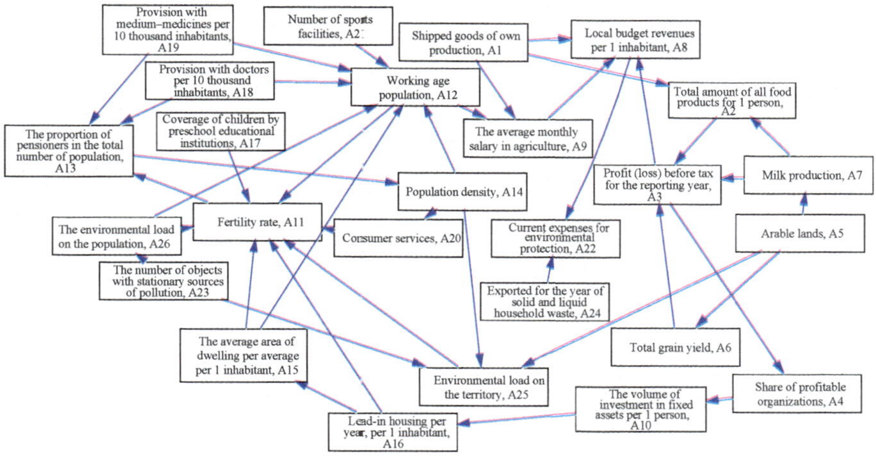

Fig. 3.2 The cognitive map

Table 3.2 Calculation of integral indexes by a rating method for pilot areas (data of 2017)

Indicators	Districts		
	Dzankoysky	Krasnogvardeysky	Saksky
Economic	0.79	0.83	0.48
Social	0.97	0.53	0.63
Ecologic	1.00	0.00	0.48
Integral index	0.91	0.76	0.53

Table 3.3 Calculation of integral indices for pilot areas by the method of comparison of paired matrices (data of 2017)

Indicators	Districts		
	Dzankoysky	Krasnogvardeysky	Saksky
Economic	0.100	0.480	0.240
Social	0.170	0.320	0.250
Ecologic	0.440	0.180	0.110
Integral index	0.189	0.300	0.187

optimized. It makes it possible to take rational decisions according to their development and correct any errors that can occur. The FCMs of rural areas were built to make an agro-ecological land assessment of pilot regions.

Acknowledgements The reported study was funded by RFBR according to the research projects #19-016-00148-a; #19-04-00939-a).

References

Averyanov AN (1985) Systemic knowledge of the world: methodological problems (Системное познание мира: Методологические проблемы) (In Russian). Politizdat, Moscow, p 263

Andreychikov AV, Andrejchikova ON (2004) Analysis, synthesis, planning decisions in the economy (Анализ, синтез, планирование решений в экономике). Finance and Statistics, Moscow, p 464

Akhmedkhanova AI, Kozhemyakina VA, Mamaev II (2015) Application of matrixes in economy (Применение матриц в экономике) (in Russian). Int Stud Sci Bull:3–4. https://docplayer.ru/50827056-Application-of-matrixes-in-economy-akhmedkhanova-a-i-kozhemyakina-v-a-mamaev-i-i-stavropol-state-agrarian-university-stavropol-russia.html. Accessed 02 Dec 2018

Bertalanffy L (1969) The general theory of systems: a critical review (Общая теория систем: критический обзор) (in Russian). Progress, Moscow, p 520

Bobylev SN (2001) Indicators of sustainable development of Russia (environmental and economic aspects) (Индикаторы устойчивого развития России (эколого-экономические аспекты) (in Russian). Moscow, p 220. http://kniga.seluk.ru/k-ekonomika/1008415-1-moskva-2001-indikatori-ustoychivogo-razvitiya-rossii-ekologo-ekonomicheskie-aspekti-pod-redakciey-bobileva-makee.php/. Accessed 02 Dec 2018

Carvalho JP (2012) Rule based fuzzy cognitive maps in humanities, social sciences and economics. In: Soft computing in humanities and social sciences. Vol. 273 of the series studies in fuzziness and soft computing. Springer, Berlin Heidelberg, pp 289–300. https://link.springer.com/chapter/10.1007/978-3-642-24672-2_16. Accessed 10 Dec 2018

Chernikov VA, Gringoff IG, Yemtsev VT (2004) Agroecology. methodology, technology, economics (Агроэкология. Методология, технология, экономика) (in Russian). Kolos, Moscow, p 400

Dunaeva EA, Golovastova ES, Vecherkov VV, Popovich VV (2018) Joint use of unmanned aircraft and earth remote sensing to estimate bioproductivity of crops (Совместное использование беспилотных летательных аппаратов и дистанционного зондирования земли для оценки биопродуктивности посевов) (in Russian). In: Regional problems of remote sensing of the earth materials of the V international scientific conference. Siberian Federal University, Institute of Space and Information Technologies, pp 302–306

Dunaeva Y, Popovych V, Melnichuk A, Mirschel W, Terleev V, Nikonorov A, Togo I, Volkova Yu, Shishov D (2018) SWAT modeling the soil properties in GIS-environment: initial calculations. In: GCCE 2018 conference "Geotechnical Construction of Civil Engineering", Submitted Jul 13

Dunaeva EA, Popovich VV, Terleev VV, Melnichuk AY (2017) Assessment of environmental risks in the water-economic complex (Оценка экологических рисков в водно-экономическом комплексе) (in Russian). Ser. "Week of Science" 2017, pp 256–258

Gazizov RM (2014) Sustainable development of rural areas: a method of assessment and typology (on the example of the Krasnoyarsk Territory) (Устойчивое развитие сельских территорий: метод оценки и типологизации (на примере Красноярского края) (in Russian). Actual problems of economics and law 3(31):34–42

Gorkovenko NE (2015) Mathematical modeling in ecology: a course of lectures (Математическое моделирование в экологии: курс лекций) (in Russian). Krasnodar, p 45

Kiryushin VI (2005) Methodical manual on agro-ecological land assessment, design of adaptive-landscape farming systems and agricultural technologies (Методическое руководство по агроэкологической оценке земель, проектированию адаптивно-ландшафтных систем земледелия и агротехнологий) (In Russian). Kiryushin VI, Ivanov AL (ed), Moscow, p 763

Kosko B (1986) Fuzzy cognitive maps. Int J Man-Mach Stud 24(1):65–75. https://www.researchgate.net/publication/230694282_Fuzzy_cognitive_maps. Accessed Dec 2018

Leon M, Rodriguez C, García M, Bello R, and Vanhoof K (2010) Fuzzy cognitive maps for modeling complex systems, part I, LNCS 6437. Springer, Berlin Heidelberg, pp 166–174. https://link.springer.com/chapter/10.1007/978-3-642-16761-4_15. Accessed 10 Dec 2018

Nikonorov A, Badenko V, Terleev V, Togo I, Volkova Yu, Skvortsova O, Nikonova O, Pavlov S, Mirschel W (2016a) Use of GIS-environment under the analysis of the managerial solutions for flood events protection measures. Proc Eng 165:1731–1740

Nikonorov A, Terleev V, Pavlov S, Togo I, Volkova Yu, Makarova T, Garmanov V, Shishov D, Mirschel W (2016b) Applying the model of soil hydrophysical properties for arrangements of temporary enclosing structures. Proc Eng 165:1741–1747

Official statistics of Department of Federal State Statistics Service for Republic of Crimea and the city of Sevastopol (2017) (Официальная статистика ОФСГС по Республике Крым) (in Russian). http://crimea.gks.ru. Accessed 05 Dec 2018

Saaty T, Kearns K (1991) Analytical planning. organization of systems (Аналитическое планирование. Организация систем) (in Russian). Radio and Telecommunications, Moscow, p 224

Saaty TL (1993) Decision making. Hierarchy analysis method (Принятие решений. Метод анализа иерархий) (in Russian). Translated from English by R. G. Vachnadze, Radio and Telecommunications, Moscow, p 278

Spitsnadel VN (2000) Basics of system analysis: textbook (Основы системного анализа) (in Russian). Business Press, Saint-Peterburg, p 326

Surmin YP (2003) System theory and systems analysis: train guide (Теория систем и системный анализ) (in Russian). MAUP, Kiev, p 368

Sovetov BY, Yakovlev SA (2007) System modeling: textbook for academic bachelor (Моделирование систем: учебник для академического бакалавриата) (in Russian). Yurait Publishing House, Moscow, p 343

Terleev V, Nikonorov A, Badenko V, Guseva I, Volkova Yu, Skvortsova O, Pavlov S, Mirschel W (2016) Modeling of hydrophysical properties of the soil as capillary-porous media and improvement of Mualem-Van Genuchten method as a part of foundation arrangement research. In: Advances in Civil Engineering 2016: Article number 8176728

Tikhomirova AN, Sidorenko EV (2012) Modification of the hierarchy analysis method by T. Saati for calculating criteria weights when evaluating innovative projects (Модификация метода анализа иерархий Т. Саати для расчета весов критериев при оценке инновационных проектов) (in Russian). Modern problems of science and education, p 2

Vasantha Kandasamy WB, Smarandache F (2003) Fuzzy cognitive maps and Neutrosophic cognitive maps. XIQUAN, Phoenix, p 213. http://fs.unm.edu/NCMs.pdf. Accessed 05 Dec 2018

Chapter 4
Integral Assessment of Condition and Sustainability of Socio-Ecological-Economic Systems

Vasiliy V. Dmitriev, Aleksandr N. Ogurtsov, Nikolai V. Hovanov, George K. Osipov, Valeriy P. Kulesh, Yuriy N. Sergeyev and Irina V. Fedorova

Abstract The methodological fundamentals and methods of integral assessment of the condition and sustainability of socio-ecological-economic systems are examined. The aim of the research was to summarize the results of the development and testing of classification models of the integrated assessment of the condition and sustainability of socio-ecological-economic systems. The research focuses on socio-ecological-economic systems which include three subsystems: 1—environmental (quality of the environment), 2—economic, 3—social. In some models, a separate component is distinguished in the third subsystem: 4—morbidity rate in the population. The research topics are the following: I—theoretical and methodological aspects and methods of generation of integral condition indicators (integral indicators of the first level of development for ecological, economic, and social subsystems, as well as the integral indicators of the second level of development or summary indicators); II—sustainability of socio-ecological-economic systems or their ability to maintain the characteristics, parameters of modes and class of the condition in which the system (or its individual subsystems) had been before the start of the impact. The study examines basic definitions, fundamentals of theory and methodology of assessment of condition and sustainability of socio-ecological-economic systems, algorithms

V. V. Dmitriev (✉) · A. N. Ogurtsov · N. V. Hovanov · V. P. Kulesh · Y. N. Sergeyev · I. V. Fedorova
St. Petersburg State University, Universitetskaya Naberezhnaya, 7/9, St. Petersburg 199034, Russia
e-mail: v.dmitriev@spbu.ru; vasiliy-dmitriev@rambler.ru

A. N. Ogurtsov
e-mail: aogurcov@yandex.ru

N. V. Hovanov
e-mail: nick@polyidea.com

I. V. Fedorova
e-mail: umnichka@mail.ru

G. K. Osipov
Military Spaces Academy Named After A. F. Mozhaisky, St. Petersburg, Russia
e-mail: Osipov-G-K-2005@yandex.ru

W. Mirschel et al. (eds.), *Landscape Modelling and Decision Support*, Innovations in Landscape Research,
https://doi.org/10.1007/978-3-030-37421-1_4

for the realization of the integral estimates, and results of the assessment research. Examples include (1)—the integral assessment of socio-ecological-economic system of Russia and comparative integral assessment of socio-ecological-economic systems of Russia and Canada, as well as Russia and Mexico; (2)—the condition of socio-ecological-economic systems of the Federal Districts of the Russian Federation; (3)—the condition of socio-ecological-economic systems of the individual regions of the Russian Federation and their subregions (the Northwest, the Center, the northern regions, etc.); (4)—the condition of socio-ecological-economic systems of the cities of the Russian Federation; (5)—the record of morbidity among the population in the integral assessment of socio-ecological-economic systems of the regions, and (6)—the study of the sustainability of regional systems in relation to the impacts on them and their subsystems. This research focuses on the development of multi-criteria and multilevel classifications (classification models) to study the condition and sustainability of socio-ecological-economic systems and the construction of integral indicators based on analysis and synthesis of Indicators in information deficit (ASPID) and Aggregated Preference Indices System (APIS) methods and models.

Keywords Socio-ecological-economic · System-integrated indicator · Regions of Russia · Impact-sustainability · Results of integrated assessment · Health of the population

4.1 Introduction

The current period time is characterized by the accumulation of methodological and practical experience in the study of condition and (to a lesser extent) sustainability of socio-ecological-economic systems and their individual subsystems. Some recent decisions in this area were presented in detail in the publication "Report on Measurement of Economic Performance and Social Progress" by Stiglitz-Sen-Fitoussi Commission (http://www.stiglitz-sen-fitoussi.fr/documents/rapport_anglais.pdf) in which the authors suggested "recommendations on the formation of statistical tools for assessment of quality of life and social sustainability" (United Nations Organization. Economic and Social Council. I.—Recommendations of Stiglitz Commission 2015). It is also necessary to mention the resolution adopted by the UN General Assembly on September 25, 2015 named "Transforming our world: The 2030 Agenda for Sustainable Development" and "Report of the Inter-Agency Expert Group on Sustainable Development Goal Indicators" at the forty-seventh session of the United Nations (March 2016) (Transforming our world. The resolution was taken by the UN General Assembly 2015; Report of the interdepartmental group of experts for indicators of objective fulfillment in the sphere of sustainable development. UNO 2016).

Modern research focuses on the following issues: 1—development of approaches to the assessment of emergent (complex, nonadditive, and integrative) properties of complex systems that characterize the system as a whole (modern or retrospective status of the system, its well-being, tension, degree of transformation of the system, stability, vulnerability of the system, integrity, etc.); 2—development of approaches to the assessment of the impact on systems and their response to the impact. In solving these problems, the researcher aims to predict the condition of a complex system (and its subsystems) on the basis of the models of system dynamics and the models of integral evaluation, as well as to develop approaches to systemic standardization of impacts on systems.

The need for the development of systemic standardization is due to the fact that the impact on a system can be judgment-based or accidental and relate to changes in one, several, or all the subsystems simultaneously (social, environmental, and economic). At present, systemic standardization of impacts on complex systems in nature and society has no alternatives in solving the problem of sustainable development of socio-ecological-economic systems. Numerous attempts to assess the state of complex systems on a component-by-component basis do not make it possible to assess the changes of emergent properties characterizing the system as a whole, and multi-criteria estimates underlying the indicator approach and indicative control can lead to a situation where a system belongs to one class of conditions (quality) based on one indicator and to another class based on another indicator. In another time interval, the system may be classified into a different class in terms of several indicators, etc. Thus, the systems of indicators that are used by management structures in evaluation studies of complex natural and social systems and quality of life of the population, which are often called sustainable development indicators, often also generate uncertainty in the results of the evaluation of the temporal dynamics of development. The process of management of such systems, based on the monitoring of factor indicators, calculation of indices and their analysis taking into account the impact on the target indicator on the basis of various approaches suggested by researchers and based on the forecast of possible changes in the selected indicators, often fails to provide the desired effect.

4.2 Methods of Assessment of Condition and Sustainability of Socio-Ecological-Economic Systems

4.2.1 Fundamentals of Methodology

In her work, Korchagina (2012) developed and proposed a definition of the concept of a system's socio-ecological-economic sustainability which characterizes it as a "subsystem in the systems of higher levels: macro-regional and national". As we understand it, the subject has to be not the "socio-ecological-economic sustainability", but the sustainability of the socio-ecological-economic system as an integrative (complex, emergent) property of the social system by which we mean "dynamic self-developing and self-regulating system "human society—nature" or "human society—environment", the dynamic balance in which (or balanced development of which) has to be supported by the reasoning of the society" (Reymers 1990; Alimov et al. 1999; Dmitriev 2008). The conventional formula of such a system is presented by us as follows: social system = biocenosis (biocenoses) + physical and geographical environment (biotopes) + population + economy + culture + politics, or social system (socio-ecosystem) = geosystem (geo-ecosystem) + economy + culture + politics.

A typical example of research in this area is the work of Lior (2015). The author conducted an analytical review of various aspects of the concept of sustainable development. References are made to 87 sources published from 1996 to 2013, which directly or indirectly consider the subject of the study. Special attention is paid to the role and current state of the scientific approach which uses quantitative criteria of sustainable development in all spheres of human activity which lead to global changes in the environment. Two examples of the methods of quantitative assessment of sustainability in their development are presented as the "working tools": 1—web diagram (the Russian term is a "rose diagram") which brings together into a single display, the information on 12 of the 20 indicators for sustainable development of the United States developed by the U.S. Interagency Working Group on Sustainable Development Indicators in 1998, and 2—method of constructing a complex sustainability indicator (*CSI*) which is represented by the unification of set of the M_i parameters used by the author into a single Composite Sustainability Indicator (*CSI*), where they are taken into consideration with their w_i weighting factors which reflect the priority of each of them. In fact, the author implements the construction of an integral indicator (with one level of convolution, rather than two or more, as in our case) without discussing the need to implement multilevel convolutions, standardize initial characteristics, and evaluate accuracy, reliability, etc. The main question that remains unanswered is why the author regards *CSI* to be the "sustainability indicator" rather than, for example, a "condition indicator". The next question is whether it is possible to produce an adequate evaluation on the basis of one level of the convolution of indicators, whether one level is able to "simulate" a study of the emergent properties of a complex system. The prospect of using the index to establish the standards of the impacts on subsystems is not clear.

Unlike most of these studies which focus on the use of estimates and indicators primarily for monitoring purposes, the author emphasizes the importance of integrating sustainability indicators into projects at the design and development stage so that the properties corresponding to the principle of sustainable development are included in the products and systems being created in advance. The author mentions the following challenges that are necessary to overcome… (a) need to make efforts to go beyond the field of specialization in order to work with the material that is unusually heterogeneous for the researcher;… (b) lack of certainty of the spatial and temporal scales of the objects of study, and their multi-scale nature; and … (c) lack of sufficiently powerful methodological and technical tools for modeling large-scale complex systems.

Tretyakova (2013) defines sustainable development as "the set of processes of positive change and the technologies that embody them aimed at harmonization of the interaction with the economic, environmental and social spheres in order to meet the needs of the socio-ecological-economic system in the long-term future". The author proposed to "measure sustainability" on the basis of two main approaches: "the indicator-based approach" and through "calculation of the integral index that makes it possible to obtain a comprehensive assessment of the sustainability of the socio-economic development". Let us note that the author is dealing with socio-economic development, not with the development of the socio-ecological-economic system. For the purpose of "development sustainability" assessment, the author suggests using the "method of dynamic standards". The method is based on a certain ordering of the dynamics indicators. The author argues that the order of the indicators of the dynamics of the observed parameters characterizes the structural changes which the system undergoes, and, therefore, by choosing the observed parameters of the system and ordering them in a certain way it is possible to build a reference model of change in its structural characteristics and, thus, to measure the effectiveness of activities through comparison of the reference and actual modes of operation of the economic system. For this purpose, a matrix with the actual ordering of the indicators of the economic component of development for the Russian Federation is constructed (the author analyzes the period from 2005 to 2011), and the conclusion is made that in terms of the economic component taken as the average annual value for this period, the measure of similarity of ratios of dynamics of sustainable development indicators in the Russian Federation was 79.2%. This means, in the author's opinion, that "during the study period the average growth rates of the basic economic indicators were more or less balanced compared to each other". Let us note the absence of evaluation scale for the resulting indicator which could be used to determine that "more or less" balanced growth rate, and the fact that the author focuses on the economic component of the socio-ecological-economic system. As a result, the paper proposes a comprehensive assessment of the dynamics of individual indicators of the country's development at a certain time interval. There is still no answer to the question as to how these indicators characterize the ability of the system as a whole to maintain its properties and parameters of the modes of operation when the system is exposed to external influences or makes a transition to another class, with better or worse conditions of the socio-ecological-economic system or the quality of life of the population.

A common interpretation of sustainable development indicators is a component-by-component analysis of the dynamics or variability of the indicators that reflect the composition and the properties of systems. At the same time, the authors do not emphasize that they are studying the sustainability of the system as a whole (or that of its subsystems). A comparison is performed with a certain reference or standard (also referred to as an "ideal image" in other definitions) for individual indices, which is compared with the indices calculated from the real statistical data for a number of years. In this case, in our experience, it is often found that one index (or criterion) of the system falls into one class, and another criterion (other criteria) to another class (other classes). As an option, the trajectory of the index change is set which characterizes, for example, only the "economic well-being" of the country.

Let us focus on the terminology and fundamental difference between our research and the studies examined above, and on the evaluation issues that exist in this sphere. The form and the initial stage of the expression of the relation of the person to the subject of estimation is the *diagnostic analysis* (*diagnostics*) of the condition of the object (the complex system). A diagnostic analysis includes exploratory identification of the advantages (positive value) and disadvantages (negative value) of the object (system), its individual properties, their natural range of variation, structure and modes of operation based on the analysis of the parameters of the condition and their critical values (Aleksandrova et al. 2000a, b; Dmitriev 2008, 2009, 2010). In our case, the complexity of diagnosing the state of the socio-ecological-economic system is due to the absence of axiometry used in evaluation practice for the selected criteria (necessary and sufficient criteria by which the condition of the system can be characterized as of high, medium, or low quality). In simple terms, the person performing the rating does not have rating scales for all or some of the evaluation parameters which could be used to develop a classification for performing integrated assessments. Less common is the opposite situation, when the researcher has several evaluation scales and needs to choose one of them and justify that choice.

The quality of life is a term widely used in human ecology and social ecology which refers to the quality of satisfaction of material and cultural needs of people, such as the quality of food, comfort of homes, quality of education, health care, services, natural environment, recreation structure; availability of fashionable clothing, degree of satisfaction of the need in objective information, level of stress, etc. In addition, the quality of life can be understood as the adequacy of the living environment to the socio-psychological attitudes of the individual. Based on the definitions of the quality of life, the main objective of the assessment may be to identify a set of natural, social, and economic conditions that determine, to a greater or lesser degree, human health, both personal and social, and human needs, i.e., the match between the life environment of a healthy person and the person's needs. The World Health Organization defines the quality of life as individuals' perception of their position in life in the context of the culture and value systems in which they live and in relation to their goals, expectations, standards, and concerns (Glukhov and Okrepilov 2008; Dmitriev and Kaledin 2016; Boblakova and Dmitriev 2014; Dmitriev et al. 2015; Osipova and Dmitriev 2014).

The existing interpretations of the concept of "quality of life" are quite numerous and ambiguous, and, therefore, researchers have fundamentally different approaches to its measurement. This is associated with different value systems in different countries (regions and ethnic groups) and different ideas about personal needs among different categories (social classes) of the population. For this reason, the differences in the existing methods of evaluation of the quality of life are manifested in dealing with issues such as selection of nomenclature of quality of life indicators, measurement of these indicators, selection of methods and assessment procedures to receive a generalized assessment of the standard of living of an individual, a group of people, a particular region, or a country as a whole. Methods or models of the quality of life are mainly built on the following: 1—taking into account the value system for subsystems in different countries (regions); 2—taking into account the ideas about personal needs of different categories (social groups) of the population, usually on the basis of questionnaires; 3—combination of 1 and 2. Thus, research papers sometimes refer to objective (type 1) or subjective (type 2) measurement of the quality of life. Apparently, both positions are not entirely appropriate. Ideally, it is necessary to combine approaches (approach 3) with the justification of the priority system within the levels (blocks and groups of characteristics) and between them.

A separate direction in the evaluation of the quality of life of the population is taking into account for personal and public health. Two directions that can be distinguished here are as follows: 1—evaluation of personal and public health within the "social conditions" component. Sometimes research papers also mention the "social determinants" of public health. Public health refers to the health of the population as a whole determined by the impact of social and biological factors. It is estimated based on the main (demographic data, morbidity rate, and physical development) and additional (medico-demographic and sociological) indicators. Human health, which includes biological, psychological, and social elements of a complex multilevel system, is the subject of interdisciplinary research and must be studied on the basis of an integrated approach (Semenova and Chistobaev 2015).

Let us clarify the terminology. The concepts of a "complex" and a "set" are close to the concept of a "system", but not identical to it. They can be seen as truncated, incomplete concepts in relation to the "system". The complex includes parts that do not necessarily have system properties, but these parts themselves can be systems, and the elements of the latter may have such properties in relation to such parts. The term "set" denotes a group of elements that have some common properties that are their essential characteristics and do not necessarily have systemic relationships or connections to each other. The "Standard" is most often based on an approved regulation or the permissible range of its variation. In modern research, the "standard" is often replaced by some of the so-called "own requirements". In this case, the "requirements" mean the need or expectation which has a compulsory nature. These requirements are established, or they are usually set by the person who performs the evaluation. This approach generates a sort of "ideal image", the proximity to which it should indicate some (often high, positive) significance of the object of the study or its properties. The arbitrary nature of such assessment is caused by the following: quality is the degree of satisfaction of a consumer who has an a priori point of view or "expected" ideas in respect of the result of the evaluation. To do this, the person performing the assessment must know the requirements of the consumer and perform the evaluation in accordance with the "ideal image" set by the "customer". In this sense, any "assessment" is subjective since it characterizes the relationship of a certain person to the object of evaluation, determination of the object's significance, either of it as a whole or of its individual properties, for this person on the basis of the selected criteria and their weight, as well as in accordance with the introduced concepts of the "standard", "requirements", and "personal preferences".

4.3 Research Methods

4.3.1 The Composite Indicators Method

The condition of the socio-ecological-economic system (SEES) is characterized by the ability of the SEES to maintain its properties and parameters of modes in the context of the system's exposure to external impacts or to internal, intra-system changes (with the impact on one of the subsystems) which characterize the quality of life of the population. To implement the approach, classes of condition (sustainability) of the system are introduced as well as components (for example, environmental, economic, and social ones), levels of convolution of indicators (the first is within the components while the second is between the components or under a more complex scheme). The necessary and sufficient criteria for evaluation are determined. Our research is based on the construction of integrated indicators of socio-ecological-economic systems (SEESs), quality of life, public health on the basis of multilevel (2–3 levels of convolution of indicators), and multi-criteria (3–5 components containing 5–10 or more evaluation criteria) indicators. We performed such studies in the first stage of our research. The goal was to obtain exploratory information about the integrative properties of the systems; to identify the effects of interconnection and interaction that are not additive to the local intra-system effects. Figure 4.1 shows the general scheme of construction of the integral indicator for one of the components (subsystem) in the method of consolidated indicators. This scale is used at the next convolution level to produce an integral indicator of the second level (the composite indicator).

After the preparation of evaluation scales for all the components, the second level of convolution is performed. The priority level (the weight) of individual components is taken into account here. An example of such convolution is shown in Fig. 4.2 and

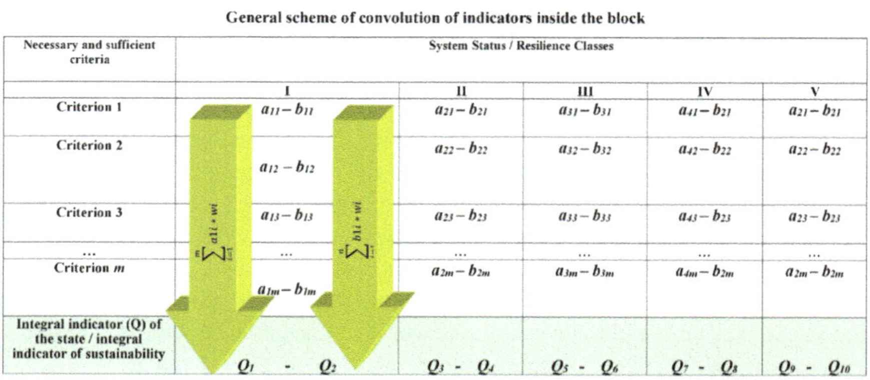

Note: a_{11}, a_{12}, a_{13}, ... a_{1m}; b_{11}, b_{12}, b_{13}, ... b_{1m} are the left and right boundaries of the estimated scales for the criteria (the result of rationing), w_i is the weight of the criterion

Fig. 4.1 The construction of the scale of the integral indicator for a separate unit

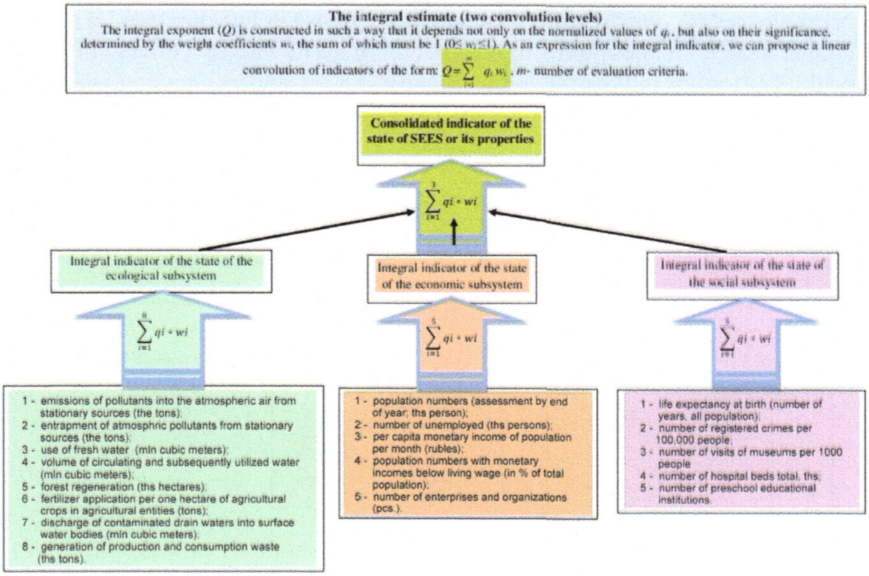

Fig. 4.2 Construction of the integral indicator of the state/stability of the socio-ecological-economic system (two levels of convolution)

in (Dmitriev and Osipov 2017; Dmitriev 2017). As a result, a researcher works with the natural evaluation scales of representative evaluation criteria of this property; introduces sustainability classes, introduces or uses evaluation scales, can introduce several levels of convolution of indicators, solves the problem of standardizing the initial data taking into account the type of correlation (direct and inverse) and its nonlinearity, sets or models the weights (priorities) for evaluation within and between groups and changes them if necessary, takes into account quantitative and qualitative information about the indicators and priorities of evaluation, simulating the change of the development priorities, and justifies the type of integral indicator.

4.3.2 The Randomized Consolidated Indicators Method

As a methodological basis for the integrated assessment in the solution of a wide range of problems in the framework of the research on the issue of integrated assessment of environmental entities, we use the idea of building randomized consolidated indicators and the principles of analysis and synthesis of indicators in information deficit methodology (ASPID) (Hovanov 1996; Hovanov et al. 2009), on which the model algorithms of the software were built by Geoexpert (Vasilyev et al. 2004). The modern development of ASPID is APIS, Aggregated Preference Indices System based on ASPID methodology (Hovanov et al. 2009).

The stages and content of the assessment are shown in Fig. 4.3. Here, it is important that there is an opportunity to select the weights on the basis of the information about the priorities of the evaluation, because in such studies the importance of the individual criteria was traditionally assessed through comparative judgments such

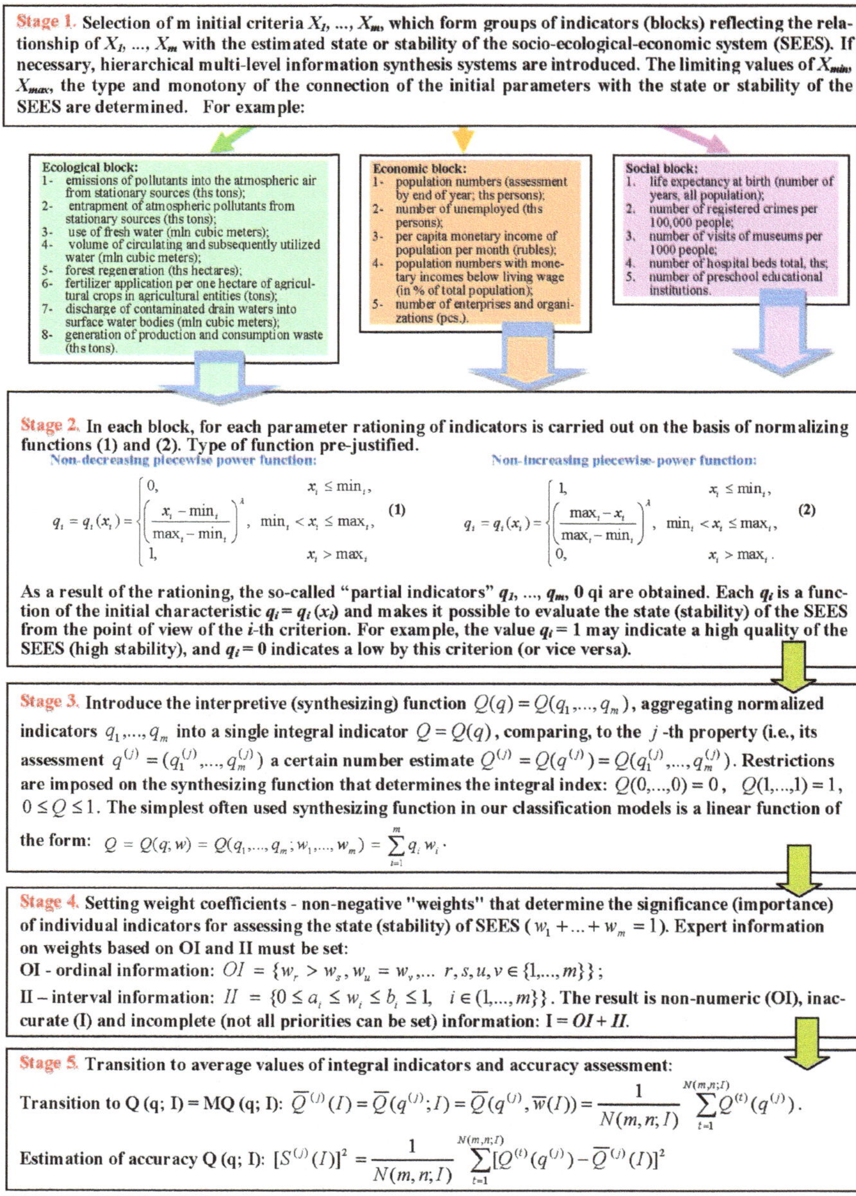

Fig. 4.3 Algorithm for obtaining integral indicators and estimates of accuracy of calculations

as "this criterion is more important for the overall evaluation than other one" or "these criteria have the same value for the integral evaluation". Thus, the significance of individual criteria is usually measured on a *non-numeric* (ordinal) scale, or all the criteria are given equal priority for the purpose of evaluation. In other cases, the researcher sets the intervals of possible variation of weight coefficients. In this connection, it becomes necessary to work with *non-numeric* (ordinal), *inaccurate* (interval) information, which is also often *incomplete* (not all the weight coefficients are given nontrivial equations and inequations corresponding to the interval and ordinal information). Non-numeric, inaccurate, and incomplete information (*nii* information) induces a whole set of admissible sets of weight coefficients for integral evaluation. To overcome this difficulty, modern approaches use the *Bayesian model of uncertainty randomization*. The idea of this model goes back to the work of Thomas Bayes (1702–1762) and represents the transition from an indefinite choice of weight coefficients to a random (randomized) choice from the set of all possible sets of weighting coefficients. Thus, the researcher gets random weights and random (randomized) integral indicators (Hovanov 1996). In this direction, it is promising to avoid formal averaging and use the methodological techniques based on the theory of modeling in the context of the deficit of information with the use of stochastic processes and fields (Hovanov 1996). This theory serves as a basis for the construction of generalized functions of the desirability of the condition and sustainability of the SEESs, taking into account the uncertainty of the definition of individual parameters and particular functions of desirability and their impact on the integrated assessment.

Figure 4.3 clarified the above statement. Upon the completion of the fifth stage, we obtained a scale of the changes in the consolidated indicator for the classes of the condition (sustainability) of the SEESs. After this stage, it is useful to analyze the resulting scale of the change of the consolidated indicator (for continuity and uniformity). If more than 50% of the total range of the changes of values of the consolidated indicator falls within the same class of the condition, it is likely that this is due to the fact that $Xi_{\min}$ or, more often, $Xi_{\max}$ was determined in haste. In this case, it is necessary to return to the previous stages and, if possible, to eliminate the drawback revealed.

At the following stages, the values of integral indicators are calculated based on the collected statistical data for individual regions. In the same way, it is possible to calculate the value of the integral indicator for another region, for other natural data or other classification models. In the examples which take into account incomplete, inaccurate, and non-numeric information, multilevel convolutions of information are introduced about the condition of the SEES. The weighting factors are set on the basis of information deficit models. The comparison of the condition of the ecosystems on an integrated basis makes it possible to quantify the spatial and temporal specifics of their dynamics, degree of their transformation, trends in their changes, and degree of permissible impact on the systems. In the latter case, it can be recommended to use, as the "maximum permissible" value of the consolidated indicator in the assessment, the value obtained on the basis of the convolution of the maximum permissible values of the initial parameters or the rightmost value of the consolidated indicator range for the selected class with the best quality (the first or the second). The second approach used by us is the assessment of the ability of the system to maintain its class of state (sustainability) when it is exposed to some impact. In this case, in hypothetical scenarios with an impact on individual components or on the whole system, the limits are revealed within which the class of the condition of the system remains the same after the impact. Some examples and results of evaluation studies are discussed below.

4.4 Results of Evaluation Studies and Their Discussion

Let us consider some examples of evaluation studies and their results. In the examples under consideration, the main attention is paid to the development of the methods for quantitative assessment of the state of the SEESs and their integrative, emergent properties (sustainability) which are accompanied by the construction of new functional units of the systems, i.e., integral indicators from several levels of information synthesis and indicators characterizing the last level of generalization (consolidated indicators). All these indicators reflect the integrity of individual subsystems and system as a whole and can serve as the basis for their systematics. They are used to determine the class of the property under evaluation and to compare the condition of the systems in space and time, identify the effects of interconnection and interaction, environmental, economic, and social determinants of social development which are not additive in relation to the local intra-system effects considered at the component level (Dmitriev and Ogurtsov 2017).

An important methodological element of the research is the need for multi-aspect visualization of the results based on the use of GIS and GIS technologies. In this case, GIS serves as a tool for spatial modeling and the analysis of integral estimates.

4.4.1 Socio-Economic Well-Being Index (SEWBI) for Russia and Mexico

To evaluate the socio-economic well-being of the countries and compare them to each other, we used eight indicators from the UN information database, which are given in Table 4.1 (Dmitriev 2010).

The above methodology was used for the assessment of the socio-economic well-being of Mexico and the Russian Federation in the time interval from 2000 to 2003. Table 4.2 shows the parameters from the UN database and the parameters selected for the calculation and the additional information needed to standardize the data and perform the convolution of indicators. Unfortunately, we did not use all the parameters stated in the UN database in our calculations due to the lack of the data for the selected countries or the time intervals under consideration.

Table 4.1 Indicators for assessing the socio-economic well-being of countries from the UN information database (www.data.un.org)

Indicator from the UN database	Indicator designation in calculations
1. Education expenditure of government, total, as percentage of GNI	X1
2. Education expenditure of government, total, as percentage of total government	X2
3. Illiteracy rates, aged 15+, %	X3
4. Consumer price index, general, 2000 = 100 (ILO)	X4
5. Unemployment, rates (%)	X5
6. Agricultural production per capita index, 1999–2001 = 100	X6
7. Energy supply (apparent consumption; Kg oil equivalent) per $1,000 (2000 PPP) GDP (World Bank)	X7
8. GDP per capita, annual growth rate, 2000 US$ (UN DPAD/Link estimates)	X8

To identify the trends and features of the socio-economic sphere of these countries, four evaluation scenarios were used (Table 4.3), in which individual criteria were given different priorities. The following are used in the table: *Qj* is the average value of the integral index for 2000–2003, *Sj* is the standard deviation in accordance with step 5 (Fig. 4.3).

The analysis of the SEWBI index calculation results (Table 4.3) showed

1. In all scenarios with the change of the values which are the evaluation priorities, w_i, synchrony is observed in the direction of changes in the index for both countries. In general, it is also possible to draw a conclusion on the quasi-constancy of the SEWBI value within the period of evaluation (from 2000 to 2003).
2. Positive dynamics of the SEWBI indices are observed for three variants of calculation. The exception is the calculation results for variant 3, where priority is given to the indicators of the “Employment” component (the only value from the component taken into consideration was “consumer price index for the core commodities”). Despite the negative dynamics of the indices, this scenario features the largest year-to-year changes in its values over the years.
3. The values of the SEWBI indices for Mexico as a whole exceed the values of SEWBI for the Russian Federation by 10–15%. Thus, the level of socio-economic well-being of the Mexican population in these years was higher than the corresponding indicator in Russia. Only variant 4, which is based on the preference of indicators of components “Agriculture”—“World indicator”, featured equal values of SEWBI. It is necessary to recognize that taking into account other indicators and giving them more weight (“the share of the undernourished in the total population”, “gross national income per capita”, “availability of drinking water”, etc.), for which in some cases there are no data available, can change the general picture and the results of the evaluation as a whole.

Table 4.2 Parameters of the UN database, selected for calculating the index of socio-economic well-being (SEWBI) and additional information necessary for data rationing and convolution of indicators

Blocks for indicators	N	Indicators	Years	Index	Rating-scale intervals, type of communication
Education	1	Book production, titles by the Universal Decimal Classification	1996	–	
	2	Education expenditure of government, total, as percentage of GNI	2000–2004	X1	0.5–12.0 increasing function
	3	Education expenditure of government, total, as percentage of total government	2000–2004	X2	3–30 increasing function
	4	Illiteracy rates by sex, aged 15+, per cent	2000–2003	X3	0.1–90.0 decreasing function
Employment	1	Youth unemployment, share of youth unemployed to total unemployed, per cent	1999	–	
	2	Youth unemployment, share of youth unemployed to youth population, per cent	1999	–	
	3	Consumer price index, general, 2000 = 100 (ILO)	1999–2006	X4	90–300 decreasing function
	4	Unemployment by sex, rates (%) and number (thousands)	2000–2006	X5	0.5–30.0 decreasing function
Agriculture	1	Agricultural production per capita index, 1999–2001 = 100	2000–2006	X6	70–150 increasing function

(continued)

Table 4.2 (continued)

Blocks for indicators	N	Indicators	Years	Index	Rating-scale intervals, type of communication
	2	Nutrition, undernourished as percentage of total population	2002–2003	–	
Health	1	Gross national income per capita (PPP international $)	2000, 2006	–	
	2	Population with sustainable access to improved drinking water sources (%) total	2000, 2006	–	
World indicator	1	Energy supply (apparent consumption; Kg oil equivalent) per $1,000 (2000 PPP) GDP (World Bank)	2000–2004	X7	0–20 increasing function
	2	GDP per capita, annual growth rate, 2000 US$ (UN DPAD/Link estimates)	2000–2004	X8	−5 to 15 increasing function

Table 4.3 Calculation results of SEWBI for Mexico (M) and the Russian Federation (R) for 2000–2003

Objects	Option 1		Option 2		Option 3		Option 4	
	Qj	*Sj*	*Qj*	*Sj*	*Qj*	*Sj*	*Qj*	*Sj*
1.2000 M	0.607	0.112	0.638	0	0.802	0.106	0.424	0
2.2001 M	0.611	0.112	0.647	7.58E-09	0.796	0.101	0.426	2.78E-09
3.2002 M	0.602	0.110	0.641	7.91E-09	0.781	0.097	0.420	0
4.2003 M	0.625	0.105	0.669	2.49E-09	0.782	0.085	0.447	0
5.2000R	0.484	0.127	0.493	0	0.681	0.107	0.337	0
6.2001R	0.530	0.113	0.528	4.05E-09	0.678	0.081	0.436	0
7.2002R	0.527	0.108	0.530	0	0.659	0.072	0.433	7.81E-10
8.2003R	0.536	0.104	0.542	0	0.634	0.054	0.460	5.19E-09
Priorities	w1 = w2 = w3 = w4 = w5 = w6 = w7 = w8		w1 = w2 = w3 > w4 = w5 = w6 = w7 = w8		w6 = w7 = w8 > w1 = w2 = w3 = w4 = w5		w6 = w7 = w8 > w1 = w2 = w3 = w4 = w5	

4.4.2 Comparative Integrated Assessment of Life Quality in Russia and Canada

Let us consider the example of the integral evaluation of the quality of life of the population of Russia and Canada for 2005, 2009, 2012, and the results of the evaluation (Lobacheva and Dmitriev 2015). The assessment is based on a three-level classification model with 5 classes, with the following names of the classes: I—"high", II—"above average", III—"average", IV—"below average", V—"low", with 25 objective criteria united in 6 groups and pertaining to 3 components: economic, social, and environmental. The proximity of the integral value to 0.0 corresponds to the high quality of life of the population, the proximity to 1.0, to the low quality. The economic component includes 4 indicators; the social one has 15 indicators, and the environmental one has 6 indicators. 3 variants of the model differing in the evaluation priorities of the second level of data generalization are proposed as follows: "model 1": equality of weights; "model 2": priority of "public health"; and "model 3": priority of "labor market". The calculation of the integral indicator of the quality of life (IIQL) at the first level (within groups) was performed on the basis of the assumption of the equality of priorities (weights) for the evaluation. In model 2, the following priorities are set for the second level: public health > personal safety = environmental quality > labor market = the media > transport infrastructure ($w_3 > w_4 = w_6 > w_1 = w_5 > w_2$). In model 3, the second level priorities are as follows: labor market > transport infrastructure = the media = public health > environmental quality > personal safety ($w_6 > w_4 = w_5 = w_3 > w_2 > w_1$). The information for the construction of the *IIQL* was taken from the websites "International economic statistics" and "The Federal State Statistics Service". In general, the quality of life in Russia is hindered by a gradual decrease in the quality of the environment, low personal safety (in terms of the integral indicator (II), it is twice lower than in Canada), and low, although improving, level of health of the population (the II is 2.5 times lower than in Canada). At the same time, Russia at the first level of convolution has better performance than Canada in the groups of "labor market" and "the media". The quality of life in Canada is due to the high level of public health, personal safety and steady trend to the improvement of the quality of the environment. It is in these groups of characteristics that Russia has had results that are one class lower than those of Canada (Russia has Class III and Canada has Class II). Both countries were found to have a tendency to the improvement of the quality of life of the population: Russia from IV to III and Canada from III to II.

4.4.3 Integral Assessment of Socio-Ecological-Economic Systems of Federal Districts of the Russian Federation

The results of the integrated assessment of the quality of life in eight Federal Districts (FD) of Russia are evaluated on the basis of ASPID modeling. The assessment of the quality of life is performed for 5 classes of quality on the basis of calculation of individual statistical indicators for three groups of the criteria of evaluation (social, economic, and ecological) on three levels of convolution of indicators (within subcomponents, within components, between components). The evaluation scales for the criteria of the first level of convolution were constructed on the basis of literature data and taking into account the regional extremities of their range. Calculations of integral indicators were performed for 2001, 2003, 2005, 2007, 2009, and 2011. Nine scenarios were considered which were based on equal weights and non-equal weights of priorities at the last level of convolution of indicators. The list of indicators, which includes 36 criteria, is presented in (Dmitriev and Boblakova 2014). The findings suggest that in terms of the value of the integral indicator of the quality of life (IIQL), the most favorable conditions for living were found in the Northwest region in 2011 (IIQL = 0.45; the third class of quality, closer to the left end of the range), and the least favorable, in the Siberian FD in 2001 (IIQL = 0.65, the fourth class of quality, the middle of the interval) (Fig. 4.4, Table 4.4). Almost all the Federal Districts were found to have positive dynamics of the quality of life of the population (except for the Volga Federal District, where in 2011 the quality of life has slightly deteriorated compared to 2009). It should be noted that, throughout the period under consideration, the quality of life in the Northwestern, North Caucasian, and Far Eastern Federal Districts did not go beyond the third class of quality and was characterized as "average". In other Federal Districts, the quality of life fell into the fourth class of quality ("below average"). In the Siberian Federal District, the QL was characterized as "below the average" throughout the entire study period, except for 2011 ("average"). Next, 8 models were built, taking into account the conditions of non-equal weights for individual components. All the weights were calculated, as well as new IIQL scales for QL evaluation by the Federal Districts of Russia, and the role of setting priorities in the construction of IIQL was assessed. In general, based on the analysis of the dynamics of IIQL over time, the positive dynamics of the QL in the Federal District was revealed in the period from 2001 to 2011 (Fig. 4.4a). In 2011, all the Federal Districts fall into the "middle" class of the QL. When different weights of the components were taken into account, it did not result in any significant changes. It can be noted that if the study is based on the model where the environmental component has the highest priority (weight) and the social component has less weight than the economic one, the quality of life in the NCFD falls into the second class of quality ("above average") (Fig. 4.4a and b). The results obtained under the condition that the environmental component has a higher priority (weight) than the social one, but lower than the economic one, are very close to the results obtained under the condition of an equal weight of all the components. The consideration of the situation in more detail within individual components makes it possible to conclude that the most favorable conditions, in terms of the set of environmental criteria, were found in the North Caucasian Federal District (for the entire study period, the QE in this Federal District falls into the second class of quality, varying only slightly). The least favorable environmental conditions were found in the Siberian and Central Federal Districts (all the values of the integral index of QE fall into the fourth class of quality, "below average"). In the other Federal Districts,

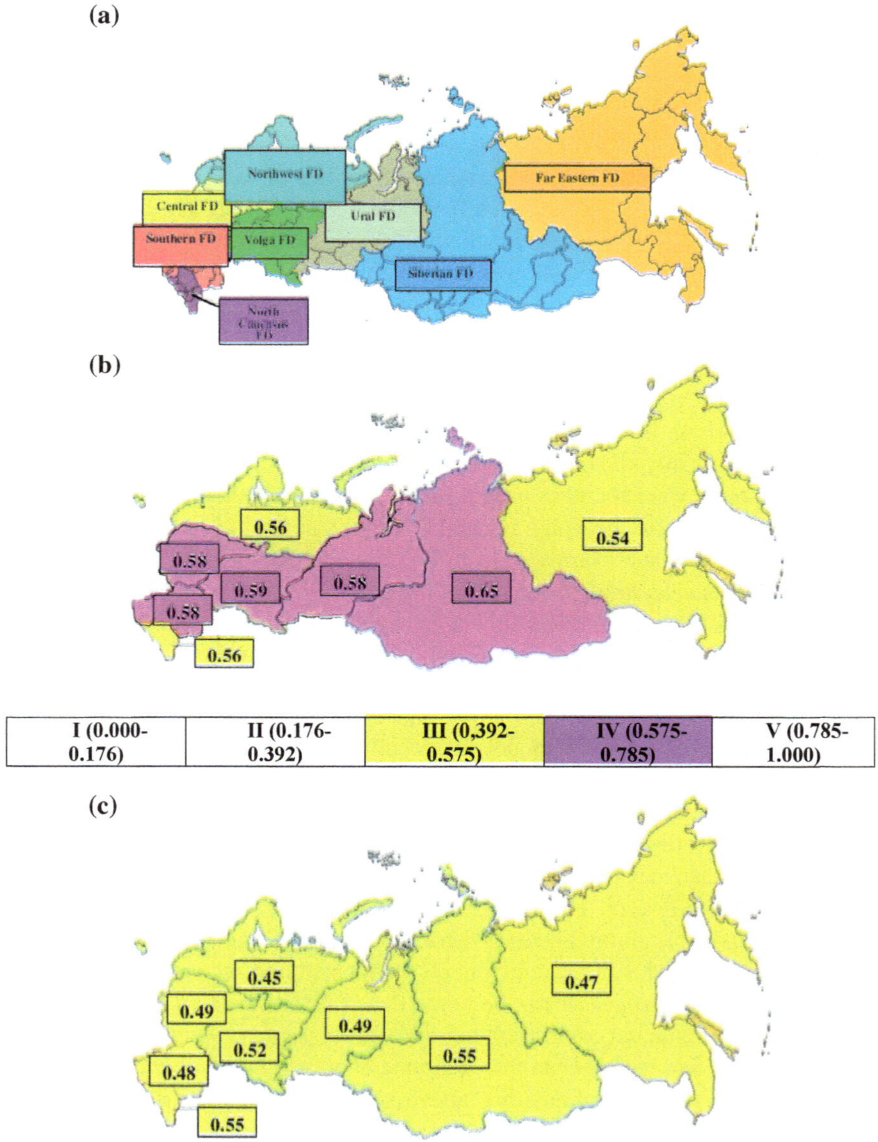

Fig. 4.4 Federal Districts of Russia (**a**). Integral indicators of the quality of life of the population in the Federal Districts of Russia in 2001 (**b**) and 2011 (**c**)

Table 4.4 Integral indicators of quality of life for the Federal districts of Russia in 2001–2011

Federal districts	Year					
	2001	2003	2005	2007	2009	2011
Central Federal District	0.58	0.57	0.55	0.52	0.49	0.49
Northwest FD	0.56	0.53	0.51	0.49	0.47	0.45
Southern FD	0.58	0.58	0.55	0.52	0.50	0.48
North Caucasus FD	0.56	0.55	0.54	0.53	0.50	0.50
Volga FD	0.59	0.58	0.57	0.54	0.50	0.52
Ural FD	0.58	0.55	0.54	0.53	0.50	0.49
Siberian FD	0.65	0.64	0.63	0.59	0.58	0.55
Far Eastern FD	0.54	0.52	0.51	0.49	0.47	0.47

the QE throughout the years under consideration falls into the third class of quality ("average"). An important conclusion is that there is no positive temporal momentum for the improvement of QE in this component. In the Central and Siberian Federal Districts, the least favorable conditions for the set of indicators were observed in 2005, in the Northwestern, in 2001, in the Southern, in 2003, in the North Caucasian, in 2007, and in Volga and Ural Districts, in 2011. In the Far Eastern Federal District, the period with the worst environmental conditions was from 2001 to 2005.

4.4.4 Integral Assessment of the Condition of Russian Regions

4.4.4.1 Integral Assessment of the Condition of the Northern Regions and Their Sustainability to Hypothetical Changes in the Condition of the Subsystems

The ecological subsystem includes eight evaluation parameters, the economic subsystem has five, and the social one has five. The integral evaluation is made for 2003–2015 (Fig. 4.5 and Table 4.5).

The composition of the parameters is discussed in (Dmitriev 2017). All the indicators are selected from the data of the Russian Federal State Statistics Service website ("the Regions of Russia"). The calculation technology is based on the convolution of indicators at the first and second levels and identification of situations in which the SEES will not be able to preserve its properties and mode parameters (state class) with the hypothetical impact on it given a priori and considered as separate systems and as a whole. When planning scenarios of the deterioration of the situation in the subsystems at the first level of the convolution, the subsystem characteristics were set with due consideration for the various directions of the parameters. The condition

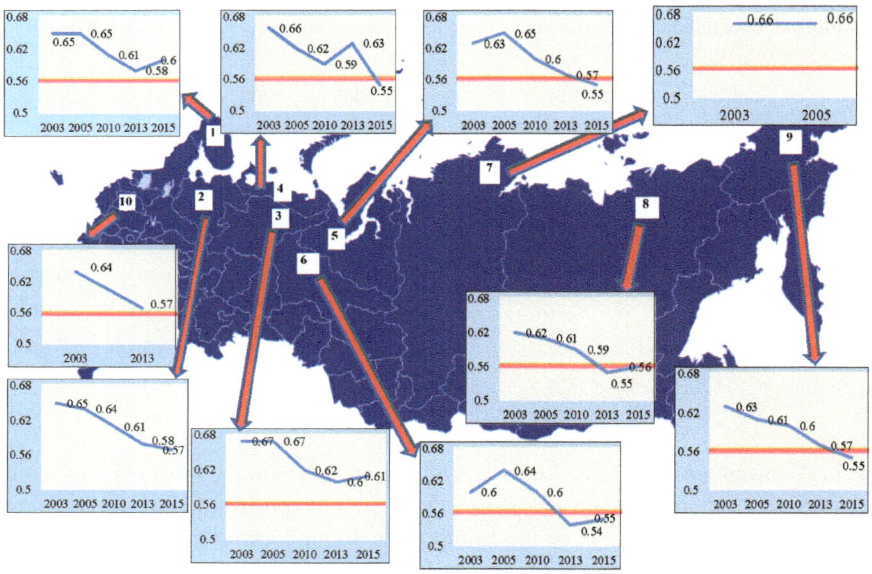

Fig. 4.5 Changes in the quality of life in the regions in terms of the integral indicator of the state of the SEES. Regions of the AZR: 1—Murmansk region; 2—Arkhangelsk region; 3—Komi Republic; 4—Nenets Autonomous district; 5—Yamalo-Nenets Autonomous district; 6—Khanty-Mansi Autonomous area; 7—Taimyr Dolgan-Nenets Autonomous Okrug; 8—Republic of Sakha (Yakutia); 9—Chukotka Autonomous region. Region of the Central Committee of the Russian Federation: 10—Tver region (Dmitriev 2018)

of the system and the quality of life of the region's population were graded as pertaining to one of the 5 classes (I—high, II—above average, III—average, IV—below average, and V—low). The proximity of the integral value to 0.0 corresponds to the high quality of life of the population, and the proximity to 1.0, to the low quality. The consequences of the impact were assessed against the condition of the regions in 2013.

4.4.4.2 Comparative Assessment of the Life Quality of Populations in Different Russian Regions and Their Dynamics in 2003–2015

In this example, we were interested in the trend of the change in the quality of life of the population from the westernmost region of the Russian Federation to the easternmost one, the pace and dynamics of indicators of individual subsystems, and regions as a whole (Fig. 4.6). The calculations of the integral indicators of quality of life (IIQL) were performed for the same 18 characteristics and 5 classes, with equal priorities, as in paragraph 3.4.1. Here are the main features of the changes in the IIQL in the regions (Fig. 4.6):

Table 4.5 Integral indicators of quality of life of people in the regions of the arctic zone of the RF for the period from 2003 to 2015 (the second level of the convolution of indicators) (Dmitriev et al. 2017)

Region/Year	2003	2005	2010	2013	2015
Arkhangelsk region	0.65 (IV)	0.64 (IV)	0.61 (IV)	0.58 (III–IV)	0.57 (III–IV)
Murmansk region	0.65 (IV)	0.65 (IV)	0.61 (IV)	0.58 (III–IV)	0.60 (IV)
Nenets Autonomous Okrug	0.66 (IV)	0.62 (IV)	0.59 (IV)	0.63 (IV)	0.55 (III)
Taymyr Dolgano Nenets Autonomous Okrug	0.66 (IV)	0.66 (IV)	–	–	–
Chukotka Autonomous Okrug	0.63 (IV)	0.61 (IV)	0.60 (IV)	0.57 (III–IV)	0.55 (III)
Republic of Sakha (Yakutia)	0.62 (IV)	0.61 (IV)	0.59 (IV)	0.55 (III)	0.56 (III–IV)
Yamalo-Nenets Autonomous Okrug	0.63 (IV)	0.65 (IV)	0.60 (IV)	0.57 (III–IV)	0.55 (III)
Republic of Komi	0.67 (IV)	0.67 (IV)	0.62 (IV)	0.60 (IV)	0.61 (IV)
Khanty-Mansiysk Autonomous Okrug Yugra	0.60 (IV)	0.64 (IV)	0.60 (IV)	0.54 (III)	0.55 (III)

Note 1—The table gives the value of integral indicator in brackets the class of quality of life; 2—for Taymyr (Dolgano-Nenets Autonomous Okrug) the data prior to 2005 are given, for by the results of the referendum held on April 17, 2005, from January 1, 2007 Taymyr (Dolgano-Nenets) Autonomous Okrug was abolished, and the municipal Taymyr Dolgano-Nenets Autonomous region was made part of Krasnoyarsk territory as an administrative territorial unit with a special status

1. In all the regions, the IIQL featured a positive trend from 2003 to 2013. The most significant growth of the consolidated indicator of the quality of life was observed in the Kaliningrad region (16.7%). The lowest value was observed in the Primorsky region (7.8%). In other regions, it was as follows: the Sakhalin region (15.2%), the Leningrad region (14.3%), the Krasnodar region (13.7%), and the Altai region (10.3%).
2. In all the regions, the improvement in the quality of life was due to the growth of the integral index of the economic subsystem (from 18.6% in the Kaliningrad region to 26.5% in the Sakhalin region). As a rule, from west to east, the integral indicator of the economic component grows.
3. Minimal changes are observed for the integral indicator of the environmental subsystem (environmental quality): from a downward trend in 2 regions (the Leningrad region and the Primorsky Krai) by 2.6% and 2.2%, respectively, or a zero change (the Kaliningrad region and the Altai Krai) to an improvement in the environmental situation in the region by 2% (the Sakhalin region) and 11.8% (the Krasnodar Krai).
4. Changes in the social subsystem in all regions, except for the Primorsky Krai, tend to improve, the increase ranging from 3.8% (the Krasnodar Krai) to 24.7%

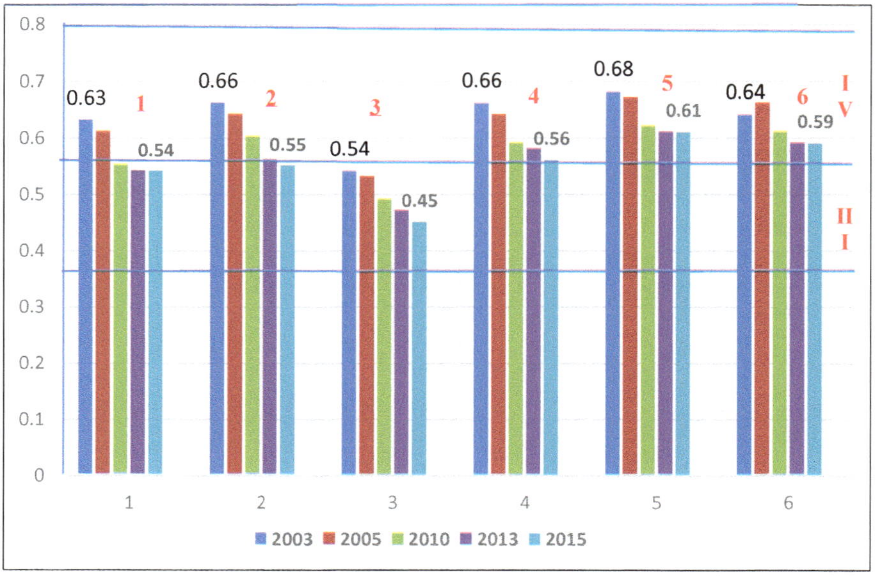

Fig. 4.6 Comparative assessment of the quality of life of the population of the Leningrad region (1), Kaliningrad region (2), Krasnodar territory (3), Sakhalin region (4), Altai territory (5), Primorsky Krai (6) and identifying the dynamics of their development from 2003 to 2015

(the Kaliningrad region). In Primorsky Krai, the social conditions in this period have deteriorated by 2.2%.

5. Assessment of the sustainability of the SEESs of the selected regions showed that two regions (the Kaliningrad region and the Leningrad region) during this time shifted from IV to III class, or rather from the middle of IV class to the right extremity of III class. The other regions have kept their previous class: the Sakhalin region, the Altai Krai, and the Primorsky Krai are Class IV, and Krasnodar Krai is Class III, with an increase in the quality of life of the population. However, the growth rate did not lead to the transition of SEES to another class, the one with a better quality of life. These rates are estimated by us as ranging from 8 to 10% (the Altai and the Primorsky Krai) to 15% (the Sakhalin region). For all the regions considered, except for the Krasnodar Krai, the main factor in the improvement of the quality of life in the coming years will be the improvement of the quality of the environment, and for the Krasnodar Krai, this factor will be the improvement of social conditions. Only in Primorsky Krai, deterioration in quality in two subsystems was observed in the 12-year period: social and environmental. This led to the fact that with an increase in the quality of the economic component by 24.6%, this region was found to have the lowest increase in the IIQL of the population for the three components among all the regions considered (7.8%).

4.5 Integral Assessment of Life Quality in Russian Cities and Regions

Figure 4.7 shows examples of integrated estimation of the quality of the environment (7e) and the quality of life of the population in the cities (a, c, d, e), constituent entities of a Federal District of Russia (b). Information for evaluation in Fig. 4.7a (Boblakova and Dmitriev 2014) was taken from the websites of the Federal State Statistics Service, the territorial body of the Federal State Statistics Service for St. Petersburg and Leningrad region, the territorial body of the Federal State Statistics

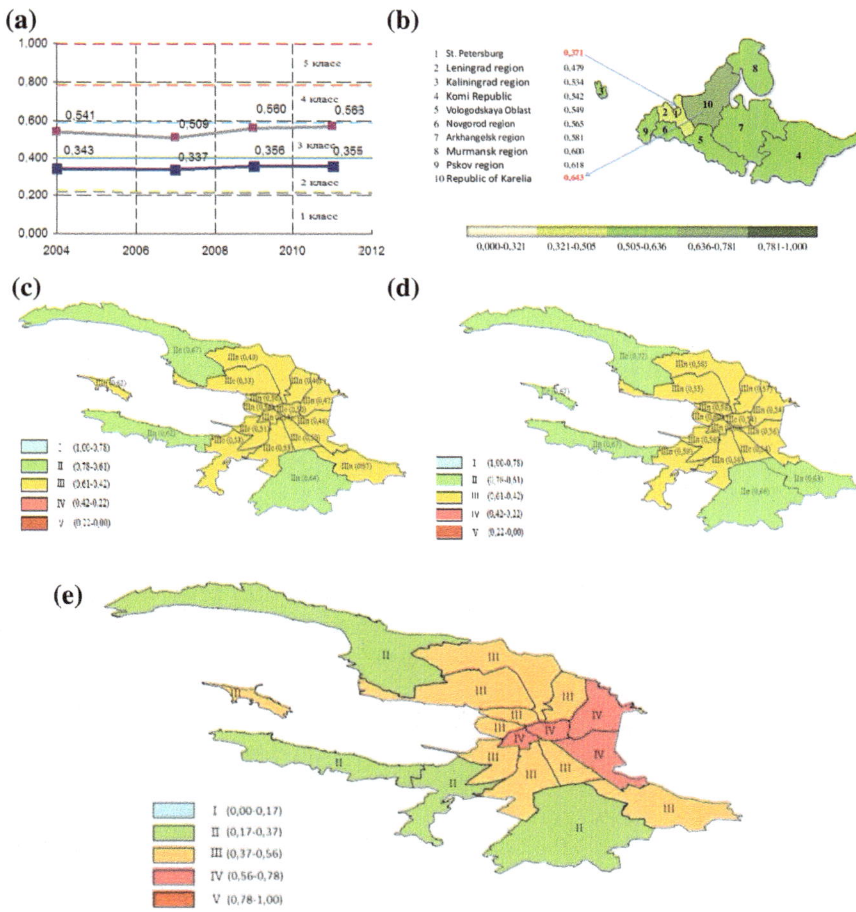

Fig. 4.7 Examples of an integral assessment of the state of socio-ecological-economic systems and the quality of life of the population in Moscow and St. Petersburg in 2004–2011 (**a**); subjects of the Northwest Federal District of Russia in 2013 (**b**); districts of St. Petersburg in 2003 and 2011 (**c**, **d**, **e**); subsystem (block) of environmental quality in 2011 (**e**) according to Boblakova and Dmitriev 2014; Dmitriev et al. 2015; Osipova and Dmitriev 2014

Service for Moscow, from bulletins "Main Indicators of Environmental Protection" for various years, "Regions of Russia, Social and Economic Indicators" for 2004, 2007, 2009, and 2011.

The system of criteria for assessing the quality of life includes three components: social, economic, and environmental (ecological) one. The social component includes 11 parameters; the economic component has 8, the component of environmental quality (environmental component) has 6 parameters for different types of environments. The analysis presented in Fig. 4.7a shows that the quality of life in St. Petersburg according to the selected parameters is, throughout the studied period, characterized by the second class of quality ("above average"). The quality of life of the population of Moscow throughout the studied years also falls into one class, the third, which is characterized as the "average" quality of life. The values of the integral indicators in the cities are close, and both cities demonstrate a tendency to move to a higher class. The evaluation of the dynamics of the integral indicator of the quality of life brings one to the conclusion that the quality of life of the population decreased slightly from 2004 to 2011 (by about 6%).

The data sources used for the assessment of the SEES and the quality of life of the constituent entities of the Northwestern Federal District of Russia were the following: State Report "On the State of the Environment and the Environmental Protection in the Russian Federation" and Annual Statistical Collection "Regions of Russia. Social and Economic Indicators" for 2013, and State Report "On the State of the Environment and the Environmental Protection in the Russian Federation" for the same year. To obtain an integrated assessment, it is proposed to use three components consisting of 9–10 indicators reflecting the quality of life of the population living in the territory of the subjects of the Russian Federation: environmental, social, and economic (Fig. 4.7b). The ecological component included 8 parameters, the social component included 11 and the economic component had 10 parameters (Dmitriev et al. 2015; Dmitriev et al. 2016). The evaluation scales of integral indicators for two levels of convolution (within the groups and between the groups) were constructed with due account for the equality and inequality of priorities (weights) for all the groups of indicators. During the implementation of the models one component, environmental, social or economic, was selected as the priority one. Four cases of correlation of the coefficients of significance were examined for the second level of convolution of indicators: 1. $p1 = p2 = p3$; 2. $p1 > p2 = p3$; 3. $p2 > p1 = p3$; 4. $p3 > p2 = p1$, where p1 is the priority of the environmental component, p2 is the priority of the social component, and p3 is the priority of the economic component. Thus, the change of the weights (priorities) between components (at the second level of convolution) was performed for four scenarios: 1—equal priority (Fig. 4.7b); 2—the priority of "quality of environment"; 3—the priority of "social conditions"; 4—the priority of "economy".

As a result, with equal weights for the first and second levels of convolution of indicators, the quality of life in 10 constituent entities was graded with classes II to IV. Class II of the quality of life was determined for two constituent entities: Saint Petersburg and Leningrad region Class III that included Kaliningrad, Vologda, Novgorod, Arkhangelsk, Murmansk and Pskov regions, and the Komi Republic.

Class IV consisted of the Republic of Karelia. With "priority 2", the Leningrad region moved down from Class II to Class III of the quality of life; the Murmansk region moved down from Class III to Class IV. With priority 4, the Kaliningrad region and the Komi Republic moved upon the quality of life scale from Class III to Class II.

The results obtained from the models are consistent with each other. After that, the influence of changes in the initial indicators within the components (social, economic, and environmental) on the value of the consolidated indicator was studied. These calculations were performed for the first scenario of the composite indicator determination (equal priority). It was shown that, for example, an increase of less than 30% of the environmental quality indicators does not lead to a change in the class of quality of life of the population, which reveals the value of the consolidated indicator which results in the transition of the socio-ecological-economic system to a higher (or lower) class of the quality of life.

Figure 4.7e shows an example of an integrated assessment of urban environment quality in St. Petersburg for one level of convolution according to seven criteria for 2011, for all the districts of the city (Osipova and Dmitriev 2014). The criteria for assessing the quality of the urban environment were the following: 1—area of green space, m^2/person; 2—level of air pollution by the main pollutants according to the data from the automated monitoring system, conventional units; 3—proportion of atmospheric air samples exceeding the maximum permissible concentration of pollutants, %; 4—equivalent level of transport noise, dBA; 5—specific combinatorial index of water pollution; 6—share of the soils, in terms of area, which have the total index of chemical pollution (Zc) higher than 16%; and 7—volume of household waste, m^3/person. The IIQI evaluation scale for the assessment of the environmental situation in individual districts of the city is presented in Fig. 4.7e. As the value of the integral index (II) increases, the quality of the environment deteriorates. The performed calculations show that the quality of the urban environment in Krasnoselsky, Kurortny, Petrodvortsovy, and Pushkinsky districts is characterized by Class II, whereas Vasileostrovsky, Vyborgsky, Kalininsky, Kirovsky, Kolpinsky, Kronshtadtsky, Moskovsky, Frunzensky, Primorsky, and Petrogradsky districts have Class III, and Admiralteysky, Tsentralny, Nevsky and Krasnogvardeysky districts have Class IV. Tsentralny district has the highest value of the II (0.64), and, therefore, the worst quality of the environment. The lowest indicator (0.23) was found in Petrodvortsovy district. The suburban areas of Saint Petersburg have lower integral indicators. In general, the quality of the urban environment in the city in 2011 is characterized by Class IIIc and has an integral index of 0.47.

The assessment of the quality of life of the population by districts of the city is shown in Fig. 4.7c and d. Based on the generalization of literary data, the criteria were selected for the evaluation of the quality of life of the population in the districts of the city. The selected 30 criteria include three components: social, economic, and environmental. Five classes of the quality of life (I—"high", II—"above average", III—"average", IV—"below average", and V—"low") were introduced, and a classification model was developed to assess the quality of life. Earlier in this work, we have considered in detail the criteria and results of the evaluation of the quality of the urban environment. The scale of the integral indicator of quality of life is presented in the legend of the figures.

The calculations show that the quality of life in 4 districts pertains to Class II, and in all the other fourteen districts, to Class III. In 3 districts, the value of the integral indicator of the quality of life was found to be in the leftmost extremity of Class III; in 8, it was in the middle of Class III; in 3, it was in the rightmost extremity of Class III. Tsentralny district has the highest value of the integral index (0.57), and, accordingly, the worst quality of life. The district with the lowest value (0.33) was Kurortny. The suburban areas of Saint Petersburg have the lower integral indicator values. Figure 4.7c and d shows examples of evaluation maps of the quality of life of the population of the city. It is obvious from the figure that the spread of indicators of the quality of life fits into 2 classes (II–III), with the worst conditions characterized by a boundary value between classes III and IV.

4.5.1 Integral Sustainability Assessment of the Social System Under Hypothetical Changes of the Environmental Quality and the Economic–Social Conditions in Russian Regions

The generalization of the results on the sustainability of SEESs of the regions was based on the examples of the Tver region (Dmitriev and Osipov 2017), entities of the NWFD (Dmitriev et al. 2015), the northern regions of Russia (Dmitriev 2018). In all cases, the research analyzed the ability of the SEES to maintain the class of the condition which it had prior to the hypothetical or predicted impact on the system as a whole or on individual subsystems. As a rule, the results of the integrated assessment of the quality of life were studied based on the assumption of the decrease of the quality in the components by 30%, by 50%, and by two times.

By the example of the Tver region, it was shown that the deterioration of the situation by 30% compared to 2013 only in the environmental component (the environmental quality component) for all 8 parameters did not move the condition of the subsystem to another class. The integrated indicator of the ecological component is increased by 7.3%. A twofold deterioration of only environmental situation in all 8 parameters changes the value of the aggregated indicator of the ecological component by 17% (0.48, III class; previous value 0.41, III class) and brings the quality of life on this component to Class IV (the width of the class interval being 0.56–0.77). The same experiments were performed with other components. The calculations of the consolidated indicator of quality of life for the situation of 30% deterioration of conditions in all the components at the same time in comparison with 2013 led to the increase of the consolidated indicator to 0.63 (IVc). Before the change, it was 0.57 (borderline value between III and IV classes). The percentage value of the change was 10.5%. This increase leads to a deterioration of the quality of life by approximately half of a class. The same experiments were performed for other values of hypothetical transformations.

Similar experiments carried out for the northern regions showed that a 30% change in the situation with the components resulted in changes only in one region out of 9; by 50%, in 3 regions out of 9 changed the class of environmental quality from III (average) to IV (below average); a twofold deterioration of the situation resulted in the transition of the social system to a lower class in 6 regions out of the 9 examined in the research. The Republic of Sakha (Yakutia), Nenets, and Chukotka or Autonomous Okrug were the regions with the highest sustainability against impacts. And the most vulnerable region in the case of the deterioration of the condition was the Khanty-Mansi Autonomous Okrug–Yugra.

The study considered the systemic effect of a record of the morbidity component in assessing the quality of life of the population of the regions along with the following components: economy, environment (quality of environment), and social aspect. For this purpose, the results of the assessment of the state of SEESs in the northern regions and the assessment of the morbidity rate there were used. The variants of calculations took into account different options for the second level of convolution (between components): type "1"—integrated assessment of four components with equal weights for each component (0.25); type "2"—integral assessment for three components +1 component (health of the population) with equal weights (0.5 and 0.5).

It was found that in general, the integral assessment of the quality of life in the period from 2005 to 2015 by type "1", for individual regions, produced the following results: the Murmansk region III-III, the Komi Republic III-IV-IV, the Khanty-Mansi Autonomous Okrug–Yugra III-III, the Republic of Sakha (Yakutia) III-III, the Arkhangelsk region III-III, the Nenets Autonomous Okrug IV-IV, the Chukotka Autonomous Okrug III-III, the Yamalo-Nenets Autonomous Okrug III-IV-III.

Convolution of the indicators of type "2" has produced the following results: the Murmansk region III-III, the Komi Republic III-III-IV, the Khanty-Mansi Autonomous OkrugYugra II-II, the Republic of Sakha (Yakutia) II-III, the Arkhangelsk region III-III, the Nenets Autonomous Okrug IV-IV, the Chukotka Autonomous Okrug III-III, and Yamalo-Nenets Autonomous Okrug III-III. The comparison of results "1" and "2" revealed minor differences in the consolidated evaluation by region. An important conclusion is the need to take into account the component of morbidity of the population and its estimated priority in the integrated assessment of the quality of life of the population.

4.6 Conclusion

The fundamentals of methodology, methods of integral assessment of the condition, and the emergent properties (sustainability) of socio-ecological-economic systems (SEES) have been analyzed. The goal of the research was to summarize the results of the development and testing of classification models of the integrated assessment of the condition and sustainability of SEESs. The subjects of the study were SEESs of individual countries, Federal Districts of the Russian Federation and their constituent entities, cities of Russia and their districts, central, northern, and other regions of Russia. The subjects of the study were theoretical and methodological aspects and methods of constructing integral indicators of the condition and the study of the sustainability of the SEESs upon the exposure to hypothetical or actual impacts. The basics of methodology for evaluation of the condition and sustainability of the SEESs and the algorithms of integral estimations were considered, as well as the findings of evaluative studies. The research was focused on the development of multi-criteria and multilevel classifications for the study of emergent properties of complex systems and the construction of integrated indicators based on ASPID and APIS methods and models.

Acknowledgements The studies are supported by RFBR grants No. 16-05-00715a; 18-05-00328-a; 18-05-60291a; 19-05-00683a.

References

Aleksandrova LV, Vasilyev VU, Dmitriev VV, Myakisheva NV, Ogurtsov AN, Tretyakov VU, Hovanov NV (2000a) Multi-criteria geographic and environmental assessments of the state and sustainability of natural and urban systems deposited manuscript no. 2342B00 01.09.2000, Dmitriev VV, Hovanov NV (ed) (In Russian). https://elibrary.ru/item.asp?id=30689943 Accessed 31 Dec 2018

Aleksandrova LV, Vasilyev VU, Dmitriev VV, Myakisheva NV, Ogurtsov AN, Terekhina NV, Tretyakov VU, Ufimtseva MD (2000b) Multi-criteria assessment of the ecological status and sustainability of geosystems based on the method of summary indicators. V. Integral assessment of the ecological state and quality of the environment in urban areas. Bulletin of St. Petersburg University. Series "Geology and Geography". SPb.: Publishing House of St. Petersburg State University, 2000, vol 4, no 34, pp 40–52 (In Russian). https://elibrary.ru/item.asp?id=24495697. Accessed 31 Dec 2018

Alimov AF, Dmitriev VV, Chistobaev AI et al. (1999) Mnogokriterial'naia otsenka ekologicheskogo sostoianiia i ustochivosti geosistem na osnove metoda svodnykh pokazatelei. V. Integral'naia otsenka ekologicheskogo sostoianiia i kachestva sredy urbanizirovannykh territorii [The integrated assessment of the environmental condition and quality of environment in urban areas]. Frolov AK (ed). St. Petersburg, SPbNTs Russian Academy of Sciences. 253 p (In Russian). https://elibrary.ru/item.asp?id=25784337. Accessed 31 Jan 2019

Boblakova LM, Dmitriev VV (2014) Comparative evaluation of integrated quality of life for Saint-Peterburg and Moscow. Int J Exp Educ (3), 91–95 (In Russian). https://elibrary.ru/download/elibrary_21168897_43175637.pdf. Accessed 31 Jan 2019

Dmitriev VV (2008) Integral estimation of ecological condition and quality of anthropogenic transforming system environment/building together our territories. Construire ensemble nos territories. Abstracts. Resume. 31st international geographical congress, Tunis 2008, August 12–15, pp 410–411

Dmitriev VV (2009) Opredelenie integral'nogo pokazatelia sostoianiia prirodnogo ob"ekta kak slozhnoi sistemy [Determination of the integral index of the natural state of the object as a complex system]. Obshchestvo. Sreda. Razvitie [Society. Wednesday. Development], 2009, no 4 (12), pp 146–165 (In Russian). http://www.terrahumana.ru/arhiv/09_04/09_04_16.pdf. Accessed 31 Dek 2018

Dmitriev VV (2010) Integral'nye otsenki sostoianiia slozhnykh sistem v prirode I obshchestve [Integrated assessment of the state of complex systems in nature and society]. Biosfera [Biosphere], 2(3):507–520 (In Russian). https://elibrary.ru/download/elibrary_15542683_14921548.pdf. Accessed 31 Dec 2018

Dmitriev VV (2018) Chapter V/4: integral assessment of the state of regional socio-ecological and economic systems, the quality of life of the population, the stability of socio-systems. Novel methods and results of landscape research in Europe, Central Asia and Siberia (in five volumes), vol 5. Landscape planning, management and rehabilitation/main editors Viktor G. Sychev, Lothar Mueller. – M.: Publishing House FSBSI «Pryanishnikov Institute of Agrochemistry», pp 23–28 (In Russian). https://doi.org/10.25680/1415.2018.75.85.350

Dmitriev VV, Boblakova LM (2014) Integral evaluation of the quality of life of the population in the regions of Russia. Information technology and systems: management, economics, transport, law, no 3, pp 37–43 (In Russian). https://elibrary.ru/download/elibrary_22600340_71704385.pdf. Accessed 31 Jan 2019

Dmitriev VV, Chistilina VS, Kravchenko VA (2015) Development and testing of integrated assessment models of socio-ecological-economic systems and the quality of life of the population in the regions of Russia. Information technology and systems: management, economics, transport, law, no 2 (16), pp 89–94 (In Russian). https://elibrary.ru/download/elibrary_25032459_70067422.pdf. Accessed 31 Jan 2019

Dmitriev V, Kaledin V and Kaledin N (2016) Theory and practice of integrated assessment of the state of complex systems in nature and society. In: 16th International multidisciplinary Scientific Geoconference SGEM 2016, Book 5, ecology, economics, education and legislation, www.sgem.org, SGEM2016 conference proceedings, 30 June–06 July, 2016, Albena, Bulgaria, vol II, pp 871–877. https://doi.org/10.5593/sgem2016/b52/s20.113

Dmitriev VV, Kaledin (2016) Russian North-west: an integral assessment of the conditions of regional social, environmental and economic systems and quality of life, vol 8, no 2. The Baltic region, Russia, pp 87–98. https://doi.org/10.5922/2079-8555-2016-2-7

Dmitriev VV, Lobacheva SV, Chistilina VS (2017) Assessment of the state of socio-ecological-economic systems of the regions of Russia: AZR regions. In: Materials of the international scientific and practical conference "Modern Ecology: Education, Science, Practice", Voronezh, pp 297–301 (In Russian). https://elibrary.ru/download/elibrary_30698447_40758645.pdf. Accessed 31 Jan 2019

Dmitriev VV, Ogurtsov AN (2017) Sustainability of complex systems in nature and society: methodology, evaluation, results. In: Scientific notes of the Russian state hydrometeorological university, no 48, pp 72–84 (In Russian). https://elibrary.ru/item.asp?id=30744920. Accessed 31 Dec 2018

Dmitriev VV, Osipov GK (2017) Integral assessment of stability osocial-ecological-economic system against changes in its functioning conditions. In: 17th international multidisciplinary Scientific Geoconference SGEM 2017. Conference proceedings, vol 17. Ecology, economics, education and legislation, no 52. Ecology and environmental protection, 29 June–5 July, 2017. Albena, Bulgaria, pp 565–572. https://doi.org/10.5593/sgem2017/52

Dmitriev VV (2017) Integral assessment of environmental quality and the quality of life of the population of the Arctic regions of Russia in the period from 2003 to 2015. In: Mathematical modeling technological and socio-economic processes, pp 147–151. https://elibrary.ru/item.asp?id=30761057. Accessed 31 Jan 2019

Glukhov VV, Okrepilov VV (2008) Managing quality of life. SPb.: Science, 2008. 484 p (In Russian). https://elibrary.ru/item.asp?id=19665032. Accessed 31 Jan 2019

Hovanov N, Hovanov K, Yudaeva M (2009) Multicriteria estimation of probabilities on basis of expert non-numeric, non-exact and non-complete knowledge. Eur J Oper Res 195(3):857–863. https://doi.org/10.1016/j.ejor.2007.11.018

Hovanov NV (1996) Analysis and synthesis of indicators in the information deficit. SPb.: SPbSU Publishers, 196 p (In Russian). https://elibrary.ru/item.asp?id=22395522. Accessed 31 Jan 2019

Korchagina EV (2012) Analiz i otsenka ustojchivogo razvitiya sotsio-ehkologo-ehkonomicheskikh sistem. Spetsial'nost' 08.00.05—ehkonomika i upravlenie narodnym khozyajstvom (regional'naya ehkonomika). Diss. na sosk. uch. st. d. eh. n. Gatchina, 2012. 377 s. (In Russian). https://elibrary.ru/download/elibrary_30058336_98490426.pdf. Accessed 31 Jan 2019

Lior N (2015) Quantifying sustainability for energy development. Energy Bull (19), pp 8–29

Lobacheva SV, Dmitriev VV (2015) Comparative evaluation of the integral quality of life in Russia and Canada. Int J Appl Basic Res (9–2):359–364 (In Russian). https://elibrary.ru/download/elibrary_24057577_53734872.pdf. Accessed 31 Dec 2018

Osipova AA, Dmitriev VV (2014) The integrated estimates of quality of life the population and quality urban environment of St. Petersburg International journal of experimental education, no 3, pp 96–102 (In Russian). https://elibrary.ru/download/elibrary_21168898_84699838.pdf. Accessed 31 Jan 2019

Report of the interdepartmental group of experts for indicators of objective fulfillment in the sphere of sustainable development (2016). UNO. Economic and Social Council. Forty-Seventh session. 8–11 March 2016. E/CN.3/2016/2, 46 p

Reymers NF (1990) Nature use. Dictionary reference. M., 488 p (In Russian). http://www.studmed.ru/download/reymers-nf-prirodopolzovanie_56ecaaca33c.html. Accessed 31 Jan 2019

Semenova ZA, Chistobaev AI (2015) Medical geography and public health. Monograph. SPb.: SPB SC RAS, SPbGU. European House Publishing House, 252 p (In Russian). https://elibrary.ru/item.asp?id=26169800. Accessed 31 Jan 2019

Transforming our World (2015) Agenda of sustainable development for the period until 2030. The resolution taken by the UN General Assembly 25 September 2015, A/RES/70/, 44 p

Tretyakova EA (2013) Evolution of research and evaluation methodology of sustainable development of social and economic systems. World Appl Sci J 25(5):756–759. https://elibrary.ru/item.asp?id=21882400. https://doi.org/10.5829/idosi.wasj.2013.25.05.13335

Vasilyev VY, Ogurtsov AN, Hovanov NV (2004) Organization of specialized applications for GIS on the principles of the ASPID-methodology. In: Proceedings of the international conference on sustainable development of territories: geo-information support and practical experience, pp. 134–143 (In Russian). https://elibrary.ru/item.asp?id=30603494. Accessed 31 Dec 2018

Chapter 5
Observation Methods and Model Approaches for Estimating Regional Crop Evapotranspiration and Yield in Agro-Landscapes: A Literature Review

Leonidas Toulios, Marios Spiliotopoulos, Giorgos Papadavid and Athanasios Loukas

Abstract The chapter presents an overview of various methods and model approaches that can be used to derive crop evapotranspiration and agricultural yield state from remote sensing data. The overview is based on an extensive literature review. The studied literature reveals that many valuable techniques have been developed both for the retrieval of evapotranspiration and crop yield from reflective remote sensing data as for the integration of the retrieved variables into crop models. However, for crop modelling and remote sensing data assimilation to be commonly employed on a global operational basis, emphasis will have to be put on bridging the mismatch between data availability and accuracy on one hand, and model and user requirements on the other. This could be achieved by the integration of images with different spatial, temporal, spectral and angular resolutions, and the fusion of optical data with data from different sources.

Keywords Earth observation · Evapotranspiration · Crop yield · Model approaches

L. Toulios (✉)
Institute of Industrial and Forage Crops, Greek National Agricultural Organization-Demeter, 41335 Larisa, Greece
e-mail: leonidastoulios@gmail.com

M. Spiliotopoulos
Department of Civil Engineering, University of Thessaly, 38334 Volos, Greece
e-mail: spilioto@uth.gr

G. Papadavid
Agricultural Research Institute of Cyprus, Nicosia, Cyprus
e-mail: papadavid@ari.gov.cy

A. Loukas
Department of Rural and Surveying Engineering, Aristotle University of Thessaloniki, 54124 Thessaloniki, Greece
e-mail: agloukas@topo.auth.gr

W. Mirschel et al. (eds.), *Landscape Modelling and Decision Support*, Innovations in Landscape Research,
https://doi.org/10.1007/978-3-030-37421-1_5

5.1 Estimation of ET Using Remote Sensing (RS) Data

5.1.1 Introduction

Evapotranspiration (ET) is a very important factor for energy, hydrologic, carbon and nutrient cycles and a key component of water balance. ET belongs to the category of secondary meteorological parameters. Traditionally, ET can be directly measured over a homogeneous surface using conventional techniques such as Bowen ratio (Prueger et al. 1997), eddy covariance (Kustas and Norman 1996) and lysimeter systems (Allen et al. 1998). However, these techniques cannot be applied to the majority of cases worldwide. The most convenient methodology for many years was the theoretical computation of ET which is grouped into (a) hydrological methods (i.e. water balance methods) (Linacre 1977), (b) micrometeorological methods (i.e. mass and energy transfer) (Kustas and Daughtry 1990) and (c) climatological methods (based on empirical relations with air temperature, computational relations on solar radiation and combined computational relations) (Allen et al. 1998).

The contribution of Earth Observation (EO) data (satellite-derived data) for estimating regional crop evapotranspiration and yield in agro-landscapes is a global issue, which must be addressed with global models and global data are needed as input to these models. EO from space has a unique capacity to provide continuously and consistently such global data sets not only on this level but also on the national and local levels and the variable estimation must be based on such data. Some of the climate and biophysical variables essential for understanding and monitoring the agro-landscapes can be efficiently observed from space since this technology enables their systematic, global and homogeneous measurement. The use of EO for agriculture and landscape research is generally based on data collected from EO satellites as well as satellites launched for other purposes, primarily for weather prediction and climate change warning purposes. To make these data useful for ET and agricultural yield studies, it is usually necessary to analyse and process the basic observational raw data and integrate them into models (Toulios et al. 2008).

5.1.2 Background

Since the 1980s, there has been an increasing effort to develop methods for estimating ET from remote sensing data. Gowda et al. (2007), Kalma et al. (2008), Li et al. (2009) and Calera et al. (2017) offer a comprehensive survey of published methods known to date, pointing out the main issues and challenges to address in the future. Advances in technology and resources, along with a significant change in the data policy by National Aeronautics and Space Administration (NASA) gave new access to satellite images in near real time giving new opportunities to Remote sensing (RS) techniques. A similar policy has been adopted recently by European Space Agency (ESA) giving free access to data from the new Sentinel-2 and improving NASA's

Landsat spatial resolution from 30 to 10 m already allowing deriving valuable information of vegetation properties (Clevers et al. 2017; Vanino et al. 2018). Additionally, a very important number of commercial sensors at very high spatial resolution (e.g. WorldView2, Pleiades, DMC and Deimos) continue to provide new capabilities to land science (Calera et al. 2017). Since RS imagery is in operational use, it is very reasonable to use RS data for the monitoring of crop biophysical parameters as well as for water resources management. RS-based techniques have the capability to estimate the spatial and temporal variation of ET from high spatial resolution catchment to low-resolution global scales (Petropoulos et al. 2018). Large area cover is a very prominent advantage of RS systems in comparison with traditional information sources as stated in Allen et al. (2011). Future research includes the development of the spatial-temporal fusion models of Landsat/Sentinel and MODIS ET to increase the temporal resolution of the final downscaled ET (Hong et al. 2011; Spiliotopoulos et al. 2013; Bhattarai et al. 2015; Ke et al. 2016).

Traditionally, Penmann–Monteith equation—known as the FAO-56 method (Allen et al. 1998; Pereira et al. 2015)—is assumed as the standard procedure for estimation of crop water requirements in irrigation scheduling and advised on procedures for calculation of the various parameters. By defining the reference crop as a hypothetical crop with an assumed height of 0.12 m having a surface resistance of 70 s m^{-1} and an albedo of 0.23, closely resembling the evaporation of an extension surface of green grass of uniform height, actively growing and adequately watered, the FAO Penman–Monteith method was developed. The method overcomes shortcomings of the previous FAO Penman method and provides values more consistent with actual crop water use data worldwide (Allen et al. 1998):

$$\mathrm{ET_o} = \frac{0.408(\mathrm{R_n} - \mathrm{G})\Delta + \gamma\left(\frac{900}{\mathrm{T_{mean+273}}}\right)u_2(e_s - e_\alpha)}{\Delta + \gamma(1 + 0.34u_2)} \tag{5.1}$$

where $\mathrm{ET_o}$ is reference evapotranspiration (mm day^{-1}), $\mathrm{R_n}$ is net radiation (MJ m^{-2} day^{-1}), Δ is slope vapour pressure curve (KPa °C), γ is psychrometric constant (KPa $°\mathrm{C}^{-1}$), $\mathrm{e_s}$ is saturation vapour pressure (kPa), e_α is actual vapour pressure (kPa), G is soil heat flux (MJ m^{-2} day^{-1}), r_α is aerodynamic resistance (s m^{-1}), $\mathrm{T_{mean}}$ is mean temperature (°C) and $\mathrm{u_2}$ is wind speed at 2 m height.

Similarly, according to ASCE (2005) reference evapotranspiration is defined as the ET from a reference crop (12 cm high clipped grass or 50 cm tall full-cover alfalfa) and therefore incorporates the effects of weather into the ET estimate. To estimate crop ET, with the use of reference ET, the ratio of a cropped and reference surface is combined into a crop coefficient according to ASCE (2005) as

$$\mathrm{ET_c} = \mathrm{ET_r} \cdot \mathrm{K_c} \tag{5.2}$$

where $\mathrm{K_c}$ is the crop coefficient, ETc is crop ET (mm) and $\mathrm{ET_r}$ is reference ET (mm). FAO-56 also promotes the concept of $\mathrm{K_c}$ in the so-called ‘two-step’ procedure (Allen et al. 1998; Calera et al. 2017). According to the ‘two-step’ procedure, ET is estimated by the product of the evaporative power of the atmosphere (ET_r) and K_c. K_c consists of three parameters. The first is the basal crop coefficient (K_{cb}) which represents transpiration when the plant is under no stress, and the second coefficient K_e is the evaporation coefficient. Allen et al. (1998), also introduced the stress coefficient for soil water limiting conditions (K_s) resulting in

$$K_c = K_{cb} * K_s + K_e \tag{5.3}$$

In contrast with ET values, K_{cb} can be represented by a smooth continuous function over time (Calera et al. 2017).

Currently, the commonly applied ET models using remote sensing data can be categorized into two types: semi-empirical methods and analytical methods. Semi-empirical methods are often accomplished by employing empirical relationships and making use of data mainly derived from remote sensing observations with minimum ground-based measurements, while the analytical methods involve the establishment of the physical processes at the scale of interest with varying complexity and requires a variety of direct and indirect measurements from remote sensing technology and ground-based instruments (Li et al. 2009).

5.1.3 *Residual Methods*

The most prevailing group of methodologies today is the Energy Balance (EB) algorithms and more specifically the residual methods (Romaguera et al. 2014). Remote sensing-based EB algorithms convert satellite sensed radiances into land surface characteristics such as albedo, leaf area index, vegetation indices, surface roughness, surface emissivity and surface temperature to estimate ET as a 'residual' of the land surface energy balance equation:

$$LE = \lambda ET = R_n - H - G \tag{5.4}$$

where LE is the latent heat flux (W/m^2), λ is the latent heat of vaporization, R_n is net radiation (W/m^2), G is soil heat flux (W/m^2) and H is sensible heat flux (W/m^2). Most recent EB models differ mainly in how H is estimated (Gowda et al. 2007). These models include the Two Source Model (TSM) (Kustas and Norman 1996), where the energy balance of soil and vegetation are separately modelled and then, combined to estimate total LE, the Surface Energy Balance Algorithm for Land (SEBAL) (Bastiaanssen et al. 1998a, b) (Fig. 5.1), the Mapping Evapotranspiration with Internalized Calibration (METRIC) (Allen et al. 2007) that uses hot and cold pixels within the satellite images to develop an empirical temperature difference equation and the Surface Energy Balance Index (SEBI) (Roerink et al. 2000; Menenti and Choudhury 1993) based on the contrast between wet and dry areas. Other variations of SEBI include the Simplified Surface Energy Balance Index (S-SEBI) (Roerink et al. 2000), and the Surface Energy Balance System (SEBS) (Su 2002). Among all these,

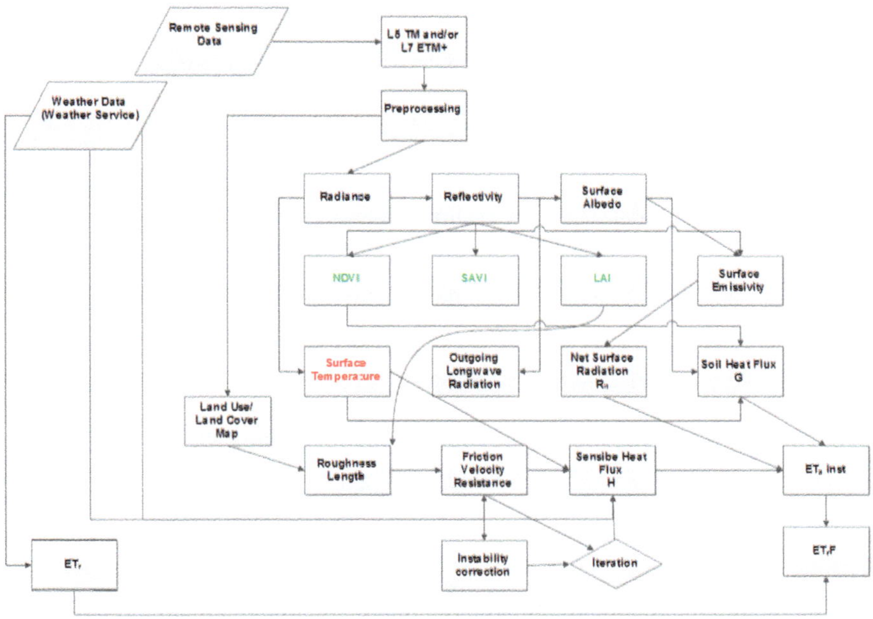

Fig. 5.1 Example of the use of remote sensing for estimating actual ET: METRIC methodology flow chart (modified from Spiliotopoulos et al. (2017), Spiliotopoulos and Loukas (2019))

SEBAL and METRIC variations seem to be the prevailing EB methodologies until now (Jaber et al. 2016; Papadavid et al. 2017; Jaafar and Ahmad 2019).

Other EB models are (Romaguera et al. 2014) as follows: the excess resistance (kB-1) (Su 2002; Kustas and Daughtry 1990), the aerodynamic temperature parameterization models proposed by Crago et al. (2004) and Chávez et al. (2005), the beta (β) Approach (Chehbouni et al. 1996), TSEB (Kustas and Norman 1996) and most recently the ET Mapping Algorithm (ETMA) (Loheide and Gorelick 2005; Gowda et al. 2007), and Soil Plant Atmosphere and Remote Sensing Evapotranspiration SPARSE (Boulet et al. 2015; Saabi et al. 2018). Finally, the Triangle Method (Goward et al. 1985; Jiang and Islam 1999), derived from a simple Crop Water Stress Index (CWSI), and the Trapezoid Method (Moran et al. 1994) derived from introducing a Water Deficit Index (WDI) for ET estimation over fully to partially vegetated surface areas (Li et al. 2009), must be mentioned. Very recently (Poon and Kinoshita 2018), the Operational Simplified Surface Energy Balance (SSEBop) was introduced for the study of ET in a postfire environment in San Diego California, USA. SSEBop utilizes RS images together with machine learning techniques, giving very promising results. Soil Vegetation Atmosphere Transfer (SVAT) models (Olioso et al. 1999) are also worth mentioning, but, though very effective, remote sensing can only have a supportive role on them.

Liou and Kar (2014) give a thorough review of basic theories, observational methods and different surface energy balance algorithms for estimating evapotranspiration

(ET). Amongst the aforementioned methodologies, some advantages and disadvantages exist (Gowda et al. 2007; Romaguera et al. 2014; Liaqat and Choi 2015; Lian and Huang 2016). For example, for METRIC, the disadvantages include the requirement for relatively high-quality weather data on an hourly or shorter time step and reliance on the accuracy of the ET_r estimate (Trezza et al. 2013). Reyes-González et al. (2017) found that ET values derived from METRIC were higher than ET values estimated with atmometers, and that difference was attributed to high wind speed values (>4 m s^{-1}) at the time of satellite image overpass. Dhungel and Barber (2018) discuss the accuracy of the selection of hot and cold pixels in METRIC, introducing an interesting automated algorithm which improves the accuracy of the methodology. With the S-SEBI method, the extreme surface temperature values are not always available for every acquired image, while, on the other hand, it is a simpler method that does not need additional meteorological data, and it does not require roughness length as for SEBS. In SEBS however, the main advantages include (1) the uncertainty from the surface temperature or meteorological variables can be limited with consideration of the energy balance in the limiting cases, (2) the new formulation of roughness height for heat transfer and actual turbulent heat fluxes are not required. Finally, one of the main disadvantages of the TSM method is the incorporation of some semi-empirical sub-models which are related to LAI and soil fraction cover, and can be easily lead to errors throughout the procedure.

5.1.4 Field Spectroscopy—Unmanned Aerial Vehicle (UAV)

Field spectroscopy for the acquisition of information about the spectral signatures of different features is another technique used for the identification of crop behaviour. Papadavid et al. (2013) used field spectroscopy for the estimation of NDVI to evaluate SEBAL methodology in Cyprus while Koksal et al. (2019) used a spectro-radiometer (model FieldSpec Pro FR, ASD, Boulder, USA) in order to calculate hyper-spectral reflectance, infrared surface temperature and net radiation over experimental plots in Turkey for the estimation of pepper crop ET in Turkey. Ortega-Farías et al. (2016) used multispectral and infrared thermal cameras placed on unmanned aerial vehicles (UAV) for the estimation of energy balance components over olive orchard fields in Chile. Results demonstrated that multispectral and thermal cameras on-board UAV could provide an excellent tool to evaluate the energy balance components. Further analysis is beyond the scope of this book.

5.1.5 Reflectance-Based Crop Coefficient

According to the 'two-step procedure', the contribution of weather on ET is described by ET_r while the properties of the crop are incorporated into K_c (Allen et al. 1998). Drought conditions and the related stress are strongly sensitive to crop development, but, standardized ET estimation methodologies generally assume fully irrigated conditions and therefore do not accurately estimate water use when soil water deficit conditions prevail. On the other hand, the basal crop coefficient K_{cb} can be used from published tabular values listed in the literature. Many researchers have been studied for the direct relationship between K_{cb} values with Vegetation Indices (VI) for corn (Bausch and Neale 1989), maize, soybean, bean and pasture (Navarro et al. 2016), wheat, oats, soybean, sunflower, cotton, sorghum and maize (Choudhury et al. 1994), alfalfa and corn (González et al. 2018). VI, generally defined as a combination of surface reflectance values between two or more wavelengths, are selected according to the desired application. Hundreds of VIs have been published in the scientific literature, but the most common RS approach for the assessment of K_c is the computation of NDVI which is called the K_c–NDVI approach (Hunsaker et al. 2005; Glenn et al. 2011; Kamble et al. 2013; González et al. 2018; Dalezios et al. 2019). NDVI utilizes red and Near-InfraRed (NIR) wavelengths, where chlorophyll reflects more NIR and green light at healthy and dense vegetation and vice versa.

The formula for the computation of NDVI is

$$\mathrm{NDVI} = \frac{(\mathrm{NIR} - \mathrm{RED})}{(\mathrm{NIR} + \mathrm{RED})} \tag{5.5}$$

The formula generates a value between −1 and +1, resulting in a standardized way to measure healthy vegetation. Crop coefficients generated from the K_c–NDVI approach could have even more reliable values than tabulated ones, because NDVI represents not only the actual crop growth conditions among different fields but also variations in canopy cover due to previous stresses (Kullberg et al. 2017).

One disadvantage of this methodology is that the use of multi-temporal crop coefficients (K_c) cannot be applied in water-limited environments without proper correction. Such correction using soil water content (SWC) measurements are discussed in Chiesi et al. (2018) among other technical issues. Kullberg et al. (2017) used several RS methodologies for the determination of K_{cb} and K_s (dual crop coefficient methodologies). K_{cb} was calculated from tables, NDVI and canopy cover while K_s were calculated using Crop Water Stress Index (CWSI) and canopy temperature measurements among others. Crop ET predicted by these methodologies was compared to observation. Piedelobo et al. (2018) introduced 'HidroMap' as a new tool for irrigation and water management utilizing NDVI derived from Landsat 8 and Sentinel-2 platforms.

One of the main tasks, in order to calculate canopy temperature indices for the later estimation of Ks, is the simplification of data demand. Taghvaeian et al. (2014) describe a CWSI methodology using only two instruments: an infrared thermometer and an air temperature/relative humidity sensor. The next idea would be the estimation of CWSI using RS data. Veysi et al. (2017) used a set of Landsat 8 satellite images for the CWSI calculation in sugarcane fields including thermal infrared data, and hot and cold pixels. Results showed a good relationship between the calculated CWSI based on field measurement and new CWSI based on satellite. Matese et al. (2018) conducted an experiment in two vineyards located in Sardinia, Italy, where CWSI was calculated from UAV images and ground infrared thermal images and, then, related to physiological measurements. Similarly, Bellvert et al. (2018) estimated CWSI and Ke using airborne remotely sensed canopy thermal-based methods for the estimation of K_c for almond and pistachio orchard fields in California.

5.1.6 Penman–Monteith RS Approaches

FAO-56 procedure can be used for the estimation of evaporation from soil and transpiration from leaves if the related surface parameters are available. These surface parameters can be related to leaf area index (LAI), surface albedo and crop height, and RS can be a perfect solution for the derivation of those parameters. A fixed value for crop height usually is assumed (e.g. 0.4 m for herbaceous crops and 1.2 m for tree crops), thus, finally, the most important remaining parameters are the surface albedo and LAI. Dhungel et al. (2014) study the estimation of latent heat flux using aerodynamic methodologies together with RS-based Penman–Monteith methodologies. Corbari et al. (2017) investigate the potential of using evapotranspiration measurements acquired by micrometeorological stations for the definition of crop coefficient functions of natural vegetated areas and extrapolation to ungauged sites through RS data. Sensitivity analysis showed that albedo and vegetation height have a low influence on crop coefficient estimates while these are highly influenced by the variability of LAI. Even if radiative transfer methodologies are used for that reason, the easiest approach remains the empirical relationships with VIs. The empirical relationships have the advantage that they can be validated easily, using portable optical analyzers (Calera et al. 2017). Saadi et al. (2018) introduced the SPARSE model computing (i) sensible heat flux (H) and (ii) daily ET over a heterogeneous semiarid landscape with complex land cover (i.e. trees, winter cereals and summer vegetables) using NDVI from TERRA and AQUA satellite platforms. SPARSE was run to compute instantaneous estimates of H and LE fluxes at the satellite overpass times with satisfactory accuracy. More recently, Fuentes-Peñailillo et al. (2018) used Shuttleworth and Wallace (SW) model to compute the intra-orchard spatial variability of actual evapotranspiration (ET) of olive trees using satellite images and ground-based climate data.

5.1.7 Using Earth Observation Data and CROPWAT Model to Estimate the Actual Crop Evapotranspiration

Since orbital sensing technologies have undergone unprecedented development, the use of multispectral satellite data in conjunction with traditional means is able to ensure the improvement of the classical determination methods of the agrometeorological parameters, greatly contributing to improved management of the agrometeorological phenomena, like drought (Stancalie et al. 2010). The European and American orbital platforms (e.g. NOAA/AVHRR, SPOT/VEGETATION, ERS, LANDSAT, EOS-Terra/Aqua, QuikSCAT and ADEOS-2) equipped with different optical or radar sensors, offer high-quality information with frequent repeat coverage. The new generation of EUMETSAT and NASA space sensor systems will certainly improve our knowledge of the surface processes for agrometeorology purposes. Remote sensing data may be used through different, simple or complex procedures (including satellite data assimilation for constructing or driving Soil Vegetation Atmosphere Transfer and crop models) for monitoring crop vegetation and plant evapotranspiration (Olioso et al. 2003).

5.1.7.1 CROPWAT Model Description and Management Variables

Satellite information can be used as input for the CROPWAT model to estimate daily evapotranspiration during the growing season from climatic data and calculate the daily actual evapotranspiration from satellite images based on the surface energy balance (Doorenbos and Pruitt 1976, 1977; Smith et al. 1991; Smith 1992, 1993; Stancalie et al. 2010). Its main functions are to calculate reference evapotranspiration, crop water requirements and crop irrigation requirements; to develop irrigation schedules under various management conditions and water supply schemes; to estimate the rain-fed production and drought effects; evaluate the efficiency of irrigation practices (Doorenbos and Pruitt 1977).

This model can be considered a useful method to support decision-making for irrigation planning and management, which is able to calculate the reference evapotranspiration, crop water requirements and irrigation requirements, to develop irrigation schedules, based on a daily soil–moisture balance using various options for the water supply and irrigation management conditions. This model also allows the development of recommendations for improved irrigation practices, the planning of irrigation schedules and the assessment of production under rain-fed conditions or deficit irrigation (Smith et al. 2002).

As input data, the model requires the following climatic, crop and soil data (Stancalie et al. 2010):

- reference crop evapotranspiration: (ET_o) values measured or calculated using the FAO Penman–Monteith equation based on monthly climatic average data of the minimum and maximum air temperature (C), relative humidity (%), sunshine duration (h) and wind speed (m/s);

- rainfall data: (daily/monthly data); monthly rainfall is divided for each month into a number of rainstorms;
- a cropping pattern: consisting of the crop type, planting date, crop coefficient data files (including K_c values, stage days, root depth, depletion fraction and K_y values) and the area planted (0–100% of the total area);
- soil type: total available soil moisture, maximum rain infiltration rate, maximum rooting depth and initial soil moisture depletion (% of the total available moisture);
- scheduling criteria: several options can be selected regarding the calculation of the application timing and application depth.

The CROPWAT model operates in two modes: computing the actual evapotranspiration using the FAO Penman–Monteith equation and using directly the evapotranspiration measurements values. The crop water requirements (CWR) or maximum evapotranspiration (ET_m) are calculated as

$$CWR = ET_o * \text{Crop } K_c \tag{5.6}$$

The average values of the crop coefficient (Kc) for each time step are estimated through linear interpolation between the Kc values for each crop development stage. The 'Crop Kc' values are calculated as

$$\text{Crop } K_c = K_c * (\text{Crop Area}) \tag{5.7}$$

where 'Crop Area' represents the area covered by the crop. Thus, if the crop covers only 50% of the area, the 'Crop K_c' values will be half of the Kc values in the crop coefficient data file.

For crop water requirements and scheduling purposes, the monthly total rainfall has to be distributed by equivalent daily values. CROPWAT for Windows achieves this in two steps. First, the month-to-month rainfall is smoothed in a continuous curve (the default curve is a polynomial curve, but other smoothing methods available in the program can be selected, e.g. linear interpolation between monthly values). Further, the model assumes that the monthly rain falls during six separate rainstorms, one every 5 days, but the number of rainstorms can be changed. The effective rainfall is calculated automatically in the files with rainfall data.

Possibilities to use the satellite-based data as input into the CROPWAT model are limited because this model was not considered to use satellite-derived information directly. This information can be useful for the comparison/validation procedures of some model input/output data, like precipitation, sunshine duration and evapotranspiration.

The CROPWAT model can use satellite-derived information in various ways:

- measured evapotranspiration may be replaced with estimations derived from satellite data, for comparison and validation procedures;
- satellite-derived evapotranspiration values may bring better accuracy for the spatial variation of the punctual computing values;

- satellite information may be used for the assessment of some reference parameters of the actual evapotranspiration (e.g. land surface temperature and vegetation indices).

5.1.8 *Method for the Assessment of the Crop Actual Evapotranspiration Using Satellite Data*

The method used for the computation of the daily crop actual evapotranspiration, ET_{cj}, is based on the energy balance of the surface. The method uses the connection between evapotranspiration, net radiation and the difference between surface and air temperatures measured around the local time of the satellite passage. The method uses a simplified linear relationship of the form:

$$ET_{cj} - R_{nj} = A - B(T_s - T_{amax}) \tag{5.8}$$

where R_{nj} is the daily net radiation; T_s and T_{amax} are the surface and air maximum temperature, respectively; A, B are coefficients which depend on the surface type and the daily mean wind speed. Coefficients A and B may be determined either analytically, on the basis of the relationships given by Lagouarde and Brunet (1991), or statistically. A and B coefficients are stable in the case of mature crop vegetation cover and in clear sky conditions. Especially the B coefficient varies considerably the function of the land vegetation cover percent.

The use of the multispectral satellite data can ensure the improvement of the classical methods applied in determining the agrometeorological parameters, including evapotranspiration. Stancalie et al. (2010), have shown that the satellite data could bring an important contribution only for comparisons and for the validation of the model outputs for some parameters, like the actual crop evapotranspiration. The analysis of the CROPWAT model results concerning the comparison of the daily actual crop evapotranspiration (ETc) calculated by using climatic data versus satellite estimations based on the surface energy balance showed that ETc values from satellite information are in general higher than those simulated by the model, the differences being from +0.45 to 1.9 mm/day. Preliminary results emphasized a good correlation between the simulated values (CROPWAT) and those derived from the satellite data, with relative errors of ±10–20%. Generally, the CROPWAT model underestimates the actual crop evapotranspiration with respect to the estimated values (Stancalie et al. 2010).

5.2 Estimation of Crop Yield Using Remote Sensing (RS) Data

5.2.1 Introduction

Environmental factors significantly affect agricultural yield and production. Agricultural yield is highly variable affected by weather influences on crop growth and development stages. Furthermore, the spatial variability of soil properties, interacting with the weather, causes spatial yield variability. The loss and variability of yield may be alleviated through crop agronomic management practices such as planting, fertilizer application, irrigation and tillage. Crop yield may be estimated, monitored and forecasted using crop simulation models and remote sensing. Both approaches are valuable for optimizing crop yield and minimizing the effects of weather influences on agricultural production.

5.2.2 Crop Models and Remote Sensing

Crop models simulate crop growth under different environmental and management conditions, taking various limiting factors (e.g. soil, weather, water, and nitrogen) into account in a dynamic way. Thus, these simulation models are good tools for diagnosing crop growing conditions or for predicting yield over large areas. They allow us to better understand the spatial variability of crop behaviour and to design tools for crop monitoring over large area scales (Launay and Guerif 2005).

Remote sensing (RS) provides extensive spatial information on the actual growth status of the crop. The use of RS data throughout the growing season is one of the methods that allow a spatial calibration of the crop model by locally estimating the missing information on model parameters (Batchelor et al. 2002). Several ways of using RS data with crop models have been explored (Delécolle et al. 1992; Fischer et al. 1997). RS is able to provide spatial and temporal distributed information on vegetation cover, as well as on state variables of the models, such as biomass and leaf area index. RS data assimilation appears as a good tool to provide more information about canopy state variables in time and space. It permits a reduction in the uncertainties in crop functioning model predictions. The combined use of the assimilation of data acquired during the crop growth, and of spatially distributed data obtained by remote sensing allows relevant localized simulations of the plant and soil state variables (Guérif et al. 2003).

The crop models are powerful tools for dealing with agro-landscape issues like land capabilities, yield estimates, impact of climatic change on crop functioning, adaptation, mitigation, impact of agriculture on soil, water and air quality and impact of agriculture on climatic change. When provided with relevant climatic, soil and crop data, the crop models allow dynamic simulation of the behaviour of the soil–plant system. Subsequently, they can give dynamic diagnostic information about soil and crop conditions.

The crop models have been developed at different spatial scales: large region (e.g. a country yield estimates), small region (yield estimates for producers/organizations), field/farm (evaluation of cropping systems, decision-making tools for crop management, precision agriculture, impact of global change), etc. Remote sensing data are used, with the models applied at different scales, but with some parameters difficult to obtain at a coarse scale. It is important to know the size of input data errors and how they propagate, and also the 'model errors', that is, the accuracy with which the models represent the real world.

5.2.3 Estimation of Crop Yield

Quantifying crop production at regional scales is critical for a wide range of applications and remote sensing offers great potential for monitoring regional production. RS techniques have provided the opportunity for optimizing and predicting crop yields in the field of agriculture. Yield prediction plays a vital role in Agricultural Policy and provides useful data to policymakers (Papadavid and Hadjimitsis 2014). The agricultural application of such techniques requires quantitative processing of remotely sensed data with high accuracy and reliability. Especially for yield prediction and estimation of crops, it is necessary to achieve very high accuracy and reliability. This is the reason why even today no routine yield estimation method for a wide range of operational applications has been developed from the scientific community. However, important steps have been done towards this direction. The crop production environment consists of many crop and meteorological parameters that can affect production (Papadavid and Toulios 2018). The models or algorithms developed for crop yield estimation usually take into account the most basic parameters which can be measured not only directly but also indirectly using remote sensing techniques. This enables potential users to use only remotely sensed data—after they correlate them to the specific parameters. Based on data obtained before harvest, crop yield can be predicted with acceptable accuracy.

The use of RS has been proven to be effective in monitoring the growth of agricultural crops and in irrigation scheduling and efforts have been made to develop various indices for different crops of different regions throughout the globe (Balaghi et al. 2005; Atkinson et al. 2012). The production and prediction of crop yield has a direct impact on year-to-year national and international economies and play an important role in food management (Hayes and Decker 1996). Satellite remote sensing offers promise as a tool to assess crop yields, with many studies demonstrating high correlations between satellite-based estimates and traditional sources in specific case studies (Clevers 1997; Shanahan et al. 2001). That is exactly the main idea behind the yield prediction using proximal remotely sensed data. Although satellites still play a limited role in most operational efforts to monitor yields, several recent developments have enabled progress towards the more routine use of satellites for yield assessment. Whereas earlier approaches tended to be specific to a given crop or region, often with the year and site-specific calibration; newer algorithms offer more scalable approaches. This fact has pushed the remotely sensed method against other point-to-point-methods.

Crop production is strongly related to many crops, meteorological and environmental factors. Field experiments from the early 1960s have shown that these factors proportionally affect the production of a crop (Quarmby et al. 1993) while new experiments have indicated the possibility to estimate a crop's yield by combining these factors along with remotely sensed data (Balaghi et al. 2005; Atkinson et al. 2012; Bregaglio et al. 2014). Crop yield, defined as the ratio of the total mass of harvested product to the cropped area, is one of the most basic and widely sought measures of productivity (Carletto et al. 2015). The utility of yield data depends on the timeliness and spatial scale of estimates relative to the needs of a particular application.

Soil humidity and crop temperature affect the volume of production in almost all crops (Metochis et al. 1995). Numerous studies have been carried out on the impact of climate variability and change on crop production as a function of water balance (Wilhite 1993; Bacsi and Hunkar 1994; Alexandrov et al. 2002). The soil humidity and its variance is proportionally related to the production, especially for the cereals cultivated in semi-arid conditions. The NDVI data have been used extensively in vegetation monitoring, crop yield assessment and forecasting (Domenikiotis et al. 2004; Prasad et al. 2006). Although vegetation development of crop fields may differ from those of natural vegetation because of human influences involved such as irrigation, use of fertilizer and pesticides, NDVI is considered as a valuable source of information for the crop conditions (Dadhwal and Ray 2000). Different methods such as neural network (Stoikos 1995), autoregressive states-space models (Wendroth et al. 2003), least-square regression (Jones 1982), exponential-linear crop growth algorithm (Oroda 2001) and numerical crop yield model (Hayes et al. 1982) have been used in NDVI to predict crop yield with moderate success.

Recently, many studies have been performed in order to derive the crop phenological stages based on satellite images (Papadavid and Hadjimitsis 2009; Papadavid et al. 2011). These studies aim to validate vegetation indices for monitoring the development of the phenological cycle from time-series data. Sakamoto et al. (2005), Minacapilli et al. (2008) and Papadavid et al. (2011) used times series of remotely sensed data in order to develop a new systematic method for detecting the phenological stages of different crops from satellite data while Bradley et al. (2007) introduced a curve-fitting procedure in order to derive inter-annual phenologies from time series of noisy satellite NDVI data. Funk and Budde (2009) have used an analogous metric of crop performance based on time series of NDVI satellite imagery. Similarly, ground spectro-radiometers have also been used in order to study the phenological cycle of potatoes. Papadavid et al. (2009, 2010, 2011) have shown that field spectroscopy and empirical modelling, when successfully integrated, can develop new models of leaf area index (LAI) and crop height, during the phenological cycle of crops which are also factors that affect crop yield and should be analysed to predict yield (Papadavid and Toulios 2018). The CY-SEBAL algorithm (Papadavid et al.

2017) provides yield values with no statistically significant difference from the real values. The model takes into account the most basic parameters which can be measured not only directly but also indirectly using remote sensing techniques. This enables potential users of the algorithm to use only remotely sensed data—after they correlate them to the specific parameters. The model shows a promising result, which can be useful for supporting policymakers and producer groups for developing sound agricultural policies based on scientific data.

5.3 Discussion and Conclusions

Satellite data offer an unprecedented potential for agriculture landscape research provided that separate sensor/satellite data are integrated into high-quality, globally integrated products. Also, climate change influence on the biosphere can be monitored with the use of satellite data. Presence of Earth Observation, meteorological and environmental satellites in space since the 1960s allows for real agro-landscape studies.

There are examples of successful use of long time satellite data series. There have been also recorded difficulties in the use of satellite data due to the limitations of this technology (Struzik et al. 2008). The main limitation issues are the accuracy and stability of satellite measurements since most of the operational satellites were created as weather platforms. As a result, long-term absolute accuracy of satellite measurements was not a crucial issue. In the measurement of the biophysical variables, it is vital for understanding processes and changes. However, it is not as necessary for determining long-term changes or trends as long as the data set has the required stability. And, when it comes to building satellite instruments, stability appears to be less difficult to achieve than accuracy. The difficulty arises because of the many known and unknown systematic uncertainties that are to be accounted for in the calibration of the instruments. Although excellent absolute accuracy is not critical for trend detection, it is crucial for understanding processes and changes. Much improved post-launch calibration of satellite instruments and intercalibration of similar instruments flying on different satellites is highly required to achieve continuity of observations. This requires overlapping periods of consecutive satellite missions. There are also problems concerning data management during processing and reprocessing because the rapid development of Earth Observations resulted in extremely huge volumes of satellite data. It is envisaged that in the future missions, new and more accurate sensors will be developed.

Despite the limitations of satellite measurements, a wide spectrum of remote sensing-based ET methods has been developed during the last three decades. However, there is no consensus on which method is the best considering that each method has both advantages and disadvantages relative to the other approaches. Future research should be based on the identification of uncertainty sources and error propagation analyses and building integrated ensemble methods combining various remote sensing ET estimation methods by taking advantage of their respective merits and compensating their limitations.

The integration of RS and crop models represents an interesting alternative in crop yield estimation, monitoring and forecasting compensating their limitations. Remote sensing can provide quantitative spatial information about the crop status at any given time during the crop growing season and the crop models can give information about the crop growth every day throughout the growing season. Remote sensing may indirectly provide spatial and temporal information about canopy state variables used by the crop models. This information may, then, be used to tune and refine the model simulation.

Concluding, it is evident that the development of remote sensing capabilities has made great contributions to operational environmental and hydrological monitoring and agricultural management. Continuation of remote sensing development through the development of better and more accurate satellite sensors will contribute to further improvements in the accuracy and capacity of ET and crop yield estimation and monitoring.

References

Alexandrov V, Eitzinger J, Cajic V, Oberforster M (2002) Potential impact of climate change on selected agricultural crops in northeastern Austria. Glob Change Biol 8:372–389

Allen RG, Pereira LS, Raes D, Smith M (1998) Crop evapotranspiration-guidelines for computing crop water requirements-FAO irrigation and drainage paper 56, vol 300. FAO, Rome, p 6541

Allen R, Tasumi M, Trezza R (2007) Satellite-based energy balance for mapping evapotranspiration with internalized calibration (METRIC)-model. J Irrig Drain Eng 133:380–394

Allen R-G, Pereira L-S, Howell T, Jensen M-E (2011) Evapotranspiration information reporting: I. Factors governing measurement accuracy. Agric Water Manag 98:899–920

ASCE (2005) The ASCE standardized reference evapotranspiration equation. ASCE-EWRI, Reston, VA

Atkinson P-M, Jeganathan C, Dash J, Atzberger C (2012) Inter-comparison of four models for smoothing satellite sensor time-series data to estimate vegetation phenology. Remote Sens Environ 123:400–417

Bacsi Z, Hunkar M (1994) Assessment of the impacts of climate change on the yields of winter wheat and maize, using crop models. J Agric Sci 149:145–157. https://doi.org/10.1017/s0021859610000808

Balaghi R, Tychon B, Eerens H, Jlibene M (2005) Empirical regression models using NDVI, rainfall and temperature data for the early prediction of wheat grain yields in Morocco. Int J Appl Earth Obs Geoinf 10:438–452. https://doi.org/10.1016/j.jag.2006.12.001

Bastiaanssen W-G-M, Menenti M, Feddes R-A, Holtslag A-A-M (1998a) A remote sensing surface energy balance algorithm for land (SEBAL) 1. Formulation. J Hydrol 212:198–212. https://doi.org/10.1016/S0022-1694(98)00253-4

Bastiaanssen W-G-M, Pelgrum H, Wang J, Ma Y, Moreno J, Roerink G-J, van der Wal T (1998b) The surface energy balance algorithm for land (SEBAL): Part 2 validation. J Hydrol 212–213:213–229

Batchelor W-D, Basso B, Paz J-O (2002) Examples of strategies to analyze spatial and temporal yield variability using crop models. Eur J Agron 18:141–158

Bausch W-C, Neale C-M-U (1989) Spectral inputs improve corn crop coefficients and irrigation scheduling. Trans ASAE 32:1901–1908

Bellvert J, Adeline K, Baram S, Pierce L, Sanden B-L, Smart D-R (2018) Monitoring crop evapotranspiration and crop coefficients over an almond and pistachio orchard throughout remote sensing. Remote Sens 10. https://doi.org/10.3390/rs10122001

Bhattarai N, Quackenbush L-J, Dougherty M, Marzen L-J (2015) A simple Landsat–MODIS fusion approach for monitoring seasonal evapotranspiration at 30 m spatial resolution. Int J Remote Sens 36:115–143. https://doi.org/10.1080/01431161.2014.990645

Boulet G, Mougenot B, Lhomme J-P, Fanise P, Lili-Chabaane Z, Olioso A, Bahir M, Rivalland V, Jarlan L, Merlin O, Coudert B, Er-Raki S, Lagouarde J-P (2015) The SPARSE model for the prediction of water stress and evapotranspiration components from thermal infra-red data and its evaluation over irrigated and rainfed wheat. Hydrol Earth Syst Sci 19:4653–4672. https://doi.org/10.5194/hess-19-4653-2015

Bradley A-B, Jacob W-R, Hermance F-J, Mustard F-J (2007) A curve fitting procedure to derive inter-annual phenologies from time series of noisy satellite NDVI data. Remote Sens Environ 106:137–145

Bregaglio S, Frasso N, Pagani V, Stella T, Francone C, Cappelli G, Acutis M, Ballaghi R, Ouabbou H, Paleari L, Confalonieri R (2014) New multi-model approach gives good estimations of wheat yield under semi-arid climate in Morocco. Agron Sustain Dev 35(1):157–167

Calera A, Campos I, Osann A, D'Urso G, Menenti M (2017) Remote sensing for crop water management: from ET modelling to services for the end users. Sensors 17:1104

Carletto C, Jolliffe D, Banerjee R (2015) From tragedy to renaissance: improving agricultural data for better policies. J Dev Stud 51:133–148. https://doi.org/10.1080/00220388.2014.968140

Chávez J, Neale C-M-U, Hipps L-E, Prueger J-H, Kustas W-P (2005) Comparing aircraft-based remotely sensed energy balance fluxes with eddy covariance tower data using heat flux source area functions. J Hydrometeorol 6:923–940. https://doi.org/10.1175/jhm467.1(2005)

Chehbouni A, Lo Seen D, Njoku EG, Monteny BM (1996) Examination of the difference between radiative and aerodynamic surface temperatures over sparsely vegetated surfaces. Remote Sens Environ 58:177–186. https://doi.org/10.1016/s0034-4257(96)00037-5

Chiesi M, Battista P, Fibbi L, Gardin L, Pieri M, Rapi B, Romani M, Maselli F (2018) A semiempirical method to estimate actual evapotranspiration in mediterranean environments. Adv Meteorol 2018, Article No. 9792609

Choudhury BJ, Ahmed NU, Idso SB, Reginato RJ, Daughtry C-S-T (1994) Relations between evaporation coefficients and vegetation indices studied by model simulations. Remote Sens Environ 50:1–17

Clevers J-G-P-W (1997) A simplified approach for yield prediction of sugar beet based on optical remote sensing data. Remote Sens Environ 61:221–228. https://doi.org/10.1016/s0034-4257(97)00004-7

Clevers J-G-P-W, Kooistra L, Van den Brande M-M (2017) Using Sentinel-2 data for retrieving LAI and leaf and canopy chlorophyll content of a potato crop. Remote Sens 9:405. https://doi.org/10.3390/rs9050405

Corbari C, Ravazzani G, Galvagno M, Cremonese E, Mancini M (2017) Assessing crop coefficients for natural vegetated areas using satellite data and eddy covariance stations. Sensors 17(11), Article No. 2664. https://doi.org/10.3390/s17112664

Crago R, Friedl M, Kustas W, Wang Y (2004) Investigation of aerodynamic and radiometric land surface temperatures. NASA Scientific and Technical Aerospace Reports (STAR) 42

Dadhwal V-K, Ray S-S (2000) Crop assessment using remote sensing—Part-II: crop condition assessment and yield forecasting. Indian J Agric Econ 55(2):55–67

Dalezios N-R, Dercas N, Spyropoulos N, Psomiadis E (2019) Remotely sensed methodologies for crop water availability and requirements in precision farming of vulnerable agriculture. Water Resour Manage. https://doi.org/10.1007/s11269-018-2161-8

Delécolle R, Maas S-J, Guérif M, Baret F (1992) Remote sensing and crop production models: present trends. 1991/01/14-18. ISPRS J Photogr Remote Sens 47(2–3):145–161

Dhungel R, Allen R-G, Trezza R, Robison C-W (2014) Comparison of latent heat flux using aerodynamic methods and using the penman-monteith method with satellite-based surface energy balance. Remote Sens 6:8844–8877. https://doi.org/10.3390/rs6098844
Dhungel S, Barber M-E (2018) Estimating calibration variability in evapotranspiration derived from a satellite-based energy balance model. Remote Sens 10:1695
Domenikiotis C, Spiliotopoulos M, Tsiros E, Dalezios NR (2004) Early cotton yield assessment by the use of the NOAA/AVHRR derived vegetation condition index (VCI) in Greece. Int J Remote Sens 25(14):2807–2819
Doorenbos J, Pruitt W-O (1976) Guidelines for predicting crop water requirements, FAO irrigation and drainage paper 24, Second ed. Rome, 156 pp
Doorenbos J, Pruitt W-O (1977) FAO irrigation and drainage paper 24, (Rev.) Rome, 156 p
Fischer A, Kergoat L, Dedieu G (1997) Coupling satellite data with vegetation functional models: review of different approaches and perspectives suggested by the assimilation strategy. Remote Sens Rev 15:283–303
Fuentes-Peñailillo F, Ortega-Farías S, Acevedo-Opazo C, Fonseca-Luengo D (2018) Implementation of a two-source model for estimating the spatial variability of olive evapotranspiration using satellite images and ground-based climate data. Water 10(3)
Funk C, Budde E-M (2009) Phenologically-tuned MODIS NDVI-based production anomaly estimates for Zimbabwe. Remote Sens Environ 113:115–125
Glenn E-P, Neale C-M-U, Hunsaker D-J, Nagler P-L (2011) Vegetation index-based crop coefficients to estimate evapotranspiration by remote sensing in agricultural and natural ecosystems. Hydrol Process 25:4050–4062
González A-R, Kjaersgaard J, Trooien T, Hay C, Ahiablame L (2018) Estimation of crop evapotranspiration using satellite remote sensing-based vegetation index. Adv Meteorol 2018, Article No. 4525021. https://doi.org/10.1155/2018/4525021
Goward S-N, Cruickshanks G-D, Hope A-S (1985) Observed relation between thermal emission and reflected spectral radiance of a complex vegetated landscape. Remote Sens Environ 18:137–146. https://doi.org/10.1016/0034-4257(85)90044-6
Gowda P-H, Chávez J-L, Colaizzi P-D, Evett S-R, Howell T-A, Tolk J-A (2007) Remote sensing based energy balance algorithms for mapping ET: current status and future challenges. Trans ASABE 50:1639–1644
Guérif M, Hollecker D, Beaudoin N, Bruchou C, Clastrel P, Houlès V, Machet J-M, Mary B, Moulin S, Nicoullaud B (2003) Conference information: 4th European conference on precision agriculture, 2003 Berlin, Germany, 253–258
Hayes J-T, O'Rourke P-A, Terjung W-E, Todhunter P-E (1982) YIELD: a numerical crop yield model of irrigated and rainfed agriculture. Publications in Climatology, p 35
Hayes M-J, Decker W-L (1996) Using NOAA AVHRR data to estimate maize production in the United States Corn Belt. Int J Remote Sens 17:3189–3200
Hong S, Hendrickx J-M-H, Borchers B (2011) Down-scaling of SEBAL derived evapotranspiration maps from MODIS (250 m) to Landsat (30 m) scales. Int J Remote Sens 32:6457–6477
Hunsaker D-J, Barnes E-M, Clarke T-R, Fitzgerald G-J, Pinter PJ (2005) Cotton irrigation scheduling using remotely sensed and FAO-56 basal crop coefficients. Trans ASAE 48(4):1395–1407. https://doi.org/10.13031/2013.19197
Jaafar H-H, Ahmad F-A (2019) Time series trends of Landsat-based ET using automated calibration in METRIC and SEBAL: the Bekaa Valley, Lebanon. Remote Sens Environ (in press)
Jaber H-S, Mansor S, Pradhan B, Ahmad N (2016) Evaluation of SEBAL model for evapotranspiration mapping in Iraq using remote sensing and GIS. Int J Appl Eng Res 11:3950–3955
Jiang L, Islam S (1999) A methodology for estimation of surface evapotranspiration over large areas using remote sensing observations. Geophys Res Lett 26:2773–2776. https://doi.org/10.1029/1999gl006049
Jones DR (1982) A statistical inquiry into crop–weather dependence. Agric Meteorol 26:91–104

Kalma J, McVicar T, McCabe M (2008) Estimating land surface evaporation: a review of methods using remotely sensed surface temperature data. Surv Geophys 29:421–469. https://doi.org/10.1007/s10712-008-9037-z
Kamble B, Kilic A, Hubbard K (2013) Estimating crop coefficients using remote sensing-based vegetation index. Remote Sens 5(4):1588–1602. https://doi.org/10.3390/rs5041588
Ke Y, Im J, Park S, Gong H (2016) Downscaling of MODIS one kilometer evapotranspiration using Landsat-8 data and machine learning approaches. Remote Sens 8(3), Article No. 215. https://doi.org/10.3390/rs8030215
Koksal E-S, Artik C, Tasan M (2019) Crop evapotranspiration estimations of red pepper using field level remote sensing data and energy balance. Polish J Environ Stud 28:165–175
Kullberg E-G, DeJonge K-C, Chavez J-L (2017) Evaluation of thermal remote sensing indices to estimate crop evapotranspiration coefficients. Agric Water Manag 179:64–73
Kustas W-P, Daughtry C-S-T (1990) Estimation of the soil heat flux/net radiation ratio from spectral data. Agric Forest Meteorol 49:205–223. https://doi.org/10.1016/0168-1923(90)90033-3
Kustas W-P, Norman J-M (1996) Use of remote sensing for evapotranspiration monitoring over land surfaces. Hydrol Sci J 41:495–516
Lagouarde J-P, Brunet Y (1991) A simple model for estimating the daily upward longwave surface radiations from NOAA–AVHRR data. Int J Remote Sens 12:1853–1864
Launay M, Guerif M (2005) Assimilating remote sensing data into a crop model to improve predictive performance for spatial applications. Agr Ecosyst Environ 111:321–339
Li Z-L, Tang R, Wan Z, Bi Y, Zhou C, Tang B, Yan G, Zhang X (2009) A review of current methodologies for regional evapotranspiration estimation from remotely sensed data. Sensors 9:3801–3853
Lian J, Huang M (2016) Comparison of three remote sensing based models to estimate evapotranspiration in an oasis-desert region. Agric Water Manag 165:153–162. https://doi.org/10.1016/j.agwat.2015.12.001
Liaqat U-W, Choi M (2015) Surface energy fluxes in the Northeast Asia ecosystem: SEBS and METRIC models using Landsat satellite images. Agric Forest Meteorol 214–215:60–79. https://doi.org/10.1016/j.agrformet.2015.08.245
Linacre E-T (1977) A simple formula for estimating evaporation rates in various climates, using temperature data alone. Agric Meteorol 18:409–424. https://doi.org/10.1016/0002-1571(77)90007-3
Liou Y-A, Kar S-K (2014) Evapotranspiration estimation with remote sensing and various surface energy balance algorithms: a review. Energies 7:2821–2849. https://doi.org/10.3390/en7052821
Loheide S-P, Gorelick S-M (2005) A local-scale, high-resolution evapotranspiration mapping algorithm (ETMA) with hydroecological applications at riparian meadow restoration sites. Remote Sens Environ 98:182–200. https://doi.org/10.1016/j.rse.2005.07.003
Matese A, Baraldi R, Berton A, Cesaraccio C, Di Gennaro S-F, Duce P, Facini O, Mameli M-G, Piga A, Zaldei A (2018) Estimation of water stress in grapevines using proximal and remote sensing methods. Remote Sens 10:114
Menenti M, Choudhury B (1993) Parameterization of land surface evapotranspiration using a location dependent potential evapotranspiration and surface temperature range. IAHS Publ 212:561–568
Metochis C, Eliades G, Papachristodoulou S (1995) Technoeconomic analysis of irrigation in Cyprus' (in Greek). Agricultural Research Institute, Nicosia, Cypru
Minacapilli M, Iovino M, D'Urso G (2008) A distributed agro-hydrological model for irrigation water demand assessment. Agric Water Manag 95:123–132
Moran M-S, Clarke T-R, Inoue Y, Vidal A (1994) Estimating crop water deficit using the relation between surface-air temperature and spectral vegetation index. Remote Sens Environ 49:246–263. https://doi.org/10.1016/0034-4257(94)90020-5
Navarro A, Rolim J, Miguel I, Catalão J, Silva J, Painho M, Vekerdy Z (2016) Crop monitoring based on SPOT-5 Take-5 and sentinel-1A data for the estimation of crop water requirements. Remote Sens 8(6), Article No. 525. https://doi.org/10.3390/rs8060525

Olioso A, Chauki H, Courault D, Wigneron J-P (1999) Estimation of evapotranspiration and photosynthesis by assimilation of remote sensing data into SVAT Models. Remote Sens Environ 68:341–356. https://doi.org/10.1016/s0034-4257(98)00121-7
Olioso A, Inoue Y, Ortega-Farias S, Demarty J, Wigneron J-P, Braud I, Jacob F, Lecharpentier P, Ottlé C, Calvet J, Brisson N (2003) Assimilation of remote sensing data into crop simulation models and SVAT models. In: Proceedings of the international workshop on use of remote sensing of crop evapotranspiration for large regions, 54th IEC meeting of the ICID, Montpellier, France
Oroda A (2001) The international archives of the photogrammetry. Remote Sens Spatial Inform Sci XXXIV:66–72, Part 6/W6
Ortega-Farías S, Ortega-Salazar S, Poblete T, Kilic A, Allen R, Poblete-Echeverría C, Ahumada-Orellana L, Zuñiga M, Sepúlveda D (2016) Estimation of energy balance components over a drip-irrigated olive orchard using thermal and multispectral cameras placed on a helicopter-based unmanned aerial vehicle (UAV). Remote Sens 8(8), Article No. 638
Papadavid G, Hadjimitsis D-G (2009) Spectral signature measurements during the whole life cycle of annual crops and sustainable irrigation management over Cyprus using remote sensing and spectro-radiometric data: the cases of spring potatoes and peas. In: Proceedings of SPIE, remote sensing for agriculture, ecosystems, and hydrology XI, vol 7472, 747215. https://doi.org/10.1117/12.830552
Papadavid G, Hadjimitsis D-G (2014) An image based method for crop yield prediction using remotely sensed and crop canopy data: The case of Paphos district, western Cyprus. In: 2nd international conference on remote sensing and geoinformation of the environment, RSCy 2014, Paphos, Cyprus, 7–10 April 2014
Papadavid G, Hadjimitsis D, Themistocleous K, Toulios L (2010) Spectral vegetation indices from field spectroscopy intended for evapotranspiration purposes for spring potatoes in Cyprus. Proc SPIE 7824:782410
Papadavid G, Hadjimitsis D, Toulios L, Michaelides S (2011) Mapping Potatoes crop height and LAI through vegetation indices using remote sensing, in Cyprus. J Appl Remote Sens 5:053526. https://doi.org/10.1117/1.3596388
Papadavid G, Hadjimitsis D, Toulios L, Michailides S (2013) A modified SEBAL modeling approach for estimating crop evapotranspiration in semi-arid conditions. Water Resour Manag 27:3493–3506
Papadavid G, Neocleous D, Kountios G, Markou M, Michailidis A, Ragkos A, Hadjimitsis D (2017) Using SEBAL to investigate how variations in climate impact on crop evapotranspiration. J Imag 3(3):30. https://doi.org/10.3390/jimaging3030030
Papadavid G, Toulios L (2018) The use of earth observation methods for estimating regional crop evapotranspiration and yield for water footprint accounting. J Agric Sci 156(5):599–617. https://doi.org/10.1017/s0021859617000594 © Cambridge University Press
Pereira L-S, Allen R-G, Smith M, Raes D (2015) Crop evapotranspiration estimation with FAO56: past and future. Agric Water Manag 147:4–20. https://doi.org/10.1016/j.agwat.2014.07.031
Petropoulos G-P, Srivastava P-K, Piles M (2018) Earth observation-based operational estimation of soil moisture and evapotranspiration for agricultural crops in support of sustainable water management. Sustainability 10(1)
Piedelobo L, Ortega-Terol D, Pozo S-D, Hernández-López D, Ballesteros R, Moreno M-A, Molina J-L, Aguilera DG (2018) HidroMap: a new tool for irrigation monitoring and management using free satellite imagery. ISPRS Int J Geo-Inf 7(6):220. https://doi.org/10.3390/ijgi7060220
Poon P-K, Kinoshita A-M (2018) Estimating evapotranspiration in a post-fire environment using remote sensing and machine learning. Remote Sens 10:1728. https://doi.org/10.3390/rs10111728
Prasad A, Chai L, Ramesh P, Kafatos M (2006) Crop yield estimation model for Iowa using remote sensing and surface parameters. Int J Appl Earth Obs Geoinf 8:26–33
Prueger JH, Hatfield JL, Aase JK, Pikul JL (1997) Bowen-ratio comparisons with Lysimeter evapotranspiration. Agron J 89:730–736

Quarmby N-A, Milnes M, Hindle T-L, Silleos N (1993) The use of multitemporal NDVI measurements from AVHRR data for crop yield estimation and prediction. Int J Remote Sens 14:199–210
Reyes-González A, Kjaersgaard J, Trooien T, Hay C, Ahiablame L (2017) Comparative analysis of METRIC model and atmometer methods for estimating actual evapotranspiration. Int J Agronomy. Article No. 3632501
Roerink G-J, Su Z, Menenti M (2000) S-SEBI: A simple remote sensing algorithm to estimate the surface energy balance. Phys Chem Earth, Part B Hydrol Oceans Atmos 25:147–157. https://doi.org/10.1016/S1464-1909(99)00128-8
Romaguera M, Toulios L, Stancalie G, Nertan A, Spiliotopoulos M, Struzik P, Calleja E, Papadavid G (2014) Identification of the key variables that can be estimated using remote sensing data and needed for water footprint (WF) assessment. In: Presented, during the second international conference on remote sensing and geoinformation of environment, RSCy 2014, Paphos, Cyprus 7–10 April 2014
Saadi S, Boulet G, Bahir M, Brut A, Delogu É, Fanise P, Mougenot B, Simonneaux V, Chabaane Z-L (2018) Assessment of actual evapotranspiration over a semiarid heterogeneous land surface by means of coupled low-resolution remote sensing data with an energy balance model: comparison to extra-large aperture scintillometer measurements. Hydrol Earth Syst Sci 22:2187–2209
Sakamoto T, Yokozawa M, Toritani H, Shibayama M, Ishitsuka N, Ohno H (2005) A crop phenology detection method using time-series MODIS data. Remote Sens Environ 96:366–374
Shanahan J-F, Schepers S, Francis D, Varvel G, Wilhelm W (2001) Use of remote-sensing imagery to estimate corn grain yield. Agron J 93:583–589
Smith M, Allen R, Monteith J-L, Perrier L-A, Segeren A (1991) Report on the expert consultation for the revision of FAO methodologies for crop water requirements. FAO/AGL, Rome
Smith M (1992) CROPWAT. A computer program for irrigation planning and management, FAO irrigation and drainage paper, p 46
Smith M (1993) CLIMWAT for CROPWAT, a climatic data base for irrigation planning and management. FAO irrigation and drainage paper 49, Rome, 113 pp
Smith M, Kivumbi D, Heng L-K (2002) Use of the FAO CROPWAT model in deficit irrigation studies. Water Reports FAO 22, ISSN: 1020-1203
Spiliotopoulos M, Adaktylou N, Loukas A, Michalopoulou H, Mylopoulos N, Toulios L (2013) A spatial downscaling procedure of MODIS derived actual evapotranspiration using Landsat images at central Greece. In: Proceedings of SPIE—the international society for optical engineering, vol 8795, pp 296–299
Spiliotopoulos M, Holden N-M, Loukas A (2017) Mapping evapotranspiration coefficients in a temperate maritime climate using the METRIC model and Landsat TM. Water 9:23
Spiliotopoulos M, Loukas A (2019) Hybrid methodology for the estimation of crop coefficients based on satellite imagery and ground-based measurements. Water 11:1364
Stancalie G, Marica A, Toulios L (2010) Using earth observation data and CROPWAT model for estimation the actual crop evapotranspiration. Phys Chem Earth Parts A/B/C 35:25–30
Stoikos G (1995) Sugar beet crop yield prediction using artificial neural networks (in Greek). In: Proceedings of the modern technologies conference in automatic control, Athens, Greece, pp 120–122
Struzik P, Toulios L, Stancalie G, Danson M, Mika J, Domenikiotis C (2008) Satellite remote sensing as a tool for monitoring climate and its impact on the environment–possibilities and limitations. In: Nejedlik, Orlandini (eds) Survey of agrometeorological practices and applications in Europe regarding climate change impacts. COST 734, ESF, pp 205–236
Su Z (2002) the surface energy balance system (SEBS) for estimation of turbulent heat fluxes. Hydrol Earth Syst Sci 6:85–99
Taghvaeian S, Chávez J-L, Bausch W-C, De Jonge K-C, Trout T-J (2014) Minimizing instrumentation requirement for estimating crop water stress index and transpiration of maize. Irrig Sci 32:53–65

Toulios L, Stancalie G, Struzik P, Danson M, Mika J, Dunkel Z, Tsiros (2008) Satellite spectral climatic and biophysical data for warning purposes for European agriculture. In: Nejedlik, Orlandini (eds) Survey of agrometeorological practices and applications in Europe regarding climate change impacts. COST 734, ESF, pp 163–203

Trezza R, Allen R-G, Tasumi M (2013) Estimation of actual evapotranspiration along the middle rio grande of new Mexico using MODIS and Landsat imagery with the METRIC model. Remote Sens 5:5397–5423. https://doi.org/10.3390/rs5105397

Vanino S, Nino P, De Michele C, Falanga Bolognesi S, D'Urso G, Di Bene C, Pennelli B, Vuolo F, Farina R, Pulighe G, Napoli R (2018) Capability of Sentinel-2 data for estimating maximum evapotranspiration and irrigation requirements for tomato crop in Central Italy. Remote Sens Environ 215:452–470

Veysi S, Naseri A-A, Hamzeh S, Bartholomeus H (2017) A satellite based crop water stress index for irrigation scheduling in sugarcane fields. Agric Water Manag 189:70–86

Wendroth O, Reuter H-I, Kersebaum K-C (2003) Predicting yield of barley across a landscape: a state-space modeling approach. J Hydrol 272:250–263

Wilhite D-A (1993) Drought assessment, management and planning: theory and case studies. Kluwer Academic Publishers, Hingham, MA, p 293

Chapter 6
Development of Landscape-Adaptive Land Use of the Upper Volga Region Based on Geostatistical Methods

Dmitry A. Ivanov and Nadezhda V. Grits

Abstract The article describes the applying of statistical methods in research in the field of agricultural geography—the science of the emergence, functioning, and development of agro-geosystems. The agro-geological systems are understood as geocomplexes, certain components of which are changed as a result of human agricultural activity. The paper shows the use of analysis of variance in determining the influence of environmental features of geocomplexes of various ranks on the yield of plants and some other production and natural parameters of farms. The characteristics of the main macro-geocomplexes (agro-ecological sections and meso-geocomplexes), genera, and types of agricultural landscapes located within the Tver region, Russia, and whose conditions are the objects of analysis. Studies have shown that crop yield variability, soil acidity, as well as the proportion of pastures in farms are less than 30% determined by the peculiarities of the natural environment of geosystems of different levels. From 30 to 50% of the content's variability of plant nutrients in the soil, their rockiness and swampiness, as well as the proportion of hay harvests in farms depend on the natural structure of geosystem. Other elements of the structural organization of agricultural enterprises are more than half determined by the nature of the interaction of macro-, meso-, and microterritories. The influence of landscape factors on plant productivity is individual for each crop. So the main part of the variability of potato yield is explained by differences at the micro-level. Only 10–13% of the variability of the yield of perennial grasses and 11–14% of grain yield are explained by the effect of macro factors. For flax, this indicator is 17–19%. The combination of agro-climatic—neotectonic and granulometric properties of agro-geosystems determines about 8% of the yield variability of annual grasses. Consequently, the process of determining the set of crops for the development of farming systems should be based on knowledge of the variability of their yields,

D. A. Ivanov (✉)
Federal State Budgetary Scientific Institution "All-Russian Research Institute of Reclaimed Lands" (FNBI VNIIMZ), Tver Region, Kalininsky District, 27, 170530 Emmaus, Russia

N. V. Grits
Federal State Budgetary Educational Institution of Higher Education "Tver State Agricultural Academy" (FSBEI HE Tver State Agricultural Academy), Tver Region, Tver, ul. Marshal Vasilevsky, 7, 170904 Sakharovo, Russia

W. Mirschel et al. (eds.), *Landscape Modelling and Decision Support*, Innovations in Landscape Research,
https://doi.org/10.1007/978-3-030-37421-1_6

primarily within the types and kinds of agricultural landscapes, and, if possible, at lower taxonomic levels. Indicators such as the degree of development of agricultural landscapes and the average size of the contour of farming land can be very confidently determined when analyzing the conditions of the macroenvironment. However, in this case, it is necessary to take into account local peculiarities which may lead to significant adjustments in each individual farm. The use of track analysis allows it to identify the factors of direct impact on the production process of crops. The comparison of its results with the data of correlation analysis allows it to highlight the "active" factors, the effect of which is described by reliable travel and correlation coefficients, and the "potential" factors whose direct influence is obscured by many reasons. The analysis showed that all the studied types of agricultural landscapes of the upper Volga region can be divided into two groups according to the number of active and potential factors, i.e., in geocomplexes with a relatively homogeneous lithogenic basis and in landscapes on two-layered sediments. The latter are distinguished by a large number of active factors and, commonly, by a wider range, exposed to their effects by crops. This allows us to conclude that the design of crop rotations in the above described groups of landscapes should take into account an individual set of factors which are actively or potentially influencing the production process. The paper shows that the methodology for assessing the productivity of agricultural landscapes can be based on the use of its integral indicators. In conjunction with geographic information systems and mathematical modeling techniques, they allow to reveal areas for measures to optimize the melioration status of geocomplexes. The paper identifies systems of environmental management and land melioration measures aimed to increase the degree of the potential productivity of agricultural landscapes for legume-grass stands. All systems can be combined into three groups: (1) adaptive placement of grasslands depending on the granulometry and geological structure of the soil; (2) adaptive placement of grassland with drainage/irrigation amelioration; and (3) adaptive placement of grassland with water amelioration and land management activities (integration of a certain number of fields for hay harvest into the crop rotation). Areas of these groups of activities were identified at the periphery of the upper Volga region.

Keywords Adaptive-landscape farming systems · Statistical analysis · Agro-geosystems · Agro-technological measures

6.1 Introduction

The synthesis of agronomy and physical geography is fraught with the emergence of a new scientific direction, which we call "agricultural geography"—the science of the emergence, functioning, and development of agro-geosystems (Ivanov 2012). Agricultural systems are geocomplexes, some components of which are changed as a result of human agricultural activity.

Agro-geographical ideas are expressed by some physical geographers (Milkov 1973; Nikolaev 1987, 1997; Prokayev 1983: Shvebs 1988) and agricultural scientists (Kashtanov and Shcherbakov 1993; Kashtanov et al. 1994; Kiryushin 1993, 1996; Shcherbakov et al. 1994). The study of the impact of the landscape environment on the productivity of cultivated plants and the characteristics of their cultivation is of great importance for agricultural geography. On the basis of the gained knowledge, it is possible to design the newest landscape-reclamation farming systems, which allow it to determine environmental-addressed agro-technical measures and, thus, adapt crop production to the environmental conditions. A special place is occupied by studies of the influence of landscapes of various hierarchical levels on the production process of plants and the natural production parameters of farms. They allow it to create models of farming systems of different levels and, thus, most fully take into account the landscape conditions of the territories in the design of land use. This paper shows the results of various geostatistical analyses of the spatial variability of crop yields and the characteristics of the natural production environment of farms in various landscapes in the Tver region, Russia.

6.2 Application Variance Analysis in Agro-Geographical Practice

The most adequate level of elements development of the farming system should be determined in the course of analyzing the hierarchical structure of environmental factors. It should show the levels at which the studied factor is most pronounced, as well as taxonomic cells in which it can be neglected. Mostfully the hierarchical structure of factors reflects the variance analysis for unorganized plans. Below are the results of a three-factor variance analysis of the crop yields dependence and some other production and natural parameters of farms on the properties of agro-ecological sections (A), genera of agricultural landscapes (B) and types of agricultural landscapes (C).

The agro-ecological section is understood as a large territory occupying a landscape province (or part of it). There are four agro-ecological sections in the Tver region (Fig. 6.1). The area of Valdai agro-ecological section coniferous broad-leaved forests within the region is equal to 27.2 thousand km^2. The proximity to the Baltic Sea causes here the softness of winter, heavy rainfall, low continental climate, and high hydrothermal coefficient (HTC). The main type of relief is the finite-moraine ridges, and the sediments are the bottom and terminal moraine of the Valdai (Wurm) glacier. Most of the farms here are located at an altitude of 193 m above sea level—there are significant elevation differences, hardened soil, small confluence of land and a significant proportion of arable land within their limits. The close occurrence of moraine loams to the surface caused significant deposits of potassium and phosphorus in the arable horizons.

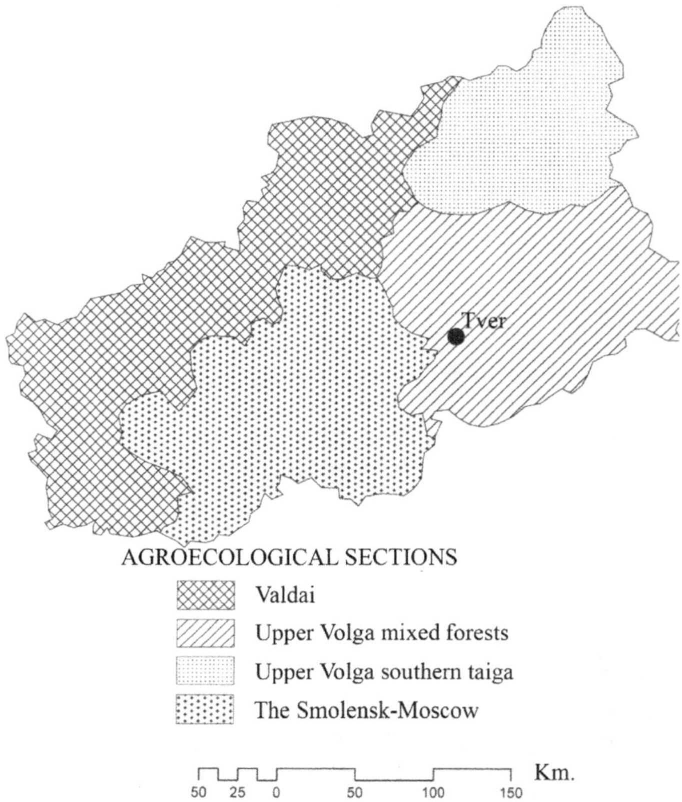

Fig. 6.1 Agro-environmental sections of the upper Volga region

The area of Smolensk–Moscow agro-ecological section of coniferous and broad-leaved forests is 21.7 thousand km^2. The nature of its climate differs from the above described in colder winters, less precipitation, and lower hydrothermal coefficient (HTC). Its orolithogenous basis was formed as a result of the stagnation of glacier waters and differs from the above described territory by a considerable age and greater homogeneity. The prevailing type of relief is the rolling plains, sediments—silty cover loam. The average height of farms is 224 m above sea level. It is characterized by erosional territory dismemberment (the average elevation difference in the farm is 69 m), a small amount of land, a significant distribution of washed away soils, and relatively large size of the contour of the site. Compared to the Valdai agro-ecological section, there is a higher soil bonitet, due to significant humus reserves and physical clay content, however, the pH of the arable layer and the content of potassium and phosphorus in it are noticeably lower. Upper Volga agro-ecological section of taiga, occupies 12.8 thousand km^2. Its climate is characterized by high continental and relatively low January temperatures. Its surface is much older than the Valdai and Smolensk–Moscow agro-ecological sections and is characterized by great homogeneity in lithological, orographic, and neotectonic relations. The dominant type of relief is the low-impressed plains, sediments—covering and moraine loams. The average height of farms above sea level is 156 m; the variability of the heights is about 22%. The agro-ecological section is characterized by a small soil stone content, but their significant erosion. The soil quality here is very high due to the large reserves of humus, phosphorus, and potassium in the arable horizons, which is reflected in the average size of the land contour.

For the Tver region, there is a maximum concerning the continental climate in the coniferous broad-leaved forests of the upper Volga agro-ecological section with an area of 23.0 thousand km^2. It mainly consists of flat moraine outwash plains, outwash plains, and alluvial outwash plains that experience an intense sinking. The average altitude level of the location of farms here is 148.4 m. The maximum elevation difference is less than 48 m. The agro-ecological section is characterized by significant soil entrainment, which prevents the creation of large land areas (the average size of the contour is 6.5 ha). The soil rating is very high due to the significant content of humus.

Within each agro-ecological section, genera of agricultural landscapes are distinguished, which, unlike the agro-geological systems described above, do not possess monolithic ranges. The characteristic features of their natural environment are determined by the granulometric composition of the soil-forming rocks. "Sandy" and "loamy" genera of agro-landscapes are distinguished (Kovalev et al. 2004). They unite in non-taxonomic aggregates, called groups (Preobrazhensky 1966). The genus of agro-landscapes consists of types of agro-landscapes. The agricultural landscapes type is a territory compatible with a specific typological physiographic landscape—a geosystem characterized by a single genesis of relief and soil-forming rocks and the maximum severity of vertical and horizontal interrelationships of the components of the geographic envelope. Its selection within the macroterritory is based on the combinations study of the most important components of the landscape and, above all, the types of the territory water supply, soil-forming rocks, soils, topography, plant associations. There are 11 non-taxonomic groups of agro-landscape types in the Tver region (Fig. 6.2, Table 6.1).

The structure of the hierarchical complex under study is presented in Table 6.2. For the purposes of variance analysis, farms located in types of landscapes belonging to four extra-taxonomic groups were selected. This is explained by the fact that only the moraine plains of Moscow (Mindel) and Valdai (Wurm) age (type 4) the sandy-loamy lake-glacial plains (type 6), the moraine-lake-glacial plains (type 7) and the moraine outwash plains (type 8) are presented in all agro-ecological sections.

Agro-geosystems that belong to the same group of types of agricultural landscapes, but are located in different agro-ecological sections, differ from each other in the agro-climatic and neotectonic relation. Types of agro-landscapes located within the same agro-ecological section, but belonging to different genera, differ from each other in the grain size characteristics of the soil. Geocomplexes belonging to one agro-ecological section and one genus of agricultural landscapes differ from each other in the degree of prevalence of moraine deposits and geomorphological parameters.

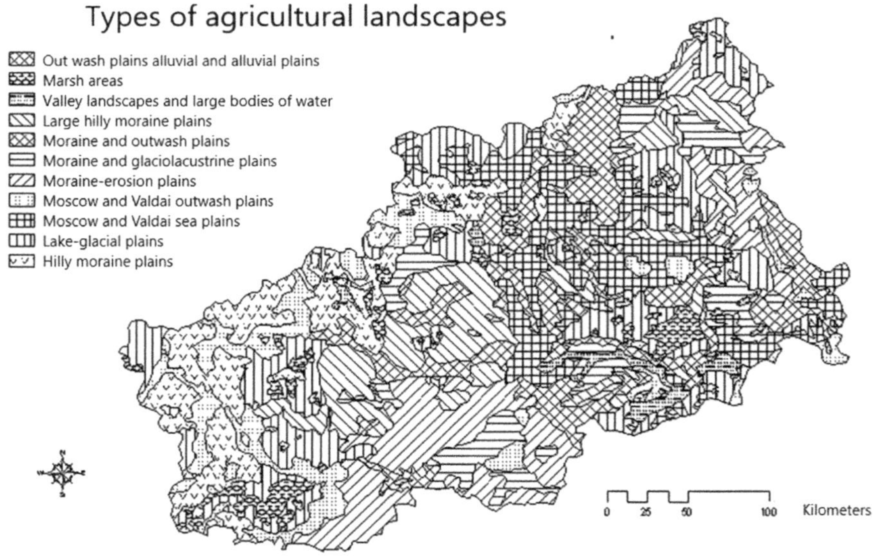

Fig. 6.2 Types of agricultural landscapes of the Tver region

The results of variance analysis are given in Table 6.3. The ranking of farm parameters according to the degree of influence on their spatial variability of the studied factors allowed us to divide them into three groups:

1. parameters with less than 30% of the variability of which is determined by the analyzed factors,
2. parameters with 30–50% of the variability and
3. parameters with more than 50% of the variability.

The first group includes such parameters as crop yields, soil acidity, as well as the proportion of pastures and deposits in farms. The second includes the content of plant nutrients in the soil, rockiness and waterlogging of the soil, as well as the proportion of haymaking. The third group includes the other elements of the structural organization of agricultural enterprises.

If we consider that the residues (values other than trend) in this case characterize the difference between farms within a particular type of agricultural landscapes, it should be assumed that they reflect the heterogeneity of agro-geosystem parameters at the level of types of agricultural landscapes-physical and geographical areas. It can be stated that the parameters of the first group mainly vary at the level of lower meso-units, while the structural organization of farms is more than half dependent on the conditions of macro- and higher meso-units.

The influence of factors of different levels of landscape environment on the productivity of plants is individually for each culture. Thus, the main part of potato yield variability is explained by differences between types of agricultural landscapes. Only 10–13% of the variability in the yield of perennial grasses and 11–14% of

Table 6.1 Characteristics of the natural conditions of the main agricultural landscape types of the Tver region (Atlas 1964; Dorofeev 1990)

Name of the agricultural landscape type (number)	Characteristics of natural conditions
Moraine-erosion plains (1)	Elevated, drained, gentle, wavy, less often wavy Moscow-era moraine-erosional plains with a cover of loess-like loam. Strongly covered, with remnants of spruce, spruce-broad-leaved forests on sod-podzolic silt-loamy soils
Large hilly moraine plains (2)	Elevated, drained, large hilly, with areas of finite-moraine-ridge relief, mainly boulder-loamy, moderately developed moraine plains of Moscow age covered with spruce, spruce-pine and spruce-small-leaved forests on sod-podzolic, mainly loamy soil
Hilly moraine plains (3)	Elevated and exalted drained undulating and hilly ridges with numerous lake basins and kames, Valdai age moraine plains with heterogeneous surface deposits, poorly covered with fir, spruce, pine and fir-aspen forests on sod-podzolic soils of different particle size distribution
Moscow and Valdai moraine plains (4)	Raised, slow-drained, wavy, with areas of hilly relief of sandy-loamy moraine plains of Moscow and Valdai age, moderately covered with spruce, spruce-broad-leaved and small-leaved forests on sod-podzolic-gley soil, mainly loamy
Outwash plains (5)	Low-lying, less elevated, slowed-drained, gently undulating sandy plains of Moscow and Valdai age, low and medium covered with pine and aspen forests on podzolic, sod-podzolic, and sod-podzolic-gley sandy soils
Lake-glacial plains (6)	Lowland, less often raised, undrained flat sandy and sandy-clayey plains, mostly poorly covered with pine, spruce-pine, and small-leaf forests on sod-podzolic-gley and peaty-podzolic-gley sandy soils
Moraine-lake-glacial plains (7)	Uneven, slow-drained gentle wavy, less often flat sandy-loamy (sometimes covered with thin layers of surface loam) moraine-lake-glacial plains, moderately covered with pine-spruce and small-leaved forests on soil of different granulometry

(continued)

Table 6.1 (continued)

Name of the agricultural landscape type (number)	Characteristics of natural conditions
Moraine outwash plains (8)	Different high-altitude, slow-drained, wavy, with areas of hilly relief moraine-sandstone plains composed of alternating sand and boulder loam, moderately covered with pine-spruce and small-leaved forests on sod-podzolic soil of different granulometry
Alluvial and outwash plains alluvial plains (9)	Low-lying flat sandy-loamy alluvial and gentle wavy, often with a finely undulating eolian relief sandy-loamy plain covered with pine forests in combination with grasses and mixed grass meadows on sod podsolic and Soddy different soil size distribution
Swamp areas (10)	Large arrays of bogs of various types on peatlands of various thickness and botanical composition
River valleys (11)	The coastal zone of large reservoirs on swampy turf of different soil granulometry

Table 6.2 Scheme of the three-factor dispersion analysis of the influence of environmental features of agro-geosystems of different hierarchical levels on the natural and production parameters of farms

First-order factors (agro-ecological section specifics)	Factors of the second-order (specificity of genera of agro-landscapes)	Factors of the third order (specificity of types of agro-landscapes)	Number of recurrences (specific farms)
Valdai	Sand	6	12
		7	3
	Loam	8	6
		4	6
Smolensk–Moscow	Sand	6	4
		7	7
	Loam	8	4
		4	7
Vernevolzhsky southern taiga	Sand	6	11
		7	8
	Loam	8	5
		4	4
Upper Volga mixed forests	Sand	6	11
		7	6
	Loam	8	6
		4	14

Table 6.3 Assessment of the natural environment's influence of agro-geological systems of different hierarchical levels on the spatial variability of agroecological parameters within the Tver region

Parameters	Factors weight						
	A	B	C	AB	AC	BC	ABC
Soil ranking			5.9		7.0		13.1
Stony land	8.2	5.0			8.3		9.6
The size of the contour land	33.6	7.7			20.4	1.4	
Soil potassium	19.0	4.0	4.0	9.9			
Phosphorus content in soil	23.0			6.0		2.8	
pH	6.9						11.6
Wetlands	16.5	3.6	3.5	7.5			
Waterlogged arable land	26.4	4.8	2.9				
Share of arable land in the economy	34.8	2.0	4.6	3.7	14.3		3.3
The proportion of hay in the farm	23.1		4.2	5.1	5.1		
The share of pastures on the farm	11.3		7.7				
Share of nonagricultural areas in the economy	28.6	2.0	4.1	5.2	11.3		
The productivity of grain	13.7						
The productivity of flax	17.4						9.7
Productivity of annual grasses				8.2			
Productivity of perennial grasses	9.5		7.2		7.9		

cereals is explained by the action of macro factors. In flax, this figure is 17–19%. The combination of agro-climatic-neotectonic and granulometric properties (ABC) agro-geosystems accounts for about 8% of the productivity variability of annual grasses. Consequently, the process of identifying a set of crops in the development of a farming system should be based on knowledge of the variability of their yields, especially within the types and types of agricultural landscapes, and if possible at lower taxonomic levels.

The parameters of the third group are mainly determined at the stage of natural conditions analysis of agro-ecological sections. Such indicators as the degree of development and plowing of agricultural landscapes and the average size of the contour of the land within them can be determined very confidently in the analysis of the conditions of the macroenvironment, but in this case, it is necessary to take into account local features that can make significant adjustments in each particular farm (Kovalev et al. 2000).

6.3 The Results of the Application of the Limit Analysis in the Development of Farming Systems

In our opinion the analysis of travel factors is very promising method of determining the factors affecting the growth of a particular culture, is the analysis of travel factors. Developed by Wright in 1932, it is an effective way to identify causes and effects in a system of interrelated traits. Its essence lies in the decomposition of the correlation of the dependent variable with each independent variable to the direct effect of one characteristic and the indirect effects of others included in the data array. The travel coefficients can be positive or negative. They, in contrast to the correlation coefficients, can be greater than one modulo (Sedlovsky et al. 1982).

Track analysis allows us to identify the factors of direct action, and the comparison of its results with the data of correlation analysis makes it possible to classify them, highlighting the "active", the action of which is described by reliable track and correlation coefficients, and "potential", the direct influence of which is obscured by many other factors.

A very simple analysis that consists in calculating the number of active and potential factors affecting the growth of crops within each studied type of agricultural landscapes showed that all the studied agro-geosystems of the upper Volga region by the number of active factors can be divided into two groups—geocomplexes with a relatively homogeneous lithogenic base and landscapes on double-stranded sediments. The latter, which include lake-glacial sandy-clay plains (6), sandy-loamy moraine-lake-glacial plains (7), and moraine outwash plains (8), are characterized by a large number of active factors and, as a rule, a wider range of exposed cultures (Fig. 6.3).

Analysis of potential impact factors reveals the same patterns. It turns out that, first, the difference between the groups within the potential factors is even more contrasting, and second, all the crops of the crop rotation in the landscapes of the second group are exposed to potential factors, and third, the distribution of factors by crops is more uniform than in the first group (Fig. 6.4).

All this leads to the conclusion that the design of crop rotations in the above groups of landscapes should take into account an individual set of factors that actively or potentially affect the production process.

It should be noted that the comparison of the conditions of growth of crops of fruit-crop rotation in different geocomplexes of the Tver region reveals their rather clear dependence on the genetic characteristics of the territory. On average, for all geocomplexes, the greatest probability of active impact on crop productivity is manifested in the factors of land organization within farms, as well as the productivity of the agricultural landscape (6.7%). Then there are orographic (5.6%) and agro-climatic (4.9%) factors, the least significant factors of soil fertility (4.4%). For geo-complexes, characterized by duality of soil-forming rocks, the most likely interaction of crops in crop rotation (17%), then there are agro-climatic factors and features of the organization of the territory of farms (10%), the least significant orographic and agro-chemical factors (7%). In the conditions of monogenicity of soil-forming rocks

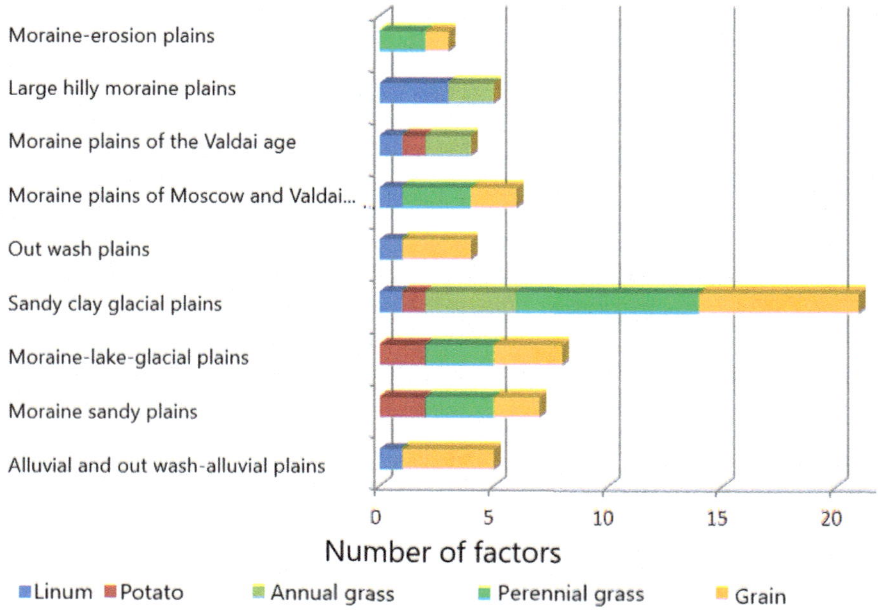

Fig. 6.3 The number of factors actively influencing the production process of plants in various agro-geosystems

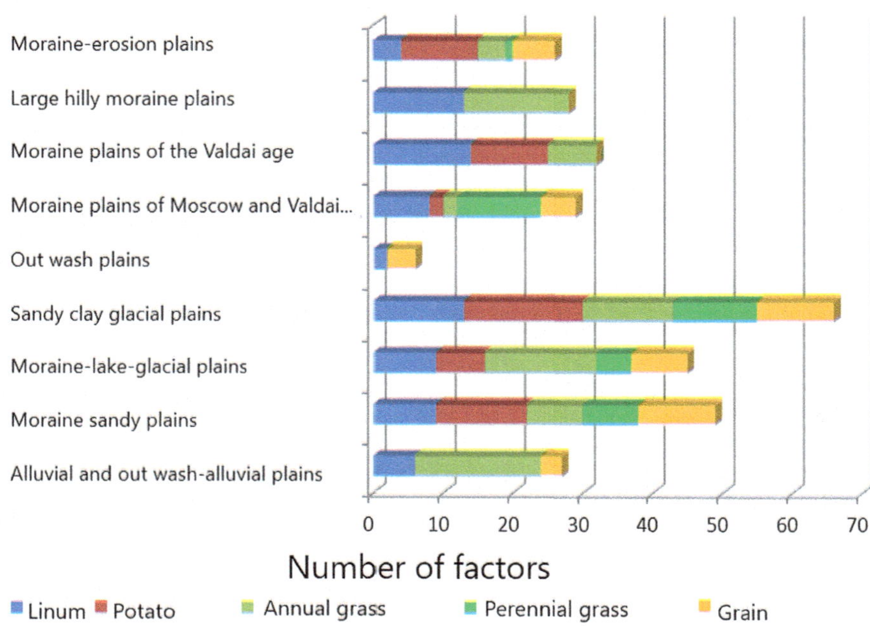

Fig. 6.4 The number of factors potentially influencing the production process of plants in various agro-geosystems

in the first place on the impact on the production process put forward factors of soil fertility (13%), then go orographic factors (10%), and the least important are agro-climatic and productive parameters of agro-geosystems (6%).

Based on this, we can say that the production process within the agro-geosystems with a fairly homogeneous lithogenic base is characterized by relative stability, since, first of all, it depends on the little changes in time orographic and agrochemical parameters, whereas in the conditions of domination of bi-membered deposits, the productivity of crops depends on less stable in time circumstances.

The duration of the stage of postglacial development of geocomplexes and the degree of homogeneity of their lithogenic basis influence the nature and depth of anthropogenic transformation of geosystems. Thus, geocomplexes of Valdai (Wurm) age, due to the small homogenization of the relief and soil-forming rocks, practically do not feel the influence of the structural organization of farms on the spatial variability of the natural environment parameters. Moraine landscapes of the Moscow (Mindel) age, the lithogenic basis of which has experienced a noticeable influence of denudation processes, are characterized by a much stronger responsiveness to anthropogenic activities. Landscapes with a sandy lithogenic base, formed for the longest time, differ even more homogeneous lithogenic base and, as a consequence, the greatest sensitivity to anthropogenic impact.

6.4 The Use of Integral Parameters in the Development of Grass Growing Enterprises

At the present stage of development of the theory of environmental management and complex melioration of agricultural landscapes important are the integral parameters of assessing the productivity of locations (Pegov and Khomyakov 1991), which allow to compare this indicator agro-landscapes of different genesis. Widely known work (Belova et al. 2008; Kireycheva et al. 2006, 2008), describing how to use integral parameters to compare the bioproductivity of different soil types. In our work, an attempt is made to use the abovementioned methodological techniques to assess the productivity of different types of agricultural landscapes of the Tver region, as well as to determine, on the basis of the results, a set of agro-meliorative measures to optimize the production process within them.

It is necessary to distinguish between the potential productivity of the type of agricultural landscapes (PTA) and its real productivity (RTA). RTA is the quantity of biomass that can be produced only in ideal circumstances when receipt of the energy balance in the agro-landscape fully goes to the formation of direct and collateral production of plant growing.

The potential productivity of agro-geosystems is estimated by the following formulas (Pegov and Khomyakov 1991):

$$\begin{aligned} \text{PTA} &= \text{S} \cdot \text{CL} \quad \text{(for natural cenoses)} \\ \text{PTA} &= \text{S} \cdot \text{ART} \cdot \text{GGR} \quad \text{(for agro-cenoses)} \end{aligned} \tag{6.1}$$

PTA is the potential productivity of vegetation biomass in these soil-climatic conditions, t ha^{-1} of air-dry matter; S—index of the soil; CL—coefficient of climate favorability; ART—an indicator of compliance of climatic conditions of this culture; GGR—coefficient depending on the characteristics of production.

Soil index—an integral indicator of soil fertility is calculated by dependence (Pegov and Khomyakov 1991):

$$\text{S} = \frac{(6.4 \cdot (Gg.a + 0.2Gf.a))}{600} + 8.5\sqrt[3]{NPK} + 5.1e^{-\frac{[Hr-1]}{4}} \tag{6.2}$$

where 6.4, 8.5 and 5.1—weighting factors; Gg.a and Gf.a—content of humate and fulvate humus, respectively, t ha^{-1}; N, P, K—the content of nitrogen, phosphorus, and potassium in the soil in fractions of their optimal value for this crop; Hr—hydrolytic acidity, mg · EQ per 100 g soil. The maximum value of the soil index is 20.

The coefficient of favorable climate (an integral indicator that takes into account the heat and moisture content of the territory) is determined by the formula (Pegov and Khomyakov 1991):

$$\text{CL} = \sqrt{\left(\text{arctg}\frac{H_f - 113}{4} + 1.57\right) \cdot \left(\text{arctg}\frac{T - 6}{2} + 1.57\right)} \tag{6.3}$$

H_f is the effective moisture index = 43.2 lg OS–T; OS is the average annual precipitation, mm; and T is the average annual temperature, °C.

For agro-cenoses, the CL coefficient cannot be applied, since each crop has its optimal range of hydrothermal conditions. Therefore, the indicator of compliance of agro-climatic conditions of the culture (ART), which is calculated by the formula (Pegov and Khomyakov 1991):

$$\text{ART} = e^{-\left[\left(\frac{H_f - H_{fo}}{DH}\right)^2 + \left(\frac{T - T_o}{DT}\right)^2\right]} \tag{6.4}$$

where the parameters H_{fo}, DH, T_o, DT are determined depending on the biological characteristics of the culture; H_{fo} and T_o characterize the optimal moisture and temperature; and DH and DT—range of appropriate conditions acceptable for the culture.

Features of agricultural production takes into account the coefficient of GGR, usually varying from 0.5 to 1.45. Using formulas 6.2–6.4 it is possible to calculate, based on the conditions of vegetation of perennial grasses, the values of PTA for agro-geosystems (Table 6.4).

The maximum value of CL (coefficient of climate favorability) is observed within the hilly moraine plains located in the west and northwest of the region. This type of agricultural landscape along with swamp massifs is also characterized by the maximum values of the s—index of the soil. The potential productivity of natural meadows is 12.2 t ha^{-1} of air-dry matter for the hilly plains and 11.44 t ha^{-1} for wetlands.

Based on the analysis of statistical data, the highest real productivity of perennial grasses was observed within the moraine-lake-glacial plains (3.23 t ha^{-1}, and the lowest on the coasts of waters (1.77 t ha^{-1}) and on hilly moraine plains (2.18 t ha^{-1}). This is partly due to the low value of ART (indicator of compliance of agro-climatic conditions of the culture), which for hilly plains is 0.86.

The contradictions between real and potential data are mainly due to factors not taken into account when working with formulas 6.2–6.4. In total, they are included in the GGR multiplier, the potential value of which can be calculated by the formula:

$$\mathrm{PotGGR} = \frac{\mathrm{PTA}}{(\mathrm{S} \cdot \mathrm{ART})} \tag{6.5}$$

where PotGGR is the potential value of GGR; PTA—potential productivity of a particular type of agricultural landscape; S—index of the soil; ART—an indicator of compliance of agro-climatic conditions of the culture.

The real value of GGR can be determined by the formula:

$$\mathrm{RealGGR} = \frac{\mathrm{RPTA}}{(\mathrm{S} \cdot \mathrm{ART})} \tag{6.6}$$

where RealGGR is the real GGR value; RTA—real productivity of a particular type of agricultural landscape; S—index of the soil; ART—an indicator of compliance of agro-climatic conditions of the culture.

The difference between the potential and actual value of GGR determines the potential of reclamation of the agricultural landscape, that is, the coefficient of the maximum possible increase in its productivity in the application of complex agronomic and agro-meliorative measures.

Share RealGGR from the value PotGGR, expressed as a percentage, determines the extent to PTA specific agro-geosystems. The cartogram of the degree of implementation of PTA in various types of agricultural landscapes of the Tver region is shown in Fig. 6.5. The data of this figure and Table 6.4 show that the maximum degree of PTA realization is observed within the limits of moraine-lake-glacial (31.0%) and large hilly moraine plains (30.5%), and the minimum in the conditions of hilly moraine plains (17.8%).

Comparison of different types of agricultural landscapes is possible only in a single system of factor coordinates. Identification of factors affecting the degree of implementation of the PTA, allows you to determine the direction of agro-meliorative

Table 6.4 PTA calculation parameters for types of agricultural landscapes within the Tver region

TA (No)	CL	S	PTA (t/ha)	RTA (t/ha)	ART	PotGGR	RealGGR	The potential of land reclamation	The degree of implementation PTA (%)
1	1.28	7.19	9.20	2.67	0.89	1.43	0.42	1.01	29.4
2	1.26	7.77	9.79	2.96	0.89	1.41	0.43	0.98	30.5
3	1.40	8.71	12.19	2.18	0.86	1.63	0.29	1.34	17.8
4	1.26	8.10	10.21	2.98	0.89	1.41	0.41	1.00	29.1
5	1.33	8.20	10.91	2.30	0.89	1.49	0.32	1.17	21.5
6	1.28	8.40	10.75	2.92	0.89	1.44	0.39	1.05	27.1
7	1.27	8.24	10.46	3.23	0.90	1.42	0.44	0.98	31.0
8	1.26	7.83	9.87	2.75	0.89	1.41	0.39	1.02	27.7
9	1.26	7.03	8.86	2.62	0.89	1.41	0.42	0.99	29.8
10	1.27	7.28	9.25	1.77	0.90	1.42	0.27	1.16	19.0
11	1.31	8.73	11.44	2.54	0.88	1.49	0.33	1.16	22.2

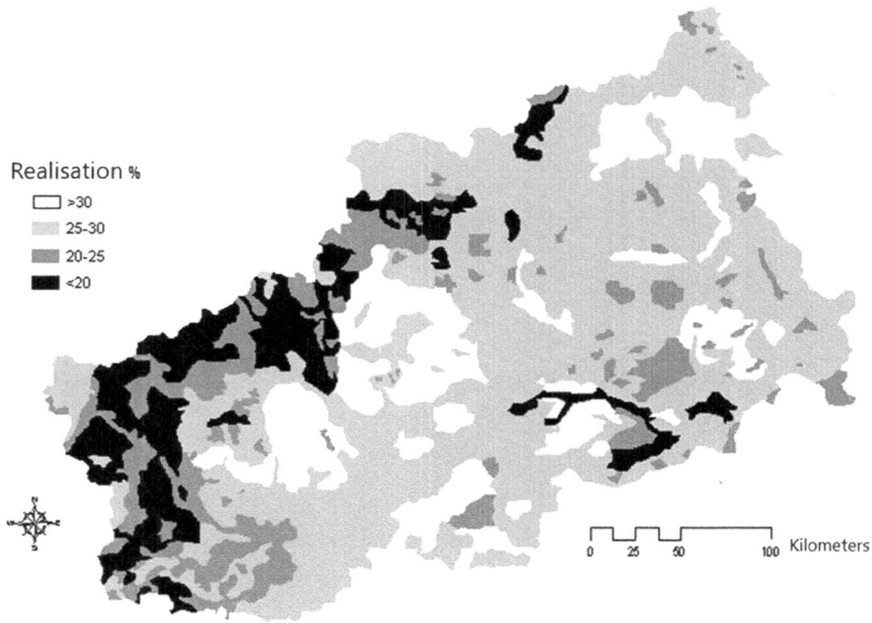

Fig. 6.5 Agro-landscape grouping by the degree of PTA implementation

effects on natural complexes. Table 6.5 presents the results of a multi-regression analysis of the impact of a single set of environmental factors of agricultural systems on the degree of implementation of various agricultural systems.

All factors affecting the degree of implementation of the PTA can be divided into three groups: 1. Strongly influencing, determining more than 10% of its variability; 2. Weakly influencing, determining from 10 to 1% of its variability; 3. Practically not influencing, defining less than 1% of its variability.

The maximum impact on the degree of realization of the potential productivity of agricultural landscapes has a ratio of meadows and arable land. Its increase leads to a decrease in the degree of implementation, which indicates the impact on her degree of biologization of crop rotation. This conclusion is confirmed by the negative impact of the growth of the share of hayfields on the degree of implementation of PTA.

The increase in productive moisture reserves in soils contributes to the increase in the degree of implementation of PTA, as it improves the water consumption of herbs, but the negative impact on it of increasing the share of wetlands in the agricultural landscape and the positive degree of erosion of arable land, which determines the self-drainage of the territory, leads to the conclusion about the need for double regulation of water–air regime of soils (Zaidelman 1991; Ministry of agriculture 1990).

Regression analysis allows it to identify the optimal and critical values of factors affecting the degree of implementation of PTA (Table 6.6). Yields of perennial grasses above the average level are obtained in agricultural landscapes with a ratio of meadow

Table 6.5 Factors affecting the degree of implementation of PTA within the Tver region

Factors	Direction of influence	The proportion of variability (%)
Ratio of meadow and arable land	–	35.1
Productive moisture reserve	+	18.9
Content of physical clay	+	9.6
Degree of wetlands	–	9.4
The share of grasslands in the agricultural landscape	–	6.7
Phosphorus content in soils	+	5.6
The potassium content in soils	–	4.7
The degree of erosion of arable land	+	3.0
Share of meadows in the agricultural landscape	+	2.1
Share of arable land in the agricultural landscape	–	1.4
pH in soils	–	0.9
Share of deposits in the agricultural landscape	–	0.8
The height of the location	–	0.5
The degree of fossil record	+	0.5
Land contour size	–	0.4

and arable land less than 0.84, which indicates a large role of seeded grass stands in the process of increasing the RTA.

Consequently, crop rotation biologization is a powerful factor in increasing the productivity of agricultural landscapes. An important condition for obtaining high yields of hay is the presence in the meter layer of soil reserves of productive moisture not less than 222 mm, physical clay not less than 26%, etc. Such indicators as the content of phosphorus and potassium in soils determine the negative impact on the growth of herbs on dense moraine loam (moraine is relatively rich in these elements). The optimal value of the share of eroded land determines the degree of self-training area.

By comparing the optimal values of the factors affecting the degree of implementation of PTA with their real values, it is possible to determine the main focus of agro-meliorative measures in various AGS. Analysis of Table 6.7 shows that some agro-geosystems, such as moraine-erosion plains or large hilly moraine plains, require little human intervention to optimize the production process of grasses, while hilly moraine plains, swampy areas, etc., require complex multifactor impact to increase the degree of implementation of PTA.

Table 6.6 Comparison of the optimal values of factors affecting the degree of implementation of PTA with their real values for different types of agricultural landscapes of the Tver region

TA (No)	Factors of the first and second groups									
	Meadow/arable land	Productive moisture	Physical clay	Bogginess	Hayfield	Phosphorus	Potassium	Erodibility	Meadows	Arable
	The optimal values of the factors									
	<0.84	>222	>26	37–67	<8	<123	<100	>16	<20	>27
	The real values of the factors									
1	0.44	234	25.9	38.0	5.4	94	94	22.1	15.6	38
2	0.53	230	25.0	42.7	6.3	112	92	23.2	14.2	30
3	1.0	210	21.6	29.2	8.7	129	106	21.9	15.6	17.3
4	0.65	222	25.2	43.2	5.9	117	103	19.1	17.3	28.9
5	0.96	207	20.5	45.2	8.0	128	99	21.5	16.7	21.4
6	0.82	224	22.8	47.0	8.6	133	99	12.7	19.0	27.6
7	0.55	228	24.3	45.3	5.4	115	99	15.3	17.1	33.9
8	0.56	219	26.9	34.4	7.2	119	103	21.2	15.5	28.9
9	0.5	224	22.3	69.8	7.6	84	77	11.1	18.6	40.4
10	0.9	223	24.8	99.9	9.9	121	102	12.8	20.9	27.5
11	0.67	212	22.0	70.7	5.0	83	69	0.0	14.0	21.0

Table 6.7 Systems of measures in different types of agricultural landscapes of the Tver region, aimed at increasing the degree of implementation of PTA for leguminous herbage

TA (No)	Event systems
1	The location of the swards on soils of heavy granulometric composition
2	The location of the swards on soils of heavy granulometric composition
3	Involvement in crop rotation of a certain number of hayfields, double regulation of water–air regime of soils, placement of grass stands on soils with heavy granulometric composition and deep moraine
4	The location of the swards on soils of heavy granulometric composition and the deep moraine
5	Involvement in crop rotation of a certain number of hayfields, double regulation of water–air regime of soils, placement of grass stands on soils with heavy granulometric composition and deep moraine
6	Involvement in crop rotation of a certain number of hayfields, soil drainage, placement of grass stands on soils with heavy granulometric composition and deep moraine occurrence
7	The soil drainage, the location of the swards on soils of heavy granulometric composition
8	Double regulation of water–air regime of soils, placement of grass stands on soils with deep moraine
9	The soil drainage, the location of the swards on soils of heavy granulometric composition
10	Involvement in crop rotation of a number of hayfields, double regulation of water–air regime of soils, placement of grass stands on soils with heavy granulometric composition and deep moraine
11	Involvement in crop rotation of a number of hayfields, double regulation of water–air regime of soils, placement of grass stands on soils with heavy granulometric composition

See Table 6.7. The systems of nature management and reclamation activities aimed at increasing the degree of implementation of PTA for grass stands in different types of agricultural landscapes of the upper Volga basin are presented.

All systems can be grouped into three groups: 1. adaptive placement of grass stands depending on the granulometric composition and geological structure of the soil; 2. adaptive placement of grass stands and drainage-irrigation melioration; and 3. adaptive placement of grass stands, water reclamation, and land management activities (involvement in the rotation of a certain number of hayfields). The scheme of zoning the territory of the region by groups of activities is shown in Fig. 6.6.

Thus, the method of assessing the productivity of types of agricultural landscapes can be based on the use of integrated indicators. At the stage of formation of methods for designing adaptive-landscape farming systems, they can be successfully used not only within the soil types but also at the level of small hierarchical units of the biosphere. Together with geographic information systems and methods of mathematical modeling, they allow to identify the areas of distribution of measures to optimize the reclamation state of geocomplexes.

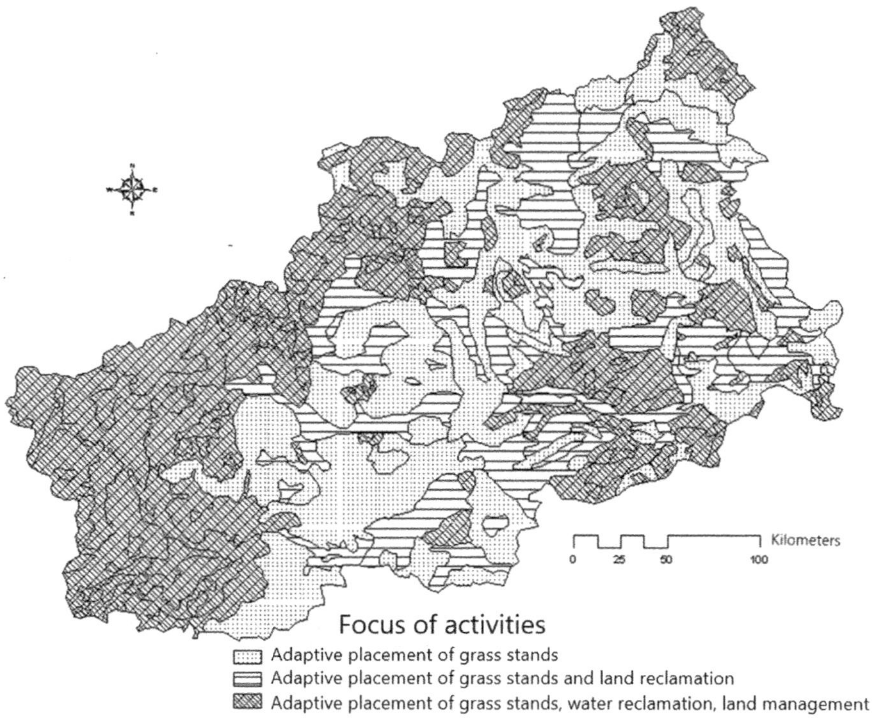

Fig. 6.6 The zoning of the Tver region in the direction of measures to optimize the production process of herbs

6.5 Conclusions and Outlook

1. Geostatistical methods are an important part of the development of adaptive-landscape farming systems. On their basis, it is possible to determine the impact of environmental conditions of geosystems of different levels on the nature of land management of farms, the structure of their acreage and crop rotation, as well as on the features of agro-technological measures of cultivation of specific crops.
2. Dispersion analysis allows to determine the "nodal" levels of typification of agricultural systems, on the basis of the characteristics of which it is possible to develop specific measures of natural-rational land use.
3. Path analysis provides information about factors that affect the growth of certain crops in different geocomplexes of the same typification level. It allows to detect the genetic relationship between the nature of the device urolithins the foundations of geo-complex and the optimal range of cultures within it.
4. Determining the integral indicator values of the productivity of agro-landscape allows us to develop sets of agronomical and melioration events achievements within the greatest possible efficiency and to identify, areas of their application, based on the GIS technology.

References

Atlas (1964) Atlas of the Kalinin region. Moscow: GUGK, 34 pp
Belova IV, Kireycheva LV, Ustinov MT (2008) forecast of productivity of agricultural lands of the Barabinsk lowland using GIS technologies. Reclam Water Manag 1:28–30
Dorofeev (1990) Legend of the landscape map of Tver region. M: GUGK MAPS, MAPS
Ivanov DA (2012) Agree (theoretical and applied aspects). Monograph, LAP LAMBERT ACADEMIC PUBLISHING Gmbh & Co. KG Heinish-BocKing, Str. Saarbrucken, Deutschland, 311 pp
Kashtanov AN, Shcherbakov AT (1993) Landscape agriculture. Guidelines for the development of landscape systems of agriculture in multicultural agriculture. Part 2: Kursk, 110 pp
Kashtanov AN, Lisetsky FN, Shvebs GI (1994) Fundamentals of landscape-ecological farming/AN Chestnut. Moscow: Kolos, p 127
Kireicheva LV, Belova IV, Khokhlova OB (2006) A methodology for predicting the production capacity and the formation of a stable, Melia-risovannogo of the agricultural landscape. Sat. "Methods and technologies of complex reclamation and ecosystem water use", M., 602 pp
Kireicheva LV, Belova IV, Karpenko NP, Agaev SB, Dedova EB, Konieva GN (2008) Technology management productivity of reclaimed agricultural lands in various regions of the Russian Federation. M., 82 pp
Kiryushin VI (1993) The concept of adaptive-landscape agriculture. Pushchino, 95 pp
Kiryushin VI (1996) Ecological bases of agriculture. M.: Kolos, 367 pp
Kovalev NG, Khodyrev AA, Ivanov DA, Tyulin VA (2004) Agrolandshaftov. Tutorial, Russian Academy of agricultural Sciences, Ministry of agriculture, TGSH, VNIIMS, Tver-Moscow, 490c
Kovalev NG, Smirnov AA, Ivanov DA et al (2000) Theoretical bases of creation of adaptive landscape-meliorative systems of agriculture and their model models (projects) for various natural and economic conditions of the humid zone. kN.I, of Tver, 119c
Milkov FN (1973) People and landscapes. M.: Thought, 224 pp
Ministry of agriculture (1990) Application of agromeliorative measures on drained mineral lands of non-Chernozem zone of the RSFSR (technological regulations). The Ministry of agriculture of the RSFSR, VNIIMS, Sevniigim M 58 pp
Nikolaev V (1987) The concept of agricultural landscape. Vestnik MGU Ser. 5, Geography, 2, pp 22–27
Nikolaev V (1997) Landscape Science and agriculture. Structure, functioning, evolution of natural and anthropogenic landscapes. Theses landscape conference M. - SPB, pp 24–28
Pegov SA, Khomyakov PM (1991) Modeling of ecological systems development. Leningrad, Gidrometeoizdat, 224c
Preobrazhensky VS (1966) Landscape research. M.: Science, 127 pp
Prokaev VI (1983) Physical and geographical zoning. M.: Education, 176s
Sedlovsky AI, Martynov SP, Mamonov LK (1982) Genetic and statistical approaches to the theory of selection of self-pollinating crops. Alma-ATA: Science, 198 pp
Shcherbakov AP, Volodin VM et al. (1994) Landscape agriculture and agrobiodiversity. Agriculture 2:8–12
Shvebs GI (1988) The concept of paragenetic landscapes and nature management. Geography and practice of science. M., pp 107–120
Zaidelman FR (1991) Ecological and soil reclamation in humid landscapes. M.: Agropromizdat, 320 pp

Chapter 7
Modelling of the Suitability of Lands to the Agrarian Use and Their Resistance to Negative Processes

Aleksey G. Osipov, Vitaly V. Garmanov, Vladimir L. Bogdanov, Viktoria A. Pavlova, Vitaly V. Terleev and Aleksandr O. Nikonorov

Abstract The authors have developed approaches for modelling the suitability of landscapes for agricultural use and their resistance to water erosion, pollution with heavy metals and acidification. The basis for modelling the suitability of landscapes for agricultural use is the methodology of multi-parameter analysis of the quality of complex objects. The basis for modelling of soil resistance to water erosion is based on empirical dependencies, which make it possible to determine the potential leaching of soil, both by rainfall and by melt runoff. The basis for modelling soil resistance to heavy metal pollution and acidification is the relationship between the geochemical potential of the soil, and their ability to accumulate chemicals coming from man-made sources.

Keywords Models of land suitability assessment · Agricultural use · Land resistance to negative processes

A. G. Osipov · V. V. Garmanov · V. L. Bogdanov
St. Petersburg State University, Universitetskaya naberezhnaya, 7/9, 199034 St. Petersburg, Russia
e-mail: osipov-g-k-2005@yandex.ru

V. V. Garmanov
e-mail: lab.naz.eco@gmail.com

V. L. Bogdanov
e-mail: garmanovv@mail.ru

V. A. Pavlova
St. Petersburg State Agrarian University, Peterburgskoye shosse, 2,
196601 St. Petersburg-Pushkin, Russia
e-mail: vikalpav@mail.ru

V. V. Terleev · A. O. Nikonorov (✉)
Peter the Great St. Petersburg Polytechnic University, St. Petersburg 29, Polytechnicheskaya, 195251, Russia
e-mail: coolhabit@yandex.ru

V. V. Terleev
e-mail: vitaly_terleev@mail.ru

W. Mirschel et al. (eds.), *Landscape Modelling and Decision Support*,
Innovations in Landscape Research,
https://doi.org/10.1007/978-3-030-37421-1_7

7.1 Introduction

The relevance of the study is caused by the fact that the organization of ecologically safe land use, which largely depends on taking into account the capabilities of natural landscapes to produce sustainable agricultural products, is becoming particularly acute at present. In this work, sustainability is understood as the ability of the landscape or its components to maintain their properties and parameters of the modes of functioning quasi-stationary under the conditions of external and internal influences. At the same time, the stability of landscapes is carried out using two mechanisms: adaptation—the ability to maintain (up to a certain limit of anthropogenic impact) its normal functioning and regeneration—the ability to restore normal functioning after the termination of man-made impact (Badenko et al. 2017; Osipov 2016a).

The need to develop models for assessing the quality of land landscapes for the production of plant-based agricultural products and their resistance to agricultural loads has repeatedly been pointed out by domestic and foreign specialists (Grodzinsky 1987; Dumanski 1997; Girardin et al. 2000; Dumanski and Pieri 2000; Petry 2001; Arefiev et al. 2003; Delbaere et al. 2004; Wood et al. 2006; Naveh 2007; Termorshuizen et al. 2007; Arefiev and Osipov 2011; Arefiev et al. 2013; Osipov 2016b).

However, at present, this task is not completely solved and continues to be highly relevant.

The study was based on the following hypothesis: the developed approaches will improve the quality of information support for land management activities, which, as a result, will ensure the effectiveness of management decisions in the field of environmentally sound agricultural land use.

7.2 Materials and Methods

The purpose of the research was to develop approaches to the qualitative assessment of the suitability of landscaped lands for agricultural use and their resistance to agricultural loads.

The methodological basis of the study was the general scientific methods of cognition. At the same time, private scientific methods were used, such as expert, analytical, descriptive and monographic. The objects of research were the problematic issues of the qualitative assessment of the land of landscapes and the sustainability of soil and vegetation cover to agricultural loads.

7.2.1 *Model for Assessing the Suitability of Land for Agricultural Use*

The model is based on a methodology for multi-parameter quality assessment of complex objects (Osipov and Timofeev 2015; Osipov 2003, 2016b). Its main principles are:

- Quality is considered as a set of properties of the evaluated object, which is represented in the form of a hierarchical structure, the so-called property tree, where the properties of higher levels are associated with the properties of lower levels, which are primary;
- The partitioning of properties is carried out until a level is formed that contains either elementary properties that cannot be further divided or quasi-elementary properties which are no longer appropriate to divide;
- Elementary and quasi-elementary properties are expressed in terms of absolute indicators. In the event that an indicator can be measured, it is assigned a quantitative value, and if not, a qualitative value is given in the form of a score. However, absolute indicators do not say anything about the quality of elementary and quasi-elemental properties from the point of view of the subject of evaluation, therefore, after determining the absolute indicators, relative indicators are calculated;
- At all levels, within each of the branches of the 'property tree' un-normalized coefficients of weight (importance) of properties are determined using expert methods. After that, for each elementary and quasi-elementary property, the normalized coefficient of its weight (importance) is calculated, which takes into account the importance of all the properties hierarchically associated with it;
- Quality is characterized in points obtained by summing the products of the normalized weighting (importance) coefficients of each elementary and quasi-elementary property to the relative values of their indicators.

The modelling process is carried out in nine stages.

The first stage includes the construction of a conceptual information model for land valuation (Nikonorov et al. 2016a, b; Terleev et al. 2016), which is understood as a structured set of information necessary and sufficient to carry out assessment work.

The construction of a conceptual information model includes the following steps:

(1) Object area investigation;
(2) Identification of graphic and semantic data and links between them;
(3) Development of local information domain models;
(4) Combing a generalized information model of the domain.

The result of the creation of a conceptual information model is its graphical representation, which makes it possible to visually analyse not only the type and content of information support but also its behaviour in case of changes in external (input of additional data) and internal (change in boundary conditions) factors. Therefore, the graphical language describing the conceptual information model should be clear, simple and informative.

The second stage includes the construction using the developed conceptual information model of the 'property tree', which is based on the following principles:

- Division within each individual group should be carried out on a single basis, i.e. on equal footing;
- Each complex property should be divided at the next higher level into such properties, the number and character of which correspond to the requirements of necessity and sufficiency;
- Within the group cannot simultaneously be generic and species properties;
- The number of levels in the property tree should be such that each group has the minimum number of properties (ideally two);
- Property splitting should be continued until the highest level is reached, at which elementary and quasi-elementary properties are located.

The third stage includes determining for elementary and quasi-elementary properties included in the 'property tree' of their absolute values. In the event that an indicator characterizing an elementary or quasi-elementary property is amenable to physical measurement methods, its values are expressed, in physical units, otherwise in points, to determine which expert methods can be used.

The fourth stage includes determining for each elementary and quasi-elementary property involved in estimating non-normalized weighting (importance) coefficients. For this, it is proposed to use the method of pairwise comparisons.

The main idea of this method is that all the analysed properties of the natural environment are compared in pairs with each other. For this, a pairwise comparison matrix is used, in which the property located in the row is compared with all the properties indicated in the columns of the matrix.

After filling in the matrix of pairwise comparisons, non-normalized priority vectors and consistency of expert estimates were calculated. The calculation algorithm was described in detail in Osipov (2016a).

This stage also includes the determination of the normalized weighting (importance) coefficients of each indicator participating in the assessment, calculated according to the following relationship:

$$P_j = \overline{P_{J1}} \cdot x \ldots x \cdot \overline{P_{jn}} \tag{7.1}$$

where P_j—normalized weight coefficient (importance) j-th elementary (quasi-elementary) properties; $\overline{P}_{J1}, \overline{P}_{jn}$—average un-normalized coefficients of weight (importance) of the i-th and n-th levels of the 'property tree' hierarchically related within the same branch with the j-th elementary (quasi-elementary) property; n—number of levels in the properties tree.

According to the given dependence, the normalized weight (importance) coefficient of the elementary (quasi-elementary) property P_j is calculated by multiplying the average un-normalized weight factors of the individual properties $P_{jl} \ldots P_{jn}$ hierarchically related to each other on the property tree.

The fifth stage includes the determination for all elementary and quasi-elementary properties of the reference values of their absolute indicators. If the assessment is carried out within one region and does not provide the comparison of results with other regions, their best values for the studied region are taken as reference values of absolute indicators.

This stage includes determining for all elementary and quasi-elementary properties the relative values of their indicators, using the following dependencies:

$$R_{pj} = \frac{w_{pj}}{w_{j\sum}} \; when \; w_{pj} < w_{j\sum} \tag{7.2}$$

$$R_{pj} = \frac{w_{j\ni}}{w_{pj}} \; when \; w_{pj} > w_{j\ni} \tag{7.3}$$

where R_{pj}—the relative value of the indicator characterizing the *j*-th elementary (quasi-elementary) property of the *p*-th land allotment; w_{pj}—the absolute value of the indicator characterizing the *j*-th elementary (quasi-elementary) property of the *p*-th land allotment; $wj_{э}$—reference absolute value of the index characterizing the *j*-th elementary (quasi-elementary) property; R_{pj} varies from 0 to 1 ($0 < R_{pj} < 1$).

The above linear dependencies should be replaced with nonlinear ones to obtain more accurate relative values of the indicators characterizing the elementary and quasi-elementary properties

The sixth stage includes the creation of digital models of basic cartographic bases, soil and landscape maps for the studied region, their entry into the geographic information system and design in the accepted conventional signs. In parallel with this, the semantic database is being filled with information necessary for assessing the suitability of land for agricultural development.

The seventh stage includes the creation of factor maps, which are understood as graphic displays of spatially associated information on the degree of manifestation of the indicator being studied. Factor maps are created in the GIS environment using standard or specially developed software.

The eighth stage includes the summation of factor maps using the 'topological overlay' operation. As a result, the study area is divided into calculated areas, which are homogeneous elementary ranges, within which each analysed indicator has only one value.

The ninth stage includes the determination of the value of a composite indicator characterizing the suitability of lands for agricultural use, for each calculated area, according to the following dependence:

$$K_p^o = \sum_{j=1}^{n} R_{pj} \times P_j \tag{7.4}$$

where K_p^0—a summary indicator characterizing the suitability of the *p*-th settlement land for agricultural development.

The most suitable is the settlement area for which the indicator K_p^0 has the greatest value. The index '0' under the indicator K_p^0 means that the estimate is given in relation to the lowest (zero) level in the hierarchy of properties, i.e. in relation to the suitability of the settlement area for agricultural development. If land plots are compared with each other, then of the plots being compared, the best in quality is the one with K_p^0 will have the greatest value. At the same time, the ratio of quality assessments will accurately reflect the reality only in the case when the evaluation of objects is performed on a full property tree, without excluding the identical properties in the compared objects.

Weighted indices obtained for the calculated plots of the study area were summarized at the landscape level using the following relationship:

$$L_m^o = \sum_{d=1}^{n} K_{dp}^0 \times Q_d \tag{7.5}$$

where L_m^0—suitability of the land of the m-th landscape for agricultural use, (point); Q_d—share in the landscape of land with d-suitability; K_{dp}^0—point, characterizing the d-th suitability of land for agricultural use.

The results of land suitability for agricultural use are presented in the form of a series of electronic maps, including a set of factor maps and a final synthetic map.

7.2.2 Model for Assessing Soil Resistance to Water Erosion

The model of soil resistance to water erosion is based on empirical dependences given in (Larionov 1993; Khrisanov and Osipov 1993), which allow determining the potential soil flushing from rainfall runoff (7.6) and from meltwater runoff (7.7):

$$G_\partial = D \times S_\partial \times P \times K_\partial \tag{7.6}$$

where G_∂—module of soil washout from rainfall runoff, t (ha year)$^{-1}$; D—erosion potential of rainfall; S_∂—soil wash-off per unit of erosion potential, t ha^{-1}; P—erosion potential of relief; K_∂—dimensionless coefficient taking into account the soil-protective properties of vegetation and agricultural engineering.

$$G_t = S_t \times K_t \tag{7.7}$$

where G_t—module of soil wash-off by meltwater runoff on slopes up to 300 m long, t (ha·year)$^{-1}$; S_t—soil wash-off by meltwater runoff, t ha^{-1}; K_t—dimensionless coefficient taking into account the soil-protective properties of the agricultural background.

The proposed dependencies are recommended for determining soil resistance to water erosion because they are most appropriate both in terms of the cost of producing the calculations, and in terms of obtaining information necessary for their implementation. The results of calculations performed on these dependencies for the control plots are in good agreement with the data obtained for the same plots according to the theoretical model of Mirtskhlava (correlation coefficient is 0.95–0.96) (Khrisanov and Osipov 1993). Consider the approaches to the definition of each of the indicators of the above empirical dependencies.

The erosion potential of rainfall is determined by the following formula:

$$D = 0.25841H \times i_{30} - 0.14921 \tag{7.8}$$

where H—layer of precipitation, mm; i_{30}—maximum rain intensity over a 30-minute interval, mm min^{-1}.

For the rains with a maximum intensity of less than 0.1 mm min^{-1}, the erosion potential should be calculated from the presented dependencies, since the formula (7.8) in this case gives underestimated results. When determining the potential, rain with a layer of precipitation of 10 mm or more is taken into account, since, according to (Larionov 1993; Khrisanov and Osipov 1993), less significant rainfall does not cause a noticeable flushing.

The calculated values of erosion potential of rainfall are plotted on a cartographic basis with reference to meteorological stations, according to which the calculations were performed, after which an isolines map of erosion potential of rainfall is compiled.

In addition, the intra-annual distribution of erosion potential of rainfall is determined, the values of which are subsequently used to calculate soil-protective coefficients, taking into account the effect of vegetation cover and agricultural background on reducing erosion processes, the need for which is related to the change in soil-protective ability of vegetation during the year.

Soil wash-off process without agricultural background and properties subject to seasonal changes depends mainly on their mechanical composition and humus content). This is determined by the schedule given in the study (Khrisanov and Osipov 1993).

After determining the wash-off process for soils of all types, the obtained values are divided into classes in increments of 0.2 units and on the soil map, the boundaries of ranges corresponding to the selected classes are drawn, and then transferred to a topographic map.

The erosion potential of the relief characterizes the intensity of soil washing off, depending on the steepness and length of the slope. It is proposed to use the following relationship to determine it (Larionov 1993):

$$P = 22.1^{-p} \frac{18.62 \sin\left[arctg\left(\frac{1}{100}I\right)\right]}{1 + 10^{0.53-0.15LI}} + 0.065 \tag{7.9}$$

$$p = 0.2 + 2.067(p_0 - 0.2)L^{-0.15} SE^{-0.45} \tag{7.10}$$

where I—incline, %; L—slope length, m.; SE—soil erodibility, t ha^{-1}; p_0 for slopes with an incline <1% equal to 0.2; 1–3%—0.3; 3–5%—0.4; > 5%—0.5.

Along with the above dependences, the erosion potential of the relief can be determined from the graph presented in study (Khrisanov and Osipov 1993).

Measurement of the steepness and length of the slope to determine the erosion potential of the relief is carried out by the point-statistical method described in (Larionov 1993; Khrisanov and Osipov 1993), which significantly reduces labour costs when performing cartometric works.

The essence of the method lies in the fact that on a large-scale topographic maps a regular network of equally spaced anchor points is plotted, for which it is convenient to use grid points. Each node measures the length of the slope, which is equal to the distance from the watershed to the node being analysed. If there is an obstacle between the watershed and the knot (road, forest, etc.), then the length of the slope is measured before this obstacle. In addition, for each analysed node, the slope is calculated according to the following dependence:

$$q = \frac{h}{d \cdot 100} \quad (7.11)$$

where h—the height difference between the watershed and the analysed node, m; d—length of the streamline from the watershed to the analysed node, m.

Using the slope and slope length values, the erosion potential of the relief is determined for each node of the grid on the graph.

After that, the distribution of agricultural land by classes of erosion potential of the relief is calculated according to the following relationship:

$$S_p = \frac{n_p}{N \cdot 100} \quad (7.12)$$

where S_p—share of arable land in the farm or morphological area in the p-th class of the erosion potential of the relief, %; n_p—the number of nodes in which the values of the erosion potential of the relief correspond to the p-th class; N—the total number of nodes in the farm or morphological area in which measurements were made.

The values of the erosion potential of the relief are divided into classes in increments of 0.25 for arable land and 1.0 for natural forage lands.

With medium-scale studies, the erosion potential of the relief is determined for individual farms, and for small-scale studies, for morphological areas. For each farm and morphological region, measurements at 400–600 points are necessary, which will ensure the accuracy of the average erosion potential of the relief, equal to 10% within the farm and 6–9% within the morphological region (Larionov 1993). In addition, for each farm and morphological area, graphs of the connection of erosion potential with a gradient are compiled.

The calculation of the coefficient of soil-protective properties of vegetation and agricultural engineering is carried out according to the following relationship:

$$K_\delta = \sum_{j=1}^{m} \left[\left(\sum_{i=1}^{n} D'_{i,j} \times P'_{i,j}/10000 \right) \times F_i \right] /100 \quad (7.13)$$

where $D'_{i,j}$—erosion potential of rainfall in the i-th period of development of the j-th group of crops; $P'_{i,j}$—coefficient of soil washout in the i-th period of development of the j-th group of crops, % (Table 9); F_i—crop area under the j-th group of cultures, ha; j—cultivated crop groups $j = 1$, m; i—periods of cultivation of crop groups $i = 1, 2, \ldots, n$.

The soil-protective properties of vegetation cover and agricultural background are determined on the basis of information about the structure of crops, the timing of sowing and harvesting of field crops, the conduct of basic soil treatment, as well as information about the area of vapours.

The soil washout from the meltwater runoff begins with the determination of the water reserves in the snow for all meteorological stations located within the study area from the climatic reference books, after which they draw up a corresponding isolinear map. Since the amount of runoff depends not only on the water supply in the snow but also on the mechanical composition of the soil, all the soil differences within the study area are combined into two groups according to the mechanical composition: sandy and clayey (loamy).

Then, using the relief information obtained in determining the erosion potential, determine its slopes, make a classification according to this indicator and distribute the areas of agricultural land according to the accepted classes.

For each class, the runoff layer is calculated according to the following relationship:

$$S_j = M_c \cdot V_i \tag{7.14}$$

where S_j—the value of the flow layer, mm; M_c—stock of water in the snow, mm; V_i—spring runoff coefficient for soils of the i-th texture; j—soil type.

In view of the fact that the agricultural background has an active influence on the soil erosion washout, besides the factors discussed above, its soil-protective coefficient is determined (Table 7.1).

At the final stage, the total soil erosion washout is determined under the influence of rainfall and meltwater and the total soil resistance to water erosion is estimated. Based on the obtained data, all values of the total soil resistance to water erosion are combined into six groups, each of which is assigned a corresponding score.

A qualification scale for assessing water resistance of soils to water erosion was given in (Arefiev 2011).

Table 7.1 Agricultural background soil-protective coefficient

Graduation indicator	Impact score
Chaff	1.0
Winter crops	0.4
Stubble and flat cut	0.5
Perennial herbs	0.1

7.2.3 Models for Assessing Soil Resistance to Heavy Metal Pollution and Acidification

In this model, soil resistance to heavy metal pollution and acidification is considered not as a potential stock of their buffering capacity, but as the speed of their 'self-purification' from the products of technogenesis, i.e. their ability to restore their normal functioning (normalization) after the cessation of human impact.

According to the data in (Dmitriev 1994), in this case, the indicator of the degree of soil resistance to anthropogenic impact is the duration of the recovery period by the soil of its initial geochemical state. Consequently, the soil resistance to heavy metal pollution and acidification, along with geochemical indicators, is also affected by their location in the landscape-geographical space.

The developed model is based on the relationship between the properties of the soil and the accumulation of chemical substances coming in with man-made streams, as well as their accessibility to biota. In the practical implementation of the model, its parameters should be selected on the basis of available in the literature data regression analysis and correlation coefficients between soil properties, chemical mobility and their accumulation in the biomass of cultivated plants. In this case, the ranking and scoring of soil parameters should be carried out using the methods of expert analysis.

Based on the foregoing, it is proposed to use the following dependencies to determine soil resistance to heavy metal pollution and acidification:

$$U_k = K \times B_k + U_p \times B_p \tag{7.15}$$

$$U_m = M \times B_m + U_p \times B_p \tag{7.16}$$

where U_k, U_m—a weighted score of a soil unit characterizing its resistance to acidification and heavy metal pollution, respectively; *K*, *M*—score characterizing the ecological and geochemical stability of the soil unit, respectively, to acidification and to heavy metal pollution; B_k, B_m—a weighting factor characterizing the influence of geochemical features of soils on their resistance, respectively, to acidification and to heavy metal pollution (in our case B_k, $B_m = 0.6$); U_p—score, characterizing the geomorphological stability of the soil unit to acidification and to heavy metal pollution; B_p—a weighting factor characterizing the influence of the geomorphological features of the territory on soil resistance to acidification and to heavy metal pollution (in our case $B_p = 0.4$).

It is proposed to use the empirical dependences given in the works to determine the ecological and geochemical stability of the soil unit, respectively, to heavy metal pollution and acidification.

$$K = [(a + al)/(o + r + ex)]/[(o + r + ex + k + c + rc)/(a + al)] \tag{7.17}$$

$$M = [(a + rc + m)/(o + r + e)]/[(o + r + e + al + k + c)/(a + rc + m)] \tag{7.18}$$

where *K*, *M*—the level of ecological and geochemical soil resistance to acidification and heavy metal pollution, respectively, and the parameters included in the right-hand sides of Eqs. (7.17) and (7.18), describing the effect on soil resistance to acid effects and heavy metal pollution such as: acid-base conditions (a); permafrost within the layer of 0–100 cm from the soil surface (m); amorphous hydroxides Fe + Al (al); horizon power A0 and A (o, r); absorption capacity of cations located in the horizons A0 and A (e); sum of exchange bases in organic and humus horizons (ex); soil carbonates (k); soil content of exchangeable Na (c); redox conditions (rc).

Baseline data for the calculation of soil resistance to heavy metal pollution and acidification are given in the study (Arefiev 2011).

Evaluation of the impact of the terrain on the intensity of pollutants and their accumulation in the soil is carried out in five stages on two scales—large (1:5000–1:25 000) or average (1:50 000–1:100 000).

The first stage includes the selection of structural lines and characteristic points of the relief on the map, defining its framework. Structural lines are used to display changes in relief, which are caused by natural and anthropogenic relief-forming processes. They distinguish between not only natural but also natural-anthropogenic and anthropogenic relief forms and act as specialized geomorphological boundaries, outlining elementary surfaces.

The second stage includes the creation of a geotopological map of elementary surfaces along structural lines and characteristic points of relief. In the course of creating a geotopological map, the selection of elementary surfaces is carried out taking into account two parameters: the angle of inclination of the relief and the change in the rate of migration of substances during the transition from one structural line to another. Along with the dividing functions, structural lines play the role of lines of communication between the characteristic points of the relief and are common parts of the adjacent elementary surfaces, thereby forming the relief carcass and determining the most characteristic topological properties of the earth's surface. The elementary surfaces identified on geotopological maps, when assessing their influence on the intensity of removal and accumulation of pollutants of anthropogenic origin, should be considered in conjunction with each other. So, for example, one cannot say that any single elementary surface is concentrating or dispersing. We can classify it in one category or another only after comparing it with the higher or lower elementary surfaces. The technique for constructing structural lines and isolating elementary surfaces is given in (Lastochkin 1995).

The third stage includes the scoring of the influence of elementary surfaces on the removal and accumulation of pollutants in the soil. After creating a geotopological map, a point estimate is made of the influence of the elementary surfaces highlighted on the map on the intensity of removal and accumulation of pollutants of anthropogenic origin.

The fourth stage involves the integration of elementary surfaces into associations according to the levels of their influence on the removal and accumulation of pollutants in the soil. To combine elementary surfaces into associations, two appraisal series are built, one of which reflects the process of accumulation of pollutants in the soil, the second—the process of their removal. Next, an appraisal matrix is formed, the axes of which are the generated grading series. According to the created matrix, the elementary surfaces are evaluated, i.e. the sequence of their location in the calibration lines is detected. According to the results of bonding, elementary surfaces are united in associations.

The fifth stage involves scoring estimation of the effect of associations on soil resistance to acidification and heavy metal pollution. Based on the cumulative effect of the terrain on the activity of the removal or accumulation of pollutants in the soil of each selected association, a score is assigned that characterizes its effect on soil resistance to acidification and contamination by heavy metals.

Qualimetric scales for assessing the influence of geochemical features of soils on their resistance to heavy metal pollution and acidification are given in (Arefiev 2011).

When conducting medium-scale studies, the slope of the terrain is used as an indicator characterizing the degree of relief effect on soil stability, and the accumulation processes in the soil are not taken into account. Estimation of the impact of relief on soil stability in this case is carried out at the level of natural-territorial complexes using the following dependency:

$$Y_p = \sum_{j=1}^{n} S_{ypj} \cdot B_{ypj} \tag{7.19}$$

where Yp—the degree of influence of the geomorphological structure of the territory on soil resistance to acidification and pollution by heavy metals; s_{ypj}—proportion of land with the j-th slope; B_{yp}—score, characterizing the degree of influence of the j-th slope on soil resistance to acidification and heavy metal pollution (Arefiev 2011).

The following dependencies are used to determine the soil resistance to acidification and contamination with heavy metals, taking into account the effect of the relief (Lastochkin 1995):

$$Y_n = N - PB \times K_n + Y_p \times K_p \tag{7.20}$$

where Y_n—the weighted score of a soil unit within the considered population characterizes its resistance to acidification; N—PB—score characterizing the ecological and geochemical stability of the soil unit to acidification (corresponds to the level of association in which the soil difference is included); K_n—a weighting factor characterizing the influence of the geochemical characteristics of soils on their resistance to acidification (in our case $K_n = 0.6$); Y_p—a score characterizing the influence of elementary surfaces within which the soil unit is located on reducing the resistance of the soil to man-made pollution; K_p—a weighting factor characterizing the influence of the geomorphological features of the territory on the soil's resistance to pollution (in our case $K_p = 0.4$).

$$Y_m = M - PB \times K_m + Y_p \times K_p \quad (7.21)$$

where Y_m—weighted score of a soil unit within the considered population characterizing its resistance to heavy metal pollution; M—PB—score characterizing the ecological and geochemical stability of the soil unit to heavy metal pollution (corresponds to the level of association which includes the soil difference); K_m—a weighting factor characterizing the influence of the geochemical characteristics of soils on their resistance to pollution by heavy metals (in our case $K_m = 0.6$); Y_p—score characterizing the influence of elementary surfaces, within which the soil unit is located on the reduction of soil resistance to man-made pollution; K_p—a weighting factor characterizing the influence of the geomorphological features of the territory on the soil's resistance to pollution (in our case $K_p = 0.4$).

7.3 Results and Discussions

The Oyat River basin, located in the northeast of the Leningrad Region, was chosen as a territory for testing and debugging models.

The soil cover of this area is relatively diverse. Here, sandy surface-podzolic and sod-podzolic soils predominate, varying degrees of gleying, as well as peaty-podzolic-gley and peat-gley soils. Closed hollows are occupied by peatlands. In the rest of the territory, medium- and strongly podzolic and sod-podzolic soils on loam are developed, as well as podzols on a sandy moraine in combination with peaty-podzolic-gley and peat soils of upland, transitional and lowland marshes.

More than 70% of the study area is occupied by forests, represented by spruce-birch green moss and grass-shrub forests. Pine green moss and lichen forests grow on the slopes of sandhills and on plains with sandy soils. On poorly drained watersheds in the lower part of the slopes, in drainage-less depressions and along the margins of marshes, swamped pine stands and sphagnum pine forests are widespread. Upland bogs are widespread within the study area.

The main types of landscapes in this area are: (1) hilly and ridge-hollow complexes with normal moisture; (2) undulating and slightly undulating plains with normal moisture; (3) weakly undulating and flat plains with normal or short-term excessive moisture; (4) flat and lowland plains with prolonged excessive moisture.

A «tree of properties» was built including three factors and nine indicators characterizing them (Fig. 7.1) to simulate the suitability of land landscapes for agrarian use. After that, a qualimetric scale of summary indicators of the boundaries of land fitness classes was developed, and using the MapInfo GIS, nine factor maps for each analysed indicator were created for the study area; examples of factor maps are shown in Fig. 7.2.

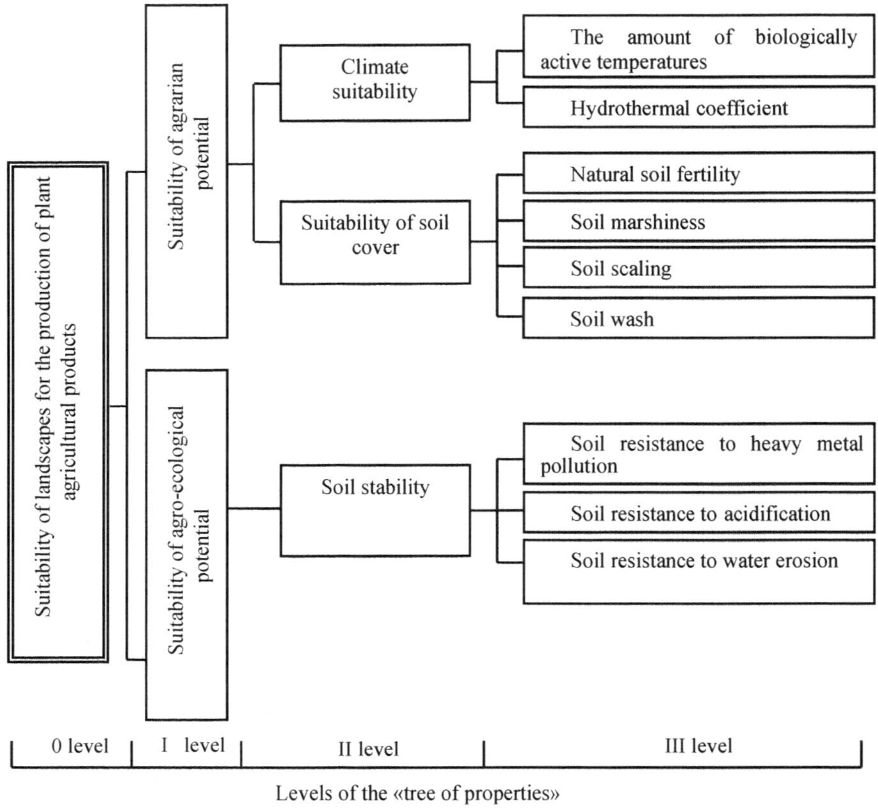

Fig. 7.1 «Tree of properties» for an integrated assessment of the suitability of the land of natural landscapes for agricultural development

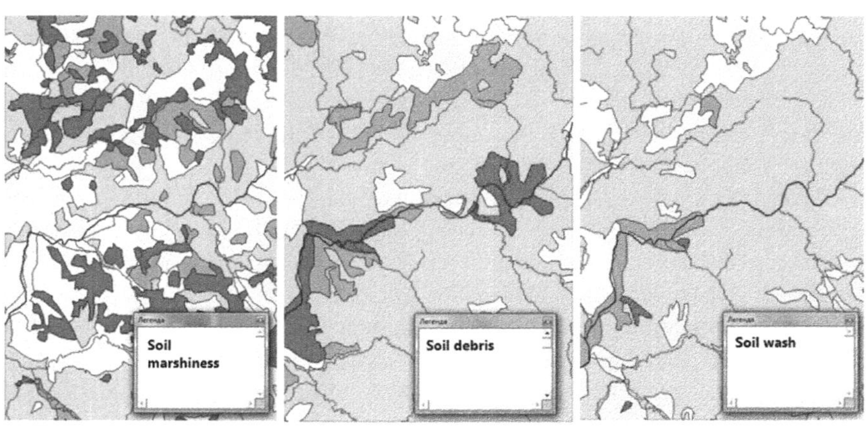

Fig. 7.2 Examples of factor maps

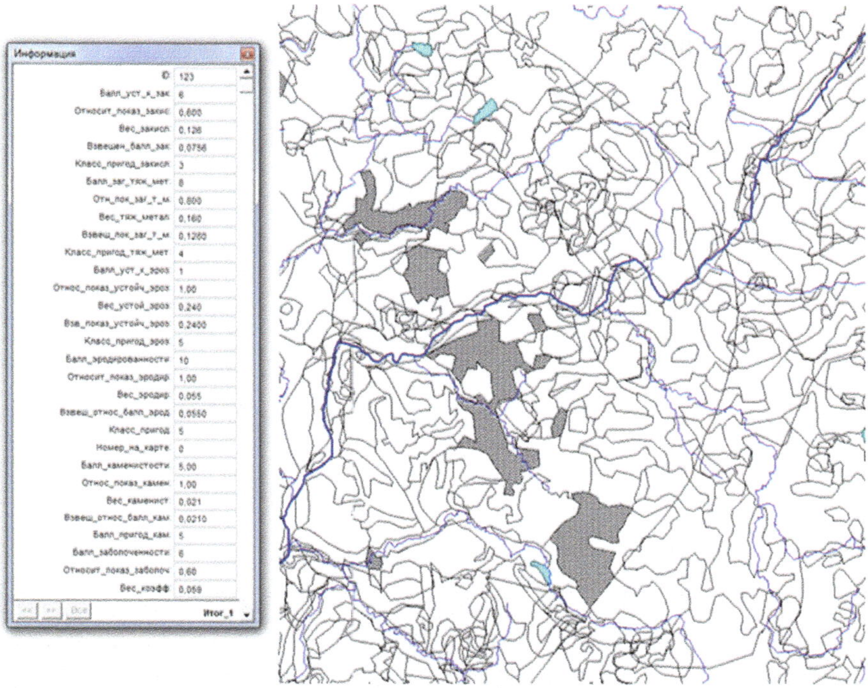

Fig. 7.3 Example of dividing the study area into calculated areas

Then, by summing the created factor maps in the GIS environment using the topological overlay operation, the study area was divided into calculated areas, which are elementary ranges, Fig. 7.3.

After that, for each calculated area, the absolute values of indicators characterizing its suitability for agricultural use were determined and weighted (aggregate) indicators were calculated by dependencies (7.2), (7.3) and (7.4) in the GIS environment. The obtained values of the weighted (summary) indicators were automatically distributed into suitability classes of calculation areas for agricultural use in accordance with the accepted qualimetric scale.

The weighted (aggregate) indicators obtained for the calculated plots were summarized at the landscape level using dependence (2.5), and each landscape was assigned a fitness class for agricultural use. After that, landscapes that fall into one class were combined and displayed by the adopted conventional symbol, Fig. 7.4. According to the conducted research, the most suitable lands for agricultural use are the lands of the following landscapes:

- undulating and slightly undulating plains with detached humid hills with normal moisture on sod-podzolic medium and light loamy soils (34.0% of the land suitable; 38.9% of the land—medium suitability);

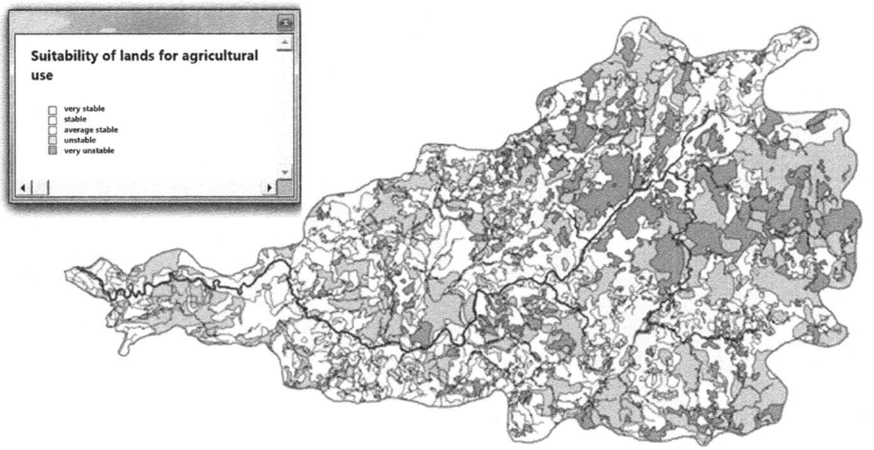

Fig. 7.4 The zoning scheme of the territory according to the suitability of land for agricultural use (on a landscape basis)

- weakly undulating and flat plains with normal or short-term excessive moisture on sod-podzolic medium and light loamy soils, in combination with peaty-podzolic-gley (38.0% of the land suitable; 37.7% of the land—medium suitability).

The most unsuitable for agricultural use are the lands of the following landscapes:

- large moraine ridges with steep slopes (5–25°) on sandy ferruginous podzols (34.9% of the land is very unsuitable; 40.7% of the land is unsuitable);
- flat and lowland plains with prolonged excessive moisture on peaty-podzolic-gley sandy soils in combination with peaty marshland (17.0% of the land is very unsuitable; 54.2% of the land is not suitable);
- flat and lowland plains with prolonged excessive moisture on sandy ferruginous podzols (17.8% of land is very unsuitable; 52.9% of land is unsuitable).

When modelling the stability of land landscapes to erosion, the erosion potential of rainfall was determined according to the following meteorological stations: Novaya Ladoga (60° 07′s; 32° 19′E); Tikhvin (59°39′N; 33°33′E); Efimovskoe (59°30′N; 34°42′E); Rooky Field (60° 43′N; 33° 33′E); Vinnitsa (60°38′N; 34°47′E); Voznesenie (61°01′N; 35°27′E). Authors used the dependence (8) to determine it. Within the study area, the maximum erosion potential of rainfall is 4.6, and the minimum is 3.7.

The percentage of fine, coarse sand and humus in the main soil types of the study area was determined according to the work (Arefiev 2011), and the soil washout value according to the schedule given in the work (Khrisanov and Osipov 1993). Within the study area, the maximum soil erosion capacity per unit of erosion potential of rainfall was 3.6 t ha^{-1} (sod-podzolic weakly glue soils), and the minimum value was 2.7 t ha^{-1} (sod-podzolic gley soils).

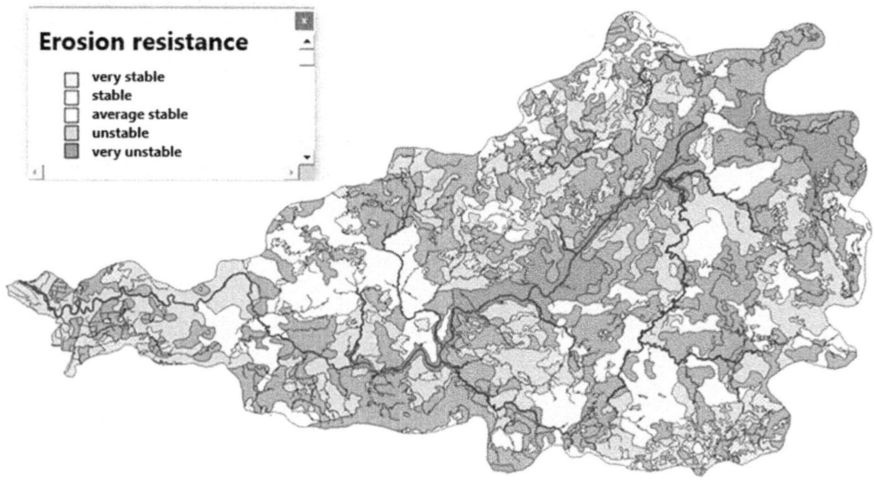

Fig. 7.5 Synthetic map of land resistance of landscapes of the study area to water erosion

Dependences (7.9) and (7.10) were used to determine the erosion potential of the relief. The slopes of the relief were calculated in the GIS environment using an elevation matrix created using a topographic map of 1: 50,000 scale. Within the study area, the maximum erosion potential of the relief was 6.5 for hilly and ridge-hollow complexes, and the minimum value was 0.2 for flat and lowland plains.

Soil washability by thawed waters was not calculated due to insignificant amounts of flushing.

A synthetic map of the land resistance of landscapes of the study area to water erosion is shown in Fig. 7.5.

Modelling the stability of land landscapes to heavy metal pollution and acidification was carried out according to dependencies (7.15) and (7.16). Authors used the qualimetric scales given in the work (Arefiev 2011) to perform the calculations.

Soil land use maps were used to characterize the soils. They were created on a scale of 1: 100,000 for the districts of the Leningrad Region by the specialists of the SevZapNIIGiproZem Institute based on materials of a large-scale soil survey, as well as generalized data on land assessments carried out by P. A. Sukhanov as part of the implementation of the regional target programme 'Fertility 2008–2013'.

Topographic maps of the scale of 1: 50,000 were used to assess the degree of influence of the geomorphological structure of the territory, on soil resistance to acidification and heavy metal pollution.

Synthetic maps of land resistance of landscapes of the study area to heavy metal pollution and acidification are shown in Figs. 7.6 and 7.7.

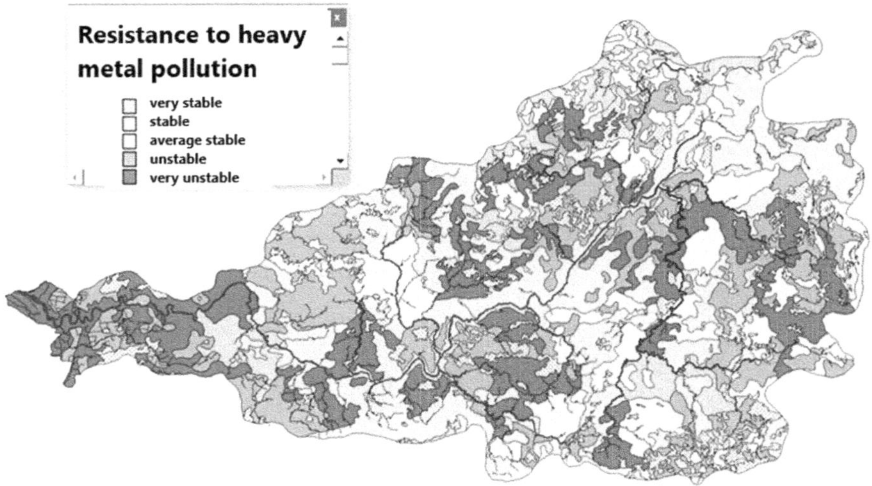

Fig. 7.6 Synthetic map of land resistance of landscapes of the study area to heavy metal pollution

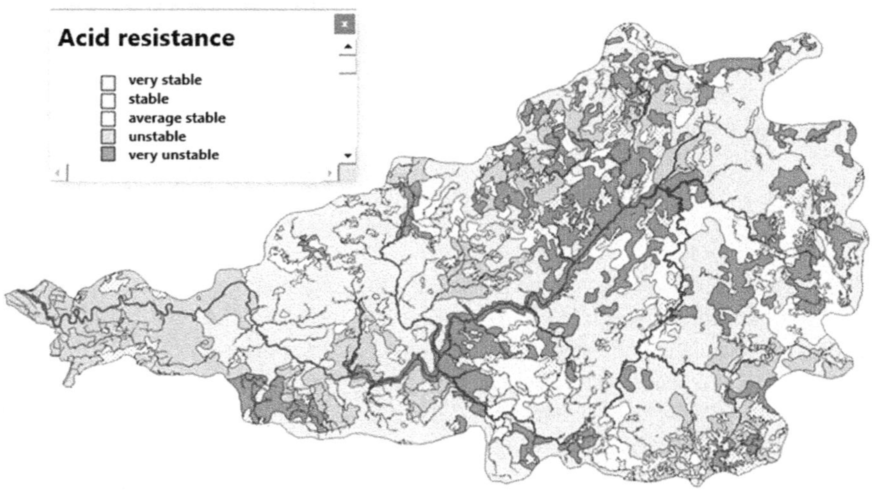

Fig. 7.7 Synthetic map of land resistance of landscapes of the study area to acidification

7.4 Conclusion

1. The basis of the model for assessing the suitability of land for agricultural use is the methodology of multi-parameter quality assessment of complex objects.
2. The model of soil resistance to water erosion is based on the empirical dependencies indicated in the above-mentioned works, which allow to determine the potential soil flushing from rainfall runoff as well as from meltwater flow.

3. The developed model for the assessment of soil resistance to heavy metal pollution and acidification is based on the relationship between soil properties and the accumulation of chemical substances entering from technogenic flows, as well as their accessibility to biota. In the practical implementation of the model, its parameters should be selected on the basis of available in the literature data of regression analysis and correlation coefficients between soil properties, chemical mobility and their accumulation in the biomass of cultivated plants. In this case, the ranking and scoring of soil parameters should be carried out using the methods of expert analysis.

Acknowledgements The reported study was funded by RFBR according to the research projects (#19-04-00939 A).

References

Arefiev NV (2011) Basics of the formation of natural-agrarian systems. Theory and practice (Основы формирования природно-аграрных систем. Теория и практика) (In Russian) Publishing house of the Polytechnic University, St. Petersburg 533

Arefiev NV, Osipov GK, Garmanov VV (2003) Optimization of the spatial structure of natural-agrarian systems on a basin-landscape basis (Оптимизация пространственной структуры природно-аграрных систем на бассейново-ландшафтной основе) (In Russian) Proceedings of the Symposium "Interregional Problems of Environmental Safety" September 17–20, 2003, Sumy 2003, Dovkillia Publishing House, pp 390–398

Arefiev NV, Osipov GK (2011) Determination of the sustainability of landscapes to land reclamation (Определение устойчивости ландшафтов к мелиоративному освоению) (In Russian) Monitoring the biodiversity of ecosystems of the steppe and forest-steppe zones. In: Proceedings of Russian scientific practical conf. with international participation under total ed. A. A. Ovcharenko. Balashov:Nikolaev, pp 7–10

Arefiev NV, Garmanov VV, Osipov AG, Trushnikov VE (2013) Landscape-ecological zoning and land monitoring of the Leningrad region (Ландшафтно-экологическое районирование и мониторинг земель Ленинградской области) (In Russian) Mining information and analytical bulletin (scientific and technical journal) 11:303–311

Badenko VL, Bogdanov VL, Garmanov VV, Osipov GK (2017) Land Management: study guide (Управление землепользованием: учеб. пособие) (In Russian) SPb, SPBSU 298

Delbaere B, Serradilla AN, Zocker C, Edwards M, Doughty M (2004) Environmental risks from Agriculture in Europe: locating environmental risc zones in Europe using agri-environmental indicators. Tilburg: European Centre for nature Conservation:184

Dmitriev VV (1994) Ecological regulation of the state and anthropogenic impacts on natural ecosystems (Экологическое нормирование состояния и антропогенных воздействий на природные экосистемы) SPbSU Bulletin 2(14):60–70

Dumanski J (1997) Criteria and indicators for land quality and sustainable land management. ITC J 3(4):216–22

Dumanski J, Pieri C (2000) Land quality indicators: research plan. Agr Ecosyst Environ 81:93–102

Girardin P, Bockstaller C, Van der Werf HMG (2000) Assessment of potential impacts of agricultural practices on the environment: the AGRO ECO method. Environ Impact Assess Rev 2(20):227–239

Grodzinsky MD (1987) the stability of geosystems: a theoretical approach to analysis and methods for quantitative assessment (Устойчивость геосистем: Теоретический подход к анализу и методы количественной оценки) (In Russian) Reports of the Academy of Sciences of the USSR. Geography series 6:5–15

Khrisanov NI, Osipov GK (1993) Management of water bodies eutrophication (Управление эвтрофированием водоемов) (In Russian) Hydrometeoizdat, St. Petersburg 278

Larionov GA (1993) Soil erosion and deflation: basic patterns and quantitative evaluations (Эрозия и дефляция почв: Основные закономерности и количественные оценки) (In Russian) Publishing House of Moscow State University 200

Lastochkin AN (1995) Landscape geoecology (environmental studies on a geotopological basis) (Геоэкология ландшафта. Экологические исследования окружающей среды на геотопологической основе) (In Russian) Publishing house of St. Petersburg University 280

Naveh Z (2007) Landscape ecology and sustainability. Landscape Ecol 22:1437–1440

Nikonorov A, Badenko V, Terleev V, Togo I, Volkova Yu, Skvortsova O, Nikonova O, Pavlov S, Mirschel W (2016a) Use of GIS-environment under the analysis of the managerial solutions for flood events protection measures. Proc Eng 165:1731–1740

Nikonorov A, Terleev V, Pavlov S, Togo I, Volkova Yu, Makarova T, Garmanov V, Shishov D, Mirschel W (2016b) Applying the model of soil hydrophysical properties for arrangements of temporary enclosing structures. Proc Eng 165:1741–1747

Osipov AG (2003) Multi-criteria assessment of land resources based on qualimetric analysis (Многокритериальная оценка земельных ресурсов на основе квалиметрического анализа) (In Russian) Regional ecology, St. Petersburg 1–2:31–39

Osipov AG Timofeev VG (2015) Scientific and methodological basis for determining the environmentally acceptable level of development of natural landscapes when creating natural-agrarian systems (Научно-методические основы определения экологически допустимого уровня освоения природных ландшафтов при создании природно-аграрных систем) (In Russian) Information and space 2:85–95

Osipov AG (2016a) Integral assessment of the sustainability of landscapes in the creation of agricultural land natural-agrarian systems (Интегральная оценка устойчивости ландшафтов при создании сельскохозяйственных угодий природно-аграрных систем) (In Russian) Bulletin of St. Petersburg University 7(3):150–162

Osipov AG (2016b) The method of geoecological assessment of landscapes in the GIS environment in the creation and reconstruction of natural-agrarian systems (Метод геоэкологической оценки ландшафтов в среде ГИС при создании и реконструкции природно-аграрных систем) (In Russian) Information and space 1: 123–130

Petry D (2001) Landscape function assessment and regional planning: creating knowledge bases for sustainable landscape development Landscape balance and landscape assessment. Springer, Berlin, pp 251–280

Terleev V, Nikonorov A, Badenko V, Guseva I, Volkova Yu, Skvortsova O, Pavlov S, Mirschel W (2016) Modeling of hydrophysical properties of the soil as capillary-porous media and improvement of Mualem-Van Genuchten method as a part of foundation arrangement research. Advances in Civil Engineering 2016: Article number 8176728

Termorshuizen JW, Opdam P, Brink AVD (2007) Incorporating ecological sustainability into landscape planning. Landscape Urban Plann 79:374–384

Wood R, Lenzen M, Dey C, Lundie S (2006) A comparative study of some of the environmental effects of traditional and organic farming in Australia. Agric Ecosyst Environ 89:324–348

Chapter 8
A GIS-Based Model for the Enhancement of Rural Landscapes: The Case Study of Valdera—Tuscany (Italy)

Massimo Rovai, Maria Andreoli and Francesco Monacci

Abstract Complexity in planning and programming applied to rural landscape and territories was increased by the new approach of the European Landscape Convention and the growing awareness on the role of rural landscapes in providing ecosystem services. While the scientific debate is still going on, one of the main challenges is how to operationalize the new attitude and knowledge about the role played by landscape in general, and not only by historical and high valued landscapes. New instruments are needed to maintain and enhance everyday landscape taking into account that it is a living and evolving system. These instruments ask for multidisciplinary and transdisciplinary approaches based on a holistic knowledge system. GIS techniques allow taking into account both spatial distribution of elements/information and their physical relations, which are paramount for the analysis of interventions about landscape. This chapter presents an expedite model for identifying policy actions aiming to protect, maintain and manage rural territories. The model has been tested on a case-study area located in Tuscany (Italy) and it identifies a set of spatialized indicators, for which it is possible to compute a cardinal or ordinal value, to be used to individuate suitable actions. While the case-study analysis is necessarily bounded by rules stated by the Tuscan administration and by the context within it was developed, the philosophy underpinning the proposed methodology may well be extended to other territories and countries. Results show that it is possible to provide simplified operative tools able to increase the effectiveness and efficiency of territorial governance.

M. Rovai
Department of Civil and Industrial Engineering (DICI), University of Pisa, Pisa, Italy
e-mail: massimo.rovai@unipi.it

M. Andreoli (✉)
Department of Agricultural, Food and Agro-Environmental Sciences (DAFE), University of Pisa, Pisa, Italy
e-mail: maria.andreoli@unipi.it

F. Monacci
University of Pisa, Pisa, Italy
e-mail: francesco.monacci@gmail.com

W. Mirschel et al. (eds.), *Landscape Modelling and Decision Support*,
Innovations in Landscape Research,
https://doi.org/10.1007/978-3-030-37421-1_8

Keywords Landscape management · Governance model · Land managers · Tuscany

8.1 Introduction

Before European Landscape Convention (ELC), only landscapes of outstanding beauty or of historical and cultural interest were deemed important and usually protected by laws setting constraints to their change. The aim was to preserve their original features, sometimes through a process of "museumification" (Magnaghi 2005) that made it increasingly difficult to maintain landscape as a living body, usually deriving from the interaction and coevolution of anthropic activities and geophysical characteristics. Moreover, standard and constraint approaches are usually opposed by land managers and risk to be ineffective in preventing negative effects deriving from "not doing", e.g. in the case of agriculture, from land abandonment (Tempesta 2014). ELC changed the above described approach stressing the need to care for "everyday" landscapes and to rule not only maintenance processes but also management, creation and enhancement of landscapes. ELC stresses also the importance of landscape perception by local population (Council of Europe 2000), thus promoting participative approaches in planning processes. This new attitude as regards landscape maintenance and management had a direct impact on planning.

Landscape can be considered as a sub-articulation of territory or, as Marson says (Marson 2012), a different way to look at territory. The ELC defines landscape as follow: a zone or area as perceived by local people or visitors, whose visual features and character are the result of the action of natural and/or cultural (that is, human) factors (Council of Europe 2000). In rural landscape farmers are key land managers, playing a very important role in landscape evolution. As a consequence, when planning for landscape sustainability (Wu 2013), all the drivers influencing farmers' decisions, such as socio-economic drivers, regulative environment, geophysical features, etc., need to be taken into account together with the opinion of local population and other stakeholders (van Zanten et al. 2013). Complexity of problems asks for new interdisciplinary and transdisciplinary approaches when dealing with landscape governance (Conrad et al. 2011; Scott 2011).

In this paper, first we provide a theoretical model of governance, and second propose a set of measurable cardinal (or ordinal) indices on which to base actions for the landscape management of a case-study area located in Tuscany, a central region of Italy. While the regulative environment is specific to Tuscany, the authors believe that the proposed approach is suitable to a wide range of contexts.

8.2 A Conceptual Model for Landscape and Territorial Governance

The ELC applies to the entire territory and concerns landscapes of outstanding beauty as well as every day or even degraded landscapes (Council of Europe 2000). Standard and constraint policies may be deemed as adequate for landscapes that either for their high value or critical situation need to be preserved or restored/reproduced. In other landscapes, management actions need to ensure an adequate provision of Ecosystem Services (ESs) without necessarily preserving the actual landscape structures or the historical actions that have contributed to their creation. According to Haines-Young (cited in Wu 2013), "sustainability should be measured or assessed by the change processes active in the landscape—not by the state the landscape is in at any one time". In other words, the focus should shift from structures to the services they can produce and, consequently, from setting constraints to adopt old management actions to enforce rules able to reproduce positive structures and functions through new management actions (Magnaghi 2005).

When dealing with policy and planning tools for landscape and territorial governance, the institutional framework cannot be overlooked. According to Grêt-Regamey et al (2017), the successful implementation of tools requires a good understanding of decision-making processes to bridge gaps in the science–policy interface.

We have previously proposed a DSS for the governance of rural landscapes (Rovai et al. 2016) as a general model for guiding and helping to design policies and individuate proper actions according to the current features and dynamic evolution of different landscape ambits. Vice versa, this chapter is based on the authors experience when dealing with the decision-makers' needs about a specific planning instrument, e.g. the drawing of the "structural plan" of Valdera,[1] an administrative body born by the union of 7 municipalities among the 14 included in the "geographical" Valdera. This means that the model had to take into account general guidelines given by the Tuscan regional administration and zonation approaches, while providing a detailed knowledge framework at local level and giving suggestions for specific area-tailored actions. The knowledge base has been mostly provided as maps and tables that could be suitable for a participative approach.

The following part of the chapter is organized as follow.

In the methodology section, we provide: (a) information on the regulatory environment of Tuscany as regards territorial and landscape planning; (b) the methodological approach and the description of the indicators chosen for the spatial knowledge framework on which guidelines and actions are based.

In the case-study section, we provide: (a) a short description of the area, (b) some maps on the main indicators used, and (c) an authors' proposal for management actions to be considered by decision-makers, based on values and criticalities characterizing each landscape unit.

[1] Valdera or Val d'Era is the plain created by Era River.

In the concluding remarks, we discuss the difficulties faced in the process and give insights on the potential of the adopted approach for steering planning and decision-making at local level.

8.3 Methodology

8.3.1 The Planning Regulatory Environment of Tuscany

Since the beginning of the nineties, the territorial approach (Magnaghi 2005) has underpinned the regulation on territorial governance of the Tuscan Region.

As regards landscape, in 2009 the Tuscan Region decided to integrate its landscape plan in the "piano di indirizzo territoriale (PIT)" or Guidance Territorial Plan (GTP) rather than ruling it by a specific plan. Territorial governance in Tuscany is currently regulated by Regional Law 65/2014 that states that the GTP is the instrument regulating territorial planning. All regional policies, sectoral plans and programmes having territorial impacts need to comply with GTP, as well as planning and urban planning tools. This decision makes the regulatory environment quite complex insofar as, as regards landscape, the Code for Cultural and Landscape Heritage states that regional landscape plans are superordinate plans with contents which are compulsory prescriptive for local authorities. Vice versa, as regards planning, the Italian Constitution assigns to municipalities the responsibility not only for the drafting but also for the approval of their territorial planning instruments, while the Region can make only specific recommendations, which do not have any binding value. This implies that the regional integrated territorial and landscape plan has a guidance role for physical planning while it gives compulsory prescriptions for the part of landscape plan (Marson 2012).

The GTP distinguishes between the structural-statutory part of planning, defining the identifying features of places, namely, their invariance and the rules for variations, and the strategic-operational part describing transformation projects (Magnaghi 2005).

The invariants that have been identified by the structural-statutory part of the current GTP are the following:

I. The hydro-geomorphological features of catchment basins and morphogenetic systems;
II. The ecosystem features of landscapes;
III. The polycentric and network characteristic of settlement, urban and infrastructural systems;
IV. The morphological and typological features of agri-environmental systems and rural landscapes.

In this chapter, we focus on rural landscapes, i.e. on the fourth invariant, and their morphotypes (deriving from their morphological ad typological features). The

Landscape Plan of Tuscany has classified rural landscapes according to the Abacus of Rural Landscape Morphotypes. Rural morphotypes have been identified through the overlapping of several informative layers, i.e. anthropic characters such as settlements, size and characteristics of cultivated plots, soil typologies and land use, and hydro-geomorphologic characters, and define rural landscape typologies (Baldeschi et al. 2017; Fastelli et al. 2018).

The regional landscape plan provides a map of morphotypes at a scale of 1:250.000 and, as a consequence, the map cannot be considered as a zoning of the rural territory, but as a broad identification of the areas within which a landscape type is prevalent on the others. Indeed, in a territorial context features characterizing morphotypes vary in intensity and boundaries are blurred. Consequently, the inclusion into a morphotype rather than into another is based on threshold values. For this reason, the classification of territory in morphotypes needs to be revised at the lower planning levels, when it is possible to have a detailed and exact description at a finer scale (Baldeschi et al. 2017). For each morphotype, the Abacus describes biophysical characteristics (elevation, acclivity, land cover, etc.), values, criticalities; gives information on its management; and provides guidelines for policy actions.

As regards rural territories, the Landscape Plan gives a central importance both to agricultural enterprises and to non-professional entities in order to ensure the maintenance and reproduction of a landscape identity which is commonly shared, without penalizing the freedom in exercising economic activities or promoting an inefficient use of the resources. The individuation of morphotypes has been based on the awareness that changes in farms structures and management techniques are necessary to maintain an adequate income in a changed market and technological scenario. For this reason, for each morphotype, it is important not only to perform a structural analysis but also functional and management analyses able to highlight the degree of "viability" of that rural landscape. Rural landscape viability depends also on the typologies of farms which are operating in the area, and their attitude towards landscape invariants and environmental protection (Fastelli et al. 2018).

At municipality level, when drafting local planning instruments such as the structural plan (Piano Strutturale) and the operational plan (Piano Operativo) it is necessary to adequately define morphotype characteristics. In urban plans, these characteristics influence rules and constraints that should be included. Vice versa, when dealing with policies for rural development or landscape maintenance and enhancement, morphotype characteristics can become targets to be reached through specific strategic projects, built through the involvement of local stakeholders.

8.3.2 The Methodological Approach

In this subsection, first we provide a general model for landscape governance inspired by the ES cascade, and then we identify and develop a specific part of the model, i.e. the role of land managers in producing landscape, according to the specific aims of the case-study analysis.

The general model proposed in Fig. 8.1 was inspired by the version of the cascade provided by the Barcelona case study (Potschin-Young et al. 2018, see Fig. 8.4). In this case, the cascade was transformed into a ring, thus solving the problem about the direction the cascade should be read. Indeed, while many authors start from structure and function to arrive at the benefits for human beings and their values, others suggest that the cascade should be read in the opposite direction, starting from the values of benefits in order to arrive to structures that are able to determine these benefits. This last option raises the question of policy instruments to be used in order to promote adequate structures. In both cases, ESs can be seen as on the boundary between the functions provided by structure and processes and the benefits that a function can represent for human beings, depending from their specific set of values. While in a strictly natural approach, policy action responses are mainly intended as a constraints aiming to limit anthropic pressure on biophysical structure and processes (Potschin-Young et al. 2018, see, e.g. Fig. 8.1), when dealing with landscape management both in its natural and anthropic components, as in the case of landscapes, the approach can change and aim to increase the value of the territorial heritage, in its immaterial and material components, both natural and anthropic.

In our general model, structure influence functions mostly as a consequence of choices operated by land managers. Vice versa, on the demand side, dwellers are the main stakeholders insofar as they decide the actual and potential uses of the territorial heritage; this by setting the values they attribute to ESs, highlighting the criticalities related to territorial depletion and contributing to setting policy goals.

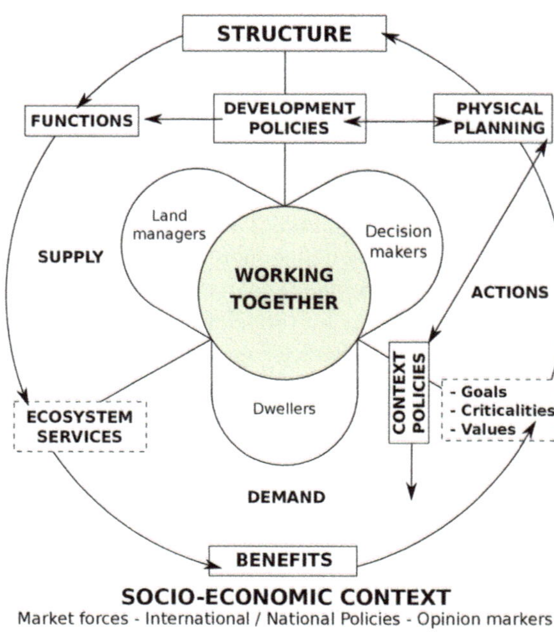

Fig. 8.1 A general model for territorial and landscape governance inspired by the ES cascade

Local decision-makers have the role to translate land managers' and dwellers' needs and aspirations into planning, programming and design instruments. They should also ensure a horizontal and vertical coordination among all the instruments that have an impact on physical planning. Besides, local public authorities have new tasks in managing local resources, insofar as planning should not only profit from the flows of information coming from stakeholders involved in supply and demand of ESs (Potschin-Young et al. 2018, see Fig. 8.4), but it should result from the joint working of all stakeholders (producers, dwellers and decision-makers) able to produce a shared vision of the territory and its development.

The described cycle is influenced by the general context. From a geographical point of view, landscape or local territory need to be seen in the framework of regional and global processes. At policy level, many interventions need to comply with rules and guidelines that are decided at international, national and regional levels, e.g. in the case of Rural Development Programmes (RDPs). Other policies that, at least in Italy, strongly affect planning are fiscal policies (Rusci 2015). Last but not least, general economic situation, global market trends and innovation in technology are influencing the context inside which land managers and dwellers operate.

GTP and landscape plan mainly refers to the identification of landscape criticalities, while the RDP mainly faces the needs of farmers. When planning mostly relates to rural landscape, as in the case of our case-study analysis, the physical planning and sectoral programming need to be coordinated and this implies that decision-makers should work together with local stakeholders, mainly represented by farmers that in rural landscape are the key land managers. This cooperation should result in a shared vision for the local territory giving rise to landscape projects, promoted by private and public bodies.

While Fig. 8.1 provides a general and comprehensive model, when dealing with farmers as key land managers it would be useful to develop the supply sector and the relationships between land managers and policy actions. Figure 8.2 focuses on the above issues.

Figure 8.2 stresses and better details the influences between structure, management and functions. The underlying hypothesis in researches aiming to provide a knowledge base for planning and landscape projects is that landscape structure partly depends on the type of land management, insofar as long-term management actions can shape structure. In their turn, management options are influenced by structural features, which provide constraints and opportunity for specific management actions. Management in rural landscape is mainly related to agricultural activities. The relation between structure and management determines the functions landscape can fulfil, and consequently also ecosystem services, or disservices provided by landscape. Indeed, while it is difficult to attribute to nature negative functions, management choices of land managers may have both positive and negative effects on community well-being, even as a consequence of the same action (Zhang et al. 2007).

According to Brunori et al. 2013, land managers' behaviour is influenced by a cultural path that starting from the intuition of the impacts their decisions can have on landscape or environmental features, evolve in awareness, which in turn modifies motivations and eventually actions. Studies dealing with the adoption of

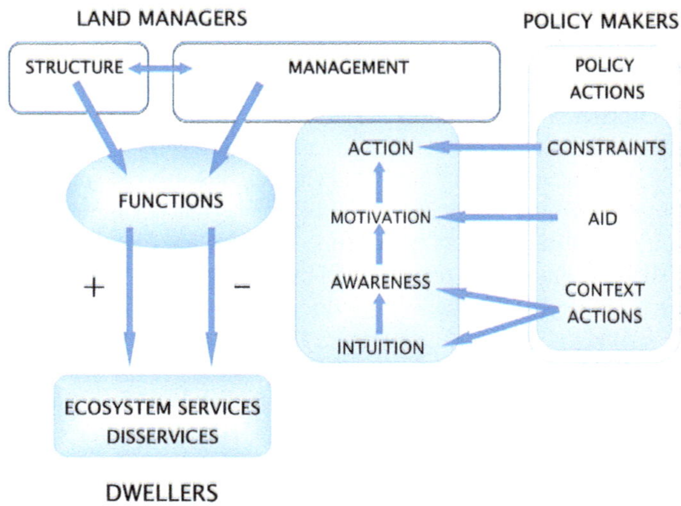

Fig. 8.2 Focusing on the supply side of the general governance model, i.e. on structure, management and policy actions and their influence on ES provision

agri-environmental schemes and their effects (see, e.g. Mills et al. 2017) have shown that in some cases the creation of motivation through a cultural change can bring about more durable effects than economic aid. From this point of view, while motivation may be modified by aid and actions constrained by rules, context policies are needed to promote intuition and awareness (Brunori et al. 2013).

A good knowledge of functions provided by a landscape as a result of structure and management may be the base for understanding the ESs landscape can provide. Among the wide range of ESs, for the case-study analysis, the following have been selected:

- Productive services (related to the economic function):
  - Agri-food production,
  - Tourism and tourism-related services,
  - Territorial marketing.
- Recreational-cultural services (related to the sociocultural function):
  - Maintenance and reproduction of the historical, cultural and identity heritage,
  - Usability of nature, environment and of the aesthetic values of landscape.
- Protective services (related to the ecological function):
  - Maintenance and reproduction of water resources,
  - Maintenance and reproduction of agroecosystem biodiversity,
  - Maintenance and reproduction of landscape heterogeneity,
  - Reduction of hydrogeological risk.

Morphotypes, according to their specific features, could be more suitable for fulfilling specific functions, while having an unsatisfactory performance for other functions, or fulfil a balanced mix of functions. A morphotype that is highly specialized in agricultural production, e.g. risks having an unsatisfactory performance as regards environmental functions. On the other hand, a morphotype characterized by land abandonment as a consequence of increasing farm economic difficulties may have a good environmental performance but be characterized by criticalities as regards aesthetic and perceptive functions or historic and identity functions. These latter, in their turn, may negatively impact on functions related to tourism and tourism-related activities.

The "state of health" of morphotypes may be analysed through a set of spatial or spatialized indicators that describe morphotype structures and the way they are currently managed.

Table 8.1 provides a list of structural and management indicators that can be computed by crossing the information of several databases provided by the Tuscan Region or other public bodies connected to the Tuscan administration (LaMMA, ARTEA). For a description of available data and the way different dataset can be integrated, see (Fastelli et al. 2018).

Table 8.1 Spatial or spatialized indicators for morphotype analysis

A—Structural indicators	B—Management indicators
A-1. Physical features	B-1. Evolutionary trends
A-1.a Elevation	B-1.a Changes in land use
A-1.b Acclivity	B-1.b Agricultural land abandonment
A-1.c Geology	B-1.b Soil consumption
A-1.d Pedology	B-2. Management aspects
A-1.e Climate	B-2.a Resident population
A-2. Naturalistic and ecologic structure	B-2.b Share of agricultural working population
A-2.a Forest vegetation	B-2.c Land managed by professional and non-professional farmers
A-2.b Ecological assets (II invariant)	B-2.d Farm characteristics
A-2.c Fragmentation and heterogeneity	B-2.d.1 Farm typology
A-2.d Protected areas	B-2.d.2 Average Utilized Agricultural Area
A-3. Settlement pattern	B-2.d.3 Number of separated parcels
A-3.a Historic / recent buildings ratio	B-2.d.4 Main crops
A.3.b Presence buildings of particular historical or architectural interest	B-2.e Presence of "Agri-tourism" activities
A-3.c Rural settlements	A-2.f Land included in denomination of origin
A-3.d Type of agrarian meshes	A-2.g Agricultural areas accessibility
A-4. Perception / visibility	

From the integrated analysis of the above listed indicators, it is possible to derive useful information on the functions a morphotype is able to fulfil as a consequence of its state of health and the way land managers operate, and recommendations for policy actions. These recommendations, when integrated in a shared vision of territorial development, may help to implement specific landscape local projects.

Few of the main indicators of management will be described and commented in the result section.

8.4 Case-Study Analysis: Area Description, Preliminary Results and Discussion

This section describes some preliminary results of the experience gained by some of the authors during a research project requested by the administration of the Valdera Municipality Union (Tuscany).

The current period is characterized by the need for municipalities and municipality unions to issue new local planning instruments after the adoption, in 2015, of the territorial and landscape plan. In particular, the research group was asked to provide a knowledge framework for the structural plan (Piano Strutturale) and the operative plan (Piano Operativo) for the case-study area, with a specific focus on rural landscapes. A structural plan includes three parts: a knowledge framework, a statutory-structural part and a strategic-operational part; these two latter should be based on the knowledge framework. At present, only a few structural plans have been issued after that the integrated territorial and landscape plan entered into force. These local plans were mainly drawn by simply reutilizing the information utilized for the regional plan, refining at a more detailed scale morphotypes boundaries, but not providing a local specific knowledge framework on which to build tailored solutions. Besides, according to the cultural attitude of local decision-makers, there is the risk for the knowledge base to be considered only a formal rule it is necessary to comply with, without exploiting its potentialities in identifying local heritage and in building landscape and territory management rules.

8.4.1 Case-Study Area Description

The geographical area called "Valdera" takes its name from the Era River, which in this area merges with Arno, the main river of Tuscany. It is mainly represented by the inland part of the ambit 8 of the Landscape Plan of Tuscany, related to the plan between Pisa, Leghorn and Pontedera.[2] Valdera includes 14 municipalities,[3]

---

[2]Only the municipality of Santa Maria a Monte is not included in ambit 8.

[3]Bientina, Buti, Calcinaia, Capannoli, Casciana Terme-Lari, Palaia, Pontedera, Santa Maria a Monte, Ponsacco, Crespina, Terricciola, Peccioli, Chianni and Lajatico.

7 of which belong to the Valdera Municipality Union.[4] The municipality union does not include the municipality of Ponsacco, although it is surrounded by the other municipalities belonging to the municipality union; for this reason, on maps there is a central area for which spatial information is not provided. Valdera includes municipalities with quite heterogeneous situations, as regards biophysical features, economic development and population size and density (Orsini 2013).

Environment and rural areas are considered as strengths for Valdera territory. Indeed, the environment still presents high quality and pristine characteristics inside which are located heterogeneous rural spaces, with highly prized characteristics. This results in a landscape with high aesthetic and perceptive value, which maintains and enhances the typical Tuscan landscape, with hilly area still unspoiled while development is mainly concentrated in flat areas and some valley floors. Hilly areas are mainly cultivated with olive groves and vineyards, whose renowned productions may ensure farm vitality. Hence, they need to be protected by containing urban expansion. In the location where rural areas have been already eroded by urban settlements, it is necessary to enhance the integration between rural space and space used for residential, productive and infrastructural uses.

At the same time, the agricultural sector is facing increasing difficulties, mostly in the more difficult areas and suffers for land abandonment that is bringing about higher risk of erosion, floods, forest fires, etc. Moreover, some of past choices about the localization/delocalization of productive activities and services are scarcely sustainable from an environmental point of view and, as a consequence, they can determine in time the depletion of stock of resources that are important for the social and economic development of these areas.

8.4.2 Preliminary Results and Discussion

The first step of the analysis was to define at local level, at a finer and more precise scale, the morphotypes characterizing the case-study area. While in the regional Abacus were included 23 general morphotypes, at local level where identified 27 morphotypes.

Figure 8.3 shows, in the left side, the distribution of rural morphotypes in the 7 municipalities of the Valdera Municipality Union. The white central part delimited in pink is the territory of the municipality of Ponsacco, which does not belong to the union, while other white parts refer to territories that are not classified as "rural".

The research has focused more on management than on structure and in particular on the causes of soil consumption. This can be caused both by land artificialisation and land abandonment, but we have privileged the last phenomenon as being more important for rural landscapes. In the following paragraphs, we describe three of the main indicators of management (B-1.b, B-2.c and B-2.d.2) included in Table 8.1.

[4]Bientina, Buti, Calcinaia, Capannoli, Casciana Terme-Lari, Palaia, Pontedera.

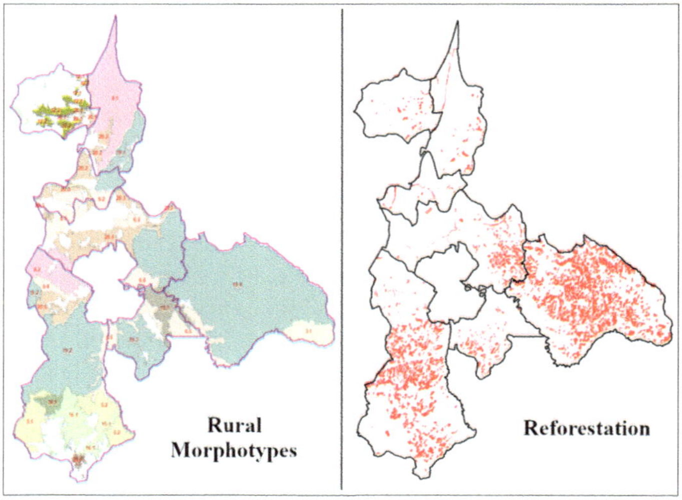

Fig. 8.3 Left side: Map of rural morphotypes identified in the territory of the Valdera Municipality Union. Right side: Map of land interested by processes of natural and artificial reforestation during the period 1954–2016

Land abandonment, especially when not adequately controlled, is deemed as having more negative effects than farming insofar as it may cause a loss of benefits for human well-being in terms of productive, protective and recreational-cultural ESs (Cooper et al. 2009; García-Ruiz and Lana-Renault 2011; Haddaway et al. 2013; Pelorosso et al. 2011) that is higher than benefits deriving from renaturalization. Sometimes land abandonment is masked by processes of reforestation, both natural and artificial, which transforms previously cultivated land into woodland, with the loss of landscape of historical interest (Agnoletti 2014). Figure 8.3, right side, shows the areas (in red) were reforestation processes affecting landscape in the period 1954–2016 were particularly strong. These processes mainly affect hill areas with the highest elevation, where cultivation is more difficult and less profitable. Territories with a high share of land cover accounted by woodland, bushes and natural or scarce cover are usually characterized by higher land abandonment, since woodland is not able to ensure an adequate income. More fertile lands are able to provide a higher per unit income and, as a consequence usually, in the case that a farmer closes its enterprise, land is not abandoned but purchased by neighbouring farmers (Rovai et al. 2016).

Indeed, several morphotypes with high share of woodlands and similar uses are also the ones that were most affected by phenomena of land abandonment from 1954 to 2016, thus confirming our previous hypothesis.

It is important to identify not only areas that are already abandoned but also areas at risk of abandonment by analysing land managed by professional and non-professional farmers. A high presence of land managed by non-professional farmers can pose questions for the future dynamics of morphotypes, insofar as non-professional farms are usually vulnerable to change in the economic context (e.g. people choice to produce its own olive oil even if it costs very much more than the purchased oil), the cultural context (when parents die, children who have grown up outside an agricultural context, are not able to carry on cultivation), etc. Besides, land managed by hobby and style farmers is more dependent on its geographical location, in terms of closeness to settlements and accessibility. Although the decision to carry on agricultural production in these areas may seem irrational from a productive-economic point of view, their abandonment could have very negative impacts as regards hydro-geologic stability and risk of forest fires.

Non-professional farmers do not respond to economic incentives, usually as a consequence of their small size that makes the economic and bureaucratic effort for applying for aid non-profitable. In this context, if these areas cannot be integrated in stronger farms, their survival may be ensured by different policy actions, e.g. promoting private or public services (e.g. pruning, harvesting, etc.) able to counteract the above-mentioned issues of high costs (in the case of style farmers) or loss of expertise (in the case of hobby farmers).

A spatialized analysis of the importance of professional farms vs non-professional ones is shown in Fig. 8.4, left side.

Figure 8.5 shows the same themes zooming on two morphotypes, i.e. morphotypes 12.1—"Terraced olive groves on the eastern side of Monte Pisano" and 8.1—"Arable land on the reclaimed land of the former Bientina Lake". Top map shows that these two rural landscapes are quite different in terms of share managed by professional farms. Indeed, while in the "Arable land on the reclaimed land of the former Bientina Lake" there is a high presence of land managed by professional farmers, in the "Terraced olive groves on the eastern side of Monte Pisano" there is a prevalence of land managed by non-professional farmers.

Usually, there is a quite strong relation between physical and economic size of farms, especially among farms with similar crop mixes. Hence, average size of farms may be considered as an important information related to farm viability and to the permanence of agricultural activities. Bottom map in Fig. 8.5 shows the spatial information about this characteristic, focusing on the same morphotypes shown in the top map. This analysis confirms the weakness of morphotype 12.1 (where about 46% of land is managed by farms with 5 ha. or less) in comparison to morphotype 8.1 where farms with average-large size are prevalent (about 73% is managed by farms whose size ranges from 50 to 500 ha.).

As stated above, in order to ensure the permanence of agricultural activities in areas that are neither suitable for self-consumption production nor close to settlements, it is necessary to promote professional farms that need to have an adequate size to be viable. In areas where small farms include fertile land, this viability and the absence of abandonment may be ensured only through the transfer of land (rent or purchase) from small to larger and stronger farms. At the same time, the growth in farm size and

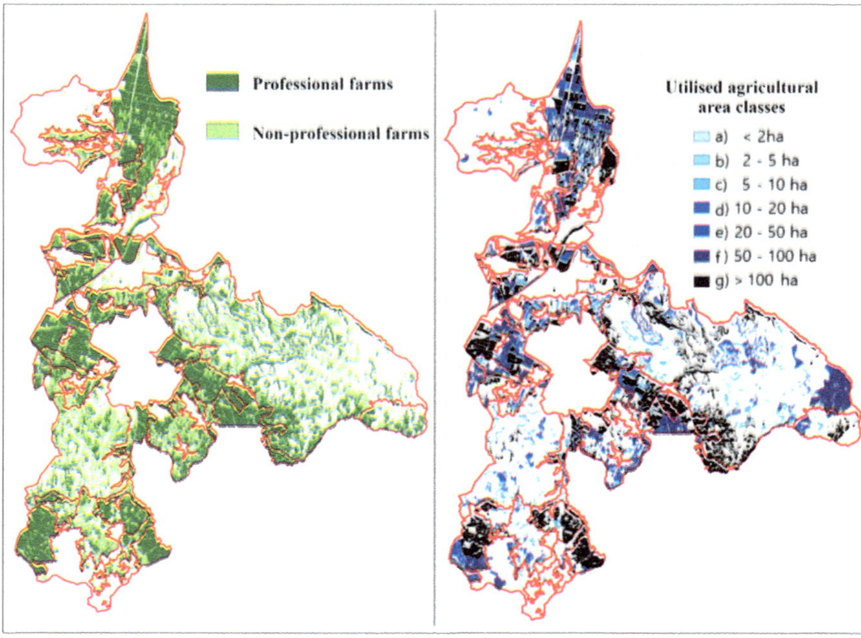

Fig. 8.4 Left side. Land managed by professional (dark green) and non-professional (light green) farmers. Right side: Farm classification according to the class of Utilized Agricultural Areas

modernization processes need not to totally destroy the typical signs characterizing farming in the Tuscan Region, namely, the agrarian meshes represented by the dense network of farm roads, field boundaries, hedges and tree lines.

The situation is more difficult to be faced in areas where land is characterized by low fertility, because in this case is more likely that the evolution will be towards abandonment. Consequently, it is necessary to ensure that this will not happen without a proper control, able to guarantee territorial stability.

Based on the variables described in Table 8.1 and of the above considerations, for each morphotype has been drawn a summary table that describes the main functions provided by each morphotype and the consequent guidelines and recommendations for policy actions. Policy actions do not have to relate necessarily to urban planning rules but may attain to sectoral policies (e.g. to Rural Development Programme—RDP) and can be at the base of a landscape strategic project.

The following Tables 8.2 and 8.3 describe functions and policy actions intended for the two morphotypes to whom we have made special reference, e.g. "Terraced olive groves on the eastern side of Monte Pisano" and "Arable land on the reclaimed land of the former Bientina Lake".

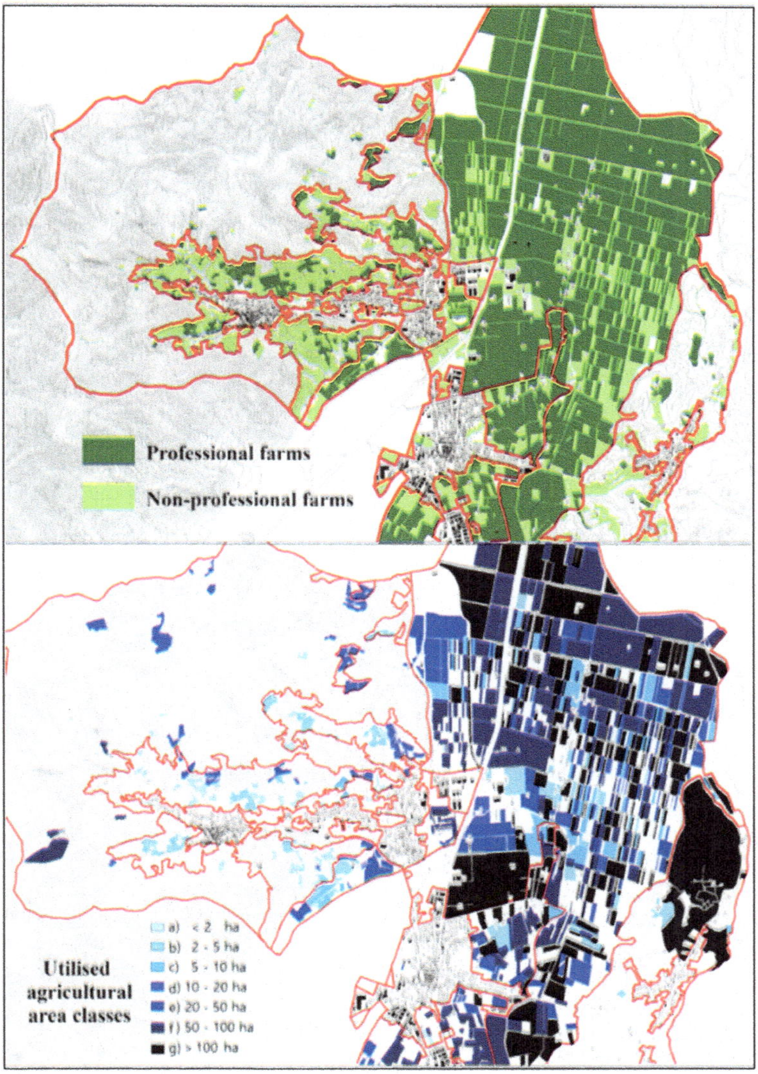

Fig. 8.5 Morphotypes 12.1—"Terraced olive groves on the eastern side of Monte Pisano" and 8.1—"Arable land on the reclaimed land of the former Bientina Lake". Top map: Land managed by professional (dark green) and non-professional (light green) farmers. Bottom map: Farm classification according to the class of Utilized Agricultural Areas

Table 8.2 A simplified framework for the analysis of morphotype "12.1—Terraced olive groves on the eastern side of Monte Pisano" and recommendations for policy actions

Function/ecosystem service	Assessment: description
Productive services	
– Agri-food production	High share of land cultivated for self-consumption or by hobby farmers
– Tourism and tourism-related services	High presence of on-farm tourism and Bed and Breakfast
– Territorial marketing	Landscape is seldom used as a leverage point for territorial enhancement
Recreational-cultural services	
– Maintenance and reproduction of historical, cultural and identity heritage	Strong identity of Monte Pisano and Tuscan landscapes
– Usability of nature, environment and of the aesthetic value of landscape	Widespread presence of roads and walking paths connecting the area with the neighbouring protected areas of Monte Pisano
Protective services	
– Maintenance and reproduction of water resources	Presence of hydraulic works that are paramount for managing surface water
– Maintenance and reproduction of agroecosystem biodiversity	Widespread presence of natural vegetation
– Maintenance and reproduction of landscape heterogeneity	Landscape specialized in olive groves, but that maintain a good level of heterogeneity thanks to the presence of ecological infrastructures
– Reduction of hydrogeological risk	Key role of terraces

Recommendations for policy actions

Policy actions need to guarantee the economic viability of productive processes insofar as they guarantee a mix of positive ecosystem services

Policy actions should aim to promote innovative management processes, such as: (a) collective services for the management of olive groves that are abandoned or at risk of abandonment, at least for the more critical phases of cultivation; (b) involvement of public bodies—e.g. municipality, land reclamation consortium—in the maintenance and management of the hydraulic and minor road networks; (c) aid to promote and maintain an adequate presence of farms in order to prevent abandonment phenomena and risks deriving from abandonment; (d) promote facilities (e.g. storage buildings) with adequate security and aesthetic features, able to make easier carrying out productive activities; (e) design regulations for common goods able to recognize the commitment of farmers in maintaining terraced olive groves and the consequent production of ecosystem services, and considering the provision of Payments for Ecosystem Services (PES)

Table 8.3 A simplified framework for the analysis of morphotype "8.1—Arable land on the reclaimed land of the former Bientina Lake" and recommendations for policy actions

Function/ecosystem service	Assessment description
Productive services	
– Agri-food production	High share of farmland belonging to professional farms that cultivate cereals and industrial crops
– Tourism and tourism-related services	Absence of on-farm tourism and other tourism services
– Territorial marketing	Landscape is seldom used as a leverage point for territorial enhancement
Recreational-cultural services	
– Maintenance and reproduction of historical, cultural and identity heritage	Reclaimed land landscape with a strong identity in Tuscany. Hydraulic work and road networks well preserved and widespread presence of farmhouses, although not so well maintained
– Usability of nature, environment and of the aesthetic value of landscape	Presence of wetlands to be maintained; widespread presence of roads and walking paths that easily allow area usability
Protective services	
– Maintenance and reproduction of water resources	Presence of hydraulic works that are paramount for managing surface water, since the area suffers for water stagnation
– Maintenance and reproduction of agroecosystem biodiversity	Widespread presence of wetlands, but low presence of ecological infrastructures
– Maintenance and reproduction of landscape heterogeneity	Landscape characterized by extensive crops, whose heterogeneity is ensured by crop diversification and fallow land (Common Agricultural Policy)
– Reduction of hydrogeological risk	Not significant, being a lowland

Recommendations for policy actions

This morphotype has a high naturalistic value but is vulnerable from the point of view of the maintenance and management of agricultural hydraulic works. It needs to be carefully preserved and enhanced.

Policy actions should aim to: (a) connect this area with other territorial systems (Monte Pisano, Cerbaie) by organizing adequate ecological corridors; (b) guarantee that agricultural hydraulic works are kept in good efficiency in order to ensure farmland production; (c) promote the restoration of abandoned farmhouses, highlighting their cultural and identity specificity. Farmhouses may be valuable resources for developing environmental tourism activities; (d) prevent the sprawl of settlements and soil consumption; (e) improve actions for territorial enhancement and territorial marketing

8.5 Conclusion

The experience gained during our research, aiming to lay down a knowledge framework for a structural plan, highlighted that in local urban planning interests are focused on land transformation, more than on rural morphotype management. Contrariwise, we believe that the role played by the rural landscape (and the territory) is as important as the role of infrastructures and residential and industrial settlements. Rural landscape role, in the coming decades, will tend to regain a new centrality, related to the organization of vital infrastructures able to provide local communities with ecosystem services directly affecting their well-being: food production, water and land management, carbon dioxide sequestration, creation and maintenance of landscape, culture and local identities, enjoyment of a balanced relationship between nature and individuals and communities (Di Iacovo et al. 2010; Rovai et al. 2010). Therefore, if we consider rural landscapes as living organisms that evolve in relation to management methods, any command and control approach will be doomed to be ineffective. For this reason, we considered the analysis of management by farmers as paramount. Through the management analysis, we reconstructed the framework of functions/services performed by each landscape and its most probable short-medium term development. Integrating the supply side of functions with the community "demands" or "hopes" for the territory/landscape, consent to identify and develop specific policies and/or projects.

Unfortunately, in the current phase, territorial planning is still guided by strong economic interests linked to land transformation and there is a scarce concern about the rural territory, about which planning tends to introduce constraints and rules, without providing aid or promoting territorial development.

Due to the fundamental role played by rural territory in providing Ecosystem Services (ESs), the need arises for a closer coordination between planning and rural development tools, in order to promote effective landscape projects with the aim to strengthen the production of ESs. Landscape projects as local development tools need to be drawn and implemented through a shared vision between farmers, as key land managers, and other stakeholders. Moreover, landscape projects may stimulate farmers to act based on a greater awareness of the role they play in producing landscape and ESs and consequently bring about more durable positive effects.

References

Agnoletti M (2014) Rural landscape, nature conservation and culture: some notes on research trends and management approaches from a (southern) European perspective. Landscape and Urban Planning 126:66–73. https://doi.org/10.1016/j.landurbplan.2014.02.012

Baldeschi P, Brunori G, Fastelli L, Gisotti M R, Rovai M (2017) I caratteri morfotipologici dei paesaggi rurali. In: Marson A (ed), La struttura Del paesaggio: una sperimentazione multidisciplinare per il Piano della Toscana, pp 202–216. Laterza

Brunori G, Fastelli L, Rovai M (2013) Politiche di sviluppo rurale e paesaggio. In: Poli D (eds), Agricoltura paesaggistica, pp 129–160. 10.978.886655/4363

Conrad E, Christie M, Fazey I (2011) is research keeping up with changes in landscape policy? A review of the literature. J Environ Manage 92(9):2097–2108. https://doi.org/10.1016/j.jenvman.2011.04.003

Cooper T, Hart K, Baldock D (2009) European Union, European Commission, Directorate-General for Agriculture and Rural Development, Institute for European Environmental Policy (2009) Provision of public goods through agriculture in the European Union. Accessed from http://www.ieep.eu/assets/457/final_pg_report.pdf

Council of Europe (2000) European Landscape Convention: CETS No.: 176. Retrieved from http://conventions.coe.int/Treaty/en/Treaties/Html/176.htm

Di Iacovo F, Rovai M, Orsini S (2010) Spazio rurale ed urbano: alla ricerca di nuovi equilibri. In: Perrone C, Zetti I (eds), I l valore della terra. Teoria e applicazioni per il dimensionamento della pianificazione territoriale, pp 105–133. Milano: FrancoAngeli

Fastelli L, Rovai M, Andreoli M (2018) A spatial integrated database for the enhancement of the agricultural custodianship role (SIDECAR)—some preliminary tests using Tuscany as a case-study region. Land Use Policy 78:791–802. https://doi.org/10.1016/j.landusepol.2018.07.025

García-Ruiz JM, Lana-Renault N (2011) Hydrological and erosive consequences of farmland abandonment in Europe, with special reference to the Mediterranean region—a review. Agr Ecosyst Environ 140(3–4):317–338. https://doi.org/10.1016/j.agee.2011.01.003

Grêt-Regamey A, Sirén E, Brunner SH, Weibel B (2017) Review of decision support tools to operationalize the ecosystem services concept. Ecosyst Services 26:306–315. https://doi.org/10.1016/j.ecoser.2016.10.012

Haddaway NR, Styles D, Pullin AS (2013) Environmental impacts of farm land abandonment in high altitude/mountain regions: a systematic map of the evidence. Environ Evidence 2(1):18. https://doi.org/10.1186/2047-2382-2-18

Magnaghi A (2005) The Urban Village. Zed Books Ltd, A charter for democracy and local self-sustainable development

Marson A (2012) Participation in landscape planning. First Careggi Seminar, Florence November 10, 2011, 4–9. Accessed from http://citeseerx.ist.psu.edu/viewdoc/download?doi=10.1.1.466.7188&rep=rep1&type=pdf

Mills J, Gaskell P, Ingram J, Dwyer J, Reed M, Short C (2017) Engaging farmers in environmental management through a better understanding of behaviour. Agric Hum Values 34(2):283–299. https://doi.org/10.1007/s10460-016-9705-4

Orsini S (2013) Explaining land management decisions to understand local landscape functions and change. Some insights from Tuscany. Local Environ 18(8):934–949. https://doi.org/10.1080/13549839.2012.748726

Pelorosso R, Della Chiesa S, Tappeiner U, Leone A, Rocchini D (2011) Stability analysis for defining management strategies in abandoned mountain landscapes of the Mediterranean basin. Landscape Urban Plann 103(3–4):335–346. https://doi.org/10.1016/j.landurbplan.2011.08.007

Potschin-Young M, Haines-Young R, Görg C, Heink U, Jax K, Schleyer C (2018) Understanding the role of conceptual frameworks: Reading the ecosystem service cascade. Ecosystem Services 29(6):428–440. https://doi.org/10.1016/j.ecoser.2017.05.015

Rovai M, Andreoli M, Gorelli S, Jussila H (2016) A DSS model for the governance of sustainable rural landscape: a first application to the cultural landscape of Orcia Valley (Tuscany, Italy). Land Use Policy, 56. https://doi.org/10.1016/j.landusepol.2016.04.038

Rovai M, Di Iacovo F, Orsini S (2010) I l ruolo degli Ecosystem services nella pianificazione territoriale. In: Perrone C, Zetti I (eds), I l valore della terra. Teoria e applicazioni per il dimensionamento della pianificazione territoriale, pp 135–161. Milano: FrancoAngeli

Rusci S (2015) Council tax policies and territorial governance: analysis and outlook of a difficult relationship. TeMA J Land Use, Mobility Environ 8(3):331–346. https://doi.org/10.6092/1970-9870/3033

Scott A (2011) beyond the conventional: Meeting the challenges of landscape governance within the European landscape convention? J Environ Manage 92(10):2754–2762. https://doi.org/10.1016/j.jenvman.2011.06.017

Tempesta T (2014) People's preferences and landscape evaluation in Italy: a review. New Medit 1:50–59

van Zanten BT, Verburg PH, Espinosa M, Gomez-y-Paloma S, Galimberti G, Kantelhardt J, Viaggi D (2013) European agricultural landscapes, common agricultural policy and ecosystem services: a review. Agron Sustain Dev 34(2):309–325. https://doi.org/10.1007/s13593-013-0183-4

Wu J (2013) Landscape sustainability science: ecosystem services and human well-being in changing landscapes. Landscape Ecol 28(6):999–1023. https://doi.org/10.1007/s10980-013-9894-9

Zhang W, Ricketts TH, Kremen C, Carney K, Swinton SM (2007) Ecosystem services and dis-services to agriculture. Ecol Econ 64(2):253–260. https://doi.org/10.1016/j.ecolecon.2007.02.024

Chapter 9
Models for Describing Landscape Hydrochemical Discharge in Mountain Countries

Yuri Kirsta, Alexander Puzanov and Tamara Rozhdestvenskaya

Abstract Using 34 river basins of the Altai-Sayan mountain country as an example, the universal mathematical models for seasonal and long-term dynamics of landscape water and hydrochemical discharges were developed. For this purpose, the system-analytical modelling and the solution of mathematical inverse problem were applied. The input factors of the models include monthly precipitation and mean monthly air temperature, spatially generalized for the country, as well as the cartographic information on the area and average altitude of individual landscapes in river basins, the altitude of the outlets, the length of river channels, and the area of arable land. The water and seven hydrochemical discharges (three nitrogen mineral forms NO_2^-, NO_3^-, NH_4^+, phosphates PO_4^{3-}, ions, total dissolved iron, suspended matter) are calculated for each of 13 specified landscape types in each river basin. The sensitivity of discharges to variations of environmental factors is evaluated. The Nash-Sutcliffe efficiency estimated as more than 0.65 for the models represent their good and very good performance. The models also make it possible to forecast the seasonal dynamics of discharges for any river basin in the country under study.

Keywords Mountain rivers · Landscapes · System-analytical modelling · Water discharge · Hydrochemical discharge · Forecast · Altai-Sayan

---

Y. Kirsta (✉) · A. Puzanov · T. Rozhdestvenskaya
Institute for Water and Environmental Problems of Siberian Branch, Russian Academy of Sciences, Molodezhnaya St. 1, Barnaul, Altai Krai 656038, Russia
e-mail: kirsta@iwep.ru

A. Puzanov
e-mail: puzanov@iwep.ru

T. Rozhdestvenskaya
e-mail: rtamara@iwep.ru

Y. Kirsta
Altai State Technical University, Lenin Ave. 46, Barnaul, Altai Krai 656038, Russia

W. Mirschel et al. (eds.), *Landscape Modelling and Decision Support*, Innovations in Landscape Research,
https://doi.org/10.1007/978-3-030-37421-1_9

9.1 Introduction

The rational use of freshwater resources needs for adequate mathematical models able to quantify the quality of these resources. Hydrochemical discharge from large and small catchments is one of the most important characteristics of ground and surface water quality affecting drinking and household water supply. Hydrochemical discharge changes markedly in different landscapes as it depends on biogeochemical processes in the catchments, water and wind erosion, atmospheric transport and deposition of chemicals (including ocean salts), and anthropogenic factors. The analysis of regularities of hydrochemical discharge is time-consuming for mountain areas due to their complex hydrogeological structure, diverse land cover, and costly experimental research of these patterns. Not surprising it is extremely difficult to quantify the impact of various environmental factors on landscape hydrochemical discharge. We solve this complex problem by the example of river basins of the Altai-Sayan mountain country through the development of mathematical models.

To do that, a system approach and an efficient system-analytical modelling (SAM) with the involvement of GIS technologies are used (Kirsta 2006, Kirsta and Kirsta 2014).

9.2 Purposes of Research

1. The developed mathematical models of water and river hydrochemical discharges should be used in environmental and water-resource monitoring of the territory, long-term forecast and management of emerging risks. In particular, the forecast of water flow is important for the control of reservoir release at mountain hydropower plans during the spring-summer flood, when the main annual flow enters the reservoirs. Models should properly take into account the key environmental factors that affect the discharge, which in mountain conditions vary significantly in space and time. To do this, it is necessary to determine the general patterns of changes in these factors in the mountains, in particular, patterns of dynamics of meteorological fields (Kirsta 2011). In this case, the models should provide a smaller discrepancy between the predicted and actual values of water and hydrochemical discharges when compared to the trivial forecast by their average annual value.
2. In mathematical modelling of complex natural systems (hydrological, hydrochemical, etc.), it is of great importance to assess the sensitivity of processes to variations of input factors. The overview of different sensitivity estimation methods can be found, for example, in (Iooss and Lemaitre 2015). The calculation error is closely related to the sensitivity (Beven and Hall 2013) and is important for the applied use of models. Thus, an adequate assessment of simulated discharge sensitivity to variations in environmental factors and calculation errors is called for.

3. For catchment basins of mountain rivers, it is necessary to specify the evolutionarily developed territorial cycles of water and chemical elements, to which natural landscapes of these basins correspond. This will allow most adequate consideration of all the processes of formation of water and hydrochemical runoffs in the models.
4. Quantitative and predictive assessment of the most important characteristics of river discharge is needed to perform environmental and water resource monitoring with advantage. It is worthwhile to characterize the main components of hydrochemical discharge, which should be evaluated as for each landscape as for the river basin as a whole.

9.3 Experimental Data and Methods of Research

The territory of the Altai-Sayan mountain country located within 49° and 56° N, 82° and 90° E is a part of the world watershed between the humid region of the Arctic Ocean and the arid drain-less region of Central Asia. The Sayan ridges reach 3000–3500 m in height, and the Altai ones – 3500–4500 m. Climate is sharply continental here, with cold winters and cold summers. Atlantic cyclones easily penetrate the territory. According to the data obtained from rare weather stations, annual precipitation varies widely. For example, the northern slopes of mountains at altitude over 3000 m get 1200–2500 mm of precipitation per year, the middle parts – up to 600 mm, and the low ones – about 200 mm. According to weather stations data (Table 9.1), the general climatic characteristics of the Altai-Sayan Mountains show that most of annual precipitation occurs during the warm period, which lasts about half a year. In all rivers, maximum discharge is observed also during this period and makes up 80–90% of the annual one. Snowmelt contribution to the formation of discharge exceeds 50%.

We selected 34 river basins (Fig. 9.1), the area of which made up from 177 to 21,000 km^2. Overall, 13 landscapes (typological groups of geosystems) were chosen for the basins (Kirsta et al. 2011) to perform SAM of landscape water and hydrochemical discharges (Table 9.2). The data from the Russian State Department of Hydrometeorology and Environmental Monitoring on water discharge (~5000 values) and the corresponding components of hydrochemical discharge (~10,000) for all 34 river basins were used in the development of the models (Kirsta et al. 2012, Kirsta and Puzanov 2016, Kirsta 2016). The input factors of the models involved monthly precipitation and average monthly air temperature for 1951–2010 (11 reference weather stations), data on the landscape structure of the basins (area and elevation of landscapes), area of arable land and other cartographic characteristics. The cartographic documents were processed in the ArcGIS 9.2 environment including 3D analyst module.

The dynamics of water and hydrochemical discharges was estimated for four specific hydrological periods/seasons, i.e. the first (winter low water, XII–III months),

Table 9.1 Long-term (1951–2010) mean air temperature and precipitation in the Altai-Sayan mountain country

Climatic characteristics	Months											
	I	II	III	IV	V	VI	VII	VIII	IX	X	XI	XII
Air temperature (°C)	−16.0	−14.6	−8.0	2.2	10.5	15.7	17.8	15.3	9.6	2.1	−7.3	−13.3
Precipitation (mm)	21.9	19.5	22.3	40.7	58.3	71.1	81.9	73.0	50.9	49.4	39.5	30.6

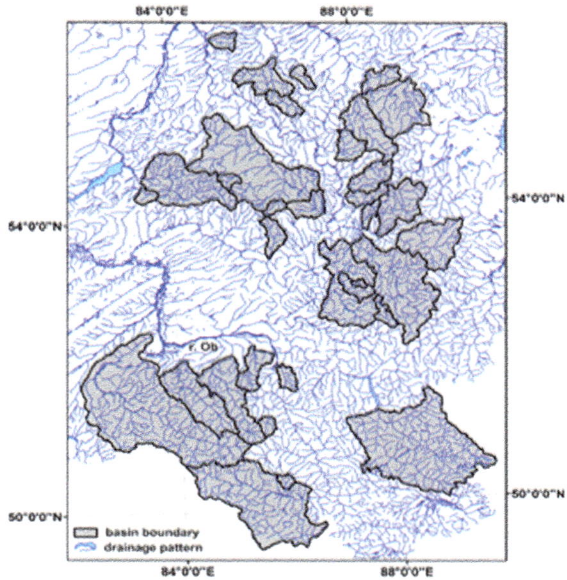

Fig. 9.1 Schematic map of 34 model river basins in the Altai-Sayan mountain country

the second (spring-summer high water, IV–VI), the third (summer low water, VII–VIII), and the fourth (autumn low water with probable floods caused by heavy rains, IX–XI). To characterize the hydrochemical discharge, we chose three separate nitrogen mineral forms (NO_2^-, NO_3^-, NH_4^+), phosphates (anions PO_4^{3-}), mineralization (ions), total dissolved iron (Fe), and suspended matter. In each season, there were one or two experimental values of analyte concentrations (mg/litre). By averaging seasonal data, we got four mean seasonal concentrations for each year of observations. For transition to dimensionless units of measurement, experimental values of river streamflow and analyte concentration were divided by their average long-term seasonal values.

For the Altai-Sayan mountain country, the spatiotemporal analysis of its meteorological fields was carried out (Kirsta 2011). Intra-annual and long-term dynamics of monthly precipitation and average monthly air temperature appeared to be uniform for the country when normalizing precipitation by its average annual value for July ‘in situ’, and temperatures – by their average annual value for January (for X–IV months) and July (for V–IX) ‘in situ’. For this reason, air temperature and precipitation were described through fraction/percent of their average long-term monthly values. Using the developed method of their spatial generalization, we obtained their long-term monthly dynamics typical to the Altai-Sayan mountain country. Such a dynamics was shown to be an adequate areal characteristic of meteorological field changes across the territory under study. The spatially generalized normalized air temperature and precipitation were independent of coordinates or altitude, were the same for all river basins, and were applied in calculation of all water and hydrochemical discharges.

Table 9.2 Mean altitude, relative area and contribution of landscapes of the Altai-Sayan mountain country to the river runoff

Landscapes (geosystem groups)	Mean altitude (m a. s. l.)	Mean relative area (%)[a]	Share of precipitation entering the river runoff for each of four hydrological seasons (XII–III, IV–VI, VII–VIII, IX–XI months)[b]			
			1	2	3	4
1. Glacial-nival high mountains with permafrost	2448	1.2	0.30	0.23	0.45	0.00
2. Goletz-alpine type high and middle mountains, pseudo-goletz low mountains with permafrost	1446	9.5	0.09	0.35	0.61	0.28
3. Tundra-steppe and cryophyte-steppe high mountains with permafrost	2329	0.4	0.30	0.56	0.02	0.00
4. Forest high middle and low mountains	742	44.1	0.29	0.33	0.69	0.91
5. Exposure-forest steppe and steppe high and middle mountains	1213	1.8	0.14	0.69	0.75	0.00
6. Forest-steppe, steppe low mountains and foothills	415	15.9	0.36	0.24	0.62	0.70
7. Intermountain depressions with different steppes and forest-steppes	720	5.2	0.33	0.31	0.85	0.44
8. Steppe and forest-steppe piedmont	268	8.8	0.06	0.31	0.42	0.31
9. Nondrainable and intrazonal landscapes with partial permafrost	1501	1.3	0.72	0.93	0.00	0.22
10. Mountain river valleys	563	9.0	0.22	0.00	0.63	0.52
11. Lowland river valleys	229	2.0	0.08	0.37	0.61	0.99
12. Forest high and piedmont plains	263	0.7	0.02	0.36	0.77	0.42
13. Aquatic landscapes	1400	<0.1	0.58	0.69	0.01	1.00

[a]In percent of total river basin area
[b]Results of calculation of b_k in Eq. (9.4a)

SAM is based on searching the structure and parameters of simulation models of complex natural processes through solving the inverse mathematical problem by optimization methods (Kirsta 2006, Kirsta and Kirsta 2014; Kirsta et al. 2012; Kirsta and Puzanov 2016). The models contain algebraic balance equations, since the use of differential equations to describe the hydrological and hydrochemical processes in mountains is unreasonable (Kirsta et al. 2012). The system of equations with the least sum of squared differences between the calculated and observed river streamflow and analyte concentrations was established via successive verification of hydrologically and physicochemically consistent dependences of different forms that describe the water and hydrochemical runoff formation under the influence of environmental factors. To carry out this work and construct the water and hydrochemical discharge models we use a MATLAB programming environment. We have developed a number of programs in MATLAB that provide the rejection of unreliable experimental data, the solution of the inverse problem for systems of more than 1500 equations with simultaneous identification of up to 100 parameters of the tested versions of models.

To describe nontrivial dependence of processes on environmental factors in SAM, we apply the universal trial function H:

$$H(X1, X2, Y1, Y2, Z1, Z2, X) = \begin{cases} Y1 + Z1(X - X1), & \text{if } X < X1 \\ \frac{Y2-Y1}{X2-X1}(X - X1) + Y1, & \text{if } \begin{matrix} X1 \le X < X2 \\ X1 \ne X2 \end{matrix} \\ Y2 + Z2(X - X2), & \text{if } X \ge X2, \end{cases} \tag{9.1}$$

where $X1$, $X2$, $Y1$, $Y2$, $Z1$, $Z2$ are the parameters selected during SAM; X is a variable of the model. Function H is a continuous piecewise-linear function of three arbitrary linear fragments, which allows to approximate a wide range of different dependencies between variables and environmental factors by changing its parameter values (Fig. 9.2).

Any mathematical model needs verification (Hauduc et al. 2011). To do that, we use the following criterion for assessing the adequacy of any calculation methods or models, which is based on the comparison of the observed and the calculated data series (Kirsta et al. 2012):

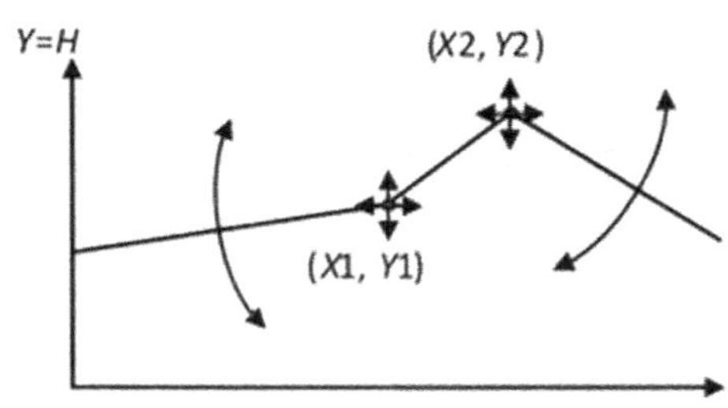

Fig. 9.2 Continuous piecewise-linear function H ($X1$, $X2$, $Y1$, $Y2$, $Z1$, $Z2$, X) of three linear fragments with arbitrarily changing parameters (see Eq. (9.1))

$$A = S_{\text{dif}}/(\sqrt{2}S_{\text{obs}}), \tag{9.2}$$

where A is the criterion of model adequacy; S_{dif} is the standard (RMS) deviation for the difference between the observed and calculated data patterns (model residuals); S_{obs} is the standard deviation for the observed pattern; $1/\sqrt{2}$ is the multiplier.

According to Eq. (9.2), criterion A is actually a model error normalized to standard deviation of the observed data. When assessing the model performance, A is similar to performance measure RSR (Moriasi et al. 2007, Koch and Cherie 2013) and Nash–Sutcliffe model efficiency coefficient NSE (Koch and Cherie 2013) related to A by equations $\text{RSR} = A\sqrt{2}$, $\text{NSE} = 1\text{–}\text{RSR}^2 = 1\text{–}2A^2$. Multiplier $1/\sqrt{2} = 0.71$ is entered (9.2) to make up a standard range 0–1 involving two meaningful intervals of A values. According to the variance sum law (applied for S_{dif}), values A can vary from 0 to 1 and more:

- The interval 0–0.71 characterizes a different degree (the best is at $A = 0$) of the identity of compared patterns and model adequacy. This interval of the adequacy corresponds to a change of model performance measure RSR between 0 and $A/0.71 = 1$. It includes RSR~1 values that lie outside the permissible limit RSR = 0.70 of model validity (Koch and Cherie 2013). At the same time, such RSR values did not prevent both the execution of SAM and the construction of adequate models of a good quality (Kirsta 1986; Kirsta and Puzanov 2015). Thus, the criterion A provides more accurate model performance measure as compared to RSR;
- The interval 0.71–1 implies that model's adequacy is low, and the regularities of the observed pattern are not taken into account properly. To make the predictive valuation, the use of the mean value of hydrological characteristic with $A = 0.71$ is recommended;
- The interval larger than 1 indicates that the calculated pattern has greater variance than the observed one. Sometimes it is essential to keep the observed pattern variance when employing the calculated data in other models/submodels. Consequently, it is reasonable to substitute the calculated data by random variations of the analysed characteristic with the same average and variance as the observed pattern has. In this case, A will be equal to 1.

Adequacy criterion A is convenient to calculate the model sensitivity to natural variations of environmental factors (Kirsta and Puzanov 2015). The heart of evaluating the sensitivity of mathematical models is criterion FS, which characterizes the sensitivity directly to natural variations of environmental factors. FS is close in meaning to the known percentage of explained variance and is calculated by the formula

$$\text{FS} = \left(A'\right)^2 - (A)^2 = \frac{\left(S'_{\text{dif}}\right)^2 - (S_{\text{dif}})^2}{2(S_{\text{obs}})^2} = \frac{2(S_{\text{fac}})^2}{2(S_{\text{obs}})^2} = \frac{(S_{\text{fac}})^2}{(S_{\text{obs}})^2}, \tag{9.3}$$

where FS is the model sensitivity to input factor; A is calculated from (9.2); A' is A value obtained from (9.2) by using the randomly mixed values of input factor instead of the initially ordered; in this case the randomly mixed observed data pattern obviously has a former statistical distribution and variance; $(S_{\mathrm{dif}})^2$ is the variance for the difference between the calculated and observed data patterns (observed runoff); $\left(S'_{\mathrm{dif}}\right)^2$ is a similar variance for the difference between the calculated and observed values of output variable after the substitution of the randomly mixed values of input factor; $(S_{\mathrm{fac}})^2$ is the contribution of natural variations of input factor to the variance of the model output variable (calculated discharge); $(S_{\mathrm{obs}})^2$ is the variance of the observed output variable.

Criterion FS is a close analogue of determination coefficient R^2 known in the variance analysis. In accordance with the variance sum law, value $\left(S'_{\mathrm{dif}}\right)^2$ is greater by two variances $(S_{\mathrm{fac}})^2$ than $(S_{\mathrm{obs}})^2$. The first $(S_{\mathrm{fac}})^2$ results from the contribution of real variations of the input factor to observed values of the output variable and the second $(S_{\mathrm{fac}})^2$ is due to the contribution of artificially created random variations of the input factor (by scrambling its observed values) into the calculated output variable. For adequate models, $(S_{\mathrm{dif}})^2$ is free of both variances $(S_{\mathrm{fac}})^2$ because of subtracting of the calculated contribution of the input factor from the observed one. At the same time, the variance derived from observational errors of input factor (or errors of spatial generalization of meteorological factors) will be present both in $\left(S'_{\mathrm{dif}}\right)^2$, and $(S_{\mathrm{dif}})^2$, and, therefore, its values in (9.3) will be subtracted from each other. Thus, the criterion FS evaluates the model sensitivity *directly to natural variations of input factor*, except for its observational errors. Based on the sensitivity FS to variations of individual environmental factors, one can assess their relative importance to the model. Since FS is expressed in proportion of $(S_{\mathrm{obs}})^2$ in (9.3), it can be also expressed in percent via multiplying by 100.

It should be pointed out that FS can be expressed as a function of RSR. Taking into account the equality $\mathrm{RSR} = A\sqrt{2}$, we have $\mathrm{FS} = \left[\left(\mathrm{RSR}'\right)^2 - (\mathrm{RSR})^2\right]/2$. Similar to A' in (9.3), RSR′ is equal to RSR obtained via using the randomly mixed values of input factor instead of the initially ordered.

9.4 Results

In the course of SAM, we created the river runoff model (Kirsta et al. 2012) that calculates water runoff for each of 13 landscapes, 34 basins, 60 years, and 4 hydrological seasons in accordance with the following balance equation:

$$
\begin{aligned}
Q^i = & \sum_k \{a_k S_k^i P_1 H(c_1, c_1, 1, 1, c_2, c_3, T_1) H\left(c_4, c_4, 1, 1, c_5, c_6, h_k^i\right)\} + \\
& + \sum_k \{b_k S_k^i P_2 H(c_7, c_7, 1, 1, c_8, c_9, T_2) H\left(c_4, c_4, 1, 1, c_5, c_6, h_k^i\right)\} + c_{10},
\end{aligned}
\tag{9.4a}
$$

where Q^i is the normalized seasonal average streamflow at the outlet of basin i, $i = 1$–34; the first and second summands in Eq. (9.4a) relate to contributions of the preceding and current seasons, respectively; but when calculating the streamflow for the first season (winter low water) these summands relate to the third and fourth seasons of the previous year since snow remains on the surface and does not affect the streamflow in winter; parameters a_k, b_k characterize the contribution of k-th landscape (geosystem group) in relevant season, $k = 1$–13; S_k^i represents relative area of k-th landscape in the basin i; h_k^i is mean altitude of the same group in the basin i, m a. s. l.; P_1, P_2 are spatially generalized normalized monthly precipitation averaged for a corresponding season; T_1, T_2 are deviations of spatially generalized normalized air temperature from 1 (1 is a long-term mean value of normalized characteristics) averaged for a corresponding season; H is the piecewise-linear function in Eq. 9.1; parameters c_1–c_9 reflect the influence of temperature T_1, T_2 and altitude h_k^i upon water runoff in the basin i; parameter c_{10} characterizes constant replenishment ($c_{10} > 0$) or depletion ($c_{10} < 0$) of both free groundwater and water of fractured rock zones.

Equation (9.4a) includes 13 parameters a_k, 13 parameters b_k, and 10 parameters c_i, i.e. 36 parameters for each hydrological season are used to calculate water runoff. All parameters were determined during SAM by the available 1300 values of river streamflow for the corresponding season and the solution of the inverse problem. In particular, Table 9.2 presents values b_k, which characterize the proportion of precipitation entering the river from landscape k in the current season. According to (4a), the seasonal water discharge Q_k^i from landscape k in basin i in a particular year is

$$Q_k^i = \left\{a_k S_k^i P_1 H(c_1, c_1, 1, 1, c_2, c_3, T_1) H\left(c_4, c_4, 1, 1, c_5, c_6, h_k^i\right)\right\} + \\ + \left\{b_k S_k^i P_2 H(c_7, c_7, 1, 1, c_8, c_9, T_2) H\left(c_4, c_4, 1, 1, c_5, c_6, h_k^i\right)\right\} + c_{10}. \quad (9.4b)$$

One can easily convert the normalized landscape and river discharges for each hydrological season into the real ones (m^3/s). To do this, the value of long-term mean river streamflow should be determined via comparison of calculated normalized streamflow and the real one for 1–2 years of observation. Subsequently, using the right side of Eqs. (9.4a)–(9.4b) one can find the desired values of discharge in m^3/s.

With the SAM, we also obtain the following universal balance equations for river hydrochemical discharge:
for the first hydrological season:

$$\text{Hydrochemical discharge} = \sum_k \{a_k Q_k^i H(c_1, c_1, 1, 1, c_2, c_3, P) \times \\ \times H\left(c_4, c_4, 1, 1, c_5, c_6, K^i\right)\} + bq^i + dS^i Q^i, \quad (9.5a)$$

for the second, third and fourth hydrological seasons:

$$\text{Hydrochemical discharge} = \sum_{k} \{a_k Q_k^i H(c_1, c_1, 1, 1, c_2, c_3, P)\times$$
$$\times H\left(c_4, c_4, 1, 1, c_5, c_6, K^i\right)\} + bq^i + d\sqrt{S^i}Q^i, \tag{9.5b}$$

where P is spatially generalized (throughout the Altai-Sayan mountain country) normalized precipitation (Kirsta 2011) for IX-XI months of the preceding year for the first season, or IV–VI, VII–VIII, IX–XI months of the current year for the second, third, fourth seasons, respectively; parameter a_k corresponds to an average long-term seasonal concentration of analyte in the calculated water discharge Q_k^i in Eq. (9.4b), the latter is formed by landscape k ($k = 1$–13) in the basin i ($i = 1$–34) due to precipitation P and other environmental factors; b is the parameter of an average long-term seasonal concentration of analyte in the estimated incoming (or outgoing) average long-term seasonal groundwater discharge q^i formed in the basin i by groundwater as well as by water from fractured rocks zones; q^i is equal to c_{10} in Eqs. (9.4a)–(9.4b); K^i is the average transverse slope of the basin i, which is calculated from map data as the tangent of slope inclination relative to the horizontal; H is the piecewise-linear function in Eq. 9.1; parameters c_1–c_6 reflect the effect of precipitation P and the slope K^i; parameter d characterizes the increase in analyte concentration in the calculated runoff Q^i by percentage of area S^i; S^i is the relative area of arable lands (in proportion/percentage of the basin i area).

In the right parts of Eqs. (9.5a) and (9.5b), the contribution of analyte from each landscape provided by surface, subsurface and groundwater runoff are summarized to form the seasonal hydrochemical discharge. Parameters a_1–a_{13} characterize the average long-term seasonal concentration of analyte in landscape water discharge. Function $H(c_1, c_1, 1, 1, c_2, c_3, P)$ makes it possible to describe the effect of precipitation P, and $H\left(c_4, c_4, 1, 1, c_5, c_6, K^i\right)$ – the effect of the slope K^i of river basins. The equations also reflect the contribution of arable lands. The summand bq^i takes into account the inflow (or outflow) of analyte with positive (or negative) q^i calculated by the water discharge model (9.4a)–(9.4b). The value q^i is practically equal to a certain percentage of water discharge Q^i, when measured in m^3/s. According to Eqs. (9.5a)–(9.5b), the seasonal and long-term dynamics of hydrochemical discharge is calculated for each landscape k in each basin i as follows
for the first hydrological season:

$$\text{Hydrochemical discharge} = \{a_k Q_k^i H(c_1, c_1, 1, 1, c_2, c_3, P)\times$$
$$\times H\left(c_4, c_4, 1, 1, c_5, c_6, K^i\right)\}, \tag{9.5c}$$

for the second, third and fourth hydrological seasons:

$$\text{Hydrochemical discharge} = \{a_k Q_k^i H(c_1, c_1, 1, 1, c_2, c_3, P)\times$$
$$\times H\left(c_4, c_4, 1, 1, c_5, c_6, K^i\right)\}. \tag{9.5d}$$

Along with the model of normalization of average monthly temperature and precipitation and their spatial generalization (Kirsta 2011), as well as the water discharge model (9.4a)–(9.4b), the Eqs. (9.5a)–(9.5c) represent a package of simulation models of climate and landscape water and hydrochemical discharges in mountain river basins. Parameters a_1–a_{13}, b, c_1–c_6, d in Eqs. (9.5a)–(9.5b) were determined in the course of SAM through the solution of the inverse problem for annual mean seasonal flows of each analyte, $Q^i C^i$, (~1200 experimental values of each analyte concentrations were used). The value Q^i characterizes the average seasonal runoff for the basin i outlet (i = 1–34) in the current year, which is calculated by the water discharge model and normalized by its average long-term value in this basin. C^i is the observed analyte concentration in river runoff for basin i, normalized by its average $\bar{C}$ for 34 basins for a given season. Average seasonal analyte concentration a_k in landscape water discharge Q_k^i can be converted from normalized values into mg/litre (Table 9.3) via multiplying by the known $\bar{C}$.

Thus, models (9.4a)–(9.4b) and (9.5a)–(9.5c) estimate the water and hydrochemical discharges from 13 dissimilar landscapes of the Altai-Sayan mountain country. For river streamflow and the flows of 7 chemical substances under study, we calculated the model performance criteria RSR (RMSE-standard deviation ratio (Moriasi et al. 2007; Koch and Cherie 2013)) and NSE (Nash-Sutcliffe model Efficiency coefficient, NSE = 1–RSR^2 (Koch and Cherie 2013)). The resulting values of RSR $\approx$ 0.4–0.5 and NSE $\approx$ 0.7–0.8 correspond to good and very good quality of all the models.

For practical use of the models, for example, in landscape planning of territories, it is important to know not only the values of landscape water and hydrochemical discharges, but the degree of environmental factors influence as well. The sensitivity of the models to environmental factors variations according to Eqs. (9.2)–(9.3) is given in Table 9.4.

The developed models can also be used to forecast water and hydrochemical discharges for the next hydrological season. This is done in Eqs. (9.4a)–(9.4b) and (9.5a)–(9.5c) by substituting the long-term mean values of meteorological factors of the forecast season instead of the actual ones. Thus, the discharge is predicted for 3–4 months in advance. Calculations have shown that the variance of discrepancy between the predicted and observed discharges is halved as compared to the discrepancy of trivial forecast using their average long-term characteristic.

It is obvious that the models can be applied for any mountain area after landscape typification of the latest and new estimation of parameters through the solution of the inverse mathematical problem in the framework of SAM. It should be noted that experimental studies of landscape hydrochemical discharge are extremely complicated because of time-consuming measurement of surface, subsurface and underground water flows with analyte concentrations in them all over the landscape area.

Table 9.3 Constant seasonal concentration of analytes (a_k, $k = 1–13$) in landscape discharge, mg/l

Landscapes	Analyte concentrations for hydrological seasons 1/2/3/4						
	Nitrite ions (NO_2^-)	Nitrate ions (NO_3^-)	Ammonium ions (NH_4^+)	Phosphate ions (PO_4^{3-})	Mineralization (Ions)	Total dissolved iron (Fe)	Suspended matter
1. Glacial-nival high mountains with permafrost	0/0/0.03/0	0/0/0.67/0	0.42/0.63/0/0	0.02/0/0.01/0	0/0/0/0	0/0/0/0	0.20/0/0/0
2. Goletz-alpine type high and middle mountains, pseudo-goletz low mountains with permafrost	0/0/0.01/0	0/0.31/0/0.09	0/0/0/0	0/0/0/0	0/0/0/0	0/0/0/0.04	0.78/0/0/0
3. Tundra-steppe and cryophyte-steppe high mountains with permafrost	0/0/0/0	0/0/0/0	0.35/1.70/0/0.50	0.02/0/0.05/0	0/494.4/0/0	0/0.20/0/0	20.0/23.2/0/0
4. Forest high middle and low mountains	0.01/0/0/0.01	0.91/0.48/0.056/0.18	0.13/0.45/0.11/0.24	0/0.01/0/0.01	124.2/73.8/43.6/36.4	0.11/0.89/0.10/0.13	4.02/74.2/10.7/7.17
5. Exposure-forest steppe and steppe high and middle mountains	0.06/0.02/0.03/0	0/0/0.79/0	0.55/0.68/0.87/0	0.02/0.01/0/0	0.52/141.0/0/0	0.01/1.08/0.01/0	29.0/230.5/18.1/0
6. Forest-steppe, steppe low mountains and foothills	0.01/0.03/0/0	1.59/1.48/0.43/0.39	0.35/0.64/0.24/0.26	0/0.14/0.01/0.01	133.7/239.8/277.5/178.9	0.33/0.34/0.14/0.13	4.16/204.5/21.8/10.2
7. Intermountain depressions with different steppes and forest-steppes	0.03/0/0/0	0.48/0/0/0	0.44/0.11/0.80/0.97	0/0.15/0.07/0.08	466.0/172.2/291.6/722.5	0/0/0/0	5.44/0/6.93/1.97

(continued)

Table 9.3 (continued)

Landscapes	Analyte concentrations for hydrological seasons 1/2/3/4						
	Nitrite ions (NO_2^-)	Nitrate ions (NO_3^-)	Ammonium ions (NH_4^+)	Phosphate ions (PO_4^{3-})	Mineralization (Ions)	Total dissolved iron (Fe)	Suspended matter
8. Steppe and forest-steppe piedmont	0.04/0.05/ 0.01/0.02	3.21/1.02/ 0.30/0.78	0.41/0.21/ 0.23/0.62	0.09/0.09/ 0.04/0.02	501.9/167.0/ 298.2/281.0	0.32/0/ 0.08/0.17	30.9/16.5/ 11.4/0
9. Nondrainable and intrazonal landscapes with partial permafrost	0/0/0.02/0.01	0/0/0.87/0.07	0/0/0/0	0/0/0/0	0/0/0/0	0/0/0/0	0/0/0/0
10. Mountain river valleys	0.05/0/0.01/0.02	2.97/0/0.18/0.78	0/0.91/0/0	0/0.07/0.07/0.02	735.3/408.0/ 1075/1206	0/0/0/0	16.1/0/34.3/20.0
11. Lowland river valleys	0.03/0.02/0.03/0	0.31/0/ 0.12/0.16	0.68/0.69/ 0.61/0	0/0.11/ 0.01/0.01	1351/188.2/ 1192/665.1	0.33/0.07/ 0.31/0.11	31.5/230.6/ 30.6/11.5
12. Forest high and piedmont plains	0.05/0/0/0	0/0/ 0.18/0.40	0.47/0/ 0.38/0	0/0/0/0	118.8/188.7/ 0/0	0/0/0/0.19	0.20/0/ 0.68/1.7
13. Aquatic landscapes	0/0/0/0.02	0/0/0/0	0.06/1.08/ 0/0	0.01/0/0/0	0/0/0/0	0/1.08/0/0	16.6/213.4/ 0/0

Table 9.4 Sensitivity of water and hydrochemical discharges of mountain rivers to environment variations

Environmental factor	Sensitivity of water and hydrochemical discharges for hydrological seasons 1/2/3/4, (%)[a]							
	Water runoff	NO_2^-	NO_3^-	NH_4^+	PO_4^{3-}	Ions	Fe	Suspended matter
Landscape structure of river basins	4 6 11 4	52 9 88 19	22 19 45 31	41 3 55 9	>100 52 55 48	8 4 7 6	12 33 64 69	98 13 21 12
Landscape altitude	0.3 0.2 ~0 0.6	–	–	–	–	–	–	–
Steepness of basin transverse slope	–	6 5 2 10	14 9 0.5 15	9 2 9 19	15 14 5 7	18 5 7 2	10 4 6 9	5 3 0.8 2
Precipitation	17 22 16 34	5 0.3 3 5	1 1 0.4 1	3 2 6 11	9 0 5 16	2 1 6 6	11 0 6 8	5 2 2 5
Air temperature	6 16 6 4	–	–	–	–	–	–	–
Area of arable land	–	4 5 0.3 3	2 0 0 0	0.1 5 0 0	4 16 16 16	12 33 7 9	0 21 0 0	1 20 9 4

[a]Expressed as a percentage of dispersion characterizing the observed values of water or hydrochemical runoff

9.5 Conclusions

1. The developed set of models allows the calculation, forecast and monitoring of seasonal and long-term dynamics of water and hydrochemical discharges for different landscapes of the Altai-Sayan mountain country.
2. To identify the models, only digital cartographic information on landscape structure of river basins, data on average monthly air temperature and monthly precipitation, as well as 1–2 years for observation of the discharge at the outlet of the river catchment are required.
3. The models can be applied to different mountain countries after anew identification of their parameters.

References

Beven K, Hall J (2013) Applied uncertainty analysis for flood risk management. Imperial College Press, London

Hauduc H, Neumann M, Muschalla D, Gamerith V, Gillot S, Vanrolleghem PA (2011) Towards quantitative quality criteria to evaluate simulation results in wastewater treatment — A critical review. In: Proceedings 8th international IWA symposium on systems analysis and integrated assessment in water management (WATERMATEX2011). San Sebastian, Spain, June 20–22, pp 36–46

Iooss B, Lemaitre P (2015) A review on global sensitivity analysis methods. In: Meloni C, Dellino G (eds) Uncertainty management in simulation-optimization of complex systems: algorithms and applications. Springer, US

Kirsta YB (1986) Modelling of desert ecosystems, Ashgabat, Ylym (in Russian)

Kirsta YB (2006) System-analytical modelling — Part I: General principles and theoretically best accuracies of ecological models. Soil-moisture exchange in agroecosystems. Ecol Model 191:315–330

Kirsta YB (2011) Spatial generalization of climatic characteristics in mountain areas. World Sci Cult Educ (Mir Nauki, Kul'tury, Obrazovaniya), 3(28):330–337 (in Russian)

Kirsta YB (2016) Modeling of hydrochemical composition of mountain river runoff: 2. Assessment of model performance for the runoff of mineral nitrogen forms. News Samara Sci Cent RAS 18(2):408–412 (in Russian)

Kirsta YB, Kirsta DY (2014) The information-physical principle of evolutionary systems formation, system-analytical modeling of ecosystems, 2nd edn, Barnaul, Altai State University Publishing House (in Russian)

Kirsta YB, Lubenets LF, Chernykh DV (2011) Typological classification of landscapes for river flow estimation in Altai-Sayan mountainous country. Sustain Dev Mnt Territ 2(8):51–56 (in Russian)

Kirsta YB, Puzanov AV, Lovtskaya OV, Lubenets LF, Kuznyak Ya E, Pakhotnova AY (2012) Simulation mathematical model of runoff for mid-size and small rivers in mountain territories. News Samara Sci Cent RAS 14(1):2334–2342. https://doi.org/10.24412/Fd1kFw4CDjU (in Russian)

Kirsta YB, Puzanov AV (2015) System-analytical modeling of mountain rivers runoff. In: Fundamental problems of water and water resources: Proceedings of IV Russian scientific conference. Moscow, Water Problems Institute RAS, pp 73–76 (in Russian)

Kirsta YB, Puzanov AV (2016) Modeling of hydrochemical composition of mountain river runoff: 1. Runoff of nitrogen mineral forms. News Samara Sci Cent RAS 18(2):96–100 (in Russian)

Koch·M, Cherie N (2013) SWAT-modeling of the impact of future climate change on the hydrology and the water resources in the upper blue Nile river basin, Ethiopia. In: Proceedings of the sixth international conference on water resources and environment research, ICWRER 2013. Germany, pp 428–523

Moriasi DN, Arnold JG, Van Liew MW, Bingner RL, Harmel RD, Veith TL (2007) Model evaluation guidelines for systematic quantification of accuracy in watershed simulation. Trans ASABE 50(3):885–900

Chapter 10
Agent-Based Modelling of a Simple Synthetic Rangeland Ecosystem

François Guerrin

Abstract The model described in this article aims at simulating free-grazing herbivores in a rangeland landscape. The first aim of the model is to find a balance between the herbivores and the vegetation dynamics guaranteeing sustainability: maintain a healthy animal population and a green landscape. Two opposite processes threaten this equilibrium: overgrazing, leading to desertification; undergrazing leading to bush invasion. Both processes may ultimately lead to the population extinction by starving and pasture invasion by bushy vegetation. The model has been implemented with the NetLogo simulation platform (Wilensky in NetLogo, http://ccl.northwestern.edu/netlogo. Northwestern University, Evanston, IL, 1999). It comprises two types of agents: cells in a spatial grid, standing for land plots, and mobile agents, standing for herbivores (here cattle). The land plots are characterized by their colour: shades of green represent grass as a function of its growth. The herbivores are characterized by attributes like their birth date, age, previous location, destination, pathway, travelled distance, ingested feed, body weight, calving dates. At each simulation time step, the herbivores iterate the following activities: find a destination, move, graze, gain and lose weight, age and, possibly, calve or die. Model simulations can be made to check variants of rangeland landscapes comprising thousands of hectares and herbivore heads. Simulation assessment criteria are the herbage biomass, herbivore population size, individual body weights, birth and mortality rates, land-use patterns and landscape fragmentation obtained after time periods of, possibly, tens of years. We give an example of simulations of a scenario giving rise to the emergence of a "grazing lawn", that is a highly productive rangeland with a low vegetation standing biomass capable of sustaining a high stocking rate of herbivores. As the model's ambition is not to mimic real specific rangelands but to offer a generic simple synthetic ecosystem to check ecological hypotheses or theories, we also show how to check the Ideal Free Distribution theory based on the simulated example.

Keywords Agent-based modelling · Landscape simulation · Grazing ecosystem · Grazing lawns · Ideal Free Distribution

---

F. Guerrin (✉)
SELMET Cirad, Inra, Montpellier SupAgro, University of Montpellier, 34398 Montpellier, France
e-mail: francois.guerrin@free.fr

W. Mirschel et al. (eds.), *Landscape Modelling and Decision Support*,
Innovations in Landscape Research,
https://doi.org/10.1007/978-3-030-37421-1_10

10.1 Introduction

This model allows the spatial dynamics stemming from the interactions between mobile agents (free-grazing herbivores) and their environment (rangeland) to be simulated. It features a herd grazing freely on a pasture. For the sake of sustainability a dynamical balance must be found between the grass intake by the animals and the herbage growth. However, two behaviours may threaten this equilibrium: overgrazing, possibly leading to desertification; undergrazing that may lead to excessive vegetation growth and landscape "closure" by a bushy cover (issue dealt with e.g. in Anselme et al. 2010; Vacquié et al. 2016). Both processes may lead the herd to extinction by starving.

This model's aim is not, actually, to mimic real specific rangelands but to offer a generic, simple, synthetic, virtual ecosystem to check variants of ecosystem's structure or animal behaviours, hypotheses and theories (e.g. Optimal Foraging, Marginal Value Theorem, Ideal Free Distribution). It has been implemented with the agent simulation platform NetLogo (Wilensky 1999) representing the landscape as a grid and herbivores as mobile agents. Although spatial grid cells are called "patches" in NetLogo we will use instead, in this article, the word "plots" to avoid confusion with the concept of "patch" in Landscape Ecology; similarly, we will not use the NetLogo term "turtle", indicating the mobile agents, but rather herbivore, animal, cattle… more relevant in the model described here. Although deliberately naive, this model has been parameterized with real rangeland and cattle values borrowed from various authors (Bayer and Waters-Bayer 1999; INRA 1988; Mayer et al. 2012; Moritz et al. 2015; Vayssières et al. 2009). If the rangeland system modelled here is not real, we believe, at least, that it be realistic enough to be fruitfully used as a synthetic rangeland ecosystem for simulation purposes.

We describe the model according to the "Overview, Design concepts, Details" (ODD) protocol (Grimm et al. 2010) which has become kind of a standard framework for agent-based model descriptions. In the overview part, we state the purpose of the model, which entities of a rangeland system are represented (herbivores, pasture) and which processes standing for the vegetation growth and the herbivores' life can be simulated with it. In the design concepts part, we give the underlying concepts of the model construction and its main functioning features. In the details part, all practical details about the model formalization (namely, all process submodels) are given. Then we demonstrate the model functioning and its relevance to simulate a rangeland system according both to its dynamical and spatial dimensions and how it can be used to check ecological theories like the Ideal Free Distribution theory taken as an example. Finally, conclusions are drawn and outlooks envisioned.

10.2 Model Overview

10.2.1 Purpose

This model aims at representing herbivores distributed over a rangeland landscape. The animals move for grazing the pasture. They gain weight while feeding but, concurrently, they lose weight to support their base metabolism and their walking and reproductive activities. The grass grazed by the herbivores grows continuously. Simulating the model, one may seek to find a good balance in the interactions between the herbivore population and the vegetation dynamics to ensure the sustainability of the rangeland system on the long run that are the maintenance of a numerous healthy population of herbivores and a green productive pasture to sustain it.

But two opposite processes threaten this balance:

- Overgrazing, due to the excessive exploitation of the pasture by the herbivores, possibly leading to desertification which is reversible but takes time to recover;
- Undergrazing, due to the under-exploitation of the pasture by the herbivores, leading to the landscape invasion by bushy vegetation hard to exploit.

These processes may lead to the extinction of the herbivore population by starving and the landscape closure by vegetation unsuitable to herbivore grazing.

This issue has been inspired by the works by Anselme et al. (2010) based on a participatory approach. However, in contrast with these authors, as well as Vacquié et al. (2016), whose goals were to represent given specific ecosystems, ours is to provide a generic testbed to check variants of the ecosystem's structure and animal behaviours as well as ecological hypotheses and theories.

10.2.2 Entities, State Variables and Scales

The model has been implemented with the agent-based simulation platform NetLogo (Wilensky 1999). It comprises thus the two types of agents[1] implemented in NetLogo:

- "Plots", partitioning the physical space in a regular grid of square cells accounting for the rangeland (called "patches" in NetLogo);
- "Herbivores", that are mobile agents (called "turtles" in NetLogo).

Plots are characterized, in addition to the spatial coordinates of their centre, by the numerical value (dimensionless) coding for their colour. We use shades of green (whose numerical value c is in the range [50, 60)) for representing the vegetation state and biomass hold by the plots. Green becoming darker and darker (c tending to 50) denotes increasing biomass and, conversely, green becoming lighter and lighter (c tending to 60) denotes a decreasing biomass (see Fig. 10.1). Whereas very dark

[1]According to the Russel and Norvig's (2010) definition: "An agent is anything that can be viewed as perceiving its environment through sensors and acting upon that environment through effectors".

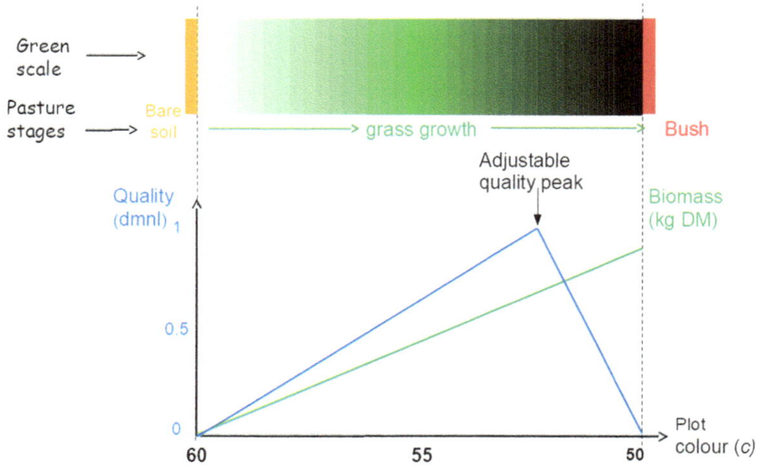

Fig. 10.1 Representation of plot vegetation as shades of green

green ($c = 50$) denotes a bushy state of vegetation (very high biomass), very light green denotes almost bare soil. The quality of the vegetation for the herbivores is characterized with respect to a colour peak value taken within the range [50, 60). The quality decreases on both sides as long as one goes away from the peak, which denotes the best quality. Using shades of green to represent the plot vegetation is illustrated in Fig. 10.1.

In addition to green plots, one can specify special plots of other colours (e.g. red in Fig. 10.2) to denote non-herbage zones endowed with special properties (e.g. places with non-edible resources by herbivores, impassable obstacles or whatever).

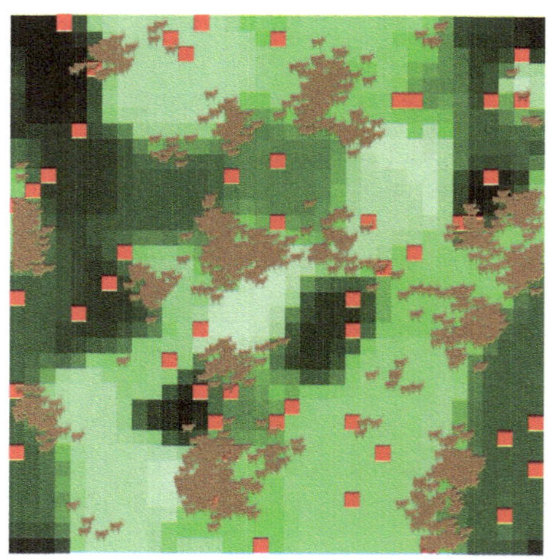

Fig. 10.2 A rangeland landscape comprising 1,225 1 ha-plots at various states of vegetation represented by different shades of green and 1,225 cattle heads forming small herds scattered all over (in brown). Red squares denote non-herbage areas

Herbivores are characterized, in addition to their id number, colour and location coordinates, by various herbivore attributes linked to their:

- Lifecycle: sex (0: female, 1: male), birth date, age, lifespan, age of sexual maturity and calving date (as days);
- Nutritional state: live weight (kg), daily live weight gain and losses (kg/day), feed intake (kg dry matter/day), body condition ratio[2] (dmnl);
- Movement in space: previous and destination plots coordinates, pathway (list of visited plots), travelled distance (either as the number of successively visited plots or as the multiple of a plot side length; 1 plot side = 100 m).

At each simulation time step each herbivore iterates the following actions: stay on the current plot or move to a new one, graze, gain and lose weight, age and, possibly, calve or die.

Simulations demonstrated here are based on a reference landscape comprising 1,225 plots (35 × 35 square grid) of 1 ha each (i.e. a plot is a 100 m × 100 m square) and 1,225 cattle heads of various ages and sizes (1 head/ha). Figure 10.2 gives an example of such a rangeland landscape. Time is discrete with constant time steps: 1 tick stands for 1 day and a year comprises 365 ticks. The simulation horizon may encompass possibly tens of years.

The main criteria of simulation assessment are the production of animal and vegetal biomasses, herbivore demography (population size, birth and mortality rates) and live weights, land-use patterns and landscape fragmentation over time.

Global variables allow one to specify agent sets endowed with specific properties or behaviours (see Sect. 10.3.9). These global variables along with other aggregating individual attributes are also used for observation purposes of the whole simulated system and to calculate assessment indicators of its functioning (e.g. herd size, birth or mortality rates, global biomasses, etc.). They are thus mentioned in Sect. 10.3.10.

10.2.3 Process Overview and Scheduling

All simulations start from time $t = 0$ and stop when the simulation horizon has been reached (possibly many years) or when no more plots with grass or no more herbivores remain. Apart from these cases and the simulation setup procedure described in Sect. 10.4.1, at each time step $t > 0$, the model executes the following functions in the order given hereafter (see details in Sect. 10.4.3):

- Reset-globals: reset some global variables to zero (e.g. the counters of deaths and births of the day), update the counters of herbivore categories (e.g. with respect to their body conditions) or plot clusters (e.g. those holding exploitable grass, bare soil, bushes, herbivores…).

[2] BCR, see Mayer et al. (2012); also called "Relative condition" in the Grazplan model (Freer et al. 1991).

- Die: eliminate herbivores too weak (i.e. with weight less than a certain percentage of the reference weight) or having reached the species longevity.
- Grow grass: make the grass grow on each plot, that is, diminishing the value of its colour attribute and concurrently updating the attribute denoting its biomass.
- Observe plots: each herbivore observes its environment within a certain scope and chooses a destination plot according to constraints defining its behaviour, e.g. its walk type (random or oriented) or its tropism towards plots matching some criteria (like maximizing the product grass biomass × quality per head of cattle); this assumes a certain knowledge of their environment by the individuals within an adjustable radius of perception.
- Stay or move: each herbivore is required to move to its destination plot if it is different from the current one, using a given walk type.
- Trample: the herbivores move on their own plot by randomly varying their location coordinates; this movement is taken into account in their total movement, for instance, in the weight loss undergone when moving.
- Graze: calculates the total grass ingested by each herbivore grazing according to its reference forage ration (kg dry matter/day), the herbage fill value (fill units/kg dry matter) depending on the forage quality, and the amount of forage offered on the plot; in addition, updates the colour of the plot.
- Gain body weight: converts the ingested grass in live weight gain using a feed conversion index (kg dry matter ingested/kg live weight gain) function of the animal live weight.
- Lose body weight: makes herbivores lose weight according to the gap between their reference forage ration and their ingested grass increased by the contribution of their daily walk.
- Adjust live weight: updates the live weight of each herbivore resulting from the daily balance between weight gain and losses.
- Age: increases each herbivore age by 1 day at each time step.
- Calve: production of a calf (possibly several) by the female herbivores above the age of sexual maturity possibly subject to a certain live weight threshold (this option allows one to make reproduce only animals in good shape).
- Tick: adds 1 day to the time clock.

10.3 Model Design Concepts

10.3.1 Basic Principles

As usual in agent-based simulation, the model includes the knowledge on the individuals not on their whole (here the herbivore population or the landscape). It is thus not based on a priori theories defined at the "macro" level (for instance, the population concept in classical population dynamics) but on the specification at the "micro" level of the biological characteristics and behaviours of the individuals themselves (here the herbivores and land plots). This approach matches the wishes expressed by Nisbet et al. (2016) and Martin et al. (2012) to base more ecological models on the physiological properties of individuals distributed over space. In contrast with classical population dynamics models, in which the demographic parameters are explicitly encoded, in agent-based models the demography (the global evolution of the whole population) "emerges" a posteriori as the result of the interplay of the local interactions of agents in a way (relatively) similar to Nature (Grimm et al. 2005).

In this perspective and to ensure the model inner coherence, the "physiological" parameter values of the herbivore and the plot agents have been chosen among the data available on cattle in rangeland systems in main France (INRA 1988), Reunion Island (Vayssières et al. 2009), Australia (Mayer et al. 2012) and Africa (Bayer and Waters-Bayer 1999; Moritz et al. 2015). This refers to the following parameters:

- For the herbivores: longevity, forage rations and intakes, reference live weights, feed conversion index, weight gain and losses related to feed intake and movement, age at calving, calving interval, birth weight;
- For the plots: growth rate and feeding quality and attractiveness of grass.

However, the model allows one to simulate scenarios also based on general theories of animal behaviour, such as Optimal Foraging, Ideal Free Distribution (see Sect. 10.5.3) or empirical models, such as various types of random walks tested to represent the movement of herbivores (see Sect. 10.4.3.3.1).

10.3.2 Emergence

The behaviours of herbivores grazing the grass produced by the land plots and moving over the space allow one to observe the emergence of collective dynamics and patterns at two levels:

- Herbivore population: demographic parameters (e.g. population size, mortality and birth rates, age structure), animal biomass, live weight structure, distance travelled, herd aggregation and distribution of animals over the landscape.
- Landscape: land cover (grass, bushes, bare soil), grass biomass distribution, landscape heterogeneity and fragmentation, pathways, spatial distribution and density of animals.

Emergence may give rise in the simulation to patterns observed in real rangeland ecosystems (e.g. "grazing lawns" as in Sect. 10.5.2).

10.3.3 Adaptation

Individual herbivores can adapt to their environment using capacities of perception and movement according to various rules or behaviours:

- Stay duration of an individual on its plot: several strategies may be implemented, like moving to the plot perceived with the best grass biomass × quality combination, or staying on the same plot until there is not enough grass to graze.
- Variable perception radius: expressed as the number of plots that can be sensed from its current location by a herbivore, the perception radius allows the individual to perceive its environment (e.g. grass available or presence of other herbivores on other plots). This perception radius may be fixed (as an input parameter) or variable (e.g. increased for thin animals to improve their chances to find a suitable amount of food to recover their nutritional gap).
- Walk types: oriented walks consisting in moving to the plots satisfying some criteria, including following a gradient, or random walks. Each type has advantages and disadvantages (e.g. some oriented walks may lead to excessive density of individuals inducing cascading effects: over grazing, mortality, bush invasion…).
- Movement steps: short-step walks (e.g. 1 plot, i.e. 100 m per day) inducing local foraging are more likely to ensure animals in good shape and to preserve herbage than long-step walks generating, in general, excess of weight losses. However, long-step walks may help in some circumstances to escape too crowded areas and better distribute herbivores over space.
- Herding behaviour: in the so-called "leader-follower" strategy (Bayer and Waters-Bayer 1999) every herbivore perceiving a leader (e.g. an individual older than oneself) in its perception radius follows it that is takes as destination plot the same as that of the leader (cf. also Dumont and Boissy 1999).
- Dying: whenever the body condition of an individual is too bad (e.g its live weight falls below a certain percentage of the reference weight) and its age too high it becomes more and more likely to die.

10.3.4 Objectives

The main criteria denoting the success of individual agents are as follows:

- For herbivores: long lifespan, high live weight, moderate distance travelled;
- For plots: good grass cover with relatively high biomass and quality to support significant herbivore stocks.
- The main criteria denoting the success of the whole landscape ecosystem are to obtain:
- For the herbivore population: high population size, birth rate, biomass;
- For the landscape: maximum cover by grass, high homogeneity and little fragmentation, high grass biomass × quality product and, at the opposite, the number of plots with bushes or bare soil as small as possible.

10.3.5 Learning

Each individual herbivore is endowed with the memory of its followed pathway (as the sequence of all visited plots) and the distance travelled during its life. Although not implemented yet in the current version of the model, it is envisaged to develop a learning function of the sites to visit or to avoid according to the past experience of the individuals. This could make the model more realistic in the light of the findings on herbivore behavioural cognition (Brooks and Harris 2008; Dumont and Hill 2004).

10.3.6 Sensing

The information perceived by the herbivores within their perception radius helps them to decide to move to another plot or not. It is relative to plots located within their perception radius. It may comprise, for example, the number of other herbivores, the nature and quality of the vegetation or any combined criteria as the value of the product grass biomass × quality.

10.3.7 Interaction

The herbivores interact with the plots located within their perception radius through the decision criterion used to choose their destination plot. By grazing the plot vegetation, the herbivores decrease its biomass which modifies its quality accordingly which, in turn, modifies its attractiveness to other herbivores and its nutritional value (see the relationship between the vegetation biomass and quality on Fig. 10.1) and, consequently the growth of the herbivores. While being consumed, the grass still regrows continuously, otherwise, it gets to its maximum level and remains constant after.

The herbivores also interact among themselves by adopting a gregarious behaviour: whenever an animal senses a "leader" within its perception range (e.g. an older individual), it can decide to follow it (i.e. to adopt the same destination plot). The result is the formation of herds as shown on the view in Fig. 10.3. An option in the model allows also each plot to interact with its neighbours by diffusing a certain proportion of an attribute value of its own. Diffusing the plot colour, leads to create smoother landscapes (see Fig. 10.4B).

Fig. 10.3 Model simulation interface

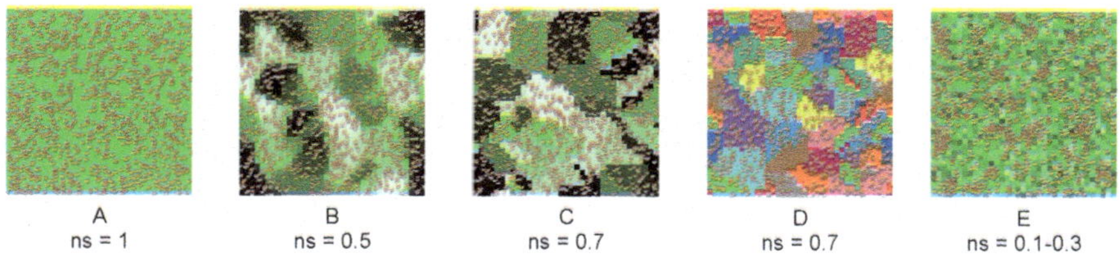

Fig. 10.4 Examples of initialization of rangeland landscapes comprising 1,225 plots of 1 ha each with 1.12 heads per ha of free-grazing cattle here randomly distributed over space (ns: landscape neighbourhood score)

10.3.8 Stochasticity

According to the built-in schedule in NetLogo, each of the processes enumerated in Sect. 10.2.3 (but the time advance) is activated for each agent (herbivore or plot) successively, the execution order being determined by a random drawing at each time step of the concerned agents.

Other random functions are used to set

- At initialization ($t = 0$): the plot colours and the position, sex, heading, age, calving date of the herbivores;
- During the simulation: the probability of dying according to the Body Condition Ratio of the animal, its age, and its daily weight change (from Mayer et al. 2012); the gestation success rate, the headings and sex of new-born calves; the location of herbivores around the centre of their destination plots.

Two-dimensional random walks (RW) implemented in the model, as their name suggests, also use randomness where the "walkers" are the herbivores. These are described in Sect. 10.5.3 and Fig. 10.7.

10.3.9 Collectives

Collectives, that are agent sets of herbivores or plots, are represented in the model in two ways:

- As an emerging property of individual behaviours: the formation of herds according to the gregarious behaviour of their members (see Sect. 10.3.3);
- As user-defined agent sets with respect to some shared common property:
  - Plot sets: encompassing plots holding a specific quality of grass, hosting or not herbivores;
  - Herbivore sets: encompassing herbivores with live weight above or below a reference weight according to their age, or with different behaviours like the way they move or feed.

10.3.10 Observation

Figure 10.3 displays the model simulation interface. In the middle lies the rangeland landscape view featuring the cattle in brown, scattered all over and grouped in little herds, and the pasture, made of square green plots of different shades (darker the green, higher the grass biomass and the other way round). Left to the landscape view are the buttons, sliders, input windows to initialize (see Sect. 10.4.1) and launch simulations. To the right of the view, are the monitors and graphs to observe the time evolution of relevant model variables set by the user (every time step are saved).

One can distinguish from Fig. 10.3 monitors and graphs related to

- Herbivores (from top to bottom): population size, % of herbivore categories (+: over-weighted, 0: standard-weighted; –: underweighted), total biomass, density (global: on all plots; local: on occupied plots only; maximal: on the highest populated plot), mean live weight, mean distance walked daily, live weights and age distributions, cumulative births and deaths.
- Plots (from top to bottom): % of plot categories and respective vegetation biomasses (G exp: total grass, G med: grass excluding bushy vegetation and bare soil; BS: bare soil; B: bushy vegetation), mean grass quality coefficient (in the range [0, 1] where 0 is very low and 1 is very high).
- Landscape indicators (top right corner): the plot colour coefficient of variation and the neighbourhood score. This score is the average proportion of neighbours similar to each individual plot with respect to its colour (we chose a Moore neighbourhood: each plot has 8 neighbours). It varies in the range [0, 1]: the closer to 1, the more homogenous the landscape and conversely (see Schwarzmueller et al. 2017).
- Feeding resource for cattle (below the landscape view): distribution of resource amount per cattle head; variance and mean of grass available among occupied and unoccupied plots; coefficient of variation of the resource available per head of cattle on occupied plots. By feeding resource, we mean in the simulation example given in Sect. 10.5, the grass biomass × quality product, used there as the main criterion to decide herbivores to move.
- Simulation report (bottom right corner): displays the final values of a set of variables chosen by the user.

10.4 Model Details

10.4.1 Initialization

Model initialization means to set up the simulation parameter values at time $t = 0$, on the one hand for the plots to create a landscape, on the other hand for the herbivores to create a population.

10.4.1.1 Plot Initialization

Plot initialization consists in assigning each plot a colour standing for the vegetation type or stage it supports. Various landscape types may be set according to the desired heterogeneity characterized by the landscape mean neighbourhood score (*ns* in Fig. 10.4):

- Totally homogenous landscape in Fig. 10.4A made of only one shade of green denoting the same type of vegetation at the same stage of growth all over.
- Semi-homogenous landscape in Fig. 10.4B made of different shades of green clustered in large patches randomly distributed with smooth boundaries due to the diffusion of colour between neighbouring plots corresponding to one type of vegetation at different stages.
- Semi-heterogeneous landscape in Fig. 10.4C made of different shades of green clustered in large homogenous patches with sharp boundaries (like 10.4B but without diffusion) corresponding also to one type of vegetation at different stages.
- Semi-heterogeneous landscape in Fig. 10.4D made of different colours clustered in large homogenous patches with sharp boundaries denoting different vegetation types.
- Heterogeneous landscape in Fig. 10.4E made of a random distribution of shades of green over space, corresponding also to one type of vegetation at different stages from one plot to the other (ns varies with the random function applied).
- Some plots may be assigned a special colour to give them specific properties (e.g. red plots in Fig. 10.2 denoting non-herbage).
- Creating complex patchy landscapes like in Figs. 10.4B, C, D is made by "sowing" a certain number of "colour seeds" over the land plots and propagate their colours to their neighbours. This implies executing three successive procedures:
- List-seeds: create the list of a finite set of colours based on the base colour range partition. E.g. in Fig. 10.4C where green has been chosen as the base colour (colour $c = 55$), 9 possible shades of green [51,..., 59] are set from the partition of the whole green value range [50, 60); from this list, a number of distinct values equal to the number of seeds set by the user is randomly drawn with equal probabilities to constitute the colours to be distributed all over the landscape with the "Sow seeds" procedure (see next item). In the case of Fig. 10.4E made of different colours, the same method applies, using instead the whole colour spectrum [0, 140) (from black to pink) and drawing from its partition a certain number of colour seeds.
- Sow seeds: each colour seed is assigned to one or several randomly drawn plots.
- Propagate: each coloured plot propagates its colour to its neighbours resulting iteratively in the whole tessellation of space giving landscapes such as those in Figs. 10.4B, C, E.

10.4.1.2 Herbivore Initialization

Herbivore initialization consists in creating a certain number of individuals (with the slider Init-Cattle-Heads in Fig. 10.3) and to parameterize their main attributes:

- Space location coordinates: whereas in Fig. 10.4 herbivores are randomly distributed they may be assigned to specific places (for example to get a totally uniform distribution over space).
- Colour (dimensionless): whereas in this article herbivores are shown all brown, several colours may be used, for example, to differentiate sub-populations endowed with specific behaviours.
- Sex (0 or 1): assigned by random drawing with equal probabilities to get a 1:1 sex-ratio; however, other settings may be relevant for some herbivore populations, e.g. in a cattle husbandry context Fust and Schlecht (2018) give a 3 males: 4 females sex-ratio and Lesel (1969) in semi-feral cattle a sex-ratio of 2 males:1 female, the females being subject to higher mortality due to calving with bad feeding conditions.
- Heading (degree): assigned by random drawing of an angle in the range [0, 360] degrees.
- Age (days): assigned by random drawing of a value in the range [0, Longevity].
- Birthdate (days): equal to (–Age) locating the birthdate in the past ($t < 0$).
- Body condition ratio (dimensionless): BCR = Live-weight/Reference weight; BCR is initialized at 1 (corresponding to the standard) or can be randomly drawn from a range [1 – BCR-min, 1 + BCR-min] where BCR-min is the minimum ratio below which the animal dies (see Sect. 10.4.3.1).
- Live weight (kg): product of the reference weight by the BCR.
- Calving date (days): similarly to the birthdate, located in the past, by random drawing in the range [−Calving-interval, 0].
- The attributes corresponding to herbivore movement (destination, pathway, origin) and nutritional stage (ingested forage, offered amount of forage) are set to 0.

10.4.2 Input Data

The only data imported from an external text file are used during the model simulation as a forcing function to determine the reference live weight of the cattle. It is generated by a model based on the logistic growth equation (taking the model used by Fellah Trade 2018, although other authors use a Gompertz-like curve equation; Ng 2001):

$$\frac{dW_{\text{ref}}}{dt} = \rho W_{\text{ref}}\left(\frac{1 - W_{\text{ref}}}{K}\right) \tag{10.1}$$

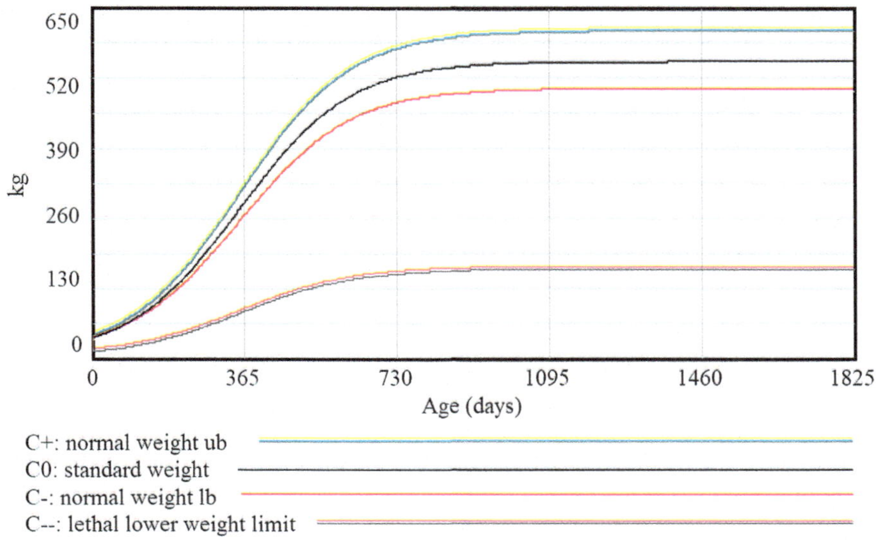

Fig. 10.5 Herbivore reference growth curves in kg (lb: lower bound; ub: upper bound)

where:

- W_{ref} is the individual live weight of reference function of animal's age; here the birthweight is $W_{birth} = 37$ kg
- K is the maximum live weight; here $K = 550$ kg
- ρ is the intrinsic growth rate; here $\rho = 0.075$.

These parameters are coherent with beef cattle like the Limousin breed (see INRA (1988)).

The resulting standard growth curve stemming from Eq. 10.1 is displayed in Fig. 10.5 (C0) along with a user-defined "normal weight" range lying within upper and lower bounds (C+ and C- resp.) arbitrarily chosen here as ± 10% of the standard weight (C0). Underweighted animals fit between the C- and C-- curves. A weight crossing below the C-- bound leads the animal to death (C-- corresponds here to-30% of the standard weight that is a body condition ratio BCR = 0.3; see Sect. 10.4.3.1).

10.4.3 Submodels

In this section are described with more details the models of all the processes listed in Sect. 10.2.3 in the sequence where they are activated at each time step during a simulation.

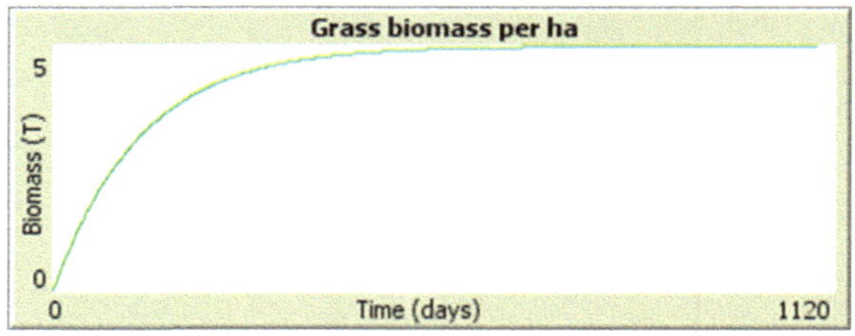

Fig. 10.6 Grass growth from bare soil to bushy cover. If the maximum biomass (here 5 t/ha) is reached after 3 years, 95% of growth is obtained in 1 year

10.4.3.1 Die

Every herbivore agent is asked to "die" (i.e. to be erased from the simulation) whenever one among the following conditions are met. First condition is when the current age of an animal exceeds its species longevity. Here, cattle longevity is fixed at 25 years since we consider free-ranging bovines which may live much beyond 20 years without human breeding intervention (Ng 2001). Second condition is based on the empirical statistical equation obtained by Mayer et al. (2012) from their data in tropical Northern Australia. According to this equation, the average probability of an animal death on an annual basis is a function of three factors: its body condition ratio (BCR), age and weight annual change (this latter factor to take account of the detrimental effect of the dry season in tropical rangelands). The probability becomes very high for an old underweighted animal and the other way round. Here is the equation:

$$P_m = \frac{1}{1 + e^{-\text{logit}}} \tag{10.2}$$

where

$$\begin{aligned} \text{logit} = &- 21.3 + 40.7 \times BCR - 24.2 \times BCR^2 + 1.05 \times Age \\ &- 0.0255 \times \Delta W - 0.893 \times Age \times BCR \end{aligned} \tag{10.3}$$

with P_m the mortality probability and ΔW is the weight balance over the year. Since this probability is set on an annual basis, we use in the model $P_m/365$ as the daily probability of death for the individuals.

In addition, as suggested by Mayer et al (2012), a third mortality condition is when the BCR drops down to a floor limit fixed, here, at 0.3 (see Sect. 10.4.2 and Fig. 10.6). Whiting et al. (2012) state for all mammals a general limit of 40% of their normal body weight likely to lead the individuals to death.

10.4.3.2 Grow Grass

Varying each land plot colour is used to represent in the model the biomass variation of the grass it holds. As grass growth increases in the opposite direction of the plot colour value (i.e. from $c = 60 \rightarrow c = 50$), making the grass grow is performed by decreasing first the plot colour value c, then by updating correspondingly its biomass attribute. At each simulation time step, everything remaining equal, the green plots become darker and darker as the grass grows. The rate of plot colour decrease, used as a proxy for the grass growth rate, is given by

$$GGR = GGR_{\max} \times \left(\frac{1}{\gamma} \times c - 5\right) \tag{10.4}$$

where $GGR_{\max}$ is the maximum grass growth rate (taken here equal to 40 kg dry matter/ha/day), γ is the extent of the colour value range and c is the colour value of the current plot (the expression in brackets takes on [0, 1)). Here $\gamma = 60–50 = 10$, where 50 denotes the darkest shade of green (denoting bushy vegetation) and ≈ 60 the lightest one (denoting very low-biomass, high-quality grass). The GGR thus is 0 for $c = 50$ and tends to $GGR_{\max}$ for c tending towards 60 (in fact since 60 is formally out of the shades of green, we use the approximate value $60 - 10^{-14}$ which is still in that range). It means that no more growth occurs when the vegetation has reached its maximal bushy state whereas the growth is maximum when the grass is regrowing from bare soil (young grass growth is much faster than old ligneous vegetation; see Bonnet (2008)). The growth curve of the grass biomass is shown in Fig. 10.6.

Biomass per colour unit (BCU), computes the conversion coefficient from the plot colour value to the corresponding grass biomass it holds. Assuming a linear distribution, it is given by the ratio $GB_{\max}/\gamma$ where $GGR_{\max}$ is the maximum grass biomass per plot (an input parameter) and γ the colour value range. Here, with $\gamma = 10$ and $GGR_{\max} = 5,000$kg $BCU = 500$ kg dry matter/unit of colour value. Therefore as the biomass increases from $c \approx 60$ (quasi-bare soil) to $c = 50$ (bushes), when the plot colour $c = 50$, its grass biomass $GB = 5{,}000$ kg dry matter, … when $c = 55$, $GB = 2{,}500$ kg,… and when $c \rightarrow 60$, $GB \rightarrow 0$.

Eventually, making the grass grow at the next time step $t + 1$ is twofold:

- Calculate plot colour (dimensionless):

$$c_{t+1} = c_t - \frac{GGR}{BCU} \tag{10.5}$$

- Calculate the corresponding biomass (kg dry matter):

$$GB_{t+1} = (60 - c_{t+1}) \times BCU \tag{10.6}$$

At $t = 0$, as the initial plot colours are directly determined from the initialization procedure, only the second step is performed.

10.4.3.3 Observe, Decide and Move

Each herbivore observes its environment within a certain scope set by its perception radius (e.g. a radius = 3 encompasses a set of 29 plots including the current one), chooses a destination plot and move to it according to specific constraints defining its behaviour (walk type). Among these constraints is the fact to be submitted (or not) to a dominant animal. This leader is often, in many herbivores, the oldest (Dumont and Boissy 1999). The rule used here is thus as follows: if in the perception radius of each animal there is an elder, then the animal destination plot will be that of the elder. Otherwise, the individual chooses its own destination by itself (as the leaders do). The existence of leaders fosters continuously the formation of herds (see Figs. 10.2 and 10.3).

Many different strategies with a quasi-infinite number of variants my be implemented, for example:

- If the current plot still holds exploitable grass, the animals stay (the destination plot is thus the current one) otherwise move.
- If they move, this can be done using a random walk or a walk oriented by some criterion (for example, like in the example given in Sect. 10.5 the aim is the plot in the perception scope maximizing the feeding resource); note that in some species like zebras, both types of walks may be observed, depending on the observation scale (Brooks and Harris 2008).

Random Walks

Although possibly looking somewhat arbitrary and loosely biologically based, the use of random walks to represent animal movement is widespread as it has been observed that many foraging animals, including herbivores (Brooks and Harris 2008), behave as if they were following these theoretical models. Random walks have given rise to many research works in mathematical biology and ecology. It is out of the scope of this article to analyse the large and interesting literature on this topic; one may report to Codling et al. (2008) and Ibe (2013).

Different types of random walks (RW) can be implemented. Figure 10.7 provides for example the traces of five of them:

- Local RW (brown trace): one of the 8 neighbouring plots of the current plot is randomly chosen with equal probabilities and a fixed step length (i.e. the distance to the target plot is, here, 1 plot per move).
- Symmetric RW (orange trace): the walker's direction is chosen between North, South, West and East with equal probabilities (here fixed step size of 1).
- Alternating RW (blue trace): the walker's direction is alternately chosen between West or East followed by North or South (here with a fixed step size of 1).
- Pearson RW (red trace): the walker's direction is chosen in a 180° range and a variable step length (here randomly chosen between 1 and 10 steps per move with equal probabilities).

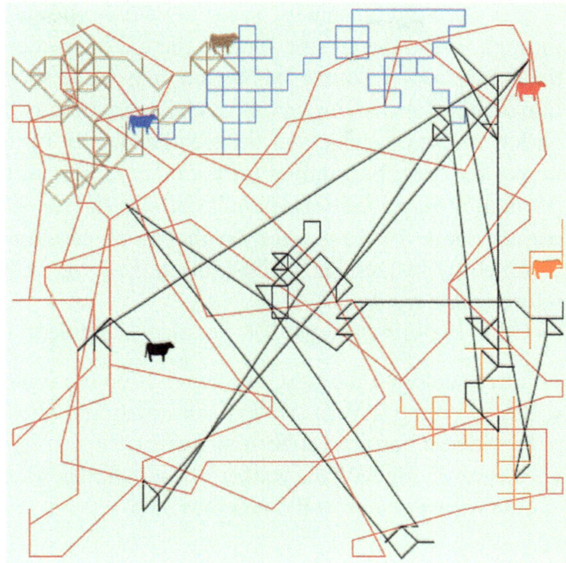

Fig. 10.7 Traces of 5 types of random walks (RW) simulated for 133 steps of various lengths (explanations in the text). Brown: local; Orange: symmetric; Blue: alternating; Red: Pearson; Black: Levy

Table 10.1 Comparison between 5 different random walks in terms of distance walked and area visited

Random walk	Distance walked	Total number of visited plots	Number of distinct visited plots
Local	132	133	67
Symmetric	132	133	65
Alternating	132	133	85
Pearson	504.97	123	113
Levy	383.32	133	9

- Levy RW (black trace): the walker's destination is chosen in its vicinity with a high probability (here 0.9) or at a remote distance with a lower probability (here 0.1).

Of course, the travelled distances and the number of visited plots are quite different from one RW to the other. Table 10.1 highlights these differences which result in different performances in terms of space exploration and, for grazing herbivores, in terms of foraging strategies: going far allows one to visit a larger extent of territory, which may be efficient to find new resources, but has the inconvenient to waste more energy than locally focused searches.

Strategies like Levy walk, alternating frequently short-step moves (local foraging) and more rarely long-step moves (remote displacement) give good results in heterogeneous patchy landscapes. But if this walk type is rather adapted when preys are sparse and searching is required (Humphries and Sims 2014), this is not always the case in rangelands as in the example given in Sect. 10.5.

If in some circumstances, long moves may allow a quick distribution of individuals through space, decreasing rapidly the excessive density of overcrowded areas and their correlative resource exhaustion, in general, short-move walks outperform long-move ones. Introducing in our model simulation two sub-populations of herbivores within the same rangeland differentiated only by their walking behaviours (long moves for one, short moves for the other) confirms this conclusion: the short-move population outperforms (in population size and survival) the other which may become extinct due to the important weight losses stemming from the excess of distances travelled by its members. Substituting the long-move walk with a Levy walk gives intermediate results.

Other two-dimensional RW are also available to model simulation:

- Correlated RW: the next step depends on the direction of the previous one.
- Self-avoiding RW: the walker's destination is chosen among all possible locations provided it has not yet been visited.
- Non-reversing RW: the walker's destination is chosen among all possible locations but the one visited at the previous step.

Oriented Walks

In contrast with random walks, oriented walks consist in responding to tropisms towards plots complying with some criterion. This implies the capacity of observation by the individuals according to their perception radius set by the user fixed or variable to give the animal some means of adaptation to its environment (cf. Sect. 10.3.3). The attraction criterion of an individual towards a plot located in its perception scope (including the current one) can be made of many factors used alone or in conjunction: maximize the grass biomass or its attractiveness, minimize the fill value of the herbage (another quality indicator, see Sect. 10.4.3.4), minimize the distance to walk, maximize the number of neighbours of the aimed plot capable of providing also adequate resources, etc. The example simulation given in Sect. 10.5 is based on such an oriented walk using as a criterion the product grass biomass × quality.

Once chosen, the herbivore moves to the destination plot while updating its attributes related to its movement:

- Path: adds the current destination plot to the list of previously visited plots.
- Origin: the current plot.
- Distances: daily distance walked and cumulative distance since the simulation start.

Trampling

In addition to the movement of herbivores from plots to plots, a model option allows one to make move the animals within the plots where they are located by randomly varying their location coordinates within the plot perimeter. This movement comes in addition to the total animal movement and is taken into account in the calculation of the weight losses due to the excess of energy wasted by the individuals.

10.4.3.4 Graze

The question is here to calculate the daily grass intake by the animals and to update concurrently the colour and the residual biomass of the plots where they are located.

The grass intake by the animals is calculated as a function of the herbage fill value (fill units/kg forage dry matter) which depends on its quality, the recommended feed ration (kg dry matter/day) and the offered amount of biomass on each considered plot (kg dry matter).

The herbage fill value (*HFV*) measures the ingestion capacity of an herbivore: 1 fill unit corresponds to 1 kg dry matter of young, good quality grass taken as the reference (Meyer 2018). In the model it is computed as a decreasing linear function of the plot colour c with two specific points: the colour value of the quality peak c_p (cf. Figure 10.1), where $HFV = 1$ and the colour corresponding to the highest biomass, that is $c = 50$ for the shades of green, for which it is maximal $HFV = HFV_{\max}$ (we took $HFV_{\max} = 1.25$ as it is a classical value for cows; cf. INRA 1988). The equation is thus:

$$HFV = a \cdot c + b \,\text{with}\, a = \frac{1 - HFV_{\max}}{c_{\text{peak}} - 50} \,\text{and}\, b = HFV_{\max} - a \times 50 \qquad (10.7)$$

We took the feeding ration values as a function of animal live weights from the data for the Limousin breed given in the feeding tables of INRA (1988). Feeding ration is the reference quantity of forage an animal fed ad libitum can ingest daily provided it has a fill value (*HFV*) of 1 unit (Meyer 2018). The ingested quantity for any kind of forage is given by the ratio *Ration*/*HFV*. Therefore when *HFV* > 1, the ingested quantity decreases and, conversely, it increases when *HFV* < 1. The actual daily ingestion by a herbivore calculated by the model at each time step is the minimum between this *Ration*/*HFV* ratio and the herbage allowance (i.e. the herbage biomass of a plot divided by the total number of animals on it, as kg dry matter/head).

The reasoning behind using fill values for determining ingestion is, basically, that all stages of grass are considered having the same intrinsic nutritive value, however, a better quality of grass with lower fill value increases the quantity ingested by the animal and thus its nutritional benefit (and the other way round for lower quality grass). The results obtained using this method to calculate the feeding ration and ingested quantity of herbivores match the practical recommendations of Chambre d'Agriculture de Charente-Maritime (2011) for different ages of cows, although they are lower than the model given by Delagarde and Pérez-Prieto (2016) for dairy cows in rotational grazing.

Updating the plot colour consecutively to the grass intake by all the herbivores having grazed on it at a time step is made by increasing this plot's colour value (i.e. fading its green shade) by its equivalent in colour units to the grass intake of each animal, that is its individual grass intake GI_i divided by the biomass per colour unit coefficient (BCU, see Sect. 10.5.2):

$$c_{t+1} = c_t + \frac{GI_i}{BCU} \tag{10.8}$$

Although Eq. 10.8, like Eq. 10.5, is used to compute a new plot colour from its value at the preceding time step and the ratio of a biomass change (here the grass intake) divided by the BCU, this ratio should this time be added to the colour since, as the biomass is decreased by grazing, the colour value is increased. With this new plot colour, its vegetation biomass is then updated using Eq. 10.6.

10.4.3.5 Gain and Lose Body Weight

Converting the grass ingested biomass (kg dry matter) into live weight gain by the animals (kg) is made with a feed conversion index (kg dry matter/kg live weight) function of the animal live weight (Boval et al. 2015). As a proxy, this index has been determined from the recommended feeding rations for heifers at different weights (cf. INRA 1988) as 4.2% of the animal live weight. Therefore, the weight gain *WG* (kg live weight) is given by:

$$WG = \frac{ING}{0.042W} \tag{10.9}$$

where *ING* is the daily ingested quantity of forage (kg dry matter) and *W* is the live weight of the animal.

The total weight losses are the sum of losses due to the animal metabolism and losses due to its movement (in kg live weight):

$$WL_{\text{tot}} = WL_{\text{met}} + WL_{\text{mvt}} \tag{10.10}$$

Metabolic weight losses: assuming the recommended feeding ration allows at least the animal to maintain its live weight, there will be weight losses only when the grass ingested will be less than the feeding ration in proportion of that lack. This loss, however, will be limited: we take from Johnson (2008) in the *GrazeMod* documentation a maximum weight loss of 1 kg/day. The equation giving the metabolic weight losses is therefore,

$$WL_{\text{met}} = \text{Min}\left(1, \frac{RAT - ING}{0.042\text{W}}\right) \text{if}\, ING - RAT < 0, \text{ otherwise}\, 0 \tag{10.11}$$

with *RAT* the feeding ration and *ING* the grass ingested daily (both in kg dry matter/day).

Movement weight losses: the energy lost due to animal movement has been taken from the *Ramdry* model (Fust and Schlecht 2018). It is dependent upon the distance walked per day, the walking speed and the animal live weight. The distance d walked is an output of the simulation expressed as a multiple of a plot side length converted into kilometres (1 plot side = 0.1 km in the model). The daily average walking speed v (km/h) is obtained by the ratio $v = d/T_w$, where T_w is the time spent walking daily in hours, and the expression taking into account the walking speed v (km/h) is an energy loss polynomial coefficient e (J/m/kg):

$$e = 6.6v^2 - 44v + 75 \tag{10.12}$$

The energy lost daily by an animal movement (J) is thus the product of the three factors cited above:

$$E = d \cdot e \cdot W \tag{10.13}$$

The conversion of energy losses (J) to weight losses due to animal movement (kg) is given by dividing E by a conversion ratio k (J/kg live weight):

$$WL_{\text{mvt}} = E/k \tag{10.14}$$

with $k = 1.9 \times 10^7$ if the animal is sexually mature and 10^7 otherwise (from Johnson 2008, in the *GrazeMod* documentation).

Finally, updating each animal body weight (kg live weight) at each time step is done by making the balance between the gain from the forage ingested and the losses due to metabolism and movement:

$$W_{t+1} = W_t + WG - (WL_{\text{met}} + WL_{\text{mvt}}) \tag{10.15}$$

The results obtained by this method are coherent with the data given by Bayer and Waters-Bayer (1999) about the weight variations of cattle in Northern Australia over the year: about 0.12% of live weight lost per day during the 6 months dry season with bad quality forage and about 0.22% per day of weight gain in the wet season with good quality forage. This latter result on daily weight gain falls also within the range [0, 0.34%] found by meta-analysis of various works in the tropics by Boval et al. (2015), depending on the climatic zone and the cattle species considered (0.22% was for high potential cattle). The weight losses due to animal movement are also in accordance with the order of magnitude given by Corson (2002) for the African buffalo according to which moving individuals spend 4–7 times more energy than still ones.

10.4.3.6 Calve

An individual may give birth to one offspring provided it complies with the following conditions:

- The individual is a cow (sex randomly determined at initialization, Sect. 10.4.1) with age above sexual maturity (here 2 years).
- Current time is greater than the last calving date plus the calving interval (here 400 days).
- A random drawing of integers on the range [0, 100] falls below the pregnancy rate.

As the model is not aimed at representing the herbivore physiology and for the sake of simplicity, the increase of the cow live weight before calving (foetus and amniotic sac) is omitted. Therefore, the decrease of weight after calving is just estimated as the difference in the cow live weights before and after gestation. As for Gauthier et al. (1984) and Petit (1979) it lies between 0 and −6% of the weight before gestation, we consider here -5% to update the cow live weight after calving.

The new-born calf attributes are initialized: age = 0 (days), birth date = current time (days), reproduction date = birth date + age of sexual maturity (days), sex = 1 or 0 (random), birth weight = 37 kg and $BCR = 1$ (the calf is assumed, in the example of Sect. 10.5, in normal reference shape at birth). Its location is inherited from its mother and its ingestion rate is calculated as described in Sect. 10.5.4. Note that, as we assume the herbivore population to be free-ranging with no human intervention, no condition is put on the shape of animals to reproduce. It means that thin females (with a $BCR < 1$) may give birth as well as fat ones ($BCR > 1$). This, of course, is avoided in a cattle farming context. To simulate that case, an additional condition may be added requiring at least a normal body condition ratio increased by the calf birthweight contribution:

$$BCR \geq 1 + \frac{W_{\text{birth}}}{SRW} \tag{10.16}$$

where SRW is the "Standard reference weight", that is the weight of a mature animal of average body condition (Mayer et al. 2012; Freer et al. 1991); here, we can take SRW as the reference body weight for a 2 years old animal (i.e. 519.82 kg).

10.5 Simulation Examples

10.5.1 Scenario Parameterization

As an example, a rangeland simulation scenario is presented based on a landscape comprising 1,225 plots of 1 ha each. Main simulation parameters for this scenario are given in Table 10.2.

Table 10.2 Parameterization of the scenario given as an example of the model simulation

Parameters	Values	Explanations	References
Herbivores			
Init-cattle-heads	1,225 heads	Initial number of herbivores	Arbitrary
Max-density	100 heads	Maximum number of herbivores allowed on a single plot	Arbitrary
Longevity	25 years	Animal dies when it becomes 25 years old	Ng (2001)
BCR-min	0.3	Minimum condition ratio below which an animal dies (i.e. animal weight 30% below standard weight). Whiting et al. (2012) give a limit of 40% of normal body weight for all mammals	Bayer and Waters-Bayer (1999) Mayer et al. (2012) Moritz et al. (2015) Whiting et al. (2012)
Marge-To	0.1	±0.1 are the bounds of BCR within which lies the BCR value considered as normal (i.e. ± 10% of standard weight)	Arbitrary
Perception radius	3	Maximal distance an herbivore can perceive its environment (i.e. plots and other herbivores). Corresponds to a 300 m radius encompassing 29 plots of 1 ha	Arbitrary
Moving time	12 h	Daily duration of animal movement. Chirat et al. (2014) give a total duration of cattle activity (grazing and moving) in the range 8–12 h/day	Fust and Schlecht (2018) Chirat et al. (2014)
Sexual maturity	2 years	Age from which a cow may calve. Ng (2001) evaluates it around 1.5 years old	Ng (2001)
Pregnancy rate	81%	The probability a cow may calve when all conditions are met (particularly its age); 81% is a good rate	Gauthier et al. (1984) Diskin (2016)

(continued)

Table 10.2 (continued)

Parameters	Values	Explanations	References
% Loss-at-birth	5%	% of cow live weight lost at calving	Gauthier et al. (1984) Petit (1979)
Birthweight	37 kg	All calves are given this live weight at birth (age = 0) considered as standard for the Limousin breed	INRA (1988)
Calving interval	400 days	Minimum period between two calvings; approximate value holding for creole cows	Gauthier et al. (1984) Diskin (2016) Vayssières et al. (2009)
Plot vegetation			
Base-grass-colour	55 (green)	The base colour standing for the vegetation	Arbitrary
Grass peak-val	60	The colour value denoting grass with most attractiveness and quality ($c = 60$; see Fig. 10.1). This quasi-linear relationship has been shown by Bonnet (2008)	Bonnet (2008)
Max-fill-val	1.25	The maximum filling value of low-quality grass holding for permanent pastures in mountainous areas	INRA (1988)
Biomax	5,000 kg dry matter/ha	The maximum grass biomass per plot. Value holding in the high plains of Reunion island (Vayssières et al.) and close to the one given for a floodplain in Africa by Moritz et al. (4,000 kg d.m./ha) in the wet season. Fresh matter has 20–30% dry matter content	Vayssières et al. (2009) Moritz et al. 2015
Max-grass-growth	40 kg dry matter/ha/day	Maximum grass growth rate. Value holding in the high plains of Reunion Island	Vayssières et al. (2009)

Herbivores are supposed to form herds based on the leader-follower pattern (younger animals follow elders; see Bayer and Waters-Bayer (1999)) and to move to the plot located in the animals perception radius (here 300 m) maximizing the product grass biomass × quality per head. We assume that low-biomass young grass is the best quality and most attractive for animals, whereas the darkest green high-biomass grass is the least. The grass peak colour value (see Sect. 10.2.2 and Fig. 10.1) has thus been set to 60 that is the lightest shade of green denoting younger grass with most attractiveness and quality. This can be viewed as a quasi-linear relationship decreasing in quality from good young grass to bad old grass as highlighted by Bonnet (2008) in Africa.

10.5.2 Emergence of a "Grazing Lawn"

Starting with any of the landscapes in Figs. 10.4A, C, E leads to very similar patterns shown in Fig. 10.8. Although all of these landscapes have quite different initial neighbourhood scores (see Fig. 10.4), after 1 year they all get a neighbourhood score in the range [0.3, 0.5] and, finally, all converge to about the same score of 0.4 from 5 years on (see Fig. 10.8). Homogenous landscapes at initialization become more heterogeneous and conversely: in a few years, due to the grazing activity of cattle they all get the same heterogeneity degree. The landscape, bushy in the first years becomes clearer and clearer to remain totally open after 10 years, offering a good pasture to the herbivore population.

Concurrently, the cattle population size grows rapidly. Starting from 1,225 heads, it reaches 1,600 heads after 1 year, 1,900 at 2 years, 2,100 at 3 years, 2,500 at 4 years, 3,000 at 5 years and finally stabilizes around 7,000 heads after 10 years, that is almost six times the initial population size (Fig. 10.10a). This is due to a strong natality, since after 10 years about 45% of the population is less than 2 years old.

Meanwhile, the vegetation tends to decrease regularly, after a short growth the first year, until stabilizing after 10 years to around 1,000–1,500 t dry matter overall, that is around 1 ton dry matter/ha (Fig. 10.9a), while the grass quality is considerably improved (Fig. 10.9b). During the first year, due to the relatively low stocking rate

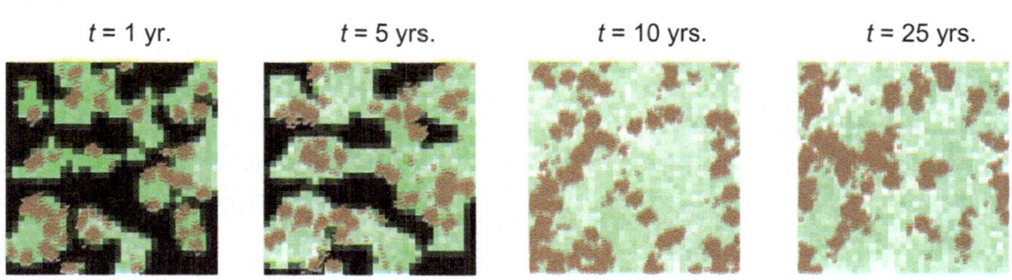

Fig. 10.8 Example of a simulated 1,225-ha landscape over 25 years with 1.12 heads/ha free-grazing cattle initialized from one of the landscapes shown in Figs. 10.4A, C, E

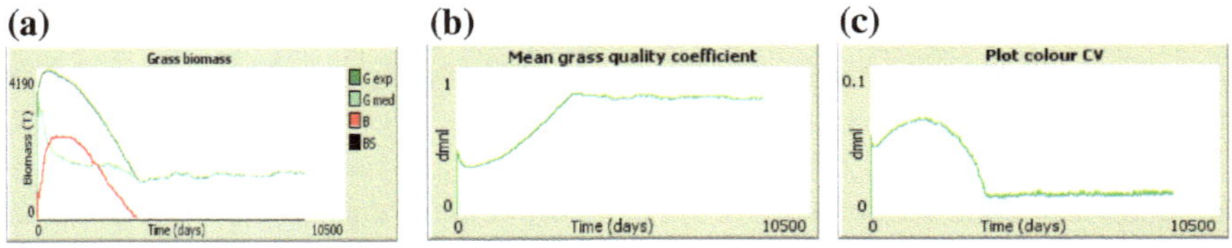

Fig. 10.9 Vegetation total biomass (**a**), mean grass quality coefficient (**b**) and vegetation coefficient of variation (**c**) for 25 years simulation

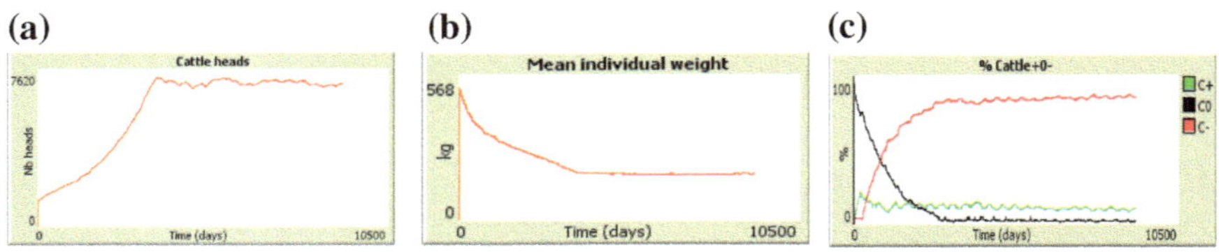

Fig. 10.10 Population size (**a**), mean individual weight (**b**) and composition (**c**) of the cattle population for 25 years simulation

with respect to the strong grass growth (40 kg dry matter/ha/day) there is the growth of bushy dark green vegetation with high biomass but low quality (Fig. 10.8). Right after 1 year, as the herd population increases considerably, the grazing pressure increases too and the bushy vegetation declines leaving all space after 9 years (Fig. 10.9a red 'B' line) to the good light green vegetation with lower biomass but high quality and attractiveness (Fig. 10.9a light green 'G med' line). Overall, landscape tends to homogenize: after an increase in the vegetation variation coefficient during the first 5 years, this coefficient drops and stabilizes at a much lower level after 10 years when the bushy vegetation leaves the room to a good pasture (Fig. 10.9c). This period of 10 years for a vegetation to change in a landscape under the pressure of grazing populations is coherent with the order of magnitude given by Buard (2013) in Zimbabwe.

Whereas the total population of herbivores has considerably increased in number and, so, the total herbivore biomass (passing from about 600 t at initialization to about 1,300 t after 10 years) the mean individual live weight has dramatically decreased: whereas it was 517 kg at initialization it drops down to 180 kg after 10 years (Fig. 10.10b). This is due partly to the population rejuvenation: the mean age of animals drops from about 12 years at initialization to 3.6 years after 10 years and stabilizes around 4.6 years after 15 years. Whereas the proportion of young individuals (<2 years old) at initialization was 12% it rises up to 45% after. Since the maximum weight is reached by individuals above 3 years old, increasing the young individuals tends to decrease the population mean weight. However, the main cause is due to the quick weight loss of individuals (Fig. 10.10c): starting from a standard-weighted population (C0), the number of underweighted individuals (C–) grows rapidly to reach about 85% after 7 years (with mean BCR = 0.4) whereas the number of standard individuals (C0) drops down to less than 5% (BCR = 1) while over-weighted herbivores (C+) represent a stable 10% of the total (mean BCR = 1.35). This result where high stocking rate of herbivores contradicts their individual growth is very classical in free-grazing contexts without feeding complementation (Bayer and Waters-Bayer 1999; Boval et al. 2015; Lesel 1969).

Summarizing, one can say this rangeland ecosystem is quite stable: pursuing the simulation up to 100 years yields the same results (the model interface presented in Fig. 10.3 corresponds to this simulation example on a temporal horizon of 25 years). It can accommodate a very high cattle biomass (5.7 heads and 1 ton live weight per ha on average) based on a dynamic productive vegetation of high quality. Cause and consequence of this, the vegetation standing biomass is relatively low (1 ton dry matter/ha versus above 3 t at initialization) as well as the animal individual weights.

This kind of rangeland is in agreement with the concept of "grazing lawns" (Bonnet 2008; Bonnet et al. 2010; Martin et al. 2015), that are short-grass, low-biomass, high-quality, high-productivity pastures, allowing for accommodation of a huge number of grazing herbivores. These grazing lawns are relatively stable provided rainfall is non-limiting (which is the case in our model where rainfall as well as seasonal variations are ignored) and worldwide distributed.

The basic mechanism leading to the creation of such a grazing lawn is as follows (see Figs. 10.8, 10.9):

- First, at the very beginning of the simulation (from t = 0 to 1 yr.), there is a relative herbivore understocking which lets bushy vegetation expand and creates a local overstocking in plots with good attractive grass.
- Due to this local overstocking and the correlative reduction of good grass, the animals alternate grazing on low-biomass/high-quality grass and high-biomass/low-quality grass plots (from $t = 1$ to 10 yrs.).
- This alternate grazing modifies the landscape by reducing the bushy plots and enlarging the good grass areas forming larger and larger lawns until they occupy the whole landscape (from $t = 5$ to 10 yrs.).
- In the meanwhile, the herbivore population increases: as the grass value and productivity increase, the animals fatten more quickly, reproduce more, graze more, and so on… according to a positive feedback maintaining dynamically the grazing lawn in the whole landscape ($t = 10$–25 yrs. and beyond).

This is very similar to the explanation of grazing lawns formation described by Bonnet (2008).

10.5.3 Theory Checking: The Ideal Free Distribution Example

Among ecological theories and hypotheses that may be checked with this model are the ones based on optimization to explain the spatial distribution and movement of animals:

- Optimal Foraging Theory (OFT) assuming that foraging behaviours tend to maximize feed intake efficiency while minimizing the cost and time spent foraging (Sinervo 1997);
- Marginal Value Theorem (MVT) based on a decreasing efficiency law aiming at deciding when to depart from an exploited resource as soon as its benefit falls below a certain average value threshold (Miller et al. 2017; Sinervo 1997);
- Ideal Free Distribution (IFD) predicting the distribution of foragers over a territory in proportion of the resource local availability (Kennedy and Gray 1993).
- We chose here to check the IFD theory whose assumptions are the following (Kennedy and Gray 1993):
- Each place of the landscape has a value in terms of food availability;
- Each forager has a perfect knowledge of the landscape and is free to reach the ideal place;
- Each place value decreases with the number of individuals located on it;
- Individuals are competitively equal.

Applied to our grazing herbivore model, we take as places the 1 ha land plots partitioning the simulated landscape and assign them, as their value, the product combining edible grass biomass and quality. The ideal plot would be for each individual the one maximizing the product grass biomass × quality divided by the number of present animals. Using model simulations we can check whether the main IFD prediction is met: the herbivores distribution is proportional to the plot values and, as consequences, occupied plots are less variable and have more value than unoccupied ones, and every herbivore is offered the same value. As indicators, like Moritz et al. (2015), we thus can try to verify whether:

- The variance of the values among occupied plots is lesser than the one among unoccupied ones;
- The mean total value of occupied plots is greater than the mean total value of unoccupied plots;
- The coefficient of variation of the value offered to each herbivore is as close to zero as possible.

Applying strictly the IFD assumptions in a simulation scenario leads to eliminate most constraints used in the above simulation scenario (Sects. 10.5.1 and 10.5.2): no more herding behaviour, no limited perception radius, no maximum animal density admitted on plots. Every individual should be free to go to their own ideal plot. This results in that the whole population goes at each time step to the same plot or, at least, to a very limited number of plots. Of course, in this scenario all the indicators comply with the theory except for very short periods of time at the beginning of the simulation (Fig. 10.11a): the variance of values in unoccupied plots is high whereas the one of occupied plots is zero (obviously, since in most cases there is only one occupied plot); the mean value per head is much higher in occupied than in unoccupied plots, and the coefficient of variation is quasi-zero. This, of course, before the population crash which occurred at $t = 1.35$ yr.

(a)

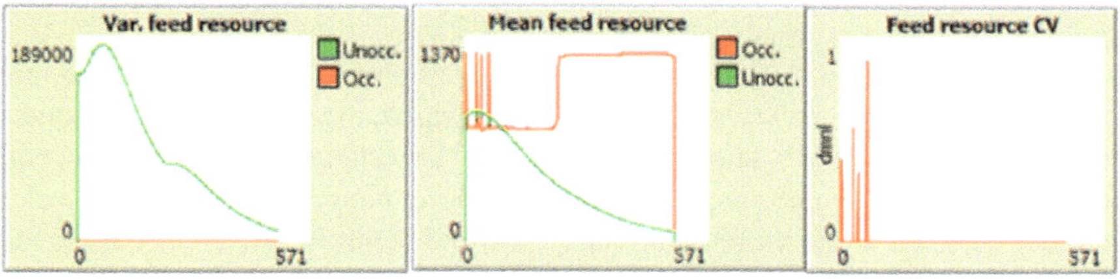

(b)

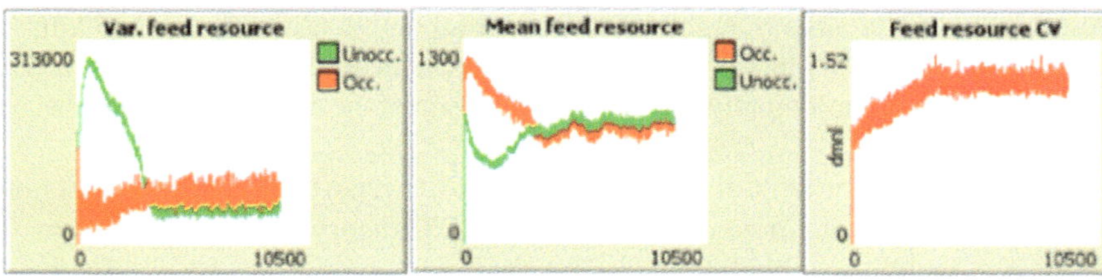

Fig. 10.11 Checking IFD theory on the variance, mean and coefficient of variation of the plot value (grass biomass × quality per head) to be maximized by individual choices: applying strict ideal free (**a**) or non-ideal non-free (**b**) assumptions (simulation in **a** is for 1.35 years whereas in **b** it is for 25 years; see explanations in text)

Because, since the population is concentrated on the same plot (or very few ones) it leaves unoccupied a large part of the landscape and the forage intake by animals on overcrowded plots is far too little. In addition, as the herbivores are limitless oriented towards the best plot they make long-distance walks: whereas in the preceding simulation (Sect. 10.5.2) their mean distance walked daily was 2–300 m/individual/day, in this case, it oscillates between 300 and 2,200 m. As a consequence, unexploited land plots become invaded by the bushy vegetation, animals lose weight and die rapidly; eventually, the whole population crashes and the landscape totally closes after 1 year and a few months. In addition to the lack of realism of IFD assumptions for herbivores, this scenario is absolutely unsustainable.

Therefore, let us restore the constraints of the preceding scenario in Sect. 10.5.2 (herding, limited perception, maximum density on plots) and check how far such a "non-ideal non-free distribution" assumptions may meet some of IFD predictions. The results on indicators are in Fig. 10.11b. Within this 25-years simulation it appears that for the first 8 years the IFD criteria are met for the variance and the mean but not after: as the landscape becomes more and more homogenous from 8 years on the oscillating curves tend to merge. About the coefficient of variation, as it remains all the time high and grows up to 10 years simulation time to oscillate later on, this criterion is not met: individuals are not offered the same value anytime. This, of course, is due to the suboptimal relaxation of the IFD constraints: as individuals are not free to go to the best plot, some get more than others. This is a confirmation that IFD is limited to work rather on patchy landscapes, where there is quite a clear difference between occupied and unoccupied plots as the former are attractive to the foragers and the latter repulsive.

Kennedy and Gray (1993) after having shown by reanalysing the data of many studies that the IFD theory was frequently not verified, explain it "as a consequence of violations of the IFD's assumptions of perfect knowledge, no interference, equal competitive abilities, and a negligible effect of travel between sites." This is exactly what we violated in the second non-ideal non-free scenario and, as a consequence, what we found also. It confirms the little interest of this theory when a non-patchy, dynamical rangeland system is considered as we did in this simulation example. This, however, has been already highlighted by several authors, like Pierce and Ollason (1987) who denied any interest to the theories based on optimality (OFT, MVT, IFD) because it is not biologically established and cannot be demonstrated in the real world.

The fact that Moritz et al. (2015) show that the repartition of pastoralists in a flooded plain in West Africa complies with the IFD theory can be explained by the strong assumptions explicitly encoded by these authors in their model where individual agents are not animals but "camps", that are compound entities encompassing multiple households and their herds. These camps have a perfect knowledge about the resource quality and quantity of each place of the whole region (ideal assumption) and are free to move anywhere (free assumption): a camp compares its energy with that of others; if it is lower than the average of all camps, then it moves to another place with more grazing resources. As it seems that most of the optimal foraging laws are embedded within the model, it is not surprising it yields the results of what has been put in it. Of course, using this explicitly optimal movement rule, they obtain for the cattle a carrying capacity 4 times higher than if the camps did not move. We get a comparable result with much lighter assumptions.

10.6 Conclusions and Outlook

- The model described here allows one to simulate dynamically and spatially the interactions between mobile agents (herbivores) and their environment (rangeland). Its aim is not to mimic real specific rangelands but to offer a generic simple synthetic ecosystem to check ecological hypotheses.
- Using model simulations, variants of the rangeland landscape features and animal behaviours may be checked in view of finding a dynamical balance between grass intake by the animals and herbage growth to avoid issues like overgrazing, leading to desertification, or undergrazing, leading to landscape "closure" by an invasive bushy vegetation.
- Implemented with the agent-based simulation NetLogo platform (Wilensky 1999) the model comprises two types of agents: spatial cells, representing the landscape as a grid of land plots, and mobile agents, standing for herbivores.

- Vegetation on land plots is represented by their colour ranging from very light to very dark shades of green: darker the shade, higher the grass biomass and lower its quality (and conversely).
- At each simulation time step each individual herbivore iterates the following actions: stay on the current plot or move to another satisfying some criteria, graze, gain and lose weight, age and, possibly, calve or die.
- Although deliberately naive, the model was parameterized with real beef cattle and rangeland data borrowed from various authors (Bayer and Waters-Bayer 1999; INRA 1988; Mayer et al. 2012; Moritz et al. 2015; Vayssières et al. 2009).
- Main simulation assessment criteria are the herbage biomass, herbivore population size, individual body weights, birth and mortality rates, land-use patterns and landscape fragmentation.
- A simulation scenario has been described and analysed, based on a reference landscape comprising 1,225 plots of 1 ha each and initialized with 1,225 cattle heads (1 head/ha) endowed with a behaviour consisting in an oriented walk towards the plot maximizing the product grass biomass × quality per head of cattle within a 300 m radius. Whatever their initial state, all landscapes converge in a few years towards a similar homogeneity degree, thanks to the grazing of herbivores. This gives rise to the emergence of what rangeland ecologists call a "grazing lawn", that is a highly productive rangeland ecosystem capable of sustaining a high stocking rate of herbivores (in this example almost 6 times the initial population size in 10 years) with a low vegetation standing biomass of high quality for many years long (in this example, the analysis was done on a 25 years simulation horizon but simulations over 100 years lead to the same conclusions).
- Ideal Free Distribution has been taken as an example to illustrate the model capacity to challenge ecological theories. In the simulation example taken in this article this theory has proven irrelevant because it leads to an unrealistic unsustainable situation; however, the system sustainability is restored when one get rid of this theory's assumptions.
- Oriented walks: a walk like the one used in the given simulation example (i.e. moving to plots with maximum grass biomass × quality per head) but not restricted to a relatively short radius (i.e. allowing long moves) tends to create too high local animal densities, inducing overgrazing, high mortality and bush extension; keeping the same rule within a limited perception range (i.e. allowing short moves only) may lead to "grazing lawns". Avoiding excess of weight losses due to movement, short-move walks (local foraging) lead to best animal and herbage production than long-move ones.
- Random walks (not demonstrated in this article): here too, short moves are most likely to lead to animal wealth and herbage preservation than long ones. "Levy walks", alternating both long and short moves, lead to intermediate results. Landscapes resulting from short walks are less heterogeneous. However, long moves, fostering a quick animal distribution in space, are better to reduce local excessive animal densities and resource depletion. Introducing two breeds of herbivores endowed with different walks confirmed this conclusion: the short walk population takes over the long-distance one because of over-mortality due to weight

losses caused by the excess of movements. However, these random walks do not comply well with the actual behaviour of most herbivores which exhibit marked preference for certain kinds of grass they graze first, leaving the less attractive for periods after the good resource has been exhausted (Dumont 2009).

- Most of these findings are in accordance with both cattle behaviour and cattle farmers' management rules. The most sustainable strategies were found to be those fostering space occupation, local foraging, short walk steps and anticipating resource exhaustion. Animal movement and grazing proved to be crucial in shaping the system.
- Simulations made on the long run (up to 100 years) allow the emergence of interesting extreme phenomena also found in real but particular ecosystems (e.g. grazing lawns). Reproducing observed patterns in ecosystems using a bottom-up modelling approach is in keeping with pattern-oriented modelling advocated by Grimm et al. (2005).
- Beyond simulating a rangeland landscape evolution, the model may also allow one to explore issues linked to ecosystem complexity (adding other trophic levels), percolation (dissemination of products over space) and resilience (assessment of disturbances through time and space on animals and land). How far this "herbage-herbivores" model, basically featuring the interaction between immobile and mobile agents (i.e. land plots and herbivores), can be used as a metaphor to represent other kinds of systems can be envisaged.
- In order to render the model more realistic to simulate real rangelands, several improvements should be integrated within the model among which: seasonality and climatic characteristics with their consequences on the grass growth and quality and the herbivore growth and reproduction; sexual dimorphism in terms of growth, live weights and, even, herding behaviours as observed by Daycard (1990) in feral cattle; varying the time spent grazing according to different vegetation covers; inter-individual attribute variations... One could also consider to improve the animal's physiological model making more extensive use of the Dynamic Energy Budget (DEB) theory, as in the NetLogo model by Martin et al. (2012), or endow the individuals with learning capabilities to feed their foraging memory (Brooks and Harris 2008; Dumont and Hill 2004). Also, introducing structuring landscape elements (e.g. water holes) known to influence herbivore movements (Valls-Fox 2015) could be done.
- In order to facilitate the landscape analysis, more effort should be devoted to use more appropriate and relevant metrics. Among the huge set of existing metrics (McGarigal 2015) we are considering, following Rütters et al. (1995), to restrict our choice to 6 univariate metrics likely to account for the six main dimensions capable of explaining most of the landscape structures.

Acknowledgements This work has been achieved while I was at the SELMET research unit in Montpellier (France). I am indebted towards several colleagues who provided me (often through informal discussions) with information, ideas, advices, data, documentation and suggestions which helped me to render my model more relevant. They are Olivier Bonnet, Jean-Marie Capron, Laurence Flori, Myriam Grillot, Lionel Julien, Mathieu Lesnoff, Marie-Odile Nozières, Eric Vall and Hugo

Valls-Fox. However, all the modelling choices, parameter assignments, model implementation and simulations presented in this article remain of my own responsibility.

References

Anselme B, Bousquet F, Lyet A, Etienne M, Fady B, Le Page C (2010) Modelling of spatial dynamics and biodiversity conservation on Lure mountain (France). Environ Model Softw 25:1385–1398
Bayer W, Waters-Bayer A (1999) La gestion des fourrages. Margraf Verlag, Weikersheim, 246 pp
Bonnet O (2008) Contraintes et avantages des «grazing lawns» en tant que ressource alimentaire chez les grands herbivores africains. Ph.D. thesis, Pierre & Marie Curie University, Paris, 175 pp
Bonnet O, Fritz H, Gignoux J, Meuret M (2010) Challenges of foraging on a high quality but unpredictable food source: the dynamics of grass production and consumption in savanna grazing lawns. J Ecol 98:908–916
Boval M, Edouard N, Naves M, Sauvant D (2015) Performances de croissance et efficacité alimentaire des bovins au pâturage en conditions tropicales: étude par méta-analyse. INRA Prod Anim 28(4):315–328
Brooks CJ, Harris S (2008) Directed movement and orientation across a large natural landscape by zebras, Equus burchelli antiquorum. Anim Behav 76:277–285
Buard E (2013) Dynamiques des interactions espèces-espace: mise en relation des pratiques de déplacement des populations d'herbivores et de l'évolution de l'occupation du sol dans le parc de Hwange (Zimbabwe). PhD thesis, Panthéon-Sorbonne University, Paris, 395 pp
Chambre d'Agriculture de Charente-Maritime (2011) Des repères pour faire son bilan fourrager. Agricultures et territoires. (Online) https://fr.scribd.com/document/362617162/Des-Reperes-Pour-Faire-Son-Bilan-Fourrager-1. Accessed 1 Dec 2018
Chirat G, Groot J, Messad S, Bocquier F, Ickowicz A (2014) Instantaneous intake rate of free-grazing cattle as affected by herbage characteristics in heterogeneous tropical agro-pastoral landscape. Appl Anim Behav Sci 157:48–60
Codling EA, Plank MJ, Benhamou S (2008) Random walk models in biology. J R Soc Interface 5:813–834
Corson PJ (2002) Le buffle d'Afrique. Biologie, chasse et mythologie. Editions du Gerfaut, collection Grandes chasses, 264 pp
Daycard L (1990) Structure sociale de bovins sauvages de l'île Amsterdam, sud de l'Océan Indien. La Terre et la Vie 45:1–35
Delagarde R, Pérez-Prieto LA (2016) Effets de la biomasse et de la quantité d'herbe offerte sur l'ingestion, les performances laitières et le comportement alimentaire des vaches laitières conduites en pâturage tournant: étude par méta-analyse. INRA Prod Anim 29(2):87–102
Diskin M (2016) Achieving high reproductive performance in beef herds. Teagasc Beef manual, Chap 22, pp 119–124
Dumont B (2009) Comportement alimentaire des herbivores et dynamique des prairies. In: Boissy A, Pham-Delègue M-H, Baudoin C (eds) Éthologie appliquée. Comportements animaux et humains, questions de société, Chap 5. Edition Quae, Versailles, pp 79–88
Dumont B, Boissy A (1999) Relations sociales et comportement alimentaire au pâturage. INRA Prod Anim 12(1):3–10
Dumont B, Hill D (2004) Spatialy explicit models of group foraging by herbivores: what can agent-based models offer? Anim Res 53:419–428
Fellah Trade (2018) Elevage bovin à l'engraissement. (On-line) https://www.fellah-trade.com/ressources/pdf/Elevage_bovins_engraissement.pdf. Accessed Dec 1 2018
Freer M, Moore AD, Donnelly JR (1991) GRAZPLAN: Decision support systems for Australian grazing enterprises-II. The animal biology model for feed intake, production and reproduction and the GrazFeed DSS. Agric Syst 54(I):17–126

Fust P, Schlecht E (2018) Integrating spatio-temporal variation in resource availability and herbivore movements into rangeland management: RaMDry–an agent-based model on livestock feeding ecology in a dynamic, heterogeneous, semi-arid environment. Ecol Model 369:13–41
Gauthier D, Coulaud G, Varo H, Thimonier J (1984) Durée de l'anoestrus post-partum et fertilité de la vache créole en climat tropical: influence de la saison de mise bas et de la variation du poids vif. Ann Zootechnie, INRA/EDP Sci 33(2):235–244
Grimm V, Berger U, DeAngelis DL, Polhill JG, Giske J, Railsback SF (2010) The ODD protocol: a review and first update. Ecol Model 221:2760–2768
Grimm V, Revilla E, Berger U, Jeltsch F, Mooij WM, Railsback SF, Thulke H-H, Weiner J, Wiegand T, DeAngelis DL (2005) Pattern-oriented modeling of agent-based complex systems: lessons from ecology. Science 310(5750):987–991
Humphries NE, Sims DW (2014) Optimal foraging strategies: Lévy walks balance searching and patch exploitation under a very broad range of conditions. J Theor Biol 358:179–193
Ibe OC (2013) Elements of random walk and diffusion processes. Wiley series in Operations Research and Management Science. Wiley, Hoboken, NJ, USA
INRA (1988) Tables de l'alimentation des bovins, ovins & caprins, 1902 pp
Johnson IR (2008) Biophysical pasture model documentation: model documentation for DairyMod, EcoMod and the SGS Pasture model. IMJ Consultants, Dorrigo, NSW, Australia, p 144
Kennedy M, Gray RD (1993) Can ecological theory predict the distribution of foraging animals? A critical analysis of experiments on the ideal free distribution. Oikos 68(1):158–166
Lesel R (1969) Étude d'un troupeau de bovins sauvages vivant sur l'île d'Amsterdam. Rev Elev Med Vet Pays Trop 22(1):107–125
Martin BT, Zimmer EI, Grimm V, Jager T (2012) Dynamic energy budget theory meets individual-based modelling: a generic and accessible implementation. Methods Ecol Evol 3:445–449
Martin J, Benhamou S, Yoganand K, Owen-Smith N (2015) coping with spatial heterogeneity and temporal variability in resources and risks: adaptive movement behaviour by a large grazing herbivore. PLoS ONE 26:1–19
Mayer DG, McKeon GM, Moore AD (2012) Prediction of mortality and conception rates of beef breeding cattle in northern Australia. Anim Prod Sci 52:329–337
McGarigal K (2015) Fragstats help. Department of Environmental Conservation, LandEco Consulting and University of Massachusetts, Amherst, USA, 182 pp
Meyer C (2018) Dictionnaire des Sciences Animales, Cirad, Montpellier, France. (On-line) http://dico-sciencesanimales.cirad.fr. Accessed 1 Dec 2018
Miller ML, Ringelman KM, Eadie JM, Schank JC (2017) Time to fly: a comparison of marginal value theorem approximations in an agent-based model of foraging waterfowl. Ecol Model 351:77–86
Moritz M, Hamilton IM, Yoak AJ, Scholte P, Cronley J, Maddock P, Pi H (2015) Simple movement rules result in ideal free distribution of mobile pastoralists. Ecol Model 305:54–63
Ng J (2001) Bos taurus. Animal Diversity Web (On-line) https://animaldiversity.org/accounts/Bos_taurus/. Accessed 29 Nov 2018
Nisbet RM, Martin BT, de Roos AM (2016) Integrating ecological insight derived from individual-based simulations and physiologically structured population models. Ecol Model 326:101–112
Petit M (1979) Effet du niveau d'alimentation à la fin de la gestation sur le poids à la naissance des veaux et leur devenir. Annales de biologie animale, biochimie, biophysique 19(1B):277–287
Pierce GJ, Ollason JG (1987) Eight reasons why optimal foraging theory is a complete waste of time. Oikos 49(1):111–117
Russel S, Norvig P (2010) Artificial intelligence. A modern approach, 3rd edn. Pearson Education Inc, New York, NJ, USA, 1132 pp
Rütters KH, O'Neil RV, Hunsaker CT, Wickham JD, Yankee DH, Timmins SP, Jones KB, Jackson BL (1995) A factor analysis of landscape pattern and structure metrics. Landscape Ecol 10(1):23–39
Schwarzmueller F, Hulthen A, Murray J, Parry H (2017) Spatially-explicit modelling of ecological processes in complex agricultural landscapes: connecting 'artificial' landscapes with 'reality'. In: 22nd international congress on modelling and simulation. Modsim 2017, Hobart, Tasmania, Australia, 3–8 Dec 2017

Sinervo B (1997) Optimal foraging theory: constraints and cognitive processes. In: Behavioral ecology, pp 105–130. University of California, Santa Cruz

Vacquié L, Houet T, Sheeren D, de Munnik N, Roussel V, Waddle J (2016) Adapting grazing practices to limit the reforestation of mountainous summer pastures: a process-based approach. Environ Model Softw 84:395–411

Valls-Fox H (2015) To drink or not to drink? The influence of resource availability on elephant foraging and habitat selection in a semi-arid savanna, 158 pp. Ph.D. thesis, University of Montpellier

Vayssières J, Bocquier F, Lecomte P (2009) Gamede: a global activity model for evaluating the sustainability of dairy enterprises. Part II–interactive simulation of various management strategies with diverse stakeholders. Agric Syst 101(3):139–151

Whiting TL, Postey RC, Chestley ST, Wruck GC (2012) Explanatory model of cattle death by starvation in Manitoba: Forensic evaluation. Can Vet J 53:1173–1180

Wilensky U (1999) NetLogo, http://ccl.northwestern.edu/netlogo. Center for Connected Learning and Computer-Based Modeling, Northwestern University, Evanston, IL

Chapter 11
The Use of Multi-temporal Spectral Information to Improve the Classification of Agricultural Crops in Landscapes

Ralf Wieland and Pablo Rosso

Abstract Machine learning opens up a wide range of possibilities for crop classification and mapping using satellite data. With the shortening of their revisit cycles, satellites are now able to provide an increasing amount of data with valuable temporal information. We propose a machine learning approach to efficiently analyze multi-temporal data for crop identification and monitoring. This methodology utilizes a Bayesian approach to gradually improve classification accuracy as the temporal resolution increases. Two multispectral satellite configurations were simulated with hyperspectral data and analyzed with a support vector machine approach and a deep learning algorithm. Results showed that both approaches are able to efficiently process information as time progresses and rapidly achieve very high accuracies. The deep learning algorithm has the advantage that the dynamic component, time, is accounted for automatically, without the need of being actively incorporated by the analyst.

Keywords Hyperspectral data · Temporal data · Machine learning · Support Vector Machine · Deep learning LSTM · Remote sensing

11.1 Introduction

Crop identification, vegetation mapping and monitoring are some of the most frequent applications in remote sensing (Gilbertson 2017; Griffiths et al. 2018). Traditionally, crop identification with remote sensing relies on spectral information collected at critical phenological stages to capture each crop's distinct development features. This approach requires a substantial amount of training data, a solid knowledge of the vegetation dynamics, and a great deal of expert involvement.

R. Wieland (✉) · P. Rosso
Leibniz Centre for Agricultural Landscape Research (ZALF), Eberswalder Str. 84, 15374 Müncheberg, Germany
e-mail: rwieland@zalf.de

P. Rosso
e-mail: Pablo.Rosso@zalf.de

W. Mirschel et al. (eds.), *Landscape Modelling and Decision Support*, Innovations in Landscape Research,
https://doi.org/10.1007/978-3-030-37421-1_11

Since the temporal resolution (cadence or revisit time) of remote sensing data has increased drastically in recent times (Ghazaryan et al. 2018), the time dimension can be added to the spectral variability to characterize and discriminate changing ground features, such as vegetation types and species (Ghazaryan et al. 2018; Zhong et al. 2016). With now up to daily image frequency at high spatial resolution and several spectral bands, optical remote sensing is able to produce large amounts of information. This frequency of data can compensate for cloud cover or low image quality and provide a more solid basis for crop classification. However, the incorporation of the time dimension and dealing with the resulting large data volume requires effective analytical tools. Traditional crop identification based solely on spectral properties has many inherent difficulties.

Figure 11.1 shows the challenge of discriminating two crops that are taxonomically and morphologically closely related, such as rye and wheat. Although it may seem that the two crops at a given time may differ sufficiently in some portions of the spectra (around 1200 nm, for example), between- and within-field variations of each crop (compare Rye 1 with Rye 2 at 1200 nm) may make these apparent differences between crops disappear.

The traditional approach based on one (or few) time observations, requires identifying the precise moments at which these differences are larger (e.g. Time 1 vs. Time 2 spectra in the figure), which may apply for a given pair of crops but not for another, or for only a particular area or year. Figure 11.1 also shows the difficulties with adding a new observation date to find more differences between crops; often trends in the development of one crop have the same pattern as the other crop. For

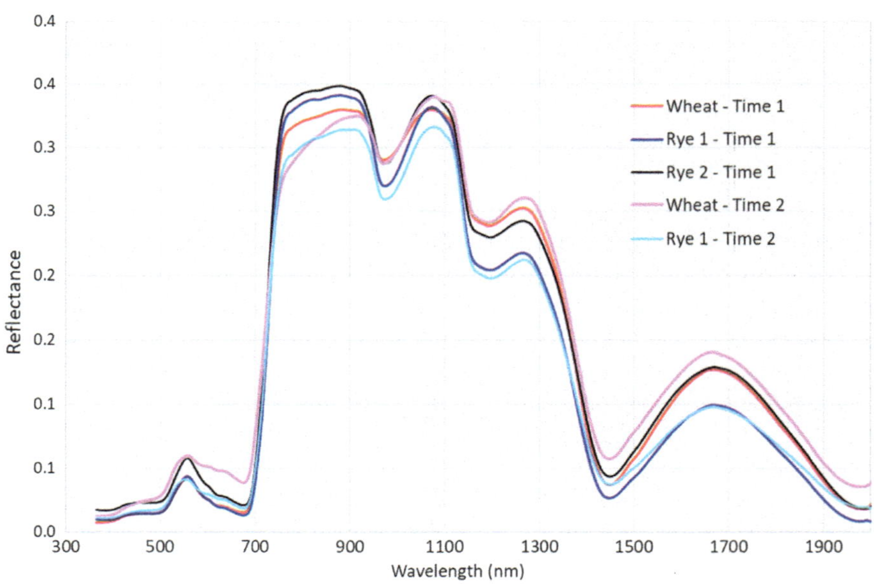

Fig. 11.1 Wheat and rye spectra from early and mid-season in 2002

example, the reflectance around 650 nm increases by late season (Time 2) in both species, which makes the separation between wheat and rye much more difficult.

Therefore, it is only the synergy of the crop spectral properties along a time continuum (several dates) that can account for differences between any pair of species or crops. An integrative analysis of spectra and time that avoids a priori decisions on specific spectral features or points in time seems to be the best approach for these complex problems. In other words, an effective time-based crop separation can be done without the need to understand or somehow characterize the temporal pattern of each crop. Such a method should be able to find the differences and accurately classify crops by only analyzing the raw information available.

Machine learning has previously been used for the classification of arable crops (Ji et al. 2018). A number of known algorithms, such as the Support Vector Machine (SVM) (Gautam et al. 2008; Jin et al. 2017), and Random Forest (Vuolo et al. 2018) can be very effective in handling multi-temporal data. A drawback of these methods is that the time dimension needs to be actively incorporated by the analyst, typically in the form of image stacks or by providing descriptive functions or statistics. Recently, 'deep learning', given its ability to manage complex data representations, has moved into the focus of resource monitoring applications. Neural networks, a method already known since the sixties, experienced a remarkable further development and a comeback as the main structure for deep learning models. A number of novel network structures, such as the convolutional network (Qayyum et al. 2017) and recursive networks, especially the Long Short-Term Memory (LSTM) (Wu et al. 2018), are widely used in complex tasks such as natural language processing and image recognition. In Sharma et al. (2018) and Ji et al. (2018) a deep learning procedure based on networks was used to classify land cover.

The aim of this paper is to show how machine learning algorithms can be applied to analyze multispectral information in a multi-temporal data framework to improve crop classification. Specifically, we want to assess the suitability of already available algorithms to handle multi-temporal information both by using them directly or by adapting them to fit the data. We also want to evaluate to what extent increasing temporal resolution can compensate for a decrease in spectral resolution, a situation often encountered by remote sensing analysts with the advent of nanosatellites.

11.2 Methods

11.2.1 Data Description and Processing

A data set with hyperspectral crop canopy measurements taken in 2002 was used to simulate satellite data. Spectra ranging from 365 to 2455 nm with a resolution of 1 nm were obtained with a field spectrometer FieldSpec Pro JR (A 110080; Analytical Spectral Devices), for the following dates: 08-05-2002, 17-05-2002, 30-05-2002, 18-06-2002, 05-07-2002 and 30-07-2002(not all crops were collected on all dates.

Details can be found in the repository: https://github.com/Ralf3/RemoteSensing). Measurements were error corrected and normalized according to the manufacturer's recommendations and software. Spectra from potato (POT), silo maize (SMA), triticale (TRI), winter rape (WRP), winter rye (WRY) and winter wheat (WWH) were used for classification.

Hyperspectral data were resampled to simulate the spectral properties of two currently available satellite sensors: a Sentinel-2 type spectrum (Mura et al. 2018), with ten bands centred at 490, 560, 665, 705, 740, 783, 842, 865, 1610 and 2190 nm; and a Planet Dove type with four bands (red, green, blue and near infrared, NIR) centred at 490, 560, 665 and 865 nm. About 20 spectra per crop type and date were used for the analyses.

11.2.2 Multi-temporal Modelling

In general, spectral information at the beginning of the growing season tends to show a few specific characteristics and therefore, little discrimination potential can be expected at that stage. As time progresses, the discrimination feasibility of the system tends to increase due to both, changes in crop development and the availability of new information.

In our case, since we assumed that no previous knowledge was available at the beginning of the growing season, an equal distribution of classes was assumed at the first modelling step. Then a Bayesian approach was adopted, by which the initial probabilities are corrected as new information becomes available. Consequently, at each new step, the classification results from the previous one were added as an a priori probability.

11.2.3 Support Vector Machine

Support Vector Machine (SVM) is a widely used machine learning algorithm particularly suited for limited training data sizes. To apply this algorithm on our classification using several time steps (Figs. 11.2 and 11.3), first a classification was performed and saved as time step 0. In the next time step (step 1), the result of time step 0 was provided as input (indicated by 'X' in the figure). Each additional step was based on the values of the spectra at the current date plus the result from the previous step. Since no previous result for the initial classification was available, column X was initially filled with an arbitrary value (here set to 1.0). The six time steps described above were applied to both, the ten- and the four-band spectral configurations. Scikit (Pedregosa et al. 2011) was used for the implementation of Support Vector Machine (SVM) and the calculation of metrics to evaluate the learning results. Recently, other algorithms

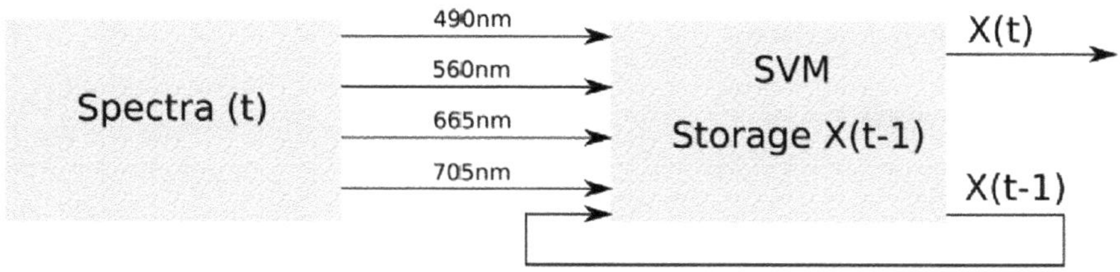

Fig. 11.2 Spectra from the current date, Spectra (t), together with the previous classification result, X(t − 1), are used for the classification at the current date, X(t)

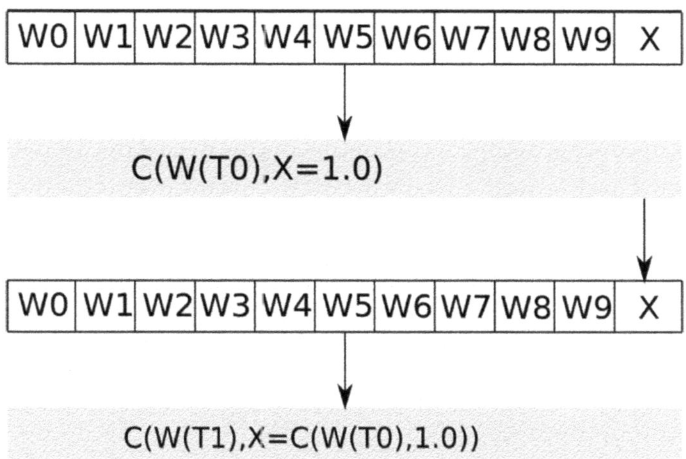

Fig. 11.3 Recursive use of the results of the previous classification (0) for the subsequent classification (1)

such as Random Forest (RF) have been used instead of SVM. An XGBoost implementation (Chen and Guestrin 2016) was built to be used as an alternative modelling tool for applying Random Forest (Vincenzi et al. 2011).

Since the goal of this study was to test the effectiveness of machine learning applied to time series, we wanted to compare an SVM method that uses the classification results from previous steps to a method that does not use previous classification results. This latter approach would represent the traditional approach in which only spectral differences at a single point in time are considered, regardless of previous experience or knowledge. To simulate this approach we applied SVM at the different time steps but obviating the results from previous classifications.

11.2.4 Deep Learning

A multi-temporal modelling procedure using deep learning may find by itself a feedback loop to use previous classification results, and the method chosen in this way can be even more effective than any other approach the analyst might have come up with. In this paper, an approach consisting of Long Short-Term Models (LSTM) in combination with a layer of traditional neuronal nodes (Dense) is presented (Fig. 11.4).

The stacked LSTM nodes (here represented by a stack of two repetition nodes, LSTM1 and LSTM2) constitutes a one-to-one sequence prediction model, by which each input representing a time step is connected with an output of the same time step. The LSTM stores the internal state of the previous time steps, t − i. In the implementation, the input D (t − 3) feeds an LSTM connected in turn to the LSTM for the D (t − 2) and so on. The inputs themselves represent a single time step but they are vectors of the spectral data for this time step. It should be underlined that the deep learning model adopts a Bayesian strategy without being explicitly instructed by the modeller to do so. In Fig. 11.4, a change in colour (pink to red) of Dense node 1 (corresponding to D (t − 3)) and Dense node 2 (connected to D (t − 2)) indicates a change in classification as new information is available.

The dense nodes represent the conditional probabilities $p(Dense_i(t-3)|D(t-3))$, $p(Dense_i(t-2)|D(t-2), Dt-3))$. Each LSTM node has six outputs (representing the six crops) to which each of the corresponding six dense nodes is connected. Since both the LSTM and the dense node have

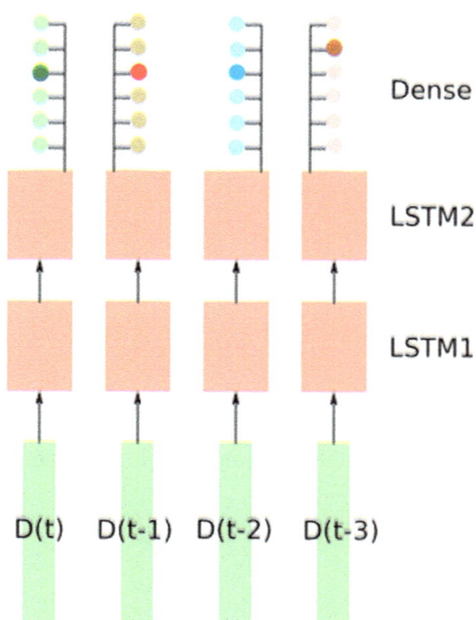

Fig. 11.4 Neural network structure with four time steps D (t − 3) … D(t), two layers of LSTM nodes and a 4 * 6 Dense nodes as outputs

parameters that need to be adjusted during the training, the total number of parameters in the example is 762, which is a rather low number for deep learning. Nevertheless, caution must be taken to avoid 'over training'. In our case this was achieved by dividing the data set into training and validation sections. The training used Adam's optimization algorithm with a batch size of 10 and a number of 800 epochs.

As for SVM, six time steps were applied to both, the ten- and the four-band spectral configurations. Tensorflow (Abadi et al. 2016) was used as basic software for deep Learning and Keras (Chollet 2015) as a development software for neural networks, in the implementation of LSTM.

11.3 Results

11.3.1 SVM

A training run on a PC (Intel(R) Core(TM) i7-6700T CPU @ 2.80 GHz, 32874 MB, Ubuntu 18.04 LTS) took less than 1 s until completion. The predicted values were calculated using a cross-validation procedure with a cv = 10 (i.e. 10 runs applied on a random selection of 90% of the data used as training data, and 10% of the total used as test data). For the Sentinel-2 data set (ten-band) simulation, both the prediction precision (fraction of correctly classified spectra over the total assigned to a given class) and the recall (fraction of correctly classified spectra over the total belonging to that class) of the classification varied between 0.95 and 0.98, and achieved an accuracy of 100% (both precision and recall = 1.0) at the fourth step (Table 11.1).

For the four-band data set simulation, both precision and recall increased steadily after the second time step and by the fourth they reached a value higher than 0.99 (Table 11.2). When comparing the four-band with the ten-band simulations, differences in initial values of precision, recall and standard deviations are consistent with the fact that the four-band case provided much less information. However, the results also suggest that after a given number of iterations, the simpler, four-band configuration can attain results comparable to the ten-band configuration.

Table 11.1 Ten-band simulation with previous classification result: precision and recall for each time step (std—standard deviation)

Time step	Precision	Std. (precision)	Recall	Std. (recall)
0	0.959	0.032	0.958	0.053
1	0.984	0.023	0.983	0.024
2	0.950	0.042	0.950	0.076
3	1.0	0.0	1.0	0.0

Table 11.2 Four-band simulation with previous classification result: precision and recall for each time step

Time step	Precision	Std. (precision)	Recall	Std. (recall)
0	0.854	0.138	0.850	0.191
1	0.851	0.174	0.850	0.180
2	0.933	0.094	0.933	0.094
3	0.992	0.018	0.992	0.019

Table 11.3 Ten-band simulation without previous classification result: precision and recall for each time step

Time step	Precision	Std. (precision)	Recall	Std. (recall)
0	0.835	0.197	0.822	0.214
1	0.903	0.105	0.889	0.092
2	0.820	0.157	0.772	0.201
3	0.868	0.131	0.872	0.083

The analysis with no input from the previous classification resulted in lower accuracies (Table 11.3) than the ten-band spectra with previous classification results. Precision and recall varied between 0.82 and 0.90, and unlike the analysis with the previous classification, both parameters fluctuated from time step to time step with no apparent trend towards stabilization, and a final accuracy closer to 1 was never achieved.

11.3.2 LSTM

A training run on the PC took about 25 s, which is substantially longer than the SVM run. The effect of time dimension on the classification accuracy can be seen at each time step. Each time step (t − 3 to t − 1) can be displayed in a confusion matrix C (Tables 11.4, 11.5 and 11.6) in which $C_{i,j}$ is equal to the number of observations

Table 11.4 LSTM on the ten-band simulation: confusion matrix for time step t − 3

	WWH	WPR	WRY	SMA	POT	TRI
WWH	18	0	5	0	0	0
WPR	0	19	0	0	0	0
WRY	4	0	20	0	0	0
SMA	0	0	0	11	0	0
POT	0	2	0	1	6	0
TRI	1	0	3	0	0	0

Table 11.5 LSTM on the ten-band simulation: confusion matrix for time step t − 2

	WWH	WPR	WRY	SMA	POT	TRI
WWH	23	0	0	0	0	0
WPR	0	19	0	0	0	0
WRY	0	0	24	0	0	0
SMA	0	0	0	11	0	0
POT	0	0	0	1	8	0
TRI	3	0	1	0	0	0

Table 11.6 LSTM on the ten-band simulation: confusion matrix for time step t − 1

	WWH	WPR	WRY	SMA	POT	TRI
WWH	23	0	0	0	0	0
WPR	0	19	0	0	0	0
WRY	0	0	24	0	0	0
SMA	0	0	0	11	0	0
POT	0	0	0	1	8	0
TRI	0	0	1	0	0	0

known to belong in group i but predicted to be in group j.

In the case of the Sentinel (ten-band) simulation, it can be seen that values outside the diagonal tend to diminish as time progresses, indicating an increase in accuracy at each time step. The last time step confusion matrix (Table 11.7) shows an almost perfect classification with only one misclassification between potato and maize.

Precision and recall values of each crop at the end of the exercise (Table 11.7) reflect the single misclassification between potato and maize, in an otherwise perfect accuracy score. The f1-score, a weighted mean of precision and recall, showed a minimum value of 0.94 for POT, but it reached 1 in most other crops, indicating that both precision and recall are equally significant.

Table 11.7 LSTM on the ten-band simulation: confusion matrix metrics of time step t − 1

	Precision	Recall	f1-score	bf support
WWH	1.0	1.0	1.0	23
WPR	1.0	1.0	1.0	19
WRY	1.0	1.0	1.0	24
SMA	0.92	1.0	0.96	11
POT	1.0	0.89	0.94	9
TRI	1.0	1.0	1.0	4
Avg./total	0.99	0.99	0.99	90

Table 11.8 LSTM on the four-band simulation: confusion matrix metrics of time step $t - 1$

	Precision	Recall	f1-score	bf support
WWH	0.93	0.87	0.90	15
WPR	1.0	0.94	0.97	18
WRY	0.96	1.0	0.98	23
SMA	1.0	1.0	1.0	16
POT	1.0	1.0	1.0	11
TRI	0.75	0.86	0.80	7
Avg./total	0.96	0.96	0.96	90

LSTM results with the colour-NIR (four-band) simulation showed a relatively low final classification accuracy (Table 11.8), with most precision and recall values between 0.93 and 1, with the exception of WWH and TRI that showed lower values, which is consistent with the fact that in previous analyses both crops showed a tendency to be confused (Tables 11.5).

11.4 Discussion

The classification of the six arable crops presented here showed that through repetition the system is able to provide sufficient information to achieve accurate results even in cases of limited spectral resolution, and all this, without the need of finding dynamic patterns, statistics or dynamic models of crop spectra.

Considering that satellite simulation from hyperspectral data represents an almost ideal situation (very little noise), we do not expect to achieve the same levels of accuracy in three or four dates when working with actual satellite information. Nevertheless, the idea is to present an approach that, according to the results shown here, has a great potential even with noisier or lower quality data.

The hyperspectral data set and the data processing model is available on GitHub, and can be used to reproduce the analyses presented in this paper. It should be noted that the training algorithms always start with random parameter settings, meaning that the initial states will never be identical to the results presented here.

When using multi-temporal data (Hütt and Waldhoff 2018) we recommend to start with a simple model, here represented by SVM. SVM is easy to implement, it has a fast processing time and it is thus suitable for larger areas requiring a large number of model runs. The default SVM parameters (e.g. C and gamma, see: Becerra-García et al. 2017) can be optimized using a grid search algorithm, which makes the SVM more powerful. This parameter optimization was not necessary in the example presented here. As an alternative to SVM, a Random Forest algorithm can be used in the form of XGBoost implementation. Any improvement on classification efficiency that reduces the required number of time steps makes the system more suitable to be used in areas with frequent cloud cover.

Deep learning can be considered a powerful alternative to more conventional approaches such as SVM. With the hardware already available the time needed for processing large amounts of information is in a similar order of magnitude to SVM and other algorithms. Once the data structure has been defined, LSTM automatically selects the optimal Bayes strategy (Xia et al. 2018) for the modeller. This can be very important in future research where much more data may be included and the simple feedback of the simulation as a result of the previous time step is no longer sufficient.

Whether in general a reduction of spectral bands can be replaced by an increase in the number of multi-temporal values, and what is the minimum number of spectral bands is still the subject of research. In any case, the availability of daily multispectral data reduces the dependency on optimal weather conditions and exact timing typical from satellites with longer revisit periods.

Another question is the use of a multi-temporal model over several years: How can a model that was trained with data from previous years be used in subsequent years? Plant development spectra should be comparable from year to year. Plant development features are affected by a number of factors, such as the time of sowing, temperature, radiation and precipitation during the development phase and management factors, such as fertilization, plant protection and soil quality. A plant growth model can estimate the stage of development of the crop and thus help to achieve better results in remote recognition. Even simple models based only on temperature sums may suffice (Weir et al. 1984; Mirschel et al. 2005). Alternatively, plant growth models that map a variety of internal parameters such as in Nendel et al. (2011) can be used. Growth models can also benefit from remote sensing by determining hard-to-control model parameters from the remote sensing data. This mutual model coupling is currently a focal point of research.

In the first time step, the SVM was trained with a constant column X = 1.0 because it was assumed that no previous information was available. In reality, however, information about crop spectra is often available and can be used as code for the X column to optimize the training process.

11.5 Conclusion

Although multispectral and multi-temporal remote sensing data has been extensively used for crop classification and monitoring, the approach presented here shows great potential to achieve very high accuracy with a minimal analyst intervention. Both the basic SVM implementation and the deep learning approach based on a recursive network (LSTM) achieved accuracies close to perfection. LSTM largely abstracts from the modeller's implementation and is able to incorporate the time component independently. Furthermore, we argue that it is only a matter of time before predefined networks are automatically applied to similar problems as in this article.

Acknowledgements This work was supported by the Federal Ministry of Food (BMELV) and Agriculture and the Ministry of Science, Research and Culture (MWKF) of the State of Brandenburg. Furthermore, I would like to thank our former colleague of the ZALF: Bernd Zbell, who did the data sampling and the preparation of the Excel-tables. My special thanks to the Python community which developed the used software and made it as Free and Open Source Software available.

References

Abadi M, Agarwal A, Barham P, Brevdo E, Chen Z, Citro C, Corrado GS, Davis A, Dean J, Devin M, Ghemawat S, Goodfellow IJ, Harp A, Irving G, Isard M, Jia Y, Jozefowicz R, Kaiser L, Kudlur M, Levenberg J, Mané D, Monga R, Moore S, Murray DG, Olah C, Schuster M, Shlens J, Steiner B, Sutskever Il, Talwar K, Tucker PA, Vanhoucke V, Vasudevan V, Viégas F, Vinyals O, Warden P, Wattenberg M, Wicke M, Yu Y, Zheng X (2016) TensorFlow: large-scale machine learning on heterogeneous distributed systems. CoRR, Volume abs/1603.04467. http://arxiv.org/abs/1603.04467

Becerra-García RA, García-Bermúdez RV, Joya-Caparrós G, Fernández-Higuera A, Velázquez-Rodríguez C, Velázquez-Mariño M, Cuevas-Beltrán FR, García-Lagos F, Rodráguez-Labrada R (2017) Data mining process for identification of non-spontaneous saccadic movements in clinical electrooculography. Neurocomputing 250:28–36

Chen T, Guestrin C (2016) XGBoost: a scalable tree boosting system. In: KDD 16 proceedings of the 22nd ACM SIGKDD international conference on knowledge discovery and data mining, San Francisco, California, USA, August 13–17, 2016. ACM DL, Pages 785–794

Collet F, others (2015) Keras. https://keras.io

Gilbertson J (2017) Machine learning for object-based crop classification using multi-temporal Landsat-8 imagery. MSc. thesis, Stellenbosch University, 102 pp

Gautam RS, Singh D, Mittal A, Sajin P (2008) Application of SVM on satellite images to detect hotspots in Jharia coal field region of India. Adv Space Res 41(11):1784–1792

Ghazaryan G, Dubovyk O, Löw F, Lavreniuk M, Kolotii A, Schellberg J, Kussul N (2018) A rule-based approach for crop identification using multitemporal and multisensor phenological metrics. Eur J Remote Sens 51(1):511–524

Griffiths P, Nendel C, Hostert P (2018) Intra-annual reflectance composites from Sentinel-2 and Landsat for national-scale crop and land cover mapping. Remote Sens Environ, January 2019:135–151. (Open access)

Hütt C, Waldhoff G (2018) Multi-data approach for crop classification using multitemporal, dual-polarimetric TerraSAR-X data, and official geodata. Eur J Remote Sens 51:62–74

Ji S, Zhang C, Xu A, Duan Y (2018) 3D convolutional neural networks for crop classification with multi-temporal remote sensing images. Remote Sens 10(75):1–17

Jin X, Song K, Du J, Liu H, Wen Z (2017) Comparison of different satellite bands and vegetation indices for estimation of soil organic matter based on simulated spectral configuration. Agric For Meteorol 244–245:57–71

Mirschel W, Wenkel K-O, Schultz A, Pommerening J, Verch G (2005) Dynamic ontogenesis model for winter rye and winter barley. Eur J Agron 23(2):123–135

Mura M, Bottalico F, Giannetti F, Bertani R, Giannini R, Mancini M, Orlandini S, Travaglini D, Chirici G (2018) Exploiting the capabilities of the Sentinel-2 multi spectral instrument for predicting growing stock volume in forest ecosystems. Int J Appl Earth Obs Geoinf 66:126–134

Nendel C, Berg M, Kersebaum KC, Mirschel M, Specka X, Wegehenkel M, Wenkel KO, Wieland R (2011) The MONICA model: testing predictability for crop growth, soil moisture and nitrogen dynamics. Ecol Model 222(9):1614–1625

Pedregosa F, Varoquaux G, Gramfort A, Michel V, Thirion B, Grisel O, Blondel M, Prettenhofer P, Weiss R, Dubourg V, Vanderplas J, Passos A, Cournapeau D, Brucher M, Perrot M, Duchesnay E (2011) Scikit-learn: machine learning in Python. J Mach Learn Res 12:2825–2830
Qayyum A, Anwar SM, Awais M, Majid M (2017) Medical image retrieval using deep convolutional neural network. Neurocomputing 266:8–20
Sharma A, Liu X, Yang X (2018) Land cover classification from multi-temporal, multi-spectral remotely sensed imagery using patch-based recurrent neural networks. Neural Netw 105:346–355
Vincenzi S, Zucchetta M, Franzoi P, Pellizzato M, Pranovi F, De Leo G, Torricelli P (2011) Application of a Random Forest algorithm to predict spatial distribution of the potential yield of Ruditapes philippinarum in the Venice lagoon, Italy. Ecol Model 222(8):1471–1478
Vuolo F, Neuwirth M, Immitzer M, Cl Atzberger, Ng W-T (2018) How much does multi-temporal Sentinel-2 data improve crop type classification? Int J Appl Earth Obs Geoinf 72:122–130
Weir AH, Bragg PL, Porter JP, Rayner JH (1984) A winter wheat crop simulation model without water or nutrient limitations. J Agric Sci Camb 102:371–382
Wu Y, Yuan M, Dong Sh, Lin L, Liu Y (2018) Remaining useful life estimation of engineered systems using vanilla LSTM neural networks. Neurocomputing 275:167–179
Xia W, Zhu W, Liao B, Chen M, Cai L, Huang L (2018) Novel architecture for long short-term memory used in question classification. Neurocomputing 299:20–31
Zhong L, Hu L, Gong P, Biging G (2016) Automated mapping of soybean and corn using phenology. ISPRS J Photogramm Remote Sens 119:151–164

Chapter 12
Global Evaluation of the Status and Sustainability of Terrestrial Landscapes and Water Bodies

Vasiliy V. Dmitriev, Vitaly V. Terleev, Aleksandr O. Nikonorov, Aleksandr N. Ogurtsov, Aleksey G. Osipov, Yuriy N. Sergeyev, Valeriy P. Kulesh and Irina V. Fedorova

Abstract The foundations of the methodology, methods of integrated assessment of the state and emergent properties (productivity, sustainability, environmental well-being) of complex natural systems are considered. The aim of the research was to summarize the results of the development and testing of models classifications of an integral assessment of the state, sustainability of terrestrial landscapes, water bodies and river systems and then, at the next stage, an integral assessment of their systemic well-being. Terrestrial landscapes, marine and lake systems and water catchment systems were selected as the objects of the study. The subjects of research were theoretical and methodological aspects and methods for constructing integral indicators

---

V. V. Dmitriev · A. N. Ogurtsov · Y. N. Sergeyev · V. P. Kulesh · I. V. Fedorova
St. Petersburg State University, Universitetskaya Naberezhnaya, 7/9, St. Petersburg 199034, Russia
e-mail: v.dmitriev@spbu.ru

A. N. Ogurtsov
e-mail: aogurcov@yandex.ru

Y. N. Sergeyev
e-mail: unsergeev36@gmail.com

V. P. Kulesh
e-mail: vpkulesh@gmail.com

I. V. Fedorova
e-mail: umnichka@mail.ru

V. V. Terleev · A. O. Nikonorov (✉)
Peter the Great St. Petersburg Polytechnic University, Polytechnicheskaya, 29, St. Petersburg 195251, Russia
e-mail: coolhabit@yandex.ru

V. V. Terleev
e-mail: vitaly_terleev@mail.ru

A. G. Osipov
A.F. Mozhaysky's Military-Space Academy, Zhdanovskaya Naberezhnaya, 13, St. Petersburg 197198, Russia
e-mail: osipov-G-K-2005@yandex.ru

W. Mirschel et al. (eds.), *Landscape Modelling and Decision Support*, Innovations in Landscape Research,
https://doi.org/10.1007/978-3-030-37421-1_12

of the state (integral indicator of productivity, integral indicator of environmental quality, etc.), integral indicators of sustainability, integral indicators of well-being (health) of land, lake and river systems. The basics of the methodology for assessing the state, stability and well-being of terrestrial and aquatic geosystems are considered, as well as the integral estimation algorithms; the results of assessment studies: an integrated assessment of the trophic status of water bodies; integral assessment of the resistance of water bodies to changes in the parameters of the natural and anthropogenic regimes; paleoreconstruction of hydroecological conditions on the Laptev Sea shelf based on integral indices; integral assessment of the ecological status and ecological well-being of reservoirs, watercourses, water catchment systems; integral assessment of the sustainability of terrestrial landscapes; integral assessment of the stability of a sociological system with a hypothetical change in the quality of the environment on the example of northern regions of Russia. Research was focused on the development of multi-criteria and multilevel classifications for the study of the emergent properties of complex systems in nature and society and the construction of integral indicators based on Analysis and Synthesis of Indicators for Information Deficiency (ASIID) and Aggregated Preference Indices System (APIS) methods and models.

Keywords Landscape · Modelling · Integral assessment · Anthropogenic impact · Sustainability · Ecological well-being

12.1 Introduction

Authors' research focuses on the development of methods for quantifying the state of complex systems in nature and society and their integrative, emergent properties, accompanied by the construction of new functional units of systems, which are considered integral indicators of several levels of generalization of information and indicators characterizing the latest level of generalization. These indicators reflect the integrity of individual subsystems and the system as a whole and play role of the basis of their systematics. They identify the class of the properties being assessed and compare the state of the systems in space and time, identify interconnection and interaction effects that are not additively with respect to local intrasystem effects considered at the component level (Dmitriev and Ogurtsov 2017).

The purpose of this research was to summarize the particular results of the development and testing of model classifications of an integral assessment of the state, sustainability of terrestrial landscapes, water bodies and river systems, and then, at the next stage, an integral assessment of their well-being. If we are talking about the state and stability of ecosystems, we first study their state (chemical and biological composition and physical properties), then the stability of ecosystems to changes in the parameters of the natural (potential resistance) and anthropogenic regimes. Then their ecological well-being or health should be investigated, in which the first two stages mentioned above are partially taken into account. If we are talking about geosystems or socio-systems, then social conditions and factors affecting the quality of the environment and the quality of life of the population, and then also the economic conditions, circumstances, factors determining the quality of the environment and the quality of life of the population, and their health are added to the listed emphasis (private and public).

Terrestrial landscapes, lake systems and water catchment systems were selected as the objects of the study. At the first stage of research, the watercourse is represented by a homogeneous transit-type aquatic ecosystem. The catchment at this stage is a homogeneous landscape, the stability and well-being of which are gradually evaluated on an integral basis. At subsequent stages of the work, the catchment area can be divided into characteristic landscapes or groups of landscapes, and the watercourse can be divided into characteristic sections of the river's transit system.

The subjects of research are theoretical and methodological aspects and methods for constructing integral indicators of the state (integral indicator of productivity, integral indicator of environmental quality, etc.), integral indicators of sustainability (ICS), integral indicators of well-being (health) of land, lake and river systems. In the assessment of the state, the following components are included: 1—system productivity (production potential, trophic status of the reservoir); 2—environmental quality and toxic pollution of geosystems; water (and bottom sediments).

The following components are included in the sustainability assessment: 1—potential resistance (resistance to changes in the parameters of the natural mode); 2—resistance to "malignant" increase in productivity (in water bodies—anthropogenic eutrophication); 3—resistance to environmental quality change (toxic pollution). Possible combinations of the types are 1 + 2 or 1 + 3. The combination of 1 + 2 + 3 is hardly advisable, although the ideologies of trophosaprobity and trophotoxicity make it possible to combine all the approaches together.

The assessment of ecological well-being (EWB) includes: 1—assessment of the ability to produce organic matter in accordance with the natural development phase of the ecosystem or geosystem; 2—assessment of the quality of the environment (the cleaner, the better); 3—assessment of species diversity in the system (the higher, the better); 4—assessment of the stability of the system (potential resistance plus an assessment of resistance to probable impacts); 5—assessment of possible acidification (and resistance to acidification); 6—assessment of the rate of contamination of the system (the higher, the lower the well-being, the system is sick); assessment of the self-cleaning rate of the system (the higher, the better); assessment of the ability to maintain these factors for a long time without moving to another (other) classes of well-being. All components at the second level form a consolidated assessment of EWB.

12.2 Materials and Methods

The theoretical and methodological foundations of the study of the stability of geosystems are outlined in a rather large list of publications. Here we indicate only some (Alimov et al. 1999; Glazovskaya 1997; Grodzinsky 1987; Aleksandrova et al. 2000;

Dmitriev 2009, 2010, 2017; Dmitriev et al. 2016, 2018a, b; Snakin et al. 1992). They discussed terminology, methods and research results. An analysis of modern approaches to assessing the stability of geosystems allowed to identify and summarize some important features of research. They were given below.

1. Some authors deny the possibility of a quantitative integral assessment of stability on the grounds that this property of the system cannot be measured, but can only be estimated indirectly, for example, in points. At the same time, the authors refuse to work with full-scale information and rating scales based on it, translating the values of the selected indicators into points and then summarize the points to obtain a total indirect assessment of sustainability. A more complex scheme is possible: indices—digits—points (Snakin et al. 1992). Points can be assigned to priorities (weights) with which they participate in the convolution. Rationing of points for conversion to a dimensionless form can also be performed (Bancheva and Alekseeva 2017). In the process of work, as a rule, the type of connection (direct–inverse, linear–non-linear) and the weight of parameters are not taken into account.

Indirectly, the weight of parameters can be taken into account when moving from indices to digits and from them to points, when, for example, a certain number of points are assigned to a certain number of digits (Snakin et al. 1992). This ideology has generated a large number of copyright points and point-index methods for assessing sustainability (Snakin et al. 1992; Fedorov 2008; Abalakov et al. 2014; Bancheva and Alekseeva 2017). The most common, widely known methods are also based on a point and point-index assessment system. For example, for a comprehensive assessment of the geochemical stability of soil cover to pollution by petroleum hydrocarbons, at least 10 characteristics are used, reflecting the processes of destruction (biochemical oxidation), accumulation, removal and dispersion of petroleum hydrocarbons. Based on the ideas and principles of the scoring system, research has been conducted by the staff of the Institute for the Problems of Northern Development of the Siberian Branch of the Russian Academy of Sciences, VNIIPrirody, Moscow State University, etc. (Snakin et al. 1992; Abalakov et al. 2014; Bancheva and Alekseeva 2017). Abroad, within the framework of the concept of "Nordicity," the scoring system proposed by the Canadian geographer L-E. Amelin was widely known (Hamelin 2008; Totonova 2016).

Let us dwell on the recently debated position, which is important for the further stages of work, which is connected with checking the adequacy of the obtained results of integral estimation (Dmitriev and Ogurtsov 2017). With the traditional approach—the point and point-index assessment of the emergent property of a natural object (for example, sustainability), the researchers, by summing up the indices and moving on to the discharges, and then to the scores, end up with a class (and a subclass) of sustainability. The assignment to a certain class of sustainability is the final stage of work and makes it possible to identify more or less stable landscapes or water bodies and analyse the spatial and temporal changes in the sustainability of water bodies and streams (Snakin et al. 1992). Such an approach, according to the authors, is the result of an indirect assessment of sustainability, since any emergent property cannot in principle be measured. At the same time, an attempt to switch to other methods initiates the above-mentioned question about the reliability of certain obtained results. Therefore, the authors prefer not to use the term "model" or "model classification" in relation to the totality of parameters and rating scales for them by state (stability) classes combined into a single (multilevel) assessment system.

It is important to note that the digital form of expression of an emergent property through an integral indicator should not confuse the researcher, since it not only helps to attribute the integral stability indicator (ICS) to a certain class indicating the accuracy of the calculations performed. On the other hand, it does not contradict the original thesis that "the emergence of a complex system cannot be measured." It does not make sense to compare the ICS of two objects by the values of integral indicators themselves, especially if they were built using a different set of characteristics (for example, different classification models) or priorities (weights) of individual indicators were set differently. The result of the assessment still remains the assignment of the object (system) to a certain class of stability. In this case, an assessment of the reliability of the results is provided. In this case, the idea of emphasizing only regional specifics in the calculations of integral indicators turns out to be not fully justified, retaining as a base, a certain set of basic characteristics and adding parameters reflecting regional specifics.

2. The absence of the author's understanding and definition of sustainability or the absolutisation of only one of the possible definitions and the rejection of other definitions or evaluative means in many works. As a result, it often turns out that the author does not evaluate the stability of the system, but, for example, the sensitivity of the organisms inhabiting the eco- or geosystem to a specific type of impact on them (Blinovskaya et al. 2012). Further, for the habitats of the most sensitive organisms by means of GIS, maps are constructed that are called the most vulnerable areas. In this case, the researcher must substantiate that the most sensitive organisms inhabiting a particular habitat characterize it as a vulnerable geosystem.
3. The authors estimate not the integrative (emergent, complex) property of the system as a whole, but the variability of the composition of the system (violation of the emergence principle of a complex system). At the same time, if the variability of the components is large, then a conclusion is drawn about the low stability of the system.
4. Analysis of the toxicological vulnerability of geosystems is carried out on the basis of constructing a system state vector based on the aggregate of the largest possible list of characteristics available in the author's database. This vector is considered as a kind of independent integral environmental characteristic of the estimated area (water area) not related to "sustainability" or it is known identifies with the stability of the system as a whole.
5. The frequent presence of several types of sustainability simultaneously and their simultaneous usage to obtain a final assessment of sustainability in the assessment approaches. As a result, in accordance with the methodical platform of the authors, some parameters characterize the ability of the system to change

smoothly with time, characterizing the adaptation of the system to new conditions (adaptation type of stability, type 1). Other parameters are considered from the point of view of their ability after short-term and significant changes due to the impact on the system, quickly enough to return to its former state, characterizing the system's ability to restore its properties and parameter mode after a temporary loss of (regenerative type of stability, type 2) (Dmitriev and Ogurtsov 2012, 2013, 2014, 2017). Apparently, this approach is possible with the following caveat: first, the researcher works with a system capable of restoring its properties and mode parameters. The impact on such a system does not lead to a strong transformation. This occurs as long as the system is capable of self-purification and restoration of its properties. After some critical threshold has been crossed, the system begins to gradually adapt to new conditions, changing its chemical and biological composition and properties. From this point on, the acquisition by the system of new properties will be unidirectional (in this case, it is assumed that the effect on the system is gradually increasing). And then, the more productive the system is, the more resilient it will be to a change in this property, or the higher the pollution level in it, the more resilient it will be to further changes in the quality of the environment in it. Here, high sustainability will no longer be a property that characterizes a healthy system, and will no longer be able to act as a synonym for a healthy system, as has sometimes been suggested without a caveat (Semenchenko and Razlutsky 2010).

6. Slow development of axiometry for assessing sustainability factors. Rarely, the authors offer informative parameters or indices of sustainability, for which, grading scales have been developed. Frequently used point scales, which reduces their value as the basis of "environmental qualimetry" or "environmental axiometry." In other cases, this is entrusted to a computer program that develops an evaluation scale for classes, taking into account the assignment of group priorities (second level of convolution). Such an approach in modern GIS is based on the Jenks' Natural Breaks algorithm.
7. The authors' research does not emphasize differences in approaches to assessing the sustainability of a natural object, the ecological status of the system, and its ecological well-being. There is still often the opinion that the most stable system is the most prosperous and most healthy.

As a methodological basis for integral assessment in solving a wide range of tasks in the framework of research and work on the problem of integrated assessment of environmental objects, authors use ideas and principles of the ASIID methodology (analysis and synthesis of indicators for information deficiency) (Hovanov 1996), on which model algorithms are built by GeoExpert programs (Aleksandrova et al. 2000). The modern development of "ASIID" is "APIS" (Aggregated Preference Indices System)—an aggregate system of preference indices based on the "ASIID Methodology" (Hovanov et al. 2009). For reconnaissance estimates, the usual method of summary indicators (SIM) is used. The research methodology is described in more detail in Dmitriev et al. (2019).

The methodological procedure of integrated assessment consists of a specific sequence of steps (steps) (Hovanov 1996; Aleksandrova et al. 2000; Dmitriev and Ogurtsov 2017). At the first stage of work, the structuring of all available information is carried out and the matrix of initial characteristics of the type is formed

$$\begin{vmatrix} x_1^{(1)}, \ldots x_i^{(1)}, \ldots x_m^{(1)} \\ x_1^{(j)}, \ldots x_i^{(j)}, \ldots x_m^{(j)} \\ x_1^{(k)}, \ldots x_i^{(k)}, \ldots x_m^{(k)} \end{vmatrix}$$

Here $x_i^{(j)}$—value of the i-th characteristic (i = 1, 2, 3, 4 … m) of qualities or properties for the j-th object.

The researcher develops measurement scales for each of the characteristics with an indication of the limit values ($x_{\min}$, $x_{\max}$) and indicates the methods for measuring the quality under investigation for each characteristic (criterion). At the same stage, if necessary, the entire set of initial characteristics is divided into subsets (groups of characteristics, which are often called blocks in publications), and then a hierarchical scheme of sequential synthesis of integral indicators of various levels is organized.

At the next step, a vector of individual indicators (indices) $q_1, \ldots, q_m$ is built and a matrix of the form is formed

$$\begin{matrix} q_1^{(1)}, \ldots q_i^{(1)}, \ldots q_m^{(1)} \\ q_1^{(j)}, \ldots q_i^{(j)}, \ldots q_m^{(j)} \\ q_1^{(k)}, \ldots q_i^{(k)}, \ldots q_m^{(k)} \end{matrix}$$

Here $q_i^{(j)}$—the value of the i-th normalized index (i = 1, 2, 3, 4 … m) of qualities or properties for the j-th object (j =1 … k).

Each such indicator, which will be called a separate indicator, since it characterizes only a separate aspect of the quality of an object and is a function of the initial characteristics in the form: $q_i = q_i(x_i) = q_i(\varphi(x_i))$, $0 \le q_i \le 1$, $i = 1, \ldots, m$, where $q_i(x_i)$—normalizing functions.

There are various options for rationing the original characteristics. In this case, depending on the nature of the behaviour, the increasing or decreasing functions are used as the normalizing function:

$$q_i = q_i(x_i)\begin{cases} = 0, & x_i \le x_{\min} \\ = ((x_i - x_{\min})/(x_{\max} - x_{\min}))^p, & x_{\min} \le x_i \le x_{\max} \\ = 1, & x_i \ge x_{\max} \end{cases} \tag{12.1}$$

$$q_i = q_i(x_i)\begin{cases} = 1, & x_i \le x_{\min} \\ = ((x_{\max} - x_i)/(x_{\max} - x_{\min}))^p, & x_{\min} \le x_i \le x_{\max} \\ = 0, & x_i \ge x_{\max} \end{cases} \tag{12.2}$$

Parameter p indicates the nature of the convexity of the normalizing function. Experience shows that it is often enough to limit to linear dependence during normalization ($p = 1$). The set of values of individual indicators q_i is a multi-criteria assessment of the quality (property) of the system (object) under investigation. It should be noted that the "incomparability problem" of multi-criteria quality assessments, to solve which at the next stage of work, the choice-construction of a synthesizing function is carried out that compares the value of an integral indicator characterizing the quality (property) being studied as a whole. As such a model function, authors use the function of the form $Q = Q(q; w) = \sum_{i=1}^{n} q_i w_i$, with the help of which the individual quality indicators are rolled up into a single integral indicator (index) (Hovanov 1996).

It should be noted that when generating weights, an algorithm is used to formalize non-numeric, inaccurate and incomplete information (so-called non-information), based on the method of sequential enumeration of the monotonic paths specified on the integer lattice and mutually uniquely defining the corresponding weight vector coefficients (Hovanov 1996).

As a result, the mathematical expectation of a randomized weighting coefficient (RWC) is taken as the numerical estimate of a specific weight coefficient vector. The accuracy of this estimate is characterized by the standard deviation of RWC. The use of the RWC vector in the convolution function of individual indicators leads to randomized integral indicators with similar statistical characteristics.

In this case, the synthesizing function depends on the vector of weight coefficients $w = (w_1, \ldots, w_n)$, determining the significance of individual indicators for integrated assessment. The range of index changes is in the range from 0 to 1. The index is constructed so that the maximum values (close to unity) indicate a high degree of quality, and close to zero, a low degree (it can be the other way around).

Since there is an uncertainty in the choice of a single vector of weights, authors often use the Bayesian randomization model in the calculations, which allows to move from the uncertainty of the choice of weights to their random selection (Hovanov 1996; Hovanov et al. 2009).

Since the objects being evaluated are of geospatial nature, an important methodological element of research is the need for multidimensional visualization of the results obtained using GIS and GIS technologies. At the same time, GIS serves as a tool for spatial modelling and analysis of integral estimates.

Consider some features of the methodological characteristics of evaluation studies.

At the **first stage** of work, the analysis, structuring of information and the formation of a matrix of initial characteristics for a key object are carried out. The estimated scales of measurements for each of the characteristics and their extreme values are determined. At the same stage classes, levels, left and right borders of each parameter for each class are introduced. It also develops a hierarchical scheme for constructing integral indicators for all levels of convolution. In our opinion, the simultaneous use of multi-criteria and several levels of convolution vividly illustrate the meaning of the integral assessment of the emergent property of a complex system. The content of the stage is revealed in Fig. 12.1.

At the **second stage**, the procedure of rationing of initial indicators of rating scales is performed and a matrix of normalized values of indicators is formed. Depending on the nature of the behaviour, the increasing or decreasing functions are used as the normalizing function.

Authors experience show that during normalization at the first stage, we can limit ourselves to a linear direct or inverse relationship between the selected parameter and the estimated property. The set of values of normalized indicators is a multi-criteria assessment of the estimated property of the key object under investigation (Dmitriev and Osipov 2017). As a model synthesizing function, we use the function of the form $Q = Q(q; w) = \sum_{i=1}^{n} q_i w_i$, with the help of which the construction of integral

(a)

Critical level separating classes

Criteria, signs	Classes of state systems or their properties									
	I		II		III		IV		V	
Criterion 1	a_{11}	b_{11}	a_{21}	b_{21}	a_{31}	b_{31}	a_{41}	b_{41}	a_{51}	b_{51}
Criterion 2	a_{12}	b_{12}	a_{22}	b_{22}	a_{32}	b_{32}	a_{42}	b_{42}	a_{52}	b_{52}
…	…	…	…	…	…	…	…	…	…	…
Criterion n	a_{1n}	b_{1n}	a_{2n}	b_{2n}	a_{3n}	b_{3n}	a_{4n}	b_{4n}	a_{5n}	b_{5n}
The integral criterion Q	Q_1 - Q_2		Q_3 - Q_4		Q_5 - Q_6		Q_7 - Q_8		Q_9 - Q_{10}	

(b)

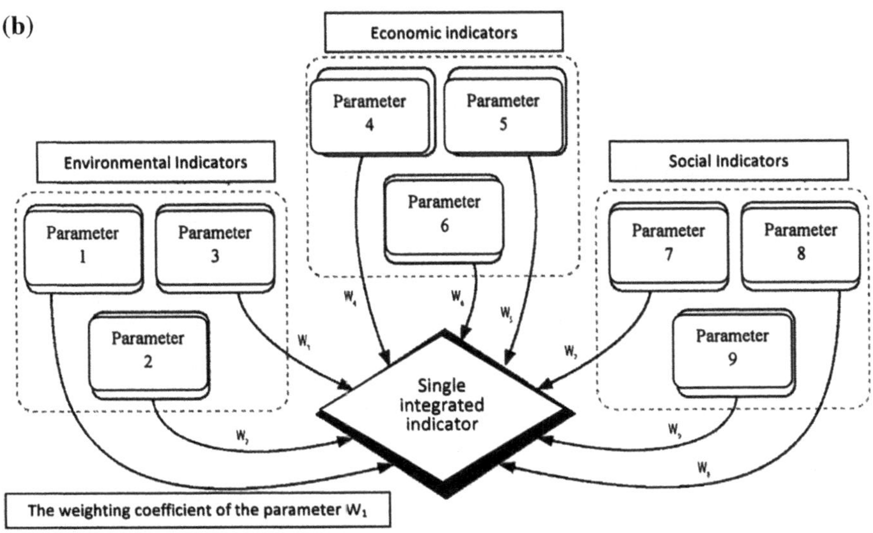

Fig. 12.1 **a** Classes, boundaries, rating scales (left and right boundaries of classes), integral indicator scale; **b** illustration of a single convolution level when building an integral indicator for economic indicators (Lior 2015)

indicators at individual levels and a composite indicator of stability at the last level of convolution (ICS) is performed. The value of the ICS depends on the weighting factors $w = (w_1, \ldots, w_n)$, determining the significance of normalized indicators for integrated assessment. The range of index changes is in the range from 0 to 1.

12.3 Results and Discussions

12.3.1 Integral Assessment of Production Properties of Reservoirs for the Lakes of the North-Western Ladoga Region

Integral assessment of production properties of reservoirs for the lakes of the north-western Ladoga region (Lake Suuri) (Fig. 12.2) in 2017 (Trushevsky et al. 2018; Dmitriev 2017) showed that this reservoir in the component evaluation had the following ratio of signs: oligo : meso : eutrophy = 4 : 3 : 2. By the value of the integral indicator (ITI = 0.28) the reservoir fell into the left boundary of the mesotrophic class with the width of the interval class 0.20–0.46. In 2016, by the largest integral trophic index (ITI), calculated for 2014–2016 the trophic status of the lake according to the average value of the integral index was assigned to the middle of the mesotrophic water class (ITI = 0.29) with class interval width from 0.14 to 0.34. In 2010–2013 by the ITI from 0.26 to 0.31, calculated by four parameters, the trophic status of the lake was defined as mesotrophic with a class interval width from 0.14 to 0.34.

Integral assessment of the trophic status of marine Arctic areas was carried out for 2005–2012. ITI of the Laptev Sea shelf area was calculated using six parameters for the period 2005–2006. By the value of ITI = 0.16 the water area was classified as oligotrophic water bodies (the interval width for the class is 0.00–0.27). The integral assessment of the trophic status for the three regions of the Kara Sea, performed on five parameters for 2007–2012, also made it possible to classify the studied areas as an oligotrophic class.

The integral assessment of the trophicity of the Northern, Central, Southern regions of the Neva Bay of the eastern part of the Gulf of Finland was made on the basis of field observations from 2000 to 2010 using four parameters. On the basis of Rosgidromet data, the average annual values of trophic indices were calculated for the Northern (by 7 stations), Central (by 8 stations) and Southern (by 7 stations) districts.

The characteristic features of changes in the integral indicators of trophicity in the areas are revealed. For the Northern District, ITI varied from 0.36 (2000) to 0.38 (2010), thus, a tendency of insignificant growth of ITI for 10 years (6.5%) within the class of mesotrophic waters was detected (the class boundaries—from 0.19 to 0.60). All ITI values are from the middle of the class. For the Central District, ITI varied from 0.36 (2000) to 0.40 (2010). The tendency of a slight increase in ITI for the period 2000–2005 and a slight decrease in ITI by 2010 was also revealed.

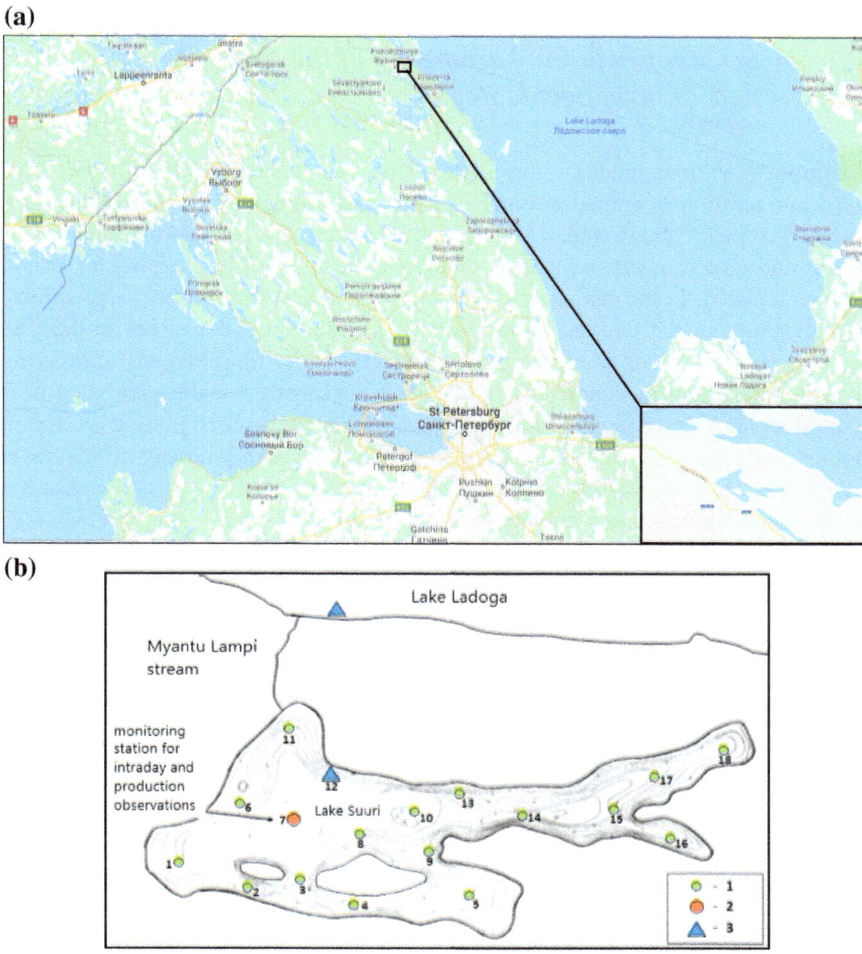

Fig. 12.2 Research area (**a**) and location of sampling stations in 2016 (**b**) [1—monitoring stations for chemical composition and hydrophysical characteristics of water (№1–18); 2—station for sampling of zooplankton, intraday and production observations (№7); 3—macrozoobenthos sampling stations (Dmitriev et al. 2017a, b)]

The obtained values are also within the mesotrophic water class. For the Southern District, ITI varied from 0.45 (2000) to 0.46 (2010). The tendency of insignificant growth of ITI for the period 2000–2005 and decrease of ITI by 2010 is revealed; all ITI values are within the mesotrophic water class.

12.3.2 *Integral Assessment of the Stability of Water Bodies to Changes in the Parameters of the Natural and Anthropogenic Regimes*

Example 1 Various methods and models have been tested for water bodies in the north and north-west of the Russian Federation according to observations from 2003 to 2004 (Dmitriev and Primak 2009). Adaptive resistance to changes in the parameters of the modes was considered. Various integral indicators of sustainability for 13 lakes were calculated. In the calculations, the high adaptation stability was characterized by the values of the integral indicators (ICS) close to 1.0, low-close to 0.0. Data were used on the group of physiographic characteristics: morphometry, climate and hydrological regime to characterize the resistance to changes in the parameters of the natural regime (Ye) (potential stability). Data on changes in the trophic status of water bodies (anthropogenic eutrophication) (Ue) or on changes in water quality (Wc) were used to characterize resistance to changes in the anthropogenic regime. At the second level of convolution, the combined effect of Ye + Ue or Ye + Wc was estimated. The most vulnerable and cleanest water bodies turned out to be the most vulnerable to changes in the parameters of the natural (six parameters) and anthropogenic (four parameters) modes with different assignments of assessment priorities: Topozero and Pyaozero—IV class of sustainability. The most stable of the considered lakes were Lake Ilmen and Lake Chudsko-Pskov, III class of sustainability. The lakes of the European north are characterized by a combination of criteria as "moderately stable" (class III) and "steady below average" (class IV). An integral assessment of the stability of individual regions (six districts) of Lake Ladoga to changes in the parameters of the natural and anthropogenic regimes showed that not a single region, nor the lake as a whole, has the maximum resistance to changes in the parameters of the natural and anthropogenic regimes. The northern coastal region of the lake (class IV) was the most vulnerable. The most sustainable was the southern coastal region (class III). The lake as a whole is estimated as "moderately stable" class III (left border). It is concluded that with this approach, the high stability of the reservoir (region) apriori does not determine its ecological well-being or the high quality of the abiotic environment and bio-resources. High adaptation stability as a whole can be caused by various groups of parameters of the natural regime (for example, the significant size of a reservoir) and the anthropogenic regime characterizing water quality or the degree of anthropogenic eutrophication of a reservoir. The study of adaptive sustainability is appropriate for long periods of development of water bodies, characterized by a gradual loss of their properties and a change in the parameters of the regimes, and a gradual transition to other classes of state and stability. When constructing the ICS of watercourses, models of classification of the second type of assessment of stability (regeneration) were used.

Example 2 Evaluation of the sustainability of the Karelian Ladoga lake. The method of aggregates (MA) was used to assess sustainability on the example of the analysis of observations 2015–2017. It is shown that an increase in the water level in the lake in 2016 led to a decrease in primary production in water bodies and an increase in the vulnerability of lakes to changes in the parameters of the natural regime and eutrophication. The analysis of the ICS revealed that an increase in the water level in the lake lowered its resistance to changes in the parameters of the natural regime and water quality (Dmitriev et al. 2017b).

Example 3 Based on the initial literature data, a model classification of the integral stability assessment for lakes-coolers of NPPs and lakes-analogues, not subject to the influence of NPPs (21 evaluation criteria) (Lubentsova et al. 2017) was constructed. The value of the integral indicator of sustainability (ICS) close to 1.0 corresponded to the maximum value of sustainability, value of ICS close to 0.0—the minimum value. The calculations were performed for the models: "M1"—potential resistance (ICS1), "M2"—resistance to changes in the natural regime and water quality (ICS1 + ICS2), "M3"—resistance to changes in the natural regime and anthropogenic eutrophication (ICS1 + ICS3). As a result, it was obtained that the cooler lakes of the key area according to "M1" belong to the III class of stability ("average") with values of 0.438 and 0.453, respectively (the interval of change of the ICS for class III is 0.645–0.355). Reservoirs analogues are assigned to the IV class of resistance with the values of 0.240 and 0.245, respectively (the interval of the ICS for class IV is 0.355–0.161). The analysis of stability according to ICS showed that the reservoirs-coolers of nuclear power plants according to "M1" are more resistant to changes in the parameters of the natural regime than their analogues.

According to "M2," the cooling pond is assigned to the left border of the III class of stability (0.500), another pond is assigned to the middle of the III class (0.453), the water bodies-analogues to the middle of the IV class with values of 0.240 and 0.245, respectively. According to "M3," the cooling pond is assigned to the III–IV class boundary with values of 0.271, another water body is assigned to the right boundary of the III class (0.283). Reservoirs analogues are attributed to the right boundary of class III (0.306) and to the right boundary of class IV (0.151).

In general, reservoirs exposed to nuclear power plants are more vulnerable to changes in the natural regime and eutrophication than changes in the natural regime and water quality. For analogous reservoirs that are not affected by nuclear power plants, on the contrary, reservoirs are more vulnerable to changes in the natural regime and water quality than to changes in the natural regime and eutrophication.

12.3.3 Paleoreconstruction of Hydroecological Conditions on the Laptev Sea Shelf Based on Integral Indices

The use of the ASIID method (analysis and synthesis of indicators for information deficiency) in palynology allows formalizing the uncertainty of information and numerically presenting it through the statistical characteristics of the integral indicator (mathematical expectation and standard deviation). The method allows sufficiently reliably to distinguish the main climatic periods of the Holocene numerically and to characterize the level of variability of conditions within these periods.

The methodological basis for the construction of the integral index was used for paleoreconstruction of advection of Atlantic waters and the formation of modern hydroecological conditions on the Laptev Sea shelf. The baseline data for the calculation of integral indices of reconstructions of paleohydrological settings were data on the content of the Laptev aquatic palynomorphs (AP) in the shelf sediments of the sea, which include cysts of marine dinoflagellate species and freshwater green algae, as well as other organic residues of aquatic microorganisms. Cysts of 10 marine dinoflagellate species were considered as assessment indicators.

Calculations by the proposed method showed that with any method of setting the degree of uncertainty, the chronology of the paleo-events in the Laptev Sea is always quite clearly identified. Option 1 (row 1) reflects the reconstruction of advection of Atlantic water masses, while option 2 reflects the reconstruction of the formation of modern hydroecological conditions.

Analysis of Fig. 12.3 (row 1) showed that at the turn of 9 thousand calendar years ago there was a significant increase in advection of Atlantic waters with short-period changes in intensity over 1600–1800 years. This time interval (9–7.2 thousand calendar years ago) is characterized by relatively large values of the integral index (more than 0.2–0.4). Starting with 7.2 thousand years ago there has been a steady decline in advection of Atlantic waters, which has not changed significantly over the next 6,000 years. With 1.2 thousand calendar years ago to date, no apparent advection processes of the Atlantic waters have been noted. The amplitude of the index does not exceed 0.05.

Option 2 (row 2) reflects the reconstruction of the formation of modern hydroecological conditions. Analysis of series 2 (Fig. 12.3) showed that this process began at the turn of 6.5–6.3 thousand calendar years ago. A number of researchers (Klyuvitkina and Bauch 2006) believe that the formation of modern hydroecological conditions began earlier (7.4 thousand years ago).

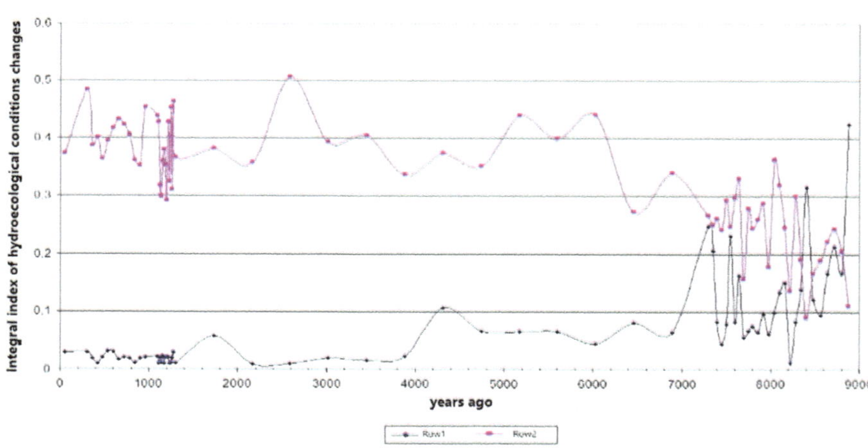

Fig. 12.3 Paleoreconstruction of hydroecological conditions in the Laptev Sea inland shelf

From Fig. 12.3 it follows that from 6.0 thousand years ago to the present, hydroecological conditions do not undergo significant changes, as evidenced by minor fluctuations in the integral index around 0.4 (steady state).

The work performed is the first detailed study of the application of the method of integral indicators, based on the principles of "ASIID" in marine paleogeographic studies. The results obtained complement and clarify the available ideas about the paleogeography of the Laptev Sea and indicate the possibility and feasibility of using model indices to analyse paleoreconstructions of advection intensity of the Atlantic water masses.

12.3.4 Integral Assessment of the Ecological Status (ES) and Ecological Well-Being (EWB) of Reservoirs, Watercourses, Water Catchment Systems

Example 4 Integral assessment of the environmental well-being of marine areas. For the integral assessment of the ecological well-being (EWB) of the key area of the Barents Sea, authors took into account: 1—benthic biomass, g/m^2; 2—species diversity of benthos by Shannon index (H); 3—degree of environmental stress experienced by benthos on the index (DE); 4—water quality according to WPI; 5—pollution of bottom sediments according to Zc pollution index; 6—degree of geomorphological risk (scoring). EWB assessment was carried out for the key area of the Barents Sea at the oil terminal design site in the area of Svyatoy Nos (Vostochniy) Cape for two priority options (weights): A—priority equality and B—priority inequality: $w2 \geq w1 = w3 \geq w5 > w4 = w6$. The results of the assessment of EWB and zoning are shown on the Fig. 12.4.

Example 5 The concepts of "ecological status" (ES) of a reservoir (water body) and ecological well-being of a water body (EWB) are introduced from the standpoint of bio- (eco) centrism and anthropocentrism. The bio- (eco) centrist definition emphasizes the use of a water body for the life of aquatic organisms or for the preservation of the ecosystem as a whole, the anthropocentric approach is the ability of the system to perform socioeconomic functions without disrupting life support functions (medium and resource reproduction). When considering ES, we took into account the trophic status of the reservoir or the ability to produce organic matter by primary producers, water quality from the standpoint of bio- (eco) centrism, system sustainability. Thus, it was assumed that one emergent property may define the other or influence it. Recently, mainly in the translation literature or among non-specialists, the definition of ES based on the term "ecological water quality" has become more frequent. There is even the term "good environmental quality." It remains to be regretted that in this form in the Russian Federation biocentrism (ecocentrism) is emphasized in assessing water quality and the state of ecosystems (Semenchenko and Razlutsky 2010). The emphasis on the methods of biological water quality control adopted in Europe used

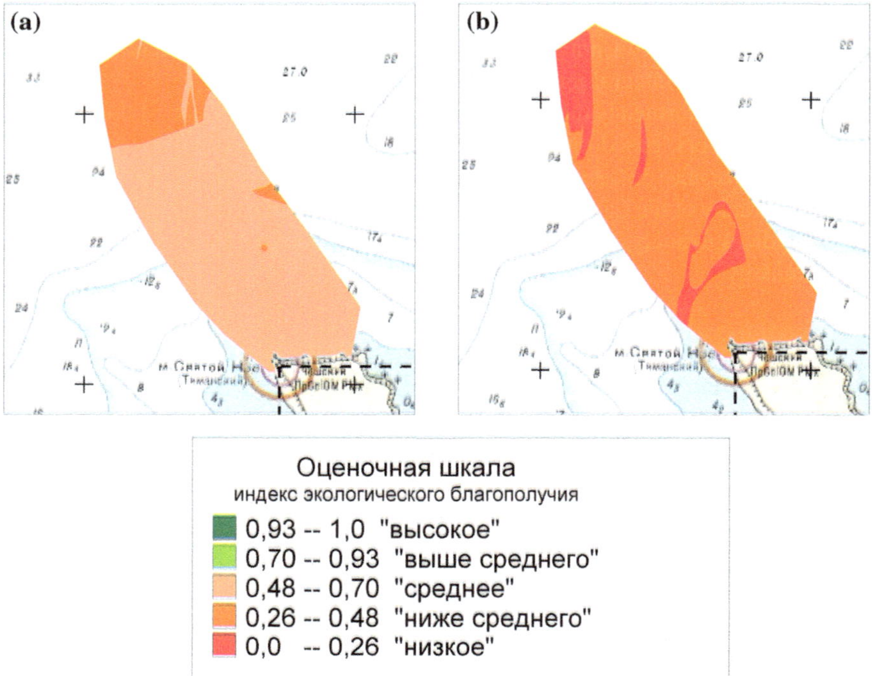

Fig. 12.4 Results of an integrated assessment of the environmental well-being of the key region of the Barents Sea (Vasilyev et al. 2009)

in these studies has been actively developed in recent years and in the Russian Federation; their simultaneous use with the methods of hydrochemical quality analysis gives results that are more reliable (Vasilyev et al. 2011a).

The classification models for the integral assessment of river system EWBs are developed. The evaluation was carried out for: 1—hypothetical scenarios that take into account the scale of the watercourse from the standpoint of the equilibrium nature of setting parameters, levels and non-equilibrium weights (priorities); 2—key watercourses based on the developed models classifications. The group of characteristics that determine potential stability includes parameters that take into account physical and geographical features, the nature of the low-water phase, the hydrological regime, the size and water content of the watercourse. Authors include in the list of characteristics the sign "change in water quality" to take into account the influence of the anthropogenic component on sustainability. At the same time, authors recommend to evaluate the quality of water on their own, based on the magnitude of hydrochemical and hydrobiological water quality indices or on a multi-criteria basis, using estimated index scales or a summary indicator of water quality for inclusion in the sustainability model classification (Vasilyev et al. 2011b).

Model 1 included three options for assessing the watercourse EWB (72, 50, and 28 parameters) related to 5 blocks. Model 2 included a river subsystem (7 blocks, 30 parameters) and a catchment subsystem (5 blocks, 33 parameters). Using the example of the equilibrium assignment of priorities for assessment, we consider hypothetical scenarios for setting the scale of the river system and the scale of impact on it: for a watercourse and for a catchment. It is shown that the developed classification can be used to assess the EWB of multi-scale objects and magnitude of impact. Accounting for disequilibrium for the second level of convolution: at the first level of indicators convolution, an equilibrium task of characteristics inside the blocks was used. At the second convolution level, 3 options for setting priorities (weights) are implemented: the equilibrium task (1) anthropocentrism (2) biocentrism (3).

For the key object (the Msta river), according to the values of the summary indicators obtained as a result of the equilibrium setting of parameters (1), the river EWB falls into the left border of class II—"EWB is above average" (the value of the composite indicator is 0.24), and the catchment basin EWB falls into the right border of class I is "Maximum EWB" (0.20). In the variant "2"—the river EWB falls into the left border of class II (0.23), the drainage collector EWB—into the right border of class I (0.18). In the variant "3"—the river EWB and river bank EWB fall into the middle of class II (0.25) and (0.33), respectively.

When developing the fundamentals of the technology for assessing the impact of anthropogenic load on the change in the EWB, a scenario of a hypothetical increase in the anthropogenic load on the watercourse and catchment area was implemented with 2 groups of parameters for each subsystem (watercourse: 2 out of 7 groups; catchment 2 out of 5 groups). With a twofold increase in the anthropogenic load on the initial parameters, the value of the aggregate indicator of the Msta River EWB goes from the class II left border (0.25) to its right border (0.32). This indicates a decrease in EWB by almost one class. The EWB class of the catchment basin, with the same load, remains unchanged, and the value of the aggregate indicator varies slightly (Dmitriev et al. 2018a).

12.3.5 Integral Assessment of the Sustainability of Terrestrial Landscapes

Example 6 The territory of the Vishersky Reserve in the Perm Territory with a reconnaissance change and the characteristic value of the sustainability assessment parameters were chosen as the first key area. The reconnaissance change of parameters in this example was set by the boundaries of the classes (from minimum to maximum within a class). As a result of the calculations, the range of changes in the integral sustainability indicator was obtained based on the use of reconnaissance data for each parameter and the characteristic value of the integral indicator of landscape sustainability (IILS), according to which the district's landscapes were classified as stability class III (right border). The study showed that adding a single parameter to the model or removing one parameter from it with their equilibrium accounting practically does not affect the final result. For example, removing the ninth parameter from the model—the climate biological efficiency index (BEI) gave a typical IILS value of 0.531, which gives a difference with the result considered above within 4% with the width of the interval of the IILS rating scale for class 0.391–0.586. At the same time, giving 2 times more weight to this parameter compared to others gives a typical IILS value equal to 0.630, which indicates that IILS approaches closely to the boundary value between the third and fourth classes (0.654 is the right third border of the third class for the rating scale taking into account the non-equilibrium nature of setting priorities).

IILS calculation was performed for the landscapes of the Baikal region in the Irkutsk region. The range of variation of the integral stability index 0.486–0.677 was obtained, which made it possible to assign landscapes to the III–IV stability classes.

The third district was selected on the territory of the Republic of Buryatia. The range of variation of the integral indicator of landscape sustainability was 0.494–0.703, which made it possible to assign them also to the III–V resistance classes.

The fourth key district is the territory of the Shushensky Bor National Park in the Shushensky District of the Krasnoyarsk Territory. The range of variation of the integral indicator of landscape sustainability was 0.459–0.575, the typical IILS value was 0.567 (right border of class III).

The next steps in working with models classifications of sustainability will be the collection of data and the refinement of baseline indicators, as well as the identification of parameters most subject to natural and man-made changes. It is also necessary to introduce into the model parameters that reflect the influence of anthropogenic impact on the system and study the stability of landscapes to changes in the natural and anthropogenic modes of their functioning. In this case, the estimated "potential sustainability" is the first step of the evaluation research. Next, a group of parameters is allocated for assessing the stability of the landscape to a change in the anthropogenic regime; its specificity is indicated. As a result, it is necessary to substantiate a set of necessary and sufficient parameters, priorities for assessing resistance to changes in the anthropogenic regime, and introduce another level of convolution of indicators to reflect the combined effect of "potential sustainability" and anthropogenic impacts on the landscape.

12.3.6 Integral Assessment of the Sustainability of a Social System with a Hypothetical Change in the Quality of the Environment in the Northern Regions of Russia

Example 7 The results of the integral assessment of the quality of life of the population of the northern regions of Russia were summarized under the assumption of a decrease in the quality of the environment in the northern regions of the Russian Federation by 30, 50% and 2 times (Fig. 12.5).

In general, the deterioration of the situation by 30% in the environmental unit (environmental quality unit) led to changes in only one out of 9 regions; by 50%—in

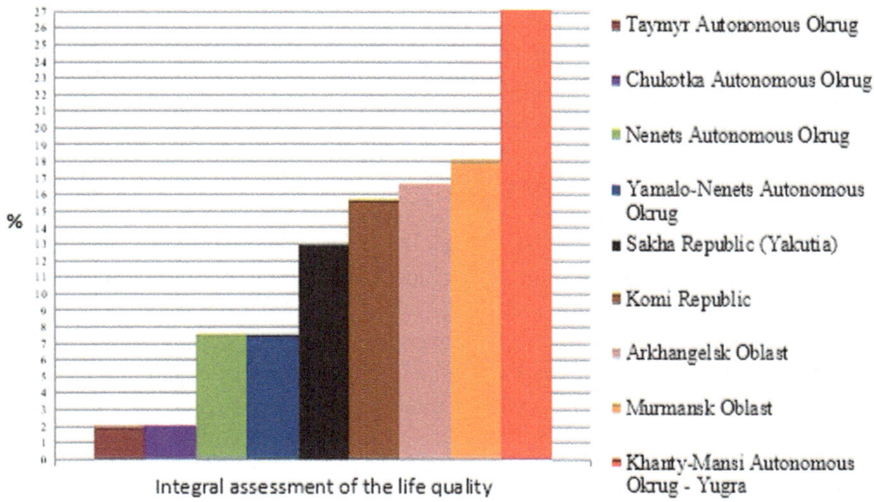

Fig. 12.5 The influence of environmental quality on the quality of life of the population deterioration (2%) to the life quality in the northern regions of Russia

3 regions out of 9 changed the class of environmental quality from III (average) to IV (below average); The deterioration of the situation by 2 times (Fig. 12.5) in 6 regions from the studied 9 entailed the transition of the region's socio-system to a lower class. Regions such as the Sakha (Yakut) Republic, Nenetsk Autonomous District and the Chukotsk Autonomous District turned out to be the most resistant to change. And the most vulnerable region in case of deterioration was the Hanty-Mansiysk Autonomous District-Yugra. It is noticeable that after a 30% deterioration of the situation economic factors has a greater influence on the consolidated assessment.

12.4 Conclusion

The foundations of the methodology, methods of integrated assessment of the state and emergent properties (productivity, sustainability, environmental well-being) of complex natural systems were considered. The aim of the research was to summarize the results of the development and testing of models classifications of an integral assessment of the state, sustainability of terrestrial landscapes, water bodies and river systems and then, at the next stage, an integral assessment of their systemic well-being. Terrestrial landscapes, marine and lake systems and water catchment systems were selected as the objects of the study. The subjects of research were theoretical and methodological aspects and methods for constructing integral indicators of the state (integral indicator of productivity, integral indicator of environmental quality, etc.), integral indicators of sustainability, integral indicators of well-being (health) of land, lake and river systems. The basics of the methodology for assessing the state, stability and well-being of terrestrial and aquatic geosystems are considered; integral estimation algorithms; the results of assessment studies: an integrated assessment of the trophic status of water bodies; integral assessment of the resistance of water bodies to changes in the parameters of the natural and anthropogenic regimes; paleoreconstruction of hydroecological conditions on the Laptev Sea shelf based on integral indices; integral assessment of the ecological status and ecological well-being of reservoirs, watercourses, water catchment systems; integral assessment of the sustainability of terrestrial landscapes; integral assessment of the stability of a sociological system with a hypothetical change in the quality of the environment on the example of northern regions of Russia Research was focused on the development of multi-criteria and multilevel classifications for the study of the emergent properties of complex systems in nature and society and the construction of integral indicators based on ASIID and APIS methods and models.

Acknowledgements The reported study was funded by RFBR according to the research projects #16-05-00715; #18-05-60291; #19-05-00683 and #19-04-00939-A.

References

Abalakov AD, Lopatkin DA, Novikova LS (2014) Mapping of landscape sustainability of the catchment area of Lake Baikal. Geopolit Ecogeodyn Reg 10(1):260–265

Aleksandrova LV, Vasilyev VU, Dmitriev VV, Myakisheva NV, Ogurtsov AN, Tretyakov VU, Khovanov NV (2000) Multi-criteria geographic and environmental assessments of the state and sustainability of natural and urban systems deposited manuscript no. 2342B00 01.09.2000 Dmitriev VV, Khovanova NV (ed). https://elibrary.ru/item.asp?id=30689943. Accessed 31 Dek 2018

Alimov AF, Dmitriev VV, Chistobaev AI et al (1999) the integrated assessment of the environmental condition and quality of environment in urban areas (Многокритериальная оценка экологического состояния и устойчивости геологических систем на основе методов сводных показателей. Интегральная оценка экологического состояния и качества средств урбанизированных территорий) (In Russian) Frolova AK (ed). St. Petersburg, SPbNTs Russian Academy of Sciences: 253

Bancheva AI, Alekseeva NN (2017) Geoecological assessment of. Hokkaido. Bulletin of Moscow University. Series 5. Geography 2:34–41

Blinovskaya JY, Gavrilo MV, Dmitriev NV, Pogrebov VB, Puzachenko AY, Usenkov SM, Knizhnikov AY, Pukhova MA, Shilin MB, Semanov GN (2012) Methodical approaches to the creation of maps of environmentally sensitive zones and areas of priority protection of waters and coasts of the Russian Federation from oil and oil products spills. Moscow, Publisher: World Wide Fund for Nature (WWF): 62

Dmitriev VV (2009) Determination of the integral index of the natural state of the object as a complex system (Определение интегрального показателя состояния природного обьекта как сложной системы) (In Russian) Obshchestvo. Sreda. Razvitie [Society. Wednesday. Development] 4(12): 146–165

Dmitriev VV (2010) Integrated assessment of the state of complex systems in nature and society (Интегральные оценки состояния сложных систем в природе и обществе) (In Russian) Biosfera [Biosphere] 2(3): 507–520

Dmitriev VV (2017) the health of a water body: axiology, axiometry, assessment. Lakes of Eurasia: problems and solutions. Proceedings of the 1st international conference (September 11–15, 2017). Karelian Research Center of the Russian Academy of Sciences, Petrozavodsk: 47–54

Dmitriev VV, Ogurtsov AN (2012) Approaches to assessment and GIS mapping of sustainability and environmental well-being of geosystems. I. Integral assessment of the sustainability of terrestrial and aquatic geosystems (Подходы к оценке и ГИС-картографии для развития устойчивости и экологического благополучия геосистем. I. Интегральная оценка устойчивости земных и водных геосистем) (In Russian) Vestnik of Saint-Petersburg University. Series 7. Geology. Geography 3: 65–78

Dmitriev VV, Ogurtsov AN (2013) Approaches to integrated assessment and GIS mapping of sustainability and environmental well-being of geosystems. II. Integrated assessment methods for sustainable land and water geosystems (Подходы к интегральной оценке и ГИС-картографии для обеспечения устойчивости и экологического благополучия геосистем. II. Методы интегральной оценки устойчивости земных и водных геосистем) (In Russian) Vestnik of Saint-Petersburg University. Series 7. Geology. Geography 3: 88–103

Dmitriev VV, Ogurtsov AN (2014) Approaches to integrated assessment and GIS mapping of sustainability and environmental wellbeing of geosystems. III. Integral assessment of the stability of the soil and land geosystems (Подходы к интегральной оценке и ГИС-картографии для обеспечения устойчивости и экологического благополучия геосистем. III. Интегральная оценка устойчивости почв и наземных геосистем) (In Russian) Vestnik of Saint-Petersburg University. Series 7. Geology. Geography 4: 114–129

Dmitriev VV, Ogurtsov AN (2017) Sustainability of complex systems in nature and society: methodology, evaluation, results. Scientific notes of the Russian State Hydrometeorological University 48: 72–84

Dmitriev VV, Osipov GK (2017) Integral assessment of stability osocial-ecological-economic system against changes in its functioning conditions. In: 17th International multidisciplinary Scientifc Geocoferrence SGEM 2017. Conference proceedings, vol 17. Ecology, economics, education and legislation. Ecology and environmental protection 52: 565–572

Dmitriev VV, Primak EA (2009) Integral assessment of the vulnerability and environmental well-being of water bodies. In: Environment and sustainable development of regions: new research methods and technologies. Proceedings of the all-Russian scientific conference with international participation. Editors in charge: Sirotkin VV, Ermolaev OP: 63–67

Dmitriev VV, Fedorova IV, Birykova AS (2016) Approaches to assessment and GIS mapping of sustainability and environmental well-being of geosystems. Part IV. Integrated assessment of ecological wellbeing of terrestrial and aquatic ecosystems. Vestnik of Saint-Petersburg University. Series 7. Geology. Geography 2: 37–53

Dmitriev VV, Zuyeva NV, Lubentsova AS, Valatin DI, Vasyakina AV, Nepomnyaschaya AV, Nikonov IA, Ostashov AA, Podshivalova DO, Prokopenya AD, Sokolova DP, Chernenko YI (2017a) Ecological state of karelian Ladoga water bodies: traditional view and contemporary accents. Scientific notes of the Russian State Hydrometeorological University 47: 126–144

Dmitriev VV, Ogurtsov AN, Morozova AS, Pilyugina AA, Sverdlov OA, Sirotina PM, Fedorova ME, Cherepanov SV, Shakurov VA (2017b) Integrated assessment of the sustainability of landscapes: models, results, prospects. International Journal of Applied and Basic Research. Publishing House Academy of Natural History. Penza 9: 110–114

Dmitriev VV, Pryakhina GV, Ogurtsov AN, Primak EA, Amaro Medina DR (2018a) Assessment of emergent properties of water bodies: trophic status, sustainability, environmental well-being In the collection: Third Grade readings. Faces of Hydrology Collection of reports of the International Scientific Conference in memory of the outstanding Russian hydrologist Yury Borisovich Vinogradov. Edited by OM Makarieva: 347–354

Dmitriev VV, Amaro Medina DR, Ogurtsov AN, Dobrynina AS (2018b) Assessment of the ecological status of the river-catchment system: approaches, methods, results. Ecology of river basins: Proceedings of the 9th Intern. Scientific-practical conf. / under total ed. prof. T.A. Trifonova; Vladimir State. unt them. A.G. and N.G. Stoletovs, Vladimir: 46–52

Dmitriev VV, Ogurtsov AN, Hovanov NV, Osipov GK, Kulesh VP, Sergeyev YN, Fedorova IV (2019) Integral assessment of condition and sustainability of socio-ecological-economic systems, in this book

Fedorov AS (2008) Soil resistance to anthropogenic influences. St. Petersburg, Publisher: Publishing House of St. Petersburg University: 115

Glazovskaya MA (1997) Methodical foundations of the ecological and geochemical resistance of soils to anthropogenic influences. M.: Publishing House of Moscow

Grodzinsky MD (1987) Stability of Geosystems: Theoretical Approach to Analysis and Quantitative Assessment Methods, Izv. Academy of Sciences of the USSR. Ser. geogr. 6: 5–15

Hamelin L-E (2008) Nordicite compare par critere en VAPO. L 'Encyclopedie canadienne

Hovanov NV (1996) Analysis and synthesis of indicators in an information deficit (Анализ и синтез показателей при информационном дефиците) (In Russian) St. Petersburg, St.-Petersburg University Press: 196

Hovanov N, Hovanov K, Yudaeva M (2009) Multicriteria estimation of probabilities on basis of expert non-numeric, non-exact and non-complete knowledge. Eur J Oper Res 195(3):857–863

Klyuvitkina TS, Bauch HA (2006) Hydrological changes in the Laptev Sea during the Holocene inferred from the studies of aquatic palynomorphs. Oceanology 46(6):859–868

Lior N (2015) Quantifying sustainability for energy development. Energy Bull 19:8

Lubentsova AS, Vereshchagina EA, Dmitriev VV (2017) Estimation of the State of Water Ecosystems, Influenced by Nuclear Power Stations on the Example of Kalininskaya NPP. Materials of the All-Russian scientific-practical conference with international participation "Ecology of reservoirs - coolers of power stations": collection of articles. Materials vseros. Scientific-practical conf. from Intern. Participation / zabaykal. State un-t; [resp. ed. G.Ts. Tsybekmitov]. Chita: ZabGU: 185–193

Semenchenko VP, Razlutsky VI (2010) Ekologicheskoe kachestvo poverkhnostnykh vod [Environmental quality of surface waters]. Minsk, Belarusian Nauka. 329 p. (In Russian)

Snakin VV, Melchenko VE, Butovsky RO, Vorontsova LI, Vasilyeva NP, Resin AL, Alyabina IO, Barinova SS, Erbanova LN, Kochetova NI, Krechetov PP, Lomakina GA, Morgun LV, Golovina LV, Barabanova GS (1992) Assessment of the status and sustainability of ecosystems. M.VNIIPriroda: 127

Totonova EE (2016) Methods of zoning analysis and zoning of the northern territories (on the example of Canada) №1 (7), Arctic XXI century. Humanitarian sciences. Publisher: North-Eastern Federal University named after MK Ammosova (Yakutsk): 28–36

Trushevsky VL, Dmitriev VV, Osipov AG (2018) Estimation of the stability of water objects: approaches and results. In the collection: ecology of River Basins Proceedings of the IX international scientific practical conference: 104–110

Vasilyev VY, Dmitriev VV, Ogurtsov AN, Fedorova IV (2011a) Paleoreconstruction hydroecological conditions on the Laptev Sea shelf based on integral indices using non-numerical, incomplete and inaccurate information. I. Statement of the problem (In Russian) Proceedings of the international conference dedicated to the 165th anniversary of the Russian Geographical Society and the 85th anniversary of the organization of the Faculty of Geography at St. Petersburg (Leningrad) State University. Under the general editorship of TA Aliyeva, VV Dmitrieva, NV Kaledina, KV Chistyakov, St. Petersburg, VVM: 145–152

Vasilyev VY, Dmitriev VV, Ogurtsov AN, Fedorova IV (2011b) Paleoreconstruction hydroecological conditions on the Laptev Sea shelf based on integral indices using non-numerical, incomplete and inaccurate information. II. In: Simulation results (In Russian) Proceedings of the international conference dedicated to the 165th anniversary of the Russian Geographical Society and the 85th anniversary of the organization of the Faculty of Geography at St. Petersburg (Leningrad) State University. VVM: 153–158

Vasilyev VY, Dmitriev VV, Ogurtsov AN, Zelenkov AI, Miskevich VM, Mashkin JL (2009) Geoinformation analysis and zoning of the Barents Sea on the degree of environmental well-being in the area of designing an oil terminal in the vicinity of Cape St. nose (East) (Устойчивое развитие территории ГИС и практический опыт) (In Russian) InterCarto / GIS Inter 15: Sustainable Development of Territories: GIS Theory and practical experience of the International Conference: 403–410

Part II
Landscape Indicators—Model Approaches for Specific Ecological and Socio-economic Indicators

Chapter 13
Spatial Estimation of Estonian Forest Landscapes' Soil Cover Humus Status: Methods, Model Samples and Assessments

Raimo Kõlli, Mait Lang, Reimo Lutter, Tõnu Tõnutare, Karin Kauer and Kaire Rannik

Abstract Humus status of soil cover is of greatest importance in the formation and functioning of landscape. On the humus status of landscape soils (soilscape) depends in a great extent the floristic composition and diversity of forest ecosystems and its functioning peculiarities (level of annual productivity and litterfall, and as well the character of soil organic matter decomposition and the fabric of formed humus profile). The substantial part of this work is devoted to the evaluation of soil organic carbon (SOC) superficial densities (in Mg per ha) and its total stocks by the dominated Estonian forest soils and their different layers, as the most important quantitative indices of soil cover. The essential findings of this work are also the estimations of total SOC stocks (pools) in whole Estonian forested soil cover and in its main sublayers (humus cover and subsoil). Much attention is paid to the ecological aspects of humus cover type (*pro* humus form) formation and its profile fabric's matching with soil properties. The humus cover type may be taken as the main qualitative index of forest soils' humus status. Relevant are also the pedo-ecological

R. Kõlli (✉) · T. Tõnutare · K. Kauer · K. Rannik
Estonian University of Life Sciences, Institute of Agricultural and Environmental Sciences, F.R. Kreutzwaldi Str., 5, 51006 Tartu, Estonia
e-mail: raimo.kolli@emu.ee

T. Tõnutare
e-mail: tonu.tonutare@emu.ee

K. Kauer
e-mail: karin.kauer@emu.ee

K. Rannik
e-mail: kaire.rannik@emu.ee

M. Lang
University of Tartu, Tartu Observatory, Observatooriumi 1, Tõravere, 61602 Tartumaa, Estonia
e-mail: lang@to.ee

M. Lang · R. Lutter
Estonian University of Life Sciences, Institute of Forestry and Rural Engineering, F.R. Kreutzwaldi Str., 5, 51006 Tartu, Estonia
e-mail: reimo.lutter@emu.ee

W. Mirschel et al. (eds.), *Landscape Modelling and Decision Support*, Innovations in Landscape Research,
https://doi.org/10.1007/978-3-030-37421-1_13

analyses on the role of humus status indexes in the characterizing of SOC cycling and annual balance, and also in the formation of ecosystems biodiversity. Totally in the soil cover of Estonian forest land, 367 Tg of SOC is sequestrated. From the total SOC amount, 66.2% is located in the humus cover and 33.8% in subsoil layers. The mineral and peat soils role in sequestration of total SOC amount are accordingly 55.0% and 45.0%. For a better understanding to international audience, the used key terms of the study are elucidated and the soil names of the local classification are juxtaposed by the names of international classifications.

Keywords Humus status · Soil cover · Humus cover · Soil organic carbon · Forest soils · Pedocentric approach · Landscape · Digital soil map

13.1 Introduction

Soil cover (SC) forms the material basis for landscape (Kõlli and Ellermäe 2001; Arold 2005). The distribution pattern, appearance and functioning of the landscape depends besides of soils' properties and diversity (pedodiversity) on the extra-soil natural agents such as climatic conditions, geology and hydrography of the area. Moreover, it depends as well on different aspects of the anthropogenic activities such as the policy of application local natural resources, kind of land use and the intensity of the land management.

In characterizing landscape type and its functioning capability (i.e. in studying ecological aspects of landscape), it is essential to follow not only SC properties. The complexity of characterization is inconceivable without studying of soil type-specific plant cover, organisms' assemblages adapted to soil condition and hydrological regimes of the ambient territory. By these above-named components, their mutual relationships are determined in general lines the landscapes functioning activity and stability, and therefore the sustainable use of available local resources (as the plant available nutrition elements and water of the SC, photosynthetically active radiation and seasonal meteorological features). With other words, the characterization of landscapes structure and functioning requires an ecosystem approach as only in such case all essential components influencing landscape functioning will be taken into account.

The fabric, diversity and functioning capacity of forest ecosystems formed in natural landscape depend much on the physical–chemical properties and hydrological conditions of soils. For the best indicators in characterizing of forest landscape's functioning intensity is the potential fertility or annual productivity of SC, which is reflected via its nutritional and humus status (HS). In this work, the main attention was paid to the HS of soils, taking it as a driving force, which determines the character of processes and evolution direction of forest landscapes. It is important to mention that HS determines partly as well the nutritional status of ecosystems as most of the nutrition elements are cycling in the composition of organic matter.

The main quantitative indexes of organic matter flux via the ecosystem (landscape) are the annual increment of phytomass, annual litterfall intensity and annual accumulation rate of new organic matter on or into the SC. In the case of deliberation (decomposition) of soil organic matter (SOM), the captured nutrition elements are switched into the new cycle of elements' turnover or into the production process of forest ecosystems. For understanding these processes, it is utmost important to know the soil organic carbon (SOC) sequestration capacities of different soil types in diverse land-use conditions and distribution of SOC by separate SC layers.

In the past decades, the databases on SOC stocks in soils and its estimation methods have been continuously increased and perfected, enfolding the generalizations of data on European and global levels (Dixon et al. 1994; Bazilevich and Titljanova 2008; Baritz et al. 2010; De Vos et al. 2015). In Estonia, the problems connected with SOC sequestration into forest SC and with its role in functioning forest ecosystems were studied versatilely and long-lastingly (Kõlli 2002; Kõlli et al. 2004, 2009a; Lutter et al. 2018). Some of our previous researches may be taken as the prerequisite or basis for the current study (Kõlli 1992; Kõlli et al. 2009b, 2010; Lang et al. 2017; Lutter et al. 2018).

The main novelty of actual study consists (1) in the generalization of SOC superficial densities (Mg ha^{-1}) distribution data on normally developed (or post-lithogenic) mineral and organic soils' united matrix table, and (2) in using of different forest land maps for extraction of forest soils data. Therefore, the main aims of the study are the following:

- to characterize Estonian forest soils' HS and pedo-ecological properties by soil groups on different levels of generalization;
- to explain the role of soils' HS in the forming and functioning of landscapes by different soil groups;
- to introduce methodological principles used for determination of SOC's superficial densities and total stocks in the SC of forest landscapes;
- to estimate the total SOC stock sequestrated in the Estonian forest landscape's SC and in its sublayers, i.e. in the top and subsoil and
- to explain some regularities of matching soil mapping units (SMU) with forest management units (compartments) in relation to dominant forest soils.

13.2 Terminology and Methodology

13.2.1 Terminology

The humus status (HS) of a soil reflects in principle the character of its SOM management or the throughout flux of SOC via the SC. This flux begins with litterfall on or into the soil and follows by variegated processes in mutual relationships with soil living, liquid, gaseous and solid phases, until to its stabilization or/and total mineralization and elimination from the SC. The main parameters for the characterization of soils' HS are the thickness and morphology of soil horizons, SOC (or SOM) concentration (g kg^{-1}) and stock densities per area (Mg ha^{-1}) in different horizons, and SOC annual turnover (Mg ha^{-1} yr^{-1}).

In the quantitative characterization of HS in relation to soil mantle the notion of soil cover (SC) or solum is used. SC embraces the superficial landscape layer influenced by soil-forming process and consists of HC and subsoil (SS). As characteristic to boreal bioclimatic belt, the thicknesses of Estonian SCs are remarkably thinner as compared with southern regions SC's thicknesses. By our previous researches (Kõlli 2002; Kõlli et al. 2004) the SC thicknesses (depth from soil surface to unchanged parent material) depend in great extent on SC moisture conditions, calcareousness and texture, reaching in automorphic pseudopodzolic and podzolic soils to 100–110 cm, but the thickness of permanently wet gley-soils is in most cases in the limits 40–65 cm. Depending on this, for the benchmark thicknesses of peat soils or histosols in the current study is taken 50 cm.

Humus cover (HC), which is known as well as epipedon and humipedon (Zanella et al. 2011), encompasses the most active superficial (topsoil) part of SC via which the dominant share of SOC (SOM) cycling takes place. The HC consists of forest floor, humus or raw humus and peat horizons and is closely coupled with plant cover. For the benchmark thicknesses of peat soils' or histosols' HCs is taken in our work 30 cm.

Subsoil (SS), which underlies the biologically active topsoil, consists in the case of mineral soils from the *eluvial* and/or *illuvial* horizons, but in the case of peat soils, the SS embraces a peat layer located in the depth from 30 to 50 cm with thickness 20 cm.

Treating of SC on the basis of large-scale soil maps (1:10,000) needs quantitative assessments using of detailed level classification taxa as (1) soil species, which is the taxon of Estonian soil classification (ESC) identified by soil-forming processes and (2) soil variety—taxon of ESC identified by soil species' texture. Soil species contours recur and they are separated on soil maps in different patterns across the landscape.

For the qualitative characterization in Estonia, the humus cover type (*pro* humus form) is in use. Humus cover type, which characterizes SOC/SOM formation ecology, is a good index as well for using in landscape level. Totally in Estonian forest humus classification, 27 HC types have been separated (Kõlli 1992).

13.2.2 Methodological Principles

For explaining SC role in the development and functioning of forest landscapes in this study, the pedocentric approach is used (Kõlli et al. 2018). For the basis of this approach are the soil species or/and soil varieties of ESC or the SMU of the large-scale (1:10,000) soil map (Estonian Land Board 2012). The main pedological information about each soil contour (as SMU) is expressed by the code of soil species. In addition to this, with each contour is connected some additional information on soil properties, such as soil texture, calcareousness/acidity and fabric of humipedon.

At the same time for the most detail mapping unit in forest management is the compartment, which, besides information on forest stands, gives some information on site forest growth conditions. Therefore, the site properties of each compartment are described indirectly by forest site type (FST), which expresses site condition mostly by forest understory plant associations and indicator species presented in plant cover (Lõhmus 2004). In connection with this, for the one of our research task was to study the matching of mapping units of soil and forest management maps (Kõlli and Köster 2018).

By our understanding, the pedodiversity depends directly on soil-forming conditions of the area, among them on soil parent material and its deposition character (relief), i.e. pedodiversity depends on areal geodiversity (Kõlli et al. 2018). For the native zonal ecosystems in Estonia are the forests formed in equilibration processes with soil conditions. We are in the opinion that the best ecological conditions are formed in the case of optimal site-specific vegetation diversity. Therefore, biodiversity should be optimal and inherent to the site (or soil). It means that the vegetation should be adequate to site conditions or should be site specific, but not to be with maximal as possible biodiversity.

Four independent data sources were used in our study to identify forest land, (1) wooded land area of 1:10,000 Estonian basic map, (2) stand map of forest management inventory database, (3) forest mask constructed using satellite images (Peterson et al. 2004, 2008), and (4) satellite images-based tree species map (Fig. 13.1; Lang et al. 2018). Spatial overlay module in GRASS GIS 7.4 was used to cut soil map objects according to forest land maps (Fig. 13.2). The tested forest land maps' acreage used to clip 1:10,000 soil map for extraction of forest soil cover's SMU data are presented in Table 13.1.

In dominating cases for SMU were soil species, only in the case of fen soils the soil species were divided into soil varieties. The results of the current study are based on the data extracted from the large-scale soil map by the compartments' map of the State Forest Register (Table 13.1). The soil map database information is in connection with normally developed mineral and organic soils' united matrix table's litho-genetic and moisture scalars, which may be taken as the coordinates for each soil patch (Fig. 13.3). For assessing SC and HC SOC stock values for each soil patch in the soil map database the lookup tables (LUTs) were used. LUTs were constructed for the SOC stock values by interpolating the data by 0.1 step resolution of soil litho-genetic and moisture coordinates (Fig. 13.5). These coordinates were used to seek SOC data from LUT that are described further in the text in more detail.

The particle size distribution was determined by Kachinsky (1965). The volume of coarse soil fractions ($\varnothing > 1.0$ mm) was determined during the field research (Astover et al. 2013). In laboratory the content of fine-earth ($\varnothing < 1.0$ mm) and of fraction 1–10 mm in soil samples was determined by sieving, but the particle size distribution of fine-earth by using the sedimentation method. SOC concentration was determined by wet digestion of SOM with acid dichromate (Tjurin 1935). The

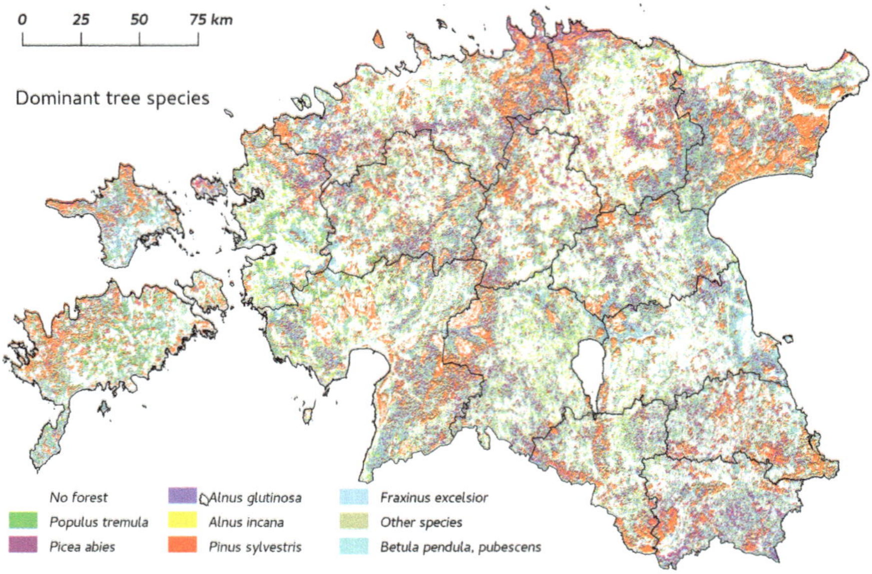

Fig. 13.1 Dominant tree species in Estonian forest landscapes. Adopted from Lang et al. (2018) with permission

SOC stocks in SC of different soil types are calculated on the basis of the SOC concentrations and bulk densities of corresponding soil horizons. In the calculation of SOM in mineral soils, the coefficient of value 1.72 was used but of peaty soils and forest floors the coefficient 2.00 was used (Astover et al. 2013).

The soil names given in the national databases by ESC were converted into World Reference Base (WRB) soil classification system (Estonian Land Board 2012; IUSS WG WRB 2015). The juxtaposition of ESC and WRB is seen also in Fig. 13.3.

13.3 Pedo-Ecological Conditions and Used Data

To the natural area of Estonia, which is located in mild and wet pedo-climatic conditions, mainly the coniferous and coniferous–deciduous mixed forests are characteristic (Laasimer 1965; Valk and Eilart 1974; Yearbook Forest 2017 2018). The list of dominant tree species found in Estonian forest landscapes contains less than 10 tree species (Fig. 13.1). As a result of intensified agricultural activity during last two centuries, the most productive areas of Estonia (soils suitable for crop cultivation and grasslands) have been turned into arable, pastured or hay-lands (Mander et al. 1995; Raukas 1995; Arold 2005).

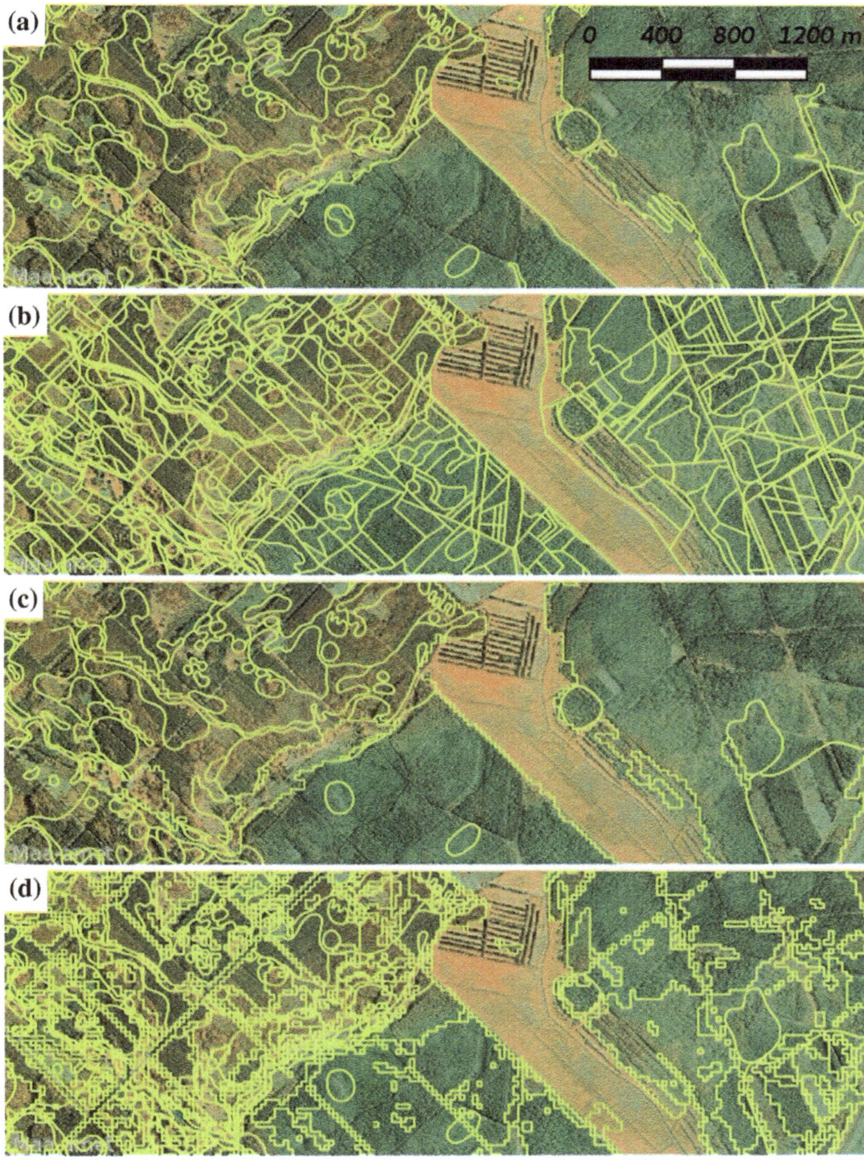

Fig. 13.2 Examples of cutting 1:10,000 digital soil map with **a** wooded land area of 1:10,000 Estonian basic map, **b** stand map of forest management inventory database, **c** forest mask constructed using satellite images (Peterson et al. 2004, 2008), and **d** tree species map based on satellite images (Lang et al. 2018)

Table 13.1 List of forest land maps used to clip 1:10,000 soil map for the extraction of forest soil cover's soil mapping units (for which is Estonian soil classification taxon 'soil species') data

No	Source of forest land data	Tested SC area, in km^2	Remarks
1	Compartments' map of the State Forest Register based on the distribution of forest site types	20,114	Forest land area by National Forest Inventory was 23,306 km^2 by the Yearbook Forest 2017 (2018)
2	Area of the basic map covered by woody plants	23,557	Exceeds National Forest Inventory (NFI) area by 251 km^2
3	Wintertime satellite images (Peterson et al. 2004, 2008)	24,002	Exceeds NFI area by 696 km^2
4	Map of forest stands' tree species composition (Fig. 13.1; Lang et al. 2018)	27,173	Exceeds NFI area by 3,867 km^2; enfolds also the bushy areas outside of forest land

The main part of parent materials of Estonian soils is derived from the glacial and aqua glacial Quaternary deposits. The parent material of half mineral soils are Pleistocene tills. The reworked tills glaciofluvial, glaciolacustrine, alluvial and aeolian sediments are distributed alternatively with tills (Raukas and Teedumäe 1997).

The integrating of soils' data into the forest landscapes management is possible thanks to the availability (1) of large-scale soil maps for forested lands and (2) of quantitative data on HS for all dominated forest soil species (Estonian Agri-Project 1983, 1985). A 1:10,000 digital soil map is provided by the Estonian Land Board (2012).

The quantitative data of soils' HS of the present study originate mainly from the soil profile horizons database (DB) *Pedon* (Kõlli et al. 2009a, b) created by us. The bulk density samples were taken from approximately one-tenth of profiles. In addition to our experimental data, the materials published on HS and productivity of mineral and peat soils of Estonia were used (Reintam et al. 2003). Overall, generalized volume weights from our own and other data sources (mainly data of Estsurvey pertaining to soil species and texture) were used.

13.4 Estimations of SOC Stocks

13.4.1 Distribution of Forest Soils by ESC Taxa and Some Essential Soil Cover Characteristics on the Level of Soil Species

For the basis of detail characterization which ever territory's HS (from one soil mapping contour to whole Estonian forest land) are the individual HS data of all

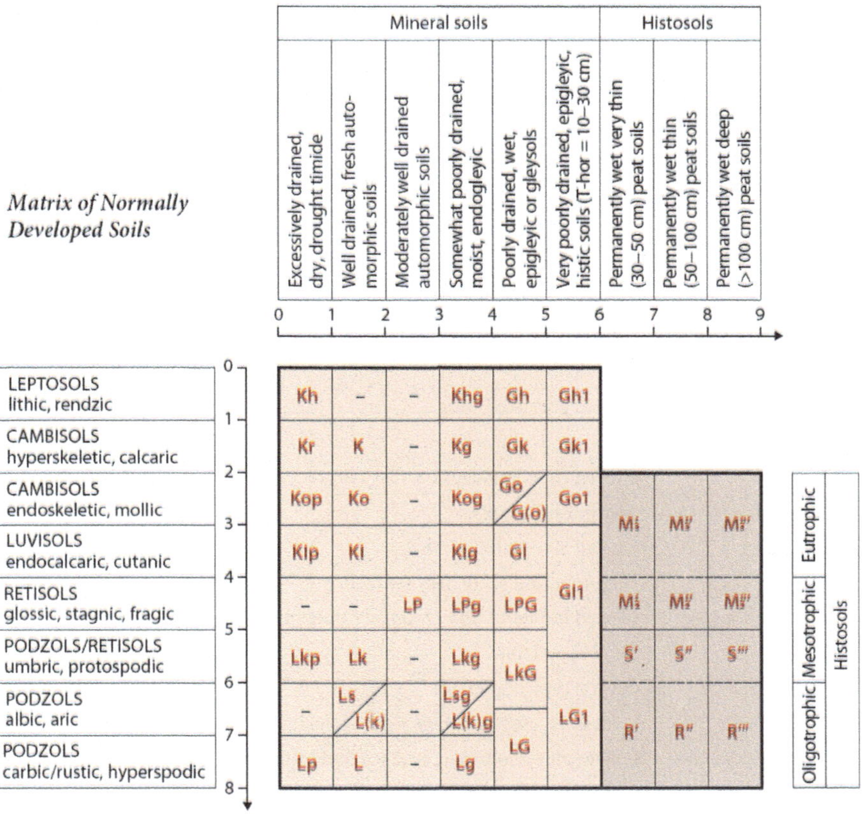

Fig. 13.3 Matrix of normally developed soils as a pedo-ecological background of SOC stocks lookup table (LUT) models with soil codes of Estonian Soil Classification (ESC). On the horizontal scalar the soils moisture conditions, but on vertical, i.e. litho-genetical scalar, the correlation with WRB reference soils are given. Additional explanation on right side characterizes the feeding water, which is drawing force of soil cover paludification processes. For soil names after their codes see Table 13.2

presented soils species and/or soil varieties and their distribution area. Totally on forested land of Estonia may be found more than 120 soil species, which enfold >300 different soil varieties. In the interest of generalization, the data on soil distribution are presented in this work in the level of small soil groups (SSG) in Table 13.2. The list of SSG is not only summarized similar by their properties soil species but also very similar to them abnormal soils. As the share of abnormal soils is relatively modest among others (Fig. 13.4), their nomenclature is not presented in actual work. In relation to tested territory, the abnormal mineral forest soils formed 2.5% from whole mineral soils and abnormal organic soils 1.0% from whole organic soils. The distribution of soil species was tested on the basis of the compartments' map of the State Forest Register (Table 13.1) in relation to territory 20,114 km^2, which forms 86.3% of total forest land (Table 13.2).

Table 13.2 Small soil groups' (SSG) codes and names by Estonian soil classification (ESC), and their areal distribution

SSG	Codes	Names by ESC[a]	Area[b]	
			ha	%
1	Kh	Limestone rendzinas	10,371	0.5
2	Khg	Gleyed limestone rendzinas	2,772	0.1
3	K Kr	Pebble and pebble-rich rendzinas	52,469	2.6
4	Kg Krg	Gleyed pebble and pebble-rich rendzinas	20,396	1.0
5	Ko Kor	Typical and pebble-rich leached soils	46,432	2.3
6	Kog Korg	Gleyed typical and pebble-rich leached soils	49,233	2.4
7	KI	Eluviated soils	29,329	1.5
8	KIg	Gleyed eluviated soils	57,968	2.9
9	LP	Pseudopodzolic soils	53,954	2.7
10	LPg	Gleyed pseudopodzolic soils	62,564	3.1
11	Lk	Sod-podzolic soils	66,457	3.3
12	Lkg	Gleyed sod-podzolic soils	41,768	2.1
13	L(k) Ls	Humuous and secondary podzols	28,409	1.4
14	L(k)g Lsg	Gleyed humuous and secondary podzols	12,748	0.6
15	L	Typical podzols	65,772	3.3
16	Lg	Gleyed typical podzols	22,881	1.1
17	Gh Gh1	Limestone gley- and peaty gley-rendzinas	2,830	0.1
18	Gk Go G(o)	Pebble gley-rendzinas; leached and saturated gley-soils	343,122	17.1
19	GI LPG	Eluviated and pseudopodzolic gley-soils	229,599	11.4
20	LkG	Sod-podzolic gley-soils	62,318	3.1
21	LG	Gley-podzols	59,767	3.0
22	Gk1 Go1	Pebble and saturated peaty gley-soils	69,456	3.5
23	GI1	Unsaturated peaty gley-soils	48,912	2.4
24	LG1	Peaty podzols	55,362	2.8
25	AM	Alluvial fen soils	5,224	0.3
25a	M3	Well decomposed lowland fen soils	252,346	12.5
25b	M2	Moderately decomposed lowland fen soils	102,923	501
26	S	Transitional bog soils	101,921	5.1
27	R	Raised bog soils	40,892	2.0
28	Tx	Technogenic eliminated soils	3,742	0.2
29	Ty	Technogenic mixed soils	5	<0.05
30	Tz	Technogenic buried soils	62	<0.05
31	Tu	Technogenic sediment soils	5,296	0.3
32	Pu	Ground sediment heaps	2,633	0.1

(continued)

Table 13.2 (continued)

SSG	Codes	Names by ESC[a]	Area[b]	
			ha	%
33	Pp	Bare ground	1,446	0.1
34	C	Artificial grounds	18	<0.05

[a]In list the codes and names of abnormal soils are absent (an exception are those of SSG 25 and 28–34), whereas for their share see Fig. 13.2

[b]Including area of similar to normal soils by their humus status different species of abnormal soils

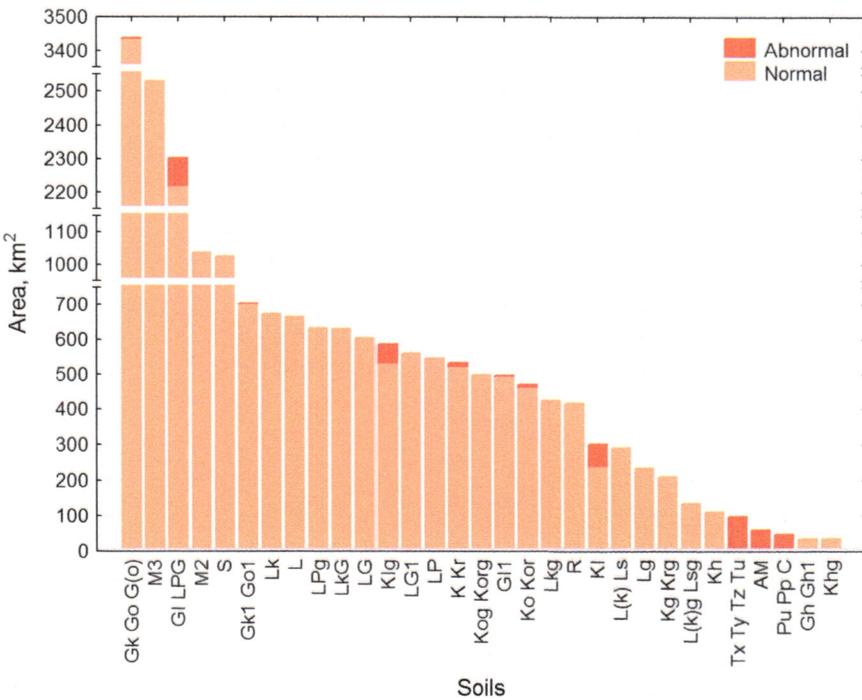

Fig. 13.4 Areas of normally developed soil species in Estonian forest lands' soil cover with these abnormal soils, which are by their properties similar to normal soils. For soil names by their codes see Table 13.2

Supporting to the data of more detail level analyses than it is presented in Table 13.2 and Fig. 13.4 was possible besides of soils HS to follow as well different other soil properties and pedo-genetic features, which are with essential importance in characterization of forest landscapes SC.

(1) By the podzolization formula—weakly (I): moderately (II): strongly (III): podzolized soils percentage, the soils with moist (gleyed) conditions are podzolized in greatest extent as compared with soils with fresh moisture conditions. If the

relationships (I:II:III) for Lk and L are accordingly 94:6:0 and 70:28:2, then the same for Lkg and Lg are accordingly 75:23:2 and 30:65:5.

(2) The ratio (in %) of pebble and pebble-rich rendzinas is an average of 73:27, but the ratio of typical and pebble-rich leached fresh and moist soils 93:7.

(3) The ratio (in %) M:S:R (fen: transitional bog: high bog) soils is an average 71:21:8.

(4) From the whole tested area, the moderately (E2) and strongly (E3) eroded soils (with slope accordingly 5°–10° and >10°) form 0.27%, whereas they are in ratio with deluvial (D) soils (E:D) as 36%:64%.

(5) If in the case of SSG 1–4 and 11–16 by the area are dominating fresh automorphic soils, then in the case of SSG 5–10 (more fertile soils) vice versa the dominating are moist or gleyed soils.

Estonian-forested landscapes SC is typical to north-eastern Europe with dominating of *Gleysols* and with a high share of *Histosols*. The share of automorphic mineral soils is approximately one-third of the forest area.

13.4.2 SOC Stocks Density Models for Estonian Normally Developed Forest Soils

The SOC superficial densities (Mg ha^{-1}) models are composed by the LUT principles on normally developed soil matrix (Fig. 13.3). The forest soils SC and HC SOC stock models are formed in the level of soils species.

The isolines of empirical SC and HC SOC stock values (Figs. 13.5 and 13.6) are too complex for the construction of mathematical equation that could represent all the variabilities with sufficient precision for all soils in the matrix. The LUT for SOC stock values were constructed by interpolating the isolines' data given in relation to soil litho-genetic and moisture scalars as coordinates (Figs. 13.5 and 13.6). The SOC stock densities, which are characterized by isolines, may be taken as decision support models (DSM). Totally two DSM are presented. One of them is elaborated in relation to whole SC (Fig. 13.5) and second in relation to HC (Fig. 13.6). By these models, it is possible to estimate whichever Estonian normally developed mineral and peat soils SC SOC stocks densities (Mg ha^{-1}). The total SOC stocks are calculated on the basis of soils' SOC densities and distribution areas.

In Fig. 13.7, it is presented total amounts of SOC, which have been sequestrated in the SC of the State Forest Register forests (20,114 km^2). In this figure, the whole SC SOC is divided into HC and SS. The SS SOC's total stocks and stocks density may be found by the formula $SOC_{SS} = SOC_{SC} - SOC_{HC}$. In connection with a modest share of abnormal soils among normal soils (see Fig. 13.3), on the basis of these two DSM were estimated also the SCs and HCs SOC total stocks of abnormal soils.

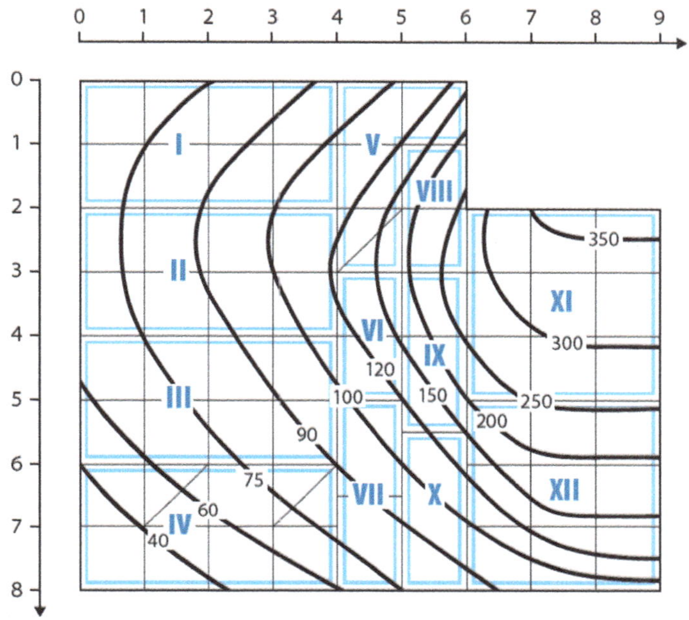

Fig. 13.5 LUT model about SOC stocks (Mg ha^{-1}) in soil cover of Estonian forest soils (given by isolines) on the background of large soil groups (LSG, I–XII). For pedo-ecological conditions of the background matrix see Fig. 13.3. The pedo-ecological characterization of LSG is given in Tables 13.4, 13.5 and 13.6

13.4.3 Estimation of SOC Stocks by Large Soil Groups (LSG)

For both tasks, generalization of the data and harmonization of calculated results by ESC with WRB, the LSG have been formed (Table 13.3). The information given by LSG was based on the same tested territory which was used in the case of SSG. For to being from the pedological aspect universally understandable to international audience, the LSG are characterized by reference soils and qualifiers of WRB (Table 13.3).

From the total amount of SC SOC (316.4 Tg) of tested area, the biggest share forms lowland fen soils (Table 13.4). Remarkable share (29.0%) belongs as well to different kinds of gley-soils (LSG V–VII). In sequestration of SOC into HC besides fen soils have formed in eutrophic conditions rich in calcium gley-soils. The subsoils of peat soils or histosols are also rich in SOC, but it should mention that this part of SOC does not participate in active cycling of SOC. At the same time, different mineral soil groups are sequestrated into their SS from 2.0 to 7.8 Tg of SOC. The highest share of SOC among mineral soils SS have different kind of podzols (as LSG II, IV, VII and X). For characterization of correlation between LSG soils and FST approximately 70% of dominated FST have been accounted in Table 13.4.

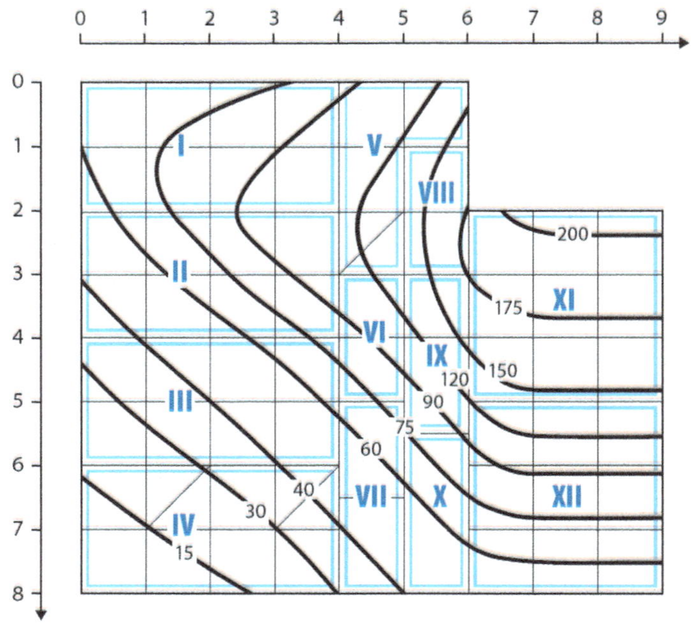

Fig. 13.6 LUT model about SOC stocks (Mg ha^{-1}) in humus cover of Estonian forest soils (given by isolines) on the background of large soil groups (LSG, I–XII). For pedo-ecological conditions of the background matrix see Fig. 13.3. The pedo-ecological characterization of LSG is given in Tables 13.4, 13.5 and 13.6

13.5 Estimations of Total SOC Stocks in the SC of Estonian Forest Land

As the tested area of Estonian forest land (20,114 km^2) enfolds 86.3%, we estimated the total SOC stocks for the whole forest area (23,306 km^2) according to three different scenarios (Table 13.5). These scenarios are based on prognoses of non-tested area's (3,192 km^2 or 13.7%) different soil cover compositions. It seems that more realistic from these three should be the first of them, where the non-tested part SC composition has been taken similar to tested one.

On the basis of forest lands, SC total SOC amounts the SOC stock densities (Mg ha^{-1}) for the different SSG were calculated (Fig. 13.8). It is important also to mention that these mean SOC sequestration capacities were calculated as mean weighted by the area in relation to whole SC and separately to mineral and peat soils. By the scenario I (Table 13.5) in the Estonian forest land SC totally 367 Tg of SOC is sequestrated, whereas 66.2% of it is located in the HC and 33.8% in SS. From the total SOC amounts, 55.0% is located in SC of mineral soils and 45.0% in peat soils.

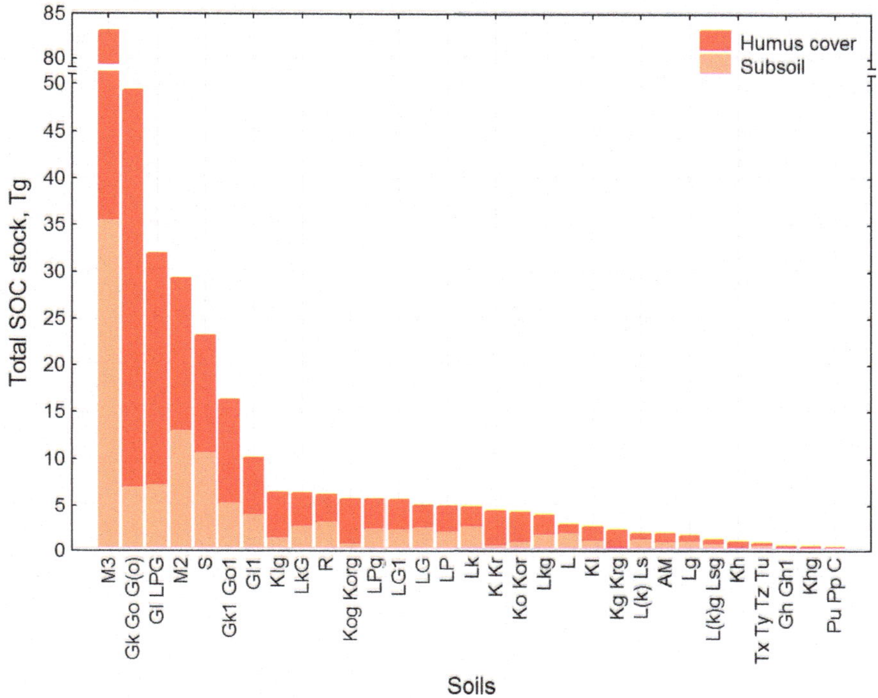

Fig. 13.7 Share of different soil species in sequestration of SOC in whole Estonian forest lands' soil cover or in solum. For soil names by their codes see Table 13.2

13.6 Essential Remarks Upon Determination Forest Soils' HS

The SC is a determining factor in the development of plant cover and its diversity. The pedodiversity of the landscape may be caused by soil texture variations (from sand to clay), mineralogical and chemical composition, calcareousness and acidity. The pattern of SC and its diversity are induced by the geodiversity and hydrological conditions. For a better understanding of mutual influences of SC and plant cover, the feedback influences of their main components (soil, plant) functioning should be studied at the ecosystem level, on typical-to-region soil types and management conditions (Photo 13.1).

In Table 13.6, besides SOC sequestration capacity into SC as well the dominant soil texture and HC types have been presented by LSG. With these characteristics are tightly connected the thicknesses of SC and HC, as well the fabric and acidity of the forest floor. The named characteristics' complex is reflected in forest ecosystems productivity (quality class) and in the composition of dominant tree species.

Table 13.3 Large groups of forest soils (LSG): composition, characterization with WRB qualifiers and distribution

LSG No	SSG	Codes by ESC	RSG of WRB	Characterization of LSG properties with WRB qualifiers	Area	
					km^2	%
I	1–4	Kh Kr K Khg Krg Kg	*Leptosols/Cambisols*	*leptic lithic skeletic rendzic calcaric hyperhumic gleyic*	860.1	4.3
II	5–8	Ko Kor KI Kog Korg KIg	*Cambisols/Luvisol*	*cambic endocalcaric luvic humic endogleyic loamic*	1,829.6	9.1
III	9–12	LP Lk LPg Lkg	*Retisols/Podzols*	*glossic stagnic fragic albic umbric epidystric endogleyic*	2,247.4	11.2
IV	13–16	L(k) Ls L L(k)g Lsg Lg	*Podzols*	*spodic albic rustic/carbic aric endogleyic arenic*	1,298.1	6.4
V	17–18	Gh Gk Gor Go G(o) Gh1	*Gleysols*	*epioxygleyic calcaric mollic eutric endoskeletic saprihistic*	3,459.5	17.2
VI	19	GI LPG	*Gleysols/Retisols*	*reductigleyic dystric glossic luvic umbric uterquic*	2,296.0	11.4
VII	20–21	LG LkG	*Podzols*	*epigleyic spodic albic dystric rustic arenic*	1,220.8	6.1
VIII	22	Gk1 Go1	*Gleysols*	*saprihistic epioxygleyic calcaric/calcic*	694.6	3.4
IX	23	GI1	*Gleysols*	*epigleyic histic reductic umbric*	489.1	2.4
X	24	LG1	*Podzols*	*epigleyic fibrihistic spodic ortsteinic arenic*	553.6	2.8
XI	25	M3 M2 AM	*Histosols*	*saprihistic/hemihistic rheic eutric fluvic*	3,604.9	17.9
XII	26–27	S R	*Histosols*	*fibrihistic ombric endohemihistic dystric*	1,428.1	7.1
XIII	28–31	T	*Technosols*	*nudic relocatic transportic cumulatic spolic*	91.0	0.5
XIV	32–34	P C	Non-soils/grounds	quarries pits dumps outcrops: ground, regolith, artifacts	41.0	0.2

Table 13.4 SOC stocks in soil cover (SC) and its distribution in SC sublayers presented by large soil groups (LSG) in relation to the tested area (20,114 km^2)

LSG No	SOC pools of SC, in Tg	Share of LSG, in %	Distribution of SOC pools by SC sublayers[a]			Dominant (>70%) forest site types
			HC, in Tg	SS, in Tg	HC:SS, in %	
I	7.3	2.3	6.8	0.5	93:7	*Calamagrotis*-alvar, *Hepatica*
II	18.0	5.7	14.9	3.1	83:17	*Hepatica*, *Aegopodium*
III	18.5	5.9	10.7	7.8	58:42	*Oxalis*, *Oxalis-Myrtillus*
IV	6.7	2.1	2.8	3.9	42:58	*Rhodococcum*, *Oxalis-Rhodococcum*, *Myrtillus*
V	49.4	15.6	42.8	6.6	87:13	*Filipendula*, *Aegopodium*, *Carex-Filipendula*
VI	31.6	10.0	24.7	6.9	78:22	*Filipendula*, *Oxalis-Myrtillus*, *Aegopodium*, *Myrtillus*
VII	10.8	3.4	6.2	4.6	57:43	*Myrtillus*, *Polytrichum-Myrtillus*
VIII	16.0	5.1	11.0	5.0	69:31	*Filipendula*, *Carex-Filipendula*, *Carex*
IX	9.8	3.1	6.1	3.7	62:38	*Filipendula*, *Oxalis-Myrtillus*, *Myrtillus*, *Carex-Filipendula*, *Carex*
X	5.3	1.7	3.3	2.0	62:38	*Myrtillus*, *Vaccinium*, *Polytrichum-Myrtillus*
XI	113.4	35.8	64.8	48.6	57:43	*Oxalis* drn[b] fen, *Alder-birch* fen, *Myrtillus* drn bog, *Filipendula*
XII	28.8	9.1	15.6	13.2	54:46	Transitional bog, Oligotrophic bog, *Myrtillus* drn bog, *Vaccinium*
XIII	0.6	0.2	0.3	0.3	50:50	Reclamationed pits, *Myrtillus* drn bog
XIV	0.2	0.1	0.1	0.1	50:50	Reclamationed pits

(continued)

Table 13.4 (continued)

LSG No	SOC pools of SC, in Tg	Share of LSG, in %	Distribution of SOC pools by SC sublayers[a]			Dominant (>70%) forest site types
			HC, in Tg	SS, in Tg	HC:SS, in %	
Total	316.4	100.0	209.9	106.5	66:34	Totally 27 forest site types may be found
Organic soils	142.2	44.9	80.4	61.8	56:44	*Oxalis* drn fen, *Alder-birch* fen, *Myrtillus* drn bog, Transitional bog
Mineral soils	174.2	55.1	129.6	44.6	74:26	Totally 21 forest site types may be found

[a]SC—sublayers: HC—humus cover, SS—subsoil; [b]drn—drained

Table 13.5 Estimation of total stocks of SOC in the soil cover of Estonian forest land, which area is by the national forest inventory data 23,306 km^2 (Forestry Yearbook 2017 2018)

No of scenario	Characterization of the scenario	Total SOC stocks of forest land,[a] in Tg		
		SC	HC	SS
I	Non-tested territory's (3,192 km^2) SC composition is taken similar to that of the whole tested area	367	243	124
II	Non-tested territory's SC composition is taken similar to the tested area's mineral soils composition	353	237	116
III	It is supposed, that from non-tested territory's SC formed 50% mineral and other 50% peat soils	380	249	131
I	Total SOC stock in mineral soils	202	150	52
I	Total SOC stock in organic soils	165	93	72

[a]SC—soil cover, HC—humus cover, and SS—subsoil

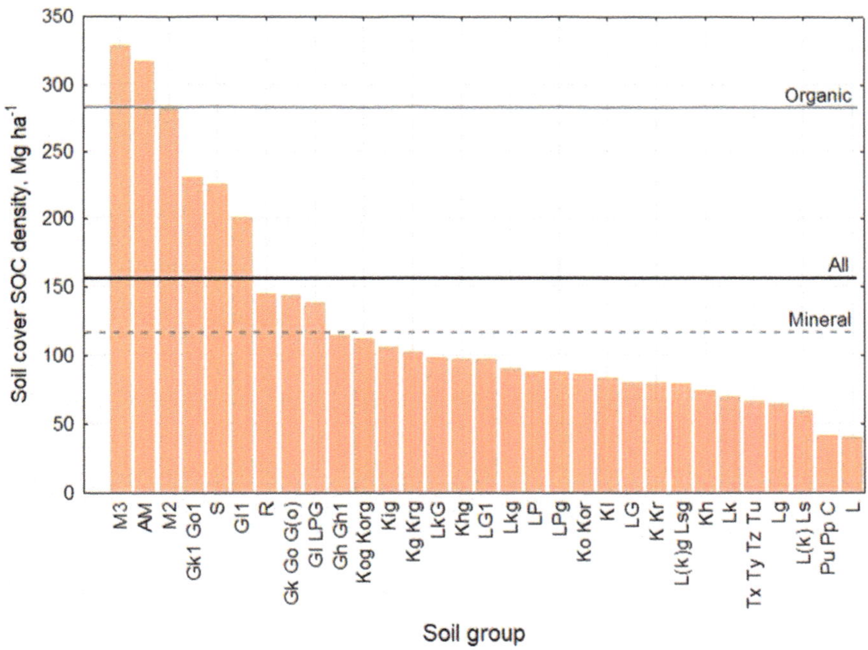

Fig. 13.8 Different soil species' SOC superficial densities (Mg ha^{-1}) per soil cover or solum. The mean density level for all soils, and separately for organic and mineral soils is given on the figure by horizontal lines. For soil names by their codes see Table 13.2

Photo 13.1 Variegated forest landscape with pine, birch, spruce and mixed-forest stands on the transitional area of Kõrvemaa and Pandivere Upland (Photo T. Kõlli). The prerequisite of the high pedodiversity of this area is its high geodiversity: lightly undulating calcareous loamy till plains are variegated here by chains of stony-rich eskers, gravelly sandy kame fields and paludified floodplain fens

The annual productivity of natural ecosystems on well-drained soils depends mainly on clay and SOM content and stocks in the soil profile but on wet soils from the moisture conditions. The matching of soil–plant systems by LSG may be followed on the basis of the data given in Tables 13.3, 13.4 and 13.6. The SC properties are characterized by WRB qualifiers (Table 13.3), by SOC sequestration capacity (Table 13.6) and its distribution in SC (Table 13.4), and as well by HC and SC thicknesses, HC types, dominating textures of SC and acidity of forest floors (Table 13.6). The plant cover of the ecosystems is characterized by the FST, dominating tree species and site quality classes (Tables 13.4 and 13.6). The maximum functioning activity of an ecosystem is observed in the presence of plant cover, which is suitable for soil properties (Kõlli 2002; Lõhmus 2004; Lutter et al. 2018).

HC type (Table 13.6) may be used as a complex landscape's indicator, which reflects the functioning of soil–plant system, among them it characterizes the intensity of biological turnover and the activity of detritus food chain organisms. For attaining ecologically sound land use or for increasing efficiency of soil resources utilization, the disharmonies in matching plant cover with SC or biodiversity with pedo(geo)diversity should be overcome. The environmental protection ability of soils

Table 13.6 Characterization of forest site conditions by the large soil groups (LSG)

LSG No	Mg SOC per 1 ha of SC[a]	Thickness, in cm		Dominant humus cover type	Dominant texture of SC	pH_{KCl} of forest floor	Dominant tree species	Quality class[c]
		HC[b]	SC					
I	84	17–23	24–56	Fresh and moist calci-mull	Ryhky loams on ryhk or limestone	4.9–5.0	*Pinus svlvestris, Picea abies*	III–IV
II	98	19–24	43–70	Fresh and moist forest-mull; fresh and moist moder-mull	Loams on ryhky (or pebble) loam	4.5–5.0	*P. abies, Betula pendula*	Ia–II
III	82	17–25	72–92	Fresh and moist moder	Loamy sands on loam	3.6–4.0	*P. abies, P. svlvestris*	Ia–II
IV	52	4–6	62–72	Fresh and moist mor; fresh and moist moder-mor	Sands	2.8–3.2	*P. sylvestris, P. abies*	II–III
V	143	23–29	34–44	Wet calci- and forest-mull	Loams on ryhky loam	5.0–5.8	*P. abies, B. pendula, Alnus incana, Populus tremula*	Ia–II
VI	138	22–28	50–60	Wet moder-mull; wet moder	Loamy sands, sandy loams, loams, clays	4.5–5.0	*B. pendula, P. tremula, A. incana*	Ia–III
VII	88	13–19	65–75	Wet moder; wet moder-mor; wet mor	Sands, loamy sands	2.7–3.1	*P. sylvestris, B. pubescens*	II–IV
VIII	230	20–26	40–50	Peaty mull	Eutric peats on loams	5.0–6.0	*B. pubescens, A. glutinosa*	II–IV
IX	200	16–20	46–56	Peaty moder	Mesic peats on loams and sands	3.5–4.8	*B. pubescens, A. glutinosa, P. svlvestris*	II–IV

(continued)

Table 13.6 (continued)

LSG No	Mg SOC per 1 ha of SC[a]	Thickness, in cm		Dominant humus cover type	Dominant texture of SC	pH_{KCl} of forest floor	Dominant tree species	Quality class[c]
		HC[b]	SC					
X	97	11–17	70–80	Peaty mor	Dystric peats on sands	2.4–3.0	*P. sylvestris*	III–IV
XI	315	30	50	Eu-and mesotrophic peat	Well and moderately decomposed peats	4.5–6.0	*B. pubescens, A. glutinosa*	III–IV (I–II)[d]
XII	202	30	50	Oligo-and mesotrophic peat	Slightly and moderately decomposed peats	2.3–4.2	*P. sylvestris*	IV–Va (III)[d]
XIII	66	<20	<50	–	Miscellaneous, mixed textures	–	*P. sylvestris*	–
XIV	41	<10	<30	–	Various mixtures	–	*P. sylvestris*	–

[a]SC—soil cover; [b]HC—humus cover; [c]By quality class scale of Lõhmus (2004): from Ia (highest) to Va (lowest); [d]In brackets—quality of drained soils

as an intrinsic property of the whole ecosystem level may be attained by the ecologically sound management of landscape (Kõlli et al. 2010). With land-use change (from natural to arable and vice versa), the more drastic changes occur in the fabric and properties of HC, whereas the SS rests in an almost unchanged state (Köster and Kõlli 2013).

The HS of natural forest landscapes or SOC throughout flux of SC is tightly connected with plant cover composition, productivity and diversity. Therefore, the awareness on the composition and properties of HC types and their relationship with plant cover and SOM decomposition potentiality are the basis of ecologically proper and sustainable management of land (soil) resources and protection of forest landscapes.

13.7 Conclusions and Outlook

- Comparative analysis of soil–plant mutual relationships on the background of pedo-ecological conditions' matrix revealed that (1) the vegetation diversity of an ecosystem depends on soil properties, being, therefore, a soil type-specific feature, and (2) the type of HC is a good ecological indicator in characterizing outlines of the biological turnover between soil and plant.
- In Estonian forest land's (23,306 km^2) soil cover totally 367 Tg of soil organic carbon is sequestrated, whereas 66.2% of it is located in biologically active humus cover and 33.8% in the subsoil.
- The pedodiversity of the landscape is an abiotic base for formation of optimal (specific to soil type) biodiversity. The ecologically sound matching of soil and plant covers is of pivotal importance in the reaching of sustainable ecosystem functioning and of good environmental status of the ambient area.
- In the formation and fabric of HC and HS of the whole forest landscape are clearly seen regional singularities, caused by soils and climatic conditions; consequently, in complex researches of landscapes a determination and analysis of humus cover types are necessary.

References

Arold I (2005) Estonian landscapes. Tartu Ülikooli Kirjastus, Tartu (in Estonian)

Astover A, Reintam E, Leedu E, Kõlli R (2013) Field survey of soils. Eesti Loodusfoto, Tartu (in Estonian)

Baritz R, Seufert G, Montanarella L, Van Ranst E (2010) Carbon concentrations and stocks in forest soils of Europe. For Ecol Manag 260:262–277

Bazilevich NI, Titljanova AA (2008) Biotic turnover on five continents: element exchange process in terrestrial natural ecosystems. Publishing House SB RAS, Novosibirsk (in Russian)

De Vos B, Cools N, Ilvesniemi H, Vesterdal L, Vanguelova E, Carnicelli S (2015) Benchmark values for forest soil carbon stocks in European: results from a large scale forest survey. Geoderma 251–252:33–46

Dixon RK, Brown S, Houghton RA, Solomon AM, Trexler MC, Wisniewski J (1994) Carbon pools and flux of global forest ecosystems. Science 263:185–190

Estonian Agri-project (1983) Metsamullad. Eesti NSV mullastik arvudes III:26–91 (in Estonian)

Estonian Agri-project (1985) Metsamullad. Eesti NSV mullastik arvudes V:3–71 (in Estonian)

Estonian Land Board (2012) Explanatory letter for the large scale digital soil map of Estonia (in Estonian). http://geoportaal.maaamet.ee/docs/muld/mullakaardi_seletuskiri.pdf

IUSS Working Group WRB (2015) World reference base for soil resources 2014, update 2015. International soil classification system for naming soils and creating legends for soil maps, world soil resources reports no 106. FAO, Rome

Kachinsky N (1965) Soil physics, vol I. University Press, Moscow (in Russian)

Kõlli R (1992) Production and ecological characteristics of organic matter of forest soils. Eurasian Soil Sci 24:78–91

Kõlli R (2002) Productivity and humus status of forest soils in Estonia. For Ecol Manag 171:169–179

Kõlli R, Ellermäe O (2001) Soils as basis of Estonian landscapes and their diversity. In: Mander Ü, Printsmann A, Palang H (eds) Development of European landscapes. Conference proceedings, vol II. Publicationes Instituti Geographici Universitatis Tartuensis, 92. IALE, Tartu, pp 445–448

Kõlli R, Köster T (2018) Interrelationships of humus cover (pro humus form) with soil cover and plant cover: humus form as transitional space between soil and plant. Appl Soil Ecol 123:451–454

Kõlli R, Asi E, Köster T (2004) Organic carbon pools in Estonian forest soils. Balt For 10(18):19–26

Kõlli R, Astover A, Noormets M, Tõnutare T, Szajdak L (2009a) Histosol as an ecologically active constituent of peatland: a case study from Estonia. Plant Soil 317:3–17

Kõlli R, Ellermäe O, Köster T, Lemetti I, Asi E, Kauer K (2009b) Stocks of organic carbon in Estonian soils. Est J Earth Sci 58(2):95–108

Kõlli R, Köster T, Kauer K, Lemetti I (2010) Pedoecological regularities of organic carbon retention in Estonian mineral soils. Int J Geosci 1:139–148

Kõlli R, Tõnutare T, Lang M, Szajdak LW, Kauer K, Rannik K (2018) Evaluation of the soils' humus status in Estonian forest landscapes. In: Sychev VG, Mueller L (eds) Novel methods and results of landscape research in Europe, Central Asia and Siberia, vol III. Landscape monitoring and modelling. Moscow, Russian Academy of Sciences, pp 60–64

Köster T, Kõlli R (2013) Interrelationships between soil cover and plant cover depending on land use. Est J Earth Sci 62(2):93–112

Laasimer L (1965) Vegetation of the Estonian SSR. Valgus, Tallinn (in Estonian)

Lang M, Kõlli R, Nikopensius M, Nilson T, Neumann M, Moreno A (2017) Assessment of MODIS NPP algorithm-based estimates using soil fertility and forest inventory data in mixed hemi-boreal forests. For Stud 66:49–64

Lang M, Kaha M, Laarmann D, Sims A (2018) Construction of tree species composition map of Estonia using multispectral satellite images, soil map and a random forest algorithm. For Stud 68:5–24

Lõhmus E (2004) Estonian forest site types. Eesti Loodusfoto, Tartu (in Estonian)

Lutter R, Kõlli R, Tullus A, Tullus H (2018) Ecosystem carbon stocks of Estonian premature and mature managed forests: effects of site conditions and over storey tree species. Eur J For Res. https://doi.org/10.1007/s10342-018-1158-4

Mander Ü, Palang H, Tammiksaar E (1995) Landscape changes in Estonia during the 20th century. Year-Book of EGS, vol 29

Peterson U, Püssa K, Liira J (2004) Issues related to delineation of forest boundaries on landsat thematic mapper winter images. Int J Remote Sens 25(24):5617–5628. https://doi.org/10.1080/01431160412331291314

Peterson U, Liira J, Püssa K (2008) Metsaga alade ning lageraiete ja nendega sarnaste häiringute kaugseire. In: Väljataga K, Kaukver K (eds) Kaugseire Eestis. Tallinn (in Estonian), Iloprint, pp 48–68

Raukas A (ed) (1995) Estonia. Nature. Valgus, Tallinn (in Estonian)
Raukas A, Teedumäe A (1997) Geology and mineral resources of Estonia. Estonian Academy Publishers, Tallinn
Reintam L, Kull A, Palang H, Rooma I (2003) Large-scale soil maps and supplementary database for land use planning in Estonia. J Plant Nutr Soil Sci 166:225–231
Tjurin IV (1935) Comparative study of the methods for the determination of organic carbon in soils and water extracts from soils. In: Materials on genesis and geography of soils, ML Academy of Science USSR, pp. 139–158 (in Russian)
Valk U, Eilart J (1974) Estonian forests. Valgus, Tallinn (in Estonian)
Yearbook Forest 2017 (2018) Keskkonnaagentuur, Tallinn
Zanella A, Jabiol B, Ponge J-F, Sartori G, DeWaal R, Van Delft B, Graefe U, Cools N, Katzensteiner K, Hager H, Englisch M (2011) A European morpho-functional classification of humus forms. Geoderma 164:138–145

Chapter 14
Dynamics of Soil Organic Matter in Agricultural Landscapes

Uwe Franko and Felix Witing

Abstract Soil organic matter is an essential key in the functioning of soils and landscapes. Not only because of its importance as a carbon sink for the mitigation of climate change but also to recommend farmers sustainable soil management, the modelling of organic matter turnover is a well-integrated part of soil science. This article provides a short review of modelling approaches for carbon and nitrogen fluxes related to organic matter turnover with an example at landscape scale where beside dedicated process simulation indicators are helpful tools to assess changes in land management. This is explained in more detail for the development of biogas production and its feedbacks to soil organic matter.

Keywords Soil organic matter (SOM) · SOM turnover · Biological active time · Organic nitrogen turnover · Modelling SOM dynamics · Climate change · SOM in agricultural landscapes

14.1 Introduction: The Relevance of Soil Organic Matter

Soil organic matter (SOM) was recently put more into focus as an important sink in the global carbon cycle (Stockmann et al. 2013, 2015) that may be relevant to reduce the increase of atmospheric CO_2 and therewith global warming. Besides this, SOM is for long known as a key component of soil quality and productivity as it controls several physical, chemical and biological soil properties (Ondrasek et al. 2019; Wiesmeier et al. 2019). Changes in the SOM content of soils can modify the soil structure and thus influence aeration, infiltration, water-holding capacity, structural

U. Franko (✉)
Department of Soil System Science, UFZ—Helmholtz Centre for Environmental Research, Halle, Germany
e-mail: uwe.franko@ufz.de

F. Witing
Department of Computational Landscape Ecology, UFZ—Helmholtz Centre for Environmental Research, Leipzig, Germany
e-mail: felix.witing@ufz.de

W. Mirschel et al. (eds.), *Landscape Modelling and Decision Support*,
Innovations in Landscape Research,
https://doi.org/10.1007/978-3-030-37421-1_14

stability and soil erodibility. From an ecological point of view, SOM is important for nutrient availability, soil biodiversity and the buffering of harmful substances. The fundamental importance of SOM for soil functioning is probably the reason for the long history of approaches developed to support the farmer's decision about a proper SOM management. In most cases, the amount or concentration of soil organic carbon (SOC) is the indicator for SOM for the experimental work as well as in modelling. Nevertheless, it is important to include the balances of carbon as well as that of nitrogen among other nutrients that may cause an environmental burden (Davies et al. 2016).

14.2 Modelling SOM Dynamics: State of the Art

From the first simple approaches to currently used more sophisticated models, it is a common understanding that SOM dynamics follow an exponential time course approaching a theoretical steady state if given management is continued under constant environmental conditions. While the first models took SOM as a homogeneous substrate—often referred to as humus—it is now generally accepted that SOM has to be subdivided into different pools with individual turnover times (Paul 1984, 2016). Furthermore, in the current understanding, it is assumed that SOM is a product of soil microbes that feed on the external organic matter from organic amendments and crop residues (Kallenbach et al. 2016). Therefore, all models include at least one pool with a fast turnover that represents this microbial biomass. The wide range of SOC concentrations that are observed at different soil types and climate conditions for fields that are under similar management led to two basic assumptions. First, the different environmental conditions control the turnover intensity in a way that under cold and wet conditions the decomposition of SOM is significantly reduced and second, that a part of the SOM is specifically protected from microbial breakdown. Only recently it became clear that this protection is controlled not only by soil texture but also by soil structure. Due to the vague nature of this protection, a part of SOM was assumed to be inert meaning that it would not change over time. The size of this inert pool was related to soil texture (Körschens et al. 1998; Rühlmann 1999) or to the overall amount of SOC (Falloon et al. 1998)—the latter related to the widely used RothC model.

Von Lützow et al. (2008) classified SOM into three pools with turnover times of 1–10 years (active pool), 10–100 years (intermediate pool) and >100 years (passive pool) where the mechanisms for stabilization were identified as aggregation and interaction with mineral surfaces beside the alteration of chemical structure. Kuka et al. (2007) provided a modelling approach for carbon turnover in arable soils based on the hypothesis that long-term stabilization of carbon in soil is mainly a function of its localization within the soil pore space. The main assumption of the CIPS (Carbon turnover In Pore Space) model is that the biological activity is not evenly distributed through the whole pore space. Because of the poor aeration in the micro-pores, they show very low biological activity leading to a strong protection of the carbon localized in this pore space. CIPS was integrated into the CANDY model providing an opportunity to link the quality-controlled primary stabilization processes (recalcitrance of organic matter) with the soil structure dependent stabilization processes (location of turnover) that may be considered as an attribute for a new generation of soil carbon turnover models. The idea to take soil structure as the main driver for SOM stabilization was further developed in the CCB model that is a simplification of CANDY working in yearly time steps. This includes the implementation of a meta-model to describe the site-specific turnover conditions in terms of the Biologic Active Time (BAT).

14.2.1 Biologic Active Time to Describe Turnover Conditions

Using the BAT concept helps to describe the impact of environmental conditions on the biologic activity in soils, which is the core driver behind the turnover of SOM (Franko et al. 1995). In a given time interval a certain biologic activity in a suboptimal environment will produce a specific turnover result. The same results occur when splitting the time into BAT intervals (optimal turnover environment) and non-BAT intervals. During the BAT interval, the microbial activity is only limited by the substrate, while during non-BAT there is no biological turnover activity at all. In the CANDY model, the calculation of the BAT interval includes the effects of soil temperature, soil water, and soil aeration.

The scheme in Fig. 14.1 demonstrates the principle of how different intensities of uniform time steps (Fig. 14.1a) are transformed into time steps of different length and uniform intensity (Fig. 14.1b). The calculated turnover, symbolized by the bar area, will be the same for both approaches, anyway. In the latter case (B) the new calculated time step ($\Delta t'$) is a product of the reduction function R(t) and the origin time step (Δt). In this case, the non-BAT time step is represented as the blank space between the BAT bars (Fig. 14.1b). The CANDY model calculates BAT in daily time steps for each of the 3 top soil layers (0–30 cm). For the CCB model, only the annual BAT sum is used as an indicator of the potential turnover under the given conditions. Based on simulation results with the CANDY model a meta-model for the annual sum of BAT was developed by Franko and Oelschlägel (1995) that considers soil texture and annual climate data (air temperature and rainfall) including the annual irrigation amount within the natural rain and an additional adaptation for conservation tillage (no mixing of soil layers). Moreover, in CCB also the effect of non-tillage on SOM is applied as a reduction of BAT (Franko and Spiegel 2016) that leads to improved accumulation rates of organic matter (OM) in soil depending on the soil type.

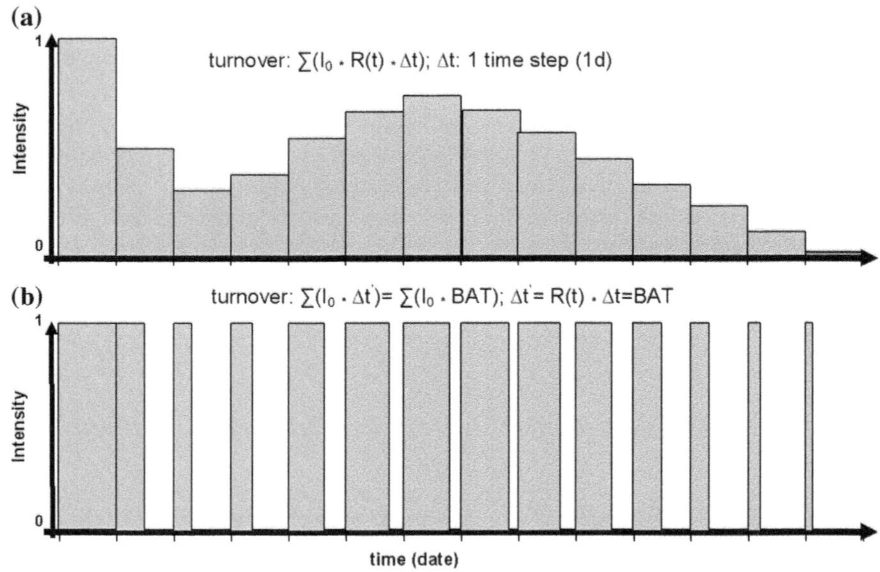

Fig. 14.1 Schematic representation of the turnover calculation by the standard approach (**a**) and the BAT approach due to the transformation of time steps (**b**)

14.2.2 Indicators Characterizing SOM Turnover

While the general understanding of turnover processes is similar in all currently used SOM models, the CCB model contains specific approaches and indicators that are usable for a generalization and upscaling of the process knowledge. As explained in the previous chapter, one of these indicators is the annual sum of BAT. It describes the impact of site-specific environmental conditions on SOM turnover in one value and enables easy comparison across different regions and scales. The federal state of Saxony in East Germany shows a large variety in soils, climatic conditions and tillage practices, which results in significant differences in regional turnover activities (Witing et al. 2016; Fig. 14.2). The arable land in the north and northeast of Saxony is characterized by sandy soils and thus high turnover rates and corresponding BAT values (45–55 days). The rather silty soils in the centre and south have lower turnover rates (10–25 days), which decreases from north to south due to the mountain regions and relating climatic conditions in the south of Saxony.

Another source of heterogeneity is connected with the properties of external organic matter from organic amendments and crop residues. Depending on the composition of these organic materials they have a very different ability for the reproduction of SOM that is continuously being decomposed by soil microbes. Generally, this quality of fresh organic matter (FOM) is connected with the amount of microbial biomass that grows upon a specific substrate and can be characterized with an efficiency factor that implicitly includes the carbon use efficiency of the microbes. Some

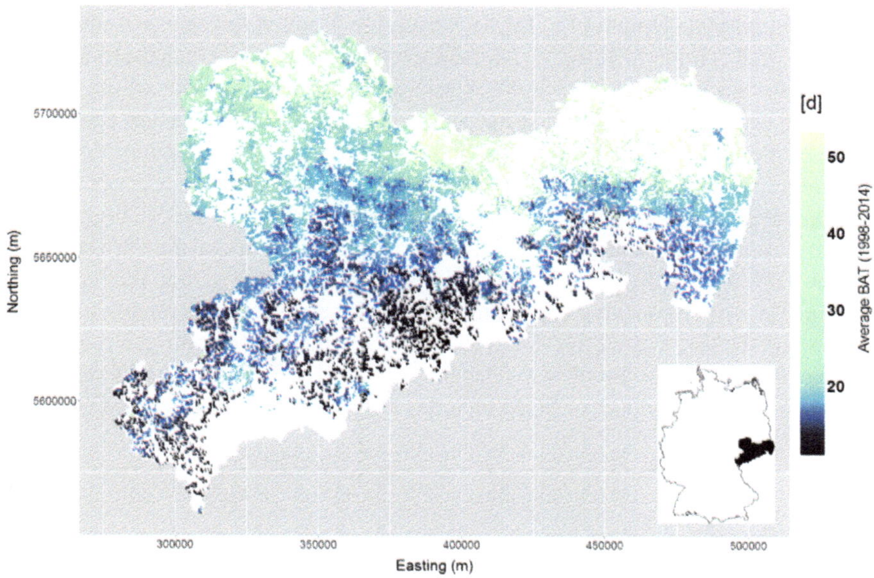

Fig. 14.2 Average annual biologic active time (BAT) of the arable land of Saxony (Germany) for the period 1998–2014 based on calculations of Witing et al. (2016)

authors relate this to the C/N ratio of the substrate and others identified the OM/C relation as a useful indicator. The best answer about the quality of FOM for SOM reproduction may come directly from the soil microbiology—by analysing incubation experiments with the help of a model. Here again, the BAT concept is very helpful to relate the laboratory conditions to the field. An example of the parameterization of biogas digestate is given by Prays et al. (2018). The model that is implemented in CCB as well as in CANDY has two characteristics that describe the turnover properties of external or fresh organic matter. The first is the turnover time, embodied in the *k*-value, which describes the stability of the substrate against a microbial attack. The second is the synthesis coefficient *eta*, which describes the quality of a substrate to build up new SOM and thus the amount of new synthesized SOM (assuming that is microbial biomass) per unit of decomposed C from FOM. The product of C_{FOM}*eta describes the amount of newly formed SOM. This carbon flux is referred to as reproducing carbon C_{rep}, a further indicator that describes how much of decomposed SOM is replaced.

14.2.3 Turnover of Organic Nitrogen

The dynamics of SOM are traditionally quantified in terms of carbon because this is usually observed from soil samplings. From an ecological point of view, nitrogen is also an important element of SOM. However, the nitrogen fluxes that connect SOM with the pool of mineral nitrogen (Nmin) are often more relevant than the total amount of organic nitrogen stored in SOM. While the dynamic behaviour of the different pools of organic matter depends mainly on the quality of the carbon compounds, the complex interaction between carbon and nitrogen fluxes is controlled by the different C/N ratios of these pools. How the individual material properties may lead to N mineralization or immobilization is schematically demonstrated in Fig. 14.3. In a first step, the decomposition of FOM results in a release of mineral nitrogen controlled by the C/N ratio of the given FOM pool γ_{FOM}. A part of the carbon that is released from FOM will be consumed by the microbial biomass that grows on this source and represents the first step of SOM reproduction. The quantity of nitrogen required for the newly formed amount of SOM depends on the C_{rep} flux and the C/N ratio of the active SOM. The turnover of SOM includes also the mineralization of SOM carbon

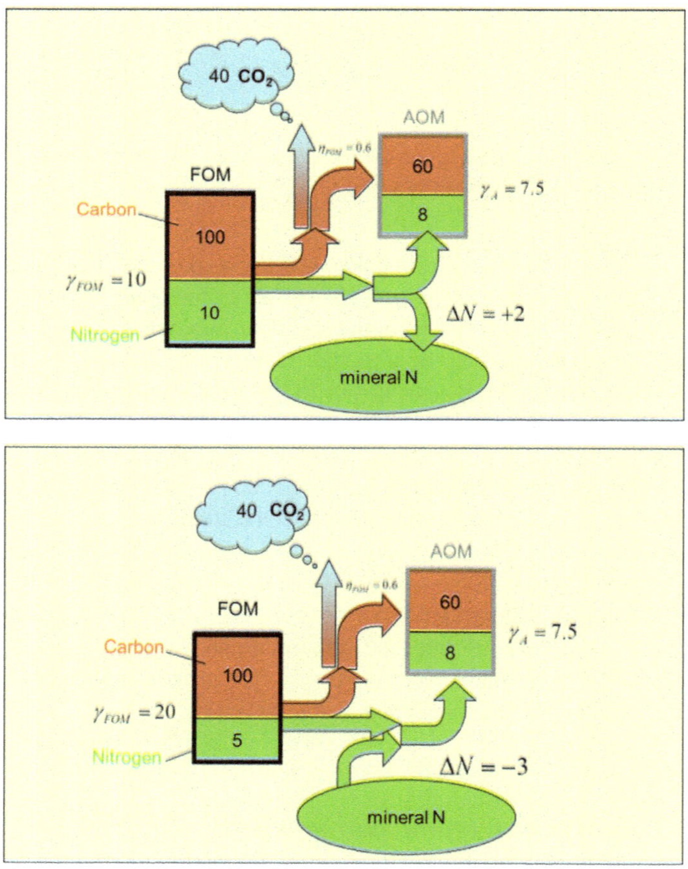

Fig. 14.3 Examples for the net N mineralization and immobilization during the turnover of fresh organic matter (FOM) into active soil organic matter (AOM) (numbers are approaches for more convenient calculation and don't exactly match model parameters)

to CO_2. Hypothesizing that nitrogen mineralization from SOM is controlled by the dynamics of carbon turnover the nitrogen release from the mineralization process is determined by the C/N ratio of the active SOM pool (AOM). The total nitrogen flux into (positive ΔN values) or from (negative ΔN values) the pool of mineral nitrogen results from the interaction of FOM decomposition and reproduction of new SOM. For better illustration Fig. 14.3 is presenting two examples with respect to FOM decomposition and nitrogen flux from FOM into AOM. Depending on the C/N ratio of the given FOM pool γ_{FOM} (10 or 20) the FOM decomposition results in a net N mineralization (+2) or immobilization (−3). Possible mineralization of the AOM pool is not considered within this example.

14.3 Controlling Soil Organic Matter on Field and Farm Scale

14.3.1 Agricultural Management and Carbon Dynamics

While the breakdown of SOM to CO_2 is mainly controlled by site conditions (with additional influence from tillage and irrigation) it is the agricultural management in terms of cropping and organic amendment that controls the build-up of new SOM. All plants leave organic residues in the soil from roots, harvest residues (like stubble) as well as root exudates and litterfall during the vegetation. Moreover, part of the harvested biomass (e.g. straw from cereals) may be removed as by-product or is left on the field as an organic input. Traditionally, farmers would apply farmyard manure too but this is more and more replaced by slurry as well as other organic amendments like compost, sewage sludge, biogas digestate, etc. Depending on the material properties all these sources have a different impact on the formation of new SOM. In Tables 14.1 and 14.2 are some data compiled that are based on the parametrization that is used in the CCB model. The amount of plant-derived organic matter may depend on the crop yield but as this relation has a high variability it is not in all cases possible to find reliable parameters.

Taking into account the crop yield as the driver for the OM input to soil it is self-evident that all management activities that control the yield—namely the application of mineral fertilizer—have an impact on SOM accumulation. This holds also for soil tillage and irrigation but these have also an effect on the turnover activity itself. Irrigation will change the water content of the soil and soil tillage is disturbing the soil structure and this way reducing the protection of SOM in the micro-pores of the soil. In the simplified approach of the CCB model, the choice between tillage or non-tillage operations results in an adaption of BAT according to the site conditions.

Table 14.1 Matter import to soil for selected crops. N_{res}: nitrogen import to soil with roots and crop residues, C_{res}: carbon import to soil with roots and crop residues, C_{rep}: newly formed carbon in SOM from matter input with roots and crop residues

Crop	Yield (t/ha)	N_{res} (kg/ha)	C_{res} (kg/ha)	C_{rep} (kg/ha)
Clover-grass	33.9	107	2515	1006
Durum wheat	4.0	10	624	343
Maize grain	11.2	40	2022	1254
Oat	3.3	13	664	365
Potato	27.7	29	1152	576
Spring barley	3.0	9	546	300
Sugar beet	26.9	20	400	140
Winter barley	6.2	18	896	493
Winter rye	4.4	15	752	414
Winter wheat	4.8	19	954	524

Table 14.2 Matter import to soil for selected organic amendments, related to the input of fresh matter (FM). C_{org}: organic carbon, N_{org}: organic nitrogen, N_{total}: sum of organic and inorganic nitrogen, C_{rep}: newly formed carbon in SOM from organic amendments

Organic amendment	C_{org} (kg/1000 kg FM)	N_{org} (kg/1000 kg FM)	N_{total} (kg/1000 kg FM)	C_{rep} (kg/1000 kg FM)
Green matter	60	4	5	21
Sewage sludge	52	7	7	41
Straw	340	6	6	153
Bio-waste compost	156	10	10	62
Beet leafs	65	3	3	37
Cattle slurry 10% dm	30	2	4	18
Manure C/N18	97	5	7	62
Manure compost	336	7	8	282

14.3.2 Humus Balance

Humus balancing methods that can be used in agricultural practice have to be simple tools for the assessment of interactions between agricultural land use and SOM (Brock et al. 2013). The term 'humus balance' covers both simple models to quantify changes of SOM (or SOC in particular) of arable soils and models that refer to the optimization of soil productivity of arable soils by calculating organic fertilizer demand, without quantifying changes in SOM or SOC. This situation naturally has caused many discussions and misunderstandings. The German VDLUFA method belongs to the second group assessing a management scenario in terms of humus equivalents (HÄQ). Crops are split into groups where each group is assigned to a certain amount of HÄQ that is required to sustain soil fertility or that is delivered to reduce the requirement of organic amendments. External sources of organic matter like straw, manure, and compost have specific HÄQ values according to their quality to reproduce SOM. The analogy to the C_{rep} concept described above is obvious and can be used to translate the parameters into each other using farmyard manure as a reference knowing that 1t manure represents about 62.2 kg C_{rep} and is equivalent to 43.2 HÄQ units.

14.4 New Challenges on Landscape Scale

14.4.1 Considering the Heterogeneity

There is a strong spatial heterogeneity in SOM concentrations within and across agricultural landscapes. Varying conditions in soil and climate are the baseline for these local SOM stocks. Policy, market but also historic preconditions drive decisions on field and farm management as well as on interactions (esp. carbon flows) between farms (Fig. 14.4). These factors strongly modify the cycle of organic matter in agricultural landscapes and as a result the local carbon input to SOM (C_{rep}, e.g. via cultivated crops, type and amount of organic amendments applied) as well as local rates of SOM turnover (e.g. via tillage system, irrigation). Furthermore, due to its 'rather slow' response to changes (long time to reach steady state), current SOM concentrations in agricultural soils are a complex result of historic changes in climate and agriculture.

Typically SOM models are applied on the pedon level where a homogeneous dataset in terms of soil, climate and management can be assumed. On higher levels from field blocks over farms to administrative units or landscapes, it is more and more difficult to handle these heterogeneous study areas as a puzzle of homogeneous parts. Especially data on land management and regional matter fluxes may be hard to retrieve, mainly because this information has to be collected from many data holders. Therefore, the use of statistical data on agricultural land-use and management aggregated within census data may be an option. The CCB model can be used in a special 'regional mode' where it is able to use aggregated management data. While the standard mode requires a complete time series of management activities and assumes a homogeneous management across the whole simulation unit, in regional mode all management activities have an additional parameter describing the area share of a spatial simulation unit to which this activity is applied. This sets some limits for the use of the model results that are now only valid for the landscape unit as a whole and not for each point within that unit. An exemplary result of this approach is shown in Fig. 14.5, where the changes in SOC stocks (1998–2014) of the arable land of Saxony

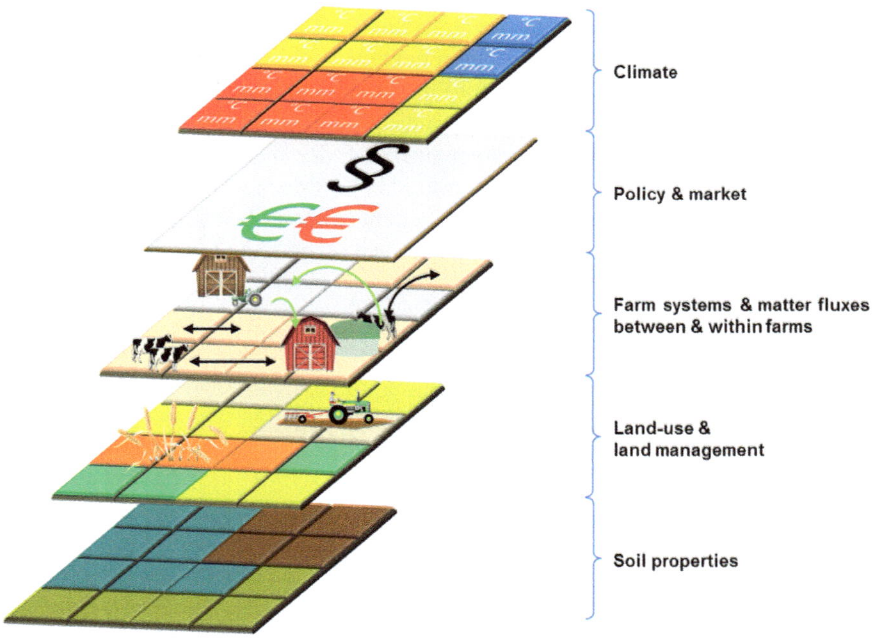

Fig. 14.4 Selected components and drivers of SOM dynamics in an agricultural landscape

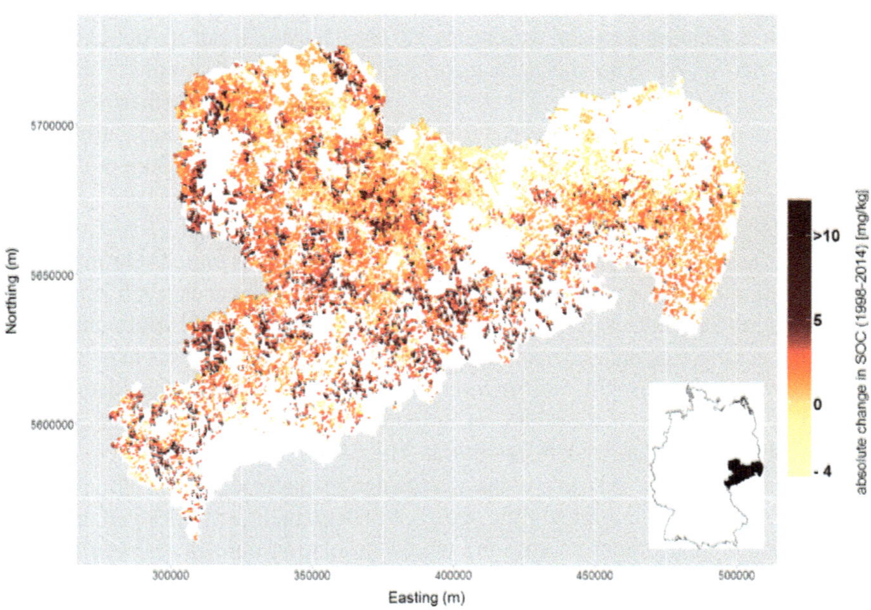

Fig. 14.5 Absolute change in the SOC concentration (mg/kg) of the arable land of Saxony (Germany) for the period 1998–2014 based on calculations using the CCB model (Witing et al. 2019)

have been quantified on a 500 m grid. For this, spatial data on soils and climate have been combined with statistical information on agricultural management on various levels (Witing et al. 2019).

Spatial fluxes or direct spatial interactions of SOM in agricultural landscapes are limited and mostly related to soil erosion. However, soil erosion aspects are rather considered in hydrologic modelling than in typical SOM models. There are some attempts trying to link both aspects, e.g. integrating SOM modules into hydrologic models (e.g. the SWAT model). However, due to their hydrologic background, current approaches to include SOM in these integrated simulated environments are addressing different research questions than typical SOM models.

SOM dynamics in agricultural landscapes is also indirectly affected by spatial fluxes, namely by fluxes related to the agricultural cycle and market of organic matter. A part of the biomass that is taken from various fields will be reapplied to agricultural soils in different forms of organic amendments (see also Fig. 14.4 component 'Farm systems and matter fluxes'). Traditionally the local cycle of agricultural biomass was quite closed, except for the products brought to the market (e.g. grain, milk, meat). Manure and slurry have been reapplied to the agricultural fields of the same farm that provided fodder and litter for the stable. However, this cycle is getting more and more complex, including the movement of organic matter between different specialized farms, the import of additional carbon (e.g. fodder) from outside of the regional agricultural system, an diversified and increased use of agricultural carbon (e.g. production of biogas from animal slurry) and by this a potentially reduced return of organic matter to soils as well as new types of organic amendments (e.g. biogas digestate). Different political, societal and environmental drivers affect these fluxes and a broader view on the agricultural systems is required to assess the impact of these drivers on SOM.

14.4.2 Changing Carbon Fluxes in Agricultural Landscapes—The Biogas Example

A prominent recent example, requiring a systematic view of the overall fluxes of agricultural carbon in a landscape, is the inclusion of biogas production into the agricultural systems of Germany. Mainly driven by political efforts, the number of biogas plants installed in Germany increased from approximate 700 (in 1999) to almost 8700 (in 2016), making Germany the largest producer of biogas in the European Union (Daniel-Gromke et al. 2017, 2018). This expansion of the agricultural system has modified agricultural management (e.g. cultivated crops, digestate application instead of slurry) and as a result the SOM cycle of the agricultural landscape. Furthermore, when biomass is used for energy purposes, carbon is taken out of the agricultural cycle (e.g. biogas-methane = CH_4) and is not available to be returned to the soil. SOM models like CCB can help to understand how the carbon input to the soil within these agricultural landscapes has changed due to bioenergy.

First results based on calculations using the CCB indicator C_{rep} for the region of Saxony indicated that the overall flux of carbon into SOM of arable land increased during the time period of the establishment of the biogas industry (Witing et al. 2018). Although biogas is consuming carbon from organic amendments (e.g. slurry) that initially would have been returned to soil, this deficit in the matter cycle has been compensated at the level of the agricultural landscape (Fig. 14.6). On one side this is an effect of changes in cultivated crops and a resulting overall increase in the carbon input to SOM supplied by crop residues and crop by-products. However, these changes in Saxony, like increased cultivation of winter wheat, maize, winter rape, field grass and sugar beet, can only partly be related to biogas production, as there have been several drivers affecting the agricultural systems in the period under study (2000–2011). On the other side, due to biogas, a part of the carbon from agricultural crop yield did not leave the agricultural cycle by fluxes into, e.g. the food market: The feedstock mix of a biogas plant typically includes plant-based substrates like maize silage and cereals. If those market-relevant, plant-based substrates are

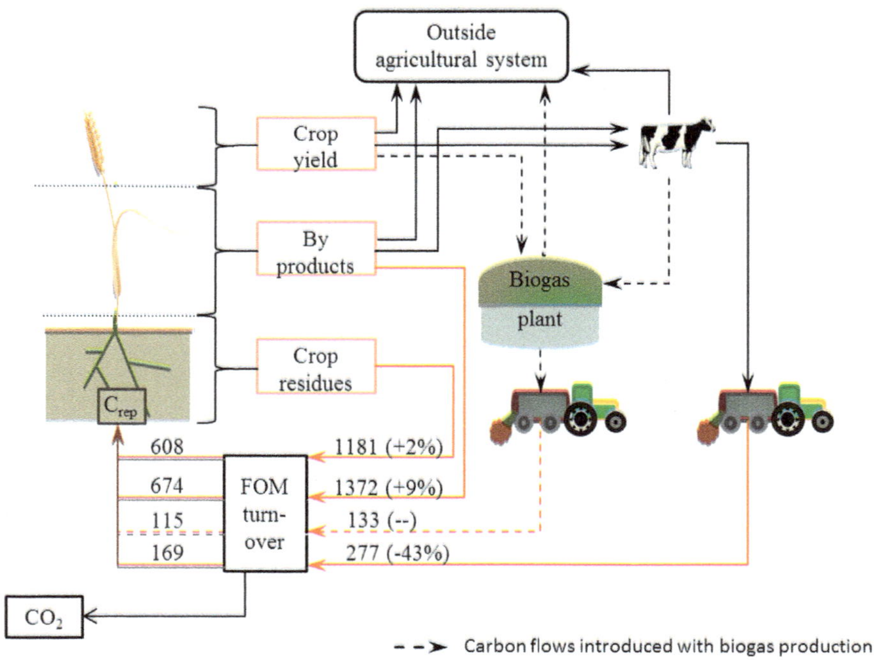

Fig. 14.6 Selected carbon flows in the regional cycling of agricultural matter and their quantification (kg C ha^{-1}) for the federal state of Saxony (Germany). The numbers represent the total carbon flux to soil from fresh organic matter (C_{FOM}) and the C_{rep} input to SOM (after FOM turnover) for 2011 and indicate the relative change compared to 2000 (% in brackets). Different pools contribute to the C_{FOM} flux: crop residues, crop by-products, biogas digestate and livestock excrement. All sources of FOM have a different quality for the formation of new SOC. The C_{rep} flux is aggregating these differences and can be used as an indicator in a given environment to characterize the land use regarding SOC storage

directly used for fermentation, a (small) part of the carbon is brought back to the regional agricultural carbon cycle and soils via biogas digestate. Nevertheless the area requirements of biogas production are high, which resulted, e.g. in a loss of fallow land for the study region Saxony. Furthermore, it is important to consider potential displacement effects, like unintended indirect land-use change around the world that could be induced by an increased import of fodder.

14.4.3 Climate Change and Soil Organic Carbon

Soil organic matter is the most important terrestrial carbon reservoir and is at the same time affecting the GHG exchange between atmosphere and biosphere from a global as well a local perspective. Globally it affects climate conditions and as result, e.g. plant growth, due to its ability to contribute to climate change mitigation. Locally SOM affects soil temperature and water conditions and as a result, also plant growth and the amount of carbon flux into SOM. Obviously, climate change will have an impact on BAT. Franko et al. (2015) did show for the region of central Germany that climate change will increase BAT. Assuming the C_{rep} flux remains constant, the steady-state level of SOM will decrease with an increase of BAT and consequently, more C_{rep} is required to keep the system on the same level. This does not mean to sustain the current SOC stock but the transient state. The required increase of C_{rep} is indicated by the relation of future BAT to past BAT and regional analysis can reveal where the most problematic regions are located.

The capabilities of soils to fight climate change are recognized by many actors. As a result of the COP 21 meeting in Paris, the initiative 4 per 1000 started with the background that a better supply of organic matter to the soil can help to improve soil fertility and mitigate climate change at the same time. If the current amount of SOC all over the world could annually be increased by 0.4% this additional storage would compensate for the CO_2 emissions from fossil fuel burning.

Generally, this aim may be reached if (i) the input of fresh organic matter to the soil is increased or (ii) the turnover of organic matter in the soil is reduced. The first option may be ecologically worthwhile if the considered changes in land-use and management also contribute to other economic and environmental objectives (like increase food production, reconstruction of landscape elements like afforestation or grassland). The second option can be used on arable land if the tillage system is changed to 'no plough'. In this case, the organic matter will have a longer turnover time in soil. The effect can be quantified by an adaptation of BAT calculation (Franko and Spiegel 2016) and is dependent on site conditions with a higher effect on heavy soils. From the current theoretical understanding of SOM turnover can be deduced that option (i) only makes sense for a limited time frame because the accumulation of SOC follows a non-linear timeline coming to a steady state with no further increase of carbon stocks while the carbon input has to take place at a constant rate requiring every year the same amount of FOM even if there is no further increase of SOC stock at the steady state of the system. Moreover, if the input rate is reduced, the soil will turn into a carbon source again which should be kept in mind when strategies for carbon sequestration are discussed.

14.4.4 Agricultural Measures to Increase SOM

We have shown that the implementation of biogas production can lead to a win-win situation where plant-derived biomass can be used for energy production that may reduce the exploitation of fossil sources and the recycling of biogas digestate still leads to an improved SOM reproduction. Of course, there are other options to generally increase the global SOC stock but especially in the case of agricultural systems, it should be kept in mind that SOC is undergone a microbial turnover and keeping a certain SOC stock at a constant level requires a steady input of FOM. Moreover, SOM is coupled with other matter fluxes that have to be kept within an environmentally acceptable range. Therefore, any increase of SOM should be related to an additional improvement of soil functions. Examples are the following:

- catch crops that increase carbon input to soil, decrease nutrient losses and reduce the risk for soil erosion,
- conservation tillage that may slow down the SOC decomposition and reduce the risk for soil erosion,
- ecological cropping systems that are adapted to the changing climate and increase the total biomass yield and conserve landscape functions.

14.5 Basic Concepts of the CCB Model

Readers that want to dig deeper into the special approaches of the CCB model on which our example was based upon find here a summary of the main features and the appropriate references:

- Quantification of the site specific turnover conditions in the BAT indicator: Franko and Oelschlägel (1995).
- Impact of conservation tillage on the turnover conditions: Franko and Spiegel (2016).
- Aggregation of different organic sources into the C_{rep} carbon flux that represents newly formed SOM: Franko (1997), Franko et al. (2011).
- Model structure to describe turnover of decomposable carbon together with the physically stabilized pool of long term stabilized OM: Franko and Merbach (2017).

14.6 Conclusions and Outlook

Any improvement of our process understanding about SOM turnover should allow for a better prediction of SOM dynamics in agricultural landscapes. Smith et al. (2018) point out that future models may require better integration of microbial interactions with organic matter to describe turnover processes with more detail and with higher resolution. This can be expanded also to mesofauna in soil. But model quality is probably not the most narrow bottleneck in this case. As model predictions depend on the quality of initial values it is very important to find quick, cheap and still reliable methods to determine the amount of SOC or SOM and, for better assessment, to relate this to soil functions. Furthermore, the amount and quality of all organic matter that enters the soil to be transformed to SOM are crucial information that is in most cases estimated using parameters in relatively general classifications. Already the estimation of the amount of OM from plant residues and roots is based on simple assumptions about its relation to crop yield that would require much more experimental work to be more reliable. On landscape scale, it is usually a problem to gather very detailed management data in a high spatial and temporal resolution. Therefore regional approaches are better to be based on indicators from the aggregated information that allow easy upscaling like in the example of biogas production within this article.

References

Brock C, Franko U, Oberholzer HR, Kuka K, Leithold G, Kolbe J, Reinhold J (2013) Humus balancing in Central Europe–concepts, state of the art, and perspectives review article. J Plant Nutr Soil Sci 176:3–11

Daniel-Gromke J, Rensberg N, Denysenko V, Trommler M, Reinholz T, Völler K, Beil M, Beyrich W (2017) Anlagenbestand Biogas und Biomethan – Biogaserzeugung und -nutzung in Deutschland., DBFZ Report Nr. 30., DBFZ - Deutsches Biomasseforschungszentrum, Leipzig

Daniel-Gromke J, Rensberg N, Denysenko V, Stinner W, Schmalfuß T, Scheftelowitz M, Nelles M, Liebetrau J (2018) Current developments in production and utilization of biogas and biomethane in Germany. Chem Ing Tec 90(1–2):1–20

Davies JAC, Tipping E, Whitmore AP (2016) 150 years of macronutrient change in unfertilized UK ecosystems: observations versus simulations. Sci Total Environ 572:1485–1495

Falloon P, Smith P, Coleman K, Marshall S (1998) Estimating the size of the inert organic matter pool from total soil organic carbon content for use in the Rothamsted carbon model. Soil Biol Biochem 30(8–9):1207–1211

Franko U (1997) Modellierung des Umsatzes der organischen Bodensubstanz. Arch Acker- Pfl Boden 41:527–547

Franko U, Kolbe H, Thiel E, Ließ E (2011) Multi-site validation of a soil organic matter model for arable fields based on generally available input data. Geoderma 166:119–134

Franko U, Merbach I (2017) Modelling soil organic matter dynamics on a bare fallow Chernozem soil in Central Germany. Geoderma 303:93–98

Franko U, Oelschlägel B (1995) Einfluss von Klima und Textur auf die biologische Aktivität beim Umsatz der organischen Bodensubstanz. Arch Agron Soil Sci 39(3):155–163

Franko U, Oelschlägel B, Schenk S (1995) Simulation of temperature-, water- and nitrogen dynamics using the model CANDY. Ecol Model 81(1–3):213–222

Franko U, Spiegel H (2016) Modeling soil organic carbon dynamics in an Austrian long-term tillage field experiment. Soil Tillage Res 156:83–90

Franko U, Witing F, Jäckel G, Volk M (2015) Large-scale identification of hot spots for soil carbon demand under climate change and bioenergy production. J Plant Nutr Soil Sci 178:199–208

Kallenbach CM, Frey SD, Grandy AS (2016) direct evidence for microbial-derived soil organic matter formation and its ecophysiological controls. Nat Commun 7:13630–13630

Körschens M, Weigel A, Schulz E (1998) Turnover of soil organic matter (SOM) and long-term balances—tools for evaluating sustainable productivity of soils. Zeitschrift für Pflanzenernährung und Bodenkunde 161(4):409–424

Kuka K, Franko U, Rühlmann J (2007) Modelling the impact of pore space distribution on carbon turnover. Ecol Model 208(2):295–306

Ondrasek G, Bakić Begić H, Zovko M, Filipović L, Meriño-Gergichevich C, Savić R, Rengel Z (2019) Biogeochemistry of soil organic matter in agroecosystems & environmental implications. Sci Total Environ 658:1559–1573

Paul EA (1984) Dynamics of organic matter in soils. Plant Soil 76(1):275–285

Paul EA (2016) the nature and dynamics of soil organic matter: plant inputs, microbial transformations, and organic matter stabilization. Soil Biol Biochem 98:109–126

Prays N, Dominik P, Sänger A, Franko U (2018) Biogas residue parameterization for soil organic matter modeling. PLoS ONE 13(10):e0204121

Rühlmann J (1999) A new approach to estimating the pool of stable organic matter in soil using data from long-term field experiments. Plant Soil 213(1):149–160

Smith P, Lutfalla S, Riley WJ, Torn MS, Schmidt MWI, Soussana JF (2018) The changing faces of soil organic matter research. Eur J Soil Sci 69:23–30

Stockmann U, Adams MA, Crawford JW, Field DJ, Henakaarchchi N, Jenkins M, Minasny B, McBratney AB, Courcelles VdRd, Singh K, Wheeler I, Abbott L, Angers DA, Baldock J, Bird M, Brookes PC, Chenu C, Jastrow JD, Lal R, Lehmann J, O'Donnell AG, Parton WJ, Whitehead D, Zimmermann M (2013) The knowns, known unknowns and unknowns of sequestration of soil organic carbon. Agr Ecosyst Environ 164:80–99

Stockmann U, Padarian J, McBratney A, Minasny B, de Brogniez D, Montanarella L, Hong SY, Rawlins BG, Field DJ (2015) Global soil organic carbon assessment. Global Food Secur 6:9–16

von Lützow M, Kögel-Knabner I, Ludwig B, Matzner E, Flessa H, Ekschmitt K, Guggenberger G, Marschner B, Kalbitz K (2008) Stabilization mechanisms of organic matter in four temperate soils: development and application of a conceptual model. J Plant Nutr Soil Sci 171(1):111–124

Wiesmeier M, Urbanski L, Hobley E, Lang B, von Lützow M, Marin-Spiotta E, van Wesemael B, Rabot E, Ließ M, Garcia-Franco N, Wollschläger U, Vogel HJ, Kögel-Knabner I (2019) Soil organic carbon storage as a key function of soils—a review of drivers and indicators at various scales. Geoderma 333:149–162

Witing F, Franko U, Gebel M (2016) Regionale Humus- und Nährstoffdynamik - Dynamische Bilanzierung von Humushaushalt und Nährstoffaustrag im regionalen Maßstab im Kontext von Landnutzungs- und Klimawandel. Sächsisches Landesamt für Umwelt, Landwirtschaft und Geologie, Schriftenreihe, Heft 20/2016: 99 S. Dresden, Germany

Witing F, Prays N, O'Keeffe S, Gründling R, Gebel M, Kurzer HJ, Daniel-Gromke J, Franko U (2018) Biogas production and changes in soil carbon input—a regional analysis. Geoderma 320:105–114

Witing F, Gebel M, Kurzer HJ, Friese H, Franko U (2019) Large-scale integrated assessment of soil carbon and organic matter-related nitrogen fluxes in Saxony (Germany). J Environ Manag 237:272–280

Chapter 15
Model-Based Assessment of Nutrient Load into Water Bodies from Different Landscape Types

Sergey A. Kondratyev, Natalia V. Ignatyeva, Marina V. Shmakova, Alexandra A. Ershova, Elena A. Minakova and Anton V. Terekhov

Abstract The overall aim of the work is the assessment of the current level of the external nitrogen and phosphorus load and its individual components at the Kuibyshev Reservoir of the Volga Cascade using the information on the landscape structure of the catchment area. A mathematical model of the average annual export of total nitrogen and phosphorus from river catchments with heterogeneous structure was developed, using the results of remote assessment of the landscape structure of the underlying surface. The nutrient load was calculated both for agricultural areas, and for areas that are not currently in agricultural use. The model is calibrated according to the state monitoring data at the pilot sites, which are the watersheds of the Kazanka River (left-bank tributary) and Sviyaga (right-bank tributary). An approximate assessment of the nutrient load on the Kuibyshev Reservoir, formed at the left-bank and right-bank parts of the catchment is presented, taking into account the background (natural) and diffuse (anthropogenic) components of the load.

S. A. Kondratyev (✉) · N. V. Ignatyeva · M. V. Shmakova · A. V. Terekhov
Institute of Limnology, Russian Academy of Sciences (IL RAS), Russian Federation, Sevastyanova St., 9, 196066, St. Petersburg, Russia
e-mail: kondratyev@limno.org.ru

N. V. Ignatyeva
e-mail: natali_ignatieva@mail.ru

M. V. Shmakova
e-mail: m-shmakova@yandex.ru

A. V. Terekhov
e-mail: tepextepex@gmail.com

A. A. Ershova
Russian State Hydrometeorological University (RSHU) Russian Federation, Malookhtensky av., 98, 195196 St. Petersbourg, Russia
e-mail: amberx19@gmail.com

E. A. Minakova
Kazan Federal University, (KFU), Russian Federation, Kremlevskaya St., 18, 420008 Kazan, Russia
e-mail: ekologyhel@mail.ru

W. Mirschel et al. (eds.), *Landscape Modelling and Decision Support*,
Innovations in Landscape Research,
https://doi.org/10.1007/978-3-030-37421-1_15

Keywords Reservoir · Catchment · External nutrient load · Nitrogen · Phosphorus · Model

15.1 Introduction

This study was conducted in the framework of the priority project "Conservation and Prevention of Pollution of the Volga River" (approved by the Presidium of the Presidential Council for Strategic Development and Priority Projects, Minutes No. 9 dated August 30, 2017) on topic 3.4. "Development of a Concept for reducing the inflow of pollutants from natural landscapes, residential areas, agricultural lands, industrial sites, enterprises of the animal-breeding complex, landfills, transport infrastructure facilities."

The goal of the study is to assess the current level of the external nitrogen and phosphorus load and its natural and anthropogenic components on the Kuibyshev Reservoir of the Volga Cascade based on the information on the landscape structure of the catchment area.

15.2 Method of Calculation of Nutrient Load Formation

The algorithm of calculation of the total nutrient export from the catchment and formation of nutrient load on the water body is presented in Fig. 15.1.

According to the accepted calculation algorithm the main components of the external load of total nitrogen and phosphorus on a water body (L) are the diffused emission of nutrients from the underlying surface not currently subject to agricultural impact (L_e); the load generated by agricultural activity (L_{agr}), discharges of point sources in the hydrographic network of the catchment area (L_{p1}) and directly into the receiving reservoir (L_{p2}), as well as the mass exchange with the atmosphere (L_a) (Kondratyev 2007; Kondratyev et al. 2011):

$$L = \left(L_e + L_{agr} + L_{P1} + L_a\right)(1 - k_r) + L_{P2} \quad (15.1)$$

where k_r—coefficient of nutrient retention by the catchment and its hydrographic net. All Eq. (15.1) components have dimensions of t/a, except for the dimensionless coefficient k_r.

An effective tool for calculating load generated on the fields of agricultural enterprises L_{agr} is the method proposed by the experts of the Institute of Engineering and Agri-Environmental Problems (IAEP), St. Petersburg (Bryukhanov et al. 2016, 2018). The advantage of the method is the calculation of the nutrient export accounting for not only the doses of fertilizer application and export of nitrogen and phosphorus with yield but also the types of soils forming the agricultural catchment, their mechanical composition, and the distance from the water body. In addition, the

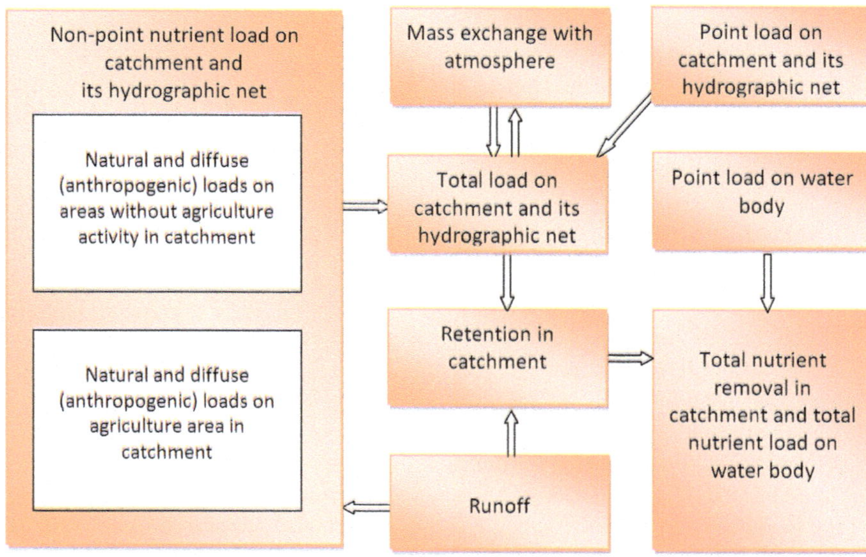

Fig. 15.1 Calculation algorithm of the total nutrient export from the catchment and nutrient load on the water body

method allows it to calculate the decrease in nutrient load using the best available technologies—BAT.

The diffused load on the catchment formed as a result of emission of chemicals from various types of underlying surface (natural and anthropogenic) not currently involved in agricultural production into the rain and meltwaters L_e is calculated as:

$$L_e = 10^{-3} \sum_i C_i \; y_i \; A_i \quad (15.2)$$

where C_i—mean concentrations of nutrients in the flow from i—type of underlying surface, mg/l, y_i—runoff depth from i—type of underlying surface, mm/a, A_i—area, km^2.

To estimate the C_i from Eq. (15.2), the results of generalization of the data of field observations on small rivers can be used, where given type of underlying surface is dominating. The runoff depth y_i with given exceedance probability is calculated based on the distribution curves built on data of long-term runoff observations on rivers of the studied region. Areas of various types of the underlying surface A_i are calculated using the cartographic materials, GIS or based on the results of aerospace images interpretation (Kondratyev et al. 2018).

As a rule, a significant portion of chemicals entering the catchment area from various sources does not reach the outlets of large rivers while they are retained by various parts of the hydrographic network. One of the methods recommended by Helsinki Commission HELCOM (Guidelines 2005) for calculating of retention of chemicals by catchments and their hydrographic network is the empirical model developed at the Institute for Freshwater Ecology and Inland Fisheries of Germany (Behrendt 1996; Behrendt and Dannowski 2007). The authors of the model suggested the following empirical relation connecting the value of the nitrogen and phosphorus retention coefficient k_r in Eq. (15.1) with the value of the specific runoff q, l/(km^2 s):

$$k_r = k^* \left(1 - \frac{1}{1 + aq^b} \right) \tag{15.3}$$

where a and b—dimensionless empirical parameters with values of 26.6 and −1.71 accordingly, k^*—calibration parameter introduced into calculations after several studies in Russia for possible accounting of local features of the nutrient retention process by the catchment and its hydrographic network. The value of specific runoff q is related to the runoff depth y (mm/a) with the ratio $q = 0.03171\ y$.

In addition to the calculation of nutrient export from the catchment and formation of the load on the water body, the model provides the calculation of the background load. In accordance with the HELCOM recommendations (Guidelines 2005) the diffuse (anthropogenic) component of nutrient export from the catchment is estimated as the difference between the total export and the background emission, while the point sources are not taken into account. Atmospheric deposition of L_a can be attributed to anthropogenic load in the case of their extremely high values in the study area.

To perform the calculations of nitrogen and phosphorus emission from river catchments according to the above algorithm the following initial information is necessary:

1. Runoff depth y, mm/a. It is set based on the assessment of the annual runoff depth or by averaging over the observation period from monitoring data.
2. Areas of various types of underlying surface A_i (residential areas, forests and natural vegetation, idle lands), km^2. Determined using GIS technology on digital maps or interpreting the satellite images' of the underlying surface.
3. Concentrations of total nitrogen and phosphorus in the flow (unfiltered water samples) from the listed types of underlying surface, mg/l. Determined as the average of the results of hydrochemical analysis of water samples taken in the primary links of the hydrographic network in homogeneous areas, or according to the available tables.
4. Coefficients used for calculation of agricultural load according to the IEAP method. Determined individually for each agricultural enterprise taking into account its location, amount of mineral and organic fertilizers applied, crops' type, soil types and mechanical composition, and methods of agricultural activity.
5. Atmospheric load on the catchment L_a, t/(a km^2). Calculated using models of atmospheric transport or estimated from the hydrochemical analysis of liquid and solid sediments or set according to the available maps and literature data.

6. Nitrogen and phosphorus discharges from point pollution sources to the hydrographic network of the catchment L_{p1} or to the receiving reservoir L_{p2}, t/a. Set according to the state statistical reporting.
7. Dimensionless parameters a and b in the retention Eq. (15.3).
8. Calibration dimensionless parameter k^* in the Eq. (15.3). Determined from calibration of the model on the data of field observations in the catchment outlets.

15.3 Study Area and Calculations

The largest in Eurasia and the third-largest in the world Kuybyshev Reservoir (6450 km^2) on the Volga River (Kuibyshev Reservoir 1983) was formed in 1955–1957 after the construction of the Volga dam (now Zhiguli Dam) for hydropower station (HPS) near the city of Stavropol (now Tolyatti). The main object of this study is an individual catchment of the Kuibyshev Reservoir with an area of $\approx$91,000 km^2, with the boundary of the Cheboksary HPS on the north, the confluence of the Vyatka River into the Kama River on the east, and the Zhiguli HPS on the south. The layout of the reservoir catchment, as well as the location of the pilot objects selected for calibration of the calculation method is shown in Fig. 15.2.

To calibrate the methodology two pilot objects were selected—the catchments of the Kazanka River (the Kazan site) and the Sviyaga River (the Buinsk site), located on the left-bank and right-bank parts of the studied basin (Fig. 15.2). Both rivers belong to the category of medium plain rivers with a catchment area from 2000 to 50,000 km^2. The selection of these rivers as pilot objects is explained by the following reasons:

1. The river catchments are located in different physical and geographical areas (Climate 1983).
2. The river catchments are located in different landscape subzones (Atlas 2005). The Sviyaga River—in the broadleaf and southern forest-steppe subzones, the Kazanka—in the subtaiga and broadleaf subzones.
3. Hydrological conditions in the river catchments vary significantly. The mean long-term value of runoff depth on the Kazanka catchment is 50% higher than on the Sviyaga.

To determine the structure of the underlying surface of the studied objects the results of the Landsat-8 satellite imagery interpretation for the summer and autumn months of 2017 were obtained using the US Geological Survey web service (http://earthexplorer.usgs.gov). The results of the classification of the underlying surface types of the watersheds are given in Table 15.1, Figs. 15.3 and 15.4 show the surfaces of pilot watersheds obtained on the basis of the interpretation of satellite images.

In order to estimate the nitrogen and phosphorus concentrations in runoff from various types of underlying surface (C_i in Eq. (15.2)) special studies were conducted on the catchments of the Sviyaga and Kazanka rivers in 2018. A series of small

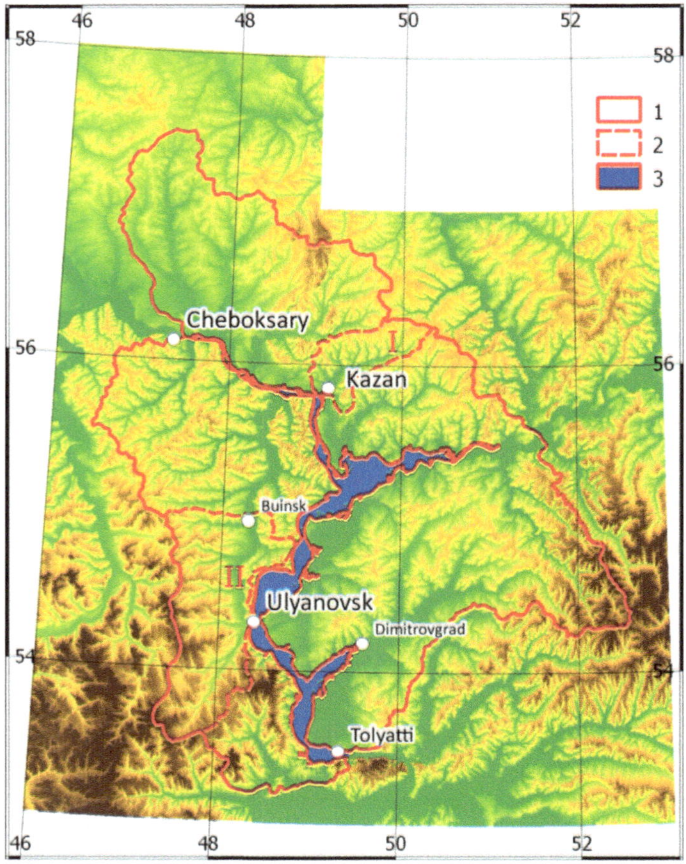

Fig. 15.2 Map of the Kuibyshev reservoir individual catchment: 1—catchment boundary, 2—boundary of pilot catchments, 3—water area of the reservoir

Table 15.1 Areas of various types of underlying surface for the studied catchments (km^2)

Landscapes	Kazanka	Sviyaga
Water area	13.6	19.6
Agricultural lands	1,720.8	5,523.2
Urban lands	151.2	299.1
Forests	415.1	1,454.2
Other natural vegetation	218.3	2,124.9
Total	2,519.0	9,421.0

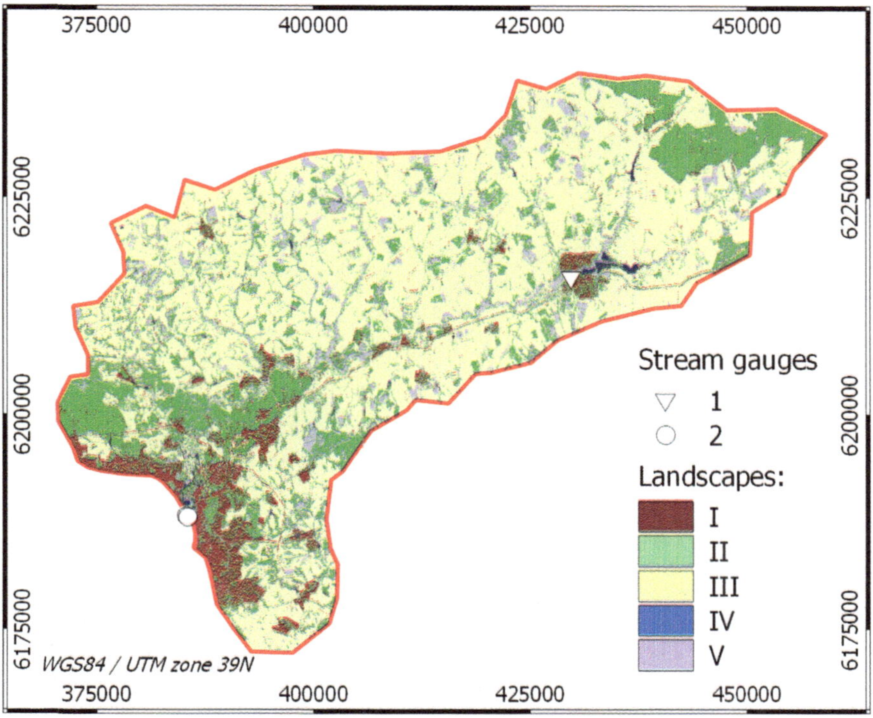

Fig. 15.3 Results of classification of the underlying surface of the Kazanka River catchment: 1—hydrological site, 2—hydrochemical site, I—urban areas, II—forests, III—open areas, IV—water surface, V—other natural vegetation

experimental catchments at primary links of the hydrographic network located on the catchment area of the Kuibyshev Reservoir were selected for fieldwork. Objects were selected based on the representative features of pilot catchments, relative surface uniformity and site accessibility. The layout of water sampling sites is shown in Fig. 15.5.

Sampling was carried out during the spring flood and at intense storm floods that occurred in the region in early June 2018, as well as in the summer (August) and autumn (October) low-flow periods. The concentrations of nitrogen and phosphorus in the runoff from various types of surface were determined based on the condition of the dominance of a particular type of surface. The obtained median concentrations are presented in Table 15.2.

The load generated as a result of agricultural activity (L_{agr}) at pilot catchments is calculated by IAEP experts (Bryukhanov et al. 2018). For this purpose, information on the activities of 87 large agricultural enterprises was collected and processed, taking into account the characteristics of their activities, the amount of mineral and organic fertilizers applied, crops grown, soil types and their mechanical composition,

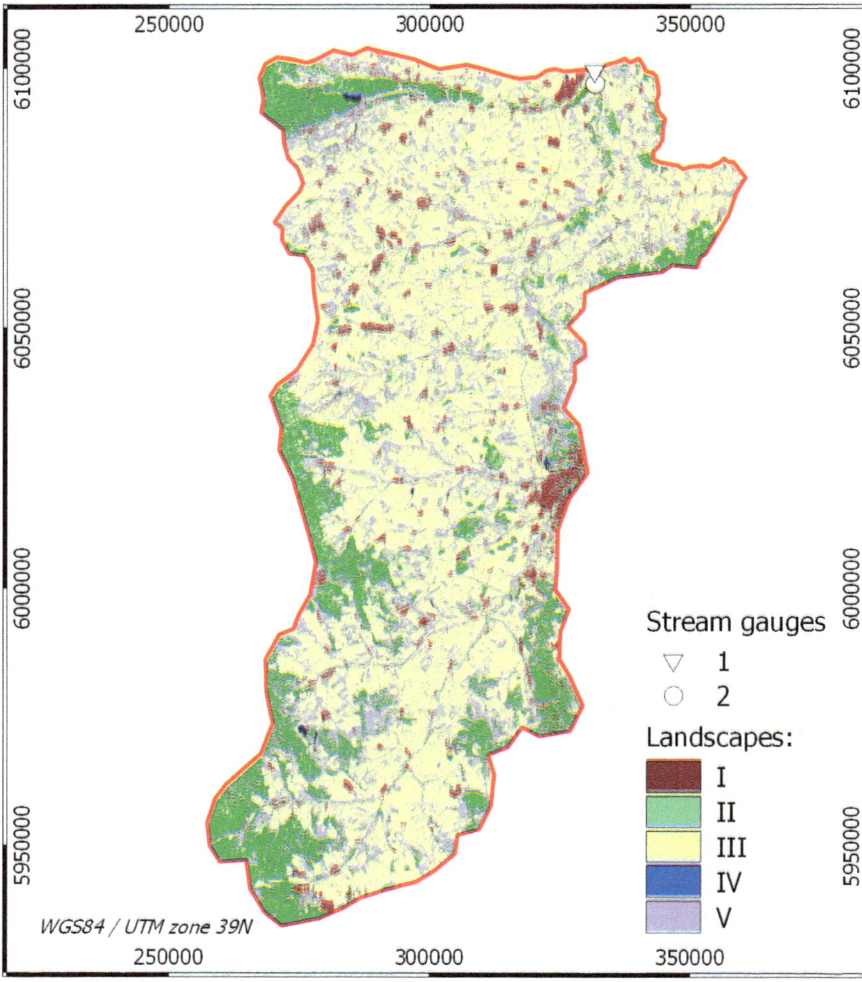

Fig. 15.4 Results of classification of the underlying surface of the Sviyaga River catchment: 1-hydrological site, 2—hydrochemical site, I—urban areas, II—forests, III—open areas, IV—water surface, V—other natural vegetation

as well as remoteness from the hydrographic network. The results of the calculations are presented in Table 15.3.

The nutrient load on the primary links of the hydrographic network of pilot watersheds, formed by discharges of pollution point sources and calculated using statistical reporting data 2TP (Russian water management accounting system), amounted to 170.2 t N/a and 0.6 t P/a for the Sviyagi catchment, and 55.6 t N/a and 3.7 t P/a for the Kazanka catchment. The values of atmospheric deposition of nitrogen and phosphorus on the surface of pilot catchments calculated by the experts of Kazan Federal University based on field observations of the chemical composition of precipitation,

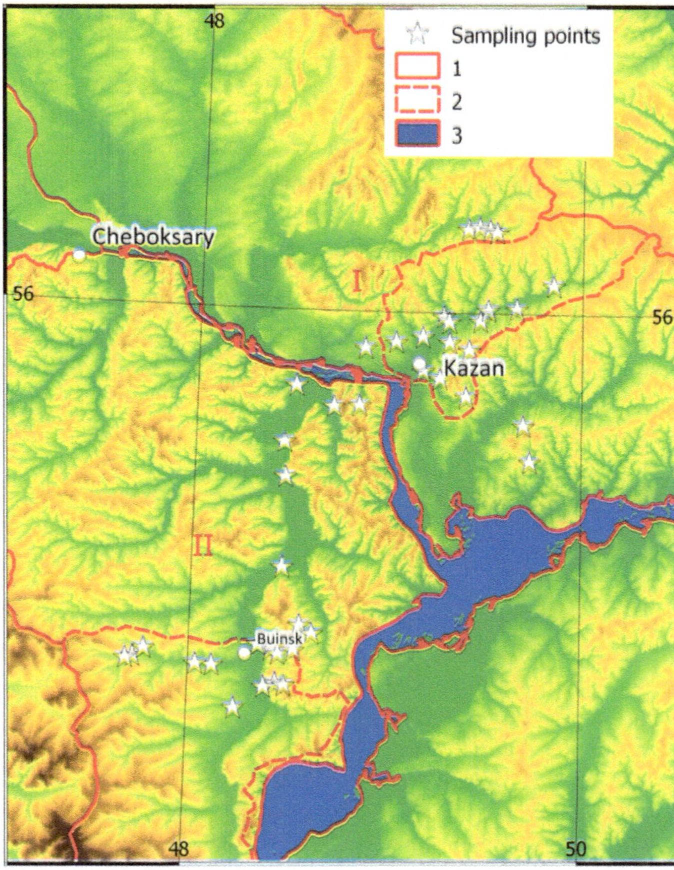

Fig. 15.5 Location of water sampling sites during the fieldworks in 2018

Table 15.2 Median values of concentrations in runoff from the underlying surface types

Nutrient, surface type	C_i, mg/l
P_{tot}, agricultural lands	0.085
P_{tot}, natural underlying surface	0.054
P_{tot}, residential areas	0.20
N_{tot}, agricultural lands	1.45
N_{tot}, natural underlying surface	1.15
N_{tot}, residential areas	3.59

Table 15.3 Agricultural load (t/a) on pilot catchments

	N_{tot}	P_{tot}
Load on the Kazanka catchment	911	86
Load on the Sviyaga catchment	4,502	306

amounted to 0.81 kg N/km^2 year and 0.035 tP/km^2 year for the Sviyaga basin, and 0.80 t N/km^2 year and 0.030 t P/km^2 for the Kazanka basin.

Calibration of the method was carried out for the period of 2008–2016 based on mean nitrogen and phosphorus export values calculated using hydrochemical monitoring data on the Sviyaga River at the Buinsk site and on the Kazanka River at the Kazan site. The resulting calibration values for the parameter k^* in the Eq. (15.3) were 1.05 for N and 0.93 for P for the Sviyaga catchment and 1.08 for N and 1.09 for P for the Kazanka catchment. The average export for the period 2008–2016 of total nitrogen and phosphorus, in tons/year, from pilot catchments, calculated according to the state monitoring of Roshydromet was 3106 t N/year and 189 t P/year for the Sviyagi catchment, and 1225 t N/year and 58 t P/year for the Kazanka catchment. The reliability of calculations was assessed based on the Student's criterion and the calculated coefficient of confidence (Kobzar 2006). The analysis of results showed that the probability of a reliable correspondence between the observed and the calculated data series is 95%.

The next step was to get an approximate estimate of nutrient load on the Kuibyshev Reservoir from the entire individual catchment area. For this the satellite imagery data was used to classify the underlying surface for the entire catchment area (shown on Fig. 15.6).

The values of the model parameters obtained for the Sviyaga were used for calculation of the load on the reservoir from the right-bank part of the catchment (Table 15.4). The export parameters from the Kazanka catchment allowed it to estimate the left-bank nutrient load on the reservoir (Table 15.4). The calculations were performed under the assumption that the intensity of the increase in agricultural load is proportional to the growth of open field areas identified by satellite imagery data. In addition to the estimation of total removal of nitrogen and phosphorus, its natural (background) and diffuse (anthropogenic) components were calculated.

15.4 Conclusion

The main results of the study aimed at assessing the nutrient load on the Kuibyshev Reservoir from the catchment area are:

1. The developed algorithm (mathematical model) allows the calculation of the average annual export of total nitrogen and phosphorus from river catchments of heterogeneous structure using the results of the remote assessment of the landscape structure.
2. The model was calibrated according to the national monitoring data on pilot sites located in the left-bank and right-bank parts of the individual catchment of the Kuibyshev Reservoir.
3. A rough estimate was obtained for the nutrient load on the Kuibyshev Reservoir, formed on the left-bank and right-bank parts of the catchment area in modern conditions. The calculated loads on the reservoir are 40,559 t N/year and 1496 t

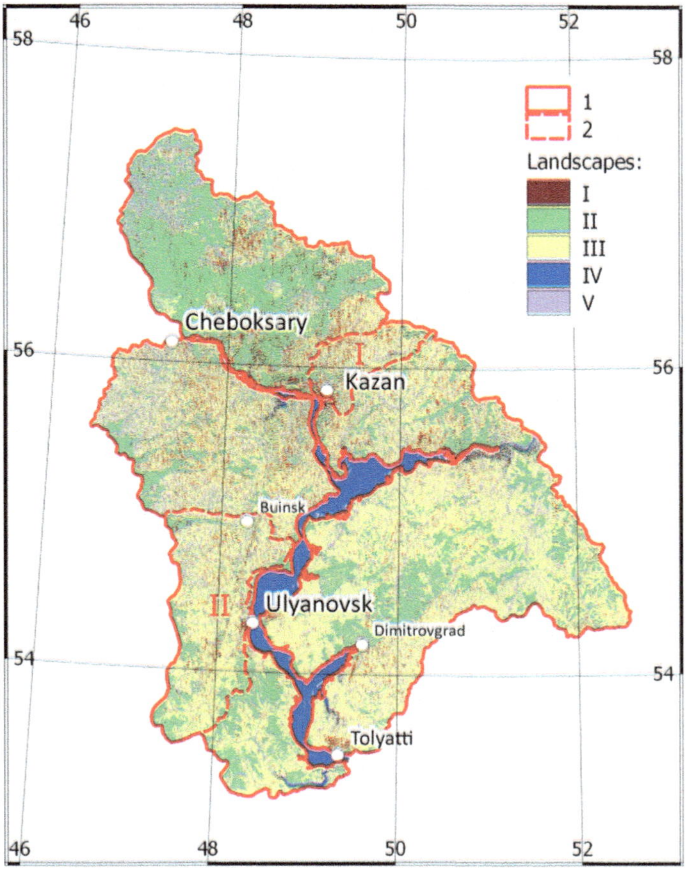

Fig. 15.6 Classification of the underlying surface of the Kuybyshev reservoir: 1—reservoir boundary, 2—pilot catchment border, 3—reservoir surface, I—urban areas, II—forest, III—open areas, IV—water surface, V—other natural vegetation

Table 15.4 Approximate estimate of the mean long-term nutrient load on the Kuibyshev Reservoir from the right bank of the catchment (area—30,878 km^2, average runoff depth 74 mm/year) and from the left bank of the catchment (area—60,207 km^2, average runoff depth 125 mm/year)

Modeling results	N_{tot}	P_{tot}
Right-bank nutrient load on the Kuibyshev Reservoir (t/a)	10,836	468
Natural (background) component	567	22
Diffuse (anthropogenic) component, including atmospheric precipitation	10,241	445
Specific export (kg/km^{2}a)	351	15
Left-bank nutrient load on the Kuibyshev Reservoir (t/a)	29,723	1,028
Natural (background) component	2,710	79
Diffuse (anthropogenic) component, including atmospheric precipitation	25,368	918
Specific export (kg/km^{2}a)	494	17

P/year for the conditions of average water content. At the same time, the contribution of the left-bank part is about 69% for phosphorus and 73% for nitrogen from the total load value.

4. The allocation of the background (natural) and diffuse (anthropogenic) components of the load showed that the contribution of the diffuse component to the total nutrient load on the reservoir from the catchment is very large (85–95% of the total load).

References

Atlas of the Republic of Tatarstan (2005) Publisher by «Kartografia», Moscow

Behrendt H (1996) Inventories of point and diffuse sources and estimated nutrient loads. a comparison for different river basins in Central Europe. Water Sci. Technol 33:99–107

Behrendt H, Dannowski R (2007) Nutrients and heavy metals in the Odra River system. Weissensee Verlag Publisher, Germany, 337 p

Bryukhanov AYu, Kondratyev SA, Oblomkova NS, Oguzdin AS, Subbotin IA (2016) Method of assessment of nutrient load of agricultural production on water bodies. Technologies and technical means of mechanized production of crop and livestock products, № 89, pp 175–183

Bryukhanov AYu, Kondratyev SA, Vassilyev EV, Minakova EA, Terekhov AV, Oblomkova NS (2018) Assessment of agricultural nutrient load formed on river catchments of the Kuibyshev Reservoir basin. Technol Tech Means Mech Prod Crop Livest Prod 3(96):175–186

Climate of the Tatar ASSR/ed by Kolobov NV (1983). Publisher by Kazan Univ-ty, Kazan, 160 p

Guidelines for the compilation of waterborne pollution to the Baltic Sea (PLC-water) (2005) HELCOM, Helsinki, 80 p

Kobzar AI (2006) Applied mathematical statistics. Moscow, pub. by Fizmatlit, 816 p

Kondratyev SA (2007) Formation of external load on water bodies: problems of modelling. St.Petersburg, publisher by Nauka, 255 p

Kondratyev SA, Kazmin MV, Shmakova MV, Markova EG (2011) Method of calculation of nutrient load on water bodies. Regionalnaya ekologiya 3–4:50–59

Kondratyev SA, Bryukhanov AYu, Terekhov AV (2018) Structure of catchment surface as a factor the structure of the catchment surface as a determining factor in the nutrient load on the reservoir (based on mathematical modeling). Vopr Geogr 145:89–108

Kuibyshev Reservoir (1983) Leningrad, publisher Nauka, 213 p

Chapter 16
Model-Based Estimation of Irrigation Water Demand for Different Agricultural Crops Under Climate Change, Presented for the Federal State of Brandenburg, Germany

Wilfried Mirschel, Ralf Wieland, Karin Luzi and Karin Groth

Abstract In the Federal State of Brandenburg, the average temperature is expected to increase by about 4.4 °C by 2100 due to climate change. In addition, precipitation is expected to decrease by about 200 mm, consequently the climatic water balance deficit will increase to 420 mm. Brandenburg is mainly characterised by morainic sandy soils with a low soil water storage capacity. All this will limit the amount of plant-available water during spring and summer months, increasing the cropping risk, especially for spring crops, and reducing crop yields as well as yield stability. This paper outlines the ZUWABE model for estimating the irrigation water demands of agricultural crops, which was developed in ZALF Müncheberg, Germany. It can be used for spatial simulation under past, present and future climate conditions. For all agricultural fields throughout the Federal State of Brandenburg, irrigation water demands were calculated for winter wheat, oats, winter barley, spring barley, sugar beets, oilseed rape, silage maize, potatoes, clover-grass-mix, lucerne-grass-mix, triticale and winter rye. The simulation runs were carried out for five climate levels (1975, 2000, 2025, 2050 and 2075) using climate data from WETTREG 2010 under the A1B emission scenario. The paper presents and discusses the averaged results for Brandenburg as well as for the five productivity classifications of arable land in Brandenburg separately. Four crops (winter wheat, sugar beets, potatoes and silage maize) are used to show and explain the spatial results for Brandenburg. At

W. Mirschel (✉) · R. Wieland · K. Luzi · K. Groth
Leibniz Centre for Agricultural Landscape Research (ZALF), Research Platform "Models & Simulation",, Eberswalder Str. 84, 15374 Müncheberg, Germany
e-mail: wmirschel@zalf.de

R. Wieland
e-mail: rwieland@zalf.de

K. Luzi
e-mail: kluzi@zalf.de

K. Groth
e-mail: kgroth@zalf.de

W. Mirschel et al. (eds.), *Landscape Modelling and Decision Support*,
Innovations in Landscape Research,
https://doi.org/10.1007/978-3-030-37421-1_16

all climate levels, the irrigation water demand was the lowest for oilseed rape, and was the highest for the clover-grass mix and silage maize.

Keywords Irrigation water demand · Agricultural crops · Climate change · Simulation model · Spatial simulation · Federal State of Brandenburg (Germany)

16.1 Introduction

The process of climate change is occurring on the global and regional scale—though every region is affected differently. For Germany, the average annual temperature has increased by 1.4 °C since 1881 (Frühauf 2017).

Over the last two decades, temperatures have increased even faster than this average. The number of hot days (Tmax $\geq$ 30 °C) has increased, and temperatures during winter are also higher. Parallel to this, annual precipitation has decreased, resulting in a decline in precipitation amounts during the main growing season of agricultural crops (Frühauf 2017).

The Federal State of Brandenburg, Germany had an annual precipitation of 550–580 mm over the last 30-year periods before 2010 (LfU 2017), which is low compared to the German average of 800 mm (see also Fig. 16.1). Additionally, Brandenburg often witnesses drought periods, which often occur in late spring and early summer (DWD 2017). The number of days with a soil water content of <50% of available field capacity for Brandenburg's arable land in April and October has increased significantly since 1971 (Frühauf 2017). Furthermore, Brandenburg is influenced by the continental climate, often featuring stable weather conditions. All this limits the crop yields of rain-fed agriculture.

In the future, Brandenburg's climate situation will be even more dramatic, which has been shown by various future climate scenario studies (LUA-BB 2010). This will lead to a serious water deficit problem in agricultural production. Coupled with the low soil water-holding capacity of the sandy soils typical for Brandenburg, it will lead to serious limitations in plant-available water during the main growing period of agricultural crops. The consequences are a greater cropping risk (especially for spring crops) on the one hand, and a decrease in crop yields and yield stability on the other.

To guarantee high crop yields and high yield stability, irrigation has been agriculture's most effective adaptation measure to climate change. Brandenburg, with the north-eastern German lowlands, is a region associated with the highest irrigation water demand in Germany (Drastig et al. 2016). The necessary irrigation water comes from three water sources: groundwater, surface water from rivers and lakes, and decontaminated wastewater from residential areas and from industry. The use of irrigation in agriculture must be in keeping with regional water availability and the landscape's water balance. In Brandenburg, irrigation water is primarily taken from groundwater. Initial land use scenario studies for Brandenburg indicate that there is sufficient groundwater recharge in most districts to provide an adequate basis for crop irrigation (Gutzler et al. 2014).

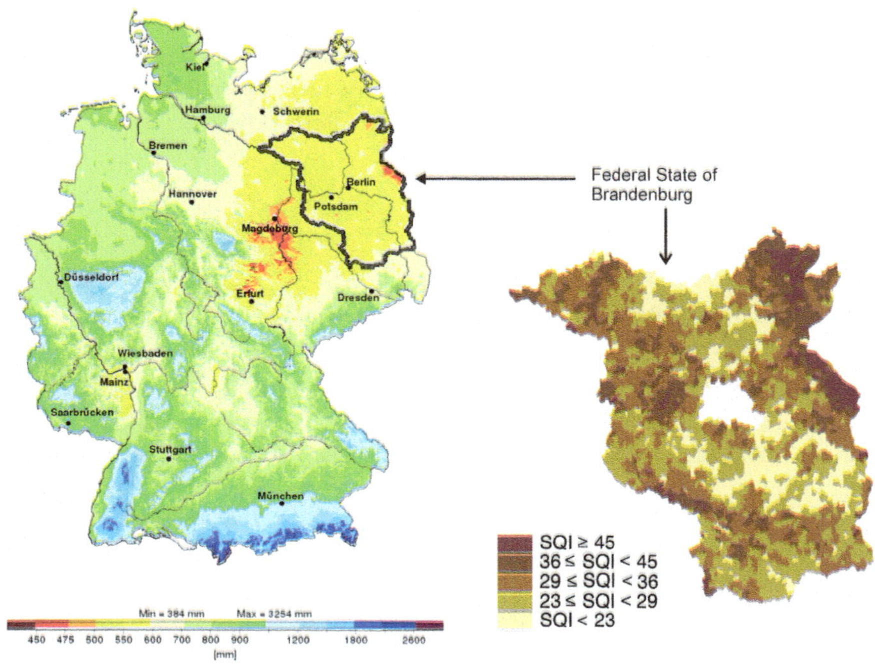

Fig. 16.1 Annual precipitation (1961–1990) for the Federal State of Brandenburg in comparison to the rest of Germany (left image; DWD 2017) and distribution of the five productivity classifications of farmland (the darker the colour, the higher the productivity of the farmland area) within Brandenburg (right image; MIL 2010)

Reliable information about the long-term irrigation water demands of agricultural crops is a necessary prerequisite for the development of climate adaptation strategies for agriculture on the regional scale. The only possibility to assess the impact of climate change on irrigation water demand is through the use of a well-validated model for calculating irrigation water demand in agriculture. To assess the climate change impacts on a spatial scale, regional models of an intermediate complexity (REMICs model; Wenkel et al. 2008) are advantageous in that they are robust, yet still have a realistic input data demand regarding geographical area and time duration.

Turning thus to the assessment of regional irrigation water demand, it makes sense to develop a specific model for the different agricultural crops grown on arable land in East German lowland regions, including Brandenburg; ZUWABE (***ZU***satz***WA***sser***BE***darf)—a statistically based, REMICs model—was developed to fill this need (Mirschel et al. 2016, 2018).

Currently, irrigation in Brandenburg is only used on 2.6% of the state's total cropland area, mainly for cropping of vegetables, potatoes, sugar beets, silage maize, winter wheat and fodder plants (Destatis 2011). Due to rising world market prices for agricultural commodities and concerns about increasing water stress related to climate change, we expect that irrigated arable land will increase significantly in the future, regardless of high investment costs for irrigation technology.

The aim of this contribution is to calculate the irrigation water demands for different agricultural crops under climate change scenarios for arable land in the Federal State of Brandenburg up to the end of this century, focusing on the five productivity classifications for farmland (MIL 2010) that exist in Brandenburg.

16.2 Materials and Methods

16.2.1 Brandenburg—A Case Study Area

The Federal State of Brandenburg is the fifth largest German state by area, with a land surface area of 29,640 km^2; 45% of this is agricultural land. Brandenburg is located in the eastern part of Germany (see Fig. 16.1). The mineral soils developed from glacio-fluvial, periglacial and glacial deposits, and river sand. Many of them have low water-holding capacity. According to the Soil Quality Index (SQI[1]), agricultural land in Brandenburg includes five productivity classifications (PC) for farmland (see Fig. 16.1). The first (PC 1), i.e. the highest productivity classification of farmland, contains soils with SQI values greater than 45, the second (PC 2) has SQI values between 36 and 45, the third (PC 3) between 29 and 35, the fourth (PC 4) between 23 and 28 and the fifth (PC 5), i.e. the lowest productivity classification of farmland, has SQI values lower than 23. In Brandenburg, arable land is only characterised by diluvial soil types (d1, d2, d3, d4, d5, d6) and alluvial soil types (al1, al2, al3). Table 16.1

Table 16.1 Percentage of productivity classifications (PC 1 … PC 5) of farmland by soil types that exist in Brandenburg (MIL 2010)

Soil type	Percentage of productivity by classification area				
	PC 1	PC 2	PC 3	PC 4	PC 5
al1	1.2	0.2	–	–	–
al2	42.3	2.9	0.1	–	–
al3	4.5	–	–	0.3	–
d1	–	–	0.5	5.5	63.3
d2	–	1.5	9.7	67.7	36.0
d3	0.7	2.7	64.2	26.2	0.7
d4	–	67.1	25.5	0.3	–
d5	48.4	25.6	–	–	–
d6	2.9	–	–	–	–

[1]The Soil Quality Index, ranging from 1 to 100, ranks agricultural soils on the basis of their parent material, pedogenetic development and hydrological boundary conditions. The lowest values are attributed to nutrient-poor diluvial sandy soils, the highest to chernozems developed from loess. The Soil Quality Index was developed during the 1930s to evaluate agricultural land in Germany.

shows the distribution of the five productivity classifications on a percentage basis across the soil types existing in the Federal State of Brandenburg.

Agricultural practice is dominated by large-farm enterprises, with an average farm size of 238 ha, which is four times the German average. The labour force of only 1.7 persons per 100 ha on average shows that the mechanisation level in agriculture of Brandenburg is high. Currently, winter wheat, winter rye, silage maize, oilseed rape, winter barley and triticale are the main crops grown in Brandenburg.

16.2.2 Model Description

ZUWABE, a statistically based model, was developed in the context of the INKA-BB project (**I**nnovation **N**etwork for c(**K**)limate **A**daptation in **B**erlin and **B**randenburg) (INKA BB 2017) to estimate site-specific irrigation water demands for a variety of agricultural crops for past and future periods; these estimates serve as a basis for determining optimal irrigation techniques and systems for future irrigation-based cropping systems, either by repairing or optimising existing irrigation systems, or by installing new systems.

The basic version of ZUWABE, developed in the 1980s and the first half of the 1990s, was based on crop-specific and site-specific irrigation water benchmarks (Roth 1991, 1993; Roth et al. 2005) for the lowlands of central East Germany. These benchmarks depended on irrigation periods (Roth 1993), rooting depth, plant-available soil water, and the long-term precipitation average for the time period 1961–1990. The benchmarks were designed to guarantee the high crop yields associated with high water use efficiency, without any water waste.

To estimate the irrigation water demands of all five federal states of East Germany at the present time as well as for future conditions as a result of climate change, this approach was expanded by additional algorithms to take into account the influence of rising CO_2 on crop evapotranspiration; rising temperatures and changing precipitation conditions on climatic water balance; the distance from the seacoast on potential evapotranspiration; and altitude on the ontogenesis-dependent irrigation period.

The recent ZUWABE model for irrigation water demand is as follows:

$$IWD_{CR,ST} = IWB_{CR,ST}\,\frac{\mathrm{Pr}_{IP,LL}}{\mathrm{Pr}_{IP,SS}} - \Delta CWB - (\mathrm{CO_2} - 380)\cdot F_{\mathrm{CO_2}}$$
$$\Delta CWB = CWB_{30ySP} - CWB_{1961-1990} \tag{16.1}$$

where $IWD_{CR,ST}$ represents crop-specific and site-specific irrigation water demand [mm]; CR is the agricultural crop; ST is the site type; $IWB_{CR,ST}$ is the crop-specific and site-specific irrigation water benchmark [mm] according to Roth (1993) (see Table 16.2); $\mathrm{Pr}_{IP,LL}$ is the precipitation average (1961–1990) for the East German lowlands calculated for IP [mm]; IP is the crop-specific irrigation period (Roth 1993) (see Table 16.4); $\mathrm{Pr}_{IP,SS}$ is precipitation average (1961–1990) for the particular site

Table 16.2 Crop- and site-specific mean irrigation water benchmark ($IWB_{CR,ST}$; mm) for the lowlands of central East Germany according to Roth (1993), exemplarily for seven agricultural crops

Agricultural crop	Classification of plant-available soil water capacity (PASWC)[a]			
	Low	Medium	High	Very high
Winter wheat	80	60	40	15
Potetoes	100	80	60	55
Sugar beet	110	90	65	55
Winter rape	25	15	0	0
Spring barley	65	50	35	20
Silage maize	100	80	60	45
Winter barley	50	30	20	5

[a]The relation between PASWC classes and the agricultural site types correspond to the Mesoscale Agricultural Site Map (MMK—Mittelmaßstäbige Landwirtschaftliche Standortkartierung) for arable land (Schmidt and Diemann 1991) is given in Table 16.3

calculated for IP [mm]; $CWB_{1961-1990}$ is the 30-year average of climatic water balance for 1961–1990 [mm]; CWB_{30ySP} is the 30-year average of climatic water balance for concrete simulation period [mm]; CO_2 is the atmospheric CO_2-content [ppm]; and F_{CO2} is a function that reflects the influence of increasing CO_2 in the atmosphere on the reduction of plant transpiration (see Fig. 16.2). The development of the F_{CO2} function bases on data from the Free Air Carbon Enrichment (FACE) Experiment in Brunswick, Germany (Manderscheid and Weigel 2012).

Table 16.3 Allocation of agricultural site types correspond to the MMK to plant-available soil water capacity (PASWC) classes

PASWC	Agricultural site type correspond to MMK
Low	D1a; D2a; D2b; D3a; D3c V3a; V3c; V5c; V7c Al3c
Medium	D3b; D4a; D4b; D4c; D5a; D5b; D5c; D6a V2a; V2c; V3b; V4a; V4c; V5a; V5b; V6b V7a; V7b; V8a; V9a Al1c; Al2c; Al3a; Al3b Lö2c; Lö3a(t); Lö4c; Lö5b(t); Lö5c; Lö6b(t)
High	D6b; D6c V1a Al1a; Al1b; Al2a; Al2b Lö1a(t); Lö1b(t); Lö1c(t); Lö2d(t); Lö3c(t); Lö4b(t); Lö5b(l); Lö6c(t)
Very high	Lö1a(l); Lö1b(l); Lö1c(l); Lö2d(l); Lö3a(l); Lö3c(l); Lö4b(l); Lö6b(l) Lö6c(l)

D - Diluvial soils, *V* - Pretertiary soils, *Al* - Alluvial soils, *Lö* - Loess soils, *t* - clayey, *l* - loamy

Table 16.4 Average crop-specific irrigation periods typically for East Germany in 1961–1990 (Roth 1993)

Agricultural crop	Irrigation period
Winter wheat	10. May–10. June
Potatoes	15. June–15. August
Sugar beet	01. July–05. September
Spring barley	15. May–05. July
Winter barley	25. April–15. June
Silage maize	10. June–10. September
Winter rape	15. April–20. May

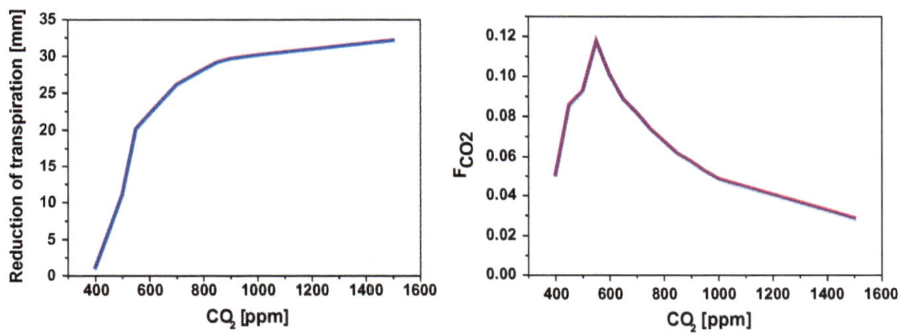

Fig. 16.2 Average reduction of crop transpiration caused by increasing atmospheric CO_2 (left) and reduction factor F_{CO2} in dependence on atmospheric CO_2 used in Eq. (16.1) (right)

Potential evapotranspiration (PET [mm]), required for the computation of climatic water balance, was calculated using the TURC-WENDLING approach (Wendling et al. 1991), which takes into account global radiation (G [J/cm^2]), temperature (T [°C]) and a correction factor (F_{Coast}), the so-called 'coast factor', which takes into account the distance to the sea (for example, to the Baltic Sea). The climatic water balance (CWB [mm]) as a difference between precipitation (Pr [mm]) and PET was calculated as:

$$CWB = Pr - \left((G + 93 \cdot F_{\text{Coast}})\frac{T + 22}{150(T + 123)}\right) \quad (16.2)$$

For a distance from a seacoast up to 50 km, F_{Coast} is 0.6, and for all other regions, F_{Coast} is 1.0. If global radiation was not available, it was estimated on the basis of sunshine duration using Goudriaan's approach (Goudriaan and van Laar 1978).

16.2.3 Regional Data Input and Simulation Framework

Regional input data for the estimation of irrigation water demands was taken from two maps: the farmland map for Brandenburg, divided into the five farmland productivity classifications for farmland according to SQI classifications, and the map of agricultural site types according to the Mesoscale Agricultural Site Map (MMK) for all arable land (Schmidt and Diemann 1991) existing for East Germany. All maps have a resolution of 1 ha grids (100 × 100 m).

For simulation runs for the past and future, ZAWABE used the WETREG 2010 climate regionalisation method (Spekat et al. 2010), taking into account the A1B climate emission scenario. This data is available for all meteorological stations of the German Weather Service with 100 equiprobable realisations for each station.

The YIELDSTAT statistics-based model (Mirschel et al. 2014a) for crop yield estimation also contains an algorithm for additional crop yield due to irrigation. This algorithm is based on the same irrigation water demand calculated using the ZUWABE model algorithm described above. ZUWABE was incorporated into YIELDSTAT and produced irrigation water demand values as output. The simulation runs for Brandenburg under climate change up to the end of this century were carried out using the interactive software solution for YIELDSTAT (Mirschel et al. 2014b).

16.2.4 Assumptions for the Simulation

The simulation runs for the calculation of irrigation water demand were carried out using the results from the ECHAM5/MPI-OM global climate model (Roeckner et al. 2004) with the A1B emission scenario by the SRES-IPCC. For the time period 1961–2100, the results were regionalised for Brandenburg using the WETTREG 2010 method (Spekat et al. 2010). The study used only every 10th of the existing 100 equally probably realisations for each year, i.e. the 1st, the 11th, the 22nd, the 33rd, the 44th, the 55th, the 66th, the 77th, the 88th and the 99th realisation. The results for each year depict the average irrigation water demand over all ten realizations.

To describe the dynamics of crop-specific irrigation water demand in Brandenburg up to 2100, the study used 25-year time intervals (1975, 2000, 2025, 2050 and 2075). The simulations were run for 30-year time periods with the aforementioned years at the halfway point (e.g. a simulation may start in 1985, have 2000 as the halfway point, and end in 2015). The averaged results were projected onto these five intervals, respectively. The same agro-management was assumed in all simulations.

Every simulation took into account the following crops: winter wheat, winter barley, oat, spring barley, triticale, winter oilseed rape, sugar beet, potato, silage maize, lucerne-grass mix and clover-grass mix.

16.3 Results

16.3.1 Climate Change by 2100

Analysis of the changing climate in Brandenburg according to the A1B emission scenario during the 1961–2100 time period can be shown using the Potsdam weather station located in the middle of Brandenburg as an example. From 1961 to 2100, annual average temperature is projected to increase by 4.4 °C, with annual precipitation decreasing by 200 mm. Based on these changes the climatic water balance deficit will increase to 420 mm.

The thermal vegetation period calculated according to Chmielewski (2003) is projected to increase by 115 days. The vegetation period will begin significantly earlier, and as a consequence, the irrigation periods of agricultural crops will also start earlier in the future.

The distribution of precipitation during the year will change significantly, i.e. more precipitation will occur during the winter, and precipitation amounts during spring and summer will decrease. Coupled with higher evapotranspiration rates during early summer and summer, the climatic water balance deficit will increase significantly. According to Adler (1987), growth temperature—mean of temperature above 5 °C—will increase by about 2.5 °C.

16.3.2 Irrigation Water Demand

The simulation results for Brandenburg showed that irrigation water demand will increase continuously until 2075 for all agricultural crops addressed in this study. Both the absolute amount of irrigation water demand and the course of its increase were crop-specific over the time between 1975 and 2075 (see Fig. 16.3).

Crop-specific irrigation water demand primarily depends on the length of the irrigation period. Since this period was short for oilseed rape, which matures relatively early compared to other agricultural crops, the irrigation water demand for oilseed rape was also low. Even up to the end of the century, the irrigation water demand for oilseed rape will remain at a low level and will not change significantly (see Fig. 16.3). For winter and spring cereals, which reach their maturity in the summer months (July–August), irrigation water demand is at a higher level; by 2075, it increases by 1.4 times for winter barley, by 1.7 times for oat and spring barley, and by 1.9 times for winter wheat and triticale compared to 1975 levels.

Agricultural crops such as clover-grass mix, lucerne-grass-mix, silage maize, sugar beets and potatoes need significantly more irrigation water for high and stable yields under climate change. Their irrigation periods include the entire summer, with higher temperatures and higher daily evapotranspiration rates. Compared to irrigated cereal crops, the irrigation water demand of fodder and root crops by 2075 will be significantly higher compared to 1975 levels: three times higher for sugar

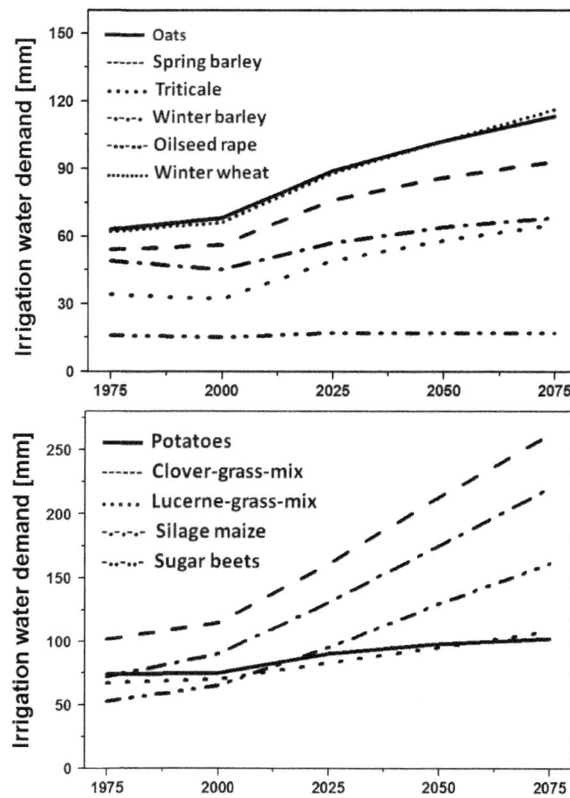

Fig. 16.3 Irrigation water demand for various agricultural crops during the period 1975–2075 as averaged over all five productivity classifications of arable land in Brandenburg

beets and silage maize; 2.6 times higher for clover-grass-mix; and 1.6 times higher for lucerne-grass-mix. Since the rooting system of lucerne reaches deep into the ground, the amount of soil water available for lucerne plants is much higher, and can compensate slightly better for the rising water demand by plants caused by climate change combined with an increased water balance deficit by 2075. Compared to fodder crops and sugar beets, the irrigation period of potatoes is about one month shorter. This is the reason for the lower irrigation water demand level for potatoes, as well as its lower rate of increase by 2075.

In the first quarter of the 100-year period taken into account here, for all agricultural crops under study, the simulation results show only a moderate increase in irrigation water demand. After the first quarter, the simulation results show a more rapid increase in irrigation water demand towards the end of the period under study. In addition to climate change, the main reason is that, at approximately the beginning of the 2000s, the frequency of occurrence of droughts in spring and early summer began to increase in Brandenburg.

Breaking down the simulation results into the five productivity classifications of arable land in Brandenburg for the different agricultural crops gives a relatively similar impression (see Fig. 16.4). In all five productivity classifications, irrigation

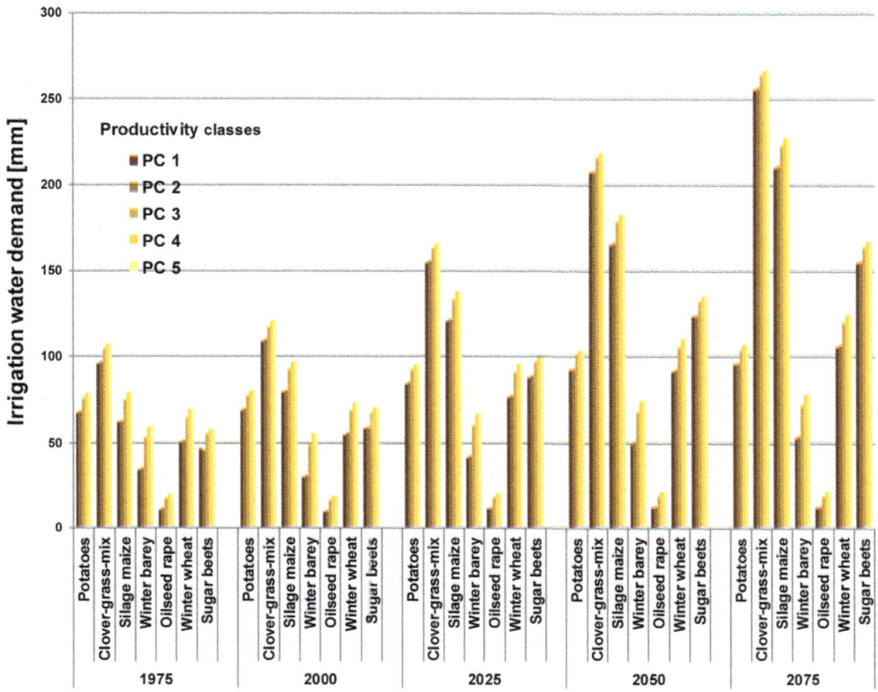

Fig. 16.4 Irrigation water demand for agricultural crops (potatoes, clover-grass-mix, silage maize, winter barley, oilseed rape, winter wheat, sugar beets) in the period 1975–2075 for the five separate productivity classifications of arable land in Brandenburg

water demand will increase until 2075, and will be more or less dramatic depending on the particular agricultural crop.

The irrigation water demands in the productivity classifications PC 1 and PC 2 were lowest compared to the other classifications. Productivity classifications PC 1 and PC 2 were dominated by better soils with higher soil water storage capacities, i.e. they could guarantee a better and longer water supply to agricultural crops. The differences between PC 1 and PC 2 were not so large. For PC 3, irrigation water demand increased significantly compared to PC 1 and PC 2, almost by 10–20 mm, i.e. one normal irrigation water application. This increase was biggest for winter barley, silage maize and winter wheat (see Fig. 16.4). In arable areas of productivity classifications PC 4 and PC 5, irrigation water demand rose as well. PC 4 and PC 5 represent arable land areas in Brandenburg with poor soils, mostly sandy soils, which result in a low capacity for soil water storage. Under climate change conditions, relying on rain-fed farming in areas of PC 4 and PC 5 carries high risks. The differences in irrigation water demand between PC 4 and PC 5 were small.

For all agricultural crops, the increase in irrigation water demand from PC 1/PC 2 to PC 3 was 2.9 times greater than the increase from PC 3 to PC 4/PC 5. This was true for all time periods between 1975 and 2075 under study (see Fig. 16.4).

The spatial distribution of the simulation results for irrigation water demand within Brandenburg for the period 1975–2075 is shown in Fig. 16.5. For all agricultural crops, irrigation water demand increases continuously to 2075. Analysis of the maps in Fig. 16.5 shows that, in most cases, the spatial distribution for irrigation water demand in Brandenburg corresponds with the Brandenburg map of productivity classifications (see also Fig. 16.1). This is a reasonable finding, because the soil map is the main basis for the map showing productivity classifications, and it is very closely connected to the distribution of soil water storage capacity as one of the main factors affecting irrigation water demand.

The spatial simulation results in Fig. 16.5 show that irrigation water demand for the crops selected as examples (winter wheat, potatoes, sugar beets, silage maize) was different depending on the particular crop, but that it nevertheless increased continuously until 2075 for all crops in the study. Looking at the maps in Fig. 16.5, it is possible to note spatial differences depending on the productivity classifications. The Oderbruch region in central eastern Brandenburg, Uckermark in the northeast, Fläming in the southwest and parts of Prignitz in the northwest (which all belong to PC 1 and PC 2 with better soils), need lower amounts of irrigation water to maintain high yields compared to the other regions in Brandenburg to compensate for drought periods induced by climate change.

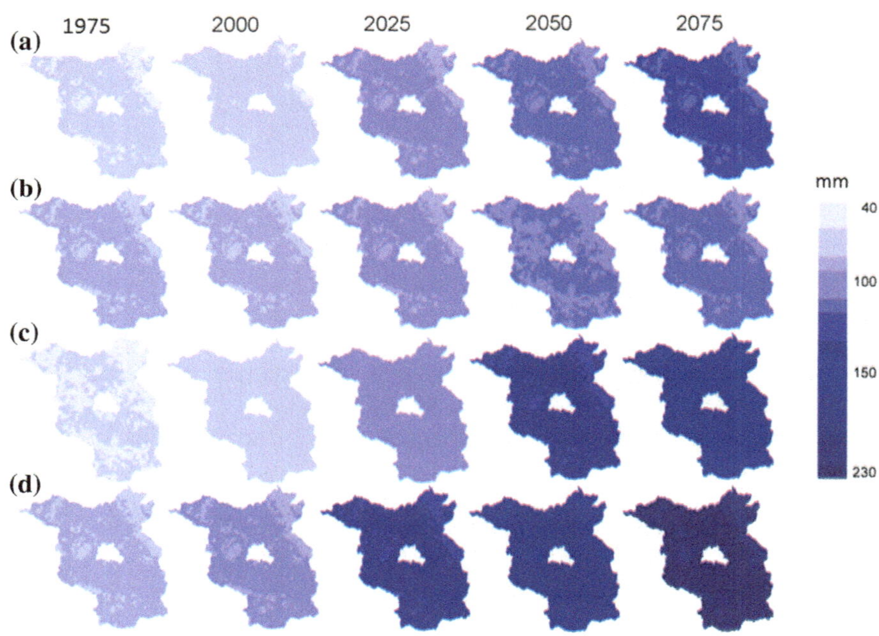

Fig. 16.5 Spatial distribution of irrigation water demand (mm) in the time period 1975–2075 for winter wheat (**a**), potatoes (**b**), sugar beets (**c**) and silage maize (**d**) in the Federal State of Brandenburg, Germany

In the region south of the Lusatia mountain ridge in southern Brandenburg, in the region south of Berlin and north of the Fläming up to the Oder river, in the region southeast of Berlin and in the region between Uckermark and Prignitz in northern Brandenburg, irrigation water demand for all agricultural crops was significantly higher. All these regions belong to the PC 4 and PC 5 productivity classifications with poor soils, associated with a low soil water storage capacity.

16.4 Discussion and Conclusions

Climate changes over the last two decades have revealed increasing temperatures and decreasing amounts of summer precipitation and, as a result, increasing water balance deficits in all regions of the Federal State of Brandenburg. According to the climate simulation, this trend will continue in the future until 2100, and will lead to a critical situation in rain-fed farming concerning the water supply of agricultural crops. For stable crop yields under future conditions, irrigation will be the most important adaptation measure of agriculture to climate change. It will be especially important for contract farming crops, such as silage maize for bioenergy production. Until 2075, irrigation water demand and consequently additional crop yields due to irrigation will increase significantly, but with major differences in the arable land of Brandenburg. Variations in irrigation water demand between years will also increase until 2075.

As a reaction to climate change effects over the last two decades, farmers working with rain-fed production in PC 4 and PC 5 productivity classifications have in practice already reduced the cultivation of spring cereals because of the increasing risk of insufficient water and associated increasing yield risk. Alternatively, farmers plant winter cereals, which have the higher yield stability.

Compared to the climate situation in 2000, irrigation water demand is projected to increase further in Brandenburg until 2075—for winter wheat by 50 mm on average (total irrigation water demand in 2075: 116 mm); for oats by 45 mm (total: 113 mm); for potatoes by 77 mm (total: 102 mm); for clover-grass-mix by 147 mm (total: 262 mm); for lucerne-grass-mix by 38 mm (total: 108 mm); for silage maize by 130 mm (total: 220 mm); for spring barley by 37 mm (total: 93 mm); for triticale by 35 mm (total: 65 mm); for winter barley by 23 mm (total: 68 mm); for oilseed rape by 2 mm (total: 17 mm); and for sugar beets by 96 mm (total: 161 mm).

The results from a simulation study concerning the irrigation water demand of spring barley, oats, winter wheat and potatoes across Germany between 1902 and 2010 by Drastig et al. (2016) using the AgroHyd Farmmodel (Drastig et al. 2012) show that, for the four crops, the results of this Brandenburg study using the ZUWABE model are in the same range. The distribution of spatial irrigation water demand in Brandenburg based on the ZUWABE model is also in accordance with the results taken from Drastig et al. (2016). Depending on the weather situation during the irrigation period, each year for the northeast German lowlands Michel and Sourell (2014) determined an irrigation water demand of 20–105 mm per year for spring barley, 15–130 mm per year for winter wheat, and 55–150 mm per year for potatoes. For the same agricultural regions with low soil water storage capacity, Hanke (1986) determined an irrigation water demand of 95 mm per year in dry years and 50 mm per year in normal years for early potatoes; 175 mm per year and 110 mm per year, respectively, for sugar beets; 240 mm and 150 mm for clover-grass-mix; 280 mm and 190 mm for field grass; 75 mm and 25 mm for winter rye; 120 mm and 75 mm for oats; 70 mm and 25 mm for winter barley; 90 mm and 60 mm for spring barley; and 110 mm per year and 70 mm per year, respectively, for winter wheat.

Compared to rain-fed farming, irrigation induces additional crop yields, which lead to yield stabilisation. The additional crop yield due to irrigation can be estimated using a crop-specific irrigation water use efficiency rate, which has emerged from various field and practical irrigation experiments (e.g. Teichhardt et al. 1984; Roth and Kachel 1989; Kachel and Roth 1990; Roth 1993; Roth and Roth 1993). Averaged over all available experimental results, irrigation water use efficiency amounts to 15 kg/ha/mm for winter wheat, 17 kg/ha/mm for oats, 12 kg/ha/mm for winter barley, 95 kg/ha/mm for sugar beets and 120 kg/ha/mm for potatoes (Mirschel et al. 2014c).

Consequently, compared to the climate situation in 2000, the additional crop yield due to irrigation is projected to increase significantly in Brandenburg until 2075—for winter wheat on average by 0.8 t/ha (total additional yield due to irrigation in 2075: 1.8 t/ha), for oats by 0.7 t/ha (total: 1.9 t/ha); for winter barley by 0.3 t/ha (total: 0.8 t/ha); for sugar beets by 9.1 t/ha (total by 15.3 t/ha); and for potatoes by 9.2 t/ha (total: 12.2 t/ha).

Intensive irrigation also means the intensive use of different water resources. Irrigation can utilise groundwater, surface water from lakes, rivers and watercourses, and decontaminated waste water (with restrictions). In Brandenburg, groundwater can only be used to irrigate agricultural crops with quota limitations based on a licence from the Official Water Agency of the Federal State of Brandenburg. Irrigation with decontaminated waste water is not allowed for any agricultural crops, i.e. its use is strictly prohibited for food crops. Under climate change scenarios with decreased amounts of precipitation especially during the irrigation period, i.e. during summer, surface water resources in lakes and rivers will also be limited.

One land use scenario study in Brandenburg for 2025 that included an irrigation scenario, in which all fields growing silage maize, oilseed rape, winter wheat and sugar beets are irrigated, showed that the irrigation water demand amounted to 667 million m^3 per year, which was 20% of the total Brandenburg annual groundwater recharge of 3,350 million m^3 per year, taken as an average of the years 1976–2005 (MUGV 2009). In Brandenburg, the irrigation water demand only amounted to more than 25% of the annual groundwater recharge in the two districts of Uckermark and Märkisch Oderland. In other districts, the irrigation water demand was between 16% and 25% of the annual groundwater recharge (Gutzler et al. 2014).

Groundwater resources in Brandenburg are limited, and will only be restored by percolation water. One possible solution might be the creation of artificial water storage facilities such as cisterns, water retention basins or small reservoirs across the Brandenburg region to retain surface water, especially from winter precipitation.

Additional, but no less important, functions of such water retention basins for irrigation are that they can guarantee better regularity of outflows (and thus can sometimes be used for flood protection), and save valuable nutrients such as nitrate and phosphorus, and sediments, both coming from agricultural fields via erosion and surface runoff. These nutrients and sediments can be returned periodically to fields as fertiliser. Here, the Free State of Thuringia, Germany can serve as an example; such a network of artificial small water retention basins for irrigation has already existed for a long time (Wegener 1988).

Another important fact in saving irrigation water, to guarantee no additional percolation under irrigation and to increase the efficiency of irrigation water use, is the use of a scheduling system for field irrigation. In the eastern part of Germany, and especially in Brandenburg, two model-based irrigation scheduling systems are currently in use: BEREST-90, an old local PC-based system from the 1990s (Wenkel and Mirschel 1991; Mirschel and Wenkel 2004), and WEBBEREST, a new, web-based system (Mirschel et al. 2014c; Berg and Mirschel 2013). The latter is based on the algorithms of the first system, but is adapted to new conditions, i.e. to new irrigation crops, new varieties, new irrigation techniques, new weather and climate data supplied by the German Weather Service (including forecast data for one week), and the modern data communication possibilities between the German Weather Service and the system user, the advisory service or the farmer himself.

Acknowledgements The authors would like to acknowledge the support received from the German Federal Ministry of Food and Agriculture and the Brandenburg Ministry of Sciences, Research and Cultural Affairs.

References

Adler G (1987) Zur mesoskaligen Kennzeichnung landwirtschaftlich genutzter Standorte von Pflanzenbau-betrieben. Zeitschrift für Meteorologie 37:291–298

Berg M, Mirschel W (2013) Anpassung des interaktiven Beregnungsberatungssystems BEREST 90 an die Möglichkeiten des Internets. In: Nguyen XT (ed) Modellierung und simulation von Ökosystemen: workshop Kölpinsee 2012. Shaker Verlag, Aachen, pp 247–256

Chmielewski FM (2003) Phenology and agriculture (Chapter 7). In: Schwartz MD (ed) Phenology: an integrative environmental science. Kluwer Academic Publishers, Boston/Dordrecht/London, pp 505–522

Destatis (2011) Land-und Forstwirtschaft, Fischerei: Bodenbearbeitung, Bewässerung, Landschaftselemente (Erhebung über landwirtschaftliche Produktionsmethoden (EKLPM)). Statistisches Bundesamt Fachserie 3, Heft 5, 132 pp

Drastig K, Prochnow A, Kraatz S, Libra J, Krauß M, Döring K, Müller D, Hunstock U (2012) Modeling the water demand on farms. Adv Geosci 23:9–13

Drastig K, Prochnow A, Libra J, Koch H, Rolinski S (2016) Irrigation water demand of selected agricultural crops in Germany between 1902 and 2010. Sci Total Environ 569:1299–1314

DWD (2017) Deutscher Klimaatlas. http://www.dwd.de/DE/klimaumwelt/klimaatlas/klimaatlas_node.html. Accessed 01 Feb 2017

Frühauf C (2017) Which challenges pose climate change for crops? J Cultiv Plants 69:35–38

Goudriaan J, van Laar HH (1978) Calculation of daily totals of the gross CO_2 assimilation of leaf canopies. Neth J Agric Sci 26:373–382

Gutzler C, Helming K, Balla D, Dannowski R, Deumlich D, Glemnitz M, Knierim A, Mirschel W, Nendel C, Paul C, Sieber S, Stachow U, Starick A, Wieland R, Wurbs A, Zander P (2014) Agricultural land use changes—a scenario-based sustainability impact assessment for Brandenburg, Germany. Ecol Ind 48:505–517

Hanke B (1986) Taschenbuch der Bewässerung—Wasser in der Pflanzenproduktion. VEB Deutscher Landwirtschaftsverlag, Berlin, p 312

INKA BB (2017) INKA BB—Innovationsnetzwerk Klimaanpassung Region Brandenburg Berlin. Müncheberg: ZALF. http://www.inka-bb.de. Accessed 13 Mar 2017

Kachel K, Roth D (1990) Beziehungen zwischen Zusatzwasser und Ertrag—Ableitung und Anwendung effizienter Beregnungsmengen. Gartenbau 37(4):107–109

LfU (2017) Klimawandel in Brandenburg—Temperatur- und Niederschlagsentwicklung im 20. Jahrhundert. http://www.lfu.brandenburg.de/media_fast/4055/Plakate-Klimawandel.pdf. Accessed 23 Feb 2017

LUA-BB (Landesumweltamt Brandenburg) (2010) Auswertung regionaler Klimamodelle für das Land Brandenburg—Darstellung klimatologischer Parameter mit Hilfe vier regionaler Klimamodelle (CLM, REMO 10, WettReg, Star2) für das 21. Jahrhundert. Fachbeiträge des Landesumweltamtes, Heft 113, 23 p

Manderscheid R, Weigel H-J (2012) Das Braunschweiger FACE-Experiment. Promet 38(1/2):20–31

Michel R, Sourell H (eds) (2014) Bewässerung in der Landwirtschaft. Klein Sachau, ERLING Verlag, p 176

MIL (Ministerium für Infrastruktur und Landwirtschaft des Landes Brandenburg) (2010) Datensammlung für die Betriebsplanung und die betriebswirtschaftliche Bewertung landwirtschaftlicher Produktionsverfahren im Land Brandenburg—Ackerbau/Grünlandwirtschaft/Tierproduktion–. Schriftenreihe des LELF, Reihe Landwirtschaft, Band 11 (H VIII), 131 p

Mirschel W, Wenkel K-O (2004) System zur operativen computergestützten Beregnungseinsatzsteuerung (BEREST 90). In: Lüttger A (ed.). Pilotprojekt Beregnung des Landes Brandenburg (4. Brandenburger Beregnungstag, Güterfelde, 27.11.2003), LVLF des Landes Brandenburg, 39–50. http://www.pilotprojektberegnung.de/tagung_03.htm

Mirschel W, Wieland R, Groth K (2014a) Regionales Ertragsmodell YIELDSTAT als interaktive Software für Modellanalysen und Simulationen. In: Nguyen XT (ed) Modellierung und Simulation von Ökosystemen: Workshop Kölpinsee 2013. Rhombos, Berlin, pp 1–17

Mirschel W, Wieland R, Wenkel K-O, Nendel C, Guddat C (2014b) YIELDSTAT—a spatial yield model for agricultural crops. Eur J Agron 52:33–46

Mirschel W, Klauss H, Berg M, Eisenhut K-U, Issbrücker G, Prochnow A, Schörling B, Wenkel K-O (2014c) Innovative Technologien für eine effiziente Bewässerung im Pflanzenbau. In: Bloch R, Bachinger J, Fohrmann R, Pfriem R (eds) Land-und Ernährungswirtschaft im Klimawandel—Auswirkungen, Anpassungsstrategien und Entscheidungshilfen. München: oekom verlag, pp 261–277

Mirschel W, Wieland R, Luzi K (2016). Simulated additional crop yields due to irrigation of agricultural fields throughout Brandenburg, Germany, for 1975–2075. In: Materialy Vserossijskoj nauchnoj konferencii, s mezhdunarodnym uchastiem: „Agroehkosistemy v estestvennykh i reguliruemykh uslovijakh: ot teoreticheskoj modeli k praktike precizionnogo upravlenija“, Sankt-Peterburg, 21–23 sentjabrja 2016 g., FGBNU AFI, Sankt-Peterburg, pp 21–31

Mirschel W, Wenkel K-O, Berg M, Wieland R, Terleev V, Luzi K (2018) ZUWABE: a model for estimation of spatial irrigation water demand for agricultural crops. In: Sychev VG, Müller L (eds) *Novel* methods and results of landscape research in Europe, Central Asia and Siberia (Monograph in 5 Volumes)—vol IV Optimising Agricultural Landscapes. Russian Academy of Sciences: FSBSI „All-Russian Research Institute of Agrochemistry named after D.N. Pryanishnikov“, Moscow, Chapter IV/74: 361–365

MUGV (2009) Mittlere Abflussspende des Landes Brandenburg für die Zeitreihe 1976–2005. http://www.mugv.brandenburg.de/lua/gis/wh_abimo.zip. Accessed 12 Dec 2009

Roeckner E, Brokopf R, Esch M, Giorgetta M, Hagemann S, Kornblueh L, Manzini E, Schlese U, Schulzweida U (2004) The atmosphere general circulation model ECHAM5. Part 2: Sensitivity of simulated climate to horizontal and vertical resolution. MPI-Report 354, Hamburg: Max Planck Institute for Meteorology

Roth D (1991) Grundlagen und Methodik zur Ermittlung des Wasserbedarfs für die Beregnung. Zeitschrift für Bewässerungswirtschaft 26(2):40–54

Roth D (1993) Richtwerte für den Zusatzwasserbedarf in der Feldberegnung. Schriftenreihe der Thüringer Landesanstalt für Landwirtschaft, Jena, H. 6:53–86

Roth D, Kachel K (1989) Zusatzwasser-Ertragsbeziehungen von Winterweizen, Sommergerste, Zuckerrüben, Kartoffeln und Welschem Weidelgras auf drei Standorten mit unterschiedlichem Bodenwasserbereitstellungsvermögen. Archiv für Acker-und Pflanzenbau und Bodenkunde 33:393–403

Roth D, Roth R (1993) Der Einfluss differenzierter Wasserversorgung auf den Ertrag von Silomais. Archiv für Acker-und Pflanzenbau und Bodenkunde 38:455–465

Roth D, Günther R, Knoblauch S, Michel H (2005) Wasserhaushaltsgrößen von Kulturpflanzen unter Feldbedingungen. Landwirtschaft und Landschaftspflege in Thüringen, Schriftenreihe der Thüringer Landesanstalt für Landwirtschaft, 1(2005), 159 pp

Schmidt R, Diemann R (eds) (1991) Erläuterungen zur Mittelmassstäbigen Landwirtschaftlichen Standort-kartierung (MMK). FZB Müncheberg, Eigenverlag, p 78

Spekat A, Kreienkamp F, Enke W (2010) An impact-oriented classification method for atmospheric patterns. Phys Chem Earth 35:352–359

Teichhardt R, Roth D, Zenner I (1984) Langjährige Untersuchungen zur Ertragswirksamkeit der Beregnung unter Produktionsbedingungen. Feldwirtschaft 25(5):195–198

Wegener U (1988) Wasserspeicher im Thüringer Becken—ihre Verteilung und landeskulturelle Verteilung. Hercynia 25(4):377–385

Wendling U, Schellin H-G, Thomä M (1991) Bereitstellung von täglichen Informationen zum Wasserhaushalt des Bodens für die Zwecke der agrarmeteorologischen Beratung. Zeitschrift für Meteorologie 41:468–475

Wenkel K-O, Mirschel W (1991) BERSIM und BEREST 90—Modelle zur Simulation der Bodenfeuchte und Evapotranspiration sowie zur operativen Beregnungseinsatzsteuerung im integrierten Landbau. Agrarinformatik (Informationsverarbeitung Agrarwissenschaft), Stuttgart: Verlag Eugen Ulmer, Bd. 21, 349–360

Wenkel K-O, Wieland R, Mirschel W, Schultz A, Kampichler C, Kirilenko A, Voinov A (2008) Regional models of intermediate complexity (REMICs)—a new direction in integrated landscape modelling. In: Jackman AJ, Voinov AA, Rizzoli AE, Chen SH (eds) Environmental modelling, software and decision support -state of the art and new perspectives-, developments in integrated environmental assessment, vol 3, Elsevier, Amsterdam, pp 285–295

Chapter 17
Forecasting Scanning Branches of the Hysteresis Soil Water-Retention Capacity for Calculation of Precise Irrigation Rates in Agricultural Landscapes Using a Mathematical Model

Vitaly V. Terleev, Wilfried Mirschel, Alex Topaj, Kirill Moiseev, Issa Togo, Yulia Volkova, Aleksandr O. Nikonorov, Roman Ginevsky and Viktor Lazarev

Abstract A mathematical model of the hysteretic soil water-retention capacity is proposed. Based on this model, a computer program called «Hysteresis» was developed. This program has options for identifying model parameters by the method of dot-fitting of experimental data, as well as for performing predictive calculations and

V. V. Terleev (✉) · I. Togo · Y. Volkova · A. O. Nikonorov · R. Ginevsky · V. Lazarev
Peter the Great St. Petersburg Polytechnic University,
Polytechnicheskaya 29, St. Petersburg 195251, Russia
e-mail: vitaly_terleev@mail.ru

I. Togo
e-mail: issatogo@mail.ru

Y. Volkova
e-mail: yv1975@mail.ru

A. O. Nikonorov
e-mail: coolhabit@yandex.ru

R. Ginevsky
e-mail: rginevski@gmail.com

V. Lazarev
e-mail: lviktor.97@mail.ru

W. Mirschel
Leibniz-Centre for Agricultural Landscape Research (ZALF) e. V.,
Eberswalder Straße 84, 15374 Müncheberg, Germany
e-mail: wmirschel@zalf.de

A. Topaj · K. Moiseev
Agrophysical Research Institute, Grazhdanskii pr. 14, St. Petersburg 195220, Russia
e-mail: alex.topaj@gmail.com

K. Moiseev
e-mail: kir_moiseev@mail.ru

W. Mirschel et al. (eds.), *Landscape Modelling and Decision Support*,
Innovations in Landscape Research,
https://doi.org/10.1007/978-3-030-37421-1_17

graphical representation of the branches of the hysteresis loop. A series of computational experiments was performed in which the possibility of identifying the parameters of the mathematical model from the data on the main (boundary) branches of soil drying and wetting was investigated, and the accuracy of the predictive calculations of the scanning branches of the hysteresis loop was estimated. Data from the literature on four soils are used. The model has been compared with three models of predecessors. A sufficiently high accuracy of forecasting the scanning branches has been achieved. The practical value of the proposed model is the possibility of calculating precise rates for crop irrigation. Application of such rates: (i) prevents the percolation of excess moisture from the root layer of the soil; (ii) minimizes the loss of irrigation water, fertilizers, ameliorants and plant protection products and (iii) reduces the risk for groundwater contamination with agrochemicals and the threat of water eutrophication.

Keywords Water-retention capacity of soil · Hysteresis loop · Identification of parameters · Main (boundary) branches · Forecasting the scanning branches · Precise rates · Irrigation

17.1 Introduction

In irrigation farming, for every crop, it is necessary to take into account the physiological threshold characterizing the minimum permissible value of soil volumetric water content θ ($cm^3\ cm^{-3}$), at which yet negative irreversible changes in the growth and development of plants do not occur. Farther, for soil *humidity before irrigation*, we will use the designation HBI. In order to maximally moisten the soil, in irrigation practices it is usually applied the method to calculate the rates according to the difference FC-HBI, where FC is the *field capacity*, which characterizes the maximum possible supply of capillary-suspended water in the soil after the excess of gravitational moisture (percolation). The soil water-retention capacity (WRC) curve characterizes the change in the quasi-equilibrium states of water in form of a dependence of θ value on the capillary pressure of moisture ψ (cm H_2O). If the initial state of water corresponds to the complete saturation of the soil with moisture, then (taking into account the phenomenon of hysteresis) this curve describes the WRC main drying branch. On this branch the FC value corresponds to a certain capillary pressure ψ (FC). The capillary pressure averaged over the set of the WRC main drying branches for different soils is equal to −330 cm H_2O. On the main drying branch, the value $\psi = -330$ cm H_2O corresponds to the θ value for which we propose to use the term «*normative field capacity*» (NFC). In order not to confuse FC with NFC, for the measured field capacity of the soil we suggest using the refining term «*potential field capacity*» (PFC).

The determining factor of soil water-retention capacity is not the volume content of moisture in the soil, but the balance between capillary sorption forces and gravity. At the ψ (PFC) value, the capillary sorption forces counterbalance the force of gravity. With allowance for hysteresis, the change in quasi-equilibrium water states is characterized by two main branches and a multitude of scanning branches (drying and wetting) of the soil water-retention capacity. Any such scanning branch starts from the inversion point on the previous branch (main or scanning) and ends with another inversion point on the subsequent branch (main or scanning). The ψ (PFC) value corresponds to a set of θ values on different branches of the hysteresis loop. In this set, the maximum θ value is equal to the PFC. All other θ values at ψ (PFC) should be considered as values of the variable for which we propose to use the term «*effective field capacity*» (EFC).

Suppose that the PFC value is measured (for example, by the field flooding method). In the opinion of the authors, in this case the irrigation rate, calculated according to the PFC-HBI formula, is overestimated. This is explained as follows: if such rate is applied, immediately after watering the θ variable reaches the PFC value, but the corresponding value of the ψ variable on the wetting scanning branch is higher than ψ (PFC). At the same time, the excess of gravitational moisture, characterized by the difference in the PFC-EFC, flows off the root-inhabited soil layer. After the run-off of excess moisture, the ψ variable reaches the ψ (PFC), but the corresponding θ variable on the scanning drying branch turns out to be lower than PFC (θ = EFC < PFC). Incidentally, we note that if the NFC is higher than the PFC, then the irrigation rate calculated from the difference NFC-HBI is even more overestimated. Therefore, in order to prevent the unproductive loss of irrigation water, it is necessary to calculate the irrigation rate not according to the difference PFC-HBI (or NFC-HBI), but by the difference EFC-HBI. The EFC index is equal to the θ value at ψ (PFC) on the scanning wetting branch, which starts from the inversion point on the previous drying branch at ψ (HBI), and corresponds to the next watering (precipitation).

In order to estimate the EFC and predict the scanning hysteresis branches, it is proposed to use a physically adequate mathematical WRC model. It should be noted that in this method it is impossible to do without using such model, since it is impossible to measure an infinite set of scanning branches. The purpose of the study is to verify the WRC model underlying the proposed method for calculating the precise irrigation rates.

17.2 Theory and Mathematical Model

The property of the soil water-retention capacity is usually described in the form of functional dependence $\theta(\psi)$ (Brutsaert 1966; Ahuja and Swartzendruber 1972; Haverkamp et al. 1977). Now, there is no generally accepted physically adequate mathematical WRC model in the literature. Among the most famous models is the Van Genuchten model (Van Genuchten 1980). Along with the merits, Van Genuchten's model has a significant drawback, which lies in the fact that its parameters (α и n) have no physical meaning.

Following Kosugi (1994, 1996) and Hopmans (Kosugi and Hopmans 1998) on the basis of the concept of the lognormal pore size distribution (D'Hollander 1979) and the phenomenon of capillarity, the authors of this study described the functional dependence of the soil differential moisture capacity on ψ. An antiderivative function is obtained for the soil differential moisture capacity. This antiderivative, by definition, is a function of the soil integral moisture capacity, which describes the soil water-retention capacity in the form $\theta(\psi)$. Using the modified Winitzki approximation (2008), the continuous approximation for $\theta(\psi)$ function, which coincides with the Haverkamp et al. model (1977), was obtained, but which has an additional parameter (Terleev et al. 2015, 2016a, b). Thus the Haverkamp et al. (1977) model is physically justified and modified. Based on the physical meaning of the parameters of this modified model and using the assumption that the function of the soil differential moisture capacity assumes only two values corresponding to the sorption and desorption quasi-equilibrium moisture states, the authors of this study describe the hysteresis inherent in the soil water-retention capacity, using two sets of parameters: for the drying branches, the other for the wetting branches.

To describe the drying branch, starting from the point «*i*» on the wetting branch, the following formulas are used

$$\left\{\begin{array}{l}\left[\begin{array}{l}S_{e,d}=\left(1+\left(\frac{\psi-\psi_{\mathrm{ae}}}{\psi_{0,d}-\psi_{\mathrm{ae}}}\right)^{n_d}\right)^{-1},\\ \theta=\theta_R+\left(\theta_S^*-\theta_R\right)S_{e,d},\\ \theta_s^*=\theta_s,\ \psi_{\mathrm{we}}\le\psi_i,\ \psi<\psi_{\mathrm{ae}};\\ \theta_s^*=\theta_i,\ \psi_{ae}\le\psi_i<\psi_{\mathrm{we}},\ \psi<\psi_{ae};\\ \theta_s^*=\frac{\theta_i-\theta_R\left(1-S_{e,d}(\psi_i)\right)}{S_{e,d}(\psi_i)},\ \psi_i<\psi_{\mathrm{ae}},\ \psi\le\psi_i;\end{array}\right.\\ \left[\begin{array}{l}\theta=\theta_s,\ \psi_{\mathrm{we}}\le\psi_i,\ \psi_{\mathrm{ae}}\le\psi\le\psi_i;\\ \theta=\theta_i,\ \psi_{ae}\le\psi_i<\psi_{we},\ \psi_{\mathrm{ae}}\le\psi\le\psi_i.\end{array}\right.\end{array}\right. \tag{17.1}$$

To describe the wetting branch, starting from the point «*j*» on the drying branch, the following formulas are used

$$\left\{\begin{array}{l}\left[\begin{array}{l}S_{e,w}=\left(1+\left(\frac{\psi-\psi_{\mathrm{we}}}{\psi_{0,w}-\psi_{\mathrm{we}}}\right)^{n_w}\right)^{-1},\\ \theta=\theta_R^*+\left(\theta_S-\theta_R^*\right)S_{e,w},\\ \theta_R^*=\theta_j=\theta_r,\ \psi_j<<\psi_{\mathrm{ae}},\ \psi_j\le\psi<\psi_{\mathrm{we}};\\ \theta_R^*=\frac{\theta_j-\theta_S S_{e,w}(\psi_j)}{1-S_{e,w}(\psi_j)},\ \psi_j<\psi_{\mathrm{ae}},\ \psi_j\le\psi<\psi_{\mathrm{we}};\end{array}\right.\\ \left[\begin{array}{l}\theta=\theta_S,\ \psi_j<\psi_{\mathrm{ae}},\ \psi_{\mathrm{we}}\le\psi;\\ \theta=\theta_j=\theta_s,\ \psi_{\mathrm{ae}}\le\psi_j,\ \psi_j\le\psi.\end{array}\right.\end{array}\right. \tag{17.2}$$

where $S_{e,d}$ and $S_{e,w}$ [dimensionless]—the effective moisture saturation of the soil for desorption and sorption quasi-equilibrium water states, respectively; θ_s (cm^3 cm^{-3})—the volumetric water content of the soil; θ_R (cm^3 cm^{-3})—minimum volumetric soil moisture at which water has the properties of a liquid; ψ_{ae} (cm H_2O) is the capillary pressure of the air entrance (bubbling pressure); ψ_{we} (cm H_2O) is the capillary pressure of the water entrance (pressure of displacement of the trapped air from dead-end pores); $\psi_{0,d}$ and $\psi_{0,w}$ (cm H_2O) are the values of capillary pressure at the most probable values of the random variable (the logarithm of the effective radius of the soil pore) for desorption and sorption quasi-equilibrium moisture states, respectively; σ_d and σ_w [dimensionless] are the values of the standard deviation of this random variable for desorption and sorption quasi-equilibrium moisture states, respectively, $n_d = 4\Big/\left(\sigma_d\sqrt{2\pi}\right)$ and $n_w = 4\Big/\left(\sigma_w\sqrt{2\pi}\right)$ [dimensionless].

17.3 Results of Verification of the Mathematical Model

Based on the model described by formulas (17.1) and (17.2), the computer program «Hysteresis» was developed. For the verification of the model, literature data for four soils were used: *White silica sand* (Huang et al. 2005), *Dune sand* (Gillham et al. 1976), *Rideau clayey loam* and *Rubicon sandy loam* (Mualem 1976). The calibration of the model (identification of parameters) was carried out using the procedure of dot-fitting of experimental data on the main (boundary) branches of the hysteresis loop (Tables 17.1 and 17.2).

Using the parameterized model, predictive calculations of the scanning hysteresis branches were performed. The results were compared with the experimental data (Table 17.3). The proposed model was compared with three models of predecessors (Scott et al. 1983; Kool and Parker 1987; Huang et al. 2005) using the errors of the approximation of experimental data on the main (boundary) branches, as well as the errors of the prediction for the investigated hysteretic scanning branches. In Tables 17.2 and 17.3, the smallest errors in the calculations were given off in bold underlined font. From the analysis of Tables 17.2 and 17.3, it follows that the model proposed by the authors differs by the greatest number of results with the least error. For this model, with the parameter shown in Table 17.1, the artificial «pumping effect» was not detected.

Table 17.1 Parameters of the model (proposed by the authors) estimated from experimental data on the hysteretic main (boundary) branches using the dot-fitting procedure

Soils	Parameters							
	θ_R	θ_S	ψ_{ae}	ψ_{we}	$\psi_{0,d}$	$\psi_{0,w}$	n_d	n_w
White silica sand	0.0861	0.3574	−12.09	−1.797	−112.2	−41.42	3.996	2.287
Dune sand	0.0934	0.3010	−19.82	−3.594	−33.68	−19.99	3.170	3.298
Rideau clayey loam	0.2896	0.4179	−20.00	6.26	−66.96	−29.44	1.951	1.999
Rubicon sandy loam	0.1688	0.3829	−13.00	16.00	−88.42	−36.32	2.911	2.993

Table 17.2 The mean absolute difference between the results of calculating the hysteretic main (boundary) branches (by the dot-fitting procedure) and the experimental data

Soils	Comparable models			
	Scott et al. (1983)	Kool and Parker (1987)	Huang et al. (2005)	Proposed
White silica sand	0.0028	0.0107	0.0030	**0.0019**
Dune sand	0.0027	0.0080	0.0031	**0.0023**
Rideau clayey loam	**0.0032**	0.0057	0.0057	**0.0032**
Rubicon sandy loam	**0.0045**	0.0130	0.0055	0.0098

On Figs. 17.1a, b, 17.2a, b, 17.3a, b and 17.4a, b the measured data are shown by points; the results of the approximation (dot-fitting) for the main (boundary) branches, as well as the results of the predictive estimation for the scanning branches of hysteresis loop (using the model proposed here) are shown by solid curves. Based on the results of the computational experiments, a comparative analysis for an accuracy of the predictive estimating the hysteretic scanning branches was carried out. The model proposed here and three models of predecessors were used (Scott et al. 1983; Kool and Parker 1987; Huang et al. 2005).

17.4 Discussion

The idea of using the Haverkamp et al. model (1977) to describe hysteresis is not new. An example is the model of the hysteresis of the soil water-retention capacity, proposed in (Scott et al. 1983), where an original solution was found for scanning branches starting from turning points. Nevertheless, the hysteresis models based on the dependencies proposed by Haverkamp et al. (1977) and Van Genuchten (1980) in the literature (Huang et al. 2005) have been criticized for the potential manifestation of the «pumping effect» . This criticism is true, but—in part. The «pumping effect» is that when the values of the capillary pressure of soil moisture oscillate in a fixed interval, the values of volumetric water content «drift» , and it is possible an intersection between the scanning and main branches, that is accompanied by «exit» of the scanning branches out of the physically acceptable area. Indeed, the dependencies $\theta(\psi)$ proposed in (Haverkamp et al. 1977; Van Genuchten 1980) contain the potential manifestation of this undesirable artificial effect in the models of hysteretic soil water-retention capacity. However, this effect does not always appear, but only with certain combinations of parameters. If the values of the parameters do not exceed the limits of the physically acceptable area, then the «pumping effect» does not appear. Of course, such boundaries can only be defined for physically adequate models. The model proposed by the authors belongs to this class of models.

Table 17.3 The mean absolute difference between the results of calculating the hysteretic scanning branches (using four models) and the experimental data

Soils	Scanning branches	Comparable models							
		Scott et al. (1983)		Kool and Parker (1987)		Huang et al. (2005)		Proposed	
		Wetting	Drying	Wetting	Drying	Wetting	Drying	Wetting	Drying
White silica sand	Primary	0.0033	0.0070	0.0035	**0.0028**	0.0035	0.0066	**0.0024**	0.0087
	Secondary	0.0029	0.0035	0.0054	0.0095	0.0050	0.0031	**0.0014**	**0.0028**
	Tertiary	0.0099	0.0128	0.0130	0.0137	**0.0082**	**0.0042**	0.0130	0.0149
Dune sand	Primary	0.0074	0.0096	0.0096	0.0151	**0.0057**	0.0096	0.0067	**0.0095**
Rideau clayey loam	Secondary	0.0038	**0.0050**	**0.0024**	0.0065	0.0034	0.0071	**0.0024**	0.0062
Rubicon sandy loam	Tertiary	0.0106	0.0141	0.0118	0.0175	0.0076	**0.0105**	**0.0052**	0.0108

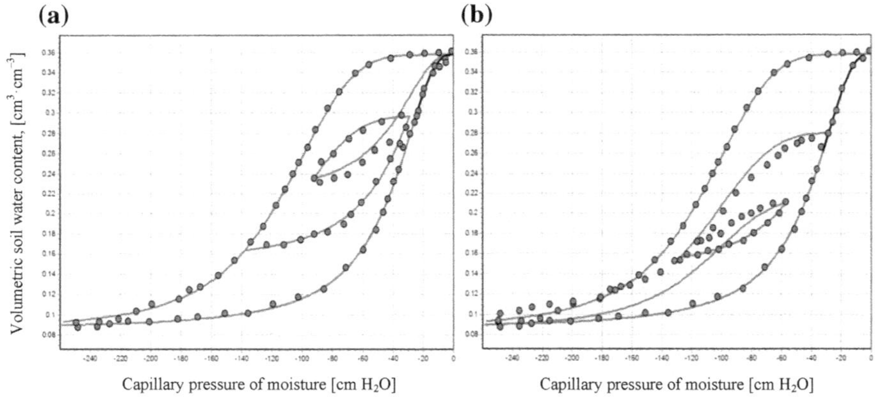

Fig. 17.1 **a** Approximation (dot-fitting) of measured data about the main branches; predictive estimation of the wetting primary branch, the drying secondary branch, the wetting tertiary branch for soil *White silica sand* using the proposed model. **b** Approximation (dot-fitting) of measured data about the main branches; predictive estimation of the drying primary branch, the wetting secondary branch, the drying tertiary branch for soil *White silica sand* using the proposed model

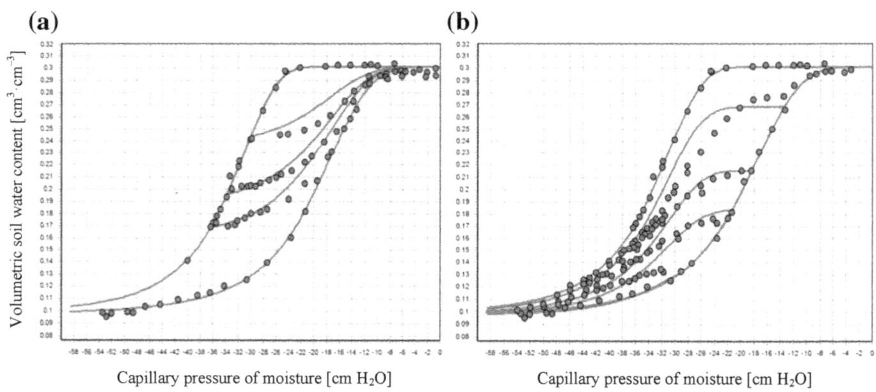

Fig. 17.2 **a** Approximation (dot-fitting) of measured data about the main branches; predictive estimation of the three wetting scanning branches for soil *Dune sand* using the proposed model. **b** Approximation (dot-fitting) of measured data about the main branches; predictive estimation of the four drying scanning branches for soil *Dune sand* using the proposed model

Thus, the obvious drawback of any model of the hysteretic soil water-retention capacity based on the functional dependence $\theta(\psi)$ proposed by Van Genuchten (1980) is the fundamental impossibility to determine for the parameters the boundaries of the physically feasible area (since the parameters α and n of the given function (1980) have no physical meaning): hence, in any such model an emergence of the «pumping effect» is highly likely. The situation becomes completely different if the functional dependence $\theta(\psi)$ proposed by Haverkamp et al. (1977) is used, or its

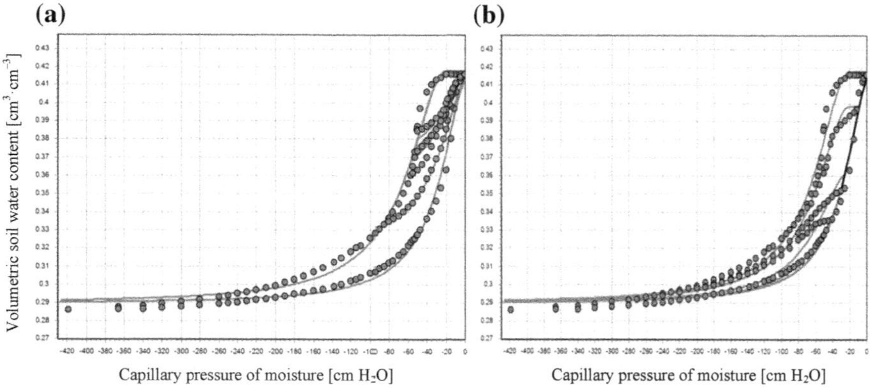

Fig. 17.3 **a** Approximation (dot-fitting) of data about the boundary branches; predictive estimation of the four wetting scanning branches for soil *Rideau clayey loam* using the proposed model. **b** Approximation (dot-fitting) of data about the boundary branches; predictive estimation of the three drying scanning branches for soil *Rideau clayey loam* using the proposed model

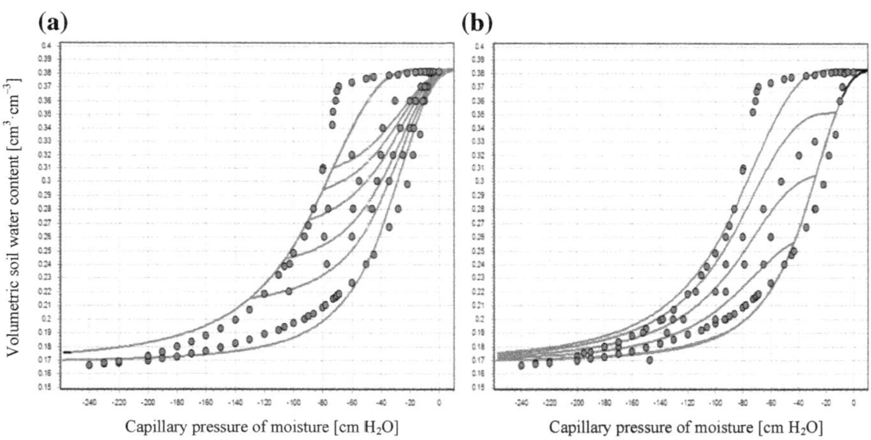

Fig. 17.4 **a** Approximation (dot-fitting) of data about the boundary branches; predictive estimation of the five wetting scanning branches for soil *Rubicon sandy loam* using the proposed model. **b** Approximation (dot-fitting) of data about the boundary branches; predictive estimation of the four drying scanning branches for *soil Rubicon sandy loam* using the proposed model

modified version described by formulas (17.1) and (17.2). In this case, calculating the hysteresis loop of the soil water-retention capacity using these dependencies reduces to the problem of an adequate estimation of the values of the parameters of these dependence. And if a solution is found for this problem, then the «pumping effect» does not appear.

The authors of this study are very sceptical of the idea that the hysteresis loops formed by the scanning branches should be artificially closed (Huang et al. 2005). It is clear that in this case there is no «pumping effect». But then we cannot expect the physical adequacy of the model. Indeed, if the different scanning (secondary) drying branches «come» to the same point, for example, on the main drying branch, then the meaning of the function of the differential moisture capacity of the soil disappears, because it has infinitely many values. According to the authors of this study, not a single scanning branch can cross the main branches of the hysteresis loop of the soil water-retention capacity. But the intersection between the scanning drying and wetting branches is possible. Oscillation of the values of the capillary pressure of soil moisture between the points of neighbouring intersections can lead to a certain «drift» of such a loop, which gradually decreases, asymptotically approaching an infinitesimal value. That is, with such oscillations, the closed loop should gradually localize in a certain area of the hysteresis loop, but not beyond the physically acceptable boundaries. Moreover, at each point on any branch of hysteresis, the function of the differential moisture capacity of the soil must take only two values: one for drying, and the other for wetting the soil (using two sets of physically interpreted parameters) (Terleev et al. 2016c, 2017a, b, 2018a, b). These ideas form the basis of the model proposed by the authors of this study.

17.5 Conclusion and Outlook

The mathematical model proposed by the authors of this paper corresponds to physical concepts in relation to the phenomenon of hysteresis of the soil water-retention capacity. Undesirable artificial «pumping effect» is not revealed. The application of irrigation precise rates calculated with the help of this model prevents the percolating the excess moisture under the action of gravity beyond the rooting zone of the soil, which minimizes the unproductive losses of irrigation water. The present model can be integrated into existing irrigation scheduling programs and systems such as BEREST90 or WEB-BEREST (Wenkel and Mirschel 1991; Mirschel et al. 2014a) used in German agriculture. It is also important for the estimation of irrigation water demand and additional crop yields due to irrigation under climate change (Mirschel et al. 2014b). The use of the proposed model in the development of farming technologies, as well as in substantiating land reclamation measures, will contribute to optimizing the water-air and nutrient regimes of the soil, as well as the rational use of water resources and agrochemicals.

Acknowledgements The research was supported by DAAD (PID: 91619700; A/10/01103) and Russian Foundation for Basic Research (#16-04-01473-a).

References

Ahuja LR, Swartzendruber D (1972) An improved form of soil-water diffusivity function. Soil Sci Soc Am Proc 36:9–14

Brutsaert W (1966) Probability laws for pore-size distribution. Soil Sci 101:85–92

D'Hollander EH (1979) Estimation of the pore size distribution from the moisture characteristic. Water Resour Res 15:107–112

Gillham RW, Klute A, Heermann DF (1976) Hydraulic properties of a porous medium: measurement and empirical presentation. Soil Sci Soc Am J 40:203–207

Haverkamp R, Vauclin M, Touma J, Wierenga PJ, Vachaud G (1977) A comparison of numerical simulation model for one-dimensional infiltration. Soil Sci Soc Am J 41:285–294

Huang HC, Tan YC, Chen CH (2005) A novel hysteresis model in unsaturated soil. Hydrol Process 19:1653–1665

Kool JB, Parker JC (1987) Development and evaluation of closed-form expressions for hysteretic soil hydraulic properties. Water Resour Res 23:105–114

Kosugi K (1994) Three-parameter lognormal distribution model for soil water retention. Water Resour Res 30:891–901

Kosugi K (1996) Lognormal distribution model for unsaturated soil hydraulic properties. Water Resour Res 32:2697–2703

Kosugi K, Hopmans JW (1998) Scaling water retention curves for soils with lognormal pore-size distribution. Soil Sci Soc Am J 62:1496–1505

Mirschel W, Klauss H, Berg M, Eisenhut K-U, Issbrücker G, Prochnow A, Schörling B, Wenkel K-O (2014a) Innovative Technologien für eine effiziente Bewässerung im Pflanzenbau. In: Bloch R, Bachinger J, Fohrmann R, Pfriem R (eds) Land- und Ernährungswirtschaft im Klimawandel—Auswirkungen, Anpassungsstrategien und Entscheidungshilfen. oekom verlag, München, pp 261–277

Mirschel W, Wieland R, Wenkel K-O, Nendel C, Guddat C (2014b) YIELDSTAT—a spatial yield model for agricultural crops. Eur J Agron 52:33–46

Mualem Y (1976) A catalogue of the hydraulic properties of unsaturated soils. Research Project 442. Technion, Israel Institute of Technology, Haifa, Israel, p 100

Scott PS, Farquhar GJ and Kouwen N (1983) Hysteretic effects on net infiltration. In: Proceeding of national conference on advances in infiltration publication 11-83 (Michigan: St. Joseph American Society of Agricultural Engineers), pp 163–170

Terleev VV, Topaj AG, Mirschel W (2015) The improved estimation for the effective supply of productive moisture considering the hysteresis of soil water-retention capacity. Russ Meteorol Hydrol 40:278–285

Terleev V, Petrovskaia E, Nikonorov A, Badenko V, Volkova Y, Pavlov S, Semenova N, Moiseev K, Topaj A, Mirschel W (2016a) Mathematical modeling the hydrological properties of soil for practical use in the land ecological management. In: MATEC web of conferences 73, Article number 03001

Terleev V, Nikonorov A, Togo I, Volkova Y, Garmanov V, Shishov D, Pavlova V, Semenova N, Mirschel W (2016b) Modelling the hysteretic water retention capacity of soil for reclamation research as a part of underground development. Procedia Eng 165:1776–1783

Terleev V, Petrovskaia E, Sokolova N, Dashkina A, Guseva I, Badenko V, Volkova Yu, Skvortsova O, Nikonova O, Pavlov S, Nikonorov A, Garmanov V, Mirschel W (2016c) Mathematical modeling of hydrophysical properties of soils in engineering and reclamation surveys. In: MATEC web of conferences 53, Article no 01013

Terleev VV, Nikonorov AO, Togo I, Volkova YuV, Ginevsky RS, Lazarev VA, Khamzin ER, Garma-nov VV, Mirschel W, Akimov LI (2017a) Hysteretic water-retention capacity of sandy soil. Mag Civ Eng 2:84–92

Terleev V, Ginevsky R, Lazarev V, Nikonorov A, Togo I, Topaj A, Moiseev K, Abakumov E, Melnichuk A, Dunaieva I (2017b) Predicting the scanning branches of hysteretic soil water-retention capacity with use of the method of mathematical modeling. In: IOP conference series: earth and environmental science 90, Article no 012105

Terleev VV, Nikonorov AO, Ginevsky RS, Lazarev VA, Togo I, Topaj AG, Moiseev KG, Pavlova VA, Layshev KA, Arkhipov MV, Yu Melnichuk A, Dunaieva IA, Mirschel W (2018a) Hysteresis of the soil water-retention capacity: estimating the scanning branches. Mag Civ Eng 1:141–148

Terleev V, Mirschel W, Nikonorov A, Ginevsky R, Lazarev V, Topaj A, Moiseev K, Layshev K, Arkhipov M, Melnichuk A, Dunaieva I, Popovych V (2018b) Five models of hysteretic water-retention capacity and their comparison for sandy soil. In: MATEC web of conferences 193, Article no 02036

Van Genuchten M Th (1980) A closed form equation for predicting the hydraulic conductivity of unsaturated soils. Soil Sci Soc Am J 44:892–989

Wenkel K-O, Mirschel W (1991) BERSIM und BEREST 90 Modelle zur Simulation der Bodenfeuchte und Evapotranspiration sowie zur operativen Beregnungseinsatzsteuerung im integrierten Landbau. Agrarinformatik 21:349–360

Winitzki S (2008) A handy approximation for the error function and its inverse (in https://sites.google.com/si-te/winitzki/sergei-winitzkis-files/erf-approx.pdf?attredirects=0)

Chapter 18
Estimation of Leaf Area Index (LAI) of Russian Forests Using a Mechanical Model and Forest Inventory Data

Michael Korzukhin and Vasily Grabovsky

Abstract We computed tree live biomass, leaf biomass and netto primary production for forests of Russian Federation using raw data of official forest inventories (7.42 million km^2 in 2004). The leaf biomass that was found makes it possible to find tree canopy leaf area index (LAI). The method applied was a physical model of growth and dying of forest biomass; the model was calibrated on 3719 individual tree biomass empirical growth curves (in carbon units). Most of the model parameters were borrowed from related literature, while four of them were left free and used for calibration. Comparison of calculated LAI values for 81 Subjects of the Russian Federation showed good agreement with LAI data found in global studies reporting LAI products for the Russian Federation.

Keywords Carbon balance · Forests of Russia · Forest LAI

18.1 Introduction

The leaf-area index LAI (m^2 m^{-2}) is the ratio of the projected leaf area to the stand (ground) area. It is among the most important functional parameters defining the intensity of matter (carbon and water) and energy flows between the atmosphere and the surface. The functional importance of LAI is illustrated by the large number of studies dealing with its estimation (e.g., Asner et al. 2003; Masson et al. 2003; Chen et al. 2012; Liu et al. 2012; Boussetta et al. 2013; Camacho et al. 2013; NASA Earth Observations 2016).

M. Korzukhin (✉)
Institute of Global Climate and Ecology, Moscow 107258, Russia
e-mail: mdkorz@gmail.com

V. Grabovsky
Center for Forest Ecology and Productivity, Russian Academy of Sciences, Moscow 117997, Russia
e-mail: wgrabo@mail.com

W. Mirschel et al. (eds.), *Landscape Modelling and Decision Support*, Innovations in Landscape Research,
https://doi.org/10.1007/978-3-030-37421-1_18

A number of methods are applied for LAI estimation; they can be sorted into three main groups: remote or satellite-derived; model calculations; ground-based observations. Each method has its own advantages and shortcomings so they complement each other in creating a more objective total picture.

With the exception of our previous study (Grabovskii et al. 2016) we are not aware of any systematic LAI estimates for Russian forests based on ground or remote data; for example, the global database of LAI compiled by Asner et al. (2003), includes only 5 locations within the territory of Russia. So, the aim of this study was to fill this gap and to calculate LAI for all Russian forests; it was realized by means of a new physical model of forest biomass growth that uses yield growth curves derived from official Russian State Forest Account (RSFA) inventory data. Besides LAI, other output variables were: tree live biomass, tree mortality and netto primary production (NPP).

Within the goal of this chapter, the model was validated on global remote and ground-based data related to LAI; output values of NPP were used to provide an additional control of the adequacy of the model.

18.2 Object and Data

According to the RSFA, Russian forests occupy $S_{\mathrm{RF}} = 7.42 \times 10^6$ km^2 (data on 2004, the year used below). The data from RSFA is broken down by Federal Subjects (FS) of the Russian Federation (the Subjects are the main self-governing administrative units of the Federation). The list can be found in Table 18.2. For each of 66 tree species (see Table 18.3 for the list of species names) within each FS, RSFA gives raw data ($k = 1, \ldots, 6$ is the age group number):

- total merchantable tree stem volume within administrative unit, V_k (m^3);
- area occupied by a species (as a dominant one), S_k (ha).

The ratio of V to S gives specific yield curves:

$$U_k = V_k/S_k\left(\mathrm{m^3ha^{-1}}\right) \tag{18.1}$$

Because of the physical nature of our model, the volume values were converted to carbon ones by multiplying by $K_{\mathrm{VC}k}$ coefficients (t C m^{-3}) which convert U_k to total carbon in live tree biomass u_k (kg C m^{-2}) (Zamolodchikov et al. 2003 updated);

$$u_k = K_{\mathrm{VC}k} \times U_k \times 10^{-1}\ \left(\mathrm{kg\,C\,m^{-2}}\right)\left(\mathrm{t\,C\,m^{-3}}\right)\left(\mathrm{m^3ha^{-1}}\right) \tag{18.2}$$

(the constant 10^{-1} converts (t) to (kg) and (ha^{-1}) to (m^{-2})). Resulting carbon growth curves (CGC) (18.2), 3719 in total, presented the main body of data used in the forest growth model that we formulate below.

Table 18.1 Original LAI values from selected sources; in bold: used for finding the averages and their final values (the bottom row)

LAI_{max}, (m^2 m^{-2})	Vegetation type	Method/Region	References
1.9	Tundra	Global biome-averages, ground-based observations	Asner et al. (2003)
2.7	Boreal needleleaf evergreen (*Abies, Picea, Pinus*)		
2.6	Boreal broadleaf deciduous (*Betula, Populus*)		
5.5	Temperate evergreen needleleaf (*Abies, Picea, Pinus*)		
5.1	Temperate deciduous broadleaf (*Acer, Betula, Populus, Quercus*)		
3.0	Global zonal average of annual LAI_{max}, 60°N		Boussetta et al. (2013)
3.3		BFB average Remote (MODIS)	
3.8	Boreal forest	BFB average	Camacho et al. (2013)
4.5–3.0		BFB gradient Remote	
4.0	Boreal forest	BFB average Model BEPS, remote	Chen et al. (2012)
4.375		BFB average	Gibelin et al. (2006)
5.25–4.0		BFB gradient model ISBA-A-gs and databases: ISLSCP-II, MODIS, ECOCLIMAP	
4.575	Needle-leaf trees		
3.5	Boreal forest	BFB average Remote (AVHRR, MODIS)	Liu et al. (2012)
3.83	Boreal forest	BFB average	Masson et al. (2003)
4.5–3.17		BFB gradient Remote plus ECOSYSTEM, POLDER, ISLSCP II databases	
4.5	Boreal forest	BFB average	NASA Earth Observations (July 2012)
3.75	Boreal forest	BFB average	Original average of the above-listed estimates
4.75–3.39		BFB gradient	
3.30	Boreal forest	BFB average	Averages corrected
4.18–3.10		BFB gradient	

18.3 Growth Model of Forest Biomass

18.3.1 The Model

We start with a tree growth equation that has the form of a mass conservation law for carbon:

$$dY/dt = (\text{GPP} - R_a) - \Sigma(\text{Loss}_k) \quad (18.3)$$

where Y (kg C tree^{-1}) is tree live biomass. Here, time step is one year, dimension of all members is (kg C tree^{-1} a^{-1}), GPP is Gross Primary Production of photosynthesis, R_a is carbon (autotrophic) respiration losses to tree maintenance, assimilate transport and growth. Loss_k is the amount of carbon lost due to the following processes: Loss_L—decay of foliage, Loss_{FR}—decay of fine roots, Loss_L—decay of seeds, Loss_{PH}—losses due to phytophags, Loss_F—losses due to non-destructive fires, Loss_B—losses due to branch decay, Loss_D—tree mortality loss (not applicable to a single tree but introduced here having in mind the future transition from per tree to per area units). Tree resources directed to the defense against phytophag attacks is difficult to estimate, but data about affected areas (Moiseev 2009) show that average Loss_{DEF} constitutes around 10^{-5} part of total NPP, so are not considered in what follows.

We simplify the model using a well-established proportion: NPP ~ CUE*GPP, where NPP (Net Primary Production of photosynthesis) is total net tree carbon gain, then CUE (Carbon Use Efficiency) is close to a constant, CUE ~ 0.50 (e.g., DeLucia et al. 2007); this constancy allows to substitute NPP for GPP-R_a:

$$dY/dt = \text{NPP} - \text{Loss}_L - \text{Loss}_{\text{FR}} - \text{Loss}_P - \text{Loss}_{\text{PH}} - \text{Loss}_F - \text{Loss}_B - \text{Loss}_D \quad (18.4)$$

18.3.2 Model Parametrization

We choose the simplest 'allometric' type of Loss_k parametrization, namely each Loss_k is taken proportional either to Y or to tree leaf biomass

$$Y_{\text{FOL}} = b_{\text{FOL}} \times Y^{q_{\text{FOL}}} \left(\text{kg C tree}^{-1}\right) \quad (18.5)$$

(b_{FOL}, q_{FOL} are allometric parameters). We can simplify the model further if we make the assumption that the net photosynthesis rate is proportional to leaf biomass (in principle, much more sophisticated approaches are available):

$$\text{NPP} = A \times Y_{\text{FOL}} \quad (18.6)$$

where A (a^{-1}) is the specific assimilation rate equal to kilogrammes of carbon assimilated by one kilogramme of leaf biomass in one year. Although the large literature on the subject has not reached a consensus, the following parametrizations of Loss_k seem reasonable:

$$\begin{aligned} \mathrm{Loss}_L &= d_L \times Y_{\mathrm{FOL}},\ \mathrm{Loss}_{\mathrm{FR}} = d_{\mathrm{FR}} \times a_{\mathrm{FR}} \times Y_{\mathrm{FOL}},\ \mathrm{Loss}_P = a_P \times Y \\ \mathrm{Loss}_{\mathrm{PH}} &= d_{\mathrm{PH}} \times Y_{\mathrm{FOL}},\ \mathrm{Loss}_F = d_F \times Y_{\mathrm{FOL}}, \\ \mathrm{Loss}_B &= d_B \times a_B \times Y,\ \mathrm{Loss}_D = d_D \times Y \end{aligned} \tag{18.7}$$

Here, d_{INDEX} (a^{-1}) are yearly decay rates, while a_{INDEX} are allometric parameters that allow to find values of corresponding biomass compartments:

$$Y_{\mathrm{FR}} = a_{\mathrm{FR}} \times Y_{\mathrm{FOL}} - \text{ fine root biomass}$$

$$Y_P = a_P \times Y - \text{ fruit biomass}$$

$$Y_B = a_B \times Y - \text{ branch biomass}$$

It is clear from formula $\mathrm{Loss}_P = a_P \cdot Y$ that omitted fruit decay rate, d_P, is set to 1.

18.3.3 Mortality Sub-model

One of the essential features of U_k curves is their non-monotonic dynamics, in particular, the frequently observed drop in the values of U_k in mature and over-mature stands. The decline is usually not very strong, but is significant enough so as not to be ruled out as an artefact of measurement (it is exhibited by around 40% of 3719 individual CGCs). This kind of non-monotonic behaviour is not specific to RSFA data but was observed in the dynamics of various stands (e.g., Whynot and Penner 1992; Lorimer et al. 2001; Wirth et al. 2002; Garet et al. 2009; Gao et al. 2017; Portier et al. 2018). Related studies do not offer a single explanatory mechanism—it can be tree senescence, litter accumulation that deteriorates site quality, susceptibility to windthrow and pathogens growing with age, or tree competition—all these mechanisms are discussed in literature.

As soon as the right-hand side of growth Eq. (18.4) depends only on Y (making the equation 'autonomous'), $Y(t)$ cannot display non-monotonic dynamics. To capture non-monotonic behaviour, we enriched the model by adding an additional equation (No 10 below). The most natural extension lies in making tree death rate to be an independent variable that will result in age-dependent $d_D = d_D(t)$. First, following to Community Land Model (Levis et al. 2004) we subdivide mortality rate into the sum of stochastic, d_{SM}, and age- and density dependent, d_{AM} components:

$$d_D(t) = d_{\mathrm{SM}} + d_{\mathrm{AM}}(t) \tag{18.8}$$

Physically, this division is somewhat conventional, since for example fire and phytophag damage contributes to both d_{SM} and d_{AM}. The stochastic component d_{SM} was fixed as

$$d_{\mathrm{SM}} = -\mathrm{Ln}(0.01)/T\mathrm{max} \tag{18.9}$$

providing 0.01 part of initial tree biomass survival to maximal tree age $T_{\max}$ (the parameter given by RSFA data). Dynamical equation for d_{AM} is taken in plausible form

$$dd_{\mathrm{AM}}/dt = g_{\mathrm{AM}} \times Y_{\mathrm{FOL}} \times (d_{\mathrm{AMmax}} - d_{\mathrm{AM}}) \tag{18.10}$$

Here, Y_{FOL} factor aims to describe the dependence of d_{AM} on tree age and its current state (c.f. Levis et al. 2004, where d_{AM} increases with NPP), while g_{AM}, d_{AMmax} are parameters.

18.3.4 Biomass Growth Per Stand Area

Meanwhile, all carbon-related variables were taken on on-tree basis. But starting from stand level and higher variables of vegetation cover are considered on an area basis (that is the case for empirical u_{k} variables (18.2)), so Eq. (18.4) has to be rewritten from (kg C tree^{-1}) to (kg C m^{-2}). If all Losses were proportional to Y, the equation would retain its form after just rewriting Y (kg C tree^{-1}) to m (kg C m^{-2}) (tree live biomass per unit area), but non-linearity in $Y_{\mathrm{FOL}} = b_{\mathrm{FOL}} \cdot Y^{q\mathrm{FOL}}$ spoils the picture.

So, we are looking for parameters a_L, p_L in allometric formula for m_L (foliage biomass per unit area) derived from

$$Y_{\mathrm{FOL}} = b_{\mathrm{FOL}} \times Y^{q_{\mathrm{FOL}}}\left(\mathrm{kg\ C\ tree}^{-1}\right), m_L = a_L \times m^{p_L}\left(\mathrm{kg\ C\ m}^{-2}\right) \tag{18.11}$$

Several ways of determining a_L, p_L are possible depending on the type of allometric and other data used. The vast majority of measurements connect a variable with the power function of the tree diameter, D (cm). Hence, the following algorithm is offered (denotations 'M_{INDEX}' relate to empirically observed components of single tree biomass, a_{INDEX}, p_{INDEX} are parameters, S_{crown} is the tree crown projection area, β_M, β_L are explained below):

$$M_{\mathrm{BEL}} = a_{\mathrm{BEL}} \times M_{\mathrm{AB}}\ \left(\mathrm{kg\ DW\ tree}^{-1}\right) \tag{18.12a}$$

$$M_{\mathrm{AB}} = a_{\mathrm{AB}} \times D^{p_{\mathrm{AB}}}\ \left(\mathrm{kg\ DW\ tree}^{-1}\right) \tag{18.12b}$$

$$M_{\mathrm{TOT}} = a_{\mathrm{AB}} \times (1 + a_{\mathrm{BEL}}) \times D^{p_{\mathrm{AB}}}$$
$$M_{\mathrm{FOL}} = a_{\mathrm{FOL}} \times D^{p_{\mathrm{FOL}}} \left(\text{kg DW tree}^{-1}\right) \tag{18.12c}$$

$$Y = \beta_M \times M_{\mathrm{TOT}},\, Y_{\mathrm{FOL}} = \beta_L \times M_{\mathrm{FOL}} \left(\text{kg C tree}^{-1}\right) \tag{18.12d}$$

$$S_{\mathrm{crown}} = a_{\mathrm{CR}} \times D^{p_{\mathrm{CR}}} \left(\text{m}^2\text{tree}^{-1}\right) \tag{18.12e}$$

$$m = Y/S_{\mathrm{crown}},\, m_{\mathrm{L}} = Y_{\mathrm{FOL}}/S_{\mathrm{crown}} \left(\text{kg C m}^{-2}\right) \tag{18.12f}$$

$$m = (\beta_M \times a_{\mathrm{AB}} \times (1 + a_{\mathrm{BEL}})/a_{\mathrm{CR}}) \times D^{(p_{\mathrm{AB}} - p_{\mathrm{CR}})}$$
$$m_{\mathrm{L}} = (\beta_L \times a_{\mathrm{FOL}}/a_{\mathrm{CR}}) \times D^{(p_{\mathrm{FOL}} - p_{\mathrm{CR}})} \tag{18.12g}$$

$$a_L = \frac{\beta_L a_{\mathrm{FOL}}}{a_{\mathrm{CR}}} \left[\frac{a_{\mathrm{CR}}}{\beta_M a_{\mathrm{AB}}(1 + a_{\mathrm{BEL}})}\right]^{p_L} \tag{18.13}$$

$$p_L = \frac{p_{\mathrm{FOL}} - p_{\mathrm{CR}}}{p_{\mathrm{AB}} - p_{\mathrm{CR}}} \tag{18.14}$$

A comment on (18.12): 'AB' and 'BEL' are above- and belowground biomasses. The three empirical relationships (18.12a, 18.12b, 18.12c) allow us to derive the two basic equations in (18.12d), which determine the full tree biomass and the leaf biomass in carbon units. The central point of the algorithm is the use of tree crown area, $S_{\mathrm{crown}}(D)$ (18.12e), division by which converts biomasses from (kg C tree^{-1}) to the area of ground surface, (kg C m^{-2}) (18.12f). Tree biomass compartments are usually given in dry weight (DW) biomass units, so parameters $\beta_M = 0.5$, $\beta_L = 0.45$ (widely accepted values) convert (DW) to (C). So (18.12a–18.12f) allow us to derive (18.12g), which shows the dependence of tree biomasses on D. Simple algebra that eliminates argument D from two equations in (18.12g) results in the final form of $m_L(m)$ in (18.11).

Finally, working form of biomass growth model used in m, LAI and other variables we arrive at:

$$dm/dt = Am_L - d_L m_L - d_{\mathrm{FR}} a_{\mathrm{FR}} m_L - a_P m - d_{\mathrm{PH}} m_L$$
$$- d_F m_L - d_B a_B m - (d_{\mathrm{SM}} + d_{\mathrm{AM}}(t))m \tag{18.15a}$$

$$dd_{\mathrm{AM}}/dt = g_{\mathrm{AM}} m_L (d_{\mathrm{AMmax}} - d_{\mathrm{AM}}) \tag{18.15b}$$

where *dm*/*dt* and all terms in right part of (18.15a) are in (kg C m^{-2} a^{-1}), and $m_L = a_L m^{p_L}$ is in (kg C m^{-2}); dd_{AM}/dt is in (a^{-2}), while the empirical nature of the second equation makes non-essential the dimension of g_{AM}.

After all, our main target variable is found as

$$\text{LAI} = \text{SLA} \times m_L \left(\text{m}^2\text{m}^{-2}\right) \left[\text{m}^2(\text{kg C foliage})^{-1}\right] \left(\left[\text{kg C foliage}\right) \text{m}^{-2}\right] \quad (18.16)$$

where SLA is one-sided (projected) specific leaf area corresponding to one kilogramme of carbon (its species-specific values are found in Table 18.3).

18.4 Parameter Values

18.4.1 Fixed Parameters from the Literature

The main allometric parameters $(a_{\text{AB}}, p_{\text{AB}})$ and $(a_{\text{FOL}}, p_{\text{FOL}})$ could be estimated from the results of numerous tree-level measurements (e.g., Cannell 1982; Ter-Mikaelian and Korzukhin 1997; Jenkins et al. 2003; Lambert et al. 2005; Zianis et al. 2005; Wang 2006). The main conclusion that can be derived from these studies and many other species-specific studies is that there is no hope of attributing a reliable value to these parameters on a species-specific level—the same species exhibit quite different values. For our purposes, there is a natural solution—to use the same values for all simulated species. Examination of the empirical 'clouds' on $(a_{\text{AB}}, p_{\text{AB}})$ and $(a_{\text{FOL}}, p_{\text{FOL}})$ planes led to selection of the following 'basic' values:

$$\begin{array}{c} \left(a_{\text{AB},} p_{\text{AB}}\right), \ (a_{\text{FOL}}, \ p_{\text{FOL}}) \\ (0.11, 2.4), \ (0.018, 1.8) \end{array} \quad (18.17)$$

Other parameters were estimated after a thorough survey of the literature that led to either species-specific (for d_L and d_{FR}—Table 18.3) or to generic values:

$a_{\text{BEL}} = 0.4$ is taken from Cannell's (1982) database (treated by Niklas 2005);
$a_{\text{CR}} = 0.0631$, $p_{\text{CR}} = 1.6$ were borrowed from Levis et al. (2004);
$a_{\text{FR}} = 0.8$—a mean over number of works;
$a_{\text{P}} = 0.0018$—estimated from (Brophy et al. 2007);
$d_{\text{FH}} = 0.07$—taken from (Utkin et al. 2008) as a mean value of foliage losses due to phytophags;
$d_{\text{F}} = 2.62 \times 10^{-4}$—estimated from RSFA data as (area covered by non-destructive fires)/S_{RF});
$a_B = 0.13$—estimated from review by (Ter-Mikaelian and Korzukhin 1997);
$d_B = 0.01$—expert estimate, literature values are rare and very variable.
Example of (a_L, p_L) calculation following (18.13), (18.14) for 'basic' parameter variant (18.17):

$$\begin{array}{ccc} a_{\text{L}} & p_{\text{L}} & q_{\text{FOL}} = p_{\text{FOL}}/p_{\text{AB}} \\ 0.1221 & 0.25 & 0.75 \end{array} \quad (18.18)$$

Note that ratio p_{FOL}/p_{AB} is precisely equal to the value predicted by the Metabolic Theory (West et al. 1999).

18.4.2 Fitted Parameters

There are four free parameters left that were appointed to fit the model to the empirical species CGCs, namely:

$$m(1) - \text{initial biomass value, } A - \text{specific assimilation rate,}$$
$$g_{AM}, d_{AMmax} - \text{parameters of ‘biomass mortality’} \tag{18.19}$$

The parameters were computed independently for each of 3719 CGCs u_k (18.2), with one technical correction: instead of fitting to 6 or less (because of missing data) values of u_k, we presented each value of u_k as a set of identical one-year values with number equal to age interval proper to a given u_k (result of this ‘spread’ in red horizontal lines is clear from Fig. 18.1, giving in total a step-wise function $w(t)$); the same presentation was made for areas S_k:

$$u_k \rightarrow w(t),\ S_k \rightarrow s(t),\ t = 1, \ldots, T_{max} \tag{18.20}$$

This substitution allowed us to avoid instabilities in the iteration process under parameter search, and to automate it almost entirely. Finally, free parameters (18.19) were found by means of minimization of standard deviation

$$\sigma = \left[\sum (m(t) - w(t))^2 / T_{max}\right]^{1/2} \tag{18.21}$$

equal to model accuracy per one year in $m(t)$ calculation.

18.4.3 Data for Validation of the Model

Evaluation of model's ability to predict LAI values for Russian forests was performed by means of comparison with a specially organized set of the observed LAI. We selected several recent large-scale studies based either on global ground observations, and/or on remote LAI products, all related to LAI_{max} (July–August) values.

Because we did not have access to digital forms of LAI maps, printed maps were used to evaluate two LAI reference parameters. At first, a «Boreal Forest Belt» (BFB) was determined as a spatial rectangle [57°N:63°N; 30°E:120°E] that encompasses approximately the core of the boreal forest range in Russia (due to accepted distribution of Plant Functional Types, e.g., Grosso et al. 2008; Beer et al. 2010). Within BFB, we evaluated:

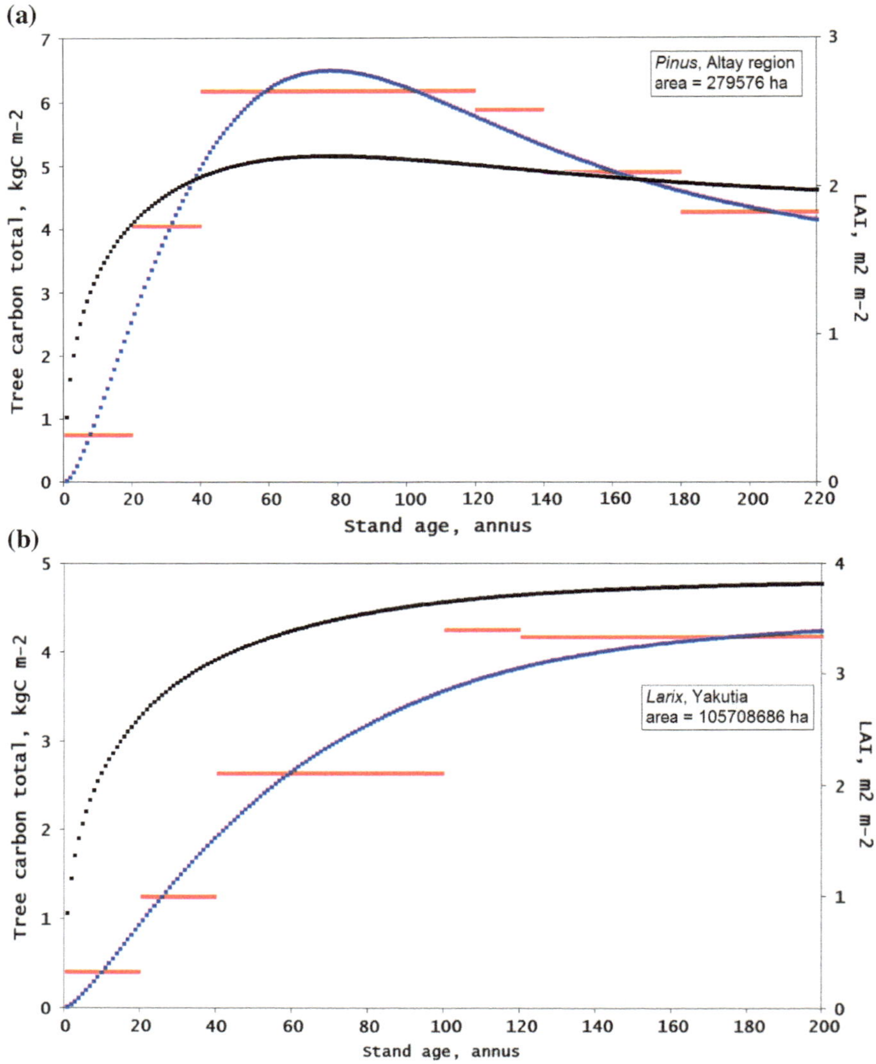

Fig. 18.1 Examples of model (18.15) fitting for 'basic' parametric case (18.17): RSFA data (▬); ■—theoretical curve $m(t)$, ■—$\mathrm{LAI}(t) = 11.33\, m_{\mathrm{L}}(t)$ where $m_{\mathrm{L}}(t) = 0.1221\, m(t)^{0.25}$

(a) average value of $\mathrm{LAI}_{\mathrm{aver}}$, and, where the map resolution allowed
(b) West-East LAI gradient, LAI_W-LAI_E.

Results are presented in Table 18.1. The most impressive feature of this data is the substantial variation in estimates of LAI and LAI_W-LAI_E among quoted studies. The reasons are hard to explain, and the authors of this chapter cannot give preference to any one set of results. In this situation, a way of getting the needed LAI reference values is just to find their averages. The method is not new, e.g., Coupled Model

Intercomparison Project-5 (CMIP5) produces a multimodel climate dataset by means of averaging the output over an ensemble of selected climatic models (Taylor et al. 2012). The result of our averaging is given at the bottom of Table 18.1. (A correction was applied to originally found $\mathrm{LAI_{aver}}$: our model actually estimates not $\mathrm{LAI_{max}}$ but an effective $\mathrm{LAI} = \mathrm{LAI_{effective}}$ during vegetation season; seasonal profiles of LAI were taken from (Heiskanen et al. 2012; Zhu et al. 2013) to find a correcting coefficient in $\mathrm{LAI} = k_{\mathrm{LAIeff}} \cdot \mathrm{LAI_{max}}$; estimation gave $k_{\mathrm{LAIeff}} \approx 0.88$.)

18.5 Results

Typical examples of CGC fitting and $\mathrm{LAI}_{\mathrm{spec}}^{\mathrm{theor}}(t)$ found for two individual dominant species are given in Fig. 18.1a, b ('basic' parametric case (18.17) was taken).

Territory-distributed LAI values were found by weighting over corresponding areas as follows (Where $\mathrm{LAI}_{\mathrm{spec,\,FS}}^{\mathrm{theor}}$ is the theoretically derived LAI value for species spec and Federal Subject FS.)

1. Average LAI values for given species in a Federal Subject:

$$\mathrm{LAI}_{\mathrm{spec,FS}}^{\mathrm{theor}} = \left[\sum \mathrm{LAI}_{\mathrm{spec}}^{\mathrm{theor}}(t) \times s_{\mathrm{spec}}(t)\right] \Big/ s_{\mathrm{spec,FS}}, \quad s_{\mathrm{spec,FS}} = \sum s_{\mathrm{spec}}(t) \tag{18.22a}$$

2. Average LAI values for all species in a Federal Subject:

$$\mathrm{LAI}_{\mathrm{FS}}^{\mathrm{theor}} = \left[\sum \mathrm{LAI}_{\mathrm{spec,FS}}^{\mathrm{theor}} \times s_{\mathrm{spec,FS}}\right] \Big/ s_{\mathrm{FS}}, \quad s_{\mathrm{FS}} = \sum s_{\mathrm{spec,FS}} \tag{18.22b}$$

3. Average LAI values for all species in all Federal Subjects (Russia):

$$\mathrm{LAI}_{\mathrm{Russia}}^{\mathrm{theor}} = \left[\sum \mathrm{LAI}_{\mathrm{FS}}^{\mathrm{theor}} \times s_{\mathrm{FS}}\right] \Big/ s_{\mathrm{Russia}}, \quad s_{\mathrm{Russia}} = \sum s_{\mathrm{FS}} \tag{18.22c}$$

In spite of the rather coarse spatial resolution, $\mathrm{LAI}_{\mathrm{FS}}^{\mathrm{theor}}$ map (Fig. 18.2) drawn in colour gradations demonstrates well a sloping West-East LAI gradient, as well as the expected steep North-South gradient. Figure 18.3 presents these spatial changes in more informative numerical form. We suppose that the model demonstrates good fidelity in respect to LAI observational data.

There is a way of independent control for $\mathrm{LAI}_{\mathrm{FS}}^{\mathrm{theor}}$ fidelity, namely comparison of model $\mathrm{NPP}_{\mathrm{FS}}^{\mathrm{theor}}$ output with observations. A comprehensive analysis of this task is far beyond the scope of this Chapter, so we just give the map of $\mathrm{NPP}_{\mathrm{FS}}^{\mathrm{theor}}$ (Fig. 18.4) from the model run that gave $\mathrm{LAI}_{\mathrm{FS}}^{\mathrm{theor}}$ in Fig. 18.2. In general, $\mathrm{NPP}_{\mathrm{FS}}^{\mathrm{theor}}$ corresponds well to the values reported in vast literature devoted to NPP examination (e.g., Running et al. 2004; Luyssaert et al. 2007; Barman et al. 2014 and many others).

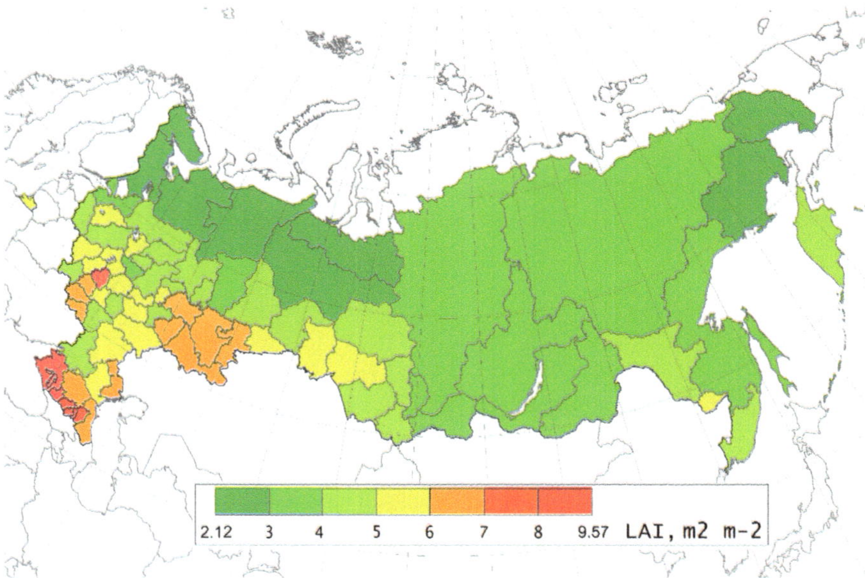

Fig. 18.2 Calculated LAI values for 'basic' parametric case (18.17), $LAI^{theor}_{Russia} = 3.56\,(m^2m^{-2})$

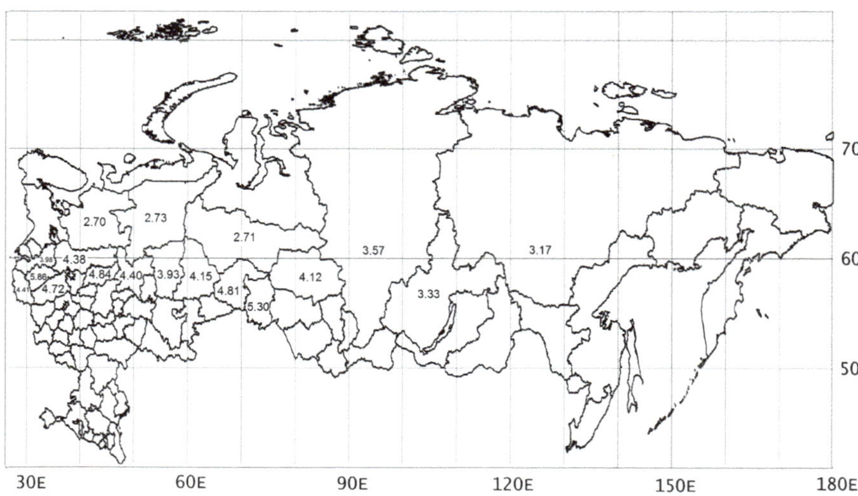

Fig. 18.3 Calculated LAI values for 'basic' parametric case (18.17) within Boreal Forest Band (approximately), c.f. to the observed values in the bottom row of Table 18.1

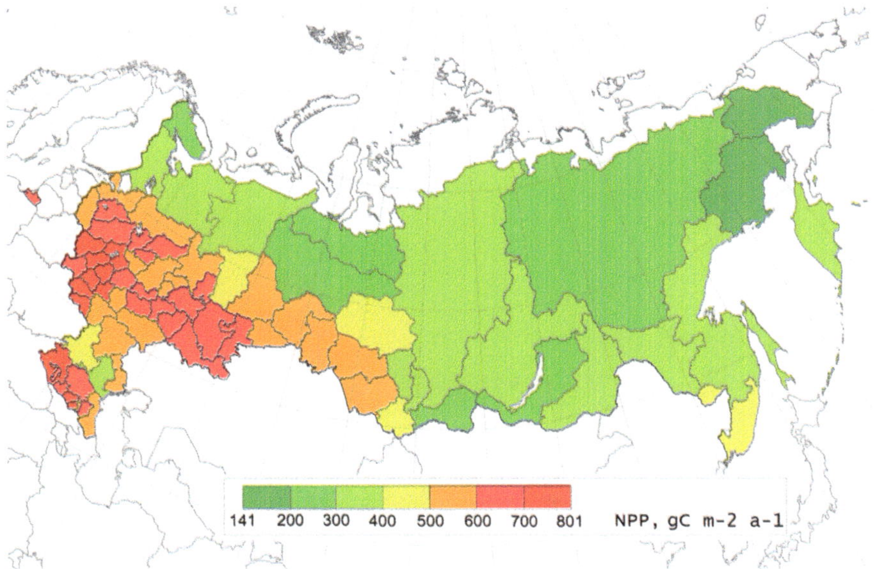

Fig. 18.4 Calculated NPP values for 'basic' parametric case (18.17), $NPP^{theor}_{Russia} = 329\,(g\,C\,m^{-2}\,a^{-1})$

18.6 Conclusions

I. The tree biomass growth Eq. (18.15a) supplemented by the tree mortality Eq. (18.15b) demonstrated good quality in description of LAI over territory occupied by forests of Russia.

II. The proposed algorithm needs essentially only empirical yield curves plus several parameters with their typical values (in agreement with the literature). The construction offered can be implemented in any encompassing scheme (model) that needs a simple LAI and NPP calculation, e.g., in global or regional climate models that need description of carbon cycle dynamics including vegetation.

III. In calculating LAI we used three independent sources: Russian State Forest Account, LAI observations, and allometric relationships for tree biomass. The fact that the resulting picture of LAI and NPP values is quite realistic confirms both the reliability of our sources and the empirical worth of the model. (It could have easily turned out that reasonable predictions of the model had to be purchased at the price of unrealistic parameter values. But in fact our parameter values are quite realistic.) We are confident that our model can be profitably used in future ecological research.

Acknowledgements The authors wish to thank Ter-Michaelian for numerous helpful suggestions and criticisms, and Theodore Korzukhin for help in preparing the English text for publication.

Appendix

Tables 18.2 and 18.3.

Table 18.2 Federal subjects of Russian Federation, their forested areas and tree species number reported by Russian state forest account

nn	fsub	FS name	Forested area (km^2)	Store specific (m^3 ha^{-1})	Tree species number	Individual yield curve number
1	1101	Altai Territory	27,114	153.4	24	68
2	1102	Republic of Altai	35,603	187.0	13	34
3	1103	Krasnodar Territory	11,222	193.7	46	105
4	1104	Krasnoyarsk Territory	1,017,876	104.5	15	39
5	1105	Primorye Territory	113,877	151.2	25	70
6	1107	Stavropol Territory	1,090	113.9	37	64
7	1108	Khabarovsk Territory	509,522	97.7	25	74
8	1109	Republic of Ingushetia	755	148.3	23	38
9	1110	Amur Region	226,566	88.1	22	60
10	1111	Arkhangelsk Region	201,327	105.1	10	39
11	1112	Astrakhan Region	1,124	73.1	20	24
12	1113	Chechen Republic	2,654	149.3	36	77
13	1114	Belgorod Region	2,403	169.3	24	35
14	1115	Bryansk Region	7,659	185.5	18	42
15	1117	Vladimir Region	9,837	178.7	15	42
16	1118	Volgograd Region	4,406	73.8	25	31
17	1119	Vologda Region	71,491	143.4	11	29
18	1120	Voronezh Region	3,711	143.8	24	54
19	1121	Jewish Autonomous Region	15,638	107.5	20	56
20	1122	Nizhny Novgorod Region	30,063	149.2	21	58
21	1123	Nenets Autonomous Area	1,898	95.0	4	4
22	1124	Ivanovo Region	7,936	179.0	16	42

(continued)

Table 18.2 (continued)

nn	fsub	FS name	Forested area (km^2)	Store specific (m^3 ha^{-1})	Tree species number	Individual yield curve number
23	1125	Irkutsk Region	605,093	143.9	15	40
24	1127	Kaliningrad Region	2,478	167.6	22	55
25	1128	Tver Region	22,534	169.1	15	40
26	1129	Kaluga Region	6,726	196.4	17	45
27	1130	Kamchatka Territory	188,425	60.8	13	31
28	1132	Kemerovo Region	40,995	114.8	17	41
29	1133	Kirov Region	59,125	145.3	17	52
30	1134	Kostroma Region	34,484	163.6	15	37
31	1135	Chukotka Autonomous Area	49,124	16.7	8	12
32	1136	Samara Region	5,137	142.1	20	76
33	1137	Kurgan Region	12,948	138.4	18	48
34	1138	Kursk Region	2,227	164.4	24	32
35	1141	Leningrad Region	35,206	176.8	16	47
36	1142	Lipetsk Region	1,855	191.9	23	28
37	1144	Magadan Region	162,598	23.2	12	22
38	1146	Moscow Region	20,576	224.9	18	40
39	1147	Murmansk Region	51,319	42.4	6	18
40	1149	Novgorod Region	33,474	172.3	17	36
41	1150	Novosibirsk Region	27,135	108.5	24	52
42	1152	Omsk Region	26,150	129.2	16	53
43	1153	Orenburg Region	4,818	136.7	23	38
44	1154	Orel Region	1,126	182.8	23	44
45	1156	Penza Region	8,298	146.5	18	57
46	1157	Perm Territory	93,579	131.8	15	48
47	1158	Pskov Region	11,131	160.7	15	39
48	1160	Rostov Region	2,245	63.1	28	61
49	1161	Ryazan Region	7,201	172.7	18	39
50	1163	Saratov Region	4,610	111.1	26	39
51	1164	Sakhalin Region	55,268	108.2	24	56
52	1165	Sverdlovsk Region	119,474	158.3	17	42
53	1166	Smolensk Region	8,382	193.9	17	42
54	1168	Tambov Region	3,215	166.1	18	48
55	1169	Tomsk Region	174,712	140.2	10	34
56	1170	Tula Region	3,170	208.7	18	29

(continued)

Table 18.2 (continued)

nn	fsub	FS name	Forested area (km^2)	Store specific (m^3 ha^{-1})	Tree species number	Individual yield curve number
57	1171	Tyumen Region	55,073	128.3	14	50
58	1172	Khanty-Mansi Autonomous Area—Ugra	278,462	104.4	11	29
59	1173	Ulyanovsk Region	9,494	172.1	16	43
60	1174	Yamal-Nenets Autonomous Area	157,390	70.5	11	17
61	1175	Chelyabinsk Region	25,637	159.1	22	73
62	1176	Trans-Baikal Territory	276,284	90.3	18	48
63	1178	Yaroslavl Region	9,703	168.6	17	34
64	1179	Republic of Adygeya	1,945	215.6	34	72
65	1180	Republic of Bashkortostan	48,164	144.7	24	75
66	1181	Republic of Buryatia	220,319	93.9	17	47
67	1182	Republic of Daghestan	3,757	107.1	34	69
68	1183	Kabardino-Balkarian Republic	1,352	164.5	31	78
69	1185	Republic of Kalmykia	142	20.0	16	21
70	1186	Republic of Karelia	92,465	98.6	11	32
71	1187	Komi Republic	283,868	95.9	10	39
72	1188	Republic of Mari EL	10,933	149.2	18	53
73	1189	Republic of Mordovia	5,235	151.1	17	44
74	1190	Republic of North Ossetia Alania	1,699	177.1	29	74
75	1191	Karachayevo-Circassian Republic	3,732	211.0	33	66
76	1192	Republic of Tatarstan	11,927	146.4	21	55
77	1193	Republic of Tuva	77,239	136.4	15	32
78	1194	Udmurtian Republic	14,938	170.2	17	44
79	1195	Republic of Khakassia	28,435	153.5	17	40
80	1197	Chuvash Republic	5,294	140.8	17	56
81	1198	Republic of Sakha (Yakutia)	1,430,617	59.8	14	22
0	1100	Russia	7,242,714	100.9	66	3719

Table 18.3 Species name list of Russian State Forest Account and species-specific parameters used in biomass growth Eq. (18.15)

nn	Species code in RSFA	Species name	Turnover rate foliage d_L, a^{-1}	Turnover rate fine roots d_{FR}, a^{-1}	SLA_{proj} (m^2 kg^{-1} C)
1	101	*Pinus* spp.	0.3825	0.61	11.33
2	102	*Picea* spp.	0.219	0.84	11.78
3	103	*Abies* spp.	0.266	0.61	21.11
4	104	*Larix* spp.	1.0	0.3	21.78
5	105	*Pinus sibirica* spp.	0.266	0.70	11.33
6	106	*Juniperus communis* L.	1.0	0.3	23.56
7	110	*Quercus* spp. <long-boled>	1.0	0.80	32.89
8	111	*Quercus* spp.<short-boled>	1.0	0.80	32.89
9	112	*Fagus sylvatica* spp.	1.0	0.80	50.89
10	113	*Carpinus betulus* L.	1.0	0.80	50.89
11	114	*Fraxinus* spp.	1.0	0.80	39.11
12	115	*Acer* spp.	1.0	0.80	50.00
13	116	*Ulmus* spp.	1.0	0.80	58.67
14	117	*Betula ermanii. paraermani* Cham.	1.0	1.22	44.44
15	120	*Haloxylon* spp.	1.0	0.80	44.89
16	121	*Robinia pseudoacacia* L.	1.0	0.80	45.33
17	124	*Betula* spp.	1.0	1.22	44.22
18	125	*Popilus tremula* L.	1.0	1.28	35.33
19	126	*Alnus incana* L.	1.0	1.22	49.11
20	127	*Alnus glutinosa* L.	1.0	1.22	46.67
21	130	*Tilia cordata* L.	1.0	1.22	61.33
22	131	*Populus alba* L.	1.0	1.22	35.33
23	132	*Salix caprea* L.	1.0	1.22	48.89
24	141	*Prunus armeniaca* L.	1.0	0.86	34.89
25	143	*Phellodéndron amurénse* Rupr.	1.0	0.86	34.89

(continued)

Table 18.3 (continued)

nn	Species code in RSFA	Species name	Turnover rate foliage d_L, a^{-1}	Turnover rate fine roots d_{FR}, a^{-1}	SLA_{proj} (m^2 kg^{-1} C)
26	144	*Carpinus orientalis* Mill.	1.0	0.86	53.33
27	145	*Gleditsia triacanthos* L.	1.0	0.86	34.89
28	147	*Pyrus communis* L.	1.0	0.86	30.00
29	150	*Zelkova carpinifolia* L.	1.0	0.86	29.56
30	151	*Quércus súber* L.	1.0	0.86	14.67
31	153	*Celtis australis* L.	1.0	0.86	34.89
32	154	*Catalpa* spp.	1.0	0.86	34.89
33	155	*Castanea sativa* L.	1.0	0.86	45.56
34	156	*Pterocarya pterocarpa* Kunth.	1.0	0.86	34.89
35	157	*Prunus* ssp.	1.0	0.86	34.89
36	160	*Júglans régia* L.	1.0	0.86	31.11
37	161	*Júglans mandshúrica* Maxim.	1.0	0.86	31.11
38	163	*Sorbus aucuparia* L.	1.0	0.86	51.11
39	164	*Buxus* L.	1.0	0.86	34.89
40	165	*Prunus sogdiana* Vass.	1.0	0.86	34.89
41	166	*Taxus baccata* L.	1.0	0.86	28.44
42	167	*Pistacia vera* L.	1.0	0.86	10.00
43	168	Without name	1.0	0.86	10.00
44	171	*Prúnus pádus* L.	1.0	0.86	34.89
45	172	*Prunus avium* L.	1.0	0.86	60.00
46	173	*Morus nigra* L.	1.0	0.86	33.33
47	175	*Eucommia ulmoides* Oliv.	1.0	0.86	34.89
48	177	*Malus sylvéstris* L.	1.0	0.86	34.67
49	180	Other tree species	1.0	0.86	34.89

(continued)

Table 18.3 (continued)

nn	Species code in RSFA	Species name	Turnover rate foliage d_L, a^{-1}	Turnover rate fine roots d_{FR}, a^{-1}	SLA_{proj} (m^2 kg^{-1} C)
50	181	*Sasa kurilensis* (Rupr.) Makino & Shibata.	0.320	0.70	41.56
51	182	*Betula fruticosa* Pallas.	0.320	0.70	45.33
52	183	*Euónymus europaea* L.	0.320	0.70	66.67
53	184	*Crataégus* ssp.	0.320	0.70	25.33
54	185	*Támarix* ssp.	0.320	0.70	41.56
55	186	*Cornus alba Sibirica Variegata* L.	0.320	0.70	57.78
56	187	*Calligonum* ssp.	0.320	0.70	41.56
57	190	*Salix* ssp.	0.320	0.70	48.89
58	191	*Pinus pumila* (Pall.) Regel.	0.266	0.70	11.33
59	193	*Alnus incana* L.	0.320	0.70	49.11
60	194	*Elaeágnus angustifólia* L.	0.320	0.70	41.56
61	195	*Juniperus communis* L.	0.320	0.70	23.56
62	196	*Hippóphaë rhamnoídes* L.	0.320	0.70	41.56
63	197	*Rhododendron ferrugineum* L.	0.320	0.70	21.11
64	198	*Ribes nigrum* L.	0.320	0.70	81.56
65	200	*Rosa* ssp.	0.320	0.70	26.67
66	204	Other bushes	0.320	0.70	41.56

References

Asner GP, Scurlock JMO, Hicke JA (2003) Global synthesis of leaf area index observations: implications for ecological and remote sensing studies. Glob Ecol Biogeogr 12:191–205

Barman R, Jain AK, Liang M (2014) Climate-driven uncertainties in modeling terrestrial gross primary production: a site level to global-scale analysis. Glob Change Biol 20(5):1394–1411

Beer C, Reichstein M, Tomelleri E, Ciais P, Jung M, Carvalhais N, Rödenbeck C, Arain MA, Baldocchi D, Bonan GB, Bondeau A, Cescatti A, Lasslop G, Lindroth A, Lomas M, Luyssaert S, Margolis H, Oleson KW, Roupsard O, Veenendaal E, Viovy N, Williams C, Woodward FI, Papale D (2010) Terrestrial gross carbon dioxide uptake: global distribution and covariation with climate. Science 329:834–838

Boussetta S, Balsamo G, Beljaars A, Krala T, Jarlan L (2013) Impact of a satellite-derived leaf area index monthly climatology in a global numerical weather prediction model. Int J Remote Sens 34(9–10):3520–3542

Brophy C, Gibson DJ, Wayne PM, Connolly J (2007) A modelling framework for analysing the reproductive output of individual plants grown in monoculture. Ecol Model 207:99–108

Camacho F, Cernicharo J, Lacaze R, Baret F, Weiss M (2013) GEOV1: LAI, FAPAR essential climate variables and FCOVER global time series capitalizing over existing products. Part 2: validation and intercomparison with reference products. Remote Sens Environ 137:310–329

Cannell MGR (1982) World forest biomass and primary production data. Academic Press, New York, NY, p 391

Chen JM, Mo G, Pisek J, Liu J, Deng F, Ishizawa M, Chan D (2012) Effects of foliage clumping on the estimation of global terrestrial gross primary productivity. Glob Biogeochem Cycles 26:GB1019

DeLucia EH, Drake JE, Thomas RB, Gonzalez-Meler M (2007) Forest carbon use efficiency: is respiration a constant fraction of gross primary production? Glob Change Biol 13(6):1157–1167

Gao B, Taylor AR, Searle EB, Kumar P, Ma Z, Hume AM, Chen HYH (2017) Carbon storage declines in old boreal forests irrespective of succession pathway. Ecosystems. https://doi.org/10.1007/s10021-017-0210-4

Garet J, Pothier D, Bouchard M (2009) Predicting the long-term yield trajectory of black spruce stands using time since fire. For Ecol Manag 257:2189–2197

Gibelin A-L, Calvet JC, Roujean JL, Jarlan L, Los SO (2006) Ability of the land surface model ISBA-A-gs to simulate leaf area index at the global scale: Comparison with satellites products. J Geophys Res 111(D18102):1–16

Grabovskii VI, Zukert NV, Korzukhin MD (2016) Estimates of leaf-area index for the territory of Russia using state forest inventory data. Contemp Probl Ecol 9(7):814–819

Grosso SD, Parton W, Stohlgren T, Zheng D, Bachelet D, Prince S, Hibbard K, Olson R (2008) Global potential net primary production predicted from vegetation class, precipitation, and temperature. Ecology 89(8):2117–2126

Heiskanen J, Rautiainen M, Stenberg P, Mõttus M, Vesanto V-H, Korhonen L, Majasalmi T (2012) Seasonal variation in MODIS LAI for a boreal forest area in Finland. Remote Sens Environ 126:104–115

Jenkins JC, Chojnacky DC, Heath LS, Birdsey RA (2003) National-scale biomass estimators for United States tree species. For Sci 49(1):12–35

Lambert M-C, Ung C-H, Raulier F (2005) Canadian national tree aboveground biomass equations. Can J For Res 35(8):1996–2018

Levis S, Bonan GB, Vertenstein M, Oleson KW (2004) The community land model's dynamic global vegetation model (CLM-DGVM): technical description and user's guide. Terrestrial Sciences Section, Climate and Global Dynamics Division, National Center for Atmospheric Research, Boulder, Colorado, NCAR/TN-459+IA, p 64

Liu Y, Liu R, Chen JM (2012) Retrospective retrieval of long-term consistent global leaf area index (1981–2011) from combined AVHRR and MODIS data. J Geophys Res 117:G04003

Lorimer CG, Dahir SE, Nordheim EV (2001) Tree mortality rates and longevity in mature and old-growth hemlock-hardwood forests. J Ecol 89(6):960–971

Luyssaert S, Inglima I, Jung M, Richardson AD, Reichstein M, Papale D, Piao SL, Schulze E-D, Wingate L, Matteucci G, Aragao L, Aubinet M, Beer C, Bernhofer C, Black KG, Bonal D, Bonnefond J-M, Chambers J, Ciais P, Cook B, Davis KJ, Dolman AJ, Gielen B, Goulden M, Grace J, Granier A, Grelle A, Griffis T, Grünwald T, Guidolotti G, Hanson PJ, Harding R, Hollinger DY, Hutyra LR, Kolari P, Kruijt B, Kutsch W, Lagergren F, Laurila T, Law BE, Le Mmaire G, Lindroth A, Loustau D, Malhi Y, Mateus J, Miglavacca M, Misson L, Montagnani L, Moncrieff J, Moors E, Munger JW, Nikinmaa E, Ollinger SV, Pita G, Rebmann C, Roupsard O, Saigusa N, Sanz MJ, Seufert G, Sierra C, Smith M-L, Tang J, Valentini R, Vesala T, Janssens IA (2007) The CO_2-balance of boreal, temperate and tropical forests derived from a global database. Glob Change Biol 13:2502–2537

Masson V, Champeaux J-L, Chauvin F, Meriguet C, Lacaze R (2003) A global database of land surface parameters at 1-km resolution in meteorological and climate models. J Clim 16:1261–1282

Moiseev B (2009) Case studies on measuring and assessing forest degradation. Results of pathological monitoring in degraded Russian forests. Forest Resources Assessment Programme of FAO, Working Paper 175, Rome, Italy, p 17

NASA Earth Observations (2016)

https://neo.sci.gsfc.nasa.gov/view.php?datasetId=MOD15A2_M_LAI

Niklas KJ (2005) Modelling below- and above-ground biomass for non-woody and woody plants. Ann Bot 95:315–321

Portier J, Gauthier S, Cyr G, Bergeron Y (2018) Does time since fire drive live aboveground biomass and stand structure in low fire activity boreal forests? Impacts on their management. J Environ Manag 225:346–355

Running SW, Nemani RR, Heinsch FA, Zhao M, Reeves M, Hashimoto H (2004) A continuous satellite derived measure of global terrestrial primary production. Bioscience 54(6):547–560

Taylor KE, Stouffer RJ, Meehl GA (2012) An overview of CMIP5 and the experiment design. Bull Am Meteor Soc 93:485–498

Ter-Mikaelian MT, Korzukhin MD (1997) Biomass equations for sixty-five North American tree species. For Ecol Manag 97(1):1–24

Utkin AI, Ermolova LS, Utkina IA (2008) Leaf area surface of forest plants: content, parameters, and use). (ed SE Vompersky), Moscow, Nauka 292 p (Площадь поверхности лесных растений: сущность, параметры, использование. Отв. ред СЭ Вомперский. Москва, Наука, 2008 292 с)

Wang C (2006) Biomass allometric equations for 10 co-occurring tree species in Chinese temperate forests. For Ecol Manag 222:9–16

West GB, Brown JH, Enquist BJ (1999) A general model for the structure and allometry of plant vascular systems. Nature 400(6745):664–667

Whynot TW, Penner M (1992) Growth and yield of black spruce ecosystems in the Ontario Clay Belt: implications for forest management. Petawawa National Forestry Institute Information Report PI-X-99, p 44

Wirth C, Schulze E-D, Kuznetova V, Milukova I, Hardes G, Siry M, Schulze B, Vygodskaya NN (2002) Comparing the influence of site quality, stand age, fire and climate on aboveground tree production in Siberian Scots pine forests. Tree Physiol 21:537–552

Zamolodchikov DG, Utkin AI, Chestnyh OV (2003) Coefficients that allow to find a phytomass of forest-forming species in Russia from stand store (Коэффициенты конверсии запасов насаждений в фитомассу основных лесообразующих пород России. Forest taxation and Management 119–127 (Лесная таксация и лесоустройство. Выпуск 1(32):2003, с 119–127)

Zhu Z, Bi J, Pan Y, Ganguly S, Anav A, Xu L, Samanta A, Piao S, Nemani RR, Myneni RB (2013) Global data sets of vegetation leaf area index (LAI)3 g and fraction of photosynthetically active radiation (FPAR)3g derived from global inventory modeling and mapping studies (GIMMS) normalized difference vegetation index (NDVI3g) for the period 1981 to 2011. Remote Sens 5(2):927–948

Zianis D, Muukkonen P, Makipaa R, Mencuccini M (2005) Biomass and stem volume equations for tree species in Europe. Silva Fennica Monographs 4:63

Chapter 19
Assessment Approach of the Spatial Wheat Cultivation Risk for the Main Cereal Cropping Regions of Russia

Vera N. Pavlova, Anna A. Karachenkova, Svetlana E. Varcheva and Nikita M. Sinitsyn

Abstract This article presents the fundamental structure and functionality of the Climate–Soil–Yield (CSY) system, contents of its database, the developed methodology for assessing climate-related risks in crop production. The authors provide analysis for climate-based risks in the territories of steppe and forest-steppe zones of Russia, evaluate the extent of the disadvantageous character of the territories, as well as draw examples of the use of CSY to evaluate current climate change and its influence on the productivity of Russia's agriculture.

Keywords Climate change · Climate–Soil–Yield (CSY) imitation system · Vulnerability of territories Climate-related risks · Crop failure · Grain crop productivity · Bioclimatic potential

19.1 Introduction

Observed climate change may result in changes in frequency and/or intensity of extreme weather events (floods, hurricanes, droughts, etc.) (IPCC 2013; Roshydromet 2014). By the end of the twenty-first century, the frequency of temperature maximums in the territory of Russia may increase by 3–7 times compared to the beginning of the century. Annual maximums and minimums of air temperature are rising throughout the most part of Russia's territory; all seasons see a greater number

V. N. Pavlova (✉) · A. A. Karachenkova · S. E. Varcheva
National Research Institute on Agricultural Meteorology (NRIAM), Lenina St., 82, Obninsk 249037, Russia
e-mail: vnp2003@bk.ru

A. A. Karachenkova
e-mail: karachenkova@gmail.com

S. E. Varcheva
e-mail: s.varcheva@mail.ru

N. M. Sinitsyn
Russian Ministry of Foreign Affairs, Smolenskaya-Sennaya pl., 32/34, Moscow 119200, Russia
e-mail: sinitsyn@gmail.com

W. Mirschel et al. (eds.), *Landscape Modelling and Decision Support*,
Innovations in Landscape Research,
https://doi.org/10.1007/978-3-030-37421-1_19

of days with anomalously high air temperature and fewer days with an extremely low night air temperature. The territory of Russia is more likely to see more "sudden" rainfall in the form of sudden showers or blizzards, increased number of strong deluges and river floods, storms, rapid changes between cold and warm weather. Most of the European part of Russia in winter witnesses more days with anomalously high precipitation, while in summer they are fewer (Roshydromet 2014).

In different regions of Russia with differing types of climate, the factors of vulnerability of agriculture also vary. For wet northern regions, relevant factors are surges of cold air from the north, excessive precipitation and frosts; in dry southern regions that would be heat waves, droughts, dust storms, dry winds, wind and water erosion. Yet for agricultural production importance lies not only with the analysis of general trends of changes in near-ground temperature or rainfall, but also the frequency of extreme stress situations that lead to substantially negative consequences.

Relevance of research for assessment of the scale of consequences of climate change for Russia's agriculture may be explained by a few reasons. Firstly, Russia's share in the world's grain market has been growing consistently in recent years (FAO 2016; USDA-FAS 2016). Secondly, most of the arable territories of Russia (around 60%) are classified as zones of high-risk and critical agriculture. Frequent and large-scale draughts are the main reason for the unparalleled yearly variability in grain crops yield in Russia (Vil'fand and Strashnaya 2011; Frolov and Strashnaya 2011; Zolotokrylin et al. 2014; Cherenkova 2017; Lioubimtseva et al. 2015; Alcamo et al. 2007).

19.2 Climate–Soil–Yield (CSY) Imitation System

19.2.1 Means of Imitation Modelling of Agroecosystems

Recent research (Wechsung et al. 2008; Ortiz-Bobea and Just 2012) shows that dynamic models for yield composition take precedence in evaluating the impact of climate change on agriculture.

Russia's system of imitation modelling AGROTOOL is a digital dynamic model of a production process of agricultural works (Poluektov et al. 2006; Mirschel et al. 2016). AGROTOOL can be used to resolve such tasks of agronomic monitoring as yield forecast, assessment of the speed of phenological development, research, and choice of agro-technical solutions. The system also allows it to study theoretical problems, including the forecast of the reaction of agroecosystems to climate change in the twenty-first century.

Notable examples of actively utilized models and decision support systems (DSS) are WOFOST (van Dipen et al. 1989), Sirius (Jamieson et al. 1998), DSSAT (Jones et al. 2003; Hoogenboom et al. 2012), CropSyst (Stöckle et al. 2003), as well as integrated interactive DSS Land Care (Mirschel et al. 2016).

Land Care-DSS allows to assess adaptation strategies and the impact of climate change on different levels: from a single farm up to a region. The modular structure of this system lowers the costs for adaptation to new countries and regions, as well as for integration with new imitation models. Interactive interface allows users to select relevant imitation models in a dialogue window, circumventing the data exchange process between the modules of the system. The geoinformation component allows to select the needed level of spatial detail in a dialogue window with subsequent presentation of results on the map of the selected territory.

The Land Care-DSS is planned for further development as a learning tool for students, consultants, and decision-makers. Using this system they will understand more easily the interconnections between the processes that form complex agro-economic landscapes. Developers are planning to adopt the Land Care-DSS to Russian-developed imitation models: AGROTOOL (Poluektov et al. 2002; Poluektov et al. 2012; Mirschel et al. 2016), an imitation model for soil nitrogen (Poluektov and Terleev 2007), and Russian information resources with data on soils (Terleev et al. 2010, 2014).

The Climate–Soil–Yield imitation system considered in this article is utilized for agroclimatic monitoring of current and forecast agro-meteorological conditions, assessment of climate-based risks in crop production, the level of unfavourability of territories, etc.

The system operates in the following modes: *retrospective* (observation data from 1892 to current time), *operational* (data by 10-day periods) and *forecast* (ensemble of middle- and long-term scenarios of expected climate change).

The toolset of the system allows it to *give a quantitative assessment to climate resources of a territory and the reaction of agriculture to input change in climate and non-climate indicators* (CO_2 concentration, concentration of ozone in the lowest atmospheric layer, degradation or enrichment of soils) relevant to the global warming phenomena. Currently, the agro-monitoring system includes main grain crops (summer and winter wheat, summer barley, winter rye) and perennial grasses.

19.2.2 Structure of CSY System

The CSY imitation system oriented for support of agricultural complex was developed under the leadership of Sirotenko (Sirotenko et al. 1994; Pavlova et al. 2019). The CSY system includes algorithmic implementation of dynamic models of agroecosystem productivity, a database on climate, soils, and crops in the territories of ex-USSR, as well as calculation, presentation, and auxiliary elements.

The foundation of the CSY imitation system is laid by the following elements:

- programmed implementation of a dynamic model of agroecosystem productivity Weather–Yield;

- database containing hydro-meteorological information for over a 100-year long period, data on physical and chemical properties of soils, agroclimatic parameters (phenology dates, soil water supply, etc.); information on climate scenarios;
- stochastic model of generation of annual variation of daily meteorological elements;
- programs of statistical processing for analysis of the imitation system output data;
- programs for presenting flat patterns of climate scenarios and visualization of input and output data.

Dynamic productivity model within the CSY system is a closed set of differential equations integrated numerically with a step value of 1 day for the length of the vegetation period of the considered crop. The model includes three interconnected modules and the dynamic is calculated within each of them via separate subsets of equations (Sirotenko et al. 1997; Pavlova et al. 2019):

- phytomasses of separate organs of plants during photosynthesis, respiration, growing, decaying, development and ageing;
- soil water supply during infiltration, evaporation, transpiration, and water uptake through roots;
- dynamic of mineral nitrogen in soil during nitrification, denitrification, uptake though roots, and leaching.

Below one may find the main equations from the growth module of the CSY system (Pavlova et al. 2019).

Crop phenology
Crop phenology is expressed in terms of the thermal time V^j [°C d], i.e., the sum of daily air temperatures T^j [°C] in excess of a base temperature T_0 [°C]. Formally

$$V^j = V^{j-1} + \Delta t(T^j - T_0) \tag{19.1}$$

where j is a time step (day), V^j is calculated for the germination-earing and earing-maturity periods.

Crop growth
The crop is considered as a whole with six structural units (compartments, p): the above-ground vegetative organs (leaves, leaf part of the stem, stem, further referred to using subscripts l, ls, s, respectively), the root system (subscript r), the grains (glume and corn, subscripts k, z). The root layer of the soil is limited to a depth of 1.5 m for water and 1.0 m for nitrogen compounds.

The equations describing the evolution of biomass, soil moisture, and nitrogen contents are as follows:

$$\frac{\mathrm{d}m_p}{\mathrm{d}t} = G_p - D_p - q_p - P_p \tag{19.2}$$

$$\frac{\mathrm{d}W_i}{\mathrm{d}t} = q_{i-1} - q_i - TR_i - \delta_i E \tag{19.3}$$

$$\frac{dN_k}{dt} = H_k + \delta_k U_N - h_k + V_{k-1} - V_k - A_k \tag{19.4}$$

where m_p ($p = \{l, s, ls, r, k, z\}$) denotes the dry biomass [mg cm^{-2}] in each of the six compartments; G_p, D_p, q_p, P_p are the rates of growth, breathing, decomposition, and plant matter fall, [mg cm^{-2}d^{-1}]; W_i is the water supply in the ith soil layer, [mm] ($i = 1{,}15$); q_{i-1}, q_i are the water flows through upper and lower layer boundaries of the ith layer, respectively [mm d^{-1}]; TR_i is the water loss on transpiration from the ith soil layer, [mm d^{-1}]; δ_i is the logic variable, $\delta_i = 1$ for the first soil layer at the day of fertilizer application and $\delta_i = 0$ in all other cases; E is the evaporation from soil surface, [mm d^{-1}]; N_k is the nitrates concentration in kth soil layer ($k = 1{,}3$), [mg cm^{-2}]; H_k is the rate of mineralization of easily hydrolyzable organic nitrogen, [mg cm^{-2}d^{-1}]; U_N is the amount of mineral fertilizer accounting for its immobilization, [mg cm^{-2}d^{-1}]; V_{k-1} and V_k are the flows of mineral nitrogen with water through upper and lower layer boundaries of the kth layer, respectively ($V_0 =$ 0), [mg cm^{-2}d^{-1}]; h_k is the rate of denitrification of mineral nitrogen, [mg cm^{-2}d^{-1}]; A_k is the term describing the nitrogen absorption by plants from the kth soil layer, [mg cm^{-2}d^{-1}].

19.2.3 CSY System Database

The database contains the data of instrumental observations of various time scales: daily average, 10-day average, monthly average.

Daily average meteorological data include air temperature, total precipitation, duration of sunshine, and air humidity deficiency. The observation time series are monthly average data on air temperature and monthly total precipitation in Russia's regions from 1892 to 2017. Average monthly data on air temperature and rainfall was collected from 455 meteorological stations (MS) in Russia and adjacent countries from 1951 to 2017.

The CSY system is also capable of processing operational data of 10-day period telegrams ("Prometei" system) for the period from 1992 to 2017 (752 stations in 76 regions of Russia); meteorological data averaged for the base period 1961–1990 (climate normal) and data of climate scenarios selected from the project for intercomparison of global climate models CMIP5 (Coupled Model Intercomparison Project) (Emori et al. 2016).

The database also contains information on prevalent soils: hydro-physical and agrochemical properties, amount of humic matter and other characteristics, as well as phenological information on grains and industrial crops, areas of cultivation and available statistics on crop productivity (data from Rosstat).

In Fig. 19.1 one can find the map with indicated meteorological stations (MS) of the observation network in the Central (black-earth centre), Volga and Southern Federal Districts.

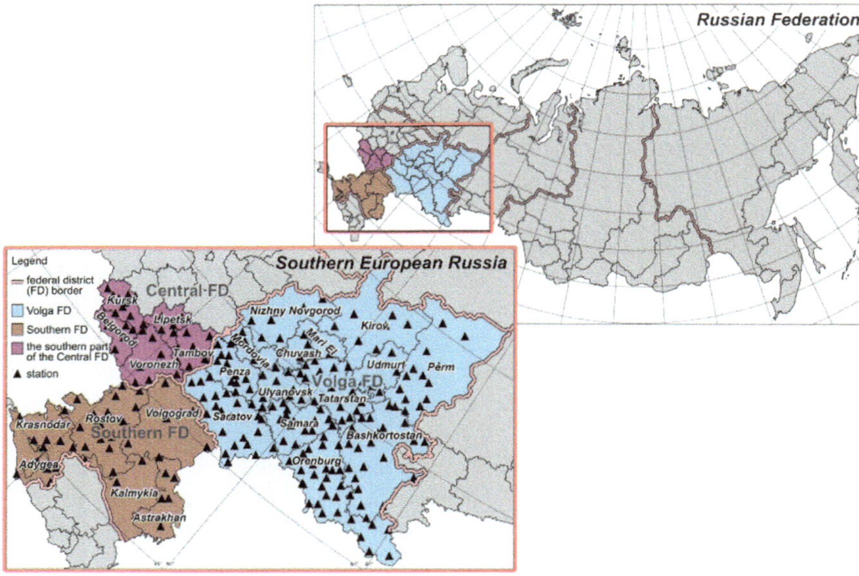

Fig. 19.1 Meteorological stations (MS) of the observation network in the Central (black-earth centre), Volga and Southern Federal Districts

19.2.4 Adequacy of the CSY System

Identification of the parameters of the CSY simulation core, verification (verification of adequacy) and applicability for agroclimatic calculations were carried out with different spatial resolution—from Federal Districts (FD), several regions (territories, regions, and republics) to municipal districts of several regions.

The series of actual crop yields were used as input for model verification (Rosstat data).

In Fig. 19.2, the results of checking the adequacy of the CSY model from observations of moisture reserves in the soil under spring wheat at 4 observation points (OP) in the steppe zone of the Volga Federal District (zone of insufficient moisture) are presented.

Years with contrasting agro-meteorological conditions were chosen for testing: 2008—favourable, high-yielding, 2010—abnormally arid, 2015—medium, close to the climatic norm.

Figure 19.2 shows that the calculated moisture reserves in 2010 in the arable and metre soil layers adequately reflect the process of the strong drying of the soil. The moisture content in the upper soil layers has decreased to a wilting moisture level (approximately 10 mm).

Such conditions evolved on the OP Chebenki and Sol-Iletsk, Orenburg Region and Balashov, Saratov Region. The CSY system adequately simulates the precipitation and gives, in general, results close to the measured values. Also, with a high degree of

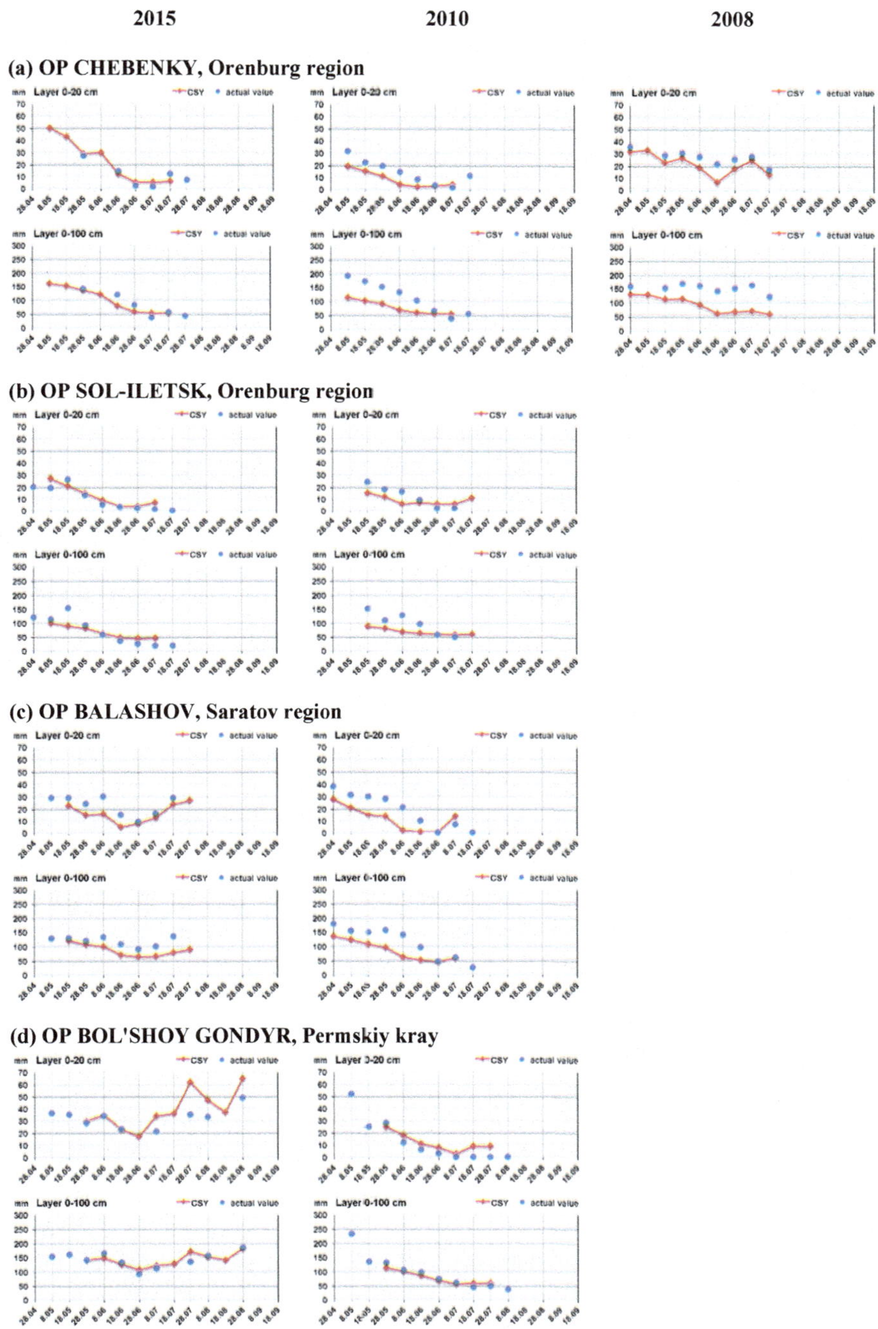

Fig. 19.2 Moisture reserves in arable (0–20 cm) and metre (0–100 cm) soil layers in 2015, 2010, and 2008 in the Volga FD: measured data (blue colour) and simulated data in the CSY system (red colour)

accuracy, it reproduced imitated moisture conditions during the 2015 growing season, according to the OP Bolshoy Gondyr—5% (Fig. 19.2d). The average accuracy of calculations is from 10 to 15%.

19.2.5 Indicators System for Agro-Climatic Monitoring

We present here the main indicators of heat and moisture supply of agricultural crops for the system of agroclimatic monitoring.

Indicators of thermal resources:

- sums of average daily air temperature values for a calendar year period with average daily temperature exceeding 0, 5, and 10 °C;
- dates of the steady transition of average daily air temperature through 0, 5, and 10 °C in spring and autumn;
- the duration of the calendar year periods with an average daily temperature exceeding 0, 5, and 10 °C;
- the average temperature of the coldest (January) and warmest (July) months of the calendar year.

Indicators of moisture resources:

- actual and potential evaporation;
- evaporation deficit or difference between evaporation and actual total evaporation (i.e., $\Delta E = E_0 - E$) during the growing season;
- reserves of productive moisture in the soil layers 0–20, 0–50, and 0–100 cm from the date of germination to maturity.

This set complements the agro-meteorological indicator: hydrothermal coefficient of Selyaninov (HTC) (Selyaninov 1958).

Climate extremes indicators:

- assessment of climate risks in crop production and vulnerability of the territory;
- assessment of the probability of large crop failures.

19.2.6 Indicators of Crops Productivity in the CSY System

The main indicators of crop productivity in the CSY is the yield conditioned by climate change (simulated yield) and the simulated bioclimatic potential of the territory (BCP).

We define and simulate BCP as the value of above-ground dry phytomass (t/ha) of conventional multi-grain grass, which is formed during the warm period of the year.

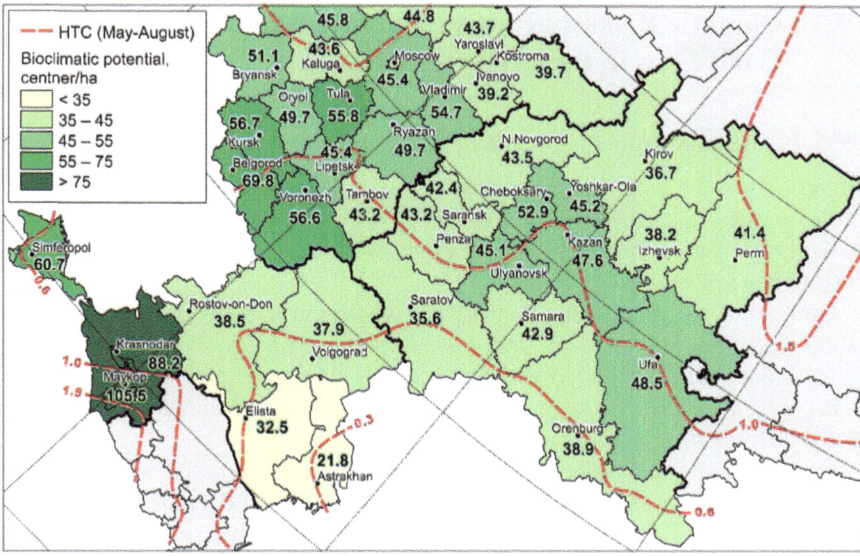

Fig. 19.3 Bioclimatic potential of the territory (centner/ha) in the Volga, Southern, and Central (only the Central black soil region) Federal Districts, simulated by CSY system for the period 1997–2017. Isolines of HTC values are plotted (red line)

This period begins on the day of the steady transition of air temperature through 5 °C in spring, and lasts until the transition through this temperature limit in autumn. During the growing season, the periodic mowing of crops is simulated, when it reaches the specified level of the leaf area index.

Figure 19.3 shows the simulation results of the BCP (according to data for 1997–2016) for the Central (the Central black soil regions only), Southern and Volga Federal Districts, which differ in soil and climatic conditions.

The spatial variability of this indicator is quite large: from the minimum values in the Republic of Kalmykia (~22 centner/ha) to the maximum—in the Krasnodar region (~ 110 centner/ha). In Fig. 19.3 you can see plotted contours (red dashed line) of HTC values, averaged over the 1997–2016. Values of HTC <=0.6 indicate highly arid areas.

The CSY system simulates four levels of HTC realization, corresponding to the four levels of agriculture intensification by optimizing two main limiting factors of the environment—the water regime of the soil and the conditions of mineral nutrition of plants.

Optimization means achieving a conditional level of moisture and/or mineral nutrition, in which these factors do not limit the growth and development of plants.

19.3 Climatic Risks and Adverse Hydro-Meteorological Conditions in the Production of Crops

19.3.1 Assessment of Climatic Risks in the CSY System

Climatic risk is manifested in a yield decrease when plants are exposed to unfavourable natural factors—weather events that do not meet the needs of plants in the provision of heat and moisture.

Until 2014, in the agricultural insurance, the insurer was obliged to pay compensation for the death of the crop, if the losses amounted to at least 30% of the planned yield, which met the requirements for entering this state support measure into the WTO "green basket". Later, in 2015, the legislator reduced this threshold to 25%, and in 2016 to 20%.

On the negative consequences in comparison with other natural disasters, droughts are among the first. Risk assessment in the agricultural production of Russia from the effects of atmospheric and soil droughts in recent years has become increasingly important.

In Russia as a whole, in the years with severe droughts, the gross grain harvest is significantly reduced. Thus, over the past decades, gross grain harvest in some years decreased by 40–50% compared with years favourable for agro-meteorological conditions.

Over the past two decades in dry years, such as 1998 and 2010, the gross yield of grain and leguminous crops was 47.5 and 61 million tons, respectively. For comparison, in 2016, the gross harvest amounted to 116 million tons.

Let us determine the climatic risk of large yield shortages of crops as the product of the drought probability and the vulnerability of agricultural production in a specific territory (Pavlova and Varcheva 2017):

$$R = p \cdot V \tag{19.5}$$

where p is the probability of a dangerous hydro-meteorological phenomenon (in %); V—vulnerability of agricultural production in a specific territory.

The lack of agroclimatic resources causing the level of vulnerability of the territory in the cultivation of crops is proposed to be assessed according to the formula:

$$V = 1 - \frac{Y}{BCP} \tag{19.6}$$

where BCP—bioclimatic potential of the territory, [centner/ha], simulated in the CSY system; Y— the average grain yield for the period under review [centner/ha].

Consider climate risk assessments according to the scheme proposed above for the territory of the Volga Federal District. The growing conditions for grain crops in the Volga Federal District are very tough.

Severe winters, short growing season, and adverse harvest conditions in the north and northeast, frequent droughts and dry winds in the southeast limit the productivity of grain crops in this region and predetermine the high yield volatility over the years.

The steppe zone of Russia is reasonably attributed to the zone of risky farming, where the level of grain production crucially depends on weather and climatic conditions.

In the simulations, data from meteorological observations from 167 meteorological stations located in the Volga Federal District were used as input information. To analyse the dynamics of yield and frequency of lean years, Rosstat data on the yield of spring and winter wheat for the past 20 years, 1994–2013, were used.

In Fig. 19.4, the spatial distribution of the risk estimates for the shortage of spring wheat yield, calculated according to Eqs. 19.5–19.6 is presented. Numbers indicate quantitative values of risk R (in % of the maximum risk equal to 100%).

The map shows that the zone of high risks of crop failure of spring wheat ($R > 30\%$) covers the territory of the south and southeast of the region belonging to the zone of insufficient moisture. The zone of low risks in the cultivation of spring wheat ($3\% < R < 10\%$) includes the central and northern regions of the Volga Federal District.

The highest risks of a shortage of spring wheat harvest occur in the Orenburg and Saratov regions, and high risks occur in the Samara region. The Republic of Bashkortostan, in which almost 6% of the total sown area of spring wheat in the Russia is located, can be arid.

In Table 19.1 averaged risk assessment of crop losses from adverse hydrometeorological conditions in the Volga Federal district, as well as the frequency

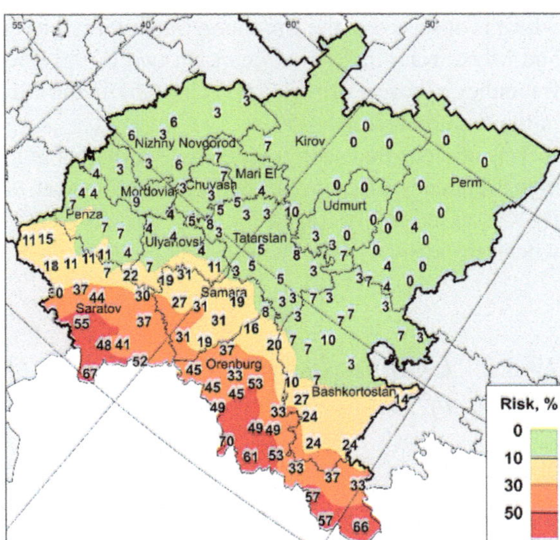

Fig. 19.4 Risk assessments (R) of spring wheat yield shortage (in %) in the territory of the Volga Federal District

Table 19.1 Estimates of the probability (P) of lean years and droughts (P_D), risks (R) of spring harvest shortages (SW) and winter wheat (WW) and the ratio of their acreage (A_{WW}/A_{SW}) in the major grain regions of the Volga Federal District for the period 1994–2013

The subject of the RF: region, republic	P ($Y < Ycr$), (%)		P_d ($HTC < 0.6$), (%)	R, (%)		A_{WW}/A_{SW}
	SW	WW		SW	WW	
Orenburg	20	15	52	42	32	0.1
Bashkiria	15	15	16	11	12	0.01
Tatarstan	20	15	8	4	7	0.3
Saratov	20	20	45	34	26	1.7
Samara	15	15	28	22	20	1.2
Ul'yanosk	15	15	12	9	11	1.1
Kirov	5	10	4	3	2	0.03
Penza	15	10	9	6	8	2.0
Nizhegorodskaya	10	10	6	4	4	1.1
Volga FD	14	12	16	23	18	0.5

of poor years and droughts, are presented. The repeatability of poor years is calculated as the ratio of the number of years with yields below a certain critical level ($Y < Y_{cr}$) to the total number of years of the period under review (1994–2013).

The critical yield level (Y_{cr}) was set as the yield of more than 40% lower than its average value for the given period. Note that on the territory of the Volga Federal District, spring wheat crops occupy twice the area of winter wheat $\left(A_{ww}/A_{sw}\right)$.

The maximum frequency of years with a significant shortfall in the yield of spring wheat is observed in the Orenburg, Saratov regions and in the Republics of Tatarstan and Mordovia—in 20% of years. In two of these regions, there is also the highest frequency of severe droughts ($HTC \leq 0.6$): in the Orenburg region—52% of the year, in the Saratov region—45%.

It is obvious that such frequency of droughts in the Saratov region led to a rather high frequency of crop failures for winter wheat too—in 4 years out of 20 years. Risk assessments, averaged for crop areas, are 23% and 18% for spring and winter wheat, respectively (see Table 19.1).

19.3.2 Changes in Large-Scale Spatio-Temporal Structure of Yield

In recent decades, significant changes have been revealed in the spatial–temporal structure of temperature and precipitation fields in Russia (Kislov et al. 2015).

What are the consequences of such changes on the productivity of agricultural crops and, as a consequence, on the food security of Russia?

Table 19.2 Correlation matrix (r_{ij}) of the yield of grain and leguminous crops with the excluded linear trend for the periods of **a** 1955–1980 and **b** 1981–2006 in the grain regions of Russia (σ_{ij}—standard deviation of yield, centner/ha)

Federal district, region		1	2	3	4	5	σ_{ij}
(a) 1955–1980							
1	Central	1.00	0.58	0.26	− 0.12	0.27	2.55
2	Volga		1.00	0.50	− 0.01	0.13	2.88
3	South			1.00	0.16	0.22	3.17
4	Siberia (West)				1.00	0.43	2.52
5	Siberia (East)					1.00	1.41
(b) 1981–2006							
1	Central	1.00	0.62	0.60	0.39	0.52	2.95
2	Volga		1.00	0.57	0.20	0.20	3.00
3	South			1.00	0.38	0.67	4.19
4	Siberia (West)				1.00	0.42	1.95
5	Siberia (East)					1.00	1.46

We calculated the correlation matrices of the yields of grain crops with the excluded technological trend. The analysis of these matrices confirms the increase in the synchronization of fluctuations in the inter-annual variability of yields induced by global warming.

In Table 19.2 estimates of the elements of the correlation matrices of yield with the excluded technological trend, calculated according to the data of 3 Federal Districts are presented: Central, Volga Southern, and two large regions: Western and Eastern Siberia for 1955–1980 and 1981–2006.

Correlation coefficients indicate a significant increase in connectivity (synchronization of fluctuations) of yield in the Russia regions. The average matrix correlation coefficient increased from 0.24 to 0.46 (Table 19.2). The degree of synchronicity of the average yield fluctuations associated with climate change in the European and Asian parts of Russia has noticeably increased.

There are 3 FDs pairs, where synchronous links of yield with the yield of other FDs (r_{ij}) increased: Eastern Siberia and Central FD (from 0.27 to 0.52), Eastern Siberia and Southern FD (from 0.22 to 0.67), Central and Southern FD (from 0.26 to 0.60). Asynchronous communications were replaced by synchronous ones for 2 FD pairs: Western Siberia and the Central and Volga Federal Districts (4th column, Table 19.2).

Also, a tendency was revealed toward an increase in the inter-annual variability of the average crop yield in the regions; so for the period from 1981 to 2006. Yield variability (σ_{ij}) on average in the territory under consideration has changed slightly (Table 19.2).

The estimates obtained allow us to conclude that the observed climate changes may lead to a decrease in Russia's food security as the mutual compensation for droughts of grain shortages in geographically remote areas decreases and the inter-annual variability of gross grain harvest increases.

The average correlation coefficient of the elements (r_{ij}) is 0.24 and 0.46; the standard deviation (σ_{ij}) is 2.50 and 2.71 centner/ha for periods (a) and (b), respectively.

To overcome the negative consequences of the observed changes in the spatial and temporal structure of the fields of agroclimatic indicators and to increase the stability of agricultural production, adaptation measures are necessary.

We recommend that the list of such measures must include: a review of the geographical structure of the acreage, adjusting the ratio between winter and spring, early and late ripening crops, the development of land reclamation, etc.

19.3.3 Hazardous Hydro-Meteorological Phenomena. Assessment of the Degree of Territories Unfavourableness for the Production of Agricultural Crops

We considered two hazardous agro-meteorological phenomena (HP) "drought" and "overwetting". To identify them, we use the meaning of G. T. Selyaninov's hydrothermal coefficient (HTC), simulated for the periods from May to August and from August to September.

Modelling is performed annually, in May, according to network observation data of 1032 meteorological stations and hydro-meteorological stations within 59 subjects in 7 Federal Districts.

The choice of climatic indicators for evaluation is determined by the "Rules for classifying territories as territories unfavourable for agricultural production" approved by the Government of the Russian Federation.

Maps with calculated contours of dry and overwetted territories are displayed using the QGIS geographic information system.

In Fig. 19.5, a map of the distribution of probability values (%) of severe droughts (HTC $\leq$ 0.6, red-brown fill) for May—August and overwetting (HTC $\geq$ 2.0; blue fill) for August—September over the territory of the agricultural zone of the Russian Federation simulated for the period from 1997 to 2016 is presented.

Upon request, customers are provided with tables containing simulated data and maps of the boundaries of areas with the probability of severe droughts of 50% or more years, and overwetting of 30% or more years.

Modelling makes it possible to identify areas and municipal districts of Federal Districts that are liable to severe droughts.

For example, the probability of severe droughts in May—August on the territory of the Saratov region ranges from 20% in the Arkady municipal district to 90% in Aleksandrovo-Gaysky. There is a high probability of droughts in Novouzensky (85%), Krasnokutsky (75%) and Yershovsky (70%) municipal districts.

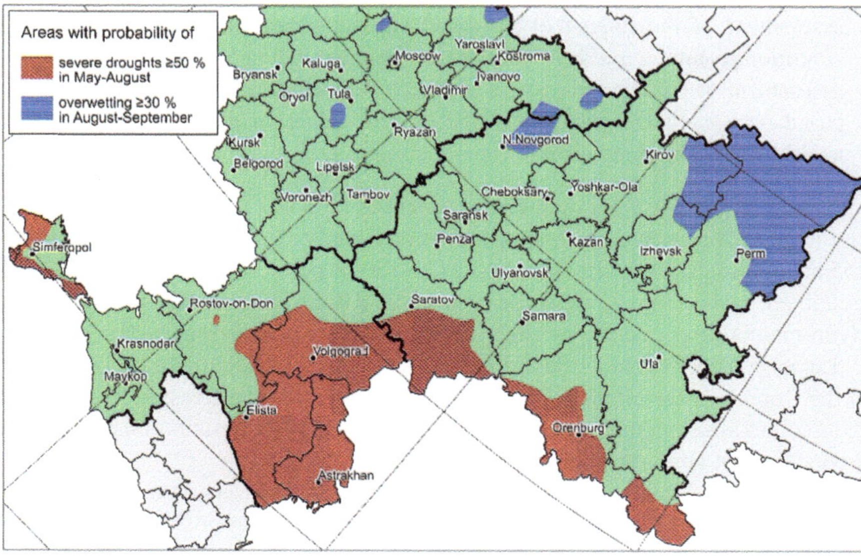

Fig. 19.5 Map with the borders of regions with unfavourable agro-meteorological conditions: severe droughts with a probability of $\geq 50\%$ and overwetting with a probability of $\geq 30\%$ (according to data from 1997–2016)

The analysis shows that a high degree of susceptibility to HP "drought" is typical for the entire Saratov region—in 50% of cases (years), 50% of the region's area is suffering from drought.

The developed technology makes it possible to identify the territories subject to one or another HP according to the specified criteria on a scale from municipal districts to large regions.

19.4 Another Example of CSY Applications

19.4.1 Implementation of Agroclimatic Monitoring on the Territory of the Agricultural Zone of Russia

Annual agroclimatic monitoring is carried out using the CSY system. In 2013, the Method for assessing the impact of climate change on agricultural productivity using the CSY was approved (Sirotenko et al. 2013). The goals of monitoring the agroclimatic component of the climate are the following:

- regular tracking of the state of the agroclimatic system, including an assessment of the degree of abnormality of the current state and the identification of extreme agroclimatic anomalies;

- assessment of the impact of observed climate change trends on agroclimatic conditions (climate trends of agro-meteorological indicators);
- assessment of the anomalousness of the current state of the agroclimatic system based on a statistical analysis of the current values of the selected parameters against the background of their long-term time series.

Monitoring of the agroclimatic component of climate is carried out (since 2010) on the basis of station data on temperature and precipitation monthly resolution for 455 meteorological stations (MS) in the CIS for the period from 1951 to the present, contained in the Climate database created in IGCE (Russian Federal Service for Hydrometeorology and Environmental Monitoring (Roshydsromet) 2018).

Let us consider the materials from the section "Agro-meteorological conditions" in the Report on the climatic features in the territory of the Russian Federation in 2017 concerning moisture availability in the territory of the agricultural zone and the productivity of spring wheat.

The analysis has shown that the growth of heat resources continues practically throughout the entire agricultural zone of the Russian Federation. The exceptions are separate regions of the Far East and Eastern Siberia.

Over the past 20 years, the growth rate of indicators such as January temperature (it is important for assessing the conditions for wintering crops) has slowed down, albeit slightly.

The negative trend (European part of Russia) towards a decrease in precipitation over the summer period of the year remains. At the same time, trends in summer precipitation for 1976–2017 are positive for the Asia part of Russian, and a positive trend towards an increase in spring precipitation throughout the entire agricultural zone of the Russian Federation also remains.

The yield of grain crops

Monitoring of agroclimatic conditions for the formation of grain yield was carried out in the territory of 16 subjects (republics, territories, regions) of the Central, Volga, Southern and North Caucasus Federal Districts from May 1 to July 20, 2017 (Fig. 19.6).

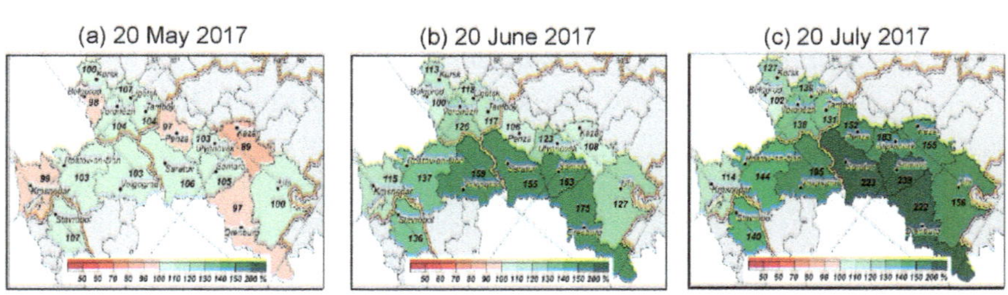

Fig. 19.6 Estimate (in %) of agro-meteorological conditions of the growing season of spring wheat in 2017 in the territory of several subjects of the European part of Russia relative to 2012–2016 at the dates: **a** May 20, **b** June 20, and **c** July 20

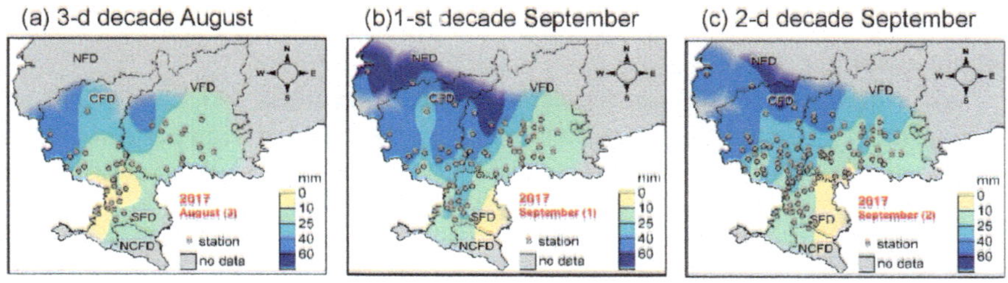

Fig. 19.7 Productive moisture reserves in the arable layer of soil (0–20 cm) under winter crops for the third decade of August and the first and second decades of September 2017

The spatial distribution of estimates of the expected yield of spring wheat in 2017 is presented at the main prognostic dates—May 20, June 20 and July 20.

Analysing the data presented here, we can conclude that in 2017 as a whole over the territory under consideration a significant increase in the gross harvest of spring wheat should be expected: in ten regions out of sixteen (63%), the yield is more than 40% higher than the average multiyear level.

In the Saratov, Samara, and Orenburg regions, the agro-meteorological conditions were favourable for the production of spring wheat: estimates of the expected yield exceeded the average long-term level (norm) by 2.2–2.4 times.

Thus, the agroclimatic conditions of 2017 in this territory contributed to the formation of a higher level of yield compared with the previous decade.

Estimates of the moisture content of agricultural fields in the pre-sowing and autumn periods show that for winter crops of the 2018 crop in the southern and southeastern regions of the Volga Federal District, there was sufficient moisture supply (Fig. 19.7).

19.5 Conclusions and Outlook

The CSY system combines a database on climate, soils, and crops on the territory of the former USSR with interpretive modules for dynamic models of energy and mass transfer and agroecosystems productivity.

This system makes it possible to quantify the climatic resources of a territory and the response of agriculture to given changes in climatic and non-climatic factors, as well as climatic risks in crop production.

The technology of monitoring the agrosphere of Russia based on the CSY system has been developed and implemented. The technology makes it possible to evaluate the current agrosphere state and the trends of its change in the future, with the calculation of estimates of anomalies and extremes of agroclimatic indicators (indices) and mapping of the results.

The archive of climatic and agroclimatic data contains observational data for a long period over a large number of points. This archive serves as the basis for obtaining reliable estimates of indicators characterizing the agro-resource potential of the regions of Russia.

Algorithms were developed for retrospective, operational, and scenario calculations in the CSY system. Estimates of trends, variability, and extremes of agroclimatic indicators of heat and moisture supply and productivity of agricultural crops by subjects in the agricultural zone of the Russian Federation since 1976 have been received and are regularly updated.

A methodology for quantitative assessment of climatic risks in the cultivation of crops, taking into account the frequency of adverse meteorological phenomena and the degree of vulnerability of the territory, has been developed and implemented. The average for Russia climatic risks of a shortage of harvest of spring and winter wheat, calculated taking into account the sown areas, are 12.5 and 10.6%. The maximum risks of obtaining low yields of spring wheat during droughts are characteristic for the Southern (37.9%) and Volga (23.0%) Federal Districts.

The obtained estimates can be used to support decision making in the field of agricultural insurance, in developing systems for supporting producers and others at various levels of management, including government agencies (the Ministry of Agriculture of the Russian Federation, regional authorities).

The development of a system of agroclimatic monitoring will reduce the negative effects and maximize the positive effects of climate change on Russia's agriculture.

Estimates of the current and expected status of agroclimatic resources and climate risk assessments in the production of crops at different levels—from Russia as a whole to republics, regions, districts, and farms should be communicated to decision-makers in farms and regional agriculture ministries.

Climate risks in crop production can only be reduced as a result of a set of measures. The main measures include technological re-equipment, design of new farming systems and agricultural technologies, extension of land reclamation, introduction of scientifically based technologies and drought-resistant varieties suitable for local conditions, etc.

To overcome the negative consequences and maximize the use of positive impacts, it is necessary to develop and adopt programs for the adaptation of Russian agriculture at various levels–from the state as a whole to subjects, districts and even farms.

These programs should include integrated regional studies to assess the risk (vulnerability) of agricultural production from the adverse effects of climatic and weather factors, as well as the development of specific recommendations for agricultural producers.

The development and implementation of such programs will reduce the likelihood of a catastrophic decline in agricultural productivity with the expected changes in climate and the environment in the twenty-first century.

Acknowledgements This study is a contribution to the ERA.net RUS Plus Project EXTRA (Impact of EXTreme events and climate change on Russian Agriculture, economic implications and adaptation). It received financial support from the Russian Science Foundation (Grant No. 16-17-00063) and (Grant 18-45-40001).

References

Alcamo J, Dronin N, Endejan M, Golubev G, Kirilenko A (2007) A New Assessment of Climate Change Impacts on Food Production Shortfalls and Water Availability in Russia. Glob Environ Change 17:429–444

Cherenkova EA (2017) Dangerous atmospheric drought in the European part of Russia in the conditions of modern summer warming. Fundam Appl Climatol 2:130–143. https://doi.org/10.21513/2410-8758-2017-2-130-143

van Diepen CA, Wolf J, van Keulen H (1989) WOFOST: a simulation model of crop production. Soil Use Manag 5:16–24

Emori S, Taylor K, Hewitson B, Zermoglio F, Juckes M, Lautenschlager M, Stockhause M (2016) CMIP5 data provided at the IPCC Data Distribution Centre Version 1 -(https://www.ipcc-data.org/docs/factsheets/TGICA_Fact_Sheet_CMIP5_data_provided_at_the_IPCC_DDC_Ver_1_2016.pdf). Visited 2019 February 15

FAO (2016) FAOSTAT Data. Food and Agric. Org. United Nations, Rome. http://www.fao.org/faostat. Accessed 15 July 2017

Frolov AV, Strashnaya AI (2011) About the 2010 drought and its impact on grain yield, Sat. reports analysis of the conditions of abnormal weather in Russia in the summer of 2010 M., Triad LTD, pp 22–31. (In Russian)

Hoogenboom G, Jones JW, Wilkens PW, Porter CH, Boote KJ, Hunt LA, Singh U, Lizaso JL, White JW, Uryasev O, Royce FS, Ogoshi R, Gijsman AJ, Tsuji GY, Koo J (2012) Decision support system for agrotechnology transfer (DSSAT) Version 4.5, Honolulu: University of Hawaii

IPCC (2013) Climate Change 2013. The Physical Science Basis. Contribution of Working Group I to the Fifth Assessment Report of the Intergovernmental Panel on Climate Change, Stocker TF, Qin D, Plattner G-K, Tignor M, Allen SK, Boschung J, Nauels A, Xia Y, Bex V and Midgley PM (eds). Cambridge University Press, Cambridge, United Kingdom and New York, NY, USA, pp 1535. https://doi.org/10.1017/cbo9781107415324

Jamieson PD, Semenov MA, Brooking IR, Francis GS (1998) Sirtus: a mechanistic model of wheat response to environmental variation. Eur J Agron 8:161–179

Jones JW, Hoogenboom G, Porter CH, Boote KJ, Batchelov WD, Hunt LA, Wilkens PW, Singh U, Gijsman AJ, Ritchie JT (2003) DSSAT cropping system model. Eur J Agron 18:235–265

Kislov AV, Varentsov MI, Tarasova LL (2015) The role of spring soil moisture in the formation of large-scale droughts of the East European Plain 2002 and 2010. Izvestiya RAN, FAO, M., Nauka 51(4):464–471 (in Russian)

Lioubimtseva E, Dronin N, Kirilenko A (2015) Grain production trends in the Russian Federation, Ukraine, and Kazakhstan in the context of climate change and international trade. In: Elbehri A (ed), Climate change and food systems: global assessments and implications for food security and trade, Food Agriculture Organization of the United Nations (FAO). Rome

Mirschel W, Wenkel K-O, Berg M, Wieland R, Nendel C, Köstner B, Topazh AG, Terleev VV, Badenko VL (2016) A spatial model-based decision support system for evaluating agricultural landscapes under the aspect of climate change. In: Mueller L et al (eds), In « Novel methods for monitoring and managing land and water resources in Siberia » , Chap. 23, pp 519–540. https://zcloud.zalf.de/index.php/s/Q8RN9zeqJtyMf6y

Ortiz-Bobea A, Just RE (2012) Modeling the structure of adaptation in climate change impact assessment. Am J Agr Econ 95(2):244–251

Pavlova VN, Varcheva SE (2017) Assessment of the degree of vulnerability of the territory and the climatic risk of major crop failures in grain sowing regions of Russia. Meteorol Hydrol 8:39–49 (in Russian)

Pavlova V, Shkolnik I, Pikaleva A, Efimov S, Karachenkova A, Kattsov V (2019) Future changes in spring wheat yield in the European Russia as inferred from a large ensemble of high-resolution climate projections. Environ Res Lett (In press). https://doi.org/10.1088/1748-9326/aaf8be

Poluektov RA, Fintushal SM, Oparina IV, Shatskikh DV, Terleev VV, Zakharova ET (2002) Agrotool—a system for crop simulation. Arch Agron Soil Sci 48(6):609–635

Poluektov RA, Smolyar EI, Terleev VV, Topazh AG (2006) Models of the production process of crops, St. Petersburg., Ed. St. Petersburg University, p 392 (in Russian)
Poluektov RA, Terleev VV (2007) Crop simulation of the second and the third productivity levels. In: Kersebaum KC, Hecker JM, Mirschel W, Wegehenkel M (eds) Modelling water and nutrient dynamics in soil-crop-systems: proceedings of the workshop on Modelling water and nutrient dynamics in soil-crop systems held from 14 to 16 June 2004 in Müncheberg, Germany. Springer, Dordrecht, pp 75–89
Poluektov RA, Topazh AG, Jakushev VP, Medvedev SA (2012) Usage of a dynamic agro-ecosystem model for impact assessment of climate change on crop yield productivity (theory and realization). Vestnik Russian Acad Agric Sci 2:7–12 (in Russian)
Roshydromet (2014) Second assessment report on climate change and its impact on the territory of the russian federation. Roshydromet, Moscow
Russian Federal Service for Hydrometeorology and Environmental Monitoring (Roshydsromet) (2018) Report on climate features on the territory of the russian federation in 2017. Roshydromet, Moscow, p 69 (In Russian)
Selyaninov GT (1958) Principles of agroclimatic zoning of the territory of the USSR. In the book: Issues of agroclimatic zoning of the USSR, M., Izd, Ministry of Agriculture of the USSR, pp 7–14 (in Russian)
Sirotenko OD, Abashina HV, Pavlova VN (1997) Sensitivity of the Russian agriculture to changes in climate, CO_2 and tropospheric ozone concentrations and soil fertility. Clim Change 36:217–232
Sirotenko OD, Pavlova VN (1994) The greenhouse effect and the food problem of Russia. Meteorol Hydrol 7:5–16 (in Russian)
Sirotenko OD, Pavlova VN, Abashina EV (2013) Technique for assessing agro-climatic conditions to monitor changes in the current climate in the Russian Federation. http://method.meteorf.ru
Stöckle CO, Donatelli M, Nelson R (2003) CropSyst, a cropping system simulation model. Eur J Agron 18:289–307
Terleev VV, Mirschel W, Schindler U, Wenkel KO (2010) Estimation of soil water retention curve using some agrophysical characteristics and Voronin's empirical dependence. Int Agrophys 24(4):381–387
Terleev VV, Narbut MA, Topazh AG, Mirschel W (2014) Modelling hydrophysical properties of soils. Agrofisica (Агрофизика) St. Petersburg 2(14):35–44) (in Russian)
USDA-FAS (2016) Grain: world markets and trade. United States Department of Agriculture, Foreign Agricultural Service. http://www.fas.usda.gov. Accessed 15 July 2017
Vil'fand RM, Strashnaya AI (2011) Climate, weather forecasts and agrometeorological support of agriculture in a changing climate. Adaptation of agriculture of Russia to the changing weather and climatic conditions, Collection of reports of the international scientific-practical conference December 7–11, 2010. Moscow (in Russian)
Wechsung F, Gerstengarbe F-W, Lasch P, Lüttger A (2008) Die Ertragsfähigkeit ostdeutscher Ackerflächen unter Klimawandel, PIK Report No. 112, Potsdam, p 3
Zolotokrylin AN, Vinogradova VV, Meshcherskaya AV, Strashnaya AI, Cherenkova YeA. (2014) Drought and desertification. The second assessment report of Roshydromet on climate change and its consequences on the territory of the Russian Federation, Moscow, Roshydromet, pp 551–587 (in Russian)

Chapter 20
Model-Based Forecasting Winter Wheat Yields Using Landscape and Climate Data

Peter A. Shary, Larisa S. Sharaya and Olga V. Rukhovich

Abstract Empirical statistical models at regional scale (resolution 600 m) were used to study links between characteristics of winter wheat yield and environmental factors that include climate, soils and topography. The study area of the size 3° by 4° is located in the western part of Oka River basin, Russia. The most strong links were found for the averaged by years addition to yield that is defined as the difference between yield with optimal fertilizers and control with no fertilizers used. The main environmental factor was insolation from southwest. The link with insolation from southwest was positive corresponding to relatively cold conditions of the region. Similar results were obtained for five other crops: perennial grasses, oats, barley, winter rye and annual grasses. A method is suggested to take into account chronological sequence of climatic factors action. Environmental factors explained 74% of variance in addition and 76% when this sequence was taken into account. When insolation increased 5% from its average (i.e., the change was 25 W/m^2), yields increased 1.86 times and additions 2.09 times. Using the climatic GISS model E, we calculated a forecast model for the year 2050 for the addition. This model demonstrated generally favourable conditions for winter wheat yields, and addition decreased southward. The role of inexpensive measures of enhancing winter wheat yields in the conditions of varying climate was evaluated and the measures include the selection of field slopes and the selection of existing cultivars. For example, the selection of mostly one of cultivars explained 81% of variance in the addition.

Keywords Winter wheat · Forecast · Topography · Multiple regression · Climate change

---

P. A. Shary (✉)
Institute of Physicochemical and Biological Problems in Soil Science, Russian Academy of Sciences, Institutskaya Str. 2, 142290 Pushchino, Moscow, Russia

L. S. Sharaya
Institute of Volga Basin Ecology, Russian Academy of Sciences, Komzina Str. 10, 445003 Toglyatti, Samarskaya Oblast, Russia

O. V. Rukhovich
All-Russian Scientific Research Institute of Agrochemistry, Russian Academy of Sciences, Pryanishnikov Str. 31a, 127550 Moscow, Russia

W. Mirschel et al. (eds.), *Landscape Modelling and Decision Support*,
Innovations in Landscape Research,
https://doi.org/10.1007/978-3-030-37421-1_20

20.1 Introduction

Process-based and statistical inventory-based models are distinguished in agriculture. The former require extensive input data on cultivars, management and soil conditions that are frequently unavailable. Even in the presence of such data these models can be very difficult to calibrate because of large numbers of uncertain parameters. Often this parameter uncertainty is ignored and a subjective decision is made to proceed with a single set of parameter values that produces acceptable agreement with observations (Lobell and Burke 2010). Inventory-based spatial models are prone to errors from omitted variables such as soil quality or fertilizer inputs that vary spatially. Statistical spatio-temporal models should assume common parameter values for all locations. The main advantage of statistical models is their transparent assessment of model uncertainties. For example, if a model does a poor job of representing crop yield responses to climate, this will be reflected in a low determination coefficient (R^2) between modelled and observed quantities. Although process-based models could in theory be accompanied with similar statistics, in practice they rarely are (Lobell and Burke 2010). Statistical inventory-based are preferred nowadays as we do here.

Influence of climate to yields of main agricultural crops is characterized by large uncertainty (Burke et al. 2015) explaining for simple regression models of linear dependence of yields on temperature and precipitation on average about 15% of variance (Lobell and Burke 2008). But more complex models usually explain also less than 50%, for example, 41% for wheat and 29% for sorghum (Lobell and Field 2007). Note in this relation that influence of microclimate (created by topography) and chronological sequence of climatic factors influence are infrequently taken into account in models. Shary (2016) have shown that microclimate described as changes in slope insolation on various fields appeared main predictor for winter wheat yields at regional scale for our study area shown in Fig. 20.1. Microclimate, climate and soil types explained 74% of variance for one of characteristics of winter wheat yield (see below).

In relation to global warming, exceeding of critical values of temperature is most important in tropical and sub-tropical regions, such as Africa and South America, where temperature for some agricultural crops is already close to critical (Jones and Thornton 2003; Adhikari et al. 2015). The response of yields to temperature may be sharply non-linear near threshold value of temperature (Schlenker and Roberts 2006, 2009). Increase of temperature in 1976–2015 in Russia was 0.45 °C/decade with the global increase 0.17 °C/decade for the same period (Rosgidromet 2016), while the known hiatus in global warming from 1998 to the present time (Tollefson 2014) was not observed in Russia. Although agricultural adaptation to climate change in Russia is currently restricted by inexpensive measures (shifting planting dates or switching to existing crop varieties). In future it may be not sufficient. Agricultural adaptation is able to reduce losses in crop yields by one third (Huang 2016), although such estimates are highly uncertain (Moore and Lobell 2014; Shrestha et al. 2013). In global estimates, the decrease in wheat yield may reach 8.3% in 2095 (Tatsumi et al. 2011), but such forecasts seem unreliable. Some authors study possible influence

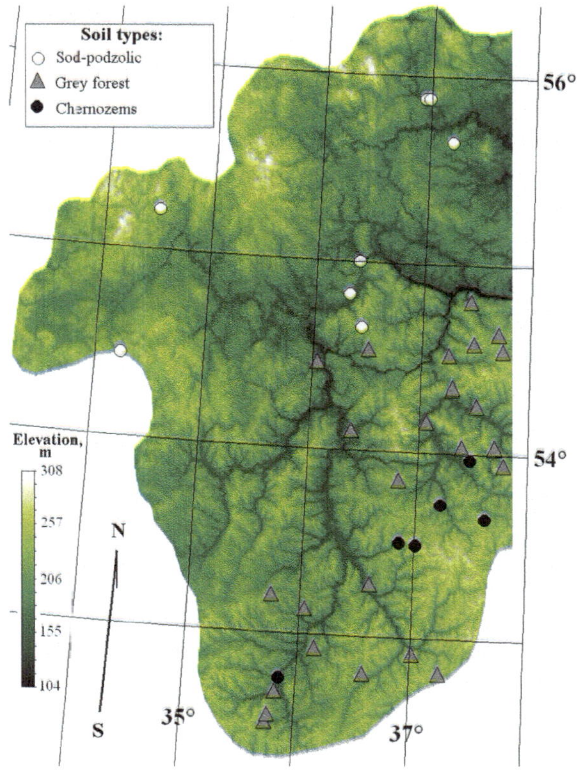

Fig. 20.1 Study area and observation points

of CO_2 fertilization to crop yields, but the reality of such fertilization is subject of debates (Müller et al. 2010). Significant increase of CO_2 is not found in Russia (Rosgidromet 2016) and we do not take it into account in our projections for 2050. Some authors try to take into account negative influence of increase in concentrations of near-surface ozone O_3 on crop yields, but extensive studies of this influence carried out in Western Europe and USA were not performed in Russia (Avnery et al. 2013). Significant increase of O_3 in Russia is not found (Rosgidromet 2016) and we do not take it into account for 2050.

Due to hiatus, current climatic models predict global increase of temperature from 1998 as large as 0.21 °C/decade, while the observed increase is 0.04 °C/decade (Tollefson 2014). However, these observations are not applicable to Russia, where this hiatus was not observed (Rosgidromet 2016), and here NASA's GISS model E may provide satisfactory results that may be considered as reliable for not too long periods of 30–60 years. Therefore we use this climatic model here.

One of important causes of low explained variance of yields is difficulty to take into account history of agricultural fields. To avoid this difficulty, we characterize yields by three averaged in time measures: (1) yield when optimal fertilizers are used, (2) control without fertilizers, and (3) addition that is the difference between

(1) and (2). The addition in the study area for winter wheat *Triticum aestivum* L. consists 60% of the control. Since history of fields enters to both yield and control, it plays essentially smaller role in the difference between them, that is, in addition. Therefore the latter is most closely related to environmental factors (see below).

To take into account microclimate at regional scale, we use relative slope insolation, $F(a, b)$. It is described as perpendicularity of solar ray incidence to land surface and is equal 100% for slopes with perpendicular ray incidence and zero for shadow slopes (Shary et al. 2002). Insolation depends on two angles that define Sun's position: angle above horizon a, and azimuth b. The dependence of yield on a at regional scale is weak, hence we fix this angle at the value of 35°. We found azimuth b from statistical comparisons. Due to finite thermal conductivity of soils, southwestern slopes are better heated in Northern hemisphere (Whittaker 1960); $b \sim 225°$ for these slopes. We used resolution 600 m that characterizes meso-topography and describes light and thermal regimes of slopes. This resolution is adequate to our task because turbulent movement of air near land surface physically averages temperature at distances about hundreds of meters (Floors et al. 2015).

Regional forecast model should be based on realistic spatial model for the base (current) period. Data on current climate were from global database WorldClim, where they are averaged for 50 years (1950–2000) and represented at resolution of 900 m (Hijmans et al. 2005). Data on features of crops were from database Agrogeos, where they are collected for 40 years (Sychev et al. 2008). Statistical methods are described in (Shary and Pinskii 2013).

Chronological sequence of influence of climatic factors may also be important. For example, small precipitation at the end of winter followed by low spring temperatures negatively influence to yield of winter wheat (Genkel 1969). This may be accounted by the use of precipitation and temperatures of different times in some combination of them (see Sect. 20.2).

The purpose of this paper is the construction of regional spatio-temporal models of yields for non-irrigated winter wheat using climate, meso-topography, soils, and finding of a technique for accounting chronological sequence of climatic variables influence to evaluate changes in yields under climate change. The main task of the paper is to reveal most influential environmental factors at regional scale.

20.2 Chronological Sequence of Climatic Factors Influence

Although the importance of chronological sequence of climatic factors influence is recognized in the literature (Genkel 1969), it was practically not taken into account in statistical models. The main reason was that corresponding climatic factors are usually interdependent and it was difficult to use them as an input to regression models because of strong effects of multicollinearity. Some authors tried to find such Earth's regions where multicollinearity effects are relatively small and to construct models for such regions (Lobell and Ortiz-Monasterio 2007). Clearly, this restricts applicability of these models by such regions.

A more general approach may consist in finding such a combination of factors *X* and *Y* that minimally varies in space, that is, has smallest coefficient of variation CV. Mathematically, this means to find some function $f(X, Y)$ from the condition CV = min. The simplest such function is a linear combination $X + aY$, where a is parameter to be determined. This combination was termed *invariant* or stable composition because it smaller varies in space than *X* or *Y* taken singly (it does not vary at all in an ideal case, hence the name). Invariant is determined from the condition $\mathrm{CV}(X + aY) = \min$. It was proven that the minimum is unique and the solution for a is given by a simple formula (Shary 2016).

$$a = \frac{\overline{Y} + A|r|\overline{X}}{A \cdot (A\overline{X} + |r|\overline{Y})} \text{ where } A = \frac{SD_Y}{SD_X} \text{ and } a > 0 \tag{20.1}$$

here r is Pearson's correlation coefficient between *X* and *Y*, their averages are $\overline{X}$ and $\overline{Y}$, and standard deviations are SD_X and SD_Y. Equation (20.1) is applicable for any sign of r, but for positive link between *X* and *Y* invariant is not composition of 'antagonists' $X + aY$, but rather a composition of 'satellites' $X + a \cdot (2\overline{Y} - Y)$ (Shary 2016).

Using Eq. (20.1) we find that the useful invariant, *INV*, for the study area is

$$INV = Pfeb + a \cdot (2 \cdot \overline{Tspring} - Tspring) \tag{20.2}$$

where *Pfeb* is precipitation of February, *Tspring* is spring temperature at a given geographical point and $\overline{Tspring}$ is its average in space. The value of a equals to 3.313 mm/°C. Map of *INV* is shown in Fig. 20.2.

So, chronological sequence of influence of climatic factors can be described by the stable composition (20.2). The statistical link between February precipitation and spring temperature is not essential because they are in a single environmental factor *INV* that does not lead to multicollinearity.

20.3 Results

According to WorldClim data for the study area, annual temperature here is 5.0 °C and annual precipitation is 637 mm. Average temperature of July, the warmest month, is 18.4 °C that is far from the critical value for winter wheat, 25–30 °C. Average temperature of January, the coldest month, is –9.4 °C, hence frost-stable cultivars of winter wheat are used, such as Mironovskaya-808 (28 observation points from 41), Ahtyrchanka, Lutescens-266 and others. Although cultivar Mironovskaya-808 is less frost-stable than Lutescens (Genkel 1969), it provides greater yields and therefore is used more frequently in the region.

For better forecasting for the year 2050, a statistical model should be well-tried for the base (current) period. The role of influential predictors is essential for this,

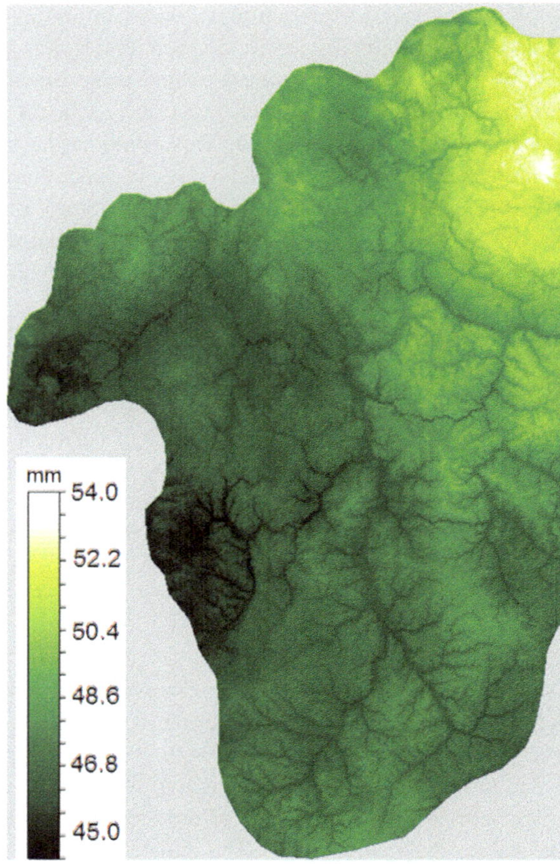

Fig. 20.2 Map of stable composition *INV* of February precipitation and spring temperature for the study area. *INV* describes repeating many years chronological sequence of influence of these factors

therefore special attention should be paid to their selection and assessment of their relative importance in the model. We consider first multiple regression model for the base period that takes into account microclimate, then take into account chronological sequence of influence of climatic factors, and then evaluate influence of cultivars of winter wheat.

To evaluate the relative role of environmental factors, we use Student's *t*-statistics (Montgomery and Peck 1982). We evaluate the relative role of environmental factors using $100|t_i|/\Sigma|t_i|$, where t_i is *t*-statistics of *i*-th predictor; these values are shown as subscripts in models below.

The model for current period with no chronological sequence and cultivars taken into account is (Shary 2016)

$$\begin{aligned} \text{Addition} =& 4.292 \cdot F(35°, 235°)_{32} + 3.411 \cdot Pfeb_{25} \\ &+ 0.202 \cdot PODZ \cdot F(35°, 235°)_{25} \\ &- 0.299 \cdot Pyear_{18} - 151.61 \quad R^2 = 0.740, P < 10^{-6} \end{aligned} \tag{20.3}$$

Here $F(35°, 235°)$ is insolation from southwest, *Pfeb* is February precipitation, *Pyear* is annual precipitation and indicator variable *PODZ* equals to 1 for sod-podzolic soils and 0 for other soil types. The main environmental factor is insolation from southwest that characterizes microclimate created by topographic conditions. Insolation is proportional to light and heat that is received by observation plots for clear sunny day. Diffused radiation in cloudiness days only diminishes the contrast of the pattern of insolation (Pierce et al. 2005). If insolation changes only 5% from its average value (the change is 25 W/m^2), winter wheat yields change 1.86 times, and additions change 2.09 times.

The positive link with February precipitation describes conservation of seedlings at the end of winter because thick snow cover prevents soils from heat loss and is favourable for yields of winter wheat. The third predictor, $PODZ \cdot F(35°, 235°)$, describes the increase of the role of insolation in colder northern part of the region in sod-podzolic soils. The negative link with the fourth predictor, annual precipitation, indicates negative influence of excess precipitation in humid climate of the region that diminishes winter wheat yields due to hypoxia.

Taken singly, these environmental factors are well known. However, revealing their relative role may be difficult when many factors act simultaneously (Genkel 1969). Multiple regression models help to solve this task and the selection of environmental factors is more substantiated for stronger links in the equation. The model (20.3) explains 74% of variance of addition to winter wheat yield, although it does not take into account cultivars and sequence of influence of climatic factors.

Model (20.3) shows the leading role of microclimate that is formed by slope insolation from southwest. Insolation describes influence of both photosyntetically active radiation, and heat from solar radiation that vary in space depending on mesotopography. Insolation depends on both slope steepness and exposure (Shary et al. 2002). If slope insolation is not taken into account, corresponding model explains essentially smaller, 48% of variance (Shary 2016).

Map of spatial distribution of addition is shown in Fig. 20.3.

We have found greater sensitivity of addition to environmental factors for a set of crops, Fig. 20.4. In all these cases insolation from southwest or similar factors played an important role.

These results confirm that links with environmental factors are stronger for addition then for yield or control.

Now we take into account chronological sequence of influence of climatic factors. We use for this invariant *INV* described in previous section. This results in the following regression model.

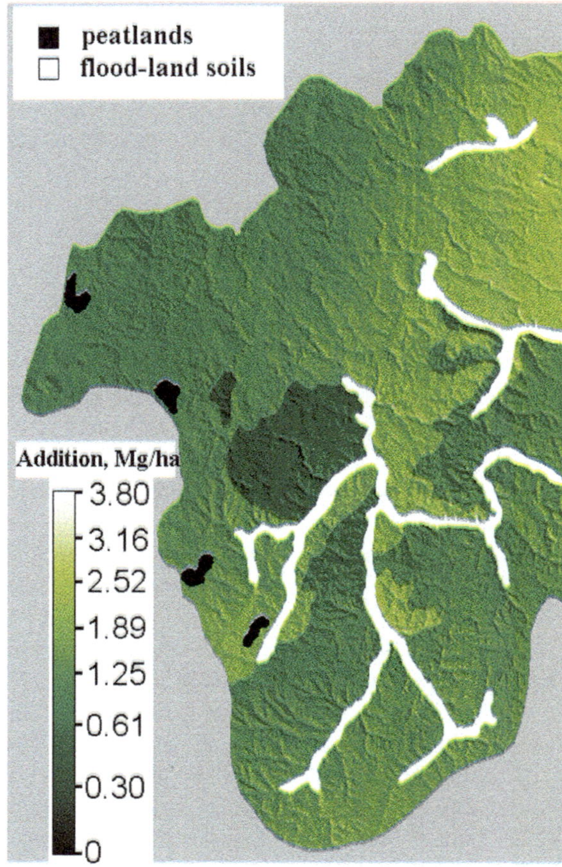

Fig. 20.3 Map of addition to yield of winter wheat calculated using model (20.3) for the base period. Peat (black) and flood-land (white) soils were excluded from consideration because of absence of observation points in them

$$\text{Addition} = 4.292 \cdot F(35°, 235°)_{36} + 10.20 \cdot PODZ \cdot T_{\text{march\_april}_{27}} - 0.6987 \cdot p_{\text{summer}_{24}} + 1.462 \cdot INV_{13} - 153.9 \quad R^2 = 0.756,\ p < 10^{-6} \tag{20.4}$$

This model explains 76% of variance, slightly greater than model (20.3). Here *PODZ*·Tmarch_april is temperature of March and April for sod-podzolic soils that are located at the north of the region. *Psummer* is summer precipitation, other predictors were explained above. As in model (20.3), slope insolation from southwest $F(35°, 230°)$ is the main predictor. However, in model (20.4) temperature plays greater role than in model (20.3). This is convenient for forecasting because temperature usually is better projected than precipitation by global climate change models (Burke et al. 2015).

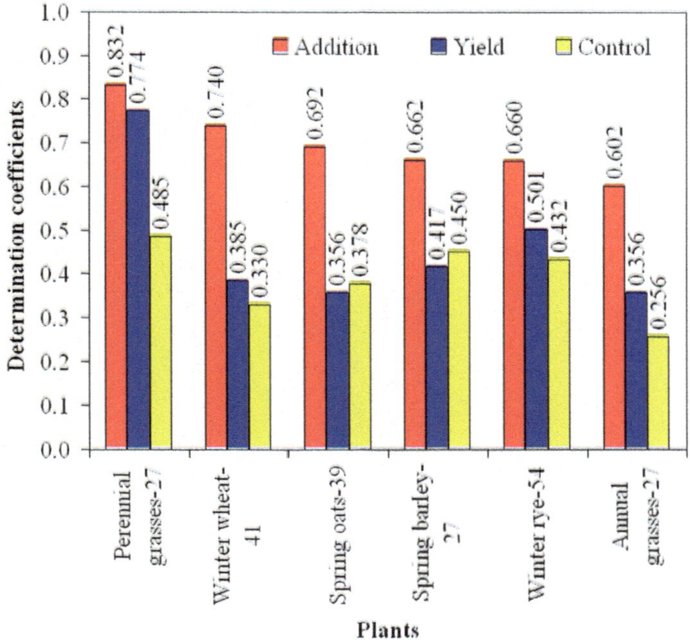

Fig. 20.4 Determination coefficients of regression models for addition, yield and control for a set of agricultural crops. The number after plant name is sample size. In all these cases, strongest links were observed for addition

February precipitation is implicitly in *INV* and the link with it remains positive. The second predictor shows that the sign of the link with spring temperature is closer to positive at the north of the region, where February precipitation is increased, so that seedlings of winter wheat are better protected from frosts at the end of winter, although they may suffer from hypoxia. Therefore high temperature of March and April results in better yields. The negative link with summer precipitation defines more exactly, precipitation of which season is more important as reflecting hypoxia.

To construct forecast map, we may substitute to model (20.4) values of temperature and precipitation that are projected for 2050 by GISS model E. According to GISS model E, temperature of March and April will increase by 2050 in the region by ~0.5 °C, summer precipitation by ~6 mm, *INV* will increase by ~3 mm. This provides map shown in Fig. 20.5.

This map predicts essential increase of addition at the north of region and some diminishing at the south (southward from ~53.5 °N) in comparison to the base period. Average projected increase of addition is 12% (average addition in base period is 1.17 ± 0.52 Mg/ha).

Now consider influence of wheat cultivars in a model for grey forest soils. These soils occupy middle part of the region (Fig. 20.1) and contain 27 observation points. We use indicator variable *MIR* that equals to 1 for cultivar Mironovskaya-808 and zero for other cultivars. The cultivar Mironovskaya-808 was used in 18 plots of 27.

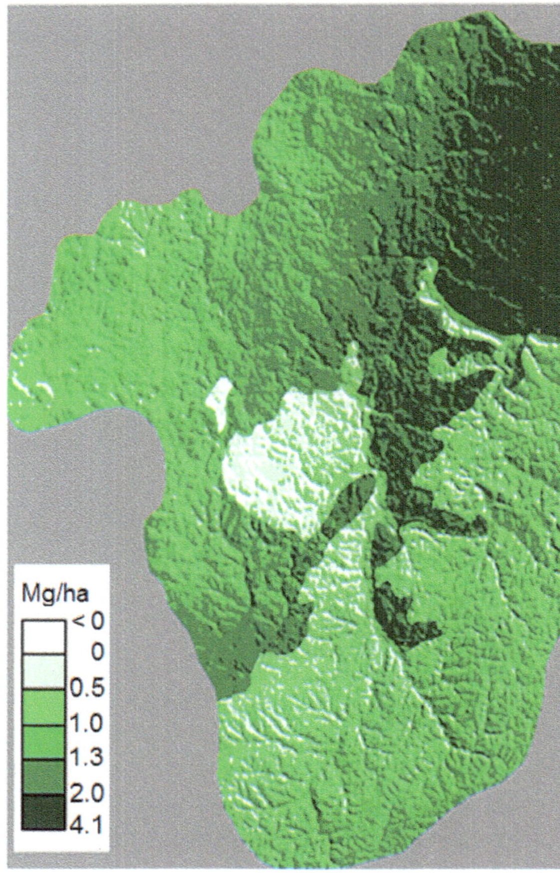

Fig. 20.5 Forecast map of addition to winter wheat yield for 2050 calculated using model (20.4)

We do not account in this model chronological sequence of climate influence because of small change in R^2. This gives the following multiple regression model.

$$\begin{aligned} \text{Addition} &= 6.530 \cdot F(35^\circ,\ 250^\circ)_{35} - 2.101 \cdot \cos A_{45}/GA_{27} \\ &+ 0.03413 \cdot MIR \cdot \left(Pyear - Pyear_{AV}\right)^2_{21} \\ &- 3.116 \cdot k\underset{17}{\overset{T}{\max}} - 363.9 \quad R^2 = 0.809 P < 10^{-6} \end{aligned} \tag{20.5}$$

The model explains 81% of variance. This is greater than in the above models reflecting positive effect of adequate cultivars choice. Here negative link with predictor $\cos A_{45}/GA$ reflects positive influence of gentle southwest slopes (Shary and Smirnov 2013) and positive link with maximal curvature *kmax* refers to slopes outside ridge landforms that correspond to areas outside water divides in flat terrains (Shary et al. 2002) The expression $kmax^{\mathrm{T}}$ means that curvature *kmax* was non-linearly transformed to diminish deviation of its distribution from the normal law as described in (Shary and Pinskii 2013).

In principle, we may choose best cultivars by testing various cultivars in models like (20.5) for the projected climate. This procedure is easy to automate. In a similar way, we might test at regional scale various other methods to diminish losses from climate change such as use of irrigation in some more distant future (e.g., for 2100). However, projection of precipitation has relatively large uncertainty which is a critical bottleneck in modern models of global climate (Burke et al. 2015).

20.4 Discussion

Finding effective environmental factors is an essential difficulty in crop yields forecasting in agriculture (Lobell and Field 2007). Therefore our study in focused primarily on the selection of effective environmental factors. We used most sensitive addition instead of yield for this. Revealing of influential environmental factors is more reliable when determination coefficients are high. It may be considered as realistic for $R^2 > 0.6$ as in our models. We have found such values of R^2 at regional scale for additions of all crops shown in Fig. 20.4. However, R^2 for yields are essentially smaller with the exception of perennial grasses, for yield of which $R^2 = 0.774$.

Low values of determination coefficients R^2 for crop yields are typical for many studies (Lobell and Field 2007). Our results show that change in insolation of 25 W/m^2 results in almost double change of winter wheat yield. Meanwhile, energy imbalance due to global climate change is 0.85 ± 0.15 W/m^2 for 2005 (Hansen et al. 2005) which is much smaller than the difference between slopes in energy (~35 W/m^2). Therefore taking into account the difference between slopes is important for assessing crop yield and especially for evaluating addition in them that is used to estimate economic effects of fertilizers application and is most sensitive to environmental factors (Fig. 20.4).

Since addition is the difference between yield and control, its dependence of field history is weaker than that of yield. Therefore account for history that was not done in our models might result in larger R^2 for crop yields. Taking into account soil properties (e.g., granulometry, pH) also might increase R^2 for yields.

According to GISS model E, climate change by the year 2050 in study area is relatively small and in general favourable for winter wheat especially in northern part of the region, where global warming will result in increase of winter wheat addition due to frost weakening. Consequently, here the link of yields with slope insolation from southwest is positive. However, largest yields come from the south of Russia where influence of global warming may be negative because of exceeding of critical temperature for winter wheat. Hence, we may expect negative link between yield and insolation from southwest in southern regions of Russia.

20.5 Conclusions

The main environmental factor for winter wheat yield in western part of Oka River in the conditions of humid climate is insolation from southwest at the resolution 600 m that characterizes perpendicularity of solar ray incidence to southwest slopes. It is advisable to include insolation from southwest into regional models of crop yields because it is able to indirectly reflect light and thermal regimes of meso-slopes. The link of yields with insolation from southwest is positive in study area. This result refers also to five other crops of the same region: perennial grasses, oats, barley, winter rye, and annual grasses. However, for sub-tropical regions of Northern hemisphere where high temperature is the limiting factor, one may expect negative link with insolation from southwest (from northwest in Southern hemisphere).

Especially strong link ($R^2 = 0.756$) was observed for addition to winter wheat yield that is defined as the difference between yield for optimal fertilizers and control with no fertilizers. The same is valid for five other crops in the region (Fig. 20.4). We explain this by smaller influence of agricultural field history to addition than to yield or control. So, history of agricultural fields may be important for finding tight links between yields and environmental factors.

If southwest meso-slopes positively influence to yields, as in our study area, preferring these slopes may result in increase 1.6–1.9 times of yields in regions with similar climatic conditions. However, in southern regions where high temperature is limiting factor, choice of colder northeast slopes may be preferable.

Taking into account chronological sequence of influence of climatic factors slightly increased determination coefficient between characteristics of yields and environmental factors (from $R^2 = 0.740$ to 0.756). Nevertheless, this allowed it to include climatic variables that are closely interrelated into spatio-temporal models. Winter wheat cultivars increased R^2 from 0.740 to 0.809.

Our regional forecast by 2050 for study area using GISS model E has shown favourable influence of warming to winter wheat yields with the addition to yield diminishing southward.

References

Adhikari U, Nejadhashemi AP, Woznicki SA (2015) Climate change and eastern Africa: a review of impact on major crops. Food Energy Sec 4:110–132

Avnery S, Mauzerall DL, Fiore AM (2013) Increasing global agricultural production by reducing ozone damages via methane emission controls and ozone-resistant cultivar selection. Glob Change Biol 19:1285–1299

Burke M, Dykema J, Lobell DB, Miguel E, Satyanath S (2015) Incorporating climate uncertainty into estimates of climate change impacts. Rev Econ Stat 97:461–471

Floors R, Peña A, Gryning SE (2015) The effect of baroclinicity on the wind in the planetary boundary layer. Q J R Meteorol Soc 141:619–630

Genkel PA (1969) Physiology of agricultural plants, vol IV. Physiology of Wheat. Moscow, Moscow State University, p 555 (In Russian)

Hansen J, Nazarenko L, Ruedy R, Sato M, Willis J, Del Genio A, Koch D, Lacis A, Lo K, Menon S, Novakov T, Perlwitz J, Russell G, Schmidt GA, Tausnev N (2005) Earth's energy imbalance: confirmation and implications. Science 308:1431–1435

Hijmans RJ, Cameron SE, Parra JL, Jones PJ, Jarvis A (2005) Very high resolution interpolated climate surfaces for global land areas. Int J Climatol 25:1965–1978

Huang K (2016) the economic impacts of global warming on US agriculture: the role of adaptation. Working Paper No. 2016–03, February 2016. The University of Adelaide

Jones PG, Thornton PK (2003) The potential impacts of climate change on maize production in Africa and Latin America in 2055. Glob Environ Change 13:51–59

Lobell DB, Field CB (2007) Global scale climate-crop yield relationships and the impacts of recent warming. Environ Res Lett 2:014002

Lobell DB, Ortiz-Monasterio JI (2007) Impacts of day versus night temperatures on spring wheat yields: a comparison of empirical and CERES model predictions in three locations. Agron J 99:469–477

Lobell DB, Burke MB (2008) Why are agricultural impacts of climate change so uncertain? The importance of temperature relative to precipitation. Environ Res Lett 3:034007

Lobell DB, Burke MB (2010) On the use of statistical models to predict crop yield responses to climate change. Agric For Meteorol 150:1443–1452

Montgomery DC, Peck EA (1982) Introduction to linear regression analysis. Wiley, New York, p 504

Moore FC, Lobell DB (2014) The adaptation potential of European agriculture in response to climate change. Nat Clim Change 4:610–614

Müller C, Bondeau A, Popp A, Waha K, Fader M (2010) Climate change impacts on agricultural yields. World Development Report. Development and Clim Change, Background note, p 11

Pierce KB Jr, Lookingbill T, Urban D (2005) A simple method for estimating potential relative radiation (PRR) for landscape-scale vegetation analysis. Landscape Ecol 20:137–147

Rosgidromet (2016) A report on climate peculiarities in Russian Federation in the year 2015. Moscow: Rosgidromet, p 67 (In Russian)

Schlenker W, Roberts M (2006) Nonlinear effects of weather on crop yields: implications for climate change. U.S. Department of Agriculture. p 41p

Schlenker W, Roberts MJ (2009) Nonlinear temperature effects indicate severe damages to US crop yields under climate change. Proc Natl Acad Sci USA 106:15594–15598

Shary PA, Sharaya LS, Mitusov AV (2002) Fundamental quantitative methods of land surface analysis. Geoderma 107:1–32

Shary PA, Pinskii DL (2013) Statistical evaluation of the relationships between spatial variability in the organic carbon content in gray forest soils, soil density, concentrations of heavy metals, and topography. Eurasian Soil Sci 46:1076–1087

Shary PA, Smirnov NS (2013) Mechanisms of the effects of solar radiation and terrain anisotropy on the vegetation of dark conifer forests in the Pechora-Ilych State Biosphere Reserve. Russian J Ecol 44:9–17

Shary PA (2016) Geomorphometric analysis of spatial variability in soils and ecosystems. Thesis of doctoral dissertation. Rostov-na-Donu, p 49 (In Russian)

Shary PA, Rukhovich OV, Sharaya LS (2016) Analytical and cartographic predictive modeling of arable land productivity. In: Mueller L, Sheudshen AK, Eulestein F (Eds), Novel methods for monitoring and management land and water resources in Siberia. Springer, Heidelberg. Chap. 21, pp 489–502

Shrestha S, Ciaian P, Himics M, Van Doorslaer B (2013) Impacts of climate change on EU agriculture. Rev Agric Appl Econ 16:24–39

Sychev VG, Rukhovich OV, Romanenkov VA, Belichenko MV, Listova MP (2008) Case study in the development of joined systematized database of field studies of Agrochimsurvey and geonetwork Agrogeos. Problems Agrochem Ecol 3:35–38 (In Russian)

Tatsumi K, Yamashiki Y, da Silva RV, Takara K, Matsuoka Y, Takahashi K, Maruyama K, Kawahara N (2011) Estimation of potential changes in cereals production under climate change scenarios. Hydrol Process 25:2715–2725

Tollefson J (2014) Climate change: The case of the missing heat. Nature 505:276–278

Whittaker RH (1960) Vegetation of the Siskiyou Mountains, Oregon and California. Ecol Monogr 30:279–338

Chapter 21
Actual and Model-Based Assessment of *Castor Fiber* Populations for Different Reserves in the European Part of Russia and Their Impact on Ecosystems

Nikolay A. Zavyalov, Lyudmila A. Khlyap, Fedor A. Osipov, Tatiana I. Oliger, Vitaly V. Osipov, Natalia N. Dergunova and Varos G. Petrosyan

Abstract The analysis of restoration of the number of Eurasian beaver (*Castor fiber* L.) in the reserves of the European part of Russia and the prediction of possible changes in ecosystems after its introduction into the reserves (Laplandsky, Darwinsky, Prioksko-Terrasny, Central Forest, Oksky, Mordovsky, Khopersky, Voronezhsky and Rdeysky) is presented. The purpose of this work is to analyse the patterns of restoration of the number of beavers based on long-term ground-based observation data and a qualitative description of possible changes in ecosystems after the introduction of beavers. The analysis of the population dynamics of the beavers in the reserves located in the north, south and in the centre of the range was carried out based on long-term monitoring data from 15 to 80 years. It is shown that the patterns of beaver population dynamics can be described using four types of patterns: eruptive; single-stage with quasi-periodic oscillations, multi-stage with quasi-periodic oscillations and a logistic trend of population change with periodic oscillations around it. It is revealed that the impact of beavers on landscapes depends on the population density, terrain features and forage resources of the territory. It is shown that in the future the beaver remains a constant component of ecosystems in the major part of

N. A. Zavyalov (✉)
Natural State Reserve "Rdeysky", Novgorod Oblast, Kholm, Russia
e-mail: zavyalov_n@mail.ru

L. A. Khlyap · F. A. Osipov · N. N. Dergunova · V. G. Petrosyan
A.N. Severtsov Institute of Ecology and Evolution, Russian Academy of Sciences, Moscow, Russia
e-mail: petrosyan@sevin.ru

T. I. Oliger
Natural State Reserve "Nizhnesvirsky", Leningradskaya Oblast, Kovkenitsy, Russia
e-mail: jghcn4351@mail.ru

V. V. Osipov
State Nature Reserve "Privolshky Forest-Steppe", Penzenskaya Oblast, Penza, Russia
e-mail: osipovv@mail.ru

W. Mirschel et al. (eds.), *Landscape Modelling and Decision Support*, Innovations in Landscape Research,
https://doi.org/10.1007/978-3-030-37421-1_21

the range. Possible changes in the biotic and abiotic characteristics of ecosystems as a result of beaver activity are analysed.

Keywords Eurasian beaver · Ecosystem engineers · Landscape · Monitoring · Modelling of population dynamics

21.1 Introduction

Beavers (*Castor fiber*, *C. canandensis*) are widely known as the forming of the environment key species, and ecosystem engineers. Currently, beavers (*Castor fiber*) occupy a huge area, within which they are distributed extremely unevenly. In Eurasia, most of the beaver populations are concentrated from the southern taiga to the forest-steppe. Beavers inhabit very different water bodies, but in the coming years it is expected that most of the populations will be concentrated on small rivers, as the most numerous hydrographic objects. Among the model sites that are convenient for studying the patterns of the influence of beavers on the small river ecosystems are Specially Protected Natural Territories (SPNT = OOPTs). Studies of beavers on the territory of Nature reserves are remarkable because the places of release, the number of individuals, sex and age of released animals are known; their security and accounting are organized; anthropogenic impact on ecosystems is significantly reduced; environmental monitoring is conducted. Studying the temporal dynamics of beaver populations in Russian reserves which are located in different natural zones with various levels of forage resources, parameters of hydrological networks, the presence or absence of large predators and other environmental factors, will provide a quantitative description of the overall vision of the recovery process of this species in Russia. The analysis of the long-term dynamics of the beaver populations will not only make a forecast about the future of the beaver populations, but also will help to understand the direction and extent of those changes in hydrology, soil, vegetation and animal populations that can be expected taking into consideration the beaver's potential for transforming its habitat.

The reserves and other OOPTs were of particular importance in the restoration of beaver populations (*Castor fiber*) in Russia. The first batches of beavers were brought to the reserves, because only there it was possible to organize reliable protection in the early stages of the restoration of beaver populations. The reserves became the primary foci of beaver resettlement. Beavers settled out of them, independently exploring new territories, and in some OOPTs, beavers were purposely caught and released into other regions. In addition, with the practice of scientific research in reserves and regular monitoring, invaluable material on the process of reintroduction of beavers, their life and the impact on various biotic and abiotic components of the environment has been accumulated.

21.2 Materials and Methods

In 2017, a questionnaire was sent to State reserves and National parks (NP) located in or near the beaver's potential range to assess the current state and to study beaver populations. Information about the Eurasian beaver was in the questionnaires sent from 63 OOPTs (Fig. 21.1). Some OOPTs have several isolated clusters. In this case, the map shows all the clusters inhabited by beavers.

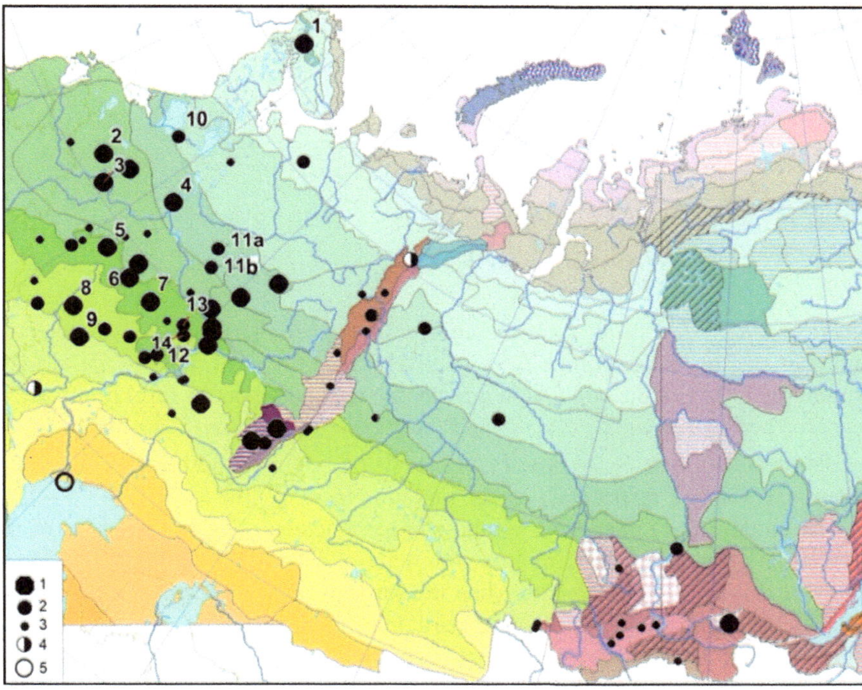

Fig. 21.1 Eurasian beaver in OOPTs of Russia (based on 2017 questionnaire data). OOPTs in which the Eurasian beaver occurs: 1—more than 10 years of observations, 2—less than 10 years of observations, 3—the availability of data on the long-term population dynamics is insufficient, 4—occurs near the boundaries of protected areas, 5—was introduced, but disappeared. OOPTs are numbered, the materials of which are presented in this study: 1. Laplandsky, 2. Rdeysky, 3. Central Forest, 4. Darwinsky, 5. Prioksko-Terrasny, 6. Oksky, 7. Mordovsky, 8. Voronezhsky, 9. Khopersky, 10. Nizhnesvirsky, 11. Kologrivsky Forest (**a**—Kologriv cluster, **b**—Manturovo cluster), 12. Privolshky Forest-Steppe, 13. Prisursky, 14. Chavash Varmane. Vegetation zones are shown in different colours (Ogureeva et al. 1999)

21.3 Modelling of Population Dynamics

To analyse the dynamics of populations, taking into account the limited resources and assumptions considered in the work (Gurney et al. 1996; Wright et al. 2004; Petrosyan et al. 2013), the following parametric time-discrete model was constructed describing the number of beavers and the proportion of active, potential and degraded resources over time.

$$P_{k+1} = P_k + \mathrm{F}\left(\frac{R_k^{(a)}}{P_k}, q_1, q_2, q_3, q_4\right) P_k, \quad R_{k+1}^{(a)} = R_k^{(a)} - q_6 P_k + q_5 R_k^{(p)}$$

$$R_{k+1}^{(p)} = R_k^{(p)} - q_5 R_k^{(p)} + q_7 R_k^{(d)}, \quad R_k^{(d)} = 1 - R_k^{(a)} - R_k^{(p)}$$

$$\mathrm{F}(x, q_1, q_2, q_3, q_4) = q_2 + (q_1 - q_2)\frac{x - q_3}{\sqrt{q_4^2 + (x - q_3)^2}};$$

$$P_k = P(t_k), \ R_k^{(a)} = R^{(a)}(t_k), \ R_k^{(p)} = R^{(p)}(t_k), \ R_k^{(d)} = R^{(d)}(t_k), \ t_{k+1} = t_k + \Delta t,$$
$$k = 1, 2, 3, \ldots; \ \Delta t = 1; \ \mathrm{k} = 1, 2, 3, \ldots; \tag{21.1}$$

where P_k, $R_k^{(a)}$, $R_k^{(p)}$ и $R_k^{(d)}$ are the number of beavers, levels of active, potential and degraded resources at time t_k, respectively; $\mathrm{F}(x, \mathbf{Q}) = \mathrm{F}(x, q_1, q_2, q_3, q_4)$ is a parametric model of the population growth rate depending on the level of the active resource per beaver where $x = R^{(a)} / P$, and $\mathbf{Q} = (q_1\, q_2\, q_3\, q_4)^T$ is a vector of the model parameters q_1, q_2, q_3, q_4; P_1, $R_1^{(a)}$ and $R_1^{(p)}$—number of beavers, active and potential resources, respectively, at the initial moment of time t_1.

The model (21.1) included the following parameters: q_1 is the threshold of the population growth rate at an unlimited amount of active resources per individual; q_2 is the population growth rate when the level of active resource per individual coincides with the optimal level under particular environmental conditions; q_3 is the level of active resource per individual necessary for the population performance; q_4 is the parameter determining steepness of the function $\mathrm{F}(x, q_1, q_2, q_3, q_4)$ at the inflection point (the value of the x derivative) which is reached at $x = q_3$ (the value of the derivative at this point is $(q_1 - q_2) / q_4)$; q_5 is the fraction of the potential resource which transforms into active resource for a year, i.e. intensity of resource restoration; q_6 is the fraction of the active resource utilized by an individual resulting in transformation of the active resource into degraded recourse; q_7 is the fraction of the degraded resource transforming into potential resource for a year. A detailed description of this model is presented in the paper (Petrosyan et al. 2013).

In current study, we follow the traditional way for analysis of correspondence of real empirical time series and model trajectories. It means that for every model we have to estimate its parameters minimizing squared deviations between time series and model trajectories (Petrosyan et al. 2013). It is rather common opinion that model gives sufficient fitting of time series if—we cannot reject the hypothesis about equivalence of average of residuals to zero, we cannot reject the hypothesis about normality of deviations (Kolmogorov—Smirnov test, Shapiro—Wilk test, etc.), we have to reject hypotheses about existence of positive or negative serial correlation in sequence of residuals (Durbin-Watson test, analysis of behaviour of autocorrelation function with testing of hypotheses about equivalence of the respective values of this function to zero).

21.4 Results and Discussion

21.4.1 Long-Term Dynamic and Prediction of Beaver Number

According to the data, the Eurasian beaver now occurs on the territory of 46 natural reserves and 14 NP, is found near the borders of NP Yugyd Va (North Ural), on the territory of the Tsimlyansky Reserve, subordinate to the Rostovsky Reserve, disappeared from the Astrakhansky Reserve (Fig. 21.1).

Data on the time of the beginning of the introduction of beavers in the OOPTs is presented in Fig. 21.2. As an aboriginal species it lived in the territory of three reserves (Voronezhsky, Malaya Sosva, Azas), and the presence of the beavers was decisive when establishing those reserves. The experience of reintroduction of beavers began with six reserves: 1934 and 1937—Laplandsky, 1936—Mordovsky and Central Forest, 1937—Oksky, 1937 and 1938—Khopersky, 1938—Pechora-Ilychsky. The following releases of beavers to reserves were in 1946 (Astrakhansky) and in 1948 (Prioksko-Terrasny, Ilmensky and again Astrakhansky). But in those years, the beavers were released in other places also. For example, in 1940, beavers were

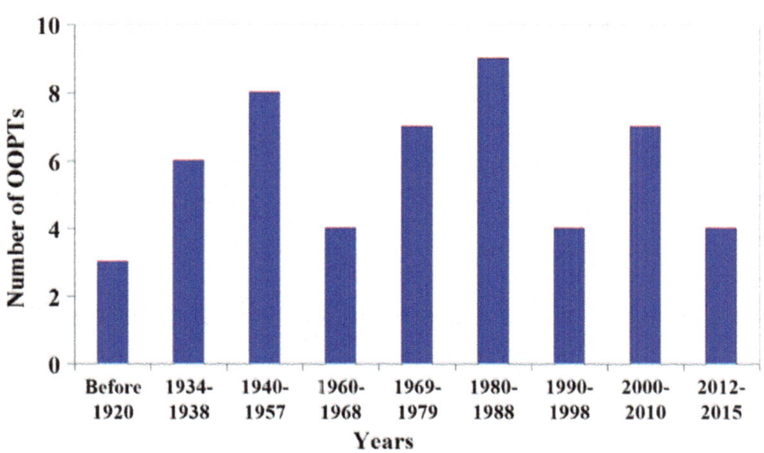

Fig. 21.2 The timing of the first beaver invasions in OOPTs based on the survey data (on the *Y*-axis—the number of OOPTs)

Table 21.1 Eurasian beavers research in OOPTs of Russia

Type of gathered data	Nature reserves	National parks
Regular counts of numbers	28	9
Long-term monitoring data	16	4
Ecosystem impact assessment	11	1

released on the territory of a hunting reserve in the middle course of the river Kerzhenets, where in 1993 Kerzhensky Reserve was established. Since the mid-1950s, the process of self-introduction of beavers began in the territory of reserves, which peaked in the 1980s (Fig. 21.2).

From the reserves of the Asian part of Russia, the beaver first appeared in Altai Reserve (1988). Until the end of the XXth century, Tigireksky (early 1990s), Kuznetsk Alatau (1990–1992) and Ubsunursky Basin (in 1994–1995 and in 1997–1998 at different clusters) were added. The introduction in the reserves continues over the last decade: 2008—Stolby Reserve, 2012—Zhigulevsky and Yugansky, 2014–2015—Ubsunurskaya Hollow (third cluster), 2014—Belogorie Nature Reserve. This reflects the general modern state of the Eurasian beaver in Russia, whose range is expanding, and the number in most regions is growing.

In those OOPTs where beavers occur, in most cases (55%) regular counts of their numbers are carried out (Table 21.1), and 16 reserves and 4 national parks have long-term (more than 10 years) observations.

In the questionnaire sent to the OOPTs, there were questions concerning the dynamics of the number of beavers at the present stage (consistently low, increasing, stabilizing and decreasing) and the intensity of the impact of beavers on the ecosystems of protected areas. Special studies of this impact were carried out only in 11 reserves and in 1 national park (Table 21.1). However, expert estimates were obtained from 57 OOPTs. According to these data, the impact of beavers on ecosystems was insignificant in 29 OOPTs (50.9%), and in the rest—noticeable or significant. Moreover, a noticeable or significant impact of the beaver was noted only with the growth or stabilization of the population size (Table 21.2).

These materials mean that the first place is occupied by OOPTs, in which the number of beavers has stabilized, and the second place—where the number is growing. Together, these OOPTs make up 86% of the questionnaires that answered this question.

Table 21.2 The role of beavers in ecosystems depending on the course of population dynamics

Impact on ecosystems	Number of OOPTs with different courses of population dynamics				
	Total	Consistently low	Increasing	Stabilized	Decreasing
Insignificant	29	6	8	13	2
Quite noticeable	18		10	8	
Substantial	10		2	8	

According to the study of changes in the number of beavers, all OOPTs can be divided into three groups.

1. Areas for which there are long-term data of beaver population counts, which allow for data analysis and making forecasts. The dynamics of the number of beavers is important not only for understanding the biology of the species itself, but also for analysing the habitats transformed by beavers (Gurney and Lawton 1996; Wright et al. 2004), and species dependent on these transformations. The patterns of dynamics and the prediction of the number of beavers in these territories are considered on the example of nine reserves: Laplandsky, Rdeysky, Central Forest, Darwinsky, Prioksko-Terrasny, Oksky, Mordovsky, Voronezhsky, Khopersky. These reserves are located in different parts of the Eurasian beaver range: from the very north (Laplandsky) to the forest-steppe (Khopersky) and the annual monitoring of the number of beavers ranges from 15 to 80 years.
2. Areas where beaver surveys have been conducted irregularly or less than 10 years, but some trends in population numbers are already clear. The dynamics of number for these reserves are considered on the example of reserves: Nizhnesvirsky, Kologrivsky Forest, Privolzhskaya Forest-Steppe, Prisurski and Chavash Varmane National Park.
3. Areas where the beavers appeared recently, or regular counts are not carried out.

The long-term dynamics of the number of beavers according to the monitoring data and model estimates of the number during the years of observation and in the future (forecast) for nine reserves are shown in Figs. 21.3, 21.4, 21.5, 21.6, 21.7, 21.8, 21.9, 21.10 and 21.11. The tests of sets of deviations between theoretical (model) trajectories and real time series showed that for selected 5% significance level we cannot reject hypotheses about equivalence of average to zero, about normality of sets of deviations, and about absence of serial correlation in sequences of residuals.

The obtained curves describing the dynamics of the population of reserves can be characterized by four main types and several types of transitions between them.

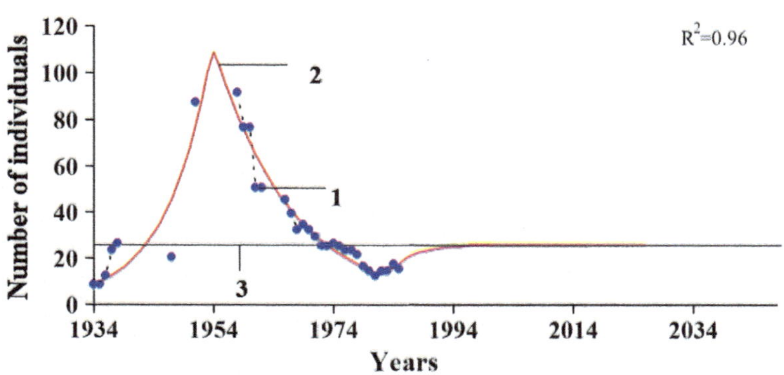

Fig. 21.3 Long-term dynamics and prediction of the beaver number for the Laplandsky reserve (1—monitoring data, 2—model estimates, 3—stationary population (I type of dynamics))

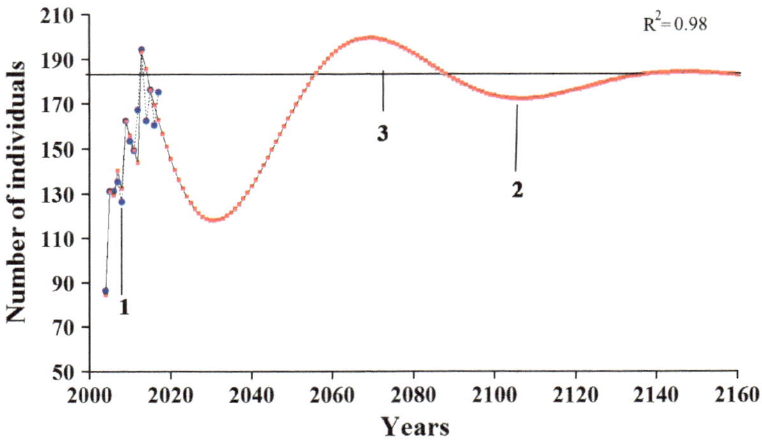

Fig. 21.4 Long-term dynamics and prediction of the beaver number for the Rdeysky reserve (version close to type II)

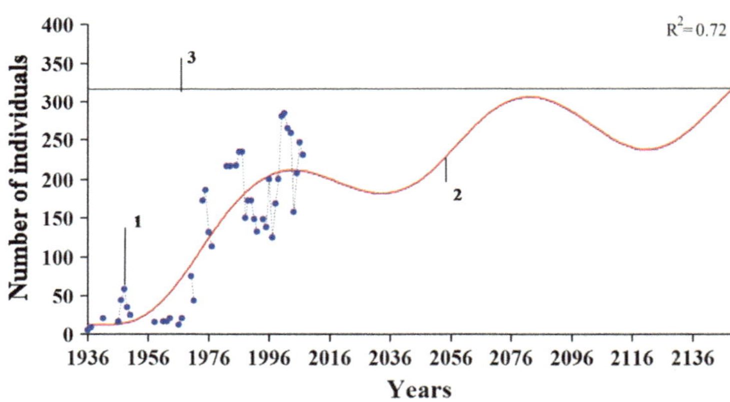

Fig. 21.5 Long-term dynamics and prediction of the beaver number for the Central Forest reserve (type III dynamics)

The first type (I) of eruptive dynamics has been noted in the Laplandsky Reserve (Fig. 21.3), where the rate of recovery of the stand removed by beaver is extremely low (Kataev 2018). It is valid for ecosystems characterized by low recovery rates and a limited amount of available feed resources. At the same time, the dynamics of number is characterized by a short-term outbreak, a further decline and a tendency towards a stationary, consistently low value.

The second type (II) of change in number is a single-stage model with quasi-periodic oscillations. It is characterized by a striving towards the level of number achieved during the first wave of a rise in the presence of a quasi-periodic component. It has been observed in the pessimal habitats of the Prioksko-Terrasny Reserve

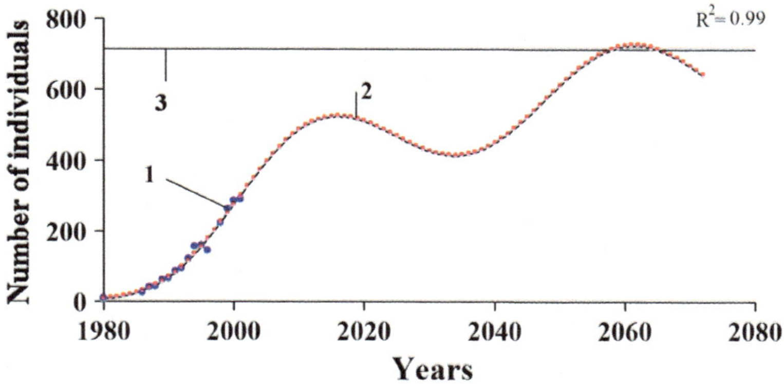

Fig. 21.6 Long-term dynamics and prediction of the beaver number for Darwinsky reserve (type III dynamics)

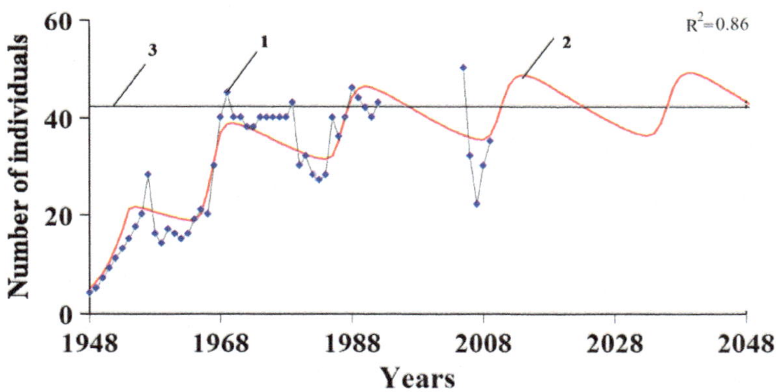

Fig. 21.7 Long-term dynamics and prediction of the beaver number for Prioksko-Terrasny reserve (type II dynamics)

(Fig. 21.7): the tree-shrub resources supply is small, and macrophytes are practically undeveloped.

The third type (III) of population dynamics is a multi-stage model with quasi-periodic oscillations and a significant increase in number. It is characteristic for beaver populations of the Central Forest (Fig. 21.5), Darwinsky (Fig. 21.6) and Khopersky (Fig. 21.11) reserves and is possible with a high rate of renewal of feed. The presence of such a multistage type of dynamics distinguishes beavers, which improve the conditions of their existence by construction activity, from most other animals.

The fourth type (IV) of dynamics is a logistic increase in number to a high level with a periodic oscillatory component. It is characterized by a desire for a stationary high state and is observed in the absence of restrictions on the available food resources, which is typical of the Oksky Reserve (Fig. 21.8).

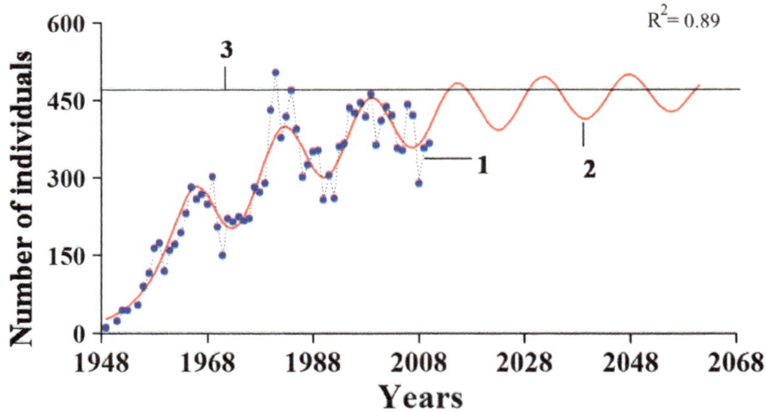

Fig. 21.8 Long-term dynamics and prediction of the beaver number for the Oksky Reserve (type IV dynamics)

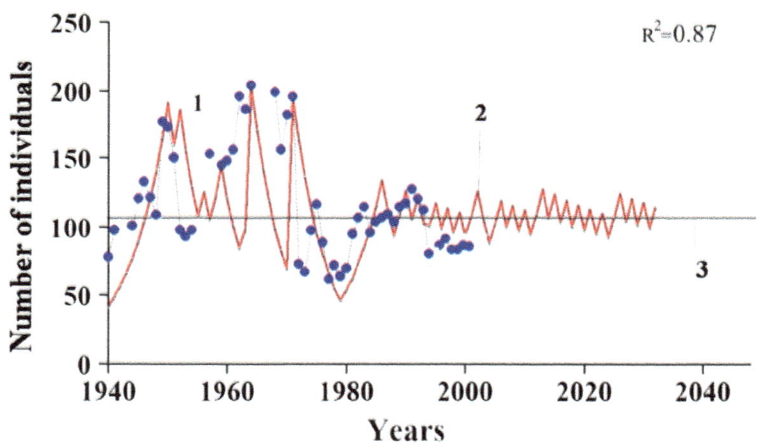

Fig. 21.9 Long-term dynamics and prediction of the beaver number for the Mordovsky reserve (transitional option between types I and II)

The intermediate position (between the first and second types of dynamics) is represented by data on the Mordovsky (Fig. 21.9) and Voronezhsky (Fig. 21.10) reserves. It should be noted that in both reserves in the first period of growth the number of beavers was caught for settlement in other regions, which could affect the nature of the dynamics of the number (Zavyalov et al. 2018; Mishin 2018). Both these reserves are also characterized by the development of black alder forests in beaver habitats. The whole complex of vital activity of beavers shifts ecological conditions to the most favourable for black alder trees direction (Zavyalov 2015), which in turn are not the best habitats for beavers themselves and, thus, forms a stable negative feedback between the number of beaver and the conditions of its habitat. But there is

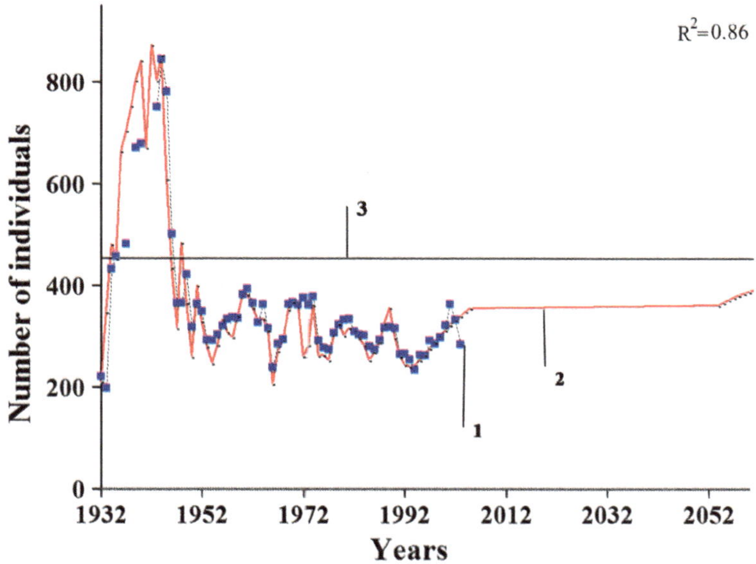

Fig. 21.10 Long-term dynamics and prediction of the beaver number for the Voronezhsky reserve (transitional option between types I and II)

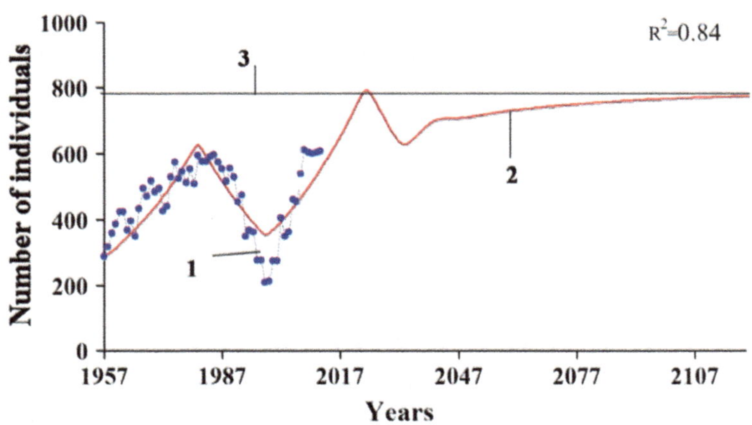

Fig. 21.11 Long-term dynamics and prediction of the beaver number for the Khopersky reserve (type III dynamics)

an important difference in the prediction of population dynamics for the Voronezhsky and Mordovsky reserves. If in the Mordovsky Reserve, population fluctuations are predicted at the stationary population level (i.e. at the level of the maximum capacity of land), then in Voronezhsky Reserve, the number is predicted at a level below the stationary (i.e. below the maximum capacity of land). These differences, apparently, are explained by the fact that in the Voronezhsky Reserve, even after the ending of the mass withdrawal of animals, the natural population dynamics is not restored, since there are obstacles to the settlement of the upper reaches of small rivers (Mishin 2018), whereas this is possible in the Mordovsky Reserve.

For the Rdeysky Reserve (Fig. 21.4), the model predicts fast and significant decline in number over the next 10–15 years, followed by restoration of number to a stationary state. Such dynamics is close to the second type of change in number. But the period of oscillation is significantly longer (76 years). The predicted decrease in the number of beavers can be explained by the depletion of tree-shrub fodder, which is now observed in many beaver settlements of the Rdeysky Reserve (Zavyalov 2018). A further increase in the number is explained by the peculiarities of the beaver habitats of the Rdeysky Reserve in large mire arrays. In beaver ponds in bogs, the maximum development of macrophytes (the main beaver feed) is observed 20–40 years after the formation of ponds (Ray et al. 2001). In addition, numerous beaver canals have a draining effect on bogs (Grootjans et al. 2014) and trees and bushes (a potential beaver feed) begin to grow along the banks of these canals. It is important that the number of habitable places for beavers in the mires is limited, and existing beaver ponds can be preserved for decades (Milbrath 2013; Karran 2018). All these processes are long-term, which, apparently, is the cause of such a long period of fluctuations in the number at the stationary level. Based on the data described in this study, our other observations and literature data, we assume that four types of population dynamics described above can apparently be considered essential for many parts of the broad range of beavers in the European part of Russia.

21.4.2 The Dynamics of Beaver Populations in Some Territories

The dynamics of the number of beavers in the territories of the second group are presented below.

Nizhnesvirsky Nature Reserve. By the time the reserve was organized in 1980, no more than a dozen beaver settlements had been observed on its territory. According to approximate estimates, the number of beavers living in the reserve, by decades, since the 1980s and ending in 2016 looks like 126, 85, 115 and 118 individuals, on average, 111 individuals annually. The greatest number of beavers lived in the late 1980s—early 1990s. This period was characterized by the presence of a relatively large number of settlements of medium size (3–5 beavers) and high-sized (more than six beavers). In the second decade (1990–2000), the number of single beavers increased several times, but the number of medium-sized settlements decreased by half, and the number of high-sized settlements decreased by four times. In the third (2000–2010) and fourth decades (2010–2018), the number of weak settlements remained almost unchanged, high-sized settlements became rare, and the basis of the number of beavers was represented by medium-sized settlements. The main factors constraining the increase in the number of beavers in the Nizhnesvirsky Reserve are low water levels in winter, ending runoff from bogs and drying of the streams of rivers and streams in summer (Oliger 2018a).

Kologrivsky Forest Nature Reserve was established in 2006. By this time, the Kostroma region, where the reserve is located, had been already inhabited by beavers, which, after a complete absence from the beginning of the 20th century, were brought back in 1958–1961. The greatest number of beavers in the region (10–12 thousand individuals) was from 1985 to 1995, i.e. about 30 years after the beginning of reintroduction. By the end of the 1990s in the Kostroma region, stabilization of the number of beavers was observed at a high level, while beavers continued to develop new, previously unpopulated areas of small water bodies. A similar process is observed now in the Kologrivsky Forest Reserve: settling new areas, beaver uses previously abandoned ones with a low intensity. In many places, food resources are already spent enough, which contributes to the search for new sites and the resettlement of previously abandoned. High-sized settlements with intensive use of resources are formed in the marginal zone of beaver penetration into new places, and within the previously developed territory there is a dispersion distribution of small settlements. In 2017, from 47 settlements, 28.3% were high-sized, 47.8% medium-sized, 10.9% weak, 13.0% abandoned. The most high-sized settlements of beavers were located on the small rivers and wetlands. The current number of beavers in the Kologrivsky Forest Reserve cluster is estimated at 250–350 individuals (Zaitsev et al. 2018).

The reserve Privolshky Forest-Steppe was established in 1989 and consists of five clusters with different dynamics of the number of beavers on each of them. On the 'Upper Sura' cluster, beavers settled before the time the reserve was founded, and the number of beavers is relatively stable here, only some annual fluctuations take place. For 2010–2017 the number of settlements was 18–23, that of individuals—68–78. In the cluster 'Borok', beaver habitat has become noticeable since 1987. Now there is an increase in the number of beavers, due to the protection measures and a good forage base. In the cluster 'Poperechenskaya Steppe', the beavers appeared in 2012 and, despite the poor forage base; the number of the beavers is now growing. In 2012, there was one settlement here, in 2017—2, in 6 years the number of beavers doubled. In the cluster 'Ostrovtsovskaya Forest-Steppe', the beaver was first observed in 2004. In 2014, there were 13 settlements with a beaver density of 3.1 individuals/km of the channel. By 2017, the number of settlements has decreased to 9, and the population density to 2.5 individuals/km of the channel. This site is characterized by a poor forage base, which with the high density of beavers and terrain features, does not suggest an increase in number in the following years (Osipov and Bashinskiy 2018).

The Prisurski Reserve is located in Chuvashia, the reintroduction of the beavers in region started in 1951 and the first beavers were released 80 km far from the modern territory of the reserve. The reserve was established in 1995, but the first registration of beavers was carried out only in 1999. The number of beavers in the reserve was grown since 1999 and reached a maximum in 2008 (47 settlements, 147–223 individuals). Then, by 2011, as a result of drought and extreme wintering, the number decreased to 43 settlements (124–188 individuals). The same number of settlements remained in 2012, but the number of individuals slightly increased (43 settlements, 133–203 individuals). Further, from 2013 to 2017 the number of settlements decreased (36–39 settlements), and the number varied from 117 to 205 individuals (Glushenkov 2018a). On the lakes of buffer zone of the Prisurski Reserve in the period of 2005–2017, the number of settlements could have increased by 3.5 times or almost halved by the following year. Peaks of numbers were noted in 2006, 2012 and 2017 (35, 34 and 33 settlements), and decline in 2005, 2009 and 2015 (10, 9 and 15 settlements). Despite the fluctuations in the number of beavers, it can be argued about the stability of the beaver population in the reserve. The presence of permanent (existing for many years) high-sized beaver settlements makes it possible to restore the number after critical periods, such as the severe drought of 2010 (Glushenkov 2018a), in a short time.

The National Park Chavash Varmane is located in the south-eastern part of the Chuvash Republic, where the beavers were reintroduced in 1963. By the beginning of establishing of the national park in 1993, beavers had already lived in all major watercourses. The first full-scale survey of beaver settlements was carried out in 2007 and since then it has been carried out continuously on all the watercourses of the park. For 10 years (from 2007 to 2017) of observations, the number remained at a relatively stable level. The greatest number was recorded in 2009—(33 settlements, 121–177 individuals), the minimum in 2010—(23 settlements, 114–142 individuals). By 2013 the number restored till 32 settlements and 114–169 individuals, and after a little decrease it reached 30 settlements and 129–185 individuals again by 2017 (Glushenkov 2018b).

Thus, if beaver surveys are irregular in the protected natural area, it is difficult to characterize the type of population dynamics, and even more so, to give a long-term forecast, but the main trends of the population state are still detected.

The results of the analysis show that for all OOPTs, located in different parts of the species range, beavers remain a constant component of ecosystems. Consequently, all changes in the biotic and abiotic characteristics of ecosystems introduced by beavers remain the same constant component.

21.4.3 The Eurasian Beaver Impact on Ecosystems of Nature Reserves

Having become a common species in many regions of the European part of Russia, the Eurasian beaver settled in the water riparian ecosystems as an ecosystem engineer—a modifier of natural regimes.

The main forms of impact as an ecosystem engineer are known in the literature and were summarized in (Zavyalov 2015). These include: the creation of new zoogenic structures—dams, burrows, canals and trails; changes in the hydrological regime of water bodies, the formation of beaver ponds and the accumulation of large volumes of water; accumulation of large amounts of dead wood; changing the mode of illumination in the riparian strip; changes in the physico-chemical characteristics of the soils of flooded beaver ponds; accumulation of large volumes of sediments above dams and their movement after dam failure. In varying degrees, changes in the environment made by beavers affect many organisms and communities, however, to trace quantitatively the influence of beavers' activity is possible only on groups of organisms for which beavers recreate or destroy already existing habitats or if the changes in the regimes are of critical importance. Let's consider the response to beaver impact on vegetation, aquatic and terrestrial invertebrates, fish, amphibians and birds. The reaction of other groups of organisms is not well studied yet.

Vegetation

The development of the flora and vegetation of beaver ponds of the Darwinsky Reserve began with the disturbance of the distinctive river and riverine vegetation and proceeded in the direction of its fast transformation. In the beaver pond, the surface area of the water increased, and the species and cenote composition, the nature of vegetation changed. Along the river banks, the meadows and forests were replaced by swamp vegetation: sedge and black alder bogs. As a result, the species genetic diversity increased, and the general appearance of the vegetation cover became similar to the small forest waterbodies (Zavyalov et al. 2005). In the Prioksko-Terrasny Reserve, a comparative analysis of the structure and dynamics of undisturbed black alder (*Alnus glutinosa*) phytocenosis and the same community after the introduction of beavers was carried out. It has been shown that sparseness of the stand has occurred—*Tilia cordata*, *Picea abies* and *Ulmus glabra* are completely excluded from its composition. The number of trunks of *Padus avium* decreased by 5 times, *Betula pubescens* by 2.5 times, *Alnus glutinosa* by 1.2 times. *Euonymus verrucosa*, *Frangula alnus*, *Lonicera xylosteum*, *Viburnum opulus* disappeared from the undergrowth, and the species of wetland group appeared in the grass cover. The long-term influence of beaver activity is associated with an increase in the area of black alder along the Tadenka River, as well as the heterogeneity and diversity of plant communities formed by grassy vascular plants in the floodplain of the river (Dgebuadze et al. 2012). The development of black alder trees in the habitats of beavers is a typical phenomenon (Fig. 21.12).

In the Mezha River floodplain (Central Forest Reserve), the grassy communities formed as a result of changes in the hydrological regime of beavers were examined. Four associations belonging to the class Pragmiti-Magno-Cariceta were revealed. These communities are wet meadows and grass mires, a significant factor in the differentiation of which is the watering level. They are widespread in Europe (Cherednichenko 2017). Thus, the activity of beavers on the Mezha River did not lead to the formation of original communities. The formation of meadow communities (beaver meadows) was also observed in other reserves (Fig. 21.13).

Aquatic Invertebrates

The reaction of aquatic invertebrates (zooplankton, zoobenthos) has been studied most fully. The small rivers of the Darwinsky Reserve have shown a fast growth in the number and biomass of invertebrates in the water column immediately after the construction of the beaver dam and the high level of these indicators in subsequent

Fig. 21.12 Black alder forest in beaver habitats of the Mordovsky reserve (Photo: N.A. Zavyalov)

Fig. 21.13 «Beaver meadow» in Nizhnesvirsky reserve (Photo: T.I. Oliger)

years, which is associated with a sharp decrease in the flow rate and an increase in the content of biogenic and organic substances (Zavyalov et al. 2005). In the zooplankton of beaver ponds of the Prioksko-Terrasny Reserve on the Tadenka River, copepods dominated, but in the total number and biomass, the proportion of rotifers was also high. New beaver ponds, regardless of the season of their formation, were characterized by a low value of the biomass of zooplankton and served as a breeding habitats for species and groups of invertebrates unusual for watercourses. Intensive water exchange, the frequent destruction of beaver dams, the periodic movements of beavers and the relatively large shade of the river contributed to the preservation in beaver ponds more characteristic of flowing water zooplankton species. In the inhabited beaver pond, the number and biomass of macrozoobenthos were lower, and the species diversity was higher than in the control non-dam sections of the river. However, in the ponds abandoned by beaver, the situation was reversed: the number and biomass of zoobenthos were greater, and the species diversity was lower than in the control plots (Dgebuadze et al. 2012).

In beaver ponds that formed on the small rivers of the Kologrivsky Forest Reserve, there was a decrease in the oxygen content and an increase in the amount of biogenic substances in the water. The development of limnophilic species was observed, which was reflected in a change in the taxonomic structure of zooplankton. The number and biomass of zooplankton species, diversity and the saprobic indexes were increased (Zaitsev et al. 2018).

In the Rdeysky Nature Reserve, aquatic beetles of beaver ponds were studied. According to species composition, occurrence and biomass, the beetles widespread in European part of Russia and characteristic for lentic ecosystems predominate here. For beaver ponds polydominant communities of aquatic beetles with a predominance of predators with wide range of food is characteristic, but in the final stages of the beaver cycle, the proportion of myxophytophages and detritivores forms increases. Shallow-water ponds with macrophyte thickets are the richest in species, small species of the genus *Hydroporus* predominate here. Species of the genera *Dytiscus* and *Cybister* dominate in deep water ponds (Sazhnev and Zavyalov 2018).

Terrestrial Invertebrates

The reaction of terrestrial invertebrates to the activity of beavers has been studied only in the Nizhnesvirsky Reserve. At the same time, the reaction of different invertebrate groups in the riparian aspen's area, which was significantly thinned by beavers as a result of foraging here (beaver logging), and in biotopes formed at places of beaver ponds that were abandoned after the beavers left, was considered separately.

The beaver logging area was superior to the control area, which was unaffected by deforestation due to a variety of environmental conditions. It was inhabited by animals preferred to dwell under the thin cover of the forest and in an open glade. The thinning (lightening) effect persisted for many years, and several years after the start of aspen removal by the beavers, formed quite stable and diverse species of invertebrate herpetobia and hortobia, which exceeded the control number of species and their numbers (Oliger 2018b).

The biotope formed on the site of the beaver pond was characterized at the first stage of the succession by the complete absence of green vegetation and an abundance of dead forest. The invertebrate colonization of this biotope began immediately after the draining of water. The first were the species that inhabited the shores of the former beaver pond and on dry dead wood: saprophages and zoophages. Then, as the soil dried out, wood waste decayed and vegetation developed, it was colonized by other invertebrate species from surrounding and distant biotopes: herbivorous, variegated or saproxil. In 2–3 years from the beginning of the succession, the species richness and the number of pioneer groups of animals exceeded the groups of the initial biotope that existed before the flooding by beavers. Pioneer invertebrate complexes were mainly represented by zoophages. The population of the new biotope included both typical eurytopic representatives of forest habitats and inhabitants of meadow or sphagnum open and semi-open spaces. Some of the species found in this biotope were not found elsewhere in the reserve or were extremely rare there. The diversity of the environment of the second and third year of succession provided the invertebrates of different taxonomic groups to dwell here. They represented a wide range of relationships to the water regime (hygro-, meso- and xerophiles). There was a large proportion of dendrofils—the inhabitants of the dead wood. As the green vegetation, mycocenosis and moss layer developed, the population of invertebrates in the place of the beaver pond was filled up with new species, their abundance increased. The community structure was unstable and unevenness, corresponding to a number of parameters for the population of animals in meadow stations, where seasonal changes in micro conditions are particularly pronounced (Oliger 2018b). Thus, the observations in the Nizhnesvirsky Reserve showed a fast response of terrestrial invertebrates to environmental changes caused by beavers, but specific invertebrate communities were unstable in time and space.

Amphibians

On the small rivers of the Rdeysky Reserve, three species of amphibians, which use beaver ponds for breeding are known. In the absence of beavers, amphibians (only two species) bred in large oxbows, sometimes in shallow floodplain water bodies. In filled beaver ponds, amphibian breeding was observed annually. It was massive and took place in the central parts of the ponds. In the abandoned beaver ponds, mass reproduction was noted in any part of the remaining water. The maximum mortality of amphibian eggs (30–40% of the total) is noted here. The survival of tadpoles was higher in the filled beaver ponds and the yield of fingerlings to dry land after the completion of metamorphosis was the greatest there. The most important feature of the influence of beaver activity on amphibians was the formation of standing bodies of water with a stable water level regime (Bashinsky 2014). In the Prioksko-Terrasny Reserve, beavers live about 70 years, and on the Tadenka River all places suitable for breeding amphibians have been altered as a result of the activities of beavers. Currently, the river has a complete set of habitats favourable for the reproduction of four species of amphibians (*Bufo bufo, Rana arvalis, R. temporaria, Pelophylax lessonae*) and the development of their larvae. These places are represented by filled ponds, various stages of pond drying, abandoned beaver pounds, and completely dry

Fig. 21.14 Beaver pond and lodge in Privolshky forest-steppe reserve (Photo: V.V. Osipov)

places where ponds once were. Large beaver ponds proved to be the most suitable habitat for amphibians (Dgebuadze et al. 2012). The age and stability of beaver ponds in the Privolshky Forest-Steppe Reserve preserve the abundance and species diversity of amphibians. Beaver rivers in the steppe are inhabited by five amphibian species (*Lissotriton vulgaris*, *Pelobates fuscus*, *Bufo viridis*, *Rana arvalis*, *Pelophylax lessonae*). The species diversity and abundance of amphibians increases as a result of the rivers damming by beavers (Fig. 21.14). In the course of the long existence of ponds, the number of amphibians increases due to an increase the shallow-water area. But when beavers leave the pond, the number of amphibians decreases. Despite some positive impact, beaver ponds in the forest-steppe conditions of the European part of Russia are not the key habitats for amphibians (Bashinskiy and Osipov 2016).

Fishes

The results of research in the Darwinsky Reserve showed that new beaver ponds lose flowage largely, the level of dissolved oxygen in the water and the availability of fish food decrease. The fish population of such ponds is extremely poor in both species and biomass. It should be noted that the level of diversity of fish, characteristic of the rivers uninhabited by beavers, is not preserved in the areas located below the beaver ponds. As the ponds age, they form lake-type ecosystems. The composition of aquatic vegetation changes, the content of dissolved oxygen increases, the diversity, abundance and biomass of fish increase substantially. In the conditions of lowland rivers, after the introduction of beavers, the structure of trophic networks changes significantly. Many species of fish (*Squalius cephalus*, *Alburnus alburnus*, *Lota lota*) disappear due to changing environmental conditions, but other species (primarily detritophages) find favourable conditions in beaver ponds. In general, communities are simplified and, apparently, become less stable (Zavyalov et al. 2005).

The ichthyofauna of the Gorelka River (Rdeysky Reserve) includes nine species of fish, of which only two species (*Esox lucius*, *Leuciscus leuciscus*) were recorded in beaver ponds. The relative density of fish per one fishing effort in beaver ponds was much less than in areas not populated by beavers. The construction of dams leads to the physical isolation of fish living in ponds from other parts of the river system (Dgebuadze et al. 2009).

For the small river of the Oksky Reserve, which is under the long-term impact of beavers, a decrease in the species richness of fish in beaver ponds has been noted. It increased after the beavers left these ponds, and when the beavers returned, it decreased again. There is a high concentration of ammonium nitrogen in ponds, populated by beavers, which limits the species diversity and abundance of aquatic organisms. The fish population is primarily affected by the fact of the beaver's habitat, but not by the length of their habitat. Under the conditions of the Oksky Reserve, beaver dams can obstruct the movement of fish only during low water periods (Ivanchev and Ivancheva 2018). The species composition and spatial distribution of fish in beaver rivers of the Privolshky Forest-Steppe Reserve are distinguished by great variability associated with seasonal and inter-annual fluctuations of water level and abundance of beavers. In the beaver ponds that emerged, the species diversity of fish falls. Most beaver ponds are small, the dams are washed out annually. Stable perennial ponds form only on rivers with low water flow, which leads to the isolation from the main watercourses of the basin. This prevents the growth of fish diversity in areas inhabited by beavers (Osipov et al. 2018).

Birds

In Nizhnesvirsky Reserve, the population of birds that formed on the site of a beaver pond, after drying of which there were many standing and fallen dead trees (Fig. 21.15), was characterized by a large number of species that have nests in hollows and, in the early years, by an almost complete absence of land nesting birds. The population density of birds here was not inferior to the density in control, and the species composition was more diverse, including not only typical forest, but also types of open spaces. The majority of species were zoophages. For some species of birds, for example, for *Tringa ochropus* or birds which have nests in hollows (*Apus apus*, *Ficedula hypoleuca*, etc.), the lowered beaver pond proved to be the preferred habitat (Oliger 2018b). Another situation was observed in the Kologrivsky Forest Reserve, the territory of which is almost entirely occupied by forest and is surrounded by equally extensive forests. This limits the colonization of habitat created by the beaver to species typical for edges, meadows and fields. There are relatively small changes in the composition of the fauna and the number of bird species. Thus, the creation of beaver ponds significantly affected the increase in the number of three species of water birds (*Anas platyrhynchos*, *Anas crecca*, *Aythya fuligula*) and

Fig. 21.15 Dead wood in beaver habitats of Nizhnesvirsky reserve (Photo: T.I. Oliger)

for some birds (*Carpodacus erythrinus*, *Emberiza schoeniclus*, *Motacilla alba*, and others), open beaver habitats became a refugium (Zaitsev et al. 2018).

21.5 Conclusion

1. The range of Eurasian beaver in Russia is expanding, and the number in most regions is growing. According to the data, the impact of beavers on ecosystems was insignificant in 29 OOPTs (50.9%), and in the rest—noticeable or significant. A noticeable or significant impact of the beaver was noted only with the growth or stabilization of the population size.
2. An analysis of the restoration of the number of beavers (*Castor fiber* L.) in the reserves of the European Russia has been made which also allows creation a forecast of possible changes in ecosystems after its introduction into the reserve. The results of the analysis of beaver populations dynamics in Laplandsky, Darwinsky, Prioksko-Terrasny, Central Forest, Oksky, Mordovsky, Khopersky, Voronezhsky and Rdeysky reserves, located in the northern, southern and central part of the species range, are presented based on the long-term monitoring data (50–80 years). The results demonstrate that the beaver population dynamics can be described using four types of patterns: eruptive, single-stage with quasi-periodic

oscillation, multi-stage with quasi-periodic oscillations, and logistic trend of population dynamics with periodic oscillations around it.

3. This work reveals that the beaver's impact on landscapes depends on the population density, relief characteristics, and territory's food resources. It has been shown that in the future in the major part of the range the beavers will remain a constant component of the ecosystems.
4. The analysis of the specifics of beaver disturbances and changes in the landscape with their constant habitat has shown that the beaver's effects include: the creation of new zoogenic structures—dams, burrows, canals and trails; changes in the hydrological regime of water bodies, the formation of beaver ponds and the accumulation of large volumes of water; accumulation of large amounts of dead wood; changing the mode of illumination in the riparian strip; changes in the physico-chemical characteristics of the soils of flooded beaver ponds; accumulation of large volumes of sediment above dams and their movement after dam failure.
5. The existing long-term data on the territory of the reserves allowed to analyse the response to beaver impact on vegetation, aquatic and terrestrial invertebrates, fish, amphibians and birds. The reaction of other groups of organisms today is not well understood and requires further research.

Acknowledgements The collection of data on the territory of nature reserves was supported by the program of the Presidium of the Russian Academy of Sciences No. 41 'Biodiversity of natural systems and biological resources of Russia'. Data analysis using a single-species was carried out with the financial support of the grant of the Russian Science Foundation No. 16-14-10323. The authors thank all the staff of protected areas that answered the questionnaire.

References

Bashinsky IV (2014) Impact assessment of European beaver reintroduction on amphibians of small rivers. Russ J Biol Invasions 5(3):134–145

Bashinskiy IV, Osipov VV (2016) Beavers in Russian forest-steppe—characteristics of ponds and their impact on fishes and amphibians. Russ J Theriology 15(1): 34–42

Cherednichenko OV (2017) The grass vegetation of the floodplain of the river Mezha in the territory of the central forest reserve. In: Contribution of the reserve system to biodiversity conservation and sustainable development: Materials of the All-Russian Scientific Conference (with international participation), dedicated to the 85th anniversary of the organization of the Central Forest State Natural Biosphere Reserve and the 100th anniversary of the Reserve System of Russia. Tver State University, Tver, pp 589–595

Dgebuadze YY, Skomorokhov MO, Zavyalov NA (2009) Preliminary materials on the fish population of the small "beaver river" of the Novgorod region. Proceedings of the State Natural Reserve «Rdeysky» (1) Veliky Novgorod, pp 173–186

Dgebuadze YY, Zavyalov NA, Petrosyan VG (ed) (2012) Eurasian beaver (*Castor fiber* L.) as a key species of a small river ecosystem (Prioksko-Terrasnyi nature biosphere reserve). Association of scientific publications KMK, Moscow, 152 p

Glushenkov OV (2018a) The number and location of the beaver (*Castor fiber*) in Prisurski state nature reserve. In: Zavyalov NA, Khlyap LA (eds) Proceedings of the State Nature Reserve «Rdeysky» (4). Beavers in the reserves of the European part of Russia. Velikiye Luki printing house, Velikiye Luki, pp 297–310

Glushenkov OV (2018b) The beavers (*Castor fiber*) in Chavash Varmane National Park. In: Zavyalov NA, Khlyap LA (eds) Proceedings of the State Nature Reserve «Rdeysky» (4). Beavers in the reserves of the European part of Russia. Velikiye Luki printing house, Velikiye Luki, pp 311–320

Grootjans A, Iturraspe R, Fritz C, Moen A, Joosten H (2014) Mires and mire types of Peninsula Mitre, Tierra del Fuego, Argentina. Mires Peat 14:1–20

Gurney WS, Lawton JH (1996) The population dynamics of ecosystem engineers. Oikos 76:273–283

Ivanchev VP, Ivancheva YY (2018) The influence of the environment-forming activity of the european beaver *Castor fiber* on the structure of the fish population of the small river of the Meshchera lowland, the Oka basin. In: Zavyalov NA, Khlyap LA (eds) Proceedings of the State Nature Reserve «Rdeysky» (4). Beavers in the reserves of the European part of Russia. Velikiye Luki printing house, Velikiye Luki, pp 442–452

Karran DJ (2018) The engineering of peatland form and function by beaver (*Castor spp.*)/A dissertation submitted to the College of graduate and postdoctoral studies in partial fulfillment of the requirements for the degree of doctor of philosophy in the Department of Geography and Planning University of Saskatchewan, Saskatoon, Canada, 2018. https://harvest.usask.ca/handle/10388/8661

Kataev GD (2018) The beavers of the Kol'sky population in terms of Laplandsky reserve. In: Zavyalov NA, Khlyap LA (eds) Proceedings of the State Nature Reserve «Rdeysky» (4). Beavers in the reserves of the European part of Russia. Velikiye Luki printing house, Velikiye Luki, pp 11–39

Milbrath JT (2013) Land-cover change within the peatlands along the Rocky mountain front, 1937–2009. Graduate Student theses, dissertations, Montana, 611 p. https://scholarworks.umt.edu/etd/611

Mishin AS (2018) Long-term number dynamics of the beavers (*Castor fiber*) of the Voronezhsky reserve and the reasons that determine it. In: Zavyalov NA, Khlyap LA (eds) Proceedings of the State Nature Reserve «Rdeysky» (4). Beavers of the European part of Russia. Velikiye Luki printing house, Velikiye Luki, pp 274–296

Ogureeva GN, Miklyaeva IM, Safronova IN, Yurkovskaya TK (1999) The map "Plant zones and their types in Russia and adjacent territories", 1: 8000000. EKOR-M, Moscow

Oliger TI (2018a) Monitoring of the beaver (*Castor fiber*) population in Nizhnesvirsky nature reserve. In: Zavyalov NA, Khlyap LA (eds) Proceedings of the 4th state nature reserve «Rdeysky» beavers in the reserves of the European part of Russia. Velikiye Luki printing house, Velikiye Luki, pp 52–84

Oliger TI (2018b) The influence of the beaver (*Castor fiber*) on the composition and structure of terrestrial animal communities of Nizhnesvirsky nature reserve. In: Zavyalov NA, Khlyap LA (eds) Proceedings of the State Nature Reserve «Rdeysky» (4). Beavers in the reserves of the European part of Russia. Velikiye Luki printing house, Velikiye Luki, pp 383–422

Osipov VV, Bashinskiy IV (2018) Beavers in Privolshky forest-steppe state nature reserve. In: Zavyalov NA, Khlyap LA (eds) Proceedings of the State Nature Reserve «Rdeysky» (4). Beavers in the reserves of the European part of Russia. Velikiye Luki printing house, Velikiye Luki, pp 337–353

Osipov VV, Bashinskiy IV, Dgebuadze YY (2018) Influence of the ecosystem engineering activity of the european beaver *Castor fiber* on the fish assemblages of the streams of the state nature reserve "Privolzhskaya forest-steppe". In: Zavyalov NA, Khlyap LA (eds) Proceedings of the State Nature Reserve «Rdeysky» (4). Beavers in the reserves of the European part of Russia. Velikiye Luki printing house, Velikiye Luki, pp 453–470

Petrosyan VG, Golubkov VV, Goryainova ZI, Zav'yalov NA, Al'bov SA, Khlyap LA, Dgebuadze YY (2013) Modeling of the Eurasian beaver (*Castor fiber* L.) population dynamics in the basin of a small Oka river tributary, the Tadenka river (Prioksko Terrasnyi nature reserve). Russ J Biol Invasions 4(1):45–53

Ray AM, Rebertus AJ, Ray HL (2001) Macrophyte succession in Minnesota beaver ponds. Can J Bot 79:487–499

Sazhnev AS, Zavyalov NA (2018) Fauna and ecology of water beetles (Insecta: Coleoptera) of the beaver ponds of Rdeysky reserve. In: Zavyalov NA, Khlyap LA (eds) Proceedings of the state Nature Reserve «Rdeysky» (4). Beavers in the reserves of the European part of Russia. Velikiye Luki printing house, Velikiye Luki, p 423–441

Wright JP, Gurney WS, Jones CG (2004) Patch dynamics in a landscape modified by ecosystem engineers. Oikos 105:336–348

Zaitsev VA, Sirotina MV, Muradova LV, Sitnikova ON (2018) Beavers in «Kologrivsky Forest» reserve. In: Zavyalov NA, Khlyap LA (eds) Proceedings of the State Nature Reserve «Rdeysky» (4). Beavers in the reserves of the European part of Russia. Velikiye Luki printing house, Velikiye Luki, pp 125–180

Zavyalov NA (2015) The environment-forming role of the Eurasian beaver (*Castor fiber* L.) in the European part of Russia. Proceedings of the State Natural Reserve "Rdeysky" (3). Veliky Novgorod, 320 p

Zavyalov NA (2018) The results of long-term monitoring of the beaver (*Castor fiber*) population of Rdeysky reserve and adjacent territories. In: Zavyalov NA, Khlyap LA (eds) Proceedings of the State Nature Reserve «Rdeysky» (4). Beavers in the reserves of the European part of Russia. Velikiye Luki printing house, Velikiye Luki, pp 85–124

Zavyalov NA, Artaev ON, Petrosyan VG (2018) Beavers (Castor fiber) of Mordovian Reserve. In: Zavyalov NA, Khlyap LA (eds) Proceedings of State Nature Reserve «Rdeysky» (4). Beavers in the reserves of the European part of Russia. Velikiye Luki printing house. Velikiye Luki, pp 253–273

Chapter 22
Modelling Biodiversity and Ecosystem Services Trade-Offs in Agricultural Landscapes to Support Planning and Policy-Making

Davide Geneletti, Blal Adem Esmail, Rocco Scolozzi, Giacomo Assandri, Mattia Brambilla and Paolo Pedrini

Abstract Agricultural areas provide non-commodity outputs besides food and fiber that can contribute to sustainability. Balancing the interest of farmers and the benefits for the society is a key challenge. Assessing the potential benefits for biodiversity and understanding spatial and temporal trade-offs among multiple ecosystem services (ES) from agricultural areas remain a key challenge, especially in mountainous landscapes. We develop an approach to assess the trade-offs and synergies between ES and biodiversity associated with agricultural areas, focusing on mountain landscapes. We first model the distribution of ES and biodiversity in seven study areas in northern Italy, aiming at providing guidance on the relationship between the intensity of use of

Text and figures of this chapter, except for Section 4, are based on a revised version of the following paper: Geneletti D, Scolozzi R & Adem Esmail B (2018) Assessing ecosystem services and biodiversity tradeoffs across agricultural landscapes in a mountain region. Int J Biodivers Sci, Ecosyst Serv Manage 14:1189–209. https://www.doi.org/10.1080/21513732.2018.1526214

D. Geneletti (✉) · B. Adem Esmail
Department of Civil, Environmental and Mechanical Engineering,
University of Trento, Via Mesiano, 77, 38123 Trento, Italy
e-mail: davide.geneletti@unitn.it

B. Adem Esmail
e-mail: blal.ademesmail@unitn.it

R. Scolozzi
Department of Sociology and Social Research, University of Trento,
Via Verdi, 26, 38122 Trento, Italy
e-mail: rocco.scolozzi@unitn.it

G. Assandri · M. Brambilla · P. Pedrini
Section of Vertebrate Zoology, Museo delle Scienze,
Corso del Lavoro e della Scienza 3, 38122 Trento, Italy
e-mail: giacomo.assandri@gmail.com

M. Brambilla
e-mail: brambilla.mattia@gmail.com

P. Pedrini
e-mail: paolo.pedrini@muse.it

W. Mirschel et al. (eds.), *Landscape Modelling and Decision Support*,
Innovations in Landscape Research,
https://doi.org/10.1007/978-3-030-37421-1_22

agricultural land and the provision of ES. We then performed a thematic aggregation of the indicators and correlation analysis followed to gain a better understanding of the spatial and temporal ES trade-offs. Finally, we discuss how the results can provide support to planning and policy-making in different sectors, with a focus on rural development and nature conservation planning.

Keywords Trade-offs · Ecosystem services · Mountain agriculture · Biodiversity conservation

22.1 Introduction

The ecosystem services (ES) concept can support policy- and decision-making promoting sustainability, from raising the awareness of stakeholders to shaping decisions (Posner et al. 2016; Cortinovis and Geneletti 2018). Greater impacts towards sustainability, i.e. achieving human wellbeing along with biodiversity and nature conservation (Kates et al. 2005; Kates 2011), can be expected when ES knowledge is deliberately used to 'generate actions' and 'produce outcomes', supporting new policies that explicitly consider effects on ES (Posner et al. 2016).

Ruckelshaus et al. (2015) defined four conceptual pathways with increasing levels of impact on decision-making, and with multiple and iterative purposes: from explorative research for systems understanding to supporting design of policy instruments. Recently, Geneletti et al. (2018) identified case studies applying ES mapping and assessment to address decision problems in real-life contexts, covering different biomes, spatial scales and different policy domains. In general, there is growing evidence of the potential benefits of integrating ES knowledge into policy and decision-making, including spatial and landscape planning, and management (Geneletti 2011; von Haaren and Albert 2011; Mckenzie et al. 2014; Adem Esmail and Geneletti 2017).

Analysing trade-offs and synergies is a key step of the ES assessment process that bears great potential to support policy and decision-making, hence is a key area of research in the field of ES (Raudsepp-Hearne et al. 2010; Geneletti 2012; Daw et al. 2015; Fichino et al. 2017). Trade-offs are defined as situations in which improvement in the provision of a service takes place at the expense of other services, while in the case of synergies the provision of multiple services can be increased simultaneously (Raudsepp-Hearne et al. 2010). Understanding spatial and temporal trade-offs or synergies of multiple services is important for informing management decisions (Bennett et al. 2015), especially when the goal is to enhance multiple ES while conserving biodiversity (Locatelli et al. 2017). Studies that addressed the spatial dimension of ES trade-offs and synergies are presented for example in Holland et al. (2011), Bennett et al. (2015), Ferrari et al. (2016).

In this chapter, we present an approach to model and assess trade-offs among biodiversity and ES in different agricultural landscapes and discuss how the results can provide support to planning and policy-making in different sectors. Particularly, we analyse seven agricultural landscapes in an Alpine region of Italy, along a gradient of land-use intensity. Our research questions are: (i) what are the main differences in the provision of ES and biodiversity across mountain agricultural landscapes characterized by different land use intensity? (ii) What are the trade-offs and synergies among individual (and aggregated) ES and biodiversity indicators, and categories of ES? (iii) How does the spatial pattern of the ES and biodiversity hotspots, i.e. areas characterized by high levels of provision of multiple ES and biodiversity, change across the selected study sites?, and (iv) How can the results of the study be used to inform planning and policy-making in different sectors (particularly rural and spatial planning, and protected area policies)?

The chapter is structured in further four parts. Section 22.2 presents the study areas, the selected ES and related indicators for assessment, as well as the methodology used for the analysis of trade-offs and synergies. Section 22.3 illustrates the results for the selected agricultural areas in mountain landscape, including an analysis of ES trade-offs and synergies. Section 22.4 discusses the potential use of the results to support planning and policy decisions. Finally, Sect. 22.5 provides some conclusions.

22.2 Methods

22.2.1 Description of Study Sites

To analyse ES associated with agricultural areas in mountain landscapes with different land-use intensity, we selected seven study sites in the Autonomous Province of Trento, in northern Italy. The ES context of this province has been described in Ferrari and Geneletti (2014) and Ferrari et al. (2016). The distribution of the seven study sites (Val di Non, Baldo, Vallarsa, Ledro, Piereni, Fiemme and Tesino) is shown in Fig. 22.1. Overall, the selected sites are representative of the variety of mountain agricultural landscapes in the province (see Table 22.1).

Operationally, we extracted the 'agricultural areas' as identified in a land cover map (an update of the CORINE 2006 by the Province of Trento), and used the related cadastral and economic data (statistics of the Autonomous Province of Trento 2013), representing the current land use.

22.2.2 Modelling Ecosystem Services and Biodiversity

We focused on the two main categories of ES associated with agricultural areas, namely: provisioning and cultural services, as identified by the Common International Classification of ES (EEA 2013). For each category, with the help of key stakeholder, we identified those ES that are considered the most significant in the context of the Autonomous Province of Trento. Similarly, we included four biodiversity indicators representing the provision of habitat for key animal and plant species

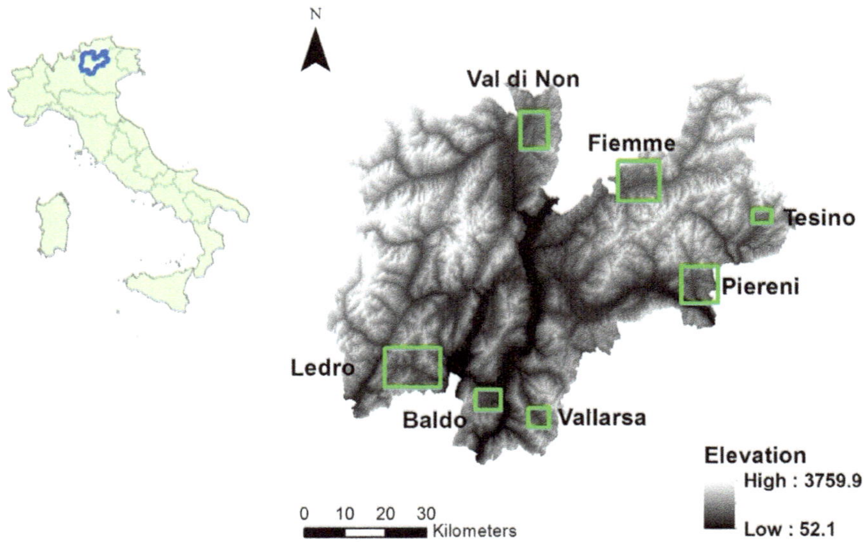

Fig. 22.1 Location of the autonomous province of Trento in Italy (left) and of the seven study sites within the province (right)

Table 22.1 Characteristics of the seven study sites

Name	Area (ha)	Average elevation (m. a. s. l.) (m)	Main crops
Val di Non	2,248	903	Fruit trees and hay meadows
Baldo	1,478	633	Mixed, i.e. vineyards, fruit trees, hay meadows, and annual crops
Vallarsa	491	784	Annual crops and hay meadows
Ledro	1,184	756	Hay meadows with some mixed crops
Piereni	371	1,134	Hay meadows, pastures and annual crops
Fiemme	2,531	1,086	Hay meadows and annual crops
Tesino	1,847	915	Annual crops, hay meadows, pastures with some mixed crops

in the region. In the following, the indicators selected for each ES and biodiversity, as well as the methods applied for their spatially explicit assessment are briefly described.

22.2.2.1 Provisioning Services

We identified agricultural production of food (i.e. 'cultivated crops' according to the CICES) and the production of forage (i.e. 'materials from plants for agricultural use') as the two most relevant provisioning services in the study areas. For each of these services, we selected an indicator combining information of biophysical (i.e. land use) and economic (i.e. market prices) nature. Respectively, we used (P1) the annual overall market value of foodstuffs (apples, small fruits, vegetables, etc.) and (P2) the annual total market value of the forage product, both expressed in Euro per hectare. Data on crop typology was obtained from the statistics of the Autonomous Province of Trento (Servizio Politiche Sviluppo Rurale, PAT, 2013). In this database, the distinction between crops is based on declaration from the individual farmers and sample verification by the administration. This information was associated to average market prices. Average productivity per hectare for each type of crop and market values were derived from provincial statistics (ISPAT, *Istituto di statistica della provincia di Trento)* or from the market values periodically collected by farmer and trade organizations; and where lacking, we relied on experts' opinion (more details in Ferrari and Geneletti 2014).

22.2.2.2 Cultural Services

We considered services related to the use (i.e. 'physical and intellectual interactions with ecosystems and land-/seascapes' according to the CICES) and the perception (i.e. 'spiritual, symbolic and other interactions with ecosystems and land-/seascapes') of agricultural areas. Accordingly, we selected the following ES indicators: density of geotagged photographs; density of cycling paths, hiking trails, horse trails, and mountain biking trails; density of elements of cultural/aesthetic interest.

Density of geotagged photographs (C1). This provides a proxy of the degree of fruition of the agricultural areas by people who appreciate its aesthetic and landscape value, as well as its recreational function. The method (adapted from the one proposed by Orsi and Geneletti 2013 and Tenerelli et al. 2017) involves the collection of the photographs available on the 'Panoramio' website (last accessed in October 2015), one of the main photo sharing sites with a geographic location system. A selection of photos uploaded by users is visible on Google Earth, where they can be downloaded along with information about their location. This allows the determination of the points from which photographs were taken, and the generation in a GIS environment of the density maps of those photographs. Density maps of photographs were constructed by considering for each cell (20 $\times$ 20 m) a radius of 500 m and reporting the number of photographs present within.

Density of elements of cultural/aesthetic interest (C2). This expresses the aesthetic and cultural value of the territory, based on the distribution of the elements of particular importance. These elements, which include areas of high landscape and environmental value, local reserves, and archeological sites, were identified based on the information contained in the Spatial Plan of the Autonomous Province of Trento.

ArcMap 10.1 (ESRI) 'Focal Statistics' module was used to calculate the number of cells (100 × 100 m) with at least one element of interest within 500 m. Hence, the ratio in percentage between the number of cells with elements of interest and the total areal extension of the study area was calculated.

Density of cycling paths, hiking trails, horse trails, and mountain biking trails (C3). This provides a quantification of the potential of the territory to be enjoyed for recreational purposes on foot and by bicycle. Based on data on pedestrian and cyclist paths (obtained from the Autonomous Province of Trento) and on hiking trails (from the local Alpine Club), the density of these elements was calculated and expressed in km/km^2.

As in the case of regulating services, the analysis for cultural services was carried out considering a buffer of 200 m beyond the boundaries of the study areas, in order to consider also the use of adjacent areas (e.g. roads between farms), which can still be traced back to the use and perception of the agricultural landscape.

22.2.2.3 Biodiversity

The aim here was to deepen aspects related to the role of agricultural areas in preserving and promoting biodiversity, building on some important modelling efforts carried out within a recent European project (LIFE11/NAT/IT000187 T.E.N—Trentino Ecological Network—http://www.lifeten.tn.it/?lang=2). Accordingly, we identified four related indicators: number of focal animal species; conservation value of priority focal species; number of endemic and sub endemic floristic species; and (total) floristic richness.

Number of focal animal species (B1). This was computed based on the habitat maps of 55 focal species. It corresponds to the total number of focal species whose potential habitat (with medium and high eligibility) falls into the analysed grid cell. The potential habitat was mapped during the above-mentioned European LIFE project through models based on the algorithm implemented in the Maxent software (Phillips et al. 2006).

Conservation value of priority focal species (B2). This was estimated by adding the conservation value of all the priority species whose habitat falls into a given location. These values were estimated according to specific importance, the ecological/functional role and the extinction risk, using the results from the mentioned LIFE project.

Number of endemic and sub endemic floristic (B3) species and Floristic richness (B4). This was derived from an existing map (1-km cell size) that shows the number of endemic and sub-endemic species found in the study area. Here only cells that are included in the selected agricultural areas for at least 20% were considered. The average number of endemic and sub-endemic species and the total number of species per study area was calculated from the chosen cell.

We added a buffer of 200 m to the extracted agricultural areas to account for 'boundary effects' linked to the production of ES that may occur outside of an agricultural area, but could actually be attributed to the area itself. An overview of the representative ES selected for comparing the study areas, classified according to the CICES, and their respective indicators is presented in Table 22.2.

Table 22.2 Overview of the selected indicators and their classification according to the CICES system

Section	ES (CICES V4.3)	Indicator	Unit
Provisioning services	Cultivated crops	**P1**. Annual overall market value of foodstuff	€/ha
	Fibers and other materials from plants, algae and animals for direct use or processing	**P2**. Annual total market value of the forage product	€/ha
Cultural services	Experiential use of plants, animals and land-/seascapes in different environmental settings	**C1**. Density of geotagged photographs	Number of photos/Km^2
	Experiential use of plants, animals and land-/seascapes in different environmental settings	**C2**. Density of elements of cultural/aesthetic interest	%
	Physical use of land-/seascapes in different environmental settings	**C3**. Density of cycling paths, hiking trails, horse trails, and mountain biking trails	km/Km^2
Biodiversity	–	**B1**. Number of focal animal species	Animal species/Km^2
	–	**B2**. Conservation value of priority focal species	Animal sp. value/Km^2
	–	**B3**. Number of endemic and sub-endemic floristic species	End. plant species/Km^2
	–	**B4**. (Total) floristic richness	Total plant species/Km^2

22.2.3 Analysis of Trade-Offs and Synergies and Identification of Hotspots

To enable the analysis of trade-offs and synergies, as a first step, the values of the selected indicators values were normalized between 0 and 1, assigning 1 to the maximum value according to the formula:

$$n_{\text{score}} = \frac{\text{score}}{\text{highestscore}} \quad (22.1)$$

where n_{score} is the normalized value; $score$ is the original value and $highestscore$ is the maximum value found for that indicator in the seven study areas considered.

The normalized indicators values were thus used to jointly display the ES provision of the seven study areas. This was done using flower diagrams (or "spider plots") in which the length of each axis is proportional to the maximum provision of the service in question (see Foley 2005). Furthermore, to gain a general understanding of the potential trade-offs and synergies among ES and biodiversity, beyond the specific study areas, an explorative correlations analysis was carried out considering the nine normalized indicators. The correlation analysis was performed using R software; specifically, the 'car' package (Fox and Weisberg 2011).

In order to simplify the reading and interpretation of the analysis of trade-offs and synergies, a two-step aggregation of the normalized indicators was undertaken. The first step consisted of calculating the arithmetic mean to obtain a single indicator for perception-related socio-cultural services (i.e. C1 and C2), a single aggregated indicator for habitat support for fauna (i.e. B1 and B2), and finally a single indicator for habitat support for flora (i.e. B3 and B4). The second-step consisted of aggregating the indicators into the two main categories of the CICES system, i.e. provisioning and cultural services, and biodiversity. As far as provisioning category was concerned, an additional aggregation step was performed in which, by way of example, the forage production was overlooked (given its values expressed in Euros per hectare were relatively much lower than the values obtained for food production). Flower diagrams were prepared, and correlation analysis performed with respect to the aggregated indicators.

As a final step, to identify spatial patterns of 'hotspots', i.e. areas characterized by high levels of provision of multiple services and biodiversity, the maps of the ES were superimposed. The rationale behind this is that identifying hotspots can offer a reference for scientifically defining boundaries for specific agro-environmental and conservation measures (e.g. Zhang et al. 2015). More specifically, the individual ES maps were first normalized and later summed through GIS Map Algebra operations. While an extensive literature has developed on the identification of 'hotspots', here, the aim was to provide a first overview of the spatial distribution of the potential trade-offs and synergies among ES and biodiversity as background for further analysis.

22.3 Results

The values of the selected biodiversity and ES indicators are shown in Table 22.3. By way of example, the spatial distribution of the provision two ES in the Baldo study areas is presented in Fig. 22.2.

22.3.1 Results for Individual ES and Biodiversity Indicators

The diagrams in Fig. 22.3 allow making a comparison of the study areas based on the selected nine ES and biodiversity indicators. The study sites are here ordered based the total value of the provisioning services (i.e. P1 + P2), here assumed to be a good

Table 22.3 Value of the selected ES and biodiversity indicator for the study areas expressed in their respective units

Indicator	Unit	Val di Non	Baldo	Vallarsa	Ledro	Piereni	Fiemme	Tesino
P1	€/ha	10,789.5	6,103.3	3,698.5	3,052.3	2,722.4	2,354.7	1,931.0
P2	€/ha	0.8	7.8	8.5	5.2	12.9	20.3	27.0
C1	Number of photos/Km^2	6.65	8.80	3.26	9.12	18.06	20.28	3.56
C2	%	1.10	3.90	8.80	20.70	3.50	11.30	0.40
C3	km/Km^2	0.89	0.19	0.42	1.13	1.83	0.38	0.13
B1	Animal species/Km^2	1.11	1.25	1.31	1.60	2.29	1.05	1.09
B2	Animal sp. value/Km^2	11.21	10.84	7.19	11.21	41.40	15.53	15.47
B3	Endemic plant species/Km^2	0.32	2.03	11.00	3.90	5.00	0.81	7.00
B4	Total plant species/Km^2	77.50	224.10	265.50	161.00	195.10	168.00	110.90

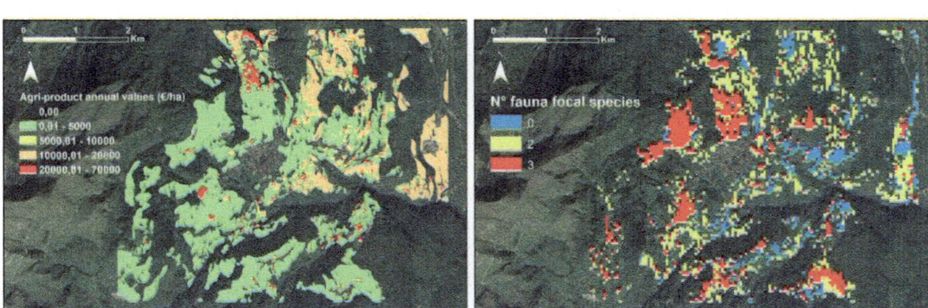

Fig. 22.2 Illustrative maps of ecosystem services associated with the Baldo study area

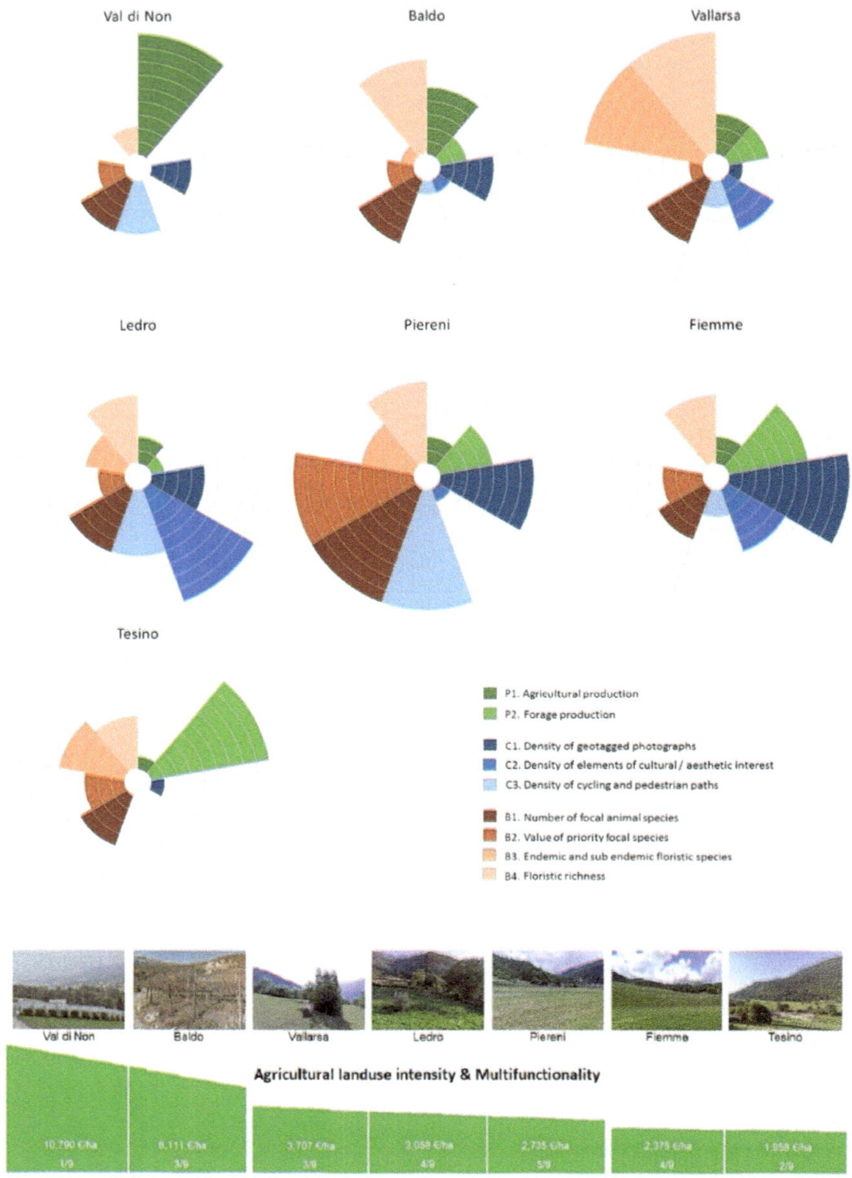

Fig. 22.3 Flower diagrams representing the normalized value of the nine selected ES and biodiversity indicator for the seven study areas. The study areas are displayed based on the actual values in €/ha of the indicators of the provisioning services (P1 + P2), and divided into three groups of agricultural land-use intensity while also specifying the degree of "multifunctionality", i.e. number of ES and biodiversity indicators exceeding the threshold value of 0.5 (see bottom)

proxy of agricultural land-use intensity. The values range from €/ha 10,790 for Val di Non to €/ha 1,958 for Tesino study area, against a national average value of €/ha 2,414 according to the Farm Accountancy Data Network (FADN).

Accordingly, the flower diagrams illustrate how a broad range of ES characterizes some agricultural areas while others have peak values only for specific ES and/or biodiversity indicators. For example, the Val di Non (€/ha 10,790) and Tesino (€/ha 1,958) study areas appear to be the most markedly 'mono-functional' ones, i.e. providing only two out of the ten selected ES (considering a threshold value of 0.5 to determine whether an area provides a given ES). In Val di Non, the maximization of agricultural productivity seems to take place at the expense of all other services, while the Tesino study site has the highest average production of forage. The Vallarsa (€/ha 3,707) study site has the highest values of biodiversity indicators, represented by the flora and the number of focal species. The Baldo (€/ha 6,111), Ledro (€/ha 3,058), and Fiemme (€/ha 2,375) sites have similar characteristics in terms of providing multiple services. All characterized by three high-value indicators belonging to two different categories, i.e. provisioning and cultural for Fiemme, and cultural services and biodiversity for Baldo and Ledro. The Piereni area shows some synergy between cultural services and biodiversity indicators against low values for the provisioning services. Overall, considering a threshold value of 0.5, the study areas can be arguably divided into three groups based on their agricultural intensity, i.e. high (Val di Non and Baldo); medium (Vallarsa, Ledro, and Piereni) and low (Fiemme and Tesino), specifying their degree of multifunctionality (as shown in Fig. 22.3).

Beyond the specific study sites, more general consideration can be made based on the results of the correlation analyses among the selected nine indicators (see Fig. 22.4). For example, it is possible to argue that the most significant positive correlations are the ones between B1-Number of focal animal species and C3-Density of cycling and pedestrian paths (0.87), between B1-Number of focal animal species and B2-Value of priority focal species (0.82); and finally between B2-Value of priority focal species and C3-Density of cycling and pedestrian paths (0.74). Similarly, it is possible to state that the highest negative correlation is registered between P1-Agrucultural production and P2-Forage production (−0.71), and between B3 Endemic and sub-endemic floristic species and C1 Density of geotagged photographs (−0.51).

22.3.2 *Results for Aggregated ES and Biodiversity Indicators*

The results based on the aggregated indicators illustrated some potential biases related to the process, originally aimed at simplifying the reading of the analysis of trade-offs and synergies. Figure 22.5 is an example of how the two-step aggregation of the ES and biodiversity indicators can actually bias the interpretation of the results. More specifically, two illustrative study sites (i.e. Val di Non and Piereni) are

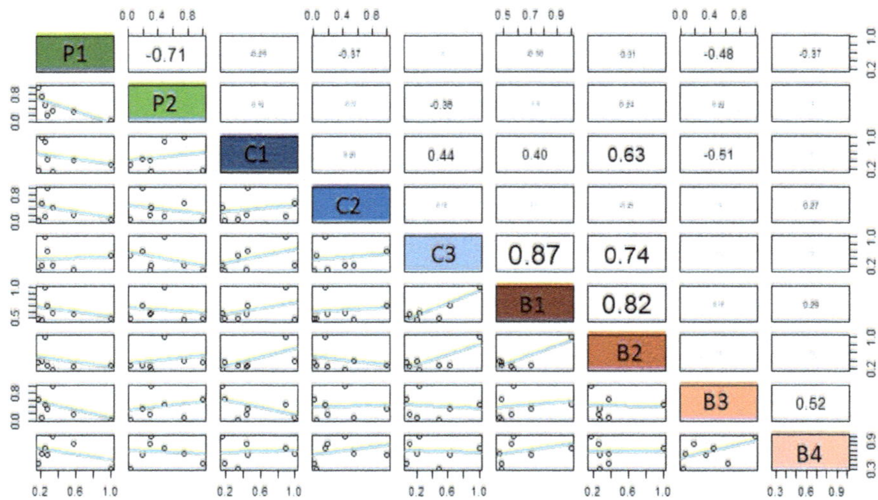

Fig. 22.4 Results of the correlation analysis among the selected indicators represented using a scatterplot matrix. The matrix contains the correlation values between the indicators expressed in the range ±1 (upper diagonal) and a graphical representation of the data (in the lower diagonal)

compared based on the aggregated indicators. With increasing level of aggregation of the indicators, some trade-offs and synergies may actually go undetected, highlighting the importance of adopting the required level of disaggregation in order to make well-informed decisions.

Another example of how the applied aggregation criteria can significantly change the results is shown Fig. 22.6. In this case, the comparison is between an aggregation that considers both food and forage production (P1 and P2) and one that considers only food production (P1) overlooking the forage production (P2) whose maximum monetary values are relatively much lower (i.e. €/ha10,790 against €/ha 27). The differences are particularly evident in the case of the Tesino study areas with respect to the provisioning services. Moreover, the different aggregation criteria result in different correlation values, which can lead to important trade-offs to be neglected (e.g. between provisioning and cultural services) or slightly underestimated (e.g. between provisioning services and biodiversity).

22.3.3 Identifications of Hotspots

The spatial distributions of hotspots, which indicate the presence of multiple ES provided by the agricultural areas, are shown in Fig. 22.7. The maps highlight that in some of the study areas the provision of ES is evenly distributed within the entire agricultural area. This is the case for example in the Fiemme and Baldo areas. In

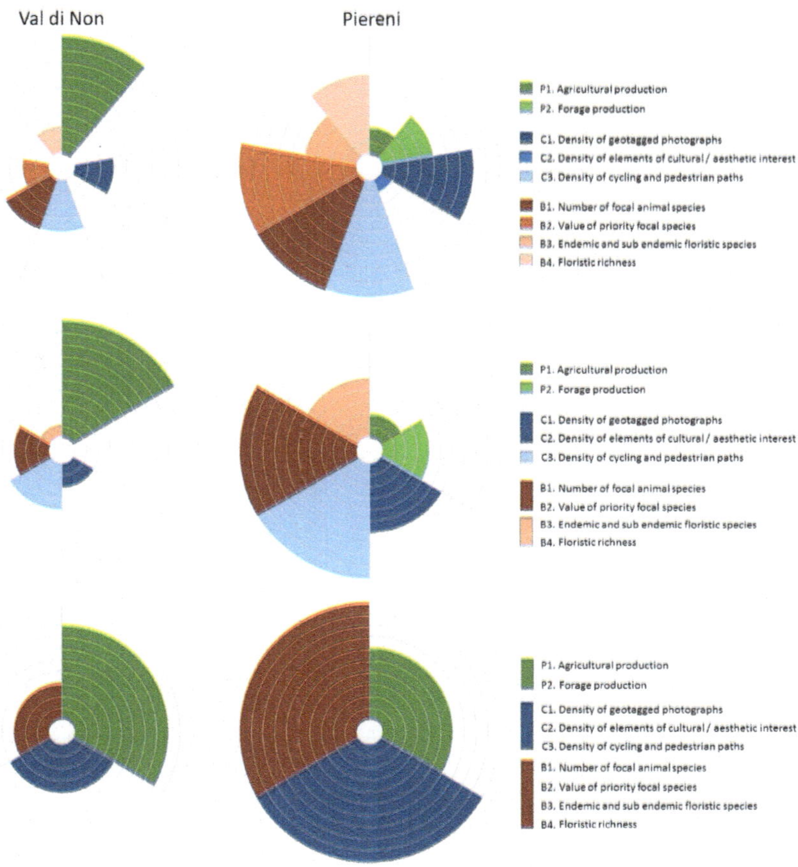

Fig. 22.5 Comparison of two study areas with respect to individual and aggregated indicators

other cases, the supply is more scattered, with a sharp contrast between the hotspots and the remaining agricultural areas (see, e.g. Tesino and Vallarsa).

22.4 Potential Applications of the Results to Support Planning and Policy-Making

Increasingly, the ES concept is being applied to support different policy and decision-making processes, by virtue of its holistic framing of the interactions between natural and human systems, and its ability to reveal synergies and conflicts between environmental and socio-economic objectives. ES provide a comprehensive framework for the analysis of trade-offs, addressing the compromises between competing land uses and assisting to facilitate planning and development decisions (Fürst et al. 2017).

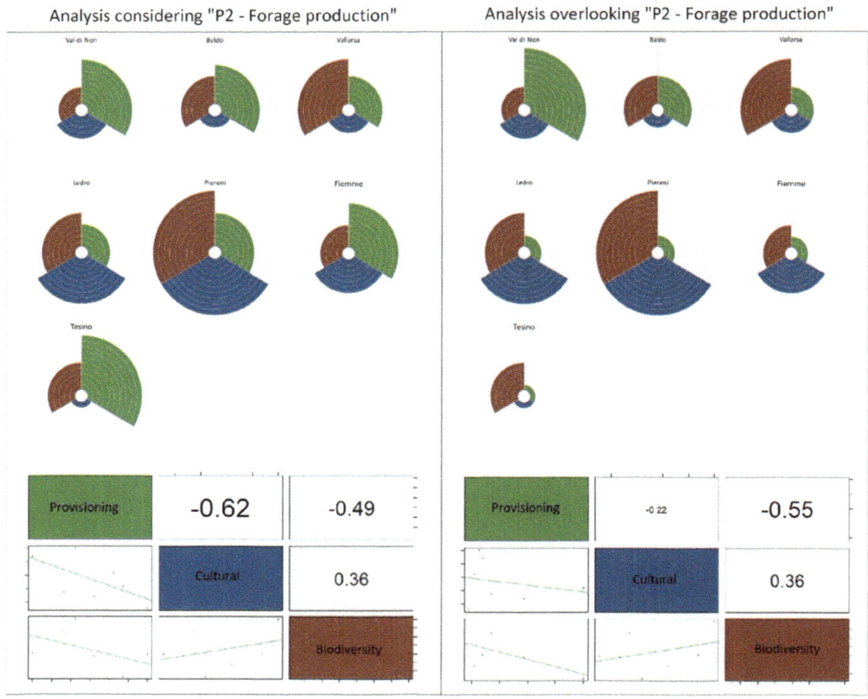

Fig. 22.6 An example of how different aggregation criteria (including both F1 and F2 indictors or only F1) can affect the different results, leading to support different conclusions (e.g. overlooking negative correlation between regulating/maintenance and cultural services)

Agriculture and forestry, spatial planning, protected area planning are some of the key sectors with a high potential for the application of ES, for example, to increase the synergies of recreation and carbon sequestration with timber production in forests, or pollination and biological control in agricultural environments (Simon et al. 2010; Buri et al. 2014). These sectors are inextricably linked to the supply of ES while depending on ES (e.g. pollination, pest and disease control, maintaining of soil fertility), and at the same time, having direct impacts on ecosystem condition and the provision of other services (e.g. maintaining habitats, chemical condition of freshwaters, global climate regulation, etc.). Furthermore, the level of supply and impacts of such ES depend directly on the applied management practices. Therefore, ES studies can be used to address the trade-offs within and between sectors, to address policy objectives and measures needed to improve the provision of ES and related payment schemes.

Considering the analysis of ES in agricultural areas presented in this chapter, in particular, it is possible to foresee a number of concrete applications to support planning and policy-making in the study region, as discussed in the next sub-sections.

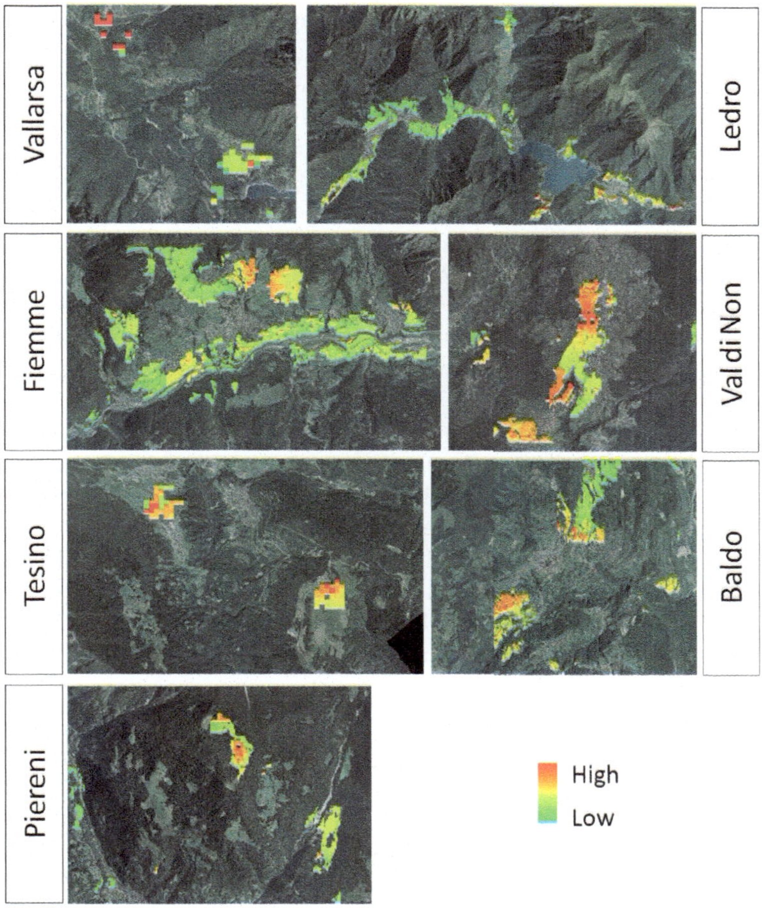

Fig. 22.7 Spatial distribution of hotspots, i.e. areas characterized by high level of ES provision and biodiversity, in the seven agricultural study areas

22.4.1 Applications in Rural Development Planning

Rural Development Programs (RDPs) represent an important tool for implementing the European policy on rural development. Funded by the European Agricultural Fund for Rural Development (EAFRD) in the framework of the Common Agricultural Policy, the RDPs are designed in cooperation between the European Commission and the Member States. They take into account the strategic guidelines for rural development policy adopted by the Council and the priorities laid down by national strategy plans as well as the needs of the local territorial context. For the 2014–2020 programming period, in particular, the three main objectives pursued by the EAFRD include (i) fostering the competitiveness of agriculture, (ii) ensuring the sustainable management of natural resources, and climate action, and (iii) achieving a balanced territorial development of rural economies and communities including the creation and maintenance of employment.

In the case of Trentino, our study region, the European Commission formally adopted the RDP in August 2015, outlining Trentino's priorities for using the € 297.5 million of public money available for the period 2014–2020. According to the proponents, the Trentino's RDP emphases actions related to restoring, preserving and enhancing ecosystems, improving the competitiveness of the farm and forestry sectors and promoting social inclusion and economic development in rural areas.

Given the case of the Trentino's RDP, and taking into account that we are approaching the end of the programming period 2014–2021, a key question to ask is "to what extent the RDP been successful in promoting a shifts towards sustainability, i.e. towards patterns of development that meet human needs while protecting life supporting ecosystems?". This is indeed a difficult question to answer, which implies looking at the actual outcomes on the ground, and not simply checking the fulfillment of the list of indicators. On the other hand, the proposed ES approach may as well be a good working alternative. By employing an adequate set of indicators of ES and biodiversity, that capture diverse stakeholders' perspectives, the approach allows assessing the cost-effectiveness of the measures funded by the RDP. This ex-post assessment can offer valuable insight for the up-coming programming period 2021–2027, in which greater emphasis will be placed on environment and climate, supporting the transition towards a more sustainable agricultural sector and the development of vibrant rural areas.

22.4.2 Implications for Conservation Planning

Even if biodiversity conservation and the optimization of the delivery of ES usually represent alternative targets or strategies in landscape planning (which is often carried out with an exclusive focus on one of the two; e.g. Mace et al. 2012), their integration into spatial planning is particularly desirable. This could be especially true for agricultural areas, where it could lead to win-win solutions (Assandri et al. 2018). Wild species often play an essential role in provisioning, regulatory and cultural ES, and the strict relationships linking biodiversity and ecosystem functions are undoubtedly fundamental. Such a key role of biodiversity in ecosystem processes, coupled with the often-shared sensitivity to human impacts and environmental modifications, make particularly important and sound the integration of the two targets into planning targeted at conservation and ecosystem management. The complex relationship between ES and biodiversity still has to shape most processes of planning and policy development, despite the importance of the subject had been highlighted since many years (Oliver et al. 2015), and thus there is an urgent need to foster interdisciplinary approaches to strategies and plans of ecosystem management (Reyers et al. 2012).

Protected areas and related management plans, ecological networks and other tools aiming at the conservation of natural resources should therefore consider what factors are likely to affect biodiversity and/or ES, and the reciprocal interactions and feedback between biodiversity and ecosystem functions. This is becoming increasingly important in a changing climate: addressing simultaneously the requirements of biodiversity and ES, and the trade-offs between the two and between different ES, is essential to develop effective adaptation plans (for protected areas as well as in general). A careful consideration of the scales involved in assessments and strategy development is also very important for planning, as the relationships occurring between biodiversity, on the one hand, and ES or functions, on the other one, are often scale-dependent. Potential synergies between biodiversity conservation and ES delivery may be found over different spatial scales and environmental contexts (Brambilla et al. 2018).

Landscape traits are crucial in driving patterns of both biodiversity and ES. They are also a key element in spatial planning, and this makes landscape planning essential for biodiversity conservation and ES delivery (Ronchi 2018). In most modern farmed landscapes, the maximization of provision services had been pursued at the expense of other services and biodiversity (Reyers et al. 2012). A particularly concerning factor for biodiversity and ES in farmed landscapes is represented by landscape simplification and homogenization (Barbaro et al. 2017). Agricultural intensification and landscape abandonment have resulted in a loss of mosaic landscapes, particularly important for several species and biocenoses (Brambilla et al. 2017) and also ES (and disservices; e.g. Persichillo et al. 2017), including cultural ones (Assandri et al. 2018).

In Trentino, grassland birds have dramatically suffered the conversion of hay meadows into intensive crops and the management intensification of the remaining ones (Assandri et al. 2019). Consequently, the correct management of landscape traits and patterns of particular relevance is a key issue for planning in agricultural areas. The conservation and restoration of marginal elements in simplified grassland landscapes in Trentino, as an example, is of basic importance for farmland birds (Ceresa et al. 2012), and at the same time may promote the aesthetic and identity values of farmed landscapes (Assandri et al. 2018). The integration of biodiversity and ES expertise into a common strategy shared with stakeholders is critical for an effective (re)design of farmed landscapes, which should consider current values and implement strategies to achieve the established ecosystem goals starting from the existing conditions (Landis 2017). In conclusion, the planning process, especially in networks of protected areas, should be followed by a careful selection of indicators to monitor biodiversity and essential ES, considering the inter-relations between them as well as the relevant scale(s) at which the monitoring efforts should take place (Feld et al. 2009).

22.5 Conclusions

In this study, we modelled the trade-offs and synergies among ES and biodiversity indicators associated with different mountainous agricultural landscapes.

Through a quantitative and spatially explicit assessment of the trade-offs and synergies among ES and biodiversity, the proposed approach can support key planning and management decisions that aim to promote multifunctionality of mountain agriculture.

An important conclusion that emerged from the application of the proposed approach is the fact that trade-offs and synergies among ES and biodiversity can be hidden or highlighted depending on how specific methodological steps are performed, such the normalization and the aggregation of the selected ES indicators. This is well illustrated by the flower diagrams based on different aggregation of indicators, which present significant differences in the results. For example, the polarization towards one or two categories of ES or a more balanced provision of ES may not appear equally clear when considering individual services. Nevertheless, individual services cannot be used to draw generalized results about a whole category. The assessment of trade-offs and synergies between ES should consider multidimensional aspects of provision and, possibly, different spatial scales (or different level of aggregation) in the data.

In line with an emerging interest in the literature (Cash et al. 2003; Clark et al. 2016), the proposed assessment of trade-offs and synergies among ES and biodiversity associated with agricultural areas implicitly addresses issues related to the usefulness and usability of the generated results in an operational setting. To start with, we here propose the use of multiple ES and biodiversity indicators that capture diverse stakeholders' perspectives: a crucial step in disentangling the different contributions to ES by different typologies of agricultural land uses with respect to different types of beneficiaries (Martín-López et al. 2012). Particularly, the different ES in the study sites are potentially used by people from different places, which is a relevant issue to consider when planning for multi-functional agriculture (e.g. in the urban-rural interface).

The assessment of the provisioning ES indicators, for example, relied on data collected routinely by the local administration by involving the farmers. In the case of cultural services, the use of geotagged photographs has been criticized for excluding a part of the population (i.e. elder population). In fact, in mountain areas like Trento, it might be the case that a reduced amount of the population uses such tools; so, it is possible to differentiate different types of users (e.g. residents from tourists). These and others, are key elements to mainstreaming ES in decision-making (e.g. Geneletti 2013), and to use this kind of analysis as a monitoring tool to assess the effectiveness of policies supporting multifunctionality. To this purpose, efforts to summarize and better communicate the results (e.g. flower diagrams and correlation analysis) are essential, particularly in participatory stakeholders setting. These are all elements that collectively contribute to enhancing the potential usefulness and usability of the results, by ensuring their scientific credibility and their saliency (e.g. Adem Esmail et al. 2017; Adem Esmail and Geneletti 2017).

References

Adem Esmail B, Geneletti D (2017) Design and impact assessment of watershed investments: An approach based on ecosystem services and boundary work. Environ Impact Assess Rev 62:1–13. https://doi.org/10.1016/j.eiar.2016.08.001

Adem Esmail B, Geneletti D, Albert C (2017) Boundary work for implementing adaptive management: a water sector application. Sci Total Environ 593–594:274–285. https://doi.org/10.1016/j.scitotenv.2017.03.121

Assandri G, Bogliani G, Pedrini P, Brambilla M (2018) Beautiful agricultural landscapes promote cultural ecosystem services and biodiversity conservation. Agric Ecosyst Environ 256:200–210. https://doi.org/10.1016/j.agee.2018.01.012

Assandri G, Bogliani G, Pedrini P, Brambilla M (2019) Toward the next common agricultural policy reform: determinants of avian communities in hay meadows reveal current policy's inadequacy for biodiversity conservation in grassland ecosystems. J Appl Ecol 56:604–617. https://doi.org/10.1111/1365-2664.13332

Autonomous Province of Trento (2013) ISPAT—Annuario on-line http://www.statweb.provincia.tn.it/annuario (accessed 12.16.19)

Barbaro L, Rusch A, Muiruri EW et al (2017) Avian pest control in vineyards is driven by interactions between bird functional diversity and landscape heterogeneity. J Appl Ecol 54:500–508. https://doi.org/10.1111/1365-2664.12740

Bennett EM, Cramer W, Begossi A et al (2015) Linking biodiversity, ecosystem services, and human well-being: three challenges for designing research for sustainability. Curr Opin Environ Sustain 14:76–85. https://doi.org/10.1016/j.cosust.2015.03.007

Brambilla M, Gustin M, Vitulano S et al (2017) Sixty years of habitat decline: impact of land-cover changes in northern Italy on the decreasing ortolan bunting *Emberiza hortulana*. Reg Environ Chang 17:323–333. https://doi.org/10.1007/s10113-016-1019-y

Brambilla M, Resano-Mayor J, Scridel D et al (2018) Past and future impact of climate change on foraging habitat suitability in a high-alpine bird species: management options to buffer against global warming effects. Biol Conserv 221:209–218. https://doi.org/10.1016/j.biocon.2018.03.008

Buri P, Humbert J-Y, Arlettaz R (2014) Promoting pollinating insects in intensive agricultural matrices: field-scale experimental manipulation of hay-meadow mowing regimes and its effects on bees. PLoS ONE 9:e85635. https://doi.org/10.1371/journal.pone.0085635

Cash DW, Clark WC, Alcock F et al (2003) Knowledge systems for sustainable development. Proc Natl Acad Sci 100:8086–8091. https://doi.org/10.1073/pnas.1231332100

Ceresa F, Bogliani G, Pedrini P, Brambilla M (2012) The importance of key marginal habitat features for birds in farmland: an assessment of habitat preferences of Red-backed Shrikes *Lanius collurio* in the Italian Alps. Bird Study 59:327–334. https://doi.org/10.1080/00063657.2012.676623

Clark WC, Tomich TP, van Noordwijk M et al (2016) Boundary work for sustainable development: natural resource management at the Consultative Group on International Agricultural Research (CGIAR). Proc Natl Acad Sci 113:4615–4622. https://doi.org/10.1073/pnas.0900231108

Cortinovis C, Geneletti D (2018) Mapping and assessing ecosystem services to support urban planning: a case study on brownfield regeneration in Trento. Italy. One Ecosyst 3:e25477. https://doi.org/10.3897/oneeco.3.e25477

Daw TM, Coulthard S, Cheung WWL et al (2015) Evaluating taboo trade-offs in ecosystems services and human well-being. Proc Natl Acad Sci 201414900. https://doi.org/10.1073/pnas.1414900112

EEA (2013) The common international classification of ecosystem services—Version 4.3 (CICES)

Feld CK, Martins da Silva P, Paulo Sousa J et al (2009) Indicators of biodiversity and ecosystem services: a synthesis across ecosystems and spatial scales. Oikos 118:1862–1871. https://doi.org/10.1111/j.1600-0706.2009.17860.x

Ferrari M, Geneletti D (2014) Mapping and assessing multiple ecosystem services in an alpine region: a study in Trentino, Italy. Ann Bot 4:65–71. https://doi.org/10.4462/annbotrm-11729

Ferrari M, Geneletti D, Cayuela L et al (2016) Analysis of bundles and drivers of change of multiple ecosystem services in an Alpine region. J Environ Assess Policy Manag 18:1650026. https://doi.org/10.1142/s1464333216500265

Fichino BS, Pivello VR, Santos RF (2017) Trade-offs among ecosystem services under different pinion harvesting intensities in Brazilian Araucaria forests. Int J Biodivers Sci Ecosyst Serv Manag 13:139–149. https://doi.org/10.1080/21513732.2016.1275811

Foley JA (2005) Global consequences of land use. Science 309:570–574. https://doi.org/10.1126/science.1111772

Fox J, Weisberg S (2011) An R companion to applied regression, second

Fürst C, Luque S, Geneletti D (2017) Nexus thinking—how ecosystem services concepts and practice can contribute balancing integrative resource management through facilitating cross-scale and cross-sectoral planning. Int J Biodivers Sci Manage 13(1):1–3. https://doi.org/10.1080/21513732.2017.1409310

Geneletti D (2011) Reasons and options for integrating ecosystem services in strategic environmental assessment of spatial planning. Int J Biodivers Sci Ecosyst Serv Manag 7:143–149. https://doi.org/10.1080/21513732.2011.617711

Geneletti D (2012) Environmental assessment of spatial plan policies through land use scenarios. A study in a fast-developing town in rural Mozambique. Environ Impact Assess Rev 32:1–10. https://doi.org/10.1016/j.eiar.2011.01.015

Geneletti D (2013) Assessing the impact of alternative land-use zoning policies on future ecosystem services. Environ Impact Assess Rev 40:25–35. https://doi.org/10.1016/j.eiar.2012.12.003

Geneletti D, Adem Esmail B, Cortinovis C (2018) Identifying representative case studies for ecosystem services mapping and assessment across Europe. One Ecosyst 3:e25382. https://doi.org/10.3897/oneeco.3.e25382

Holland RA, Eigenbrod F, Armsworth PR et al (2011) The influence of temporal variation on relationships between ecosystem services. Biodivers Conserv 20:3285–3294. https://doi.org/10.1007/s10531-011-0113-1

Kates RW (2011) What kind of a science is sustainability science? Proc Natl Acad Sci 108:19449–19450. https://doi.org/10.1073/pnas.1116097108

Kates BRW, Parris TM, Leiserowitz AA (2005) What is sustainable development? Goals, indicators, and practice. Environment 47:8–21

Landis DA (2017) Designing agricultural landscapes for biodiversity-based ecosystem services. Basic Appl Ecol 18:1–12. https://doi.org/10.1016/j.baae.2016.07.005

Locatelli B, Lavorel S, Sloan S et al (2017) Characteristic trajectories of ecosystem services in mountains. Front Ecol Environ 15:150–159. https://doi.org/10.1002/fee.1470

Mace GM, Norris K, Fitter AH (2012) Biodiversity and ecosystem services: a multilayered relationship. Trends Ecol Evol 27:19–26. https://doi.org/10.1016/j.tree.2011.08.006

Martín-López B, Iniesta-Arandia I, García-Llorente M et al (2012) Uncovering ecosystem service bundles through social preferences. PLoS ONE 7:e38970. https://doi.org/10.1371/journal.pone.0038970

Mckenzie E, Posner S, Tillmann P et al (2014) Understanding the use of ecosystem service knowledge in decision making: lessons from international experiences of spatial planning. Environ Plan C Gov Policy 32:320–340. https://doi.org/10.1068/c12292j

Oliver TH, Heard MS, Isaac NJB et al (2015) Biodiversity and resilience of ecosystem functions. Trends Ecol Evol 30:673–684. https://doi.org/10.1016/j.tree.2015.08.009

Orsi F, Geneletti D (2013) Using geotagged photographs and GIS analysis to estimate visitor flows in natural areas. J Nat Conserv 21:359–368. https://doi.org/10.1016/j.jnc.2013.03.001

Persichillo MG, Bordoni M, Meisina C (2017) The role of land use changes in the distribution of shallow landslides. Sci Total Environ 574:924–937. https://doi.org/10.1016/j.scitotenv.2016.09.125

Phillips SJ, Anderson RP, Schapire RE (2006) Maximum entropy modeling of species geographic distributions. Ecol Modell 190:231–259. https://doi.org/10.1016/j.ecolmodel.2005.03.026

Posner SM, McKenzie E, Ricketts TH (2016) Policy impacts of ecosystem services knowledge. Proc Natl Acad Sci 113:201502452. https://doi.org/10.1073/pnas.1502452113

Raudsepp-Hearne C, Peterson GD, Bennett EM (2010) Ecosystem service bundles for analyzing tradeoffs in diverse landscapes. Proc Natl Acad Sci 107:5242–5247. https://doi.org/10.1073/pnas.0907284107

Reyers B, Polasky S, Tallis H et al (2012) Finding common ground for biodiversity and ecosystem services. Bioscience 62:503–507. https://doi.org/10.1525/bio.2012.62.5.12

Ronchi S (2018) Ecosystem services for spatial planning. Springer International Publishing, Cham

Ruckelshaus M, McKenzie E, Tallis H et al (2015) Notes from the field: lessons learned from using ecosystem service approaches to inform real-world decisions. Ecol Econ 115:11–21. https://doi.org/10.1016/j.ecolecon.2013.07.009

Simon S, Bouvier J-C, Debras J-F, Sauphanor B (2010) Biodiversity and pest management in orchard systems. A review. Agron Sustain Dev 30:139–152. https://doi.org/10.1051/agro/2009013

Tenerelli P, Püffel C, Luque S (2017) Spatial assessment of aesthetic services in a complex mountain region: combining visual landscape properties with crowdsourced geographic information. Landscape Ecol 32:1097–1115. https://doi.org/10.1007/s10980-017-0498-7

von Haaren C, Albert C (2011) Integrating ecosystem services and environmental planning: limitations and synergies. Int J Biodivers Sci Ecosyst Serv Manag 7:150–167. https://doi.org/10.1080/21513732.2011.616534

Zhang L, Fu B, Lü Y, Zeng Y (2015) Balancing multiple ecosystem services in conservation priority setting. Landscape Ecol 30:535–546. https://doi.org/10.1007/s10980-014-0106-z

Chapter 23
Simulating the Effects of Agrochemicals and Other Risk-Bearing Management Measures on the Terrestrial Agrobiodiversity: The RISKMIN Approach

Matthias Trapp, Mark Deubert, Lucas Streib, Björn Scholz-Starke, Martina Roß-Nickoll and Andreas Toschki

Abstract The RISKMIN Model was a research project funded by the German Federal Office of Consumer Protection and Food Safety between 2012 and 2016 to support the development of a landscape-based mitigation approach for assessing different measures and their effects on agrobiodiversity. Conducted by an interdisciplinary cooperation between different research groups in Germany two representative landscapes were chosen as pilot regions. Based on a very high-resolution landscape analysis and a comprehensive survey of the terrestrial agrobiodiversity in these regions a simulation of different risk mitigation measures was realized by reference to potential scenarios. The resulting effects could be analysed, quantified and visualized with the help of geographic information systems (GIS). Main results are that the most effective measure on landscape level is the extensification of the

---

M. Trapp (✉) · M. Deubert · L. Streib
Institute for AgroEcology (IfA), RLP AgroScience, Breitenweg 71, 67435 Neustadt, Germany
e-mail: matthias.trapp@agroscience.rlp.de

M. Deubert
e-mail: mark.deubert@agroscience.rlp.de

L. Streib
e-mail: lucas.streib@agroscience.rlp.de

B. Scholz-Starke
Darwin statistics, Aachen, Germany
e-mail: scholz-starke@darwin-statistics.de

M. Roß-Nickoll
Institute for Environmental Research (Biology V), RWTH Aachen University, Worringerweg 1, 52074 Aachen, Germany
e-mail: ross@bio5.rwth-aachen.de

A. Toschki
Gaiac - Research Institute for Ecosystem Analysis and Assessment, Kackertstraße 10, 52072 Aachen, Germany
e-mail: Toschki@gaiac-eco.de

W. Mirschel et al. (eds.), *Landscape Modelling and Decision Support*,
Innovations in Landscape Research,
https://doi.org/10.1007/978-3-030-37421-1_23

land use, but on the other hand, that the combination of in crop and off crop measures does have the most effects for the landscape elements with the highest ecological values.

Keywords Landscape classification and analysis · Agrobiodiversity · Ecological values · Risk assessment · Mitigation measures · GIS

23.1 Introduction

Biodiversity is the basic resource maintaining and supporting ecosystem services and functions. Biodiversity is realized at different levels in the complexity of a given landscape. Therefore when dealing with biodiversity related aspects it is important to consider the whole landscape instead of only analysing single fields or elements. EFSA published different opinions and scientific papers in the context of recovery and recolonization of populations. In these opinions a clear relationship between landscape and risk is formulated. "There is, therefore, an important interplay between homogeneity of agricultural practices over spatial scales, and the potential for recovery from stresses. Landscape features therefore may need to be assessed when assessing the potential for external recovery. This is clearly problematic as it indicates that not the assessed product or species per se may be decisive for the recovery from impact, but the properties of the environment in which these products or species are having an effect. This is challenging from a regulatory perspective" (EFSA 2016).

It is repeatedly proven that species diversity and habitat quality is dramatically decreasing in Germany. The loss of habitats and intensity of agricultural land-use are one of the main accepted reasons for this. In recent years indeed a constant decline of the biodiversity of agricultural landscapes is observed, indicated by, e.g. birds and butterflies (Aichi targets adopted by the EU) (Hallmann et al. 2017; EEA 2013). Environmental stressors (like pesticides) act on the landscape level (Jahn et al. 2014). This adverse effect is relevant within national as well as EU legislation.

Important questions in landscape risk assessment are (1) whether there is a state or a known situation, to which the quality of biodiversity can be compared (reference state) (2) how to quantify and assess the impact of land-use scenarios and (3) how to effectively protect and recover biodiversity at landscape level.

Biodiversity depends on the combination of the environmental conditions of a landscape area and the amount and quality of landscape elements and structures (Tscharntke et al. 2005). The ecological value of a specific biotope can be deduced from "biotope valences" according to the environmental assessments within the German "impact mitigation regulation".

In the German multidisciplinary project RISKMIN, sponsored by the German Federal Office of Consumer Protection and Food Safety these approaches were investigated and tested for two pilot regions in Germany (Scholz-Starke et al. 2016). Based on a previous surveying publication (Trapp et al. 2018) this chapter explains the methods and results of the RISKMIN project in detail.

23.2 Materials and Methods

RISKMIN, the acronym for the present study, is an interdisciplinary cooperation between geographers, ecologists, ecotoxicologists and risk managers. At first a conceptual approach (Fig. 23.1) of the RISKMIN model was developed, in which six modules are pointed out. This modular approach offers the opportunity to hand over data and information between the work packages.

The module **Field** as practical in-field survey covers the important, regionally relevant types of biotopes of agricultural landscapes containing in crop like cropland and meadows and off-crop like hedges, fallow cropland and fallow meadows. Because the occurrence of species and species communities at a site is not solely dependent on one factor, but rather determined by a factor combination, an integrative approach to biodiversity assessment is necessary (Toschki 2008; Roß-Nickoll et al. 2004). The results of the module **FIELD** are not described in detail here.

The module **Meta** is based on a comprehensive literature research and covers the most important factors affecting the agrobiodiversity like farming practices, pesticide and fertiliser use.

In the module **Geodat** geographic information systems (GIS) are used because of their ability to detect, analyse and visualize geodata in general and landscapes with their elements in particular. Landscape metrics are used to quantify and describe landscape elements (patches) and finally to make landscapes comparable (Lang and Blaschke 2007). This module delivers a very high-resolution landscape classification following the approach of 'NatFlo', published by Tintrup gen Suntrup et al. (2014), Trapp et al. (2015) as input for the indicator based mapping of the structural diversity in the agricultural landscape as landscape elements and land cover.

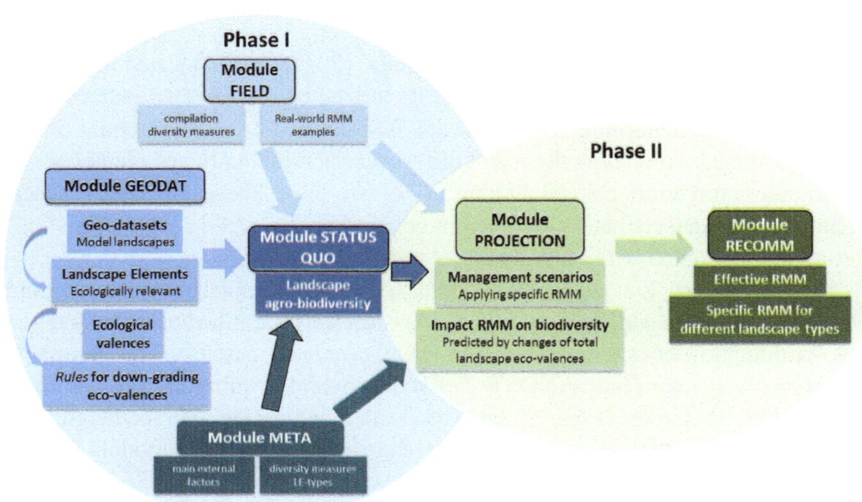

Fig. 23.1 Conceptual approach of the RISKMIN model

In the modules **Status Quo** as well as in the modules **Projection** and **Recomm** the linkage of the semi-automated analyses of landscape structures with measures of biodiversity, mobility of populations, and impacts of (pesticide) risk management measures is conducted.

An integrated landscape ecological value is calculated as the sum of the values of all single landscape elements (LE) and of in-crop land use by the module **Status Quo**, which is then corrected for intensive agricultural use for LE nearby the in-crop areas. This value is the basis for mitigation on landscape level. The module **Projection** then identifies risk mitigation measures (RMM) of different mechanisms, defines most promising RMM and implements rules for mechanisms of melioration on landscape level. The effects of different RMM and the most effective ones are then calculated and interpreted in the module **Recomm**. The modules **Field** and **Meta** deliver information and data whenever necessary and lacking for the main thread of the project.

23.2.1 Module Geodat: Landscape Classification and Modelling

The technical concept of the RISKMIN approach (described in detail in the following section) was developed based on two pilot landscapes according to the official landscape classification of the Federal Agency for Natural Conservation (BfN 2011). However, the concept is transferable to any kind of agricultural landscape. The pilot landscape HB 'Horbacher Börde' (2,200 ha) is a densely wooded, grassland-dominated agricultural landscape located in West Germany. It is comparable well equipped with structural landscape elements that host a typical and structure specific biodiversity.

The second pilot landscape VP 'Vorderpfalz' (20,000 ha) is located in the Upper Rhine Valley in South-West Germany characterized by intensive viticulture as well as cropland and orcharding. The landscape is structured by a high amount of small fields combined with a high degree of different structural landscape elements.

To assess the total ecological value of the two pilot landscapes their landscape elements were detected and converted in comprehensive geo-datasets consisting of geo-objects assigned with individual ecological values.

For that purpose, first landscape elements (LE) with an object height above ground equal or greater 1 m and vegetation characteristics were identified using remote sensing techniques. Geo-referenced objects were generated based on the Normalized Difference Vegetation Index (NDVI), derived from high-resolution (true) multispectral digital Orthophotos (Figs. 23.2 and 23.3), in combination with the height above ground (nDSM), derived from digital surface and digital terrain models based on photogrammetrical methods by using the true orthoimages (Fig. 23.3), using object based image analysis (OBIA) (Tintrup gen Suntrup et al. 2014). Therefore the NDVI as well as nDSM-values based on the same data source and without any time lag

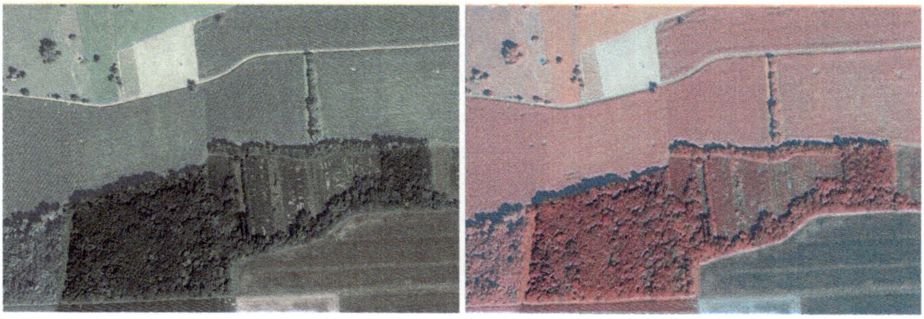

Fig. 23.2 RGB-orthoimages and false colour near infrared (NIR)

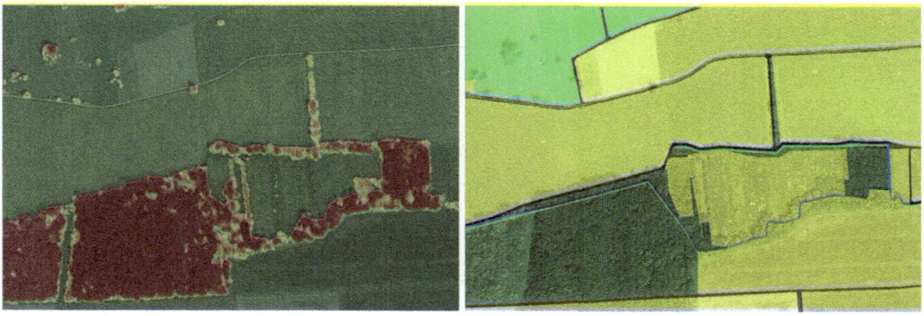

Fig. 23.3 Photogrammetrical based normalized digital surface model (nDSM) and ALKIS object classes

between. A segmentation process provided by the software eCognition (Trimble) to create vector objects with homogenous attributes (NDVI and nDSM) was used. In a further step it were classified context sensitive based on their object attributes in preliminary LE classes, e.g. Bushes, Forests, Trees. Based on shape parametres (height, width, length and area) the classification was further refined by linking additional information regarding the ecological status, which was determined in the modules META and FIELD and results, amongst others, from age, area extent and the assumed proportion of typical plants (Fig. 23.4).

Second LE were derived from the German Authoritative Real Estate Cadastre Information System (ALKIS 2015). Therefore, the areas of the ALKIS objects covering the model landscapes were reduced by the previously generated objects to avoid overlapping LE objects (Fig. 23.3). To link the necessary information regarding the possible value range of the ecological status to those objects the ALKIS actual usage attribute was used, e.g. Settlement, Agriculture, Viticulture, and classified the objects in, again, preliminary LE classes. In this context, the possible ecological status results from intensity of agricultural use and, in turn, the estimated applied amount of plant protection products.

The resulting main land cover classes are vineyards, cropland, grassland, copses, mixed orchards, hedges, trees, tree lines, woods and wood fragments.

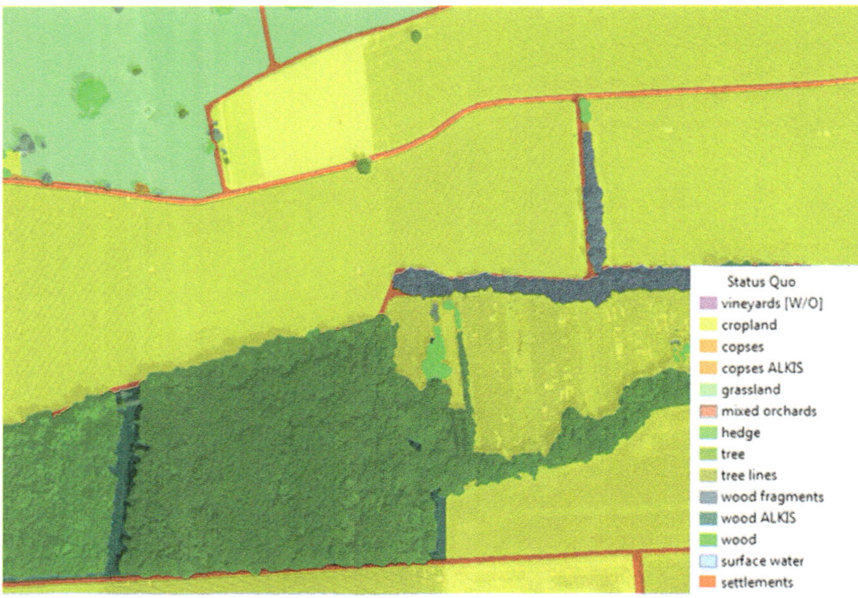

Fig. 23.4 Resulting 3D-landscape features and 2D land cover classes

The whole landscape classification process can be seen in the following figures, where one zoom into the landscape VP is shown to illustrate the basic input geodata and the resulting land cover classification in detail.

Based on this very high-resolution land cover classification combined with the ALKIS information at the moment of the creation time of the underlying data models the status quo of the land use/land cover classification of the pilot landscapes was captured and statistics were calculated.

In the following Fig. 23.5 the distribution of 3D-LE is shown for both pilot areas.

Hedges are the dominant 3D-LE type in the pilot region VP, whereas wood is the dominant 3D-LE type in the pilot region HB.

Next the comprehensive geo-objects per pilot landscape were further classified into 56 biodiversity-relevant LE types typical for agricultural landscapes. The definition of the LE types was deduced on the basis of the type list of the well-established numerical biotope value method of North Rhine-Westphalia (Adam et al. 1986; Ludwig and Meinig 1991; Biedermann et al. 2008). Here, each of the LE types a numerical ecological value in the range of 0–100 is assigned, which is basis for ecological compensation in case of a negative alteration of a LE, e.g. due to construction measures. The criteria for valuation are 'naturalness', 'substitutability', 'completeness' and 'particular hazards' of a LE.

The methodology of this approach is shown basically in Fig. 23.6.

Furthermore, for selected LE types spatial relationships in-between LE were included and adjusted the valuation thereby, e.g. the degree of isolation. It was not possible to classify the objects according to the biodiversity-relevant LE types solely

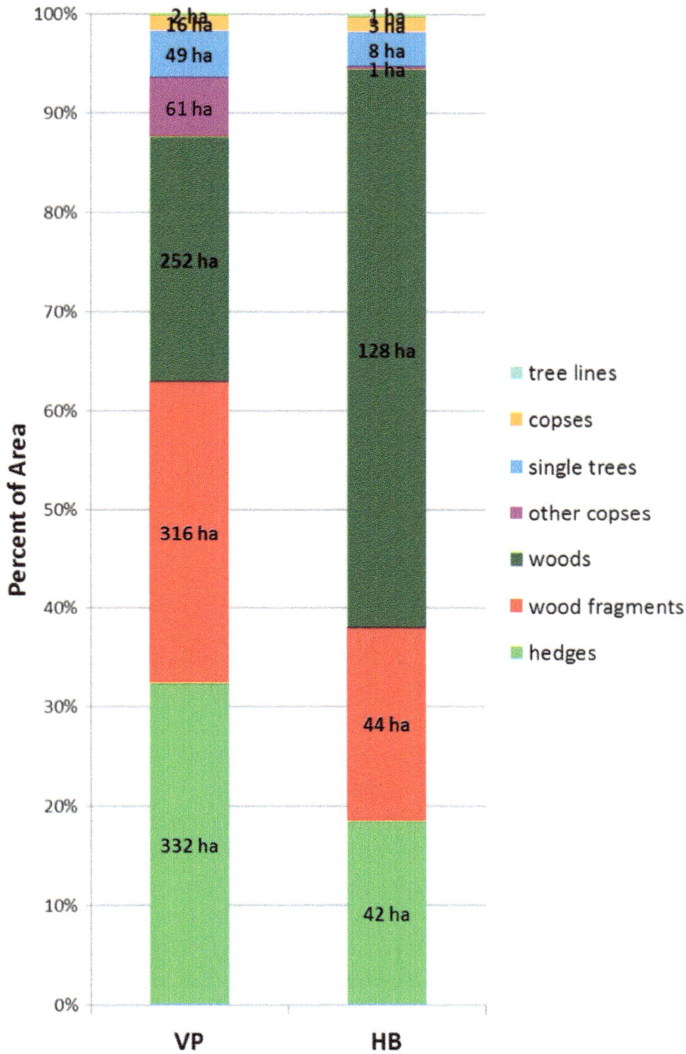

Fig. 23.5 Status quo percentage of area of the main LE categories in both pilot regions

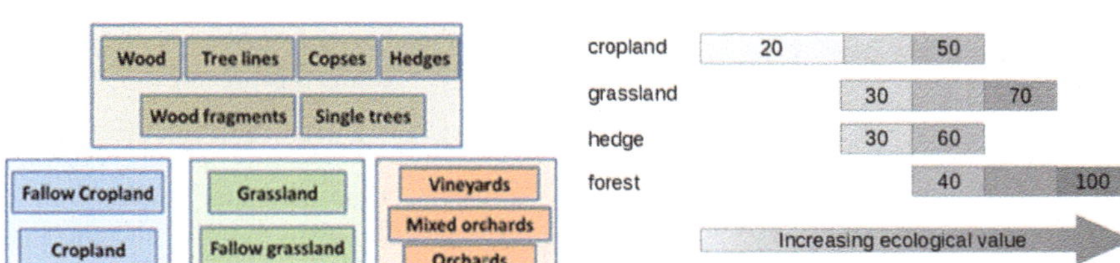

Fig. 23.6 Methodology of determining "eco-values" to the main LE categories in both pilot regions

using remote sensing techniques or ALKIS usage information, since these procedures cannot adequate deduce the criteria for valuation. Therefore, based on known or expert estimated statistical distributions of the 56 LE types per model landscapes each geo-object was stochastically further differentiated according to their preliminary classification. This approach resulted in a random distribution of LE in the considered model landscapes.

The differentiation of the main classes into the 56 LE types is based mostly on the intensity of land use and the usage of pesticides. It is characterized through the number: 1 means the most intensive type and higher numbers lower intensities. In the following Fig. 23.7 the distribution is shown for the pilot region VP.

This classification approach finally enabled the assignment of numerical ecological values to the geo-objects according to their specific classification within the biotope value method.

However, to include the negative effects of pesticide use on areas adjacent to intensively used agricultural fields LE were further systematically devaluated in buffer area of 5 m (Fig. 23.8) to integrate drift-related effects and overspray scenarios. The 5 m buffer is based on the buffer requirements in the context of pesticide use based on the approaches of risk mitigation measurements of the authorities (Enzian and Gutsche 2005).

The current ecological status of the model landscapes ('Status quo') was then calculated as the sum of ecological values of all LE within the model landscapes. To prevent random effects resulting from the spatial distribution of LE the stochastically assignment of LE types has been applied 100 times. The median ecological status value of the 100 replicates was then used as a reference to monitor the effectiveness of landscape-related risk mitigation measures (RMM).

Exemplary RMM were simulated on the basis of current measures in scientific discussion as well as existing regulations. In the module PROJECTION appropriate RMM landscape scenarios—in terms of varying distributions of the 56 LE types—were developed, the specific ecological status was calculated and finally compared to the Status quo. Again, to prevent random effects the stochastically assignment of LE types has been applied 100 times. The landscape specific ecological effectiveness of a RMM corresponds to the respective percentage deviation of the median ecological status value.

23.3 Results and Discussion

In the module **geodat** a holistic landscape classification based on very high-resolution geodata was implemented. In connection with the other modules some potential scenarios of landscape-related risk-bearing management measures were simulated and compared afterwards within the two study areas.

The 56 LE types were differentiated into 21 two- and 35 three-dimensional types depending on their height (Fig. 23.7). The 'landscape-wide eco-value' is then calculated as the status quo sum of all eco-values of the LE in the respective landscape

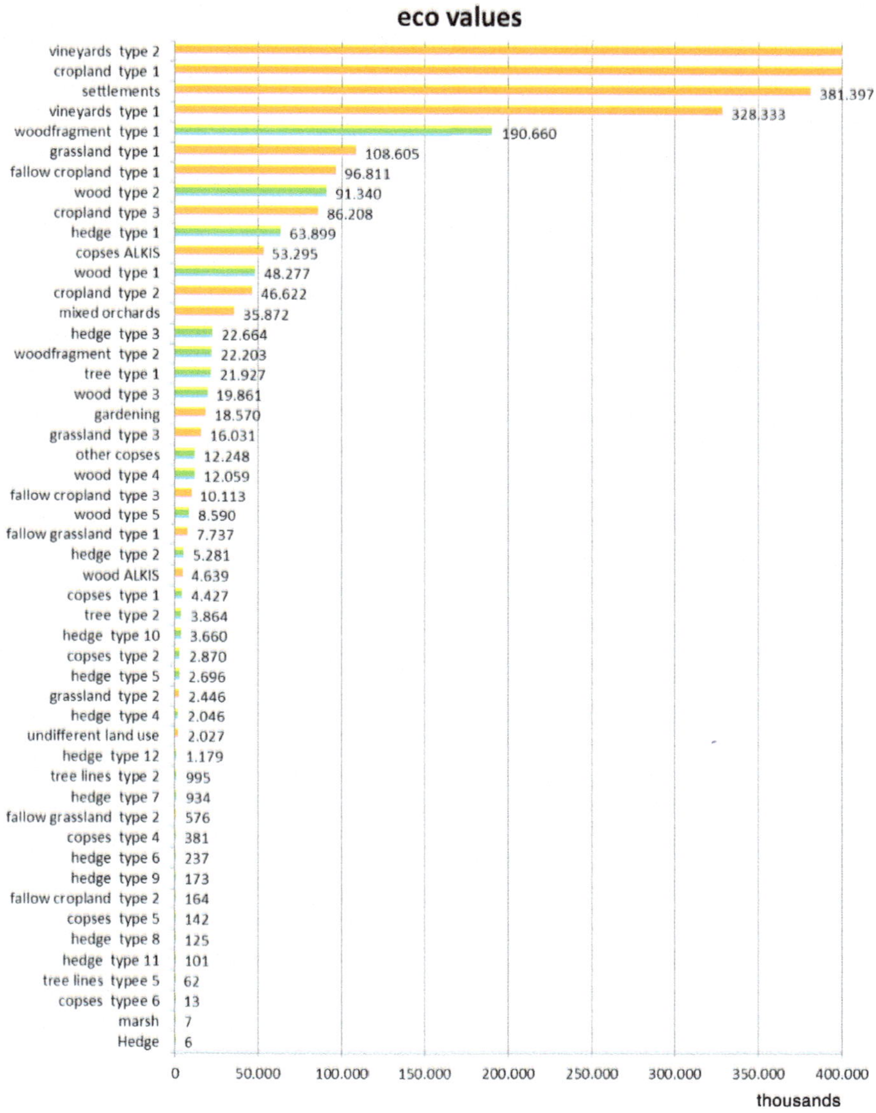

Fig. 23.7 Resulting distribution of eco-values in the pilot region VP (median)

section. In the case of model landscape 'HB' the total sum of an eco-value is about 625 million (Ø 25.9/m^2) and in 'VP' the total sum is about 5977 million (Ø 29.8/m^2), initially depending on their area sizes, because each pixel was assigned with the respective eco-value.

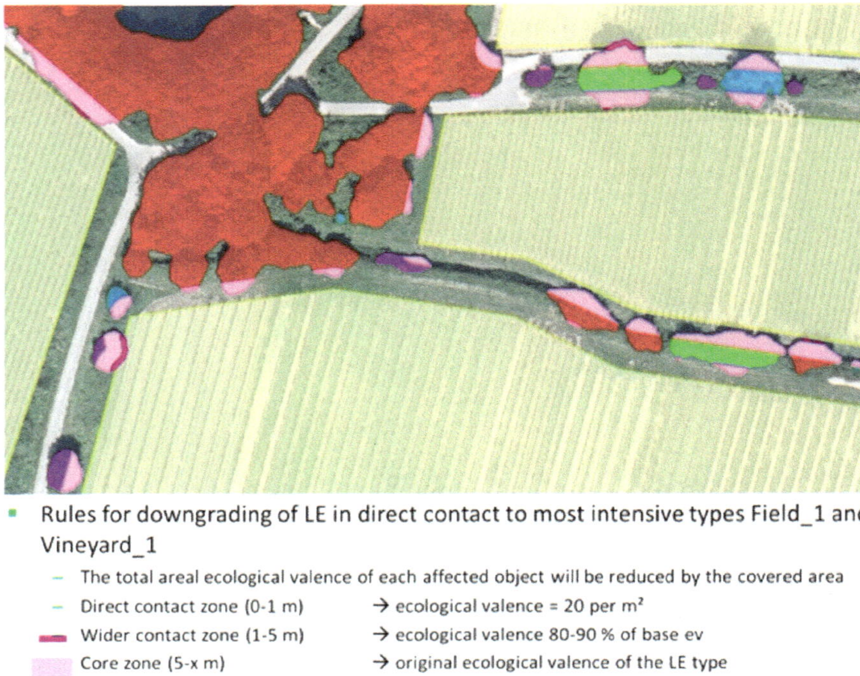

Fig. 23.8 Conceptual approach of the RISKMIN model for integrating vicinity aspects

In the module **field** comprehensive ground truth campaigns were conducted during the whole research project concerning the on-site mapping of flora, fauna and the automated landscape classification.

A comprehensive validation of the semi-automated generated land cover classes in the pilot region HB was done, too.

The results show, comparable to validations in Rhineland-Palatinate (Trapp et al. 2015), a good accuracy of 3D-features, and some discrepancies concerning the current 2D-land cover classes, depending on the intensity of land use by farmers.

Overall, the geometrical accuracy of the semi-automated land cover classification was sufficient for the purposes in this research project.

The methodology is considered as being well-established but needs calibration by data from different landscapes. The model that was derived from the RISKMIN project puts the discussion on landscape-based RMM a good step forward. The RISKMIN concept is an option for risk managers especially to optimize RMM on a local to regional level alongside with national programs for sustainable land use. The following Figs. 23.9 and 23.10 visualize an example of one of the landscape-related simulations and compares the Status Quo and the Scenario 1 (=extensification of cropland). In detail in scenario 1 we realized this extensification of the cropland use by randomly assigning 47.5% of the total cropland the value extensive use in

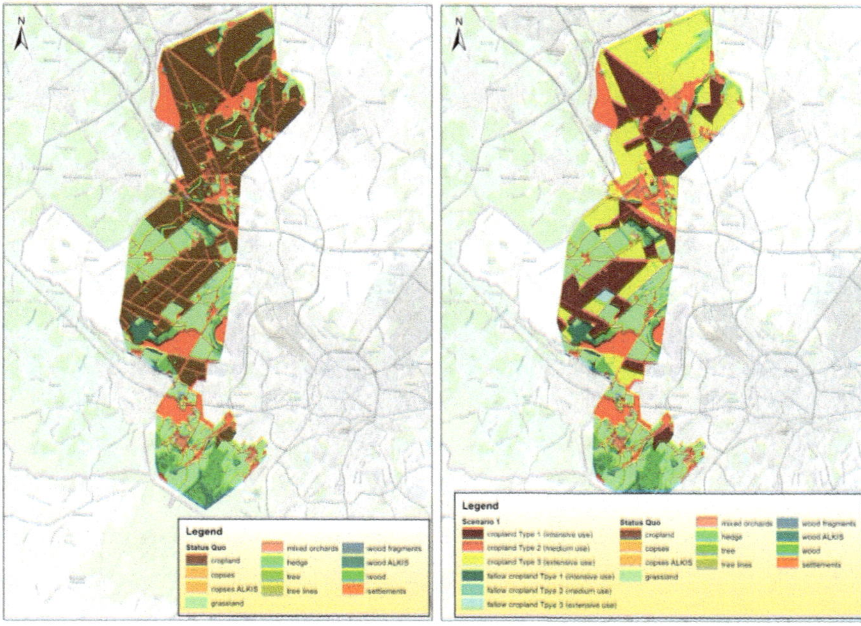

Fig. 23.9 Visualisation of one landscape-related simulation: comparison between the Status Quo and Scenario 1 in the pilot region HB

comparison to the status quo land-use definition, where we assigned 97.5% of the cropland as intensive use and 2.5% as extensive use.

Therefore you can see in detail how many fields would be extensified and with the help of geo-referenced data you can quantify the effect by calculating the sums of the overall ecological values.

The following tables show exemplarily the results of the Status Quo for both pilot regions in comparison to the Scenario 1 (=extensification of cropland) and additionally the integration of a 5 m buffer strip (DR drift reduction), here for all land cover classes and the two main LE categories coppices and hedges.

The buffer strips were calculated inside the cropland and including borders as well as landscape features as visualized in Fig. 23.11.

In the following Tables 23.1, 23.2, 23.3, and 23.4 the results of the landscape-related simulations are shown. Based on the Status Quo description the effects of the RMM of the scenario 1 (extensification of cropland use as well as the integration of a 5 m buffer zone) are described explicitly for the both pilot regions HB and VP.

The main mitigation effect is based on the extensification scenario, whereas buffers strips do not have important influence regarding the whole landscape.

Differences between the two pilot regions are resulting from a different amount of cropland.

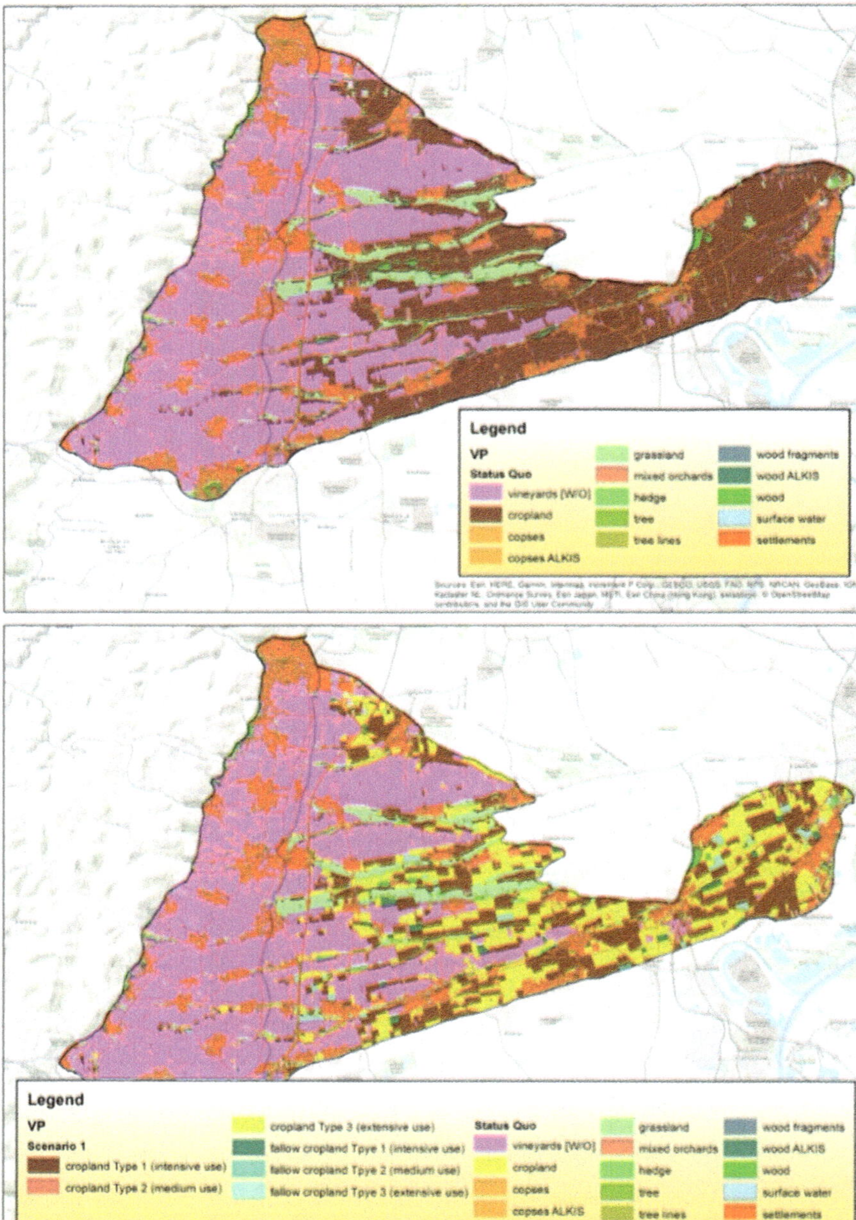

Fig. 23.10 Visualisation of one landscape-related simulation: comparison between the Status Quo and Scenario 1 in the pilot region VP

Fig. 23.11 Establishing in crop buffer zones of 5 m and off-crop field margins as mitigation measures

Table 23.1 Results of the Status quo in comparison to the Scenario 1 for all land cover classes

Total	HB			VP		
	ECO_PPTs	Change %	Effect DR %	ECO_PPTs	Change %	Effect DR %
Status quo	627,948,185	0.0		5,998,351,169	0.0	
Status quo DR	629,393,830	0.2	0.23	6,027,369,360	0.5	0.48
Scenario 1	785,075,915	25.0		6,868,880,078	14.5	
Scenario 1 DR	785,915,169	25.2	0.13	6,887,985,377	14.8	0.32

Table 23.2 Results of the Status quo in comparison to the Scenario 1 for the main LE class coppice

Coppice	HB			VP		
	ECO_PPTs	Change %	Effect DR %	ECO_PPTs	Change %	Effect DR %
Status quo	1,743,566	0.0		7,879,446	0.0	
Status quo DR	1,825,742	4.7	4.71	8,280,634	5.1	5.09
Scenario 1	1,780,097	2.1		8,050,313	2.2	
Scenario 1 DR	1,824,556	4.7	2.55	8,302,317	5.4	3.20

Regarding only the LE category coppice the buffer strips show higher effects (4.7% in HB and 5.1 in VP). Here the extensification scenario is not as evident as when regarding the whole landscape.

Table 23.3 Results of the Status quo in comparison to the Scenario 1 for the main LE class hedges

Hedge	HB			VP		
	ECO_PPTs	Change %	Effect DR %	ECO_PPTs	Change %	Effect DR %
Status quo	14,760,519	0.0		103,111,034	0.0	
Status quo DR	15,187,846	2.9	2.90	105,661,450	2.5	2.47
Scenario 1	14,938,683	1.2		104,120,621	1.0	
Scenario 1 DR	15,175,852	2.8	1.61	105,622,147	2.4	1.46

Table 23.4 Results of the in crop buffer zones

Measurement	Pilot region HB [% area]	Pilot region VP [% area]
−3 m puffer	−4.5	−6.9
−5 m puffer	−7.4	−11.2

Even lower, but similar to the LE feature coppice are the results regarding the main LE class hedges. Here the influence of buffer strips varies between 2.9 in HB and 2.5 in VP.

Regarding the loss of application area when integrating such buffer zones as mitigation measure we calculated a 5 m in crop buffer around the parcel borders and landscape features as visualized in Fig. 23.11.

In these pilot regions a 5 m buffer zone around the parcel borders including landscape features leads to a loss of 7.4 (HB) resp. 11.2% (VP) of the total area and could be an appropriate mitigation measure.

The analysis of the data from the field surveys (module Field) found some generalizable rules that are valuable to consider in risk mitigation. The biodiversity in the agricultural landscape on the one hand arises from the diversity of landscape elements in a landscape area, and on the other hand from the different characteristics and quality of a specific landscape element. Furthermore, the agricultural landscape possesses a regionally specific biodiversity due to different climatic, edaphic, land-use dependent and other regional factors (e.g. species areas of plants and animals). Species are adapted to the habitat in the competitive structure of biocoenosis. The assessment and management of biodiversity must therefore be adapted to the corresponding area, region or landscape type.

The present study showed clear and specific patterns of animal species groups and plant species. One part of the biodiversity is interchangeable across the different landscape elements. These species have a wide ecological amplitude and therefore they have no specific ties to a particular habitat type. Populations of these species are thus in contact across different landscape elements. Another part of biodiversity is not interchangeable across different landscape elements. These species have a narrow ecological amplitude. Their occurrence is adapted to one or a few landscape elements or specific conditions (e.g. humidity). Populations of these species are therefore not in direct contact with other landscape elements. The management within the agricultural landscape must therefore be adapted to the different effects of the measures on the different species groups.

Not described in this publication are the effects of introducing more, new or better equipped LE types that host biodiversity into a given landscape. Even recommendations which measures would be best suited considering the specific boundary condition in a landscape and where they would be placed best for effectiveness are not part of this publication in detail.

23.4 Conclusions

One important result of the RISKMIN project is that the 'regulations' concerning buffers strips do not impact the total landscape ecological valences much due to the small amount of area of 'high-value off-crop structures' compared to in-field area. Therefore the 'extensification' of large in-field areas has the highest absolute impact on the total landscape ecological valences.

On the other hand a further important result can be described by 'type-related affectedness'. This means that the combination of in-field and off field mitigation measures linked in landscape contexts have more effects than spatial singular measures, e.g. narrow margins would benefit most.

1. With respect to landscape level, different RMM need to be calibrated to different landscape types. Landscape-related mitigation measures are more effective than measures without respect to the environmental conditions and the neighbourhood relations in spatiotemporal context.
2. Most effective is a more extensive land use, but implementing different kinds of mitigation measures in crop and off crop to support synergistic effects and building up a network could lead to similar results.
3. In future, the effects could be validated by on-site monitoring of the biodiversity or of the quality of landscape elements and a region specific assessment should be applied by defining 'general principles' for different types of landscapes.
4. Heat-maps of biodiversity hotspots or sinks could be helpful for risk managers deciding on the most efficient measures to be implemented.

Acknowledgements We thank the German Federal Office of Consumer Protection and Food Safety for funding the research project called "Landschaftsbezogene Risikominderungsmaßnahmen zur Förderung der terrestrischen Agrobiodiversität (RISKMIN)". The full document can be downloaded by https://www.bvl.bund.de/DE/04_Pflanzenschutzmittel/01_Aufgaben/09_GesundheitNaturhaushalt/02_SchutzNaturhaushalt/psm_SchutzNaturhaushalt_RiskMin_basepage.html. We also thank the Land Survey Authorities in Rhineland—Palatinate and in North Rhine—Westphalia for supporting this research project by giving the license of using geodata: © LVermGeo 2019, www.lvermgeo.rlp.de Open.NRW—Das Open Government Portal für das Land Nordrhein-Westfalen, "www.open.nrw".

References

Adam K, Nohl W, Valentin W (1986) Bewertungsgrundlagen für Kompensationsmaßnahmen bei Eingriffen in die Landschaft. In: Ministerium für Umwelt, Landwirtschaft, Natur- und Verbraucherschutz Nordrhein-Westfalen (ed) Naturschutz und Landschaftspflege in Nordrhein-Westfalen. Düsseldorf

Biedermann U, Werking-Radtke J, Woike M (2008) Numerische Bewertung von Biotoptypen für die Eingriffsregelung in NRW. In: Landesamt für Natur, Umwelt und Verbraucherschutz Nordrhein-Westfalen (LANUV NRW) (ed.) Recklinghausen. https://www.lanuv.nrw.de/fileadmin/lanuv/natur/lebensr/Num_Bew_Biotyp_Sept2008.pdf

Enzian S, Gutsche V (2005) GIS - gestützte Berechnung der Ausstattung von Agrarräumen mit naturnahen terrestrischen Biotopen auf der Basis der Gemeinden – 2. Ausgabe des Verzeichnisses der regionalisierten Kleinstrukturanteile Biologische Bundesanstalt für Land- und Forstwirtschaft, Institut für Folgen-ab-schätzung im Pflanzenschutz, Kleinmachnow. https://www.julius-kuehn.de/sf/ab/raeumliche-analysen-und-modellierung/verzeichnis-der-regionalisierten-kleinstrukturanteile/

European Environment Agency (EEA) (2013) the European grassland butterfly indicator: 1990–2011. EEA Technical report No 11/2013, p 36

European Food Safety Authority (EFSA) (2016) Recovery in environmental risk assessments (ERA). EFSA J 14(2):4313, 85. https://doi.org/10.2903/j.efsa.2016.4313

Federal Agency for Natural Conservation (BfN) (2011) Landscapes of conservation importance. https://www.bfn.de/en/activities/protecting-habitats-and-landscapes/landscapes-of-conservation-importance.html

Hallmann CA, Sorg M, Jongejans E, Siepel H, Hofland N, Schwan H et al (2017) More than 75 percent decline over 27 years in total flying insect biomass in protected areas. PLoS ONE 12(10):e0185809. https://doi.org/10.1371/journal.pone.0185809

Jahn T, Hötker H, Oppermann R, Bleil R, Vele L (2014) Protection of biodiversity of free living birds and mammals in respect of the effects of pesticides, Federal Environment Agency (Umweltbundesamt), 06844 Dessau-Roßlau. https://www.umweltbundesamt.de/publikationen/protection-of-biodiversityof-free-living-birds, ISSN 1862-4804

Lang S, Blaschke T (2007) Landschaftsanalyse mit GIS. Ulmer, Stuttgart

Ludwig D, Meinig H (1991) Methode zur ökologischen Bewertung der Biotopfunktion von Biotoptypen. Bochum

Roß-Nickoll M, Lennartz G, Fürste A, Mause R, Ottermanns R, Schäfer S, Smolis M, Theissen B, Toschki A, Ratte HAT (2004) Die Arthropodenfauna von Nichtzielflächen und die Konsequenzen für die Bewertung der Auswirkungen von Pflanzenschutzmitteln auf den terrestrischen Bereich des Na-turhaushaltes. Umweltbundesamt (UBA), Berlin. FKZ 20063403, p 148

Scholz-Starke B, Trapp M, Streib L, Deubert M, Oellers J, Fürste A, Luther S, Peters S, Toschki A, Roß-Nickoll M (2016) Landschaftsbezogene Risikominderungsmaßnahmen zur Förderung der terrestrischen Agrobiodiversität (RISKMIN). Federal Office of Consumer Protection and Food Safety (BVL). https://www.bvl.bund.de/DE/04_Pflanzenschutzmittel/01_Aufgaben/09_GesundheitNaturhaushalt/02_SchutzNaturhaushalt/psm_SchutzNaturhaushalt_RiskMin_basepage.html

Tintrup gen Suntrup G, Jalke T, Streib L, Keck N, Nieland S, Moran N, Kleinschmit B, Trapp M (2014) New methods in acquisition, update and dissemination of nature conservation geodata – implementation of an integrated framework. In: The 36th international symposium ion remote sensing of environment, 11–15 May 2015, Berlin, Germany

Toschki A (2008) Eignung unterschiedlicher Monitoring-Methoden als Grundlage zum Risk-Assessment für Agrarsysteme - Am Beispiel einer biozönologischen Reihenuntersuchung und einer Einzelfall-studie. Phd thesis Institute for Environmental Research, RWTH-Aachen, p 162

Trapp M, Jalke T, Tintrup gen Suntrup G (2015) Automatisierung von Verwaltungsabläufen am Beispiel von Landwirtschaft und Umwelt, LSA VERM 1/2015, pp 25–30

Trapp M, Deubert M, Streib L, Roß-Nickoll M, Scholz-Starke B, Toschki A (2018) Simulating the effects of agrochemicals and other risk-bearing management measures on the terrestrial agrobiodiversity: the RISKMIN approach. J Plant Sci Nov Methods Results Landsc Res Eur Cent Asia Sib 3(2018):228–342. https://doi.org/10.25680/5187.2018.80.87.264
Tscharntke T, Klein AM, Kruess A, Steffan-Dewenter I, Thies C (2005) Landscape perspectives on agricultural intensification and biodiversity – ecosystem service management. Ecol Lett. 8:857–874. https://doi.org/10.1111/j.1461-0248.2005.00782.x
Working Committee of the Surveying Authorities of the States of the Federal Republic of Germany (AdV) (2015) Documentation on the modelling of geoinformation of official surveying and mapping. http://www.adv-online.de/Publications/AFIS-ALKIS-ATKIS-Project/binarywriterservlet?imgUid=8e1708ee-765f-8551-2357-c133072e13d6&uBasVariant=11111111-1111-1111-1111-111111111111

Part III
Landscape Decision Support—Systems, Tools and Frameworks

Chapter 24
LandCaRe-DSS—An Interactive Model-Based Decision Support System for Assessing the Impacts of Climate Change on Agriculture and Agricultural Landscapes

Wilfried Mirschel, Michael Berg-Mohnicke, Karl-Otto Wenkel, Ralf Wieland and Barbara Köstner

Abstract The current and future climate change, which is still superimposed by global change, has a serious influence on landscapes in general and agricultural landscapes in particular. To meet the challenges for society and agriculture, well-planned and sustainable adaptation measures to climate change are needed. For agriculture, the following question has to be answered: How will climate change affect regional agriculture and ecosystems and what could be possible adaptation strategies, taking into account local site potentials and the specific structure of farms and agricultural enterprises? To answer this question, both farmers and other stakeholders might need the help of computer-based decision support systems. The information and decision support system described here, called LandCaRe-DSS, is such a helping instrument. The LandCaRe-DSS is designed as a user-friendly, interactive, model-based and spatial-oriented information and decision support system. It can be used on different spatial scales while being fully interactive. It supports long-term spatial scenario simulations, multi-ensemble and multi-model simulations at the regional level as well as complex impact assessments of potential adaptation strategies for

W. Mirschel (✉) · M. Berg-Mohnicke · K.-O. Wenkel · R. Wieland
Leibniz-Centre for Agricultural Landscape Research (ZALF) e.V,
Eberswalder Str. 84, 15374 Müncheberg, Germany
e-mail: wmirschel@zalf.de

M. Berg-Mohnicke
e-mail: michael.berg-mohnicke@zalf.de

K.-O. Wenkel
e-mail: kowenkel@web.de

R. Wieland
e-mail: rwieland@zalf.de

B. Köstner
Department of Meteorology, Technical University Dresden,
Pienner Straße 23 and 21, 01737 Tharandt, Germany
e-mail: barbara.koestner@tu-dresden.de

W. Mirschel et al. (eds.), *Landscape Modelling and Decision Support*,
Innovations in Landscape Research,
https://doi.org/10.1007/978-3-030-37421-1_24

land use at the local level. All simulations are carried out with a high spatial resolution and include coupled climate and agro-economic scenarios. The easy-to-use software is controlled via a zoomable user interface known to many users from for instance Google Maps®. The system can be extended with further modules for different tasks.The article describes the structure and use of the LandCaRe-DSS in detail, including the components and modules of the system, the system framework, the operation principles, the climate and geodatabases, the scale-specific ecological impact models and the IT realisation. The LandCaRe-DSS offers different data analysis and visualisation tools, a help system for users and a farmer information system for adaptation of agriculture to climate change. Using concrete examples, the different fields of application are presented: the analysis of climate data, the analysis of phenology and ontogenesis and the climate change impact assessment on national, regional and local resp. farm scales. In particular, the LandCaRe-DSS supports strategic planning in agriculture and the sustainable development of rural areas (regions) and provides answers to the impacts and costs of possible adaptation measures to climate change. The LandCaRe-DSS was developed as a prototype and is parameterised and validated for the region Uckermark (dry lowland, 2,600 km^2) in the Federal State of Brandenburg and for the region Weisseritzkreis (wet mountain area, 400 km^2) in the Free State of Saxony. In the last 10 years, the system underwent slow albeit continuous changes and has been adapted to further regions and extended for the needs of research projects.

Keywords Decision support system · Climate change · Regional · Impact assessment · Simulation model · Agricultural productivity · Agricultural adaptation strategies

24.1 Introduction

The agricultural landscapes are particularly affected by climate change and its variability. Agricultural production and agricultural farm management are primarily dependent on the weather conditions whose patterns are determined by annual fluctuations. Depending on the natural conditions of an agricultural landscape—the associated regional climate characteristics, the agricultural cropping systems and the economic framework conditions—global climate change can have both negative and positive impacts on land use in agricultural landscapes. Agricultural landscapes can amplify climate change and its consequences through inadequate use of resources and the emission of climate-impacting trace gases, but can also mitigate climate change and its consequences through stable agricultural ecosystems, the production of bioenergy and the promotion of CO_2 sinks.

On the one hand, there is the demand for climate protection measures (Kyoto 1997; COP24 2018) and, on the other hand, there is a need for agriculture to adapt to climate change. Adaptation to climate change requires knowledge on the potential regional and local impacts of climate and weather extremes but also knowledge about the expected long-term consequences of land use activities. Adaptation must consider sustainability with respect to high plant production without degrading other ecosystem services like soil protection, purification and recycling of water or maintenance of biodiversity. Further, adaptation should not enhance but reduce greenhouse gas emissions. This implies that excellent decision-making has to consider both socio-economic and ecological consequences of adapted land use management. From the agricultural side, various options are possible for adapting to climate change, such as the cultivation of new crop species and drought-resistant varieties, extended crop rotations, increasing soil humus content, adaptation of agro-management timetables, conservation tillage, as well as adapted fertilisation, irrigation and plant protection regimes. In practice, however, the number of options is limited to just a few alternatives. Each agricultural enterprise or farmer works under different soil–climate and management conditions. Therefore farmers have to examine separately, how they can reduce the vulnerability to climate change and increase desirable outcomes with the lowest costs under consideration of the different regional and local economic and soil–climate conditions.

Climate change represents a change in the boundary conditions for many initiatives and actors in the agricultural landscape. Adapting to climate change in an agricultural region is a multifaceted task and cannot be solved by agriculture alone. Since the agricultural region itself is a limited resource for natural and economic services, it is necessary to develop integrated solutions that enable and honour the diverse functions of the agricultural landscape and its sustainable development.

For the identification of correct and sustainable decisions, it is required to assess the complex regional impacts of climate change and different land use systems on the most important ecological and economic landscape indicators, the risks for agricultural enterprises or farmers and nature as a whole, and their changes over time. These assessments, which must also take into account future changes, can only be made with the help of spatial data and robust and well-validated regional models for different landscape indicators. For decisions to be made based on many scenario simulations, these must be part of interactively usable decision support systems (DSS). The development of a DSS for this purpose, however, remains a major challenge. Such a DSS needs to address agricultural regions as a whole, providing up-to-date scientific knowledge and regionalised climate information derived from the most recent global climate scenarios. At the same time, it has to meet users' demand for transparency, interactivity and user-friendliness without any loss of information (Wenkel et al. 2013). Due to the complexity of such a challenge, only a few spatially applicable DSS and similar tools have been developed and made available, such as LADSS (LADSS 2005), ADSS (Prased et al. 2008), GPFARM (Ascough et al. 2001), EET (Elbe-Expert Toolbox)/GLOWA Elbe (Kaden et al. 2010; Wechsung et al. 2005, 2014) and Elbe-DSS (Berlekamp et al. 2005; Kofalk et al. 2004), which were developed for specific regions or water catchments and which as prototypes are not available to the public.

To close the gap in this field, LandCaRe-DSS was developed by the former Institute of Landscape Systems Analysis of the Leibniz-Centre for Agricultural Landscape Research (ZALF) Müncheberg, Germany. The aim of the LandCaRe-DSS was to integrate the latest results from the fields of climate regionalisation, plant ecology, ecosystem and landscape modelling and farm economics into an interactive usable spatial information and decision support system. This system should provide potential users with the currently available knowledge about regional climate change and its probable consequences. It should enable interactive scenario simulations in order to develop regionally adapted and cost-effective climate adaptation strategies for agriculture, taking into account existing uncertainties using mainly regional models of intermediate complexity (REMICs, Wenkel et al. 2008, 2019), but also detailed dynamic process-based agroecosystem models such as MONICA (Nendel et al. 2011) for the field plot or farm levels.

For the LandCaRe-DSS, this paper will present the conceptual framework, the basic structure, the functionality, the IT realisation, the systems use for scenario studies at different spatial scales and ways in which the system can be adapted for other regions or countries.

24.2 LandCaRe-DSS—The Information and Decision Support System

The LandCaRe-DSS (Fig. 24.1) is a modular model-based system that allows information and climate impact assessment models to be used interactively to develop cost-effective climate adaptation strategies for agriculture. It contains both information and a model-based simulation system. The LandCaRe-DSS was developed within the collaborative project LandCaRe 2020 (Köstner et al. 2012) as a part of the German research programme 'klima*zwei*' (Mahammadzadeh et al. 2009). The main philosophy of LandCaRe-DSS is that there is not only one single solution to climate change-related problems. The aim of the system is to provide all the necessary expertise to solve various issues related to climate change in agricultural landscapes. Potential users of the system, such as agricultural consultants, agronomists, decision

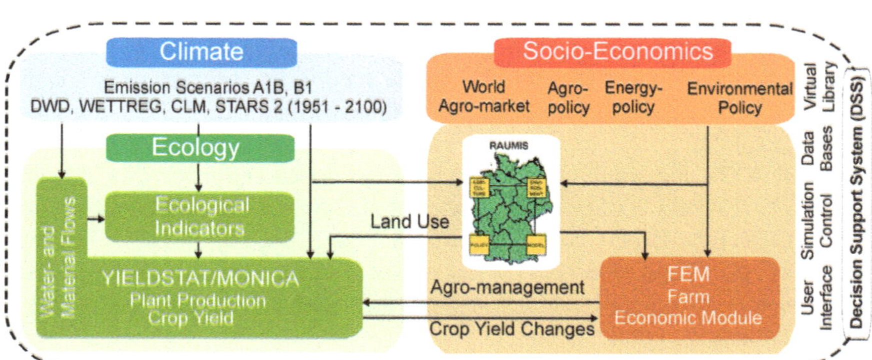

Fig. 24.1 Conceptual framework and integration of different modules in the LandCaRe-DSS

makers of public regional and landscape planning bodies, water management boards and agricultural insurance companies, can find out about the extent of the expected impact of climate change at different spatial scales. They also receive information on the variability of regional climate projections, which go along with the application of different global climate regionalisation methods and scenarios. In addition, users are given the opportunity to evaluate various adaptation options at farm level with regard to their medium- to long-term impact on ecology and socio-economy.

Unlike other decision support systems, the LandCaRe-DSS facilitates a quantitative assessment of the consequences of potential climate adaptation strategies, including information on the variability of the respective output quantities, as a basis for risk assessments. The LandCaRe-DSS offers the opportunity to also include changes to framework conditions, such as agricultural policy and markets, thus making it a powerful tool for sustainable investigations. In general, LandCaRe-DSS provides

- greater knowledge about past and future climate change in the respective region,
- an evaluation of the productivity of agricultural landscapes,
- an estimation of the potential ecological and economic consequences of climate change,
- the ability to prepare and visualise all available knowledge on potential agricultural climate adaptation strategies and
- suitable tools for interactive simulations in order to compare alternative land use systems and options for action.

24.2.1 System Components and Framework

The LandCaRe-DSS consists of five basic system components:

- Information and advisory system related to climate and climate change,
- Analysis of climate data and climate impacts on crop phenology and ontogenesis,
- Climate change impact assessment for agriculture on national scale,
- Climate change impact assessment on regional scale,
- Climate change impact assessment and integrated assessment of different agricultural adaptation strategies to climate change on farm scale.

The conceptual framework and the integration of different modules within the LandCaRe-DSS are represented in Fig. 24.1.

Special focus has been put on the following features of the LandCaRe-DSS:

- **Interactivity**
  The system user decides on its own which analyses, simulations and calculations are to be performed and which models are to be used.
- **Dynamicity**
  Depending on the desired level of detail, the user can run a large variety of model-based simulations for analysis and comparison. The preconditions chosen by the user will affect the simulation results.

- **Spatiality**
  The spatial level (national, regional or farm level) for which the model-based simulations are to be run can be freely selected via zooming on a Google Maps® like view. Different models can be activated for execution.
- **Web-integration**
  Due to the central maintenance and care of the LandCaRe-DSS software (control and update) and the data required, the web-based system could be used from everywhere via the internet.
- **Extensibility**
  The system is open to extension regarding data or content, as well as continuous improvements to the information, knowledge and the model base.

The LandCaRe-DSS includes a set of statistical and process-based simulation models for different spatial scales, various databases centred around and a zooming user interface (ZUI).

24.2.2 System Related Available Impact Models

The LandCaRe-DSS contains a wide range of models to support decision-making for adapting regional and local land use management to climate change. These models include biophysical system models, conceptual models, statistical models, a farm economy module, tools for data analysis and visualisation (elementary statistics, distribution probabilities, and others) and an information and help system that supports the user in handling the system and interpreting the simulation results. The most important models of the current LandCaRe-DSS prototype are listed in Table 24.1. Their possible uses and the most important model components are shown in Fig. 24.2.

Before being included in the LandCaRe-DSS, all models were revised and tested as independent versions under conditions of East German agricultural landscapes and adapted to the needs and structure of LandCaRe-DSS. Exemplarily for YIELDSTAT and MONICA model, test results are presented in Mirschel et al. (2010, 2012) and Nendel et al. (2011), respectively.

Not all LandCaRe-DSS models can be used interactively as they either operate with a very high temporal resolution, like the big-leaf model SVAT-CN for the primary production of non-agricultural vegetation, or operate at a higher spatial scale, like RAUMIS for a Germany-wide yield forecast. RAUMIS is a mathematical programming approach with a non-linear goal function. The model reflects regional adaptations of agriculture in Germany to agricultural and agro-environmental policy measures. Different climate and market development scenarios are taken into account. For these reasons, the analyses of the SVAT-CN and the RAUMIS models have been preprocessed for many predefined scenarios and users can assemble the necessary maps themselves from the precalculated results of these models.

Table 24.1 List of models integrated into the current LandCaRe-DSS prototype (modified from Wenkel et al. 2013)

Model name	Model description	Application scale	Reference
Climate analysis for actors from administration, agriculture, forestry and water management, nature conservation, tourism, associations, insurance companies and training institutions			
TREND	Long-term climate data analysis and trend calculation (temperature, precipitation, climatic water balance, heating and cooling degree days, Ellenberg index for forest trees, Huglin index for vine, Schwärzel index for fruit trees)	Climate stations, grid points	Franke and Köstner (2007), Huglin (1986), Schwärzel (2000)
SAISON	Seasonal data analysis of long-term climate data (temperature, precipitation, frost and ice days, summer, hot and tropical days, heating and cooling days)	Climate stations, grid points	
FREQUENCY	Frequency and value distribution for temperature and precipitation for long-term climate data	Climate stations, grid points	
Phenology/ontogenesis for actors from agriculture, viticulture, fruit-growing and horticulture, plant breeding and education institutions			
VEGPER	Calculation of begin, end and length of vegetation period	Climate stations, grid points	Chmielewski (2003)
ONTO	Calculation of development stages for agricultural crops between sowing and harvest	Climate stations, grid points	Mirschel (2010)
PHENO	Calculation of phenology phases of wild indicator plants for the estimation of seasons between pre spring and late autumn	Climate stations, grid points	Chmielewski and Henniges (2007)
Agricultural productivity, farm economics, need in irrigation for actors from agricultural enterprises, consultants, insurance companies and training institutions			
YIELDSTAT	Estimation of yield, biomass and carbon fixation for 15 agricultural crops considering site, climate, agrotechnology, plant breeding, and agro-management	Region, farm, field	Mirschel et al. (2014a)

(continued)

Table 24.1 (continued)

Model name	Model description	Application scale	Reference
FEM-YIELDSTAT	Farm economy module coupled with statistic yield estimation model to capture climate change effects on yield, cross margin and revenues less direct variable costs	Farm	Münch and Gocht (2006), Münch et al. (2014), Mirschel et al. (2014a)
MONICA	Agroecosystem model with a daily time step for all relevant crop and soil processes (water, nitrogen and carbon dynamics, biomass and crop yield formation)	Field	Nendel et al. (2011, 2014)
FEM-MONICA	Farm economy module coupled with dynamic agroecosystem model to capture climate change effects on yield, evapotranspiration, nitrate leaching, humus balance, cross margin and revenues less direct variable costs	Farm	Münch and Gocht (2006), Münch et al. (2014), Nendel et al. (2011, 2014)
GLPROD	Calculation of grassland productivity and forage quality in dependence on grassland type, site, climate, use intensity level, water and nutrition supplies	Region, farm	Käding et al. (2005), Kaiser et al. (2005)
IRRINEED	Identification of the general site-specific irrigation need	Region, farm, field	Roth (1993)
ZUWABE	Calculation of the site-specific irrigation water demand	Region, farm, field	Roth (1993), TGL (1990)
EROSION	Calculation of potential erosion risk dependent on site, farm management and climate using the modified Revised Universal Soil Erosion Equation (RUSLE)	Region, farm, field	Wischmeier and Smith (1978), DIN19708 (2005)

(continued)

Table 24.1 (continued)

Model name	Model description	Application scale	Reference
Regional land use, regional water balance and evapotranspiration, and primary production for forests and grasslands for actors of resource protection and regional planning			
LANUDIS	Computation of a stochastic distribution of agricultural crops on arable land (land use map) dependent on scenario-given crop ratios, soil types, and the economic viability	Region	Wieland (2010), Mirschel et al. (2016)
BAGLUVA	Calculation of the regional water balance (actual evapotranspiration, groundwater recharge, surface runoff) dependent on climate, vegetation and land use structure	Region	ATV-DVWK-Regelwerk (2002)
SVAT-CN	Calculation of soil–vegetation–atmosphere transfer (big-leaf-model) of energy, water vapour, carbon and nitrogen for primary production and water use efficiency of non-agricultural vegetation	Region	Kuhnert and Köstner (2008, 2009)
Information on net value added and the need for irrigation on agricultural land for actors of the Federal and State Ministries of Agriculture and nationally active companies and associations			
RAUMIS	Description of impacts of changes of crop yield, adaptation of agriculture to climate change and price–cost relations (baseline, pessimistic, optimistic) on future agricultural land use, productivity and revenues	Country	Henrichsmeyer et al. (1996), Anter et al. (2009), Gömann et al. (2005, 2009)

As part of LandCaRe-DSS, RAUMIS was also used to develop basic scenarios for future cost–price relations, which can then be used in the models FEM-MONICA and FEM-YIELDSTAT. A detailed description of the impact models integrated in LandCaRe-DSS is given in Wenkel et al. (2010a, 2010b).

The time resolution of process-based simulations within LandCaRe-DSS is 1 day. Climate impact simulations and the comparison of simulation results always refer to a period of 30 years, which corresponds to the minimum length of climatic periods considered for the analysis of climate trend signals in climatology. The maximum

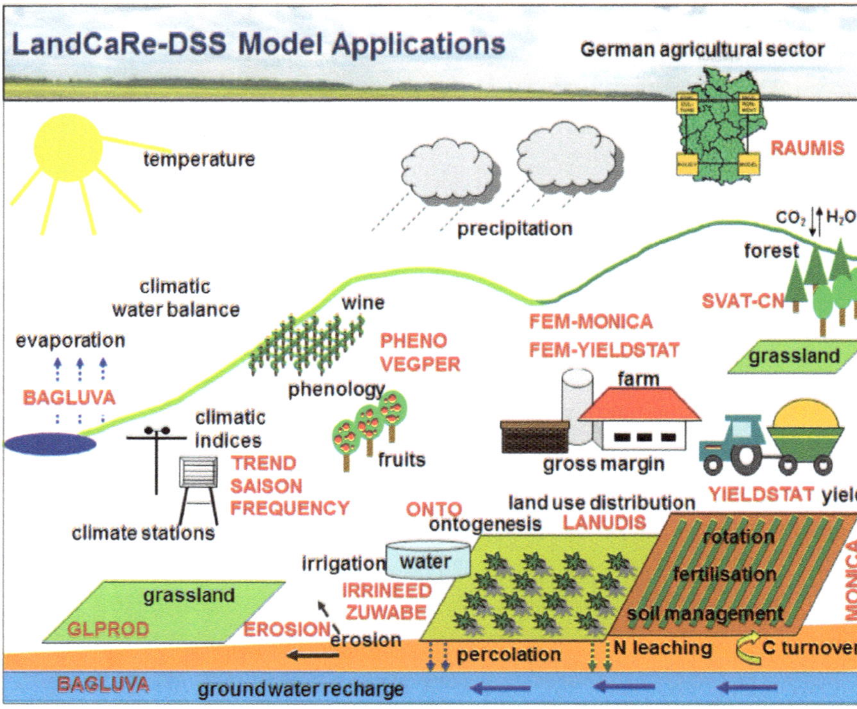

Fig. 24.2 Overview of model applications usable within LandCaRe-DSS in a virtual landscape (models are highlighted in red)

spatial resolution of the model results is 100 m × 100 m, which corresponds to the resolution of the soil map used.

24.2.3 System Related Database

The LandCaRe-DSS database comprises several local SQLite databases, mainly containing model parameters and system support data. A remote MySQL database hosts climate data. Geodata (ASCII grids) such as georeferenced data on land use, topography, hydrology, inclination, stoniness and soil quality as well as soil characteristics and soil profiles are stored in a local file-based database. For the whole territory of eastern Germany, these spatial soil data are derived from the German medium-scale soil map MMK (Schmidt 1975). In addition, crop yield models MONICA and YIELDSTAT have their own parameter databases.

Within LandCaRe-DSS, a central database contains historical climate data and regionalised future climate projections (different emission scenarios, different regionalisation methods). Climate data are visualised in order to highlight regional climate trends and weather statistics from the recent past, recorded by climate stations of the German Weather Service (DWD) since 1950. The future climate scenarios are based on simulations of the ECHAM5/MPI-OM Global Circulation Model by the Max Planck Institute for Meteorology Hamburg (Roeckner et al. 2004). Each scenario (here: A1B and B1) has been scaled down to the regional level using the statistical regionalisation methods WETTREG (Enke et al. 2005), STARS2 (Orlowsky et al. 2008) and REMO (Jacob et al. 2001), which produce weather station-based datasets. It is also possible to use climate data created using the process-based regionalisation method CLM (Böhm et al. 2006), which produces weather data in grids of 5 to 18 km length. Because of the systems' modular structure, it is possible to link to further regionalised climate data from different emission scenarios of other global climate models to LandCaRe-DSS with relatively little effort.

Furthermore, as a basis for economic calculations with the farm economic module (FEM), the local database contains parameters for different agricultural crops, detailed management data as well as prices for all crops and the costs of 188 individual agricultural production processes and cropping procedures, which can be adjusted by users interactively to match their specific local farm management techniques (Münch et al. 2014).

24.2.4 System Operation Principles

As already mentioned above, LandCaRe-DSS is a model-based knowledge platform for decision-making. It allows it to analyse interactively scenarios and thus offers adaptation and utilisation options for rural landscapes or individual farms under the influence of regional climate change and socio-economic framework conditions. A short demonstration of how LandCaRe-DSS works can be watched on YouTube (https://youtu.be/tbUHaeA6-dl) for application examples in Germany and Brazil.

The basic principle of operation is an iterative procedure (Fig. 24.3). The first step is the scenario definition by the user. The second step is the simulation and the impact assessment of the climate change. In the third step, the ecological and economic impacts of the chosen agricultural adaptation strategy at the regional and/or local level are investigated. In the fourth step, the simulation results are analysed, graphically visualised and stored. If the users are not satisfied with the result of the chosen adaptation strategy, new scenarios can be defined and analysed. This process is repeated by the user until an adaptation strategy has been found that seems economically sustainable in terms of adaptation to climate change. Special tools help users to analyse and interpret the result data.

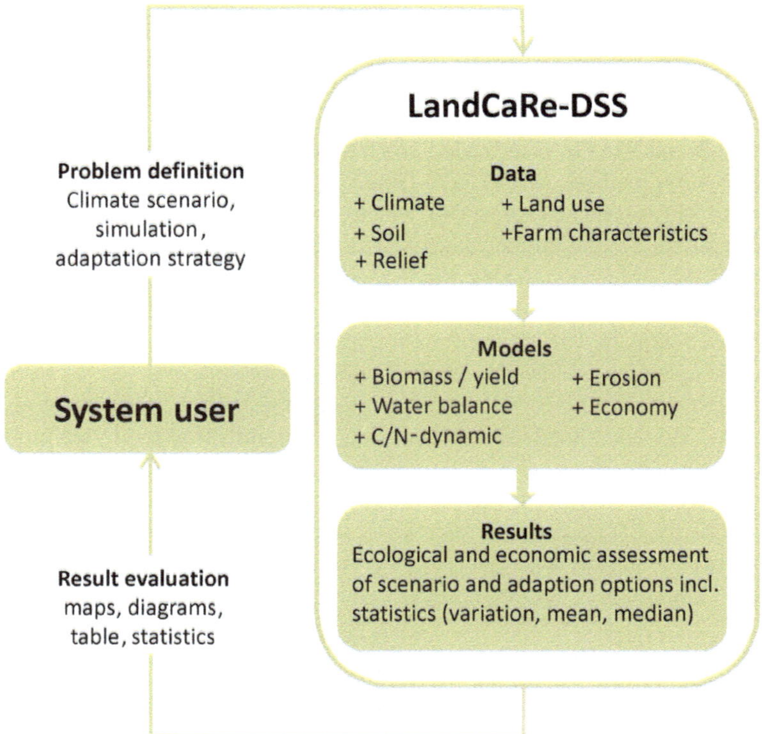

Fig. 24.3 Basic principle of operation of LandCaRe-DSS for decision-making

24.2.5 The Systems' IT Realisation

The LandCaRe-DSS has been realised as two distinct systems, a desktop application and a web-application. The former had originally been designed as a prototypical application in order to develop the latter. In the end, the collaborative project LandCaRe 2020 ended up with two systems offering different functionality. At present, only the desktop application is available and maintained.

24.2.5.1 The Desktop-LandCaRe-DSS

The desktop LandCaRe-DSS is a hybrid C++ application based on the QT framework (QT 2019). It merges web-based with traditional user interface elements. The application is centred around a web-based map view (currently Google Maps®) which takes the job of a simplified GIS system. Overlaid to the map view is a zooming user interface (ZUI), which is the user's main way to interact with the simulated landscape, the model inputs and results. In general, the user will see predefined rectangular regions, which visualise the availability of spatial data. A user of the system will then select one of these regions or create a subregion using the existing ones, to tell the system where a regional simulation will be performed. The LandCaRe-DSS will then create all necessary user interface elements to execute the simulation, change input data and visualise results in and around the chosen region. The system's user interface is designed to be mostly non-modal, so users can work on multiple regions at once or run multiple models in the same region. This property of the ZUI enables for instance a natural way to compare model results or merge results of multiple models in the same region.

Models and their outputs are available in two categories, as regional models whose results are presented to the user as a set of maps overlaid to the simulated region and as point models where multiple results are visualised in a column view. Both result visualisations offer ways to show the users the uncertainty of multiple simulation runs, usually from a set of climate scenario realisations. Initially, the LandCaRe-DSS was a pure desktop application, which means all models have been built into the system itself. The system is able to utilise all the processors available on a system, by running models in parallel where applicable. More current updates to the DSS moved models out of the system and enabled cloud computing scenarios where the system communicates with remotely running models, e.g. on a high-performance computer (HPC).

The Desktop-LandCaRe-DSS was designed to be extensible with new models. Based on the existing visualisation, it is comparably easy to integrate new C++ based models. With the later cloud computing updates to the system also, the integration of non-C++ models is possible. On the data side, it is very easy to add new regions into the system. By obeying some conventions, the user can add new ESRI-ASCII Grid formatted files and a SQLite soil database into a special data folder on the system's installation. After restarting the LandCaRe-DSS, the newly supported region is recognised by the system and can be found at the according geolocation on the world map. Via naming convention and given the appropriate rectangular projection PROJ4 parameters (https://proj4.org) in a SQLite database, a region can appear everywhere on the planet.

24.2.5.2 The Web-LandCaRe-DSS

The web-based LandCaRe-DSS was designed to be a traditional, albeit somewhat interactive, server-based web-application. It utilised the desktop application's core components to run the actual simulations, but added features on its own, which were out of reach for the prototypical desktop application. Also, in the Web-DSS, the user was able to choose a region on a Google Maps® interface. Only a reduced set of models, compared to the desktop version, was available, but all simulation runs were stored on the system's server. This sets up the possibility to search for simulations that were done in the past and simply display the results, something not possible and implemented with the desktop version. Because potentially many users might do simulations in parallel, the web-based LandCaRe-DSS queued simulation requests and notified the users later if their simulation run had been finished.

24.2.6 System Validation

Within the project LandCaRe 2020, the LandCaRe-DSS prototype was parameterised and validated for two contrasting model regions within the Eastern part of Germany—the catchment area of the river Ucker (Uckermark district) and the Weißeritz river catchment. The Uckermark district (0–169 m a. s. l.; 2,600 km^2) is located in the north-east of the German lowlands in the Federal State of Brandenburg and has very heterogeneous soils due to its location in the end-moraine landscape. Historically, the Uckermark district is a typical agricultural region with a wide range of agricultural crops and a small grassland area and is dominated by a continental climate, characterised by hot and dry summers (high risk of drought), cold winters and little rainfall. The Weißeritz catchment (102–904 m a. s. l.; 400 km^2) is located in the Free State of Saxony (west of Dresden, South-East Germany) in the eastern Ore Mountains and is characterised by a transition climate from maritime to subcontinental types with a relatively high level of precipitation. At the arable areas of the Weißeritz catchment, mainly winter wheat, winter oilseed rape, winter barley, silage maize and winter rye are cropped. At higher elevations, mainly grassland is cultivated.

24.3 LandCaRe-DSS—Levels of Application

24.3.1 Information System

The LandCaRe-DSS has an information system both concerning the system itself and general climate change topics. The information system includes the following aspects:

- Introduction to the topic ‘Climate change and agriculture’
- Various questions about climate change and agriculture, such as
  - What is climate change?
  - Do humans cause climate change?
  - How does the greenhouse effect work?
  - How does climate change affect the soil? and
  - Do all plants react identically to more CO_2?
- General information on the LandCaRe-DSS
- Model short descriptions and model documentation
- Current application possibilities and limitations of LandCaRe-DSS prototype
- Potential adaptation measures of agriculture to climate change, structured in
  - Crop production (variety selection/sowing, agricultural crop/crop rotation, soil tillage, fertilisation, irrigation, plant protection),
  - Horticulture (fruit and wine growing and vegetable growing),

- Grassland and forage cultivation, and
- Animal production (animal husbandry, feeding, animal breeding).

Using the models made available in the LandCaRe-DSS, the user can examine which probable effects (yield, ecology, economy) individual adaptation measures would produce.

It is possible to navigate between the individual information levels of the information system. Links are also included to further information that is available via the Internet.

24.3.2 Analysis of Climate Data

In order to perform a quick climate data analysis, user-friendly algorithms can be used in LandCaRe-DSS to undertake a trend analysis, an inner-yearly analysis (daily and monthly) and a frequency analysis. Trend analysis can be performed for the temperature (minimum, mean and maximum), annual precipitation, climatic water balance, heating days, cooling days, the start and end of vegetation period, the thermal sum, the chill sum and the climatic index for fruit trees according to Schwärzel (2000). Using the climate scenario A1B and the climate regionalisation WETTREG, Fig. 24.4 shows the climate levels comparison 1975 versus 2035 for the meteorological station Angermünde, Germany, for the annual course of the mean temperature and the corresponding 30-year variation ranges, the monthly precipitation sums with their 30-year fluctuation ranges and the monthly number of ice, frost, hot and tropical days.

With LandCaRe-DSS, it is also possible to compare different climate regionalisation methods and climate scenarios. The comparison for the Müncheberg climate region illustrated as an example in Fig. 24.5 shows that the use of different regionalisation methods and different climate scenarios are already two sources for the uncertainties in climate impact assessment.

24.3.3 Analysis of Phenology and Ontogenesis

Temperature is the most important variable that controls plant development. Plant processes are thereby accelerated or slowed down, switched on or switched off. What the phenological phases of special wild plants are for the temporal positioning of the beginning of seasons, the ontogenesis of agricultural plants is the most important variable for planning the site- and farm-specific agro-management within the agricultural cropping systems. The user of the LandCaRe-DSS can choose different climate scenarios, different time periods up to 2100 and different sowing dates for running the ONTO model. As a result, the DSS user gains information on the crop reaction and draws consequences for the farmer´s own agro-management.

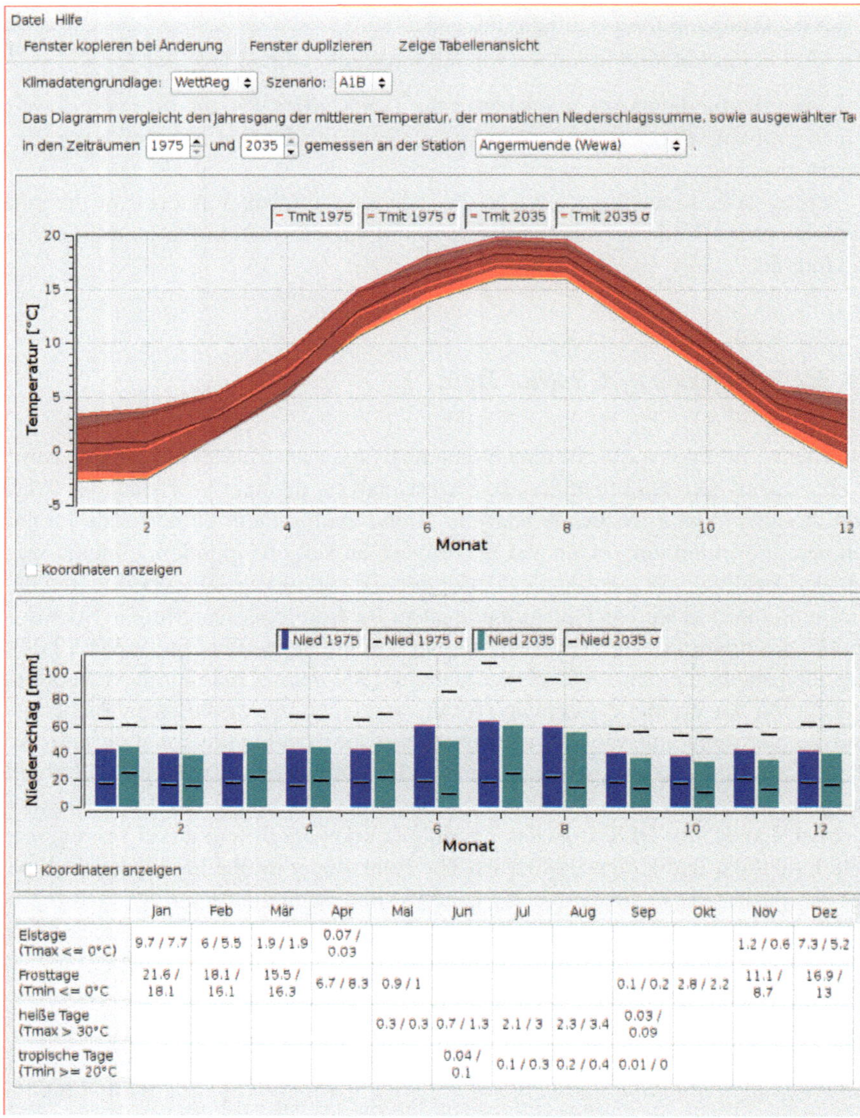

	Jan	Feb	Mär	Apr	Mai	Jun	Jul	Aug	Sep	Okt	Nov	Dez
Eistage (Tmax <= 0°C)	9.7 / 7.7	6 / 5.5	1.9 / 1.9	0.07 / 0.03							1.2 / 0.6	7.3 / 5.2
Frosttage (Tmin <= 0°C	21.6 / 18.1	18.1 / 16.1	15.5 / 16.3	6.7 / 8.3	0.9 / 1				0.1 / 0.2	2.8 / 2.2	11.1 / 8.7	16.9 / 13
heiße Tage (Tmax > 30°C					0.3 / 0.3	0.7 / 1.3	2.1 / 3	2.3 / 3.4	0.03 / 0.09			
tropische Tage (Tmin >= 20°C						0.04 / 0.1	0.1 / 0.3	0.2 / 0.4	0.01 / 0			

Fig. 24.4 Climate level comparison 1975 versus 2035 for the annual mean temperature course and the corresponding 30-year variation ranges (top), for the monthly precipitation sums with their 30-year fluctuation ranges (centre) and for the monthly number of ice days ($T_{max} \leq 0$ °C), frost days ($T_{min} \leq 0$ °C), hot days ($T_{max} \geq 30$ °C) and tropical days ($T_{min} \geq 20$ °C) (bottom, 1975/2035) at the meteorological station Angermünde, Germany (regionalisation method: WETTREG, climate scenario: A1B)

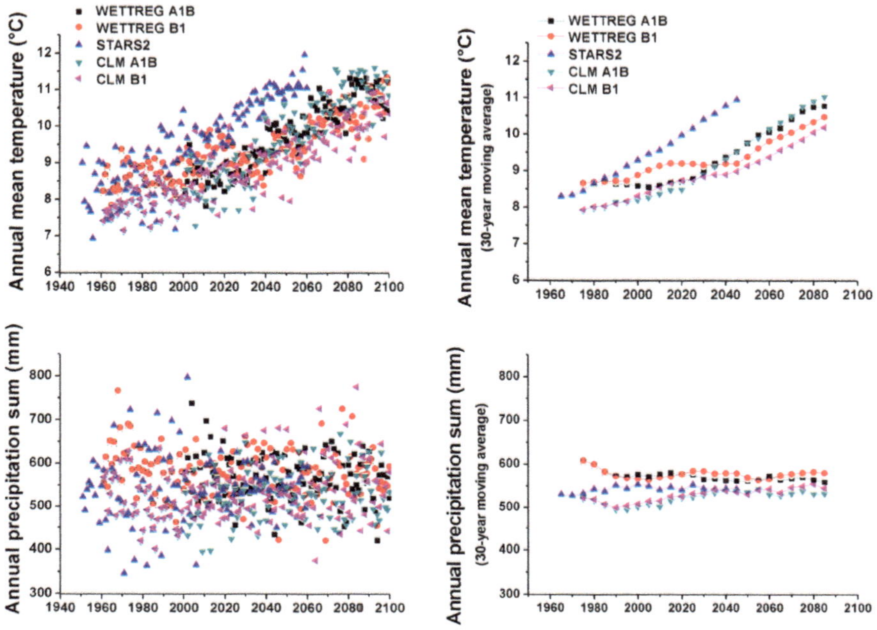

Fig. 24.5 Comparison of the climate regionalisation methods WETTREG, CLM and STARS2 with two climate scenarios each (A1B, B1) in the period 1950–2100 for the annual mean temperature (top) and the annual precipitation sum (bottom) (annual values (left), 30-year moving average (right), meteorological station Müncheberg)

For winter wheat in the Müncheberg region, Germany, Fig. 24.6 shows the reaction of ontogenesis between sowing and ripeness due to changing temperature conditions for different time slices, calculated with the model ONTO, for the climate scenarios A1B and B1 as well as three climate regionalisation methods (WETTREG, CLM, STARS2). The simulation results show that the different regionalisation methods have a greater impact on the length of individual development phases in winter wheat compared to different climate scenarios. Apparently, both the choice of regionalisation methods and the choice of possible climate scenarios contribute significantly to the uncertainty of the results. Comparing all variants, the user can see that in the future, the entire cultivation period between sowing and harvesting will be shortened and that development stages relevant to agro-management will occur earlier in the year and thus lead to a temporally changed management in the cultivation system of winter wheat.

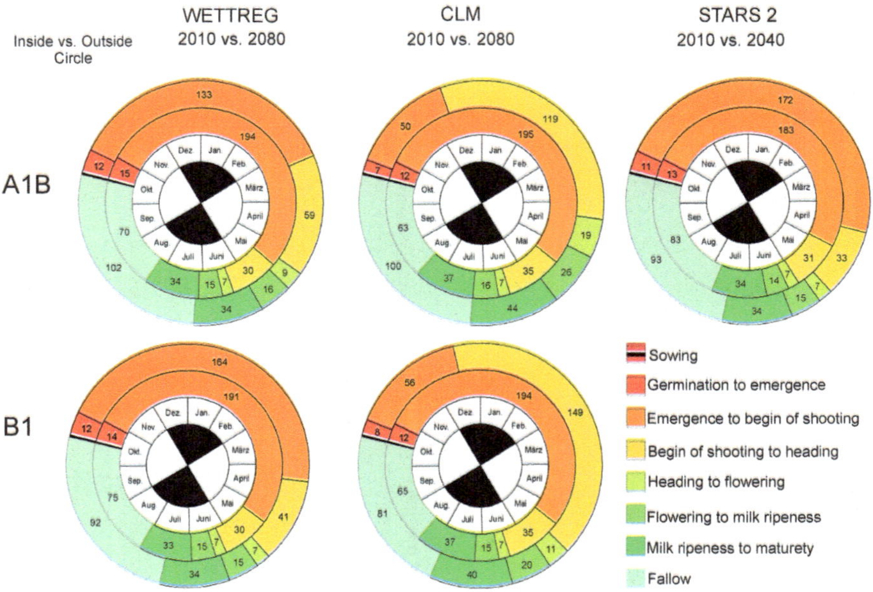

Fig. 24.6 Length of ontogenesis stages for winter wheat (in days) between sowing and harvest maturity in comparison of different climate scenarios (A1B, B1), different climate regionalisation methods (WETTREG, CLM, STARS2) and different time slices (Müncheberg region, sowing date: 15 October , model ONTO)

24.3.4 Climate Change Impact Assessment on National Scale

At the national scale, the result of various scenario runs is presented to the stakeholders and LandCaRe-DSS users in form of maps. The result maps based on the RAUMIS model contain information on changes in crop yields, cropping structure, farm economies and irrigation demand for different future time periods as a consequence of climate change. Because of the complexity of the RAUMIS model, the result maps were preprocessed by the RAUMIS team of the Thünen Institute (TI) Brunswick and incorporated into the LandCaRe-DSS. On this basis, the LandCaRe-DSS user can carry out statistical analyses. Figure 24.7 shows an example of the winter wheat yields expected in Germany at the national scale in 2025. The smallest spatial resolution here is the municipality.

The map clearly shows a much greater depression of winter wheat in the north-east of Germany, due to the continental climate influence, the expected winter wheat yields due to higher temperatures, and insufficient precipitation in the main vegetation period (and thus an increasing evapotranspiration), than in the north-western regions of Germany, which are primarily influenced by a maritime climate.

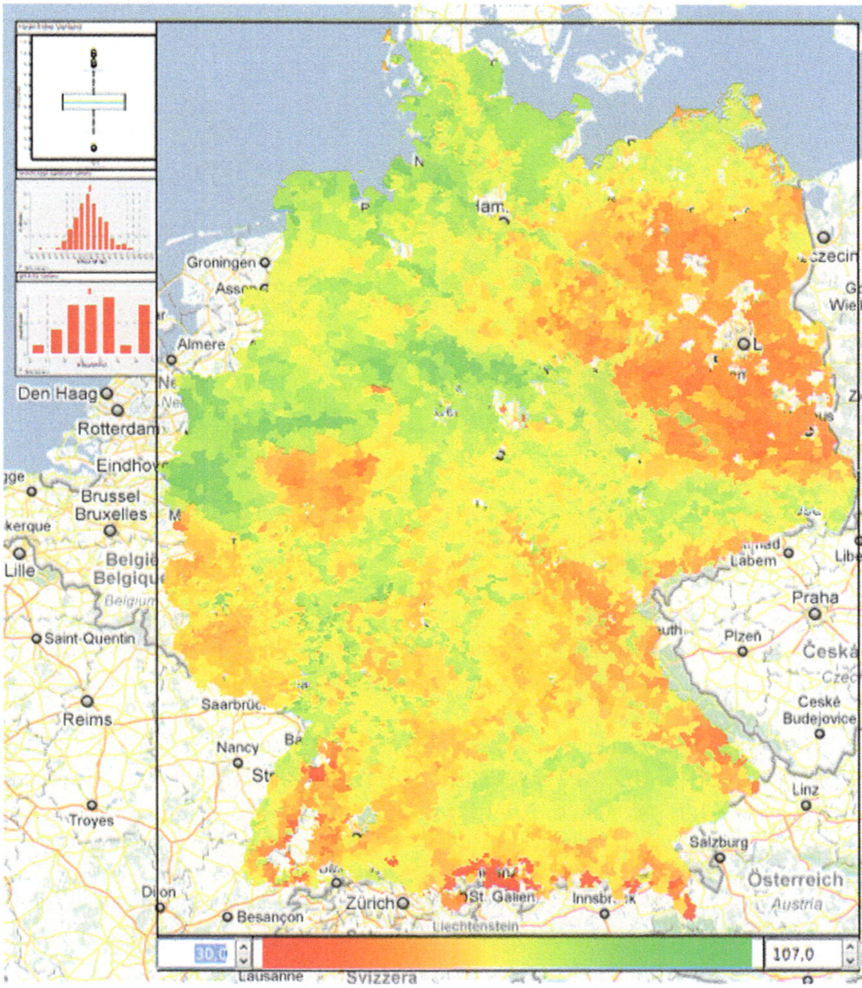

Fig. 24.7 Regional winter wheat yields for Germany expected in 2025 with a variation range between 3 t ha^{-1} and 10.7 t ha^{-1} (simulation model: RAUMIS, regionalisation method: WETTREG, climate scenario: A1B)

24.3.5 Climate Change Impact Assessment on the Regional Scale

On the regional scale, the ecological impact assessment of climate and land use changes are realised on a high spatial resolution, i.e. on a minimum pixel size of 1 ha (100 × 100 m). Most of the models mentioned above can be activated on this scale, but without coupling to the economic module. Using different models, calculations are possible for the expected impacts of climate change on yields for arable and grassland,

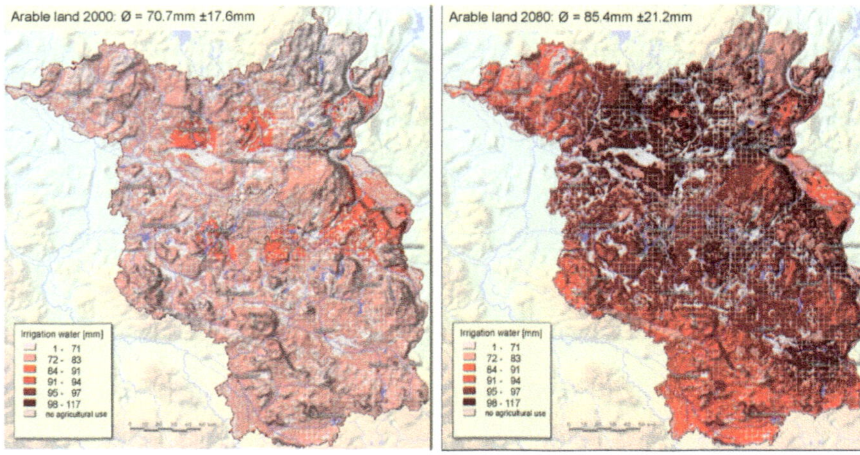

Fig. 24.8 Distribution of the irrigation water demand in 2000 (left) compared with the situation in 2080 (right) for the Federal State of Brandenburg (regionalisation method: WETTREG, climate scenario: A1B, simulation model: ZUWABE)

on the potential erosion risk, on the real evapotranspiration and groundwater recharge, on the irrigation water demand and others. At this regional scale for a statistical analysis, the average, median, histogram and others are automatically calculated and presented to the LandCaRe-DSS user.

In Fig. 24.8 for the Federal State of Brandenburg (BB), the irrigation water demand in 2000 (average for BB: 70.7 mm) is compared to the situation in 2080 (average for BB: 85.4 mm). It is seen that in 2080, the potential irrigation water demand is significantly larger than in 2000 because of higher temperatures, higher evapotranspiration and lower precipitation during the growing period of the irrigated agricultural crops.

With the statistic-based model YIELDSTAT, the crop yields can be estimated on the regional scale for 15 different agricultural crop types and two grassland intensities. Figure 24.9 shows the results of a simulation with the model YIELDSTAT for the Quillow catchment for the 30-year climatic period 1991–2020 for a given cropping ratio. In order to make the yields of the different agricultural crops comparable, they are converted into grain units (GU).

The LandCaRe-DSS screenshot (Fig. 24.9) shows the spatial distribution of yields in the Quillow catchment (centre) and on the left, the results of statistical calculations such as spatial variance or yield distribution. The yields range from 7.1 GU ha^{-1} (red) to 129.2 GU ha^{-1} (green). For the user of LandCaRe-DSS, the initial conditions for the simulation and the spatial inputs used for it are shown in map form in the lower two bands.

The LandCaRe-DSS allows scenario simulations to be carried out in order to estimate expected regional yield developments of agricultural crops under climate change and to gain information on the uncertainty of these yields. Simulations with different climate regionalisation methods and different climate scenarios can be carried out for different climatic time slices. The results can then be presented as a

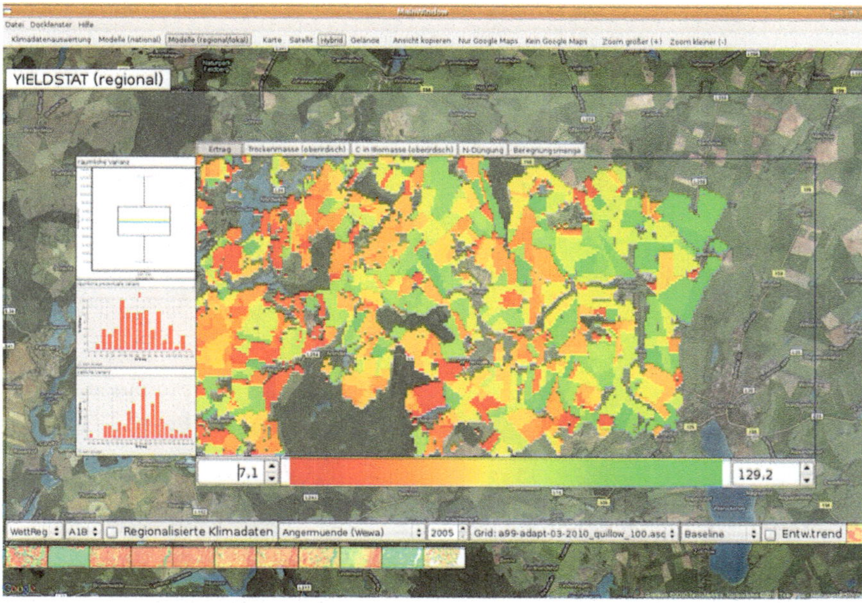

Fig. 24.9 Spatial yield (GU ha^{-1}) distribution for the Quillow catchment within the Uckermark region, Germany, for the climatic period 1991–2020 using a cropping ratio given by the LandCaRe-DSS user (regionalization method: WETTREG, climate scenario: A1B, simulation model: YIELDSTAT)

spatial difference map or as a general summary of a region. Figure 24.10 shows the yield development of different agricultural crops in the period 2005–2075 in the Uckermark region by comparing two different climate regionalisation methods and two different climate scenarios.

24.3.6 Climate Change Impact Assessment on Local or Farm Scale

At the local or farm level, the LandCaRe-DSS offers interactive simulation and integrated impact assessment of agricultural adaptation strategies to climate change (crop rotation, tillage, fertilisation, irrigation, price and cost changes, …) depending on different price–cost relationships. The user of the system is informed about changes in plant productivity (yield, yield quality), soil fertility (water, carbon and nitrogen content), water erosion and farm economy. At the farm level, the dynamic process-based agroecosystem model MONICA and the statistic-based hybrid model YIELDSTAT for yield forecasting are coupled with the farm economy module (FEM). The user of the LandCaRe-DSS gets information on several economic parameters, fertiliser amounts and costs, irrigation water needs and costs and finally on crop yields and

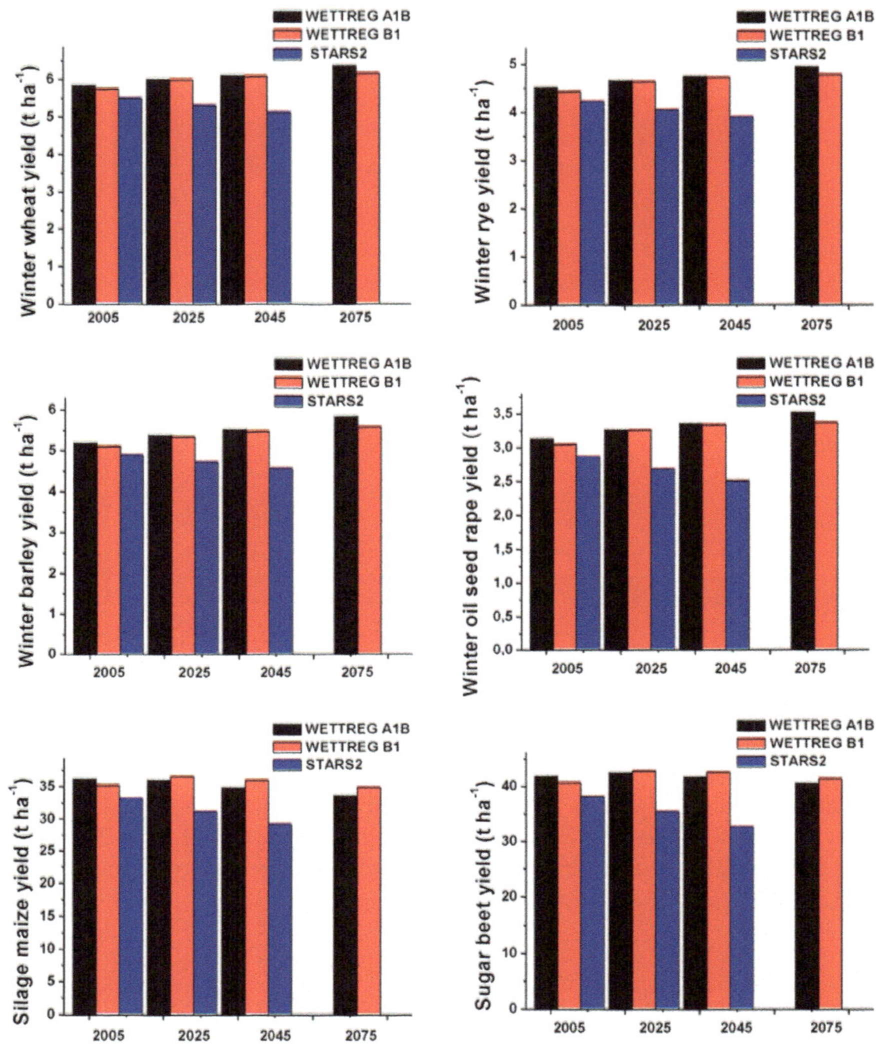

Fig. 24.10 Yield development of agricultural crops in the Uckermark region in comparison to different climate regionalisation methods (WETTREG, STARS2) and different climate scenarios (A1B, B1) for the time period 2005–2075 (assumptions: current cropping ratio (winter wheat: 33.9%, winter rye: 8.5%, winter barley: 11.5%, winter oil seed rape: 20.8%, silage maize: 8.9%, sugar beet: 2.7%), soil tillage with plough, no irrigation, no trend consideration for agro-management and plant breeding)

sales revenues. For all output information, the variances of the results as basis for risk assessment are given, based on up to 90 simulation runs. The results are visualised using normalised bar graphs to allow a better comparison between different scenario runs. On the background of a Google Maps® satellite image, Fig. 24.11 shows the

Fig. 24.11 Visualisation of simulation results for the combined FEM-MONICA model (top) and the combined FEM-YIELDSTAT model (bottom) at local level for a small part of farms within the Uckermark region, Germany (regionalisation method: WETTREG, climate scenario: A1B; green bars: ecological parameters, pink bars: economic parameters, yellow bars: crop yields, grey bars: fertilisation/nutrition parameters, blue bars: water-related parameters)

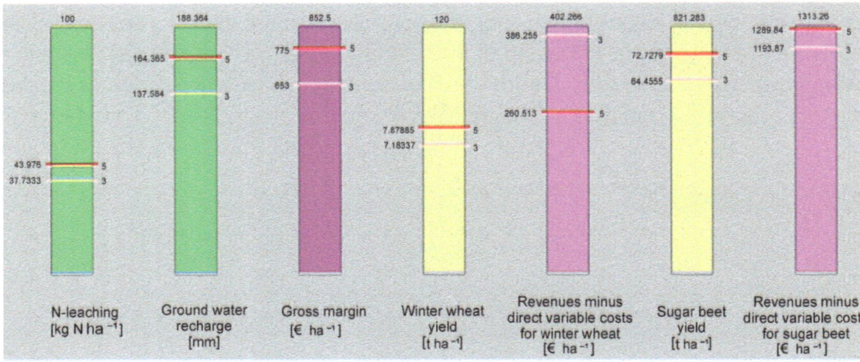

Fig. 24.12 Zoomed view on the result bars (here a result bar excerpt of an example with winter wheat and sugar beet) for two simulation runs (3—without irrigation, 5—with irrigation)

visualised simulation results for both FEM-YIELDSTAT and FEM-MONICA for the local resp. farm scale. The results for ecological, economic, nutrient- and water-related as well as yield variables are arranged as standardised bars around a part of the farm or around the farm (see also Fig. 24.12). The area considered for simulations is subdivided into grid cells of 1 ha (100 m × 100 m). For FEM-MONICA, additional information concerning the annual dynamics of the C, N and water contents in the soil as well as groundwater recharge and N leaching is provided on the left-hand side. The respective results are presented as 30-year mean as well as the variance around the mean. For FEM-YIELDSTAT, only additional information concerning the spatial and temporal variances is provided on the left. In the upper part, the data and assumptions of the actual simulation scenario are shown.

Figure 24.12 shows an enlarged view of a part of the result bars. In the result bars, result lines are plotted for all simulations, with the simulation number on the right and the simulation result on the left. If the LandCaRe-DSS user goes to a result line with the cursor, the distribution of the results around the mean of a 30-year simulation is plotted. Parallel to the graphical result presentation, a table with all necessary simulation-related information is created in the background of the system. The simulation number refers to the results documented in tabular form which can be enabled in a different part of the user interface.

24.4 LandCaRe-DSS—Further Use and Development

Because of the modular architecture of LandCaRe-DSS, there is a relatively low effort

- to adapt the system to geodatabases that cover other regions or countries,
- to include other impact models,

- to switch to other climate and emission scenario data and
- to also integrate remote models using ZeroMQ (http://zeromq.org) message-based communication.

Up to now, LandCaRe-DSS was developed, adapted and used in connection with a number of other research projects such as REGKLAM (Regional Adaptation Programme to Climate Change for the Dresden Region, Saxony, Germany (REGKLAM 2013a, b)), CARBIOCIAL (Carbon sequestration, biodiversity and social structures in Southern Amazonia: models and implementation of carbon-optimised land management strategies for regions within the Mato Grosso and Pará states of Brazil (Carbiocial 2014) and KIT LandCaRe-DSS (Model-based tools for strategic and operational irrigation measures under climate change for the model region Uelzen, Lower Saxony, Germany (KIT LandCaRe-DSS 2015)).

Models integrated in LandCaRe-DSS were used in climate scenario studies to assess the impacts of climate change on crop productivity and irrigation water demand of arable land for three German states: Saxony (Mirschel et al. 2009; Mirschel 2010), Brandenburg (Mirschel et al. 2016) and Thuringia (Mirschel et al. 2012, 2013).

Further developments of the LandCaRe-DSS concentrate now on applying the lessons learned during the last decade of using and developing the system to a model and simulation infrastructure currently being developed at ZALF. While the design principles the LandCaRe-DSS was built with in mind still hold true today, the demands to simulate large regions, countries or even continents have risen considerably. These new needs can't be catered for by locally installed stand-alone software running on a single personal computer or laptop. A contemporary decision support system must be able to use all available computational resources, no matter if available locally or remotely. Some first steps towards this goal have been taken by enabling the LandCaRe-DSS to use remote models via a ZeroMQ (http://zeromq.org)-based messaging interface. That way the system can use every model if it can be equipped with the necessary communication interface. In general, a message-driven approach to model integration and communication can be done at the level of the system itself, thus integrating models into the system, but also for the communication of models amongst each other. In the former case, the model MONICA has recently been moved from an DSS-integrated model to a loosely integrated remote model. An example for the latter case is the coupling at the daily time step level of the core computational components of the irrigation advisory system WEB-BEREST (Mirschel et al. 2014b) to the model MONICA in order to calculate the irrigation water amounts by WEB-BEREST instead of using the simplified algorithms by MONICA.

The main research and development question for a scaled-up successor version to the LandCaRe-DSS is however how to create a large powerful system of many loosely coupled components? Things that are easy at a single application, single computer level, become a challenge when being distributed. But there is no way to ignore today's networked world. A modern information and decision support system has to use resources on the network, wherever they are, do so securely and safely and still maintain an interactive and user-friendly interface. So, all goals initially set up for the first prototype of the desktop LandCaRe-DSS still apply today, but our task now is to transfer them into a system that runs in a twenty-first century networked world.

24.5 Conclusions and Outlook

Climate change will transform agriculture and the use of agro-landscapes in many ways. There is a need for better information on the expected climate changes at the regional scale and its impacts on agricultural yields, farm economy and on environmental indicators for farmers, local stakeholders and politicians. In the first decades of the twenty-first century, the expected impacts of climate change on agriculture in Germany are relatively low, but the second half of the century is expected to bring more severe impacts. The most important agricultural adaptation strategies in Germany to cope with climate changes are irrigation, flexible multi cropping crop rotations, change to low and no tillage farming systems and conservation of soils on a good fertility status. Until now, farmers and regional stakeholders have had only a few appropriate tools available to support decision-making processes in adapting to climate change. The presented information and decision support system LandCaRe-DSS—a new generation of interactive decision support systems—helps to close this gap.

LandCaRe-DSS supports interactive spatial scenario simulations, multi-ensemble and multi-model simulations using scenario techniques, uncertainty analysis, and easy-to-understand visualisations. LandCaRe-DSS offers the advantages of high simulation speed, broad functionality and open system architecture. The LandCaRe-DSS

- offers visualised regionalised data on expected changes to important climate parameters, or changes that have already occurred, for freely selected climate stations and periods,
- enables the simultaneous consideration of several climate scenarios from various regionalisation methods (new scenarios can be integrated with little effort),
- provides background information on climate change, as well as information on potential adaptation measures that could be taken in agriculture,
- makes available knowledge of climate-related changes in the yield and cultivation structures of main agricultural crops as well as national irrigation requirements,
- enables an assessment of the ecological consequences of potential climate and land use changes at the regional level,
- permits an analysis and integrated assessment (ecological, economic) of possible climate adaptation strategies (crop rotation, soil tillage, fertilisation, irrigation, price and cost changes and others) at farm level by virtually testing potential adaptation options and different climate projections,
- is characterised by a free choice of spatial region and the interactive simulation models to be applied based on modern information technologies,

- primarily supports strategic planning and decision-making concerning the adaptation of agriculture, regional water and land management to climate change (examination of action via iterative simulation),
- requires little effort to extend to other regions or countries because of its modular structure and can be adapted by incorporating region-specific geodata and static and/or dynamic models that are valid for these regions,
- enables the use of external models in later versions by the use of ZeroMQ message-based communication.

Despite all the progress made in modelling climate and climate impacts in recent decades, many uncertainties remain. There are uncertainties in the regional distribution of future climate changes, uncertainties about global and regional economic developments but also uncertainties in the understanding of dynamic key processes that describe the interactions of increased CO_2 with other climate variables (including extreme values). However, there are also uncertainties in the description of soil processes, in the assessment of water quality and in the description of water cycles, as well as in the description of ecosystem interactions with pests, weeds and diseases and with the vulnerability of the ecosystem as a whole. All these uncertainties influence the accuracy of the conclusions and results generated by LandCaRe-DSS. For LandCaRe-DSS, it means that the system should be open for further developments and integration of new knowledge, new models and new data resources. Likewise, it should be easy enough to support stakeholders in the complex process of decision-making in an interactive way.

Until now, LandCaRe-DSS has been advanced, updated and used in a number of research projects such as REGKLAM (project for agricultural regions of Saxony, Germany), CARBIOCIAL (project for regions within the Mato Grosso and Pará states of Brazil) and KIT LandCaRe-DSS (project for strategic and operational irrigation measures under climate change for the model region Uelzen, Lower Saxony, Germany).

For the future, concerning the further development of LandCaRe-DSS, it is planned

- to take the lessons learned during developing and using LandCaRe-DSS and scale it up into a ZALF model and simulation infrastructure which can cope with the demands arising more than a decade after its original inception,
- to incorporate the Müncheberg Soil Quality Rating (MSQR, Müller et al. 2014) for assessing the quality of global farmland,
- to incorporate tested regional models describing natural, scientific and socio-economic landscape indicators such as degree of biodiversity, soil fertility, trace gas emission, labour demand and labour potential, landscape fragmentation and others,
- to adapt the system to Russian conditions, i.e. to their special soil classification system (Shishov et al. 2004), to the Russian climate and emission scenario data, to Russian as system interface language, and to take into account models developed in Russia for Russian conditions, such as the agroecosystem model AGROTOOL (Poluektov et al. 2012), the Climate–Soil–Yield (CSY) Imitation System

for assessment of spatial wheat cultivation risk (Pavlova et al. 2019) or models for soil nitrogen (Poluektov and Terleev 2010) and different soil properties (Terleev et al. 2017).

It is also planned to apply the system to practical case studies in order to test its performance, practicability and limitations at local/farm and regional levels. For the future, it is intended to use the system as a tool for learning and knowledge exchange to help students, consultants, authorities and other interested stakeholders better understand the processes, interactions and feedbacks within such complex systems like an agricultural landscape.

Acknowledgements This contribution was funded by the German Federal Ministry of Education and Research (BMBF) within the 'klima*zwei*' research programme (grant: 01 LS 05109), the German Federal Ministry of Food, Agriculture and Consumer Protection and the Ministry of Infrastructure and Agriculture of the Federal State of Brandenburg (Germany).

References

Anter J, Gömann H, Kreins P, Richmann A (2009) Einfluss sich wandelnder ökonomischer Rahmenbedingungen auf die Beregnung landwirtschaftlicher Kulturen in Deutschland, Landbauforschung, Sonderheft 328 - Wasser im Gartenbau, pp 21–28

Ascough JC, Shaffer MJ, Hoag DL, Mc Master GS, Dunn GH, Ahuja LR, Weltz MA (2001) GPFARM: an integrated decision support system for sustainable great plain agriculture. In: Scott DE, Mohtar RH, Steinhardt GC (eds) Sustaining the global farm. Selected papers from the 10th international soil conservation organization meeting held from 24 to 29 May 1999 at Purdue University and the USDA-ARS National Soil Erosion Research Laboratory, pp 951–960

ATV-DVWK-Regelwerk (2002) Verdunstung in Bezug zu Landnutzung, Bewuchs und Boden. ATV-DVWK-Regelwerk: Merkblatt; M 504, p 144

Berlekamp J, Lauterbach S, Graf N, Matthies M (2005) A decision support system for integrated river basin management of the German Elbe. In: Proceedings MODSIM-2005, pp 1518–1524

Böhm U, Kücken M, Ahrens W, Block A, Hauffe D, Keuler K, Rockel B, Will A (2006) CLM—the climate version of LM: brief description and long-term applications. COSMO Newsl 6:225–235

Carbiocial (2014) Carbon sequestration, biodiversity and social structures in Southern Amazonia: models and implementation of carbon-optimized land management strategies. http://www.uni-goettingen.de/de/211024.html. Accessed 4 Apr 2019

Chmielewski FM (2003) Phenology and Agriculture (chapter 7). In: Schwartz MD (ed) Phenology: an integrative environmental science. Kluwer Academic Publishers, Boston, pp 505–522

Chmielewski FM, Henniges Y (2007) Phänologische Modelle als Grundlage zur Abschätzung des Klimaimpakts. Berichte Meteorologisches Institut Freiburg 16:229–235

COP24 (2018) COPE24 Katowice – United Nations climate change conference. http://cop24.katowice.eu/de/. Accessed 7 Mar 2019

DIN19708 (2005) Bodenbeschaffenheit – Ermittlung der Erosionsgefährdung von Böden durch Wasser mit Hilfe der ABAG (Soil quality – Predicting soil erosion by water by means of ABAG). DIN 19708:2005-02, Normenausschuss Wasserwesen (NAW) im DIN, pp 25

Enke W, Schneider F, Deutschländer T (2005) A novel scheme to derive optimized circulation pattern classifications for downscaling and forecast purposes. Theoret Appl Climatol 82:51–63

Franke J, Köstner B (2007) Effects of recent climate trends on the distribution of potential natural vegetation in Central Germany. Int J Biometeorol. 52(2):139–147

Gömann H, Kreins P, Herrmann S, Wechsung F (2005) Impacts of global changes on agricultural land-use in the German Elbe region: results of an operational modelling tool for planning, monitoring and agri-environmental policy counselling [CD-ROM]. In: 21st European regional conference—ERC 2005—integrated land and water resources management: towards sustainable rural development, 15–19 May 2005—Frankfurt (Oder), Germany—Slubice, Poland, p 11

Gömann H, Offermann F, Kleinhanss W, Kreins P, von Ledebur O, Osterburg B, Pelikan J, Salamon P, Sanders J (2009) vTI-Baseline 2009–2019: agro-economic projections for Germany. Special Issue No. 33

Henrichsmeyer W, Cypris C, Löhe W, Meudt M, Sander R, von Sothen F, Isermeyer F, Schefski A, Schleef KH, Neander E, Fasterding F, Helmcke B, Neumann M, Nieberg H, Manegold D, Meier T (1996) Entwicklung eines gesamtdeutschen Agrarsektormodells RAUMIS96. Endbericht um Kooperationsprojekt. Forschungsbericht für das BML (94 HS 021)

Huglin P (1986) Biologie et écologie de la vigne. Lavoisier (Edition Tec & Doc), Paris, p 371

Jacob D, Van den Hurk BJJM, Andrae U, Elgered G, Fortelius C, Graham LP, Jackson SD, Karstens U, Chr K, Lindau R, Podzun R, Rockel B, Rubel F, Sass BH, Smith R, Yang X (2001) A comprehensive model intercomparison study investigating the water budget during the PIDCAP period. Meteorol Atmos Phys 77(1–4):19–44

Kaiser T, Käding H, Kiesel J, Müller L, Hierold W, Behrendt A (2005) the derivation of grassland vegetation types on the basis of site and land use characteristics. Arch Agron Soil Sci 51(4):405–416

Kaden S, Kaltofen M, Timmermann R, Lüllwitz T, Roers M (2010) Die Elbe-Expert-Toolbox - ein Entscheidungshilfesystem für das integrale wasserwirtschaftliche, (öko-)hydrologische und sozioökono-mische Mangement eines Flusseinzugsgebietes. Tagungsband "Angewandte Geoinformatik"—22. AGIT-symposium, 07–09 July 2010, Salzburg Austria, pp 290–299

Käding H, Kaiser T, Werner A (2005) Model for calculating grassland yields and forage quality in North-East Germany on the basis of site and management characteristics. Arch Agron Soil Sci 51(4):417–431

KIT LandCaRe-DSS (2015) KMU-innovativ—joint project—LandCaRe-DSS: model-based tools for strategic and operational irrigation measures under climate change. http://www.zalf.de/en/forschung/projekte/Pages/detail.aspx?fpid=1543&tpid=832

Kofalk S, Boer S, Scholten M, de Kok, J-L, Matthies M, Hahn B (2004) Ein decision support system für das Flusseinzugsgebiets-Management der Elbe. In: Möltgen J, Petry D (eds) Interdisziplinäre Methoden des Flussgebietsmanagements, Workshopbeiträge 15./16. März 2004, IFGIprints, Schriftenreihe des Instituts für Geoinformatik, Westfäische Wilhelms-Universität Münster, Band 21, S.1–10

Köstner B, Wenkel K-O, Bernhofer Ch (2012) Neue modellbasierte Informationssysteme für Klimafolgen-szenarian in der Landwirtschaft. Promet 38(1/2):42–52

Kuhnert M, Köstner B (2008) Effects of different climatic conditions on the site potential of Norway spruce, common beech and grassland assessed by modelling photosynthesis. In: Geophysical research abstracts, vol 10, EGU General Assembly, Vienna, Austria

Kuhnert M, Köstner B (2009) Regional modelling of water and CO_2-fluxes with a one-dimensional SVAT model. In: Geophysical research abstracts, vol 11, EGU General Assembly, Vienna, Austria

Kyoto (1997) Kyoto-Protokoll. https://www.lpb-bw.de/kyoto_protokoll.html. Accessed 7 Mar 2019

LADSS (2005) Land allocation decision support system. http://www.macaulay.ac.uk/LADSS. Accessed 4 Apr 2019

Mahammadzadeh M, Biebler H, Bardt H (2009) Klimaschutz und Anpassung an die Klimafolgen – Strategien, Maßnahmen und Anwendungsbeispiele. Institut der deutschen Wirtschaft Köln Medien GmbH, 2009, Köln, 310 p

Mirschel W (2010) Modell ONTO zur Beschreibung der Ontogenese von landwirtschaftlichen Fruchtarten im LandCaRe DSS (Modellbeschreibung). In: Wenkel KO, Berg M, Wieland R, Mirschel W (ed) Vorsorge und Gestaltungspotenziale in ländlichen Räumen unter regionalen

Wetter- und Klimaänderungen (LandCaRe 2020) PT-DLR 01 LS 05104: Modelle und Entscheidungsunterstützungssystem zur Klimafolgenab-schätzung und Ableitung von Adaptationsstrategien der Landwirtschaft an veränderte Klimabedingungen (AGROKLIM-ADAPT); Teilprojekt 6 (DSS); Schlussbericht (Anlage 3); Müncheberg (Leibniz-Zentrum für Agrarlandschaftsforschung)

Mirschel W, Wenkel K-O, Wieland R, Albert E, Köstner B (2009) Klimawandel und Ertragsleistung: Auswirkungen des Klimawandels auf die Ertragsleistung ausgewählter landwirtschaftlicher Fruchtarten im Freistaat Sachsen - eine landesweite regionaldifferenzierte Abschätzung -. In: SÄCHSISCHES LANDESAMT FÜR LANDWIRTSCHAFT, UMWELT UND GEOLOGIE (ed) Schriftenreihe des Landesamtes für Umwelt, Landwirtschaft und Geologie (LfULG Sachsen), Issue 28/2009, Dresden, 61 p

Mirschel W, Wenkel K-O, Wieland R, Albert E, Köstner B (2010) Impact assessment of climate change on agricultural crop yields using the hybrid model YIELDSTAT—a case study for the Free State of Saxony, Germany. In: Materialy vserossiskoi konferencii "Matematiceskie modeli i informachionnye tekhnologii v sel´skokhozaistvennoi biologii: itogi i perspektivy" (14.-15.10.2010, ARI St. Petersburg, Russia), St. Petersburg, pp 200–203. http://www.agrophys.ru/upload/poluektov/sbornik.pdf

Mirschel W, Wieland R, Guddat C, Michel H, Wenkel K-O (2012) Crop yield estimation on arable land under climate change for the Free State of Thuringia, Germany, using the model YIELDSTAT. In: Tendencii razvitija agrofiziki v uslovijakh izmenjajushhegosja klimata (k 80-letiju Agrofiziceskogo NII): Materialy Mezhdunarodnoj konferencii, Sankt-Peterburg, 20–21 sentjabrja 2012 - SPb: Ljubovich, pp 169–180. http://www.agrophys.ru/Thesys-collection. Accessed 4 Apr 2019

Mirschel W, Wieland R, Wenkel K-O, Guddat CH, Michel H (2013) Modellgestützte Abschätzung der Auswirkungen des Klimawandels auf Ertrag und Zusatzwasserbedarf im Freistaat Thüringen bis 2050. In: Nguyen Xuan Thin (Ed.): Modellierung und Simulation von Ökosystemen: Workshop Kölpinsee 2012. (Berichte aus der Umweltinformatik), Shaker Verlag Aachen, pp 1–19

Mirschel W, Wieland R, Wenkel K-O, Nendel C, Guddat C (2014a) YIELDSTAT—a spatial yield model for agricultural crops. Eur J Agron 52(2014):33–46

Mirschel W, Klauss H, Berg M, Eisenhut K-U, Ißbrücker G, Prochnow A, Schörling B, Wenkel K-O (2014b) Innovative Technologien für eine effiziente Bewässerung im Pflanzenbau. In: Bloch R, Bachinger J, Fohrmann R, Pfriem R (eds) Land- und Ernährungswirtschaft im Klimawandel – Auswirkungen, Anpassungsstrategien und Entscheidungshilfen. oekom verlag, München, S. 261–277

Mirschel W, Wieland R, Gutzler C, Helming K, Luzi K (2016) Auswirkungen landwirtschaftlicher Anbauszenarien auf Ertrag und Zusatzwasserbedarf im Land Brandenburg im Jahr 2025. In: Nguyen XT (ed) Modelling and simulation of ecosystems: workshop Kölpinsee 2015. Rhombos-Verlag, Berlin, S. 1–19

Müller L, Schindler U, Shepherd TG, Ball BC, Smolentseva E, Pachikin K, Hu C, Hennings V, Sheudshen AK, Behrendt A, Eulenstein F, Dannowski R (2014) The Muencheberg soil quality rating for assessing the quality of global farmland. In: Mueller L, Saparov A, Lischeid G (eds) Novel measurement and assessment tools for monitoring and management of land and water resources in agricultural landscapes of Central Asia. Springer, Cham, pp 235–248

Münch Th, Gocht A (2006) Farm Boss - ein Instrument zur strategischen Beratung landwirtschaftlicher Betriebe, In: GIL (ed) "Land- und Ernährungswirtschaft im Wandel - Aufgaben und Herausforderungen für die Agrar- und Umweltinformatik - Referate der 26. GIL Jahrestagung in Potsdam 2006", Bonn, pp 189–192

Münch T, Berg M, Mirschel W, Wieland R, Nendel C (2014) Considering cost accountancy items in crop production simulations under climate change. Eur J Agron 52:57–68

Nendel C, Berg M, Kersebaum KC, Mirschel W, Specka X, Wegehenkel M, Wenkel KO, Wieland R (2011) The MONICA model: testing predictability for crop growth, soil moisture and nitrogen dynamics. Ecol Model 222(9):1614–1625

Nendel C, Kersebaum KC, Mirschel W, Wenkel K-O (2014) Testing farm management options as a climate change adaptation strategy using the MONICA model. Eur J Agron 52(2014):47–56

Orlowsky B, Gerstengarbe F-W, Werner PC (2008) a resampling scheme for regional climate simulations and its performance compared to a dynamical RCM. Theoret Appl Climatol 92:209–223

Pavlova VN, Karachenkova AA, Varcheva SE, Sinitsyn NM (2019) Assessment of the spatial wheat cultivation risk for the main cereal cropping regions of Russia. (In this book)

Poluektov RA, Terleev VV (2010) Computer model of nitrogen dynamics in the rooting zone. Agrokhimiya 10:68–74

Poluektov RA, Topazh AG, Jakushev VP, Medvedev SA (2012) Usage of a dynamic agro-ecosystem model for impact assessment of climate change on crop yield productivity (theory and realization). (Использование динамической модели агроекосистемы для оценки влияния климатических изменений на продуктивность посевов (теория и реализация). Вестник Российской академии сельскохозяйст-венных наук 2:7–12)

Prased JR, Prasad RS, Kulkami UV (2008) A decision support system for agriculture using natural language processing (ADSS). In: Proceedings of the international multi conference of engineers and computer scientists (IMECS 2008, 19–21 March 2008, Hong Kong), vol I, pp 1–5

QT (2019). https://www.qt.io/developers/. Accessed 4 Apr 2019

REGKLAM (2013a) REGKLAM – Regionales Klimaanpassungsprogramm Modellregion Dresden. http://www.regklam.de. Accessed 4 Apr 2019

REGKLAM (2013b) Integriertes Regionales Klimaanpassungsprogramm für die Region Dresden: Grundlagen, Ziele und Maßnahmen, REGKLAM-Publikationsreihe 7. Rhombus-Verlag, Berlin, 344 p

Roeckner E, Brokopf R, Esch M, Giorgetta M, Hagemann S, Kornblueh L, Manzini E, Schlese U, Schulzweida U (2004) The atmosphere general circulation model ECHAM5. Part 2: Sensitivity of simulated climate to horizontal and vertical resolution. MPI-Report 354, Max Planck Institute for Meteorology, Hamburg

Roth D (1993) Richtwerte für den Zusatzwasserbedarf in der Feldberegnung. In: Schriftenreihe LUFA Thüringen, Heft 6 (Richtwerte Pflanzenproduktion), pp 53–86

Schwärzel H (2000) Untersuchungen zu Kriterien der Standort- und Gehölzauswahl bei extensiven Anpflanzungen von Obstbäumen. Dissertation, Humboldt-Universität zu Berlin

Schmidt R et al (1975) Grundlagen der Mittelmaßstäbigen Landwirtschaftlichen Standortkartierung. Arch. Acker-u. Pflanzenbau u. Bodenkd, Berlin 19, 8:533–543

Shishov LL, Tonkonogov VD, Lebedeva II, Gerasimova MI (2004) Classification and diagnostic of Russian soils. (Классификация и диагностика почв России. Смоленск: Ойкумена)

Terleev VV, Mirschel W, Badenko VL, Guseva IJ (2017) An improved Mualem-van-Genuchten method and its verification using data on Beit Netofa Clay. Eurasien Soil Sci 504:445–455

TGL (1990) Verfahren der Pflanzenproduktion, Beregnung – Ermittlung des Zusatzwasserbedarfs. TGL 46200(03):30

Wechsung F, Becker A, Gräfe P (2005) Auswirkungen des globalen Wandels auf Wasser, Umwelt und Gesellschaft im Elbegebiet. Weißensee Verlag Berlin, 407 p

Wechsung F, Hartje V, Kaden S, Venohr M, Hansjürgens B (2014) Die Elbe im globalen Wandel. Eine integrative Betrachtung. Weißensee Verlag Berlin, 630 p

Wenkel K-O, Wieland R, Mirschel W, Schultz A, Kampichler C, Kirilenko A, Voinov A (2008) Regional models of intermediate complexity (REMICs): a new direction in integrated landscape modelling. In: Environmental modelling, software and decision support: state of the art and new perspective. Elsevier, Amsterdam, pp 285–295

Wenkel K-O, Berg M, Wieland R, Mirschel W (2010a) Modelle und Entscheidungsunterstützungssystem zur Klimafolgenabschätzung und Ableitung von Adaptionsstrategien der Landwirtschaft an veränderte Klima-bedingungen (AGROKLIM-ADAPT)—decision support system (DSS). Forschungs-Abschlußbericht: BMBF 01 LS 05104, Leibniz-Zentrum für Agrarlandschaftsforschung (ZALF), Selbstverlag, Müncheberg, 2010, 51 S. + 133 p (in 14 suppmements)

Wenkel K-O, Nendel C, Kersebaum KC, Mirschel W, Specka X (2010b) Modelle und Entscheidungsunter-stützungssystem zur Klimafolgenabschätzung und Ableitung von Adaptionsstrategien der Landwirtschaft an veränderte Klimabedingungen (AGROKLIM-ADAPT) - Modellbasis. Forschungs-Abschlußbericht: BMBF 01 LS 05104, Leibniz-Zentrum für Agrarlandschaftsforschung (ZALF), Selbstverlag, Müncheberg, 2010, 26 + 61 p (in 3 supplements)
Wenkel K-O, Berg M, Mirschel W, Wieland R, Nendel C, Köstner B (2013) LandCaRe DSS—an interactive decision support system for climate change impact assessment and the analysis of potential agricultural land use adaptation strategies. J Environ Manag 127:168–183
Wenkel K-O, Wieland R, Mirschel W (2019) Challenges and perspectives for integrated landscape modelling to support sustainable land use management in agricultural landscapes. (In this book)
Wischmeier WH, Smith DD (1978) Predicting rainfall erosion losses—a guide to conservation planning. In: Agriculture Handbook, vol 537. Department of Agriculture, Washington D.C. U.S.
Wieland R (2010) Modell LANUVER zur Generierung einer Landnutzungsverteilung im LandCaRe DSS (Modellbeschreibung). In: Wenkel K-O, Berg M, Wieland R, Mirschel W Vorsorge und Gestaltungs-potenziale in ländlichen Räumen unter regionalen Wetter- und Klimaänderungen (LandCaRe 2020) PT-DLR 01 LS 05104: Modelle und Entscheidungsunterstützungssystem zur Klimafolgenabschätzung und Ableitung von Adaptationsstrategien der Landwirtschaft an veränderte Klimabedingungen (AGROKLIM-ADAPT); Teilprojekt 6 (DSS); Schlussbericht (Anlage 10); Müncheberg (Leibniz-Zentrum für Agrarlandschaftsforschung)

Chapter 25
A Spatial Analysis Framework to Assess Responses of Agricultural Landscapes to Climates and Soils at Regional Scale

Edmar Teixeira, Anne-Gaelle Ausseil, Eric Burgueño, Hamish Brown, Rogerio Cichota, Marcus Davy, Frank Ewert, Jing Guo, Allister Holmes, Dean Holzworth, Wei Hu, John de Ruiter, Ellen Hume, Linley Jesson, Paul Johnstone, John Powell, Kurt Christian Kersebaum, Hymmi Kong, Jian Liu, Linda Lilburne, Sathiyamoorthy Meiyalaghan, Roy Storey, Kate Richards, Andrew Tait and Tony van der Weerden

Abstract This chapter describes the structure, datasets and processing methods of a new spatial analysis framework to assess the response of agricultural landscapes to climates and soils. Georeferenced gridded information on climate (historical and climate change scenarios), soils, terrain and crop management are dynamically integrated by a process-based biophysical model within a high-performance computing environment. The framework is used as a research tool to quantify productivity and

---

E. Teixeira (✉) · E. Burgueño · H. Brown · R. Cichota · M. Davy · W. Hu · J. de Ruiter · E. Hume · L. Jesson · P. Johnstone · H. Kong · J. Liu · S. Meiyalaghan · R. Storey · K. Richards
New Zealand Institute for Plant & Food Research Limited, Lincoln, New Zealand
e-mail: edmar.teixeira@plantandfood.co.nz

A.-G. Ausseil · J. Guo · L. Lilburne
Manaaki Whenua - Landcare Research, Wellington, New Zealand
e-mail: AusseilA@landcareresearch.co.nz

J. Powell · A. Tait
National Institute of Water and Atmospheric Research (NIWA), Wellington, New Zealand
e-mail: John.Powell@niwa.co.nz

A. Holmes
Foundation for Arable Research (FAR), Christchurch, New Zealand
e-mail: allister.holmes@far.org.nz

F. Ewert · K. C. Kersebaum
Leibniz Centre for Agricultural Landscape Research (ZALF), Müncheberg, Germany
e-mail: wiss.direktor@zalf.de

D. Holzworth
Commonwealth Scientific and Industrial Research Organisation (CSIRO), Toowoomba, Australia
e-mail: Dean.Holzworth@csiro.au

T. van der Weerden
AgResearch, Hamilton, New Zealand
e-mail: tony.vanderweerden@agresearch.co.nz

W. Mirschel et al. (eds.), *Landscape Modelling and Decision Support*,
Innovations in Landscape Research,
https://doi.org/10.1007/978-3-030-37421-1_25

environmental aspects of agricultural systems. An application case study using New Zealand spatial datasets and silage maize cropping systems illustrates the current framework capability and highlights key areas for enhancement in future gridded modelling research.

Keywords APSIM · Climate change · Crop · GIS · Modelling

25.1 Introduction

Large-scale agricultural systems are highly sensitive to climate change. It is therefore critical to develop tools that can assess crop responses to climate by accounting for multiple combinations of soils, plant genotypes and managements that characterise agricultural systems. The productivity and environmental footprint of agricultural systems are strongly influenced by interactions among genotypes (G; e.g. crop species and cultivars), environments (E; e.g. climate and soils) and management (M; e.g. sowing dates and input resource amounts) conditions (Ewert et al. 2015a). The nature of GEM interactions can largely differ spatially and temporally (Teixeira et al. 2016), as farmers tactically combine G and M into technology bundles to respond to a changing E at both short- (i.e. inter-annual weather variability) and long-term (i.e. climate change). These responses can be estimated by process-based biophysical models such as the Agricultural Production Systems sIMulator (APSIM) framework (Holzworth et al. 2014). Biophysical models quantitatively consider GEM effects on underlying crop and soil processes which enables understanding mechanisms of response. For example, by representing water and nitrogen fluxes in crops and soils, biophysical models enable a quantitative exploration of crop performance under future scenarios of land use, farmer adaptation, resource availability and other research and policy aspects relevant to agricultural landscapes (Ewert et al. 2015a; Rosenzweig et al. 2013). Historically, biophysical models were developed and calibrated for local (point-based) simulations in which input data are assumed to be known. This contrasts with large-scale studies, particularly for climate change, in which biophysical models require gridded input data on climates and soils that need to be estimated by different methods (Ewert et al. 2015b). This fundamental change implies an additional layer of uncertainty being carried from input datasets into model structures and ultimately into output simulations (Grosz et al. 2017; Wallach and Thorburn 2017). This has been the topic of investigation of many recent gridded modelling studies which assembled different datasets and models to address specific research questions at different scales (Hoffmann et al. 2014; Müller et al. 2017; Zhao et al. 2016). In this chapter, we present a comprehensive methodological description and application of a nationwide gridded analyses framework under development for New Zealand as a case study. The need for analytical solutions to evaluate large-scale agricultural systems is particularly relevant for countries such as New Zealand. The nation's high reliance on the agricultural sector can be illustrated by the economic and environmental profile that largely differs from most nations with a similar degree of development. Specifically, around 6.5% of New Zealand's gross domestic product (GDP) and 50% of greenhouse gas emissions are derived from primary sector activities (MFE 2018; Stats-NZ 2017). This chapter describes the development of a spatial analysis framework to operate as a research tool for interdisciplinary science teams working on agricultural landscapes. The approach encapsulates a dynamic biophysical model within a High-Performance Computing (HPC) environment to serve as the core engine for integration of georeferenced information on climate (historical and future change projections), management, soils and terrain nationally. The technical challenges and solutions to implement the spatial analysis framework are illustrated with an example application using a prototype version. These results provide original insights on the development of gridded modelling frameworks which can be applied to other situations and locations.

25.2 Data Flow and Input Datasets

The flow of information across the framework data pipeline is illustrated in Fig. 25.1. The method can be described in three stages: an input database stage, which holds the base input data which is harmonised to a common spatial resolution; the modelling stage, which encompasses the formatting of the data to serve as input for APSIM and running the simulations in a HPC environment; and the output database and visualisation stage, where final outputs are stored and end users can explore georeferenced simulated results.

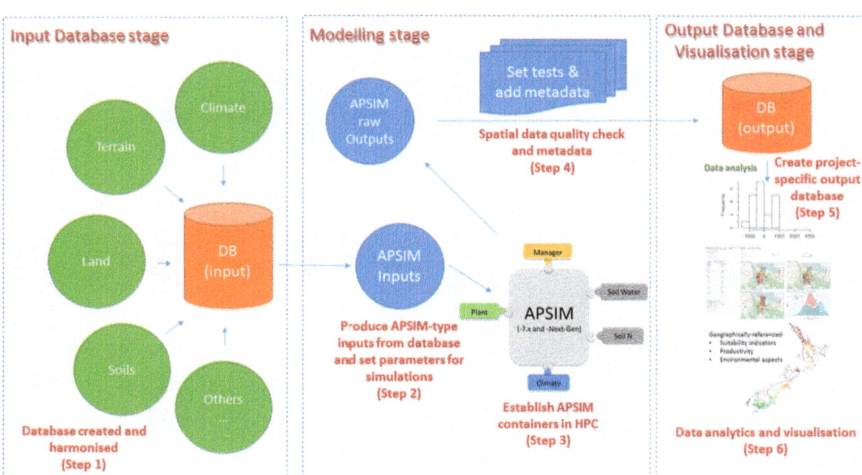

Fig. 25.1 Schematic representation of flow of data across the spatial analysis framework. Input databases (DB) use data from Manaaki Whenua—Landcare Research (MWLR) for soils and terrain and from the National Institute of Water and Atmospheric Research (NIWA) for climate. APSIM is the Agricultural Production Systems sIMulator

Input datasets are harmonised to a spatial resolution of 3 arc-minutes (~5 × 5 km). This resolution is defined by the gridded datasets for daily climate variables provided in netCDF format by the National Institute of Water and Atmospheric Research (NIWA) of New Zealand. The climate raster resolution is based on NIWA's Virtual Climate Station Network (VCSN) grid system (NIWA 2019). The netCDF climate datasets are converted into standard APSIM weather files (.met files in ASCII format) at daily time resolution.

Each file refers to an individual VCSN grid cell containing all weather variables required by the model. These daily climate data were bias-corrected and downscaled across New Zealand (Tait et al. 2016). Two different climate datasets are currently considered in this version of the framework. An historical period (1971 to 2000) with interpolated weather station data from the ERA-40 dataset (Sood 2015) is used for simulating absolute values of model outputs. For climate change assessments, climate data simulated by six General Circulation Models (GCMs) from the fifth phase of the Coupled Model Intercomparison Project (CMIP5) is used, available for baseline (1972 to 2005) and future (2006 to 2100) periods (Tait et al. 2016). The GCMs available from NIWA are HadGEM2-ES, BCC-CSM1.1, NorESM1-M, CESM1-CAM5, GISS-EL-R and GFDL-CM3. For each GCM, four different Representative Concentration Pathways (RCPs) are considered (RCP 2.5, RCP 4.5, RCP 6.0 and RCP 8.5) to characterise climate change scenarios (Meinshausen et al. 2011). The minimum set of weather variables necessary to run APSIM are air temperature (maximum and minimum, °C), rainfall (mm) and solar radiation (MJ/m^2) as illustrated for historical climate in Fig. 25.2.

For the CMIP5 data, vapour pressure (mbar) and wind speed (m/s) are also included to enable alternative evapotranspiration calculations. In addition, the daily mean atmospheric CO_2 concentration from CMIP5 is included in the weather files. This is used as input for APSIM to adjust radiation use efficiency (RUE) and transpiration efficiency (TE), being necessary to account for the "CO_2 fertilisation effect" depending on the photosynthetic pathway of specific crops (Vanuytrecht et al. 2016).

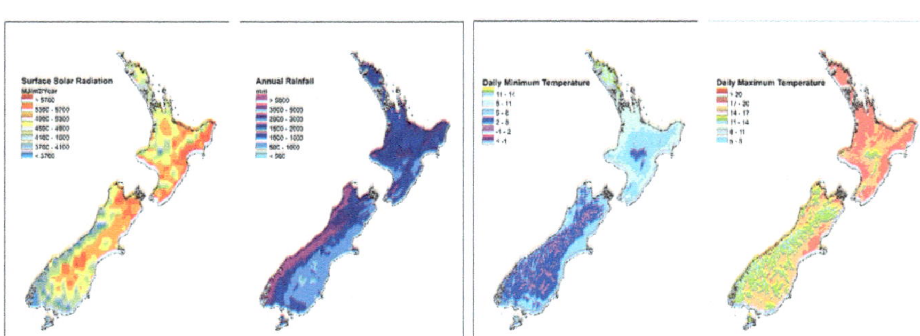

Fig. 25.2 Maps illustrating the spatial variability in historical weather data (ERA-40; 1971–2000) across New Zealand. Minimum inputs required by APSIM include daily temperature (°C; maximum and minimum), annual rainfall (mm) and total surface solar radiation (MJ/m^2 per year)

The soil dataset is derived from the S-map database (S-Map 2019), a new digital soil spatial information system for New Zealand (Lilburne et al. 2004; Webb and Lilburne 2011). The aim of S-map, under development by Manaaki Whenua – Landcare Research (MWLR), is to provide consistent and comprehensive national coverage of digital data at 1:50000 scale. Soils in S-map are characterised by "families" which are subdivided into "siblings" on the basis of any unique combination of classes of drainage, stoniness, depth and texture (Webb and Lilburne 2011). For input into APSIM, information on soil siblings is translated into a library (XML or JSON format) in which the required soil parameters are listed for layers in the soil profile (Fig. 25.3). Soil parameters are generated by pedotransfer functions (PTFs) derived from observed and measured field data and expert knowledge. The PTFs range from simple look-up tables linked to soil classes to more complex algorithms considering various soil, land use, vegetation, climate or topographic attributes (Webb and Lilburne 2011). The soil parameter library is automatically generated by a web processing service developed by MWLR using 52° North infrastructure (52north.org). Information in the soil library is structured to meet the requirement of a cascading-type water balance module in APSIM (APSIM-SoilWat; Probert et al. 1998). Parameters are separated into physical, organic matter and the soil–crop attribute groups. The physical attributes, for each functional horizon, include BD (Mg/m^3, bulk density), air-dry (mm/mm, the water content at air-dry condition), LL15 (m^3/m^3, water content

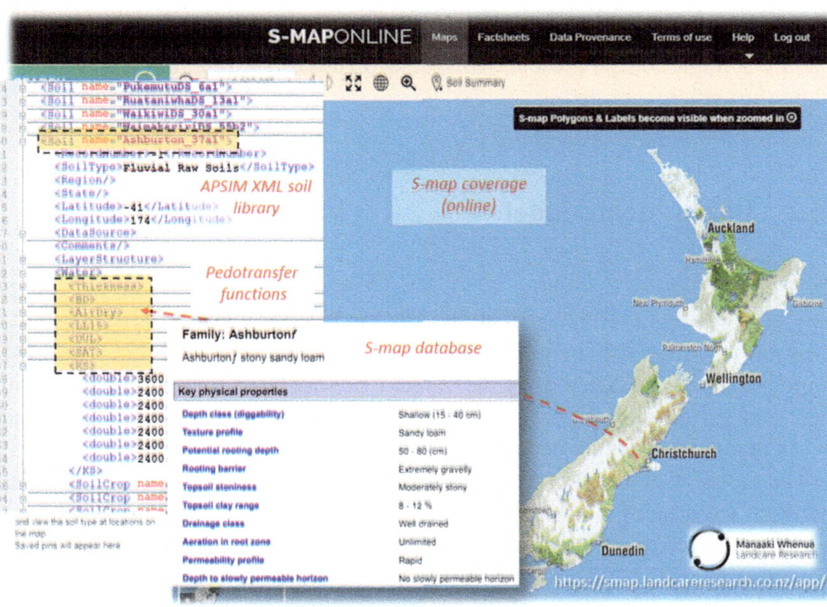

Fig. 25.3 Illustration of the national soils database in S-map used as input by APSIM in the spatial model framework. Data by soil sibling name is available in the S-map database. Soil data are translated into XML format (in this example) used in APSIM, with parameters estimated across the soil profile, by a variety of pedotransfer functions

at the −1500 kPa), DUL (m^3/m^3, the soil water content at field capacity, assumed to be at −10 kPa), SAT (m^3/m^3, the saturated soil water content) and KS (mm/day, the hydraulic conductivity of the soil at saturation). The organic matter attributes are OC (%, the organic carbon content), CNR (kg/kg, the carbon-to-nitrogen ratio of the soil), FBIOM (kg/kg, the proportion of OC that is in the microbial biomass pool, fast turnover) and FINERT (kg/kg, the proportion of the remaining OC that is inert). The soil–crop attributes are LL (m^3/m^3; the effective wilting point for the crop), KL (m^3/m^3, the maximum proportion of available water that a crop can take up each day) and XF (dimensionless, the root exploration factor that controls the rate at which roots can penetrate a soil layer).

The analysis framework uses logical rules to select which soil sibling(s) to consider within a VCSN grid. Soil siblings available per VCSN grid cell are defined by spatially overlapping S-Map to generate metadata containing a list of S-map siblings per each VCSN grid. These metadata also include the share of area occupied by each sibling within a given grid cell. This is used to select siblings for simulation based on their spatial representativeness. Additional soil metadata hold biophysical characteristics of individual siblings, such as average depth, drainage, texture, water holding capacity (WHC) and topsoil stoniness. This enables selection of siblings by their biophysical attributes (e.g. most contrasting WHC). For other terrain attributes (elevation, slope, aspect and land capability class), summary statistics (minimum, maximum, mean, median, standard deviation and 10th and 90th percentile) are included as an additional database, for each VCSN grid, based on LRIS portal datasets (LRIS-Portal 2019). Terrain characteristics enable the selection of grid cells that are suitable for the presence of specific crops, regarding topography and other land information.

25.3 Biophysical Model

The APSIM model

The APSIM model (Holzworth et al. 2014) is used as the core integrator of GEM information within the spatial analysis framework (Fig. 25.4). In brief, APSIM is a biophysical process-based dynamic model that simulates crop growth and development and corresponding carbon, water and nitrogen dynamics in the plant and the soil in response to daily weather input data.

The model is composed of multiple modules that represent specific processes at the plant (e.g. canopy development and biomass growth), soil (e.g. drainage and N mineralization) and farmer management decisions (e.g. sowing time and cultivar selection). Currently, the model exists as (i) APSIM, hereafter referred to as APSIM-7.x (Holzworth et al. 2014), where "x" refers to a version number; and (ii) APSIM Next Generation, hereafter referred to as APSIM-Next-Gen (Holzworth et al. 2018). Both APSIM models are being currently calibrated and applied for a wide range of crops and environmental conditions in New Zealand (e.g. Brown et al. 2018; Khaembah et al. 2017) and worldwide (e.g. Gaydon et al. 2017).

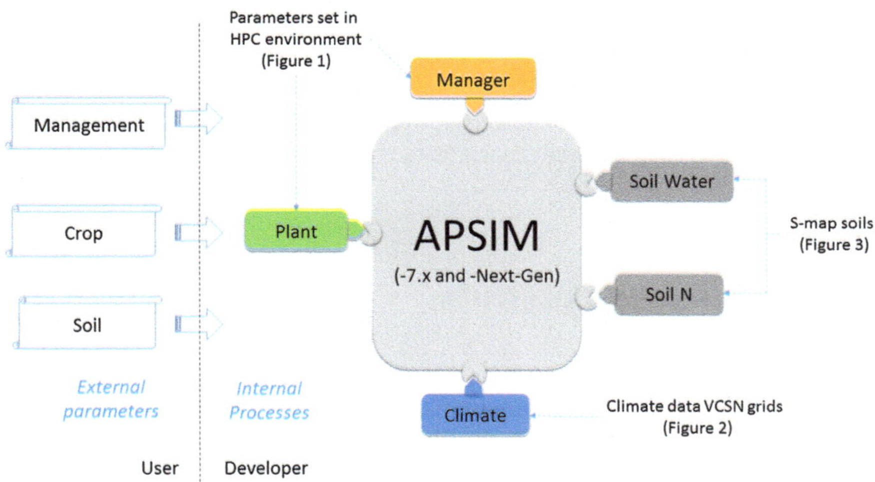

Fig. 25.4 Representation of the modularity aspect of the APSIM model. APSIM is used as the core engine to dynamically integrate georeferenced information on climates, soils, crops and managements in the analysis framework. Users can calibrate the model by providing parameters for simulations in the High-Performance Computing (HPC) environment to characterise specific agricultural systems. Developers can select plant and soil processes and define how these respond to input parameters. In APSIM-Next-Gen, the distinction between users and developers is less strict than APSIM-7.x because the model user interface enables developing new model representations. See http://www.apsim.info/

The spatial analysis framework enables both APSIM-7.x and APSIM-Next-Gen to be run in an HPC Linux environment. For APSIM-7.x, the codebase complexity presents several challenges. The application achieves portability between Operating Systems (Microsoft Windows and GNU/Linux) by leveraging the Mono Project framework, requiring a number of external dependencies and workarounds to run in a Linux environment. These issues were solved by introducing containerization technologies which permit creating a stand-alone, self-contained *artifact* that can be run without changes in any HPC environment that supports running containers. An advantage of introducing software containers is that the project becomes more robust from a "reproducible research" perspective: software containers can be created in a *reproducible* way, and once created the resulting container images are both *portable* and *immutable*. These properties ensure that the exact state of the software stack used during the analysis is captured and distributed alongside the raw data to replicate the analysis in a different environment. The analysis framework had two container engines implemented: Docker (Docker Inc.) and Singularity (Sylabs Inc.). Docker containers, however, were found to be unsuitable for multi-tenant HPC environments due to three primary concerns. Specifically, (i) the security of the computer host could be compromised because operators must have *root* access, which allows full administrative control of the machine; (ii) by default, the data created from a Docker container are owned by the container user instead of the operator running it; and (iii) container access to the persistent storage, where the raw data exist, must be explicitly declared by the operator at run time. The singularity container engine, which were developed for APSIM-7.x and APSIM-Next-Gen, addressed these limitations by restricting end users to run commands inside a container using only their own user identity, as well as automatically mapping the storage endpoints globally configured by the HPC administrator.

25.4 Analysis and Visualisation

The volume of georeferenced outputs generated by the framework is potentially large, and inherently complex, due to the multiple combinations of G (e.g. crop species and cultivars) and M (e.g. use of irrigation, rate of fertiliser application and sowing dates) scenarios within thousands of E (e.g. grid cells with many soils and climates). Grid-cell output data are visualised as raster images overlaid on maps by an application developed in R-Shiny (R-Shiny 2019). The main functionality of the Shiny app is to spatially enable the comparison of two user-defined scenarios (*reference* versus *alternative*) in the main page and to further analyse rasterised outputs (Fig. 25.5). The scenarios are constructed by selecting levels of factors that characterise climates (e.g. GCMs, RCPs and time slices), soils (e.g. S-map sibling name), crop species/cultivars and any other experimental treatment (e.g. irrigation or fertiliser treatment) specific to a project. Original model results are produced in ASCII format for APSIM-7.x and

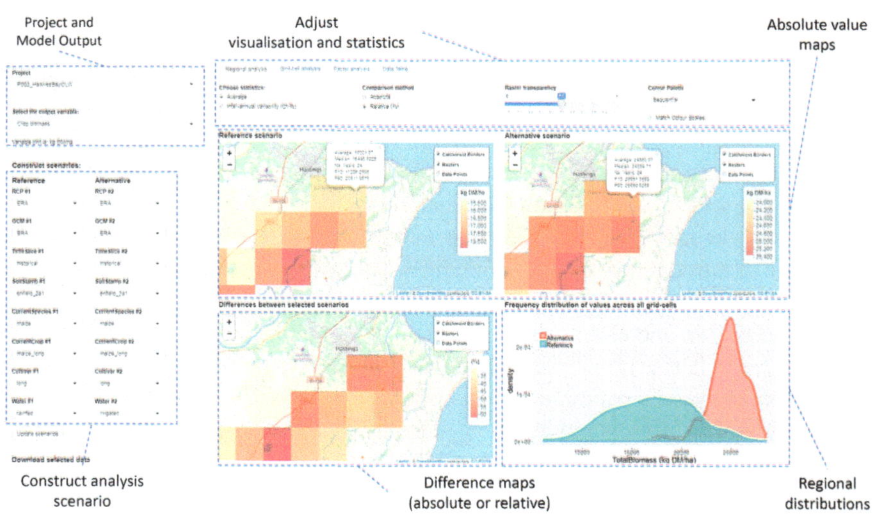

Fig. 25.5 Illustrative example of the main user interface page of the Shiny app for analysis of large areas. Grid cells illustrate the 3 arc-minute spatial resolution for the climate data in southern Hawkes Bay region. Scenarios show a comparison between rain-fed (Reference) and irrigated (Alternative) scenarios for biomass of silage maize crops using historical climate (ERA-40 dataset 1971–2000)

SQLlite relational databases for APSIM-Next-Gen. During the processing, results are combined with metadata (e.g. variables units, names and acceptable ranges) and loaded into a project-specific PostgreSQL database accessed by the Shiny app. The Shiny app user interface adjusts its configuration to include project-specific factors and factor levels in the scenario menu (left side in Fig. 25.5). Results are displayed as raster images of multi-year averages or coefficients of variation and differences between the two scenarios per grid cell. Various libraries from the R statistical package (R Core Team 2017) are used to generate additional analysis capability such as graphs of regional distributions, variability within grid cells and the relationship between selected output variables.

25.5 Application of the Spatial Analysis Framework

This section illustrates a case study application of the spatial analysis framework prototype adapted from Rutledge et al. (2017). In this example, we investigate the influence of climate change impacts on irrigated silage maize yields across arable lands in New Zealand. APSIM-7.9 runs used interpolated weather station data for historical climate (ERA-40 database 1971 to 2000) to simulate absolute silage yields in each VCSN grid cell. Relative climate change impacts were then calculated based on model runs for baseline (1985–2005) and future (2080–2099) periods from the CMIP5 dataset for the RCP 8.5 with the GCM HadGEM2-ES. Simulations assumed fully irrigated conditions and a single hypothetical soil type with high water holding capacity (160 mm/m), so the effect of high spatial resolution soil databases was not considered in this case study. Model runs were done with or without tactical farmer adaptation of using long-cycle genotypes and sowing earlier in response to the early onset of warmer spring temperatures. Simulation results showed a consistent north–south decreasing gradient of absolute maize silage yields for the historical climate (Fig. 25.6). Silage biomass estimates ranged from 7 t dry matter (DM)/ha in southern regions to up to 27 t DM/ha in northern regions. Silage quality was characterised by the percentage of grains in total biomass (i.e. Harvest Index HI) which relates to the forage metabolisable energy. The HI declined from 46% in the warmer regions of the North Island to 17% in New Zealand's cooler southern regions. These differences were mainly caused by a reduction in the length of time available for maize to grow due to low temperatures, which shortened the grain filling period.

Simulated climate change impact on irrigated maize silage yield is shown in Fig. 25.7. Negative impacts occurred mainly in the North Island. In contrast, southern regions showed potential for an increase in climatic suitability for the growth of maize silage. Adaptation of sowing dates and maize genotypes partially counteracted negative impacts in northern and central regions. In southern regions, there was a further increase in positive yield responses when adaptation options were considered.

The main implication of these results is that adaptation to climate change is possible through a shift or expansion of cropping areas to southern regions in New Zealand. This is because low-temperature stress that currently limits maize growth is

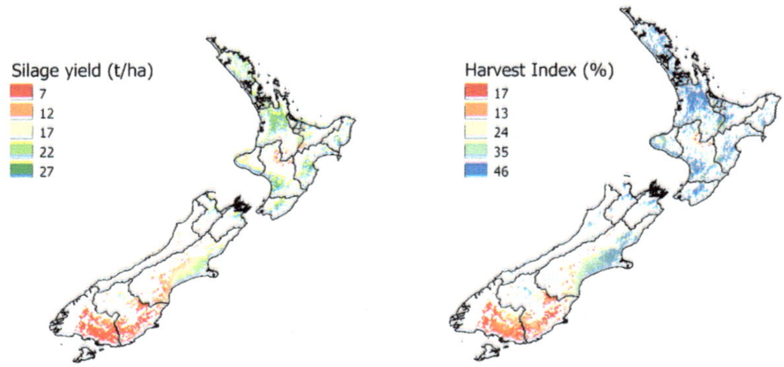

Fig. 25.6 Simulated total biomass yield of silage maize (t dry matter DM/ha) and grain content (% yield) of silage maize in arable lands for the period 1971–2000 (ERA-40 historical climate) using the Agricultural Production Systems sIMulator (APSIM). Adapted from Rutledge et al. (2017)

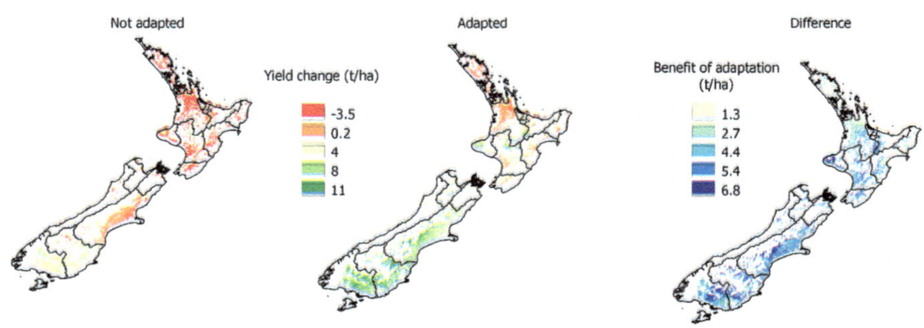

Fig. 25.7 Simulated climate change impact on silage maize yields for end-of-century climate (2080–2099, Representative Concentration Pathway RCP 8.5 using the HadGEM2-ES climate model) and the effect of adapting sowing dates and genotypes across New Zealand's arable lands. Adapted from Rutledge et al. (2017)

minimised in the climate change scenario considered. However, such analysis does not consider logistical and economic aspects that might play an important role on the viability of industry expansion at regional scales. Also, access to irrigation and soil types were not considered in the study, as these would require georeferenced information across all arable lands, not yet available and incorporated into the framework. The expansion of S-map databases (Fig. 25.3) across New Zealand in the future will enable more accurate and comprehensive investigations for multiple soils types, which is particularly important for rain-fed production systems. Finally, our results indicate that adaptation of current agronomic practices is likely to be required to minimise yield losses and exploit yield increase potentials under climate change. The spatial variability in the magnitude of adaptation effects suggest that local studies on management interventions are necessary to properly identify those adaptive measures that confer resilience to climate change. The biophysical aspects considered in this framework require further integration with socio-economic elements for a more comprehensive evaluation of adaptive strategies.

25.6 Outlook and Conclusions

This chapter describes the development of a spatial analysis framework to investigate crop and soil processes in agricultural landscapes, with New Zealand as a case study. It uses a biophysical model as the core engine to dynamically integrate national georeferenced datasets on climates, soils, terrain and simulate response to contrasting crop genotypes and management interventions. A key strength of the framework is the combination of the spatial resolution (5 $\times$ 5 km grids) able to address regional-scale questions and the high temporal resolution of simulations (daily time steps, reported annually). This enables sufficient depth of analysis to investigate inter-annual variability and also plant/soil processes that operate in short timescales to influence crop productivity (e.g. canopy development) and environmental aspects (e.g. water and N uptake). There are limitations and aspects that require attention in the prototype stage of the analysis framework. For instance, the trade-off of high spatio-temporal resolution is the large demand for data storage and computing power, which requires the HPC environment to run APSIM. Also, the use of large georeferenced data as model input (climate, soil and terrain) may introduce significant uncertainty because such datasets are often created by other modelling approaches such as interpolations, downscaling methods and transfer functions. This uncertainty permeates through the data pipeline to the model calculations and model results (Wallach and Thorburn 2017). Our approach to reduce this risk is to connect biophysical modellers and data providers who all work using a single cloud source-code repository to develop the framework, so input data issues are quickly reported and made visible across developers. Additional uncertainty comes from the use of a single biophysical model, instead of model ensembles that account for differences in model structure (Rodríguez et al. 2019). The current strategy to deal with this uncertainty source is to intensively calibrate APSIM-7.x and APSIM-Next-Gen for local conditions (crops, soils and management) which minimises the need for extensive multi-model expertise. The model testing and calibration process is largely facilitated with APSIM-Next-Gen because the version control system evaluates model performance in an automated way, at each request for model update, by comparing results against worldwide datasets in the APSIM-Next-Gen repository (Holzworth et al. 2018). Finally, although the current spatial framework uses georeferenced information to generate APSIM simulations that can cover an entire region or country, the simulations themselves are not spatially aware and do not interact with each other. This implies that some questions, such as nutrient transfer within large farms or catchments, cannot yet be addressed with the analysis framework.

In conclusion, this chapter provides a comprehensive description of the development and application of a spatial analysis framework, still in the prototyping stage, to address research and policy questions in agricultural landscapes. The technical solutions to enable the framework to remain useful as an analysis tool rely heavily on a multi-developer environment, with collaboration within and across research organisations, due to the interdisciplinary nature of the research and the requirements on specialised georeferenced datasets. The insights and solutions provided can contribute to advances in state-of-the-art gridded modelling of agricultural landscapes.

Acknowledgements This work was completed under Plant & Food Research's Discovery Science project "Spatial modelling of crops under climate change" (DS17-19) and Sustainable Agro-Ecosystems (SAE) programme, both funded from the Strategic Science Investment Fund. Additional funding was provided as an output for the Suitability programme of the Our Land and Water and Deep South National Science Challenges (Ministry of Business, Innovation and Employment contracts C10X1507 and C01X1445).

References

Brown H, Huth N, Holzworth D (2018) Crop model improvement in APSIM: using wheat as a case study. Eur J Agron 100:141–150. https://doi.org/10.1016/j.eja.2018.02.002

Ewert F, Rötter RP, Bindi M, Webber H, Trnka M, Kersebaum KC, Olesen JE, van Ittersum, MK, Janssen S, Rivington M, Semenov MA, Wallach D, Porter JR, Stewart D, Verhagen J, Gaiser T, Palosuo T, Tao F, Nendel C, Roggero PP, Bartošová L, Asseng S (2015a) Crop modelling for integrated assessment of risk to food production from climate change. Environ Model Softw 287–303. http://dx.doi.org/10.1016/j.envsoft.2014.12.003

Ewert F, van Bussel L, Zhao G, Hoffmann H, Gaiser T, Specka X, Kurt-Christian K, Sosa C, Lewan E, Yeluripati J, Kuhnert M, Tao F, Roetter R, Constantin J, Raynal H, Wallach D, Teixeira EI, Grosz B, Bach M, Doro L, Roggero PP, Zhao Z, Wang E, Kiese R, Haas E, Eckersten H, Trombi G, Bindi M, Klein C, Biernath C, Heinlein F, Priesack E, Cammarano D, Asseng S, Elliott J, Glotter M, Basso B, Baigorria GA, Romero CC, Moriondo M (2015b) Uncertainties in scaling up crop models for large area climate change impact assessments. In: Rosenzweig C, Hillel D (eds) Handbook of climate change and agroecosystems: the agricultural model intercomparison and improvement project (AgMIP)

Gaydon DS, Singh Balwinder, Wang E, Poulton PL, Ahmad B, Ahmed F, Akhter S, Ali I, Amarasingha R, Chaki AK, Chen C, Choudhury BU, Darai R, Das A, Hochman Z, Horan H, Hosang EY, Kumar PV, Khan ASMMR, Laing AM, Liu L, Malaviachichi MAPWK, Mohapatra KP, Muttaleb MA, Power B, Radanielson AM, Rai GS, Rashid MH, Rathanayake WMUK, Sarker MMR, Sena DR, Shamim M, Subash N, Suriadi A, Suriyagoda LDB, Wang G, Wang J, Yadav RK, Roth CH (2017) Evaluation of the APSIM model in cropping systems of Asia. Field Crops Res 204:52–75. https://doi.org/10.1016/j.fcr.2016.12.015

Grosz B, Dechow R, Gebbert S, Hoffmann H, Zhao G, Constantin J, Raynal H, Wallach D, Coucheney E, Lewan E, Eckersten H, Specka X, Kersebaum K-C, Nendel C, Kuhnert M, Yeluripati J, Haas E, Teixeira E, Bindi M, Trombi G, Moriondo M, Doro L, Roggero PP, Zhao Z, Wang E, Tao F, Rötter R, Kassie B, Cammarano D, Asseng S, Weihermüller L, Siebert S, Gaiser T, Ewert F (2017) The implication of input data aggregation on up-scaling soil organic carbon changes. Environ Model Softw 96. https://doi.org/10.1016/j.envsoft.2017.06.046

Hoffmann H, Zhao G, Specka X, Nendel C, Kersebaum KC, Sosa C, Lewan E, Eckersten H, Yeluripati J, Kuhnert M, Tao F, Rötter R (2014) Effects of climate input data aggregation on modelling regional crop yields. In: FACCE MACSUR mid-term scientific conference, April 1–3 (+ 4), 2014, University of Sassari, Sardinia, Italy

Holzworth D, Huth NI, Fainges J, Brown H, Zurcher E, Cichota R, Verrall S, Herrmann NI, Zheng B, Snow V (2018) APSIM next generation: overcoming challenges in modernising a farming systems model. Environ Model Softw 103:43–51. https://doi.org/10.1016/j.envsoft.2018.02.002

Holzworth DP, Huth NI, deVoil PG, Zurcher EJ, Herrmann NI, McLean G, Chenu K, van Oosterom EJ, Snow V, Murphy C, Moore AD, Brown H, Whish JPM, Verrall S, Fainges J, Bell LW, Peake AS, Poulton PL, Hochman Z, Thorburn PJ, Gaydon DS, Dalgliesh NP, Rodriguez D, Cox H, Chapman S, Doherty A, Teixeira E, Sharp J, Cichota R, Vogeler I, Li FY, Wang E, Hammer GL, Robertson MJ, Dimes JP, Whitbread AM, Hunt J, van Rees H, McClelland T, Carberry PS, Hargreaves JNG, MacLeod N, McDonald C, Harsdorf J, Wedgwood S, Keating BA (2014) APSIM—evolution towards a new generation of agricultural systems simulation. Environ Model Softw 62. https://doi.org/10.1016/j.env-soft.2014.07.009

Khaembah EN, Brown HE, Zyskowski R, Chakwizira E, de Ruiter JM, Teixeira EI (2017) Development of a fodder beet potential yield model in the next generation APSIM. Agric Syst 15. https://doi.org/10.1016/j.agsy.2017.08.005

Lilburne L, Hewitt A, Webb TH, Carrick S (2004) S-map—a new soil database for New Zealand. Reg Inst Online Publ

LRIS-Portal (2019) LRIS-Portal [WWW Document]. lris.scinfo.org.nz/data/. Accessed 15 Jan 2019

Meinshausen M, Smith SJ, Calvin K, Daniel JS, Kainuma MLT, Lamarque JF, Matsumoto K, Montzka SA, Raper SCB, Riahi K, Thomson A, Velders GJM, van Vuuren DPP (2011) The RCP greenhouse gas concentrations and their extensions from 1765 to 2300. Clim Change 109:213. https://doi.org/10.1007/s10584-011-0156-z

MFE (2018) No Title [WWW Document]. New Zealand's Greenh. Gas Invent. http://www.mfe.govt.nz/climate-change/state-of-our-atmosphere-and-climate/new-zealands-greenhouse-gas-inventory. Accessed 2 Jan 2019

Müller C, Elliott J, Chryssanthacopoulos J, Arneth A, Balkovic J, Ciais P, Deryng D, Folberth C, Glotter M, Hoek S, Iizumi T, Izaurralde RC, Jones C, Khabarov N, Lawrence P, Liu W, Olin S, Pugh TAM, Ray DK, Reddy A, Rosenzweig C, Ruane AC, Sakurai G, Schmid E, Skalsky R, Song CX, Wang X, de Wit A, Yang H (2017) Global gridded crop model evaluation: benchmarking, skills, deficiencies and implications. Geosci Model Dev 10:1403–1422. https://doi.org/10.5194/gmd-10-1403-2017

NIWA (2019) No Title [WWW Document]. www.niwa.co.nz/climate/our-services/virtual-climate-stations. Accessed 15 Jan 2019

Probert ME, Dimes JP, Keating BA, Dalal RC, Strong WM (1998) APSIM's water and nitrogen modules and simulation of the dynamics of water and nitrogen in fallow systems. Agric Syst 56:1–28. https://doi.org/10.1016/S0308-521X(97)00028-0

R-Shiny (2019) No Title [WWW Document], 15 Jan 2019

R Core Team (2017) R: a language and environment for statistical computing. R Foundation for Statistical Computing, Vienna, Austria

Rodríguez A, Ruiz-Ramos M, Palosuo T, Carter TR, Fronzek S, Lorite IJ, Ferrise R, Pirttioja N, Bindi M, Baranowski P, Buis S, Cammarano D, Chen Y, Dumont B, Ewert F, Gaiser T, Hlavinka P, Hoffmann H, Höhn JG, Jurecka F, Kersebaum KC, Krzyszczak J, Lana M, Mechiche-Alami A, Minet J, Montesino M, Nendel C, Porter JR, Ruget F, Semenov MA, Steinmetz Z, Stratonovitch P, Supit I, Tao F, Trnka M, de Wit A, Rötter RP (2019) Implications of crop model ensemble size and composition for estimates of adaptation effects and agreement of recommendations. Agric For Meteorol 264:351–362. https://doi.org/10.1016/j.agrformet.2018.09.018

Rosenzweig C, Jones JW, Hatfield JL, Ruane AC, Boote KJ, Thorburn P, Antle JM, Nelson GC, Porter C, Janssen S, Asseng S, Basso B, Ewert F, Wallach D, Baigorria G, Winter JM (2013) The agricultural model intercomparison and improvement project (AgMIP): protocols and pilot studies. Agric For Meteorol 170:166–182. https://doi.org/10.1016/j.agrformet.2012.09.011

Rutledge D, Ausseil A-G, Baisden T, Bodeker G, Booker D, Cameron M, Collins D, Daigneault A, Fernandez, Frame B, Keller E, Kremser S, Kirschbaum M, Lewis J, Mullan B, Reisinger A. Sood A, Stuart S, Tait A, Teixeira E, Timar L, Zammit C (2017) Identifying feedbacks, understanding cumulative impacts and recognising limits: a national integrated assessment. Synthesis

Report RA3. Climate changes, impacts and implications for New Zealand to 2100. MBIE contract C01X1225

S-Map (2019) No Title [WWW Document]. smap.landcareresearch.co.nz/

Sood A (2015) Improved bias corrected and downscaled regional climate model data for climate impact studies: validation and assessment for New Zealand. Unpublished

Stats-NZ (2017) No Title [WWW Document]. National accounts (industry production and investment) year end, Mar 2017. https://www.stats.govt.nz/information-releases/national-accounts-industry-production-and-investment-year-ended-march-2017. Accessed 2 Jan 2019

Tait A, Sood A, Mullan B, Stuart S, Bodeker G, Kremser S, Lewis J (2016) Updated climate change projections for New Zealand for use in impact studies. Synthesis Report RA1. Climate changes, impacts and implications (CCII) for New Zealand to 2100. MBIE contract C01X1225 18

Teixeira EI, Zhao G, Ruiter JD, Brown H, Ausseil A-G, Meenken E, Ewert F (2016) The interactions between genotype, management and environment in regional crop modelling. Eur J Agron. https://doi.org/10.1016/j.eja.2016.05.005

Vanuytrecht E, Raes D, Willems P (2016) Regional and global climate projections increase mid-century yield variability and crop productivity in Belgium. Reg Environ Change 16:659–672. https://doi.org/10.1007/s10113-015-0773-6

Wallach D, Thorburn PJ (2017) Estimating uncertainty in crop model predictions: current situation and future prospects. Eur J Agron 88:A1–A7. https://doi.org/10.1016/j.eja.2017.06.001

Webb TH, Lilburne LR (2011) Criteria for defining the soil family and soil sibling. The fourth and fifth categories of the New Zealand soil classification. Landcare research science series, no 3, 2nd edn. http://digitallibrary.landcareresearch.co.nz/cdm/ref/collection/p20022co

Zhao G, Hoffmann H, Yeluripati J, Specka X, Nendel C, Coucheney E, Kuhnert M, Tao F, Constantin J, Raynal H, Teixeira E, Grosz B, Doro L, Kiese R, Eckersten H, Haas E, Cammarano D, Kassie B, Moriondo M, Trombi G, Bindi M, Biernath C, Heinlein F, Klein C, Priesack E, Lewan E, Kersebaum KC, Rötter R, Roggero PPPP, Wallach D, Asseng S, Siebert S, Gaiser T, Ewert F (2016) Evaluating the precision of eight spatial sampling schemes in estimating regional means of simulated yield for two crops. Environ Model Softw 80:100–112. https://doi.org/10.1016/j.envsoft.2016.02.022

Chapter 26
An Integrated Modelling Approach for Land Use Changes on Different Scales

Rüdiger Schaldach, Jan Göpel and Jan Schüngel

Abstract In the scientific literature, the relation between human land use activities and the environment is often described as a coupled human–environment system. Integrated land system models aim to portray the structure and functioning of these systems. One approach to simulate the spatial and temporal dynamics of land systems on the regional to global scale level is the LandSHIFT modelling framework. It has a modular structure that supports the integration of various sub-models representing the key components of land systems. In this chapter, we describe and discuss the design of the modelling framework and present a case study in Southern Amazonia (Brazil) that illustrates how it can be applied in context of a scenario analysis to provide information on the interlinkages between land use change, its socio-economic drivers and resulting environmental impacts.

Keywords Land systems · Integrated modelling · Land use change · Southern Amazonia

26.1 Introduction

Land use describes how humans modify and manage the Earth's surface. It is regarded as an important process of global environmental change. In the year 2010, more than 37% of the global land surface was used for crop cultivation and livestock grazing (Alexander et al. 2015). The produced biomass is used for food, feed, bioenergy and biomaterials. During the last decades, the expansion of settlement and agricultural

---

R. Schaldach (✉) · J. Schüngel
Center for Environmental Systems Research (CESR), University of Kassel, Wilhelmshöher Allee 47, 34130 Kassel, Germany
e-mail: schaldach@usf.uni-kassel.de

J. Schüngel
e-mail: schuengel@usf.uni-kassel.de

J. Göpel
University Koblenz-Landau, Campus Landau, Fortstraße 7, 76829 Landau (Pfalz), Germany
e-mail: goepel@uni-landau.de

W. Mirschel et al. (eds.), *Landscape Modelling and Decision Support*, Innovations in Landscape Research,
https://doi.org/10.1007/978-3-030-37421-1_26

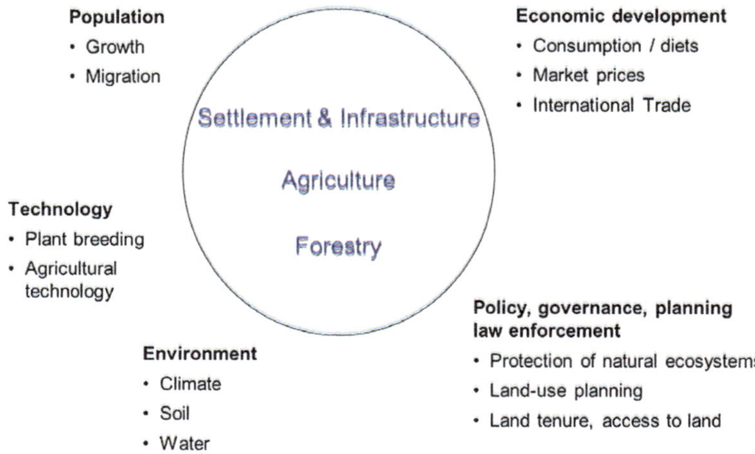

Fig. 26.1 Overview of factors that drive land use change

area was a major cause for the conversion of natural ecosystems and in consequence for the loss of biodiversity and release of carbon dioxide stored in soils and vegetation cover, especially in tropical countries (Foley et al. 2005; Newbold et al. 2016). At the same time, the intensification of agriculture with an increasing use of fertilizers and pesticides is recognized as a serious problem, e.g. for water quality of aquatic ecosystems and biodiversity (e.g. Geiger et al. 2010; Nejadhashemi et al. 2011). In the light of a growing world population and a further increasing biomass demand, competition for land is likely to become a critical problem that demands for strategies and actions to sustainably manage land resources in order to provide food security, to protect remaining ecosystems and to contribute to climate change mitigation efforts (Lambin and Meyfroidt 2011; Verburg et al. 2015). Drivers of land use change are manifold (Fig. 26.1).

They include interlinkages between population and economic development, technological advancements as well as the setting of policies, governance, spatial planning and law enforcement (e.g. Geist and Lambin 2002). It is important to note that different drivers operate on different scale levels. For example, a policy that designates land as nature protection area influences the local and regional land use by prohibiting the use of this land for crop production, whereas consumption of palm oil or soy in Europe may trigger the conversion of natural ecosystems in biomass exporting countries like Brazil or Indonesia (Bruckner et al. 2015).

In the scientific literature, the relation between human land use activities and the environment is often described as a coupled human–environment system with socio-economic as well as ecological and geochemical components (Turner et al. 2007). These "land systems" form the terrestrial part of the Earth System and are, therefore, critical components of global biogeochemical cycles and energy fluxes. The temporal and spatial dynamics of these systems are determined by the interaction of the aforementioned system components.

26.2 Integrated Modelling

The modelling of these land systems and the simulation of their dynamics in space and time requires the coupling of human and environmental drivers and processes. The aim of integrated models is the representation of these processes and their inter-linkages on different spatial and temporal scale levels (Rotmans and van Asselt 2001; Stanton et al. 2009). On the technical level, this integration can be implemented either by formulating the model with a consistent mathematical approach or by coupling different model components that can use quite different mathematical approaches or concepts internally (Hamilton et al. 2015). The latter case requires the definition of interfaces and data formats for communication and data exchange between the model components.

A mathematically consistent model, using coupled differential equations, is World-3, which is based on the research conducted by members of the Club of Rome in the early 1970s (Meadows et al. 2004). The model operates on the global scale and integrates, among others, sub-models for population development, food consumption and use of natural resources. Using carefully designed simulation experiments, the model was used to illustrate future threats of the overuse of natural resources for human societies and ecosystems.

A prominent example for the second category of integrated models is the IMAGE modelling framework (Stehfest et al. 2014), which couples different process-based environmental models, e.g. models for climate and plant growth with econometric and land use modelling approaches. Also aiming at analyses at the global scale, IMAGE has a higher spatial resolution than World-3 and subdivides the globe in different world regions as well as a global geographical raster. A major field of model application is global change-related research with an emphasis on providing policy support for the development of climate change mitigation and adaptation strategies (e.g. Doelman et al. 2018).

Integrated modelling approaches that concentrate on the representation of land systems typically belong to the second category of integrated models and combine different model paradigms within a common conceptual and software framework (Schaldach and Priess 2008). While human decision-making can be represented by optimization or rule-based approaches derived from economic or decision theory, environmental processes such as plant growth and water fluxes are often simulated with dynamic models based on physical, chemical and biological principles.

Of particular importance are spatially explicit land system models that calculate land use pattern within a study region on a geographic raster with (in most cases) uniform cells. This allows incorporating spatial variability of environmental and planning factors such as biomass productivity, soil type or the location of nature conservation areas, respectively. In their case study for the Brazilian Amazon, Chaplin-Kramer et al. (2015) demonstrate impressively how different spatial patterns of land use change can determine impacts on biological carbon storage and biodiversity.

Conceptually, two groups of land system models can be distinguished—top-down models and bottom-up models. Top-down models use macro-economic approaches in combination with scenarios to determine drivers of land use change (e.g. agricultural production) and translate this information into land use patterns by allocating production or area to a geographic raster. Examples include the CLUE model family (e.g. van Asselen and Verburg 2013; Verburg and Overmars 2009), MagPie (Lotze-Campen et al. 2008), LandSHIFT (Schaldach et al. 2011) and GLOBIOM (Havlík et al. 2011) as well as the land use component of the IMAGE model. In contrast, bottom-up models use agent-based approaches to portray decision-making processes of individual land users or groups of land users, e.g. regarding the choice of the cultivated crop type and agricultural management. On this level, decision-making is often represented by profit maximization, i.e. based on micro-economic theory. The articles from An (2012) and Groeneveld et al. (2017) give a good overview of recent research in this domain. While top-down approaches are applied on the global and regional scale level, bottom-up models are in most cases developed for studies on farm or landscape level.

Important objectives of integrated land system models are to gain a better scientific understanding of land use change processes and to providing information to support decision-making processes, e.g. by planners and political actors. In this context, also the analysis of future developments within the land system in form of scenarios is an important aspect. Applications include, among others, the analysis of deforestation patterns, effects of land use change on greenhouse gas emissions as well as competition for land between food and bioenergy production.

In the following, we describe the conceptual framework of the LandSHIFT model as an example of an integrated land system model that combines human and environmental processes on different spatial scale levels. A regional case study in Brazil demonstrates how the model can be applied in context of a scenario analysis.

26.3 The LandSHIFT Model

26.3.1 Purpose and Model Output

LandSHIFT is an integrated model to simulate the spatial and temporal dynamics of land systems on the regional up to the continental and global scale. It has been developed as a flexible tool for exploring the effects of changes in socio-economic and environmental conditions on the spatial distribution and intensity of land use as well as for the analysis of interactions between the changing land use pattern and environmental processes. The model was adapted to case studies in Europe (Schaldach et al. 2012), Jordan (Koch et al. 2008, 2018), Brazil (Lapola et al. 2010; Göpel et al. 2018a) and Eastern Africa (van Soesbergen et al. 2016) as well as for global analyses (Thrän et al. 2016).

Main model outputs are raster maps (micro-level) that depict the spatial and temporal changes of land use within the respective study region. In addition, statistical data that portrays the main characteristics of changes in the study region (macro-level) such as expansion of agricultural area or deforestation rates is calculated.

We differentiate between two main fields of application. First, the model serves as a tool to formalize knowledge of the structure and processes of land systems and to develop a better scientific understanding of their functioning, e.g. by testing hypotheses about human decision-making and interlinkages between the different system components. Second, the model is applied to simulate future development pathways of land systems as done in numerous scenario studies that were combining qualitative and quantitative elements. In this context, the model results are used to provide scientific support for the evaluation of environmental impacts of land use change. The generated maps can help to identify hotspots of change can also serve as basis for further analyses of environmental impacts, such as greenhouse gas emissions or loss of biodiversity.

26.3.2 State Variables and Spatial Scale Level

LandSHIFT is a spatially explicit model that operates on a set of interlinked scale levels. On each scale level, different state variables are specified that can be manipulated by land use processes (Sect. 26.3.3). The following two scale levels form the basis of each simulation experiment:

- **Macro-level**: This scale level is used to specify variables that describe the socio-economic drivers of land use change. Depending on the model setup, the elements of the macro-level comprise a single or a set of countries or regions, respectively. Typical variables include population development, biomass consumption, agricultural and forest production, trade of biomass-based commodities, technological change and assumptions regarding the implementation and effectiveness of environmental and spatial planning policies.
- **Micro-level**: The geographic area of each element of the macro-level is represented by a raster with uniform cells. In current case studies, the cell size varies between 5 arcminutes (~9 × 9 km at the Equator) in global and 900 × 900 m in regional applications. On this level, the spatial land use dynamics are calculated. The most important state variables are the dominant land use type, fraction of settlement area, biomass productivity (e.g. crops, pasture and forest) but also information on the intensity of land use in the form of livestock densities and actual crop production. In addition, the micro-level is used to describe landscape characteristics (e.g. terrain slope, soil type) and zoning information, for example, the presence of nature conservation or indigenous area as well as preference areas for urban development.

Additional scale levels can be included to provide information from external sources or to couple other model components with LandSHIFT. The continental and global studies described in Schaldach et al. (2011) and Thrän et al. (2016) define an additional raster level with the spatial resolution of 30 arcminutes (~50 × 50 km at the Equator) in order to provide information of potential crop yields, generated with the LPJmL model (Bondeau et al. 2007), to the simulation. A second example is the definition of a spatial level that includes watersheds to exchange data with the global water model WaterGAP (Alcamo et al. 2003). Information exchange between scale levels is facilitated by the definition of a clear spatial relationship between the different levels. For example, each cell of the micro-level is associated to a macro-level element, a cell of the coarser raster level and a watershed.

26.3.3 Modelled Processes

Conceptually, LandSHIFT is sub-divided into the Environment module, the Socio-economy module and the Land use change module (LUC module) that implement different processes of the modelled land system (Fig. 26.2). Each process is associated to one or more state variables on different scale levels that it can read and manipulate. This mechanism allows for information exchange between processes and modules. A simulation run is conducted in discrete time steps (annual to 5-year intervals). In each step, first the Environment module and the Socio-economy module are executed to update their state variables. Subsequently, the updated information is used as input by the LUC module to calculate changes of the land use pattern on the micro-level.

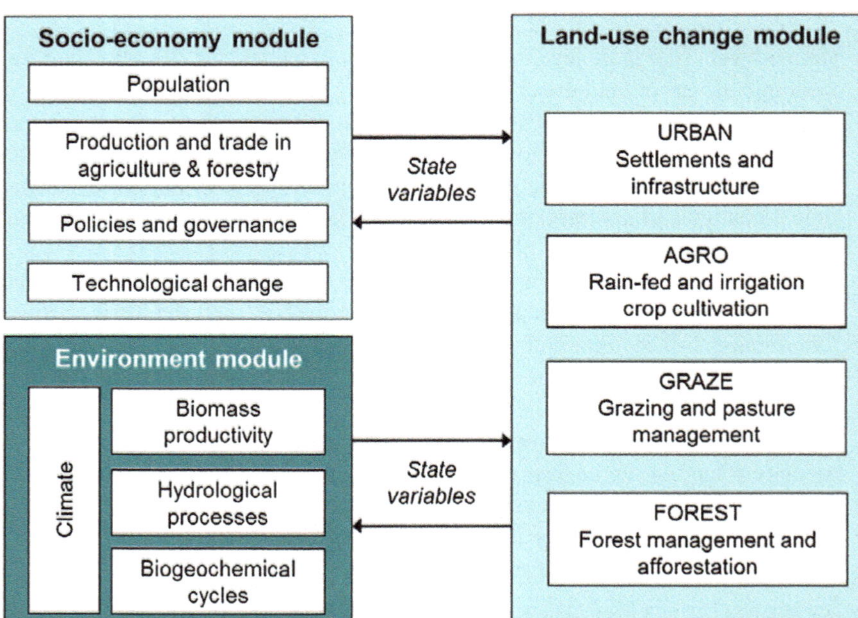

Fig. 26.2 Conceptual design of the LandSHIFT model

The **Environment module** is responsible for the representation of processes such as biomass productivity (cropland, pasture, forest), water availability and irrigation water requirements. While water availability is handled as a state variable on the spatial level of watersheds, biomass productivity is calculated either on the coarse raster level or on the micro-level. Technically, these processes are realized as C++ code in the LandSHIFT model. They either implement specific algorithms for calculating their state variables directly or access information that was generated a priori by other models (e.g. by a crop or a hydrological model) and stored in a separate database.

The **Socio-economy module** and the **LUC module** represent the human subsystem of the land system. The Socio-economy module is responsible for calculating changes of the macro-level state variables (see Sect. 26.3.2). This task can be realized either by using predefined scenarios or by exchanging data with an external macro-economic model such as MAGNET (van Meijl et al. 2006) or IMPACT (Rosegrant et al. 2017). The LUC module calculates the extent and location of land use and land cover changes in each country or region, as specified on the macro-level. For this purpose, the implemented processes distinguish different land use activities that can manipulate the state variables of the raster cells on the micro-level. Current applications of LandSHIFT include the activities settlement and infrastructure (URBAN), cropland cultivation (AGRO), livestock grazing (GRAZE) and forest use in form of wood extraction (FOREST).

In each simulation step, the land use activities are executed in a user-defined order that, for example, represents their respective economic importance or relative competiveness. Raster cells newly occupied by a land use activity within the actual time step are unavailable for the subordinate land use activities. Typical model applications start with allocating new settlement area, followed by cropland, grazing land and forest use. Each land use activity executes three functions: **demand processing**, **preference ranking** and **spatial allocation**. In the first step, demand processing translates the land use change drivers to macro-level demands for services (e.g. housing) and commodities (e.g. amount of crop or round-wood production). In the second step, the preference ranking calculates the suitability of each raster cell for the respective land use activity and constructs a ranking list of these cells. Finally, within the allocation routine, the land use type and other state variables, e.g. on population or stocking density, of the most suitable cells taken from the ranking list are changed until the demand for services or commodities of the land use activity is fulfilled. The location and extent of land use changes are determined by the demand for the service or commodity on the one hand and by the supply, e.g. the biomass productivity, of the raster cells on the other hand.

As an example in Fig. 26.3, the functioning of the cropland cultivation activity (AGRO) is depicted. Driving factors for the quantity of land use change include macro-level information on crop production in metric tonnes for different crop types and assumptions about crop yield increase by technological improvements, e.g. in form of plant breeding or better farm management. Drivers for the location of land use change are specified on the micro-level. They include factors such as terrain slope, soil type, access to road infrastructure and the location of protected areas

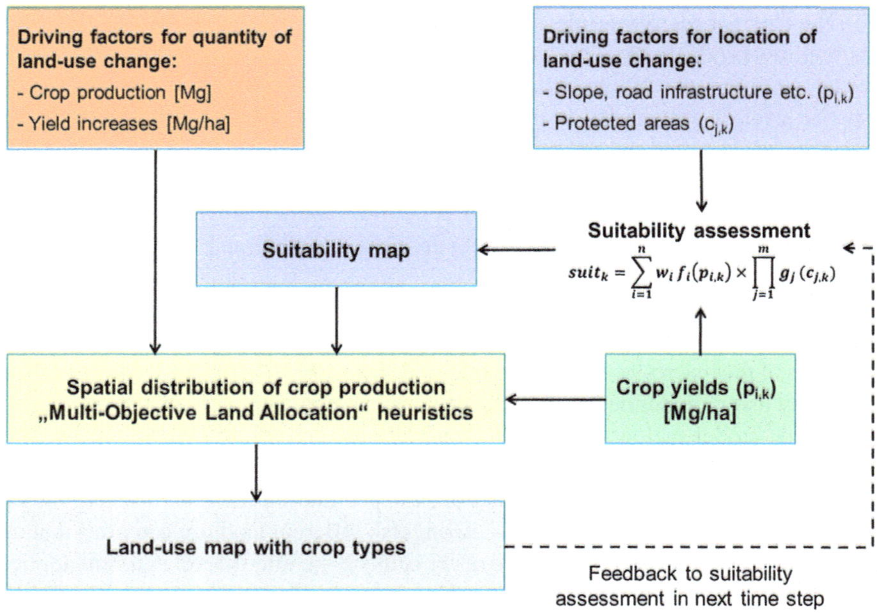

Fig. 26.3 Flow diagram of the functioning of the AGRO land use activity

that are excluded from agricultural land use. Additionally, data on achievable crop yields are provided on this level based on crop yield simulations in combination with the assumptions of technological yield changes. Based on this information, the preference (suitability) of the raster cells for each crop type is determined by multi-criteria analysis (Göpel et al. 2018a). The grid cells are then ranked according to their suitability. In the next step, the crop production is allocated to the most suitable cropland cells. In order to find a near-optimal crop distribution, the spatial allocation is calculated with the "Multi-Objective Land Allocation" algorithm (Eastman et al. 1995).

26.4 Case Study: Regional Land Use Change in Southern Amazonia

26.4.1 Background

The case study was conducted as part of the research project CarBioCial, which was funded by the German Ministry of Education and Research (BMBF) from 2011 to 2016. In this context, LandSHIFT was applied to explore land use change until the year 2030 in the Brazilian federal states Pará (PA) and Mato Grosso (MT), located in Southern Amazonia. In total, the study area comprises 2,159,971 km^2

with 1,253,165 km^2 in PA and 906,807 km^2 in MT. Agricultural production in MT is highly mechanized. Important sectors are soybean production and cattle ranching. The natural vegetation cover is predominantly characterized as Cerrado. In contrast, the dominant land use sector in PA is cattle ranching. The vegetation in PA is dominated by dense rainforest inhabited by many endemic plant and animal species. In the past decades, both states were deforestation hotspots (Lapola et al. 2014).

For the analysis, a set of four scenarios that portray different plausible development pathways of the region was constructed. Each scenario consists of a storyline, which is a short narrative of the respective future world, a set of quantitative drivers of land use change derived from the storylines, and simulated maps that depict the spatial patterns of land use change (Schönenberg et al. 2017).

26.4.2 Adaptation of LandSHIFT to the Study Region

LandSHIFT was used to simulate land use change on a micro-level raster with the cell size of 900 × 900 m that covers the territories of MT and PA. Cell-level information include the state variables "land use type", "human population density" and "livestock density" as well as a set of parameters that describe its landscape characteristics, road infrastructure and zoning regulations (e.g. indigenous and nature protection areas). Accordingly, the land use change module implements the activities URBAN, AGRO and GRAZE. Regional crop yield and biomass productivity maps that were calculated with the vegetation model LPJmL (Bondeau et al. 2007) constitute the Environment module. The Socio-economy module specifies land use change drivers derived from the scenario storylines for the regional level (see Sect. 26.4.3). Model output in the scenario analysis includes raster maps of the land use pattern in the base year 2010 and in the year 2030 under the four scenarios (Göpel et al. 2018a). The resulting land use change was analysed with a Geographic Information System (GIS) and used as the starting point for further analyses of impacts on greenhouse gas emissions (Schaldach et al. 2017) and biodiversity (Göpel 2018c). The process of model initialization, a comprehensive analysis of model uncertainties and the model validation process are described in detail by Göpel et al. (2018b).

26.4.3 Specification of Scenarios

The storylines were initially developed by a group of scientists with high expertise in Amazonia and have a long-term focus on society, economy and environment in the study region. They depict agricultural and socio-economic developments until 2030 as well as assumptions of land use policies and governance. Another important element is the linkage between regional and global markets of agricultural commodities. In the second step, the storylines were reviewed by regional stakeholders and

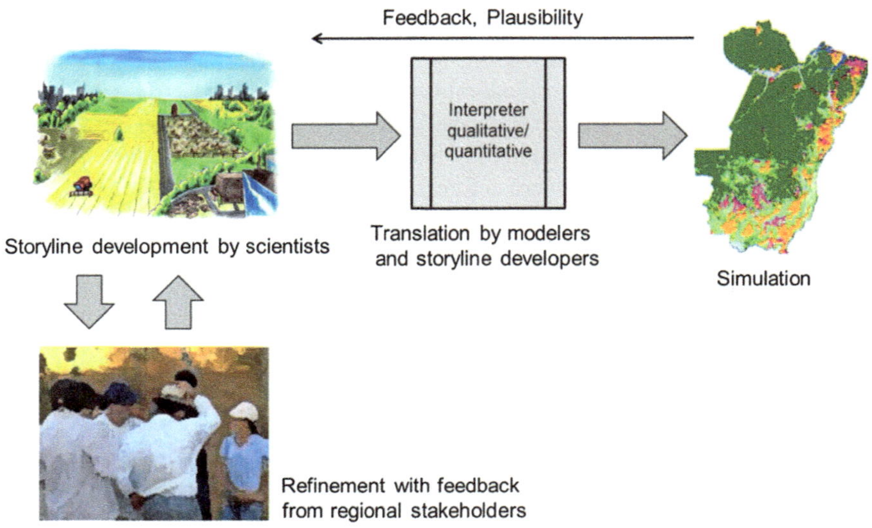

Fig. 26.4 Overview of the CarBioCial scenario development process

successively refined (Fig. 26.4). Based on the storylines, for each scenario, numerical input data for the land use models was derived that describes the evolution of key land use drivers. This includes human population, crop production, livestock numbers and crop yield increases due to technological change as well as legislation for the conversion of ecosystems to agricultural land, and assumptions regarding the effectiveness of environmental laws for natural area protection (Göpel et al. 2018a; Schönenberg et al. 2017). In the following, the key characteristics of the storylines are summarized.

The *Trend* scenario assumes a growing production of agricultural commodities in the study region. Crop yields are increased by further agricultural intensification, e.g. by cultivation of new crop varieties and improved farm management. Natural ecosystems that are not located in protected areas are still converted into cropland and pasture. Migration processes lead to a strong population increase in both states.

The *Legal Intensification* and the *Illegal Intensification* scenarios are characterized by an even stronger increase of crop production and livestock numbers mainly due to growing exports to Asian markets. In addition to growing crop yields also cattle ranching is intensified. The two scenarios differ in respect of environmental law enforcement. Under *Legal Intensification*, the conversion of protected areas of any kind is prohibited. In contrast, the *Illegal Intensification* scenario assumes weak law enforcement. In this case, only military and indigenous areas are protected, while areas under ecological protection status are de facto available for agricultural use.

The *Sustainable Development* scenario describes a society with a social model based on participation and an inclusive economic system with clear land titles and strong law enforcement. Natural resources are well protected. Due to a global shift towards a more vegetarian diet that is oriented on WHO recommendations (e.g.

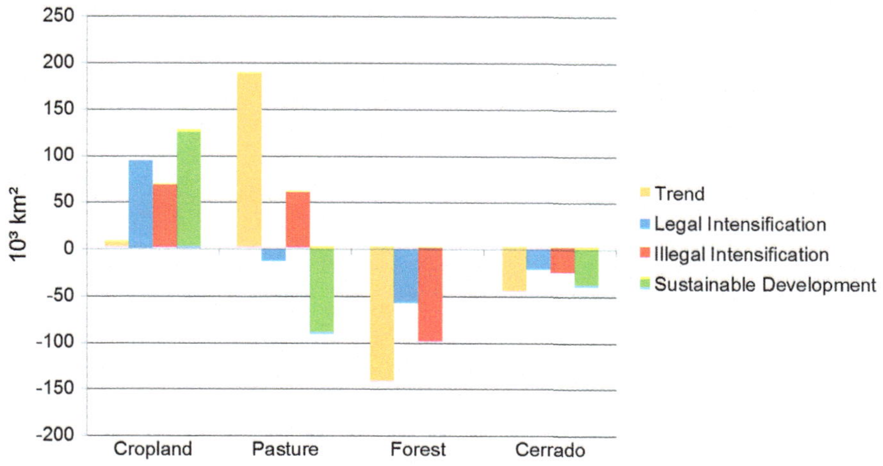

Fig. 26.5 Summary of calculated land use changes under the four scenarios

Srinivasan et al. 2006; Amine et al. 2002), the scenario is characterized by a strong decrease of livestock numbers and a significant increase of crop production for substituting calorie intake from animal-based products. Due to less immigration from other parts of Brazil, population increase is lower than in the other scenarios.

26.4.4 Simulation Results and Main Findings of the Analysis

Figure 26.5 summarizes the simulated land use changes. The maps in Fig. 26.6 depict the land use pattern in the base year 2000 and the year 2030 for the *Sustainable Development* and *Illegal Intensification* scenarios.

In all four scenarios, an expansion of agricultural area (cropland and pasture) that replaces natural ecosystems could be observed. The strongest increases were calculated for the *Trend* and the *Illegal Intensification* scenarios with pasture expansion being the dominant driver of land use change. These results confirm that the growing demand for agricultural commodities from world markets will continue to account for a significant share of deforestation in Southern Amazonia if effective government policies and multinational agreements will not be implemented. In both scenarios, the assumed improvements of agricultural management in terms of crop yields and biomass productivity of pasture could not compensate the increasing production of agricultural commodities. The role of conservation policies, depicted in the *Legal Intensification* scenario, is crucial for mitigating the pressure of the growing agricultural production on natural ecosystems. While cropland expansion has a similar order of magnitude as in the *Illegal Intensification* scenario, pasture area slightly decreases. Only under *Sustainable Development*, deforestation could

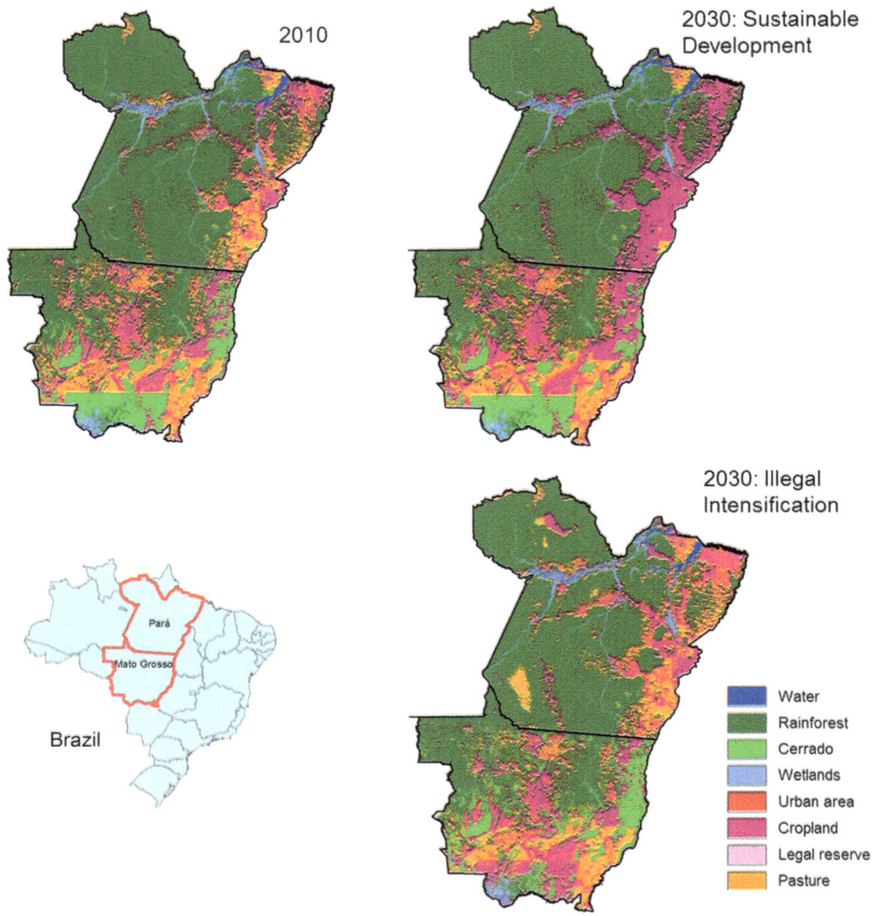

Fig. 26.6 Base map and simulated maps for the *Sustainable Development* and *Illegal Intensification* scenarios in 2030

be stopped completely, but still more than 30,000 km^2 of Cerrado vegetation is converted into cropland. Nevertheless, in total more than 65% of the new cropland is allocated on former pastureland that is abandoned due to the decreasing livestock numbers. Consequently, the scenario shows the smallest conversion rates of natural ecosystems.

The results underline that agricultural intensification and conservation policies are essential for decreasing the loss of natural ecosystems in Southern Amazonia. Therefore, economic instruments that encourage intensification of agricultural production and nature conservation will have to play a key role in Brazil's attempts to stop deforestation. But even these measures might not be effective enough if demands from global food and energy markets trigger an increase of agricultural production within the region that cannot be compensated by increasing yields and improved pasture management, thus leading to an additional loss of rainforest and Cerrado. At this point, the *Sustainable Development* scenario indicates that also changing human consumption patterns might play an essential role for decreasing pressures on natural ecosystems (Yu et al. 2013; Kastner et al. 2012).

26.5 Conclusion

The LandSHIFT modelling framework is a tool for the simulation-based spatially explicit analysis of land use change processes from the regional up to global level. It allows the integration of different models in a common software framework. Due to its modular design, information and processes on different spatial scale levels can be linked with each other. The central component is the LUC module. It distinguishes between different land use activities that define individual strategies in evaluating and allocating land resources and compete with one another for available land resources. Moreover, it supports the modelling of land use intensities as an important element for detailed environmental impact assessments, e.g. related to biodiversity loss or greenhouse gas emissions. Different studies illustrate how effects on greenhouse gas emissions and biodiversity can be assessed. In the case study in Southern Amazonia, we could demonstrate how the model can be used in the context of a scenario analysis to develop a better understanding of the mechanisms that drive land use change. The simulation results provide new insights into the interplay between different drivers of land use change, agricultural development and loss of natural vegetation. It is important to note that the model-based scenario analysis is not a method to predict concrete future events but rather a powerful tool to systematically explore plausible constellations of social and economic drivers and the emerging trajectories of land use change. As such, our results can be a valuable contribution for identifying elements that will be crucial for the development of strategies for a more sustainable land management. Current research efforts that are related to model improvements concentrate on two aspects: First is the implementation of bidirectional linkages between model components in order to explore feedback mechanisms, e.g. between land use change, economy and climate processes. Second aspect is the further refinement of the representation of human decision-making processes within the land use activities.

Acknowledgements The case study was conducted in the framework of the integrated project CarBioCial funded by the German Ministry of Education and Research (BMBF) under the grant number 01LL0902K. We thank all involved stakeholders, farmers and our Brazilian scientific colleagues for their support and CNPq, Embrapa and FAPEMAT for co-funding of Brazilian counterpart projects.

References

Alcamo J, Döll P, Henrichs T, Kaspar F, Lehner B, Rösch T, Siebert S (2003) Development and testing of the WaterGAP 2 global model of water use and availability. Hydrol Sci J 48(3):317–337

Alexander P, Rounsevell MD, Dislich C, Dodson JR, Engström K, Moran D (2015) Drivers for global agricultural land use change: the nexus of diet, population, yield and bioenergy. Glob Environ Change 35:138–147

Amine E, Baba N, Belhadj M, Deurenbery-Yap M, Djazayery A, Forrester T et al (2002) Diet, nutrition and the prevention of chronic diseases: report of a joint WHO/FAO expert consultation. World Health Organization, Geneva. ISBN: 924120916X

An L (2012) Modeling human decisions in coupled human and natural systems: review of agent-based models. Ecol Model 229:25–36

Bondeau A, Smith PC, Zaehle S, Schaphoff S, Lucht W, Cramer W et al (2007) Modelling the role of agriculture for the 20th century global terrestrial carbon balance. Glob Change Biol 13(3):679–706

Bruckner M, Fischer G, Tramberend S, Giljum S (2015) Measuring telecouplings in the global land system: a review and comparative evaluation of land footprint accounting methods. Ecol Econ 114:11–21

Chaplin-Kramer R, Sharp RP, Mandle L, Sim S, Johnson J, Butnar I (2015) spatial patterns of agricultural expansion determine impacts on biodiversity and carbon storage. Proc Natl Acad Sci 112(24):7402–7407

Doelman JC, Stehfest E, Tabeau A, van Meijl H, Lassaletta L et al (2018) Exploring SSP land-use dynamics using the IMAGE model: regional and gridded scenarios of land-use change and land-based climate change mitigation. Glob Environ Change 48:119–135

Eastman JR, Jin W, Kyem PK, Toledano J (1995) Raster procedures for multi-criteria/multi-objective decisions. Photogramm Eng Remote Sens 61:539–547

Foley JA, DeFries R, Asner GP et al (2005) Global consequences of land use. Science 309(5734):570–574

Geiger F, Bengtsson J, Berendse F et al (2010) Persistent negative effects of pesticides on biodiversity and biological control potential on European farmland. Basic Appl Ecol 11(2):97–105

Geist HJ, Lambin EF (2002) Proximate causes and underlying driving forces of tropical deforestation: tropical forests are disappearing as the result of many pressures, both local and regional, acting in various combinations in different geographical locations. Bioscience 52(2):143–150

Göpel J, Schüngel J, Schaldach R, Meurer KH, Jungkunst HF et al (2018a) Future land use and land cover in Southern Amazonia and resulting greenhouse gas emissions from agricultural soils. Reg Environ Change 18(1):129–142

Göpel J, Hissa LDBV, Schüngel J, Schaldach R (2018b) Sensitivity assessment and evaluation of a spatially explicit land-use model for Southern Amazonia. Ecol Inform 48:69–79

Göpel, J (2018c) A scenario approach to modeling land-use changes and assessing associated environmental impacts in Southern Amazonia. Dissertationsschrift Universität Kassel. https://doi.org/10.17170/kobra-2018112826

Groeneveld J, Müller B, Buchmann CM, Dressler G, Guo C (2017) Theoretical foundations of human decision-making in agent-based land use models—a review. Environ Model Softw 87:39–48

Hamilton SH, El Sawah S, Guillaume JH, Jakeman AJ, Pierce SA (2015) Integrated assessment and modelling: overview and synthesis of salient dimensions. Environ Model Softw 64:215–229

Havlík P, Schneider UA, Schmid E, Böttcher H, Fritz S et al (2011) Global land-use implications of first and second generation biofuel targets. Energy Policy 39(10):5690–5702

Kastner T, Rivas MJI, Koch W, Nonhebel S (2012) Global changes in diets and the consequences for land requirements for food. Proc Natl Acad Sci 109(18):6868–6872

Koch J, Schaldach R, Köchy M (2008) Modeling the impacts of grazing land management on land-use change for the Jordan River region. Glob Planet Change 64:177–187

Koch J, Wimmer F, Schaldach R (2018) Analyzing the relationship between urbanization, food supply and demand, and irrigation requirements in Jordan. Sci Total Environ 636:1500–1509
Lambin EF, Meyfroidt P (2011) Global land use change, economic globalization and the looming land scarcity. Proc Natl Acad Sci 108(9):3465–3472
Lapola DM, Schaldach R, Alcamo J, Bondeau A, Koch J, Koelking C, Priess JA (2010) Indirect land-use changes can overcome carbon savings from biofuels in Brazil. Proc Natl Acad Sci 107(8):3388–3393
Lapola DM, Martinelli LA, Peres CA, Ometto JP, Ferreira ME, Nobre CA, Vieira IC (2014) Pervasive transition of the Brazilian land-use system. Nat Clim Change 4(1):27–35. https://doi.org/10.1038/nclimate2056
Lotze-Campen H, Müller C, Bondeau A, Rost S, Popp A, Lucht W (2008) Global food demand, productivity growth, and the scarcity of land and water resources: a spatially explicit mathematical programming approach. Agric Econ 39(3):325–338
Meadows D, Randers J, Meadows D (2004) Limits to growth: the 30-year update. Chelsea Green Publishing
Nejadhashemi AP, Woznicki SA, Douglas-Mankin KR (2011) Comparison of four models (STEPL, PLOAD, L-THIA, and SWAT) in simulating sediment, nitrogen, and phosphorus loads and pollutant source areas. Trans ASABE 54(3):875–890
Newbold T, Hudson LN, Arnell AP, Contu S, De Palma A et al. (2016) Has land use pushed terrestrial biodiversity beyond the planetary boundary? A global assessment. Science 353(6296):288–291
Rosegrant MW, Sulser TB, Mason-D'Croz D, Cenacchi N, Nin-Pratt A et al. (2017) Quantitative foresight modeling to inform the CGIAR research portfolio. Project Report, International Food Policy Research Institute (IFPRI), Washington DC, USA
Rotmans J, van Asselt MB (2001) Uncertainty management in integrated assessment modeling: towards a pluralistic approach. Environ Monit Assess 69(2):101–130
Schaldach R, Priess JA (2008) Integrated models of the land system: a review of modelling approaches on the regional to global scale. Living Rev Landsc Res 2
Schaldach R, Alcamo J, Koch J, Kölking C, Lapola DM, Schüngel J, Priess JA (2011) An integrated approach to modelling land-use change on continental and global scales. Environ Model Softw 26(8):1041–1051
Schaldach R, Koch J, Aus der Beek TA, Kynast E, Flörke M (2012) Current and future irrigation water requirements in pan-Europe: an integrated analysis of socio-economic and climate scenarios. Glob Planet Change 94:33–45
Schaldach R, Goepel J, Klingler M (2017) The role of future land-use change in Southern Amazonia to reach the aims of Brazil's National Climate Plan. Erdkunde 71(3):213–230
Schönenberg R, Schaldach R, Lakes T, Göpel J, Gollnow F (2017) Inter-and transdisciplinary scenario construction to explore future land-use options in southern Amazonia. Ecol Soc 22(3)
Srinivasan CS, Irz X, Shankar B (2006) An assessment of the potential consumption impacts of WHO dietary norms in OECD countries. Food Policy 31(1):53–77
Stanton EA, Ackerman F, Kartha S (2009) inside the integrated assessment models: Four issues in climate economics. Clim Dev 1(2):166–184
Stehfest E, van Vuuren D, Bouwman L, Kram T (2014) Integrated assessment of global environmental change with IMAGE 3.0: model description and policy applications. Netherlands Environmental Assessment Agency (PBL)
Thrän D, Schaldach R, Millinger M, Wolf V, Arendt O et al (2016) The MILESTONES modeling framework: an integrated analysis of national bioenergy strategies and their global environmental impacts. Environ Model Softw 86:14–29
Turner BL, Lambin EF, Reenberg A (2007) The emergence of land change science for global environmental change and sustainability. Proc Natl Acad Sci 104(52):20666–20671
Van Asselen S, Verburg PH (2013) Land cover change or land-use intensification: simulating land system change with a global-scale land change model. Glob Change Biol 19(12):3648–3667
Van Meijl H, Rheenen T, Tabeau A, Eickhout B (2006) The impact of different policy environments on agricultural land use in Europe. Agric Ecosyst Environ 114:21–38

van Soesbergen A, Arnell AP, Sassen M, Stuch B, Schaldach R (2016) Exploring future agricultural development and biodiversity in Uganda, Rwanda and Burundi: a spatially explicit scenario-based assessment. Reg Environ Change 17(5):1409–1420

Verburg PH, Overmars KP (2009) Combining top-down and bottom-up dynamics in land use modeling: exploring the future of abandoned farmlands in Europe with the Dyna-CLUE model. Landsc Ecol 24(9):1167

Verburg PH, Crossman N, Ellis EC, Heinimann A, Hostert P (2015) Land system science and sustainable development of the earth system: a global land project perspective. Anthropocene 12:29–41

Yu Y, Feng K, Hubacek K (2013) Tele-connecting local consumption to global land use. Glob Environ Change 23(5):1178–1186

Chapter 27
Assessment of Soybeans Crop Management Strategies Using Crop Growth Models for Central Brazil

Rafael Battisti, Derblai Casaroli, Jéssica Sousa Paixão, José Alves Júnior, Adão Wagner Pêgo Evangelista and Marcio Mesquita

Abstract The assessment of crop management can help to improve yield across different climate and soil conditions. Soybean is the main crop in Central Brazil, where sowing date, maturity group, and irrigation management are an important decision need to be taken by farmers to get higher yields. This way, the aim of this study was to assess the total production in the region in function of crop management (sowing date, maturity group and irrigation), considering gridded weather data ($0.5 \times 0.5°$), local total plant-available soil water capacity and current production intensity of soybean by county. The yield was simulated using three crop models considering four sowing dates, two maturity groups under rainfed and irrigated conditions. The total production in the region was obtained combining yield for each management simulated to the local soil and the production intensity by county. The higher uncertainty was observed for growing seasons (coefficient of variation, mean CV = 7.13%) under rainfed, and for maturity group (mean CV = 4.97%) under irrigation. The use of irrigation reduced considerably the CV for management of sowing dates and soil types (mean CV < 1.55%). The use of irrigation resulted in a yield gain higher than 1200 kg ha^{-1} with irrigation requirement in most of the area above 51 and lower than 200 mm cycle^{-1}. The total production in the region can be increased

R. Battisti (✉) · D. Casaroli · J. S. Paixão · J. Alves Júnior · A. W. P. Evangelista · M. Mesquita
Research Group on Climate and Water Resources of the Cerrado Biome, College of Agronomy, Federal University of Goiás, Avenida Esperança Street, s/n. Campus Samambaia, Goiânia, GO, Brazil
e-mail: battisti@ufg.br

D. Casaroli
e-mail: derblai@ufg.br

J. S. Paixão
e-mail: jsousa.paixao@hotmail.com

J. Alves Júnior
e-mail: josealvesufg@yahoo.com.br

A. W. P. Evangelista
e-mail: awpego@gmail.com

M. Mesquita
e-mail: marciomes@gmail.com

W. Mirschel et al. (eds.), *Landscape Modelling and Decision Support*, Innovations in Landscape Research,
https://doi.org/10.1007/978-3-030-37421-1_27

around 12 million tons by using supplemental irrigation. Maturity groups (MG) 7.2 and 8.4 had a higher production occurring for sowing date on 20 Oct under the rain-fed condition, totalizing 65.12 and 68.74 million tons, respectively. For irrigation, MG 7.2 had 76.82 million tons occurring for sowing date on 20 Oct, while for MG 8.4, the total production was most stable across sowing date, ranging from 81.14 to 82.34 million tons. The uses of local soil and weather, current production intensity and different crop management based on multiple crop models to simulate soybean yield help to identify the best management to obtained higher total production in the region, indicating the best strategies to put efforts to promote these best management through agricultural extension and public policy.

Keywords Landscape decision · Production intensity · Total soybean production · Sowing date · Irrigation · Maturity group · Cerrado biome

27.1 Introduction

Soybean is the fourth biggest crop in the world with 121 million hectares (FAO 2018), being a source of oil and high-protein fibers (EMBRAPA 2015). It is the main grown crop in Brazil, with 35 million hectares grow during 2017/18 growing season and total production of 119 million tons, resulting in a mean yield of 3,364 kg ha^{-1} (CONAB 2018). The yield can be limited by weather during growing season and by crop management, where water deficit reduced potential yield from 46 to 74%, and crop management reduced attainable yield from 26 to 54% (Sentelhas et al. 2015; Battisti et al. 2018a), depending on the reference yield.

The close of yield gap is important due to Brazil has been received attention to be an important global supply of soybean, with a required projection of increment soybean production at the rate of 2.6% per year until 2026 in a limited area for expansion (OECD 2017). In this context, Central Brazil has around 15 million hectares grown with soybean (CONAB 2018), where reference farmers had been exceeded 6,000 kg ha^{-1} (Battisti et al. 2018a). The higher yield was reached by reducing yield gap by water deficit through deeper root system (Battisti and Sentelhas 2017), and better management through improving sowing and seed quality, reducing chemical and physical soil constraints and better cultivar allocation (Teixeira et al. 2019).

The yield improvement can be achieved defining strategies of sowing dates, cultivar maturity groups and irrigation based on local climate and total plant-available soil water capacity (Chauhan et al. 2013; Heinemann et al. 2016). After water availability, the sowing date was the most important management that defined soybean yield in the South Brazil, where soybean yield reduced 0.5% for each day of delay in the sowing after the best one (4 Nov) (Zanon et al. 2016). In this region, the lower yield was associated with short crop cycle duration due to changes in sowing dates from the best one (Battisti et al. 2018b), where the adaptation of maturity group could improve yield across the sowing window.

The crop management is in constant interaction between them to define yield, as the case of maturity group and irrigation, where Wegerer et al. (2015) verified that maturity group 5 had a better performance than early maturity group under rainfed conditions, while in the irrigated treatment, there was no consistency of optimal maturity group in Arkansas, USA. In Central Brazil, these analysis are essential due to climate pattern characterized by lower rainfall at the beginning and end of soybean growing season, which can limit yield for early sowing and longer cycles (Battisti and Sentelhas 2014), and irrigation is a possible management when considered the production system (Almeida et al. 2018).

The crop management strategies need to consider the spatial variability of soil characteristics and weather variability, which changes across the landscapes (Teixeira et al. 2017; Battisti and Sentelhas 2019). The assessment of better management can be done using crop models, associated with spatial input data (Ewert et al. 2015; Maharjan et al. 2019). The multi-model approach has been used to quantify the uncertainties in the simulation associated with model structure and parametrization (Bassu et al. 2014; Martre et al. 2015), users and calibration (Confalonieri et al. 2016) and input data (Battisti et al. 2017a) to improve the models answer to the interaction between management, soils, and weather.

This way, the hypothesis of this study is that defining sowing date, cultivar maturity group and irrigation strategies based on local climate and soil can improve total production in the region using current areas where soybean is grow. Based on that, the aim of this study was to propose an approach to obtained total soybean production in the region by management in Central Brazil, evaluating the crop yield and production by changing crop management of sowing dates, maturity groups and irrigation.

27.2 Methodologies

27.2.1 The Region, Weather, and Soils Inputs

The study was developed for Central Brazil, encompassing the states of Goiás, GO, and part of the Mato Grosso, MT, Mato Grosso do Sul, MS, and Minas Gerais, MG, states (Fig. 27.1). Soybean was grown in around of 15 million hectares during 2017/18 growing season (Conab 2018), being the main crop grown in the region, sowed from the end of September until November, followed by a grown of a second crop (maize or cotton). The region has a predominance of Tropical zone with dry winter (Aw), when considered the Köppen's climate classification (Alvares et al. 2013), with most of the rainfall occurring from October to April.

Gridded weather data were obtained from Xavier at el. (2015) in a grid of $0.5 \times 0.5°$, totalizing 563 points (Fig. 27.1). The data were obtained daily for the period from 1980 to 2015, enabling the simulation of 35 growing seasons. The data included maximum and minimum air temperature, relative humidity, solar radiation, rainfall and wind speed at 2 m. The gridded weather data were compared with surface weather

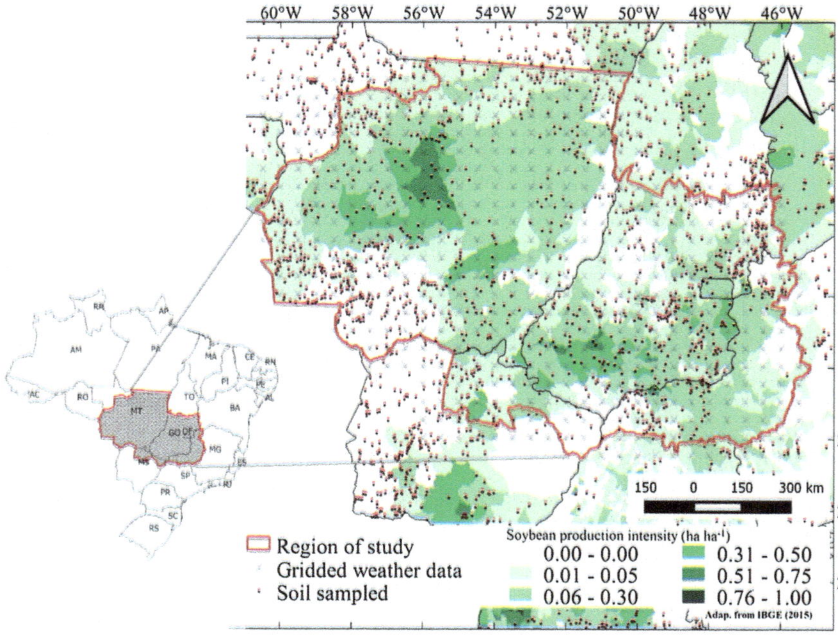

Fig. 27.1 Region of study, gridded weather data, soil samples and soybean production intensity by the county for soybean in Central Brazil

stations by Battisti et al. (2018c), being recommended gridded weather data for simulating soybean yield in Brazil.

The total plant-available soil water capacity was quantified for each soil sample, shown in Fig. 27.1. The samples were obtained from RADAMBRASIL (1974), where sand, silt and clay content, and maximum soil depth were measured. Pedotransfer function was used to quantify the total plant-available soil water capacity (Lopes-Assad et al. 2001), limiting the maximum root depth to 80 cm, as proposed by Battisti et al. (2017b), or by the maximum soil depth if shallower than 80 cm. Total plant-available soil water capacity was interpolated in the region using the inverse distance weighting (P = 10) in the QGIS v. 3.0.1 software (QGIS Development Team 2018) with a pixel size of around 20000 ha, which was grouped in three classes, lower than 60 mm, between 61 and 90 mm, and higher than 90 mm (Fig. 27.2).

27.2.2 Crop Models

Soybean yield was simulated using three crop models: CSM-CROPGRO-Soybean, present in the software Decision Support System for Agrotechnology Transfer (DSSAT) (Jones et al. 2003; Boote et al. 2003), referred to as CROPGRO; Model for

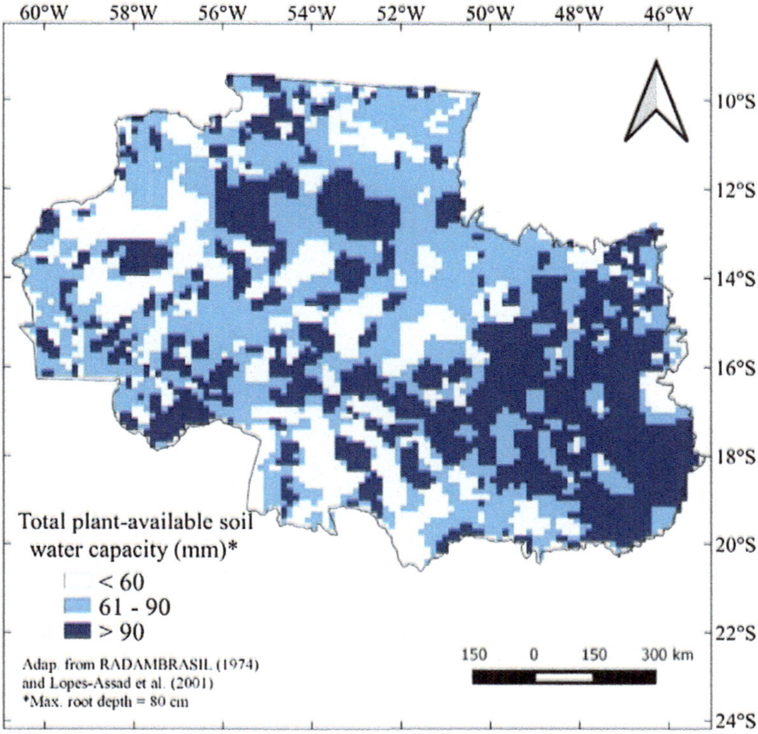

Fig. 27.2 Total plant-available soil water capacity obtained based in soil texture and pedotransfer function across de study region in Central Brazil

Nitrogen and Carbon in Agroecosystems (Nendel et al. 2011), referred to as MONICA; and FAO—Agroecological Zone (Kassam 1977; Doorenbos and Kassam 1979; Rao et al. 1988; Battisti et al. 2018a), referred to as FAO. These crop models were calibrated and tested for Brazilian conditions, showing improvement of performance when considered the model's ensemble (Battisti et al. 2017a, b, 2018a). The inputs for the crop models were soils profiles, daily weather, cultivars coefficients, and crop management. More details about the crop models can be found in the cited references and in Battisti (2016).

27.2.2.1 Soils Profiles

Three soils profiles were building based on the classes showed in Fig. 27.2 and data from Battisti and Sentelhas (2017). The soils inputs were saturation, field capacity, permanent wilting point, bulk density, and soil saturated hydraulic conductivity (Table 27.1). Soils were classified based on total plant-available soil water capacity in sandy-loam (<60 mm), sandy-clay (60–90 mm) and clayey (>90 mm).

Table 27.1 Texture and hydraulics characteristics of clay, sandy-clay and sandy-loam soils used for soybean crop simulations in Central Brazil

Layer (cm)	Clay	Silt	Sand	SAT[a]	FC	PWP	BD	Ksat
	%			$cm^3\ cm^{-3}$			$Mg\ m^{-3}$	$mm\ h^{-1}$
Clayey								
0–15	68	12	20	0.523	0.340	0.220	1.16	0.6
15–30	68	12	20	0.506	0.340	0.220	1.22	0.6
30–200	68	12	20	0.483	0.340	0.220	1.30	0.6
Sandy-clay								
0–15	41	11	48	0.438	0.260	0.160	1.40	1.2
15–30	41	11	48	0.438	0.260	0.160	1.40	1.2
30–200	41	11	48	0.389	0.260	0.160	1.56	1.2
Sandy-loam								
0–15	19	2	79	0.387	0.130	0.070	1.55	25
15–30	19	2	79	0.387	0.130	0.070	1.55	25
30–200	19	2	79	0.343	0.130	0.070	1.69	25

[a]SAT = soil water saturation; FC = field capacity; PWP = permanent wilting point; BD = bulk density; Ksat = saturated hydraulic conductivity. Maximum root depth was limited to 80 cm or maximum soil depth if lower than 80 cm. *Source* RADAMBRASIL (1974), Battisti and Sentelhas (2017)

27.2.2.2 Weather Data

The weather data of maximum and minimum air temperature, relative humidity, solar radiation, rainfall and wind speed at 2 m were used to build complementary files for the crop models. The data were used in a daily frequency from 01/01/1980 to 31/12/2015. Based on that, CROPGRO requires the daily minimum relative humidity, which was estimated using Teten's equation, assuming partial vapor pressure equal saturation vapor pressure at minimum air temperature and constant during the day, and minimum relative humidity occurring at maximum air temperature. Solar radiation was used to obtain sunshine hours per day required by FAO crop model, following Angström-Prescott approach. The potential evapotranspiration was obtained by Penman-Monteith method for all crop models.

27.2.2.3 Cultivars and Crop Models Parameters

The parameters were obtained for CROPGRO (Battisti et al. 2017b; Teixeira et al. 2019), MONICA (Battisti et al. 2017a) and FAO (Battisti et al. 2018a). The parameters were related to the crop cycle, using two cultivars maturity groups, being 7.2 and 8.4, and root growth profile calibrated by the cited authors. The parameters calibrated are shown in Table 27.2.

Table 27.2 Parameters used in the soybean crop model for simulating soybean yield considering maturity group 7.2 and 8.4 in Central Brazil

FAO

Trait	Phase duration (days)		Water deficit sensitivity index (Ky)	Coefficient of cultivation (Kc)
	MG 7.2	MG 8.4		
Sowing—second trefoil (S-V2)	10	12	0.05	0.56
Second trefoil—beginning of flowering (V2-R1)	30	35	0.15	1.00
Beginning of flowering—beginning of grain filling (R1-R5)	25	30	0.4	1.50
Beginning of grain filling—beginning of maturity (R5-R7)	32	38	0.75	1.50
Beginning of maturity—maturity (R7-R8)	8	10	0.1	1.10
Maximum leaf area index	4.40	5.00		
Maximum root depth (MRD)	0.80	0.80		

CROPGRO

Trait	Definition	Value	
		MG 7.2	MG 8.4
CSDL	Critical short day length below which reproductive development progresses with no day length effect (for short day plants) (h)	12.10	12.00
PPSEN	Slope of the relative response of development to photoperiod with time (positive for short day plants) (1/h)	0.345	0.350
EM-FL	Time between plant emergence and flower appearance (R1) (PTD[1])	25.0	22.5
FL-SH	Time between first flower and first pod (R3) (PTD)	5.5	5.0
FL-SD	Time between first flower and first seed (R5) (PTD)	10.0	9.0
SD-PM	Time between first seed (R5) and physiological maturity (R7) (PTD)	27.0	27.0
FL-LF	Time between first flower (R1) and end of leaf expansion (PTD)	20.0	18.0
LFMAX	Maximum leaf photosynthesis rate at 30 °C, 350 vpm CO_2, and high light (mg CO_2 m^{-2} s^{-1})	1.03	1.03
SLAVR	Specific leaf area of cultivar under standard growth conditions (cm^2 g^{-1})	340	320

(continued)

Table 27.2 (continued)

CROPGRO			
Trait	Definition	Value	
		MG 7.2	MG 8.4
SIZELF	Maximum size of full leaf (three leaflets) (cm^2)	180	180
XFRT	Maximum fraction of daily growth that is partitioned to seed-shell	1	1
WTPSD	Maximum weight per seed (g)	0.21	0.21
SFDUR	Seed filling duration for pod cohort at standard growth conditions (PTD)	23	23
SDPDV	Average seed per pod under standard growing conditions (no. Pod^{-1})	2.3	2.3
PODUR	Time required for cultivar to reach final pod load under optimal conditions (PTD)	12	12
THRSH	Threshing percentage, the maximum ratio of (seed/(seed + shell)) at maturity	74	74
SDPRO	Fraction protein in seeds (g (protein)/g (seed))	0.4	0.4
SDLIP	Fraction oil in seeds (g (oil)/g (seed))	0.2	0.2
Soil-root growth factor			
Layer 1 (0–5 cm)		1.00	
Layer 2 (5–15 cm)		1.00	
Layer 3 (15–30 cm)		0.42	
Layer 4 (30–40 cm)		0.34	
Layer 5 (40–40 cm)		0.20	
Layer 6 (60–100 cm)		0.16	
Layer 7 (100–150 cm)		0.04	

MONICA			
Trait	Definition	Value	
		MG 7.2	MG 8.4
stage_temperature_sum	Sowing—emergence	140	140
	Emergence—end of the juvenile	50	70
	End of juvenile—beginning of flowering	575	645
	Beginning of flowering—first pod	300	345
	First pod—last pod	940	1,000
	Last pod—harvest	370	430
	Senescence	25	25

(continued)

Table 27.2 (continued)

MONICA			
Trait	Definition	Value	
		MG 7.2	MG 8.4
day_length_requirement	Limit inferior photoperiod requirement that under this value the rate is 1 (h)	−11.8	−11.5
base_day_length	Limit superior photoperiod requirement that over this value rate is 0 (h)	−14.8	−14.7
crop_specific_max_rooting_depth	Maximum root depth (m)	0.75	

27.2.2.4 Crop Management

Crop management included the interaction between maturity group, sowing dates and irrigation strategy. Maturity group 7.2 and 8.4 were used, being recommended for the region with different cycle duration. The sowing dates assessed were 05 Oct, 20 Oct, 05 Nov, and 20 Nov, being the main sowing window for the region, when considered the climate and production system. The irrigation management was one under the rainfed condition, where yield was a result of natural rainfall, and the second, where the irrigation of 10 mm day^{-1} was applied when soil water content was below 80% of field capacity. This way, soybean yield was obtained for 2 maturity groups, 4 sowing dates, and 2 irrigation treatments, including 3 soils types for 563 weather points, and 35 growing seasons, totalizing 945,840 runs.

27.2.3 Data Analysis

The management effects on soybean yield were analyzed considering the mean results from three crop models and 35 growing seasons, applying a local soil based on the total plant-available soil water capacity (TASW) from Fig. 27.1, and the yield simulated using soil texture sandy-loam, sandy-clay and clayey. The soybean yield was interpolated in the region using the inverse distance weighting (P = 5) in the QGIS v. 3.0.1 software (QGIS Development Team 2018) with a pixel size of 2500 ha. The total production in the region was obtained considering the yield and production intensity (Fig. 27.1) by pixel, following Eq. 1. The steps showed in Fig. 27.3 were used to obtain the total production in Central Brazil in function of management considered in the simulations.

$$Tp_j = \sum_{i=1}^{n} \{Ay_i \times PA \times PSA_i\} \quad (1)$$

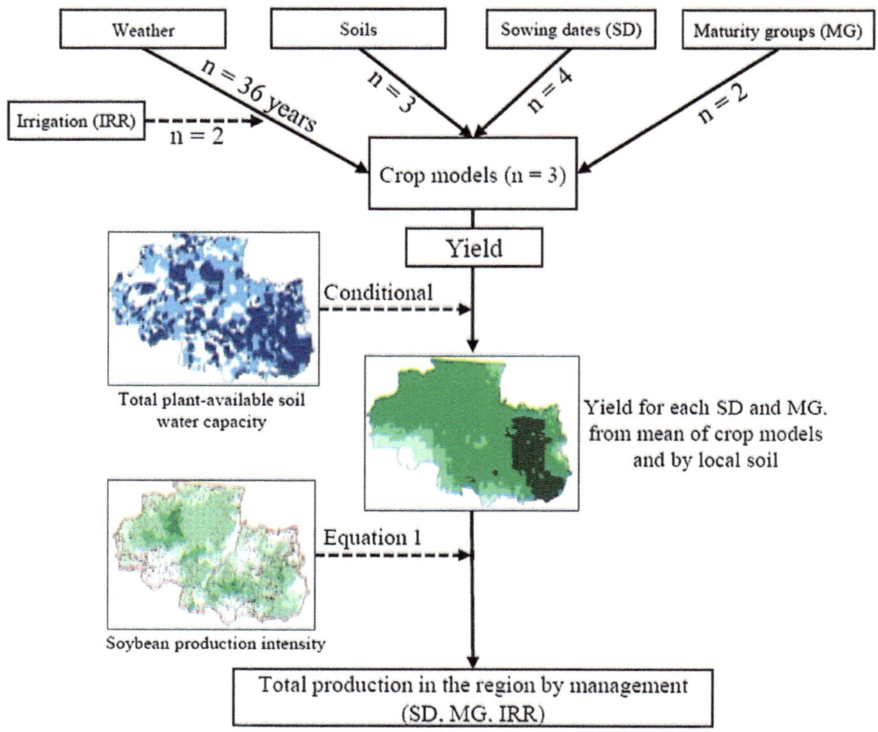

Fig. 27.3 Steps used in the simulation of total production in the region by the management of sowing dates, maturity groups, and irrigation, considering local total plant-available soil water capacity and soybean production intensity in Central Brazil

where Tp is the total production in the whole area (kg); Ay is the average yield estimated by the mean from three crop model corrected to 13% of grain water content (kg ha^{-1}) for each pixel (i) in function of crop management (j); PA is the pixel area (2,500 ha); PSA is the proportion of soybean grown area in each pixel (i) (Fig. 27.1) (ha ha^{-1}); and n is the number of pixels in the region. PSA was obtained by dividing the average soybean production area in the pixel for growing season 2016/2017 (IBGE 2018), by total pixel area (PA).

The combination of sowing dates and maturity groups were used to show the total production and soybean yield for rainfed and irrigated condition; the yield gain by using irrigation through the different between the mean yield from irrigated and rainfed; and the total water demand for irrigation. The uncertainties from growing seasons, sowing dates, soils type and maturity group were analysed considering the coefficient of variation from the mean yield of three crop models for each weather point and growing seasons for rainfed and irrigated treatment.

27.3 Crop Management and Yield Predictions

27.3.1 Weight of Uncertainties

Soybean yield changed in function of weather across the region and growing seasons (GS), sowing dates (SD), maturity groups (MG), and soil types (SO) (Fig. 27.4). They are an important source of uncertainty for soybean production simulation (Battisti et al. 2017a). GS had the higher coefficient of variation (CV) under the rainfed condition, with a mean of 7.13%, followed by MG (4.62%), sowing dates (4.57%) and soils types (3.59%) (Fig. 27.4). Sowing dates had the highest difference between the mean and superior limit 95% under rainfed condition (Fig. 27.4), which is associated with the climate condition across sowing dates and the variability between growing seasons (Battisti et al. 2018a).

The CV reduced considerably when irrigation was considered in the crop management (Fig. 27.4). The MG had the highest mean (4.97%) because the change of

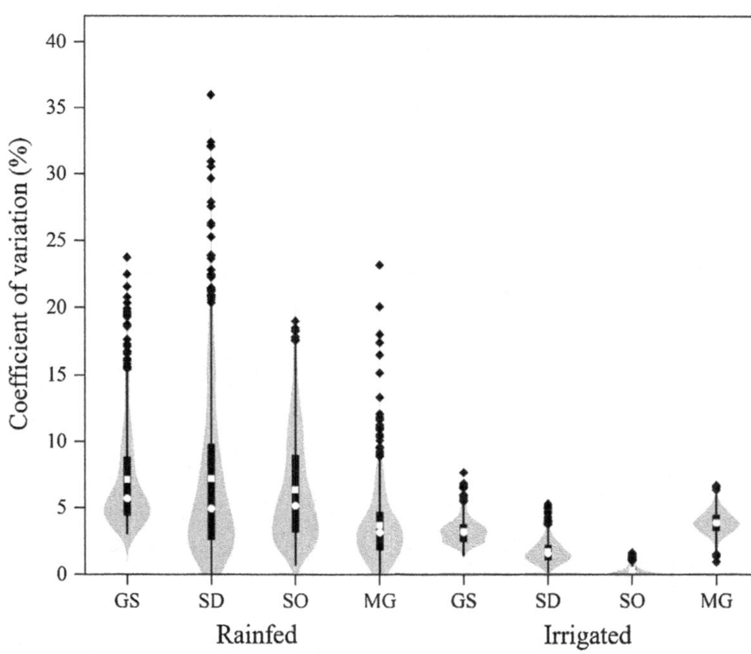

Fig. 27.4 Coefficient of variation for soybean grain yield from growing seasons (GS), sowing dates (SD), soils type (SO) and maturity groups (MG) under rainfed and irrigated management in Central Brazil. For boxplot, the square is the median, the circle is the mean, and the extremes box are the percentiles 25% and 75%, the end of the line are maximum and minimum limit (1.5 times the values percentiles 25% and 75%, respectively) and the points refer to outliers. The gray area is the data distribution using the Kernel Smooth method. Variability is from 35 growing seasons and 563 weather points

cycle length affected the maximum yield that soybean can reach, especially under the irrigated condition, where total water requirement in the cycle is not a limiting factor (Battisti and Sentelhas 2014). Following MG, GS had a mean CV of 3.25%, while for SD and SO, the CV dropped to lower than 1.55% (Fig. 27.4), showing that SD and SO had a lower weight in changes yield in the region under the irrigated system. The irrigation is important to crop management to improve soybean yield and to reduce yield variability in Central Brazil with interesting financial income (Almeida et al. 2018).

27.3.2 Yield by Sowing Date and Maturity Group

Soybean grain yield had a lower change in most of the region when sowing occurred from 05 Oct to 20 Nov for both maturity group (Fig. 27.5). The area with a yield above 4,500 kg ha^{-1} reduced when sowing date had a delay from 5 Oct to 20 Nov from Central for maturity group 7.2, but with reduction of the area in the South-West with a yield lower than 3000 kg ha^{-1} (Fig. 27.5a–d).

Current, the South-West area has a lower soybean production intensity (Fig. 27.1), contributing less for total soybean grain production in Central Brazil. The change of maturity group from 7.2 to 8.4 improved soybean yield to above 4,500 kg ha^{-1}, with large area when sowing date occurred after 20 Oct (Fig. 27.5e–h), due to a longer cycle (Zanon et al. 2016; Battisti et al. 2018a). This demonstrated that the choice of the longer cycle with sowing date after 20 Oct can improve soybean yield,

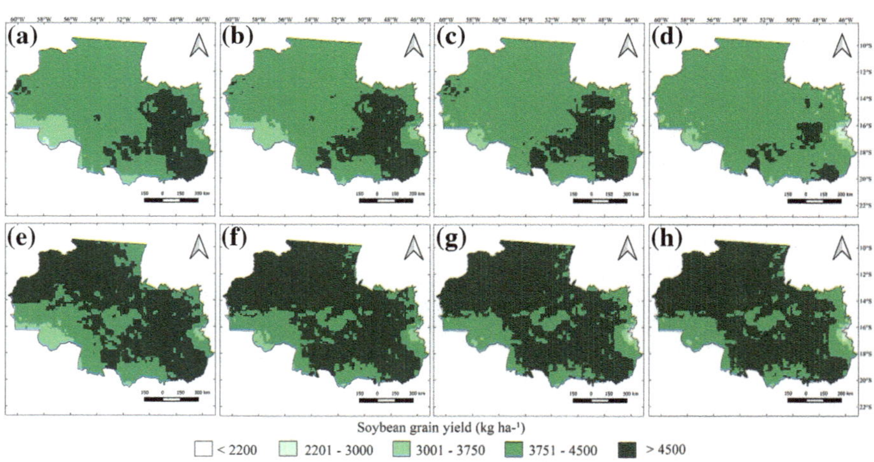

Fig. 27.5 Soybean grain yield for sowing dates on 05 Oct (**a** and **e**), 20 Oct (**b** and **f**), 05 Nov (**c** and **g**) and 20 Nov (**d** and **h**) for maturity groups 7.2 (**a**, **b**, **c** and **d**) and 8.4 (**e**, **f**, **g** and **h**) under rainfed management in Central Brazil. Mean value obtained from 3 crop models and 35 growing seasons considering local soil type (Fig. 27.2)

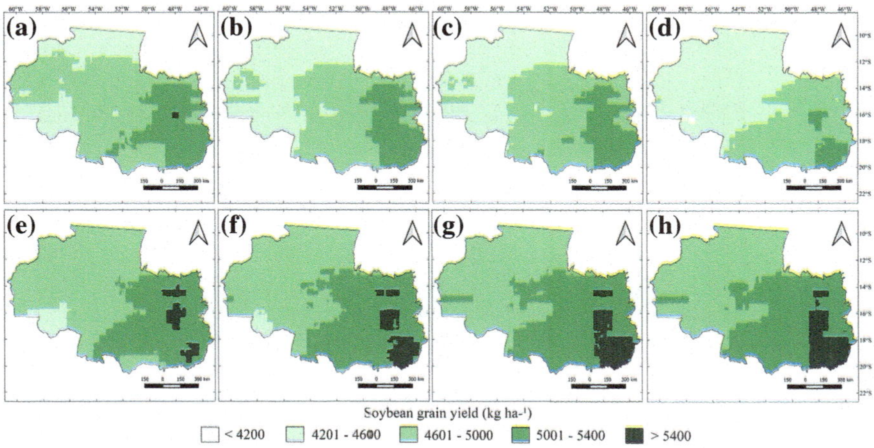

Fig. 27.6 Soybean grain yield for sowing dates on 05 Oct (**a** and **e**), 20 Oct (**b** and **f**), 05 Nov (**c** and **g**) and 20 Nov (**d** and **h**) for maturity groups 7.2 (**a**, **b**, **c** and **d**) and 8.4 (**e**, **f**, **g** and **h**) under irrigated management in Central Brazil. Mean value obtained from 35 growing seasons considering local soil type (Fig. 27.2)

especially in the region were soybean has a higher production intensity (Northwest and Southeast) (Fig. 27.1).

The yield pattern along the region was similar under irrigation than rainfed condition (Fig. 27.6), where the delay of sowing date reduced yield for maturity group 7.2 (Fig. 27.6a–d), and the delay increase yield for maturity group 8.4 (Fig. 27.6e–h). The use of irrigation created more stable yield across the region, from 4,200 to above 5,400 kg ha^{-1}, while rainfed system had values ranging from 3,000 to 4,500 kg ha^{-1} (Fig. 27.5). The highest yield was observed in the Southeast, with values above 5,000 kg ha^{-1} (Fig. 27.6), with large area occurring for maturity group 8.4. This interaction between maturity groups and irrigation need is to be considered to improve yield, increasing the water use efficiency and requiring a local analysis (Wegerer et al. 2015).

27.3.3 Irrigation: Yield Gain and Requirement

The yield gain by irrigation ranged from less than 300 kg ha^{-1} to more than 1,200 kg ha^{-1} (Fig. 27.7a). The lower yield gain occurred in the Northwest, resulting in less than 300 kg ha^{-1} (Fig. 27.7a), with an irrigation requirement amount lower than 100 mm cycle^{-1} (Fig. 27.7c). This region has a more stable rainfall during the soybean growing season, resulting in a lower yield gap by water deficit (Sentelhas et al. 2015). The yield gain by use of irrigation was increasing from Northwest to Southeast to higher than 1,200 kg ha^{-1} (Fig. 27.7a). In the same direction, the irrigation requirement increased from the range of 51-100 mm cycle^{-1} to more than

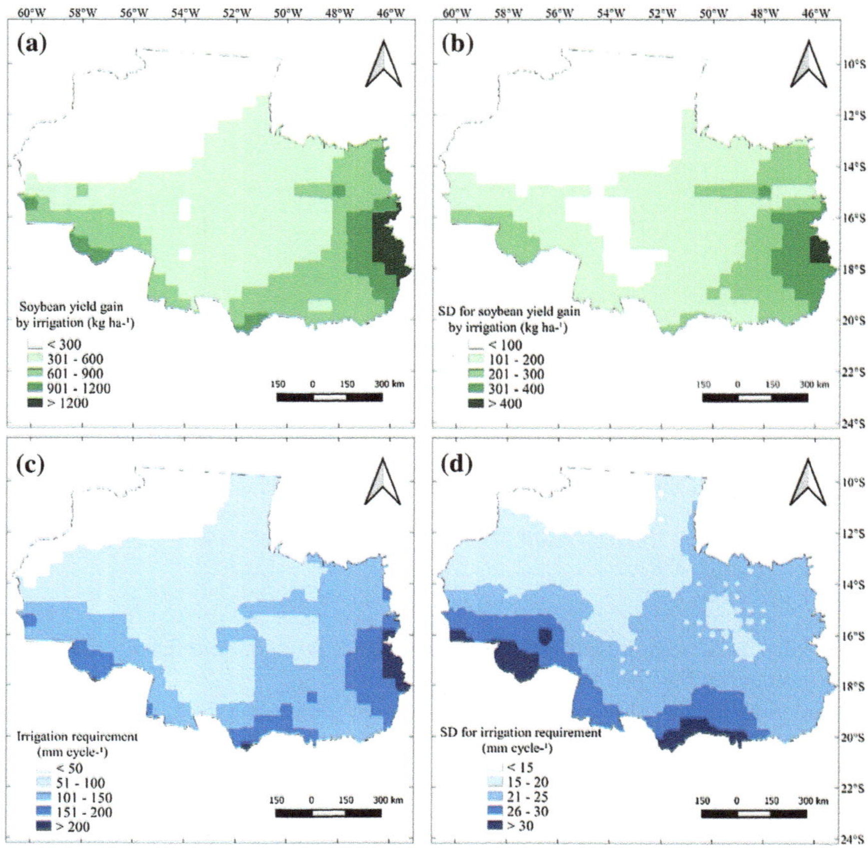

Fig. 27.7 Soybean yield gain (**a**) and standard deviation (SD) (**b**) through applying irrigation, and irrigation amount requirement (**c**) and standard deviation (SD) (**d**) in Central Brazil. Mean value obtained from 3 crop models, 35 growing seasons, 4 sowing date, 2 maturity group and 3 soil type

200 mm cycle^{-1} (Fig. 27.7c), where Southeast is a current area that uses irrigation in a crop rotation system (Soybean-Maize-Tomato) (Almeida et al. 2018). Higher yield gain and irrigation requirement amount also occurred in the South-West. The areas with higher yield gain and irrigation requirement showed higher standard deviation (Fig. 27.7b, d).

The higher yield gain occurred for the soil sandy-loam, with a mean gain of 645 kg ha^{-1}, reducing for sandy-clay and clayey, mainly for growing seasons with higher yield gain (higher water deficit) (Fig. 27.8a). The irrigation requirement amount had a mean of 117, 95, and 85 mm cycle^{-1}, respectively, for sandy-loam, sandy-clay and clayey (Fig. 27.8b). Maturity group 7.2 and 8.4 had a mean yield gain, respectively, of 460 and 560 kg ha^{-1} (Fig. 27.8c), while the irrigation requirement was 93 and 105 mm cycle^{-1} (Fig. 27.8d). Sowing dates had a similar yield gain by using irrigation (470–524 kg ha^{-1}) (Fig. 27.8e), otherwise, mean irrigation

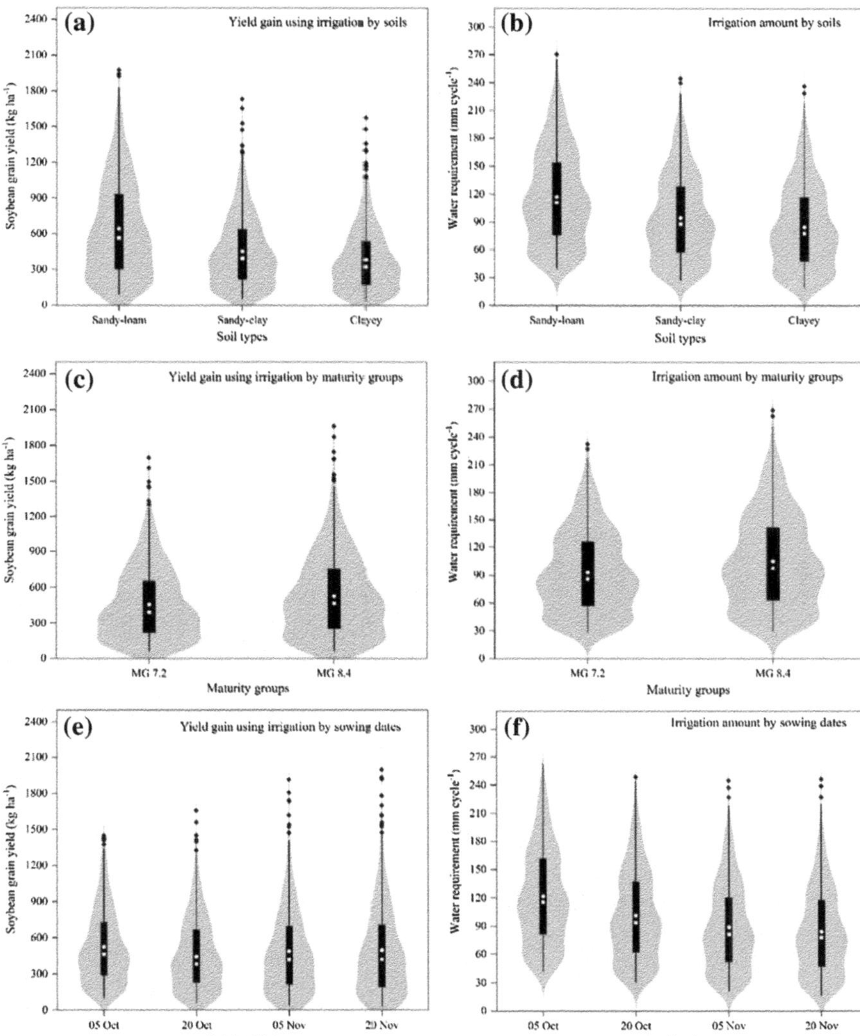

Fig. 27.8 Soybean yield gain (**a**, **c** and **e**) through applying irrigation and irrigation amount requirement (**b**, **d** and **f**) by soils type (**a** and **b**), maturity groups (**c** and **d**) and sowing dates (**e** and **f**) in Central Brazil. For boxplot, the square is the median, the circle is the mean, and the extremes box are the percentiles 25 and 75%, the end of the line are maximum and minimum limit (1.5 times the values percentiles 25% and 75%, respectively) and the points refer to outliers. The gray area is the data distribution using the Kernel Smooth method. Variability is from 35 growing seasons and 563 weather points

requirement were 121, 101, 89, and 84, respectively, for 05 Oct, 20 Oct, 05 Oct, and 20 Nov (Fig. 27.8f). The choose of sowing dates under irrigation condition can help to improve yield saving water for the system (Almeida et al. 2018; Battisti et al. 2018b), due to yield gain is the same but with lower water applied during the cycle.

27.3.4 Total Production by Crop Management

The total production was obtained considering yield from sowing dates and maturity groups, total plant-availability soil water capacity and current soybean production intensity. A similar approach was used by Battisti et al. (2017c) to define the best soybean traits to be selected in the breeding against water deficit under climate change in Southern Brazil. The total production had a mean value of 65.18 million tons for rainfed and 77.77 million tons for irrigated condition (Table 27.3), a production gain of 12.59 million tons by using irrigation. A production of 61.69 million tons was obtained for maturity group 7.2 and sowing on 05 Oct for the rainfed condition, while a delay to 20 Oct resulted in 65.12 million tons, a production gain of 3.43 million of tons in the region (Table 27.1). For MG 8.4, the highest production occurred on 05 Oct and 20 Oct, above 67 million tons, but reducing to around of 65 million tons if sowing date delay to 05 Nov and 20 Nov (Table 27.1). The total production had a similar pattern for MG 7.2 under irrigation than occurs in rainfed condition, while for MG 8.4, total production was almost constant across sowing dates under irrigation management. The results indicated the interaction between irrigation and maturity group choice (Wegerer et al. 2015).

Table 27.3 Total production by sowing dates, maturity groups and irrigation management for current soybean area production in Central Brazil

Sowing date	05 Oct	20 Oct	05 Nov	20 Nov	Mean
	Million tons				
Maturity group	Rainfed				
MG 7.2	61.69	65.12	64.86	62.74	63.60
MG 8.4	67.71	68.74	65.31	65.26	66.76
Mean	64.70	66.93	65.09	64.00	65.18
Maturity group	Irrigated				
MG 7.2	66.77	76.82	76.69	74.27	73.64
MG 8.4	81.14	81.85	82.31	82.34	81.91
Mean	73.96	79.34	79.50	78.31	77.77

27.4 Conclusion and Outlook

- It is required to include different growing seasons, sowing dates, soil types, and maturity group to simulate soybean yield under rainfed treatment due to the higher coefficient of variation for soybean yield; while for irrigated treatments, sowing dates and soil types had a lower effect on yield, resulting in a lower weight of uncertainty in the soybean yield.
- Sowing dates and maturity groups interacted to define the higher soybean yield, where maturity group 7.2 was benefited by early sowing, while maturity group 8.4 by the delay, under rainfed and irrigation management.
- Through the spatial input data was possible to identify the main strategies to improve total production of soybean in Central Brazil, highlighting the interaction between maturity groups and sowing dates, and the use of supplemental irrigation and longer cycle.
- The approach proposed showed to be capable to identify preferential management for the region based on total production, which was not clear to identify if only the soybean yield was analysed.
- The uses of local soil to define total plant-availability soil water content, gridded weather data and soybean production intensity in each county improved the capacity to define the best management at the landscape level.

References

Almeida V, Alves J Jr, Mesquita M, Evangelista AWP, Casaroli D, Battisti R (2018) Comparison of the economic viability of agriculture irrigated by central pivot in conventional and no-tillage systems with soybean, maize and industrial tomato crops. Glob Sci Technol 11:256–273

Alvares CA, Stape JL, Sentelhas PC, Gonçalves JLM, Sparovek G (2013) Köppen's climate classification map for Brazil. Meteorologische Zeitschirift 22:711–728. https://doi.org/10.1127/0941-2948/2013/0507

Bassu S, Brisson N, Durand J-L, Boote KJ, Lizaso J, Jones JW, Rosenzweig C, Ruane AC, Adam M, Baron C, Basso B, Biernath C, Boogaard H, Conijn S, Corbeels M, Deryng D, Sanctis GS, Gayler S, Grassini P, Hatfield J, Hoek S, Izaurralde C, Jongschaap R, Kemanian AR, Kersebaum KC, Kim S-H, Kumar MS, Makowski D, Müller C, Nendel C, Priesack E, Pravia MV, Sau F, Shcherbak I, Tao F, Teixeira E, Timlin D, Waha K (2014) How do various maize crop models vary in their responses to climate change factors? Glob Change Biol 20:2301–2320. https://doi.org/10.1111/gcb.12520

Battisti R (2016) Calibration, uncertainties and use of soybean crop simulation models for evaluating strategies to mitigate the effects of climate change in Southern Brazil. Thesis (Phd in Agricultural Engineering Systems)—ESALQ, University of São Paulo, Piracicaba, SP, Brazil, p 188

Battisti R, Sentelhas PC (2014) new agroclimatic approach for soybean dates recommendation: a case study. Revista Brasileira de Engenharia Agrícola e Ambiental 18:1149–1156. https://doi.org/10.1590/1807-1929/agriambi.v18n11p1149-1156

Battisti R, Sentelhas PC (2017) Improvement of soybean resilience to drought through deep root system in Brazil. Agron J 109:1612–1622. https://doi.org/10.2134/agronj2017.01.0023

Battisti R, Parker PS, Sentelhas PC, Nendel C (2017a) Gauging the sources of uncertainty in soybean yield simulations using the MONICA model. Agric Syst 155:9–18. https://doi.org/10.1016/j.agsy.2017.04.004
Battisti R, Sentelhas PC, Boote KJ (2017b) Inter-comparison of performance of soybean crop simulations models and their ensemble in southern Brazil. Field Crop Research 200:28–37. https://doi.org/10.1016/j.fcr.2016.10.004
Battisti R, Sentelhas PC, Boote KJ, Câmara GMS, Farias JRB, Basso CJ (2017c) Assessment of soybean yield with altered water-related genetic improvement traits under climate change in Southern Brazil. Eur J Agron 83:1–14. https://doi.org/10.1016/j.eja.2016.11.004
Battisti R, Sentelhas, PC, Pascoalino, JAL, Sako, H, Dantas, JPS, Moraes MF (2018a) Soybean yield gap in the areas of yield contest in Brazil. Int J Plant Prod 1–10. https://doi.org/10.1007/s42106-018-0016-0
Battisti R, Sentelhas PC, Parker PS, Nendel C, Câmara GMS, Farias JRB, Basso CJ (2018b) Assessment of crop-management strategies to improve soybean resilience to climate change in Southern Brazil. Crop Pasture Sci 69:154–162. https://doi.org/10.1071/CP17293
Battisti R, Bender, FD, Sentelhas PC (2018c) Assessment of different gridded weather data for soybean yield simulations in Brazil. Theor Appl Climatol 1–11. https://doi.org/10.1007/s00704-018-2383-y
Battisti R, Sentelhas PC (2019) Characterizing Brazilian soybean-growing regions by water deficit patterns. Field Crop Res (under review)
Boote KJ, Jones JW, Batchelor WD, Nafziger ED, Myers O (2003) Genetic coefficients in the CROPGRO-Soybean model: Link to field performance and genomics. Agron J 95:32–51. https://doi.org/10.2134/agronj2003.3200
Chauhan YS, Solomon KF, Rodriguez D (2013) Characterization of north-eastern Australian environments using APSIM for increasing rainfed maize production. Field Crops Research 144:245–255. https://doi.org/10.1016/j.fcr.2013.01.018
CONAB (2018) Survey of crop season: soybean. http://www.conab.gov.br/conteudos.php?a=1253&. Accessed 15 Nov 2018
Confalonieri R, Orlando F, Paleari L, Stella T, Gilardelli C, Movedi E, Pagani V, Cappelli G, Vertemara A, Alberti L, Alberti P, Atanassiu S, Bonaiti M, Cappelletti G, Ceruti M, Confalonieri A, Corgatelli G, Corti P, Dell'Oro M, Ghidoni A, Lamarta A, Maghini A, Mambretti M, Manchia A, Massoni G, Mutti P, Pariani S, Pasini D, Pesenti A, Pizzamiglio G, Ravasio A, Rea A, Santorsola D, Serafini G, Slavazza M, Acutis M (2016) Uncertainty in crop model predictions: what is the role of users? Environ Model Softw 81:165–173. https://doi.org/10.1016/j.envsoft.2016.04.009
Doorenbos J, Kassam AM (1979) Yield response to water. Irrigation and drainage paper, 33. FAO, Rome, p 300
EMBRAPA (2015) Soybean in numbers. https://www.embrapa.br/en/soja/cultivos/soja1/. Accessed 10 July 2015
Ewert F, Rötter RP, Bindi M, Webber H, Trnka M, Kersebaum KC, Olesen JE, van Ittersum MK, Janssen S, Rivington M, Semenov MA, Wallach D, Porter JR, Stewart D, Verhagen J, Gaiser T, Palosuo T, Tao F, Nendel C, Roggero PP, Bartošová L, Asseng S (2015) Crop modelling for integrated assessment of risk to food production from climate change. Environ Model Softw 72:287–303. https://doi.org/10.1016/j.envsoft.2014.12.003
FAO (2018) FAOSTAT: FAO statistical databases. http://www.fao.org/faostat/en/#data/QC. Accessed 10 Dec 2018
Heinemann AB, Ramirez-Villegas J, Souza TLPO, Didonet AD, Di Stefano JG, Boote KJ, Jarvis A (2016) Drought impact on rainfed common bean production areas in Brazil. Agric For Meteorol 225:57–74. https://doi.org/10.1016/j.agrformet.2016.05.010
IBGE (2018) Agricultural production. In Portuguese. http://www.sidra.ibge.gov.br/bda/pesquisas/pam. Accessed 1 Aug 2018
Jones JW, Hoogenboom G, Porter CH, Boote KJ, Batchelor WD, Hunt LA, Wilkens PW, Singh U, Gijsman AJ, Ritchie JT (2003) The DSSAT cropping system model. Eur J Agron 18:234–265. https://doi.org/10.1016/S1161-0301(02)00107-7

Kassam AH (1977) Net biomass production and yield of crops. FAO, Rome, 29 p
Lopes-Assad ML, Sans LMA, Assad ED, Zullo J Jr (2001) Relações entre água retida e conteúdo de areia total em solos brasileiros. Revista Brasileira de Agrometeorologia, Passo Fundo 9:588–596
Maharjan GR, Hoffmann H, Webber H, Srivastava AK, Weihermüller L, Villa A, Coucheney E, Lewan E, Trombi G, Moriondo M, Bindi M, Grosz B, Dechow R, Kuhnert M, Doro L, Kersebaum K-C, Stella T, Specka X, Nendel C, Constantin J, Raynal H, Ewert F, Gaiser T (2019) Effects of input data aggregation on simulated crop yields in temperate and Mediterranean climates. Eur J Agron 103:32–46. https://doi.org/10.1016/j.eja.2018.11.001
Martre P, Wallach D, Asseng S, Ewert F, Jones JW, Rötter RP, Boote KJ, Ruane AC, Thorburn PJ, Cammarano D, Hatfield JL, Rosenzweig C, Aggarwal PK, Angulo C, Basso B, Bertuzzi P, Biernath C, Brisson N, Challinor AJ, Doltra J, Gayler S, Goldberg R, Grant RF, Heng L, Hooker J, Hunt LA, Ingwersen J, Izaurralde RC, Kersebaum KC, Müller C, Kumar SN, Nendel C, O'Leary G, Olesen JE, Osborne TM, Palosuo T, Priesack E, Ripoche D, Semenov MA, Shcherbak I, Steduto P, Stöckle CO, Stratonovitch P, Streck T, Supit I, Tao F, Travasso M, Waha K, White JW, Wolf J (2015) Multimodel ensembles of wheat growth: many models are better than one. Glob Change Biol 21:911–925. https://doi.org/10.1111/gcb.12768
Nendel C, Berg M, Kersebaum KC, Mirschel W, Specka X, Wegehenkel M, Wenkel KO, Wieland R (2011) The MONICA model: testing predictability for crop growth, soil moisture and nitrogen dynamics. Ecol Model 222:1614–1625
OECD (2017) Oilseeds and oilseed products. In: OECD-FAO agricultural outlook 2017–2026. Organization for economic co-operation and development (OECD) Publishing, Paris. http://dx.doi.org/10.1787/agr_outlook-2017-8-en
QGIS Development Team (2018) QGIS geographic information system. Version 3.0.1. 2018. http://qgis.osgeo.org. Accessed 10 Jan
RADAMBRASIL (1974) Levantamento de recursos naturais. Rio de Janeiro, 4 (in Portuguese)
Rao NH, Sarma PBS, Chander S (1988) a simple dated water-production function for use in irrigated agriculture. Agric Water Manag 13:25–32
Sentelhas PC, Battisti R, Câmara GMS, Farias JRB, Hampf A, Nendel C (2015) The soybean yield gap in Brazil—magnitude, causes and possible solution. J Agric Sci 158:1394–1411. https://doi.org/10.1017/S0021859615000313
Teixeira EI, Zhao G, Ruiter JD, Brown H, Ausseil A-G, Meenken E, Ewert F (2017) The interactions between genotype, management and environment in regional crop modelling. Eur J Agron 88:106–115. https://doi.org/10.1016/j.eja.2016.05.005
Teixeira WWR, Battisti R, Sentelhas PC, de Moraes MF, de Oliveira Jr A (2019) Uncertainties assessment of soybean yield gaps using DSSAT-CSM-CROPGRO calibrated by cultivar maturity groups. J Agron Crop Sci (under review)
Wegerer R, Popp M, Hu X, Purcell L (2015) Soybean maturity group selection: irrigation and nitrogen fixation effects on returns. Field Crop Res 180:1–9. https://doi.org/10.1016/j.fcr.2015.05.002
Xavier AC, King CW, Scanlon BR (2015) Daily gridded meteorological variables in Brazil (1980–2013). Int J Climatol 36:2644–2659. https://doi.org/10.1002/joc.4518
Zanon AJ, Streck NA, Grassini P (2016) Climate and management factors influence soybean yield potential in a subtropical environment. Agron J 108:1447–1454. https://doi.org/10.2134/agronj2015.0535

Chapter 28
Estimation of Agro-Landscape Productivity in Regional Scale Using Dynamic Crop Models in a GIS-Environment

Vladimir Badenko, Alex Topaj, Sergey Medvedev, Elena Zakharova and Ielizaveta Dunaeva

Abstract Investigation and management of the productivity of agricultural landscapes in regional scale play a significant role in solving the most important problems of sustainable development of the region under interest. A modern wide-distributed and powerful scientific tool for analysis, monitoring, and decision support in agronomy and agroecology is crop simulation modelling. At the same time, the simplest calculation methods (empirical regression models) are traditionally used for the predictive estimation of the productivity of agroecosystems over rather large areas (region-wide). The article intends to fill such gap, presenting the author's view on the specifics of using dynamic models for regional (mesoscale) calculations. The issues of information support of models in the regional projects as well as the aspects of program implementation of the infrastructure of corresponding computer experiments are considered. We also discuss several prospective use cases for crop-model-based regional projects and propose the methods of processing and presentation the obtained results in a GIS-environment. The practical applicability of the described methodology is demonstrated on the example of calculations of spring barley production processes in the Republic of Crimea.

V. Badenko
Peter the Great St.Petersburg Polytechnic University, 29, Polytechnitcheskaya Street, 195251 Saint Petersburg, Russia
e-mail: v.badenko@spbu.ru

A. Topaj (✉)
Bureau Hyperborea Ltd, 40-2-87, Podvoyskogo Street, 193312 St. Petersburg, Russia
e-mail: aleksandr.topazh@bureauhyperborea.ru

S. Medvedev · E. Zakharova
Agrophysical Research Institute, 14, Grazhdansky Saint Petersburg 195220, Russia
e-mail: glorguin@yandex.ru

E. Zakharova
e-mail: elenazaharova.afi@gmail.com

I. Dunaeva
Federal State Budget Scientific Institution "Research Institute of Agriculture of Crimea", 150, Kievskaya Street, Simferopol 295543, Russia
e-mail: dunaeva_e@niishk.ru

W. Mirschel et al. (eds.), *Landscape Modelling and Decision Support*, Innovations in Landscape Research,
https://doi.org/10.1007/978-3-030-37421-1_28

Keywords Crop model · GIS-environment · Regional agro landscape · Spatial–temporal weather generator

28.1 Introduction

The study of the productivity of agricultural landscapes plays a significant role in solving the most important global and regional problems of hunger and sustainable development. Many countries pay attention to food security, and this serious challenge is formulated and researched on the scale of individual countries, regions, continents and the whole world (Najafi et al. 2018). However, many researchers rightly point out that increasing the productivity of agricultural landscapes should not be in conflict with the principles of sustainable development (Inwood et al. 2018). The relationship between farming methods and landscape dynamics is still poorly understood and many problems require the development of new methods, including those based on dynamic (biophysical) crop models (Rounsevell et al. 2012; Badenko et al. 2015). The conceptual decision-making models of the dynamics of agricultural landscapes for the integration of agricultural practices during regional planning management have been already developed (Benoît et al. 2012; Wenkel et al. 2013). The choice of crop rotation and the associated strategy play an important role in supporting of sustainable development of agro-landscapes (Badenko et al. 2017a; Inwood and Dale 2019). It entails that crop rotation description needs to be incorporated into crop-model-based regional decision support projects (Dury et al. 2012; Badenko et al. 2017a). In (Chopin et al. 2015), the authors were able to characterize the changes in the agricultural landscape and the provision of ecosystem services by analysing the influence of determinants on the changes in farm cropping plans.

Estimation of agro-landscape productivity in regional scale is also connected to agricultural monitoring systems (including the following steps: observations, analysis and forecasts), which provided information regarding food production to different actors and decision-makers to support food security (Fritz et al. 2019; Van der Velde et al. 2019). Obviously, the resolving of forecast task by means of eco-physiological crop models has good prospects (Medvedev et al. 2015; Badenko et al. 2017b). Although in (Jones et al. 2017), it is stated that data scarcity and limitations are more important than gaps in theory and approaches in crop models. The main problems of applying crop models for regional planning are related to the fact that such models have been developed and can be applied for individual locations with location-specific, spatially homogeneous input data (Van Bussel et al. 2011). Crop models are able to account for the variation in climatic conditions and management activities and their effects on crop productivity on a regional scale, but, the spatial heterogeneity in the timing of all phenological stages in a region can have the sufficient variance (Badenko et al. 2017a, b). However, there are some examples of successful application of crop models on a regional scale, including the territories of the former USSR (Dunaeva 2018; Popovich and Molyar 2018). In (Nendel et al. 2011), it has been shown that crop model MONICA proved to be a suitable tool for long-term predicting crop growth and yield formation for cereals, sugar beet and maize under today's and future climatic conditions, even though the simulation of single crops in a rotation failed for different seasons. Expanding this conclusion for spatial scale authors mean that MONICA can be regarded as a suitable simulation model for use in regional applications also.

The main feature of crop models is that they are site-specific models ("point models"), but they are often also successfully used for regional planning tasks (Wenkel et al. 2013; Köstner et al. 2014; Badenko et al. 2015, 2017b). For example, in (Nendel et al. 2013) it is apparent from crop modelling simulation study that the good result obtained for the reproduction on regional scale for winter wheat yields using single weather station and nearest neighbour simulation is based on compensation of errors that operate at the lower scale and affect yield estimates in different directions. The model overestimates yield at the local scale due to the calibration with data from optimum-conditioned field experiments, but on the other hand, the weather information provided with inadequate spatial distribution leads to considerable underestimation of yields in some regions. Currently, the expansion of the scope of application of crop models to the regional level involves integration with geographic information systems (GIS), which should allow resolving some mentioned contradictions. Moreover, such a combination of information systems and technologies of different functional orientations can provide in the future a unique opportunity to study agro-landscape productivity in regional scale in time and space. In general, based on the analysis of recent publications, it can be concluded that there is now a certain tendency to use single-point models to assess the productivity of agricultural landscapes at the regional level (Ceglar et al. 2019; Lecerf et al. 2019).

During the development and implementation of the decision-making system LandCaRe DSS (Wenkel et al. 2013; Köstner et al. 2014), it has been proposed and implemented the idea to use different crop model with its own accuracy for each scale. This approach was applied due to the fact that crop models require detail data, and this always difficult to ensure on a regional scale. Nevertheless, data acquisition technologies are developing rapidly (Wolfert et al. 2017) and the scalability and the universalization of relevant software and methodologies can become a solution to this increasingly critical question (Mirschel et al. 2004). The project MACSUR is a pan-European information portal on the use of simulation models for the information support of agricultural adaptation processes in Europe for the consequences of anthropogenic climate change highlighted in a special working group «Scale-It!». This working group supports and performs ongoing research about modelling cropping systems at various scales, typically from field to regional ones (Ewert et al. 2011). Ensembles of soil-crop models are applied for this purpose. The site gathers information about the people, models and partners involved, presenting challenges and advances of this research.

Therefore, the objective of the current paper is to present a method that will allow the use of the same crop model in a long time scale and in an extended spatial one for estimation of agro-landscape productivity in regional scale in a GIS-environment.

28.2 Materials and Methods

Analysis of landscape productivity and sustainability under extended agricultural use for the particular geographic region relates to so-called mesoscope spatial level of crop model application. Unlike micro- (point, field or farm level) and macroscaling of model application (country or continental level) mesoscopic model application has a number specific features and requirements. Firstly, regional-intended simulation necessitates the use of peculiar sources and needed detail of the input data. Secondly, the set of practical and theoretical problems solved by computer experiments is also determined by spatial resolution. And, finally, it entails the use of proper tools and methods for adequate interpretation of simulation results (GIS-environments mainly).

28.2.1 Regionalized Crop Models: Input Data

Let's consider the problem of input data supply for region-oriented crop modelling in comparison with model use in micro- and macro special scales (Table 28.1). We use the standard classification of input data domains for dynamic crop models, which

Table 28.1 Sources and types of input data for crop-model-based computational projects of various spatial detail

Input factor	Spatial detail level		
	Microscale (field, farm)	Mesoscale (geographic region)	Macroscale (country, region, continent)
Soil	Site-specific detailed field test data	Reference soil map	Reference soil map
Site	Coordinates + relief micro-peculiarities	Coordinates	Coordinates
Technology	Precision farming	Common regionalized technology	Zone-specific approved technologies
Culture	Actually cultivated variety	Basic regionalized ideotype	Zone-specific recommended cultivars
Weather (actual)	Local automated meteostation	Nearest WMO meteostation + short-term forecast	Nearest WMO meteostation + short-term forecast
Weather (forecast)	One-point calibrated weather generator	Spatial–temporal weather generator	A set of benchmark years-analogues or one-point weather generator
Initial state	Rotation (predecessor culture)	Spin up computation	Conditional values

corresponds to the classification of the main factors influencing crop production process (Medvedev et al. 2015). The only difference here is that we specifically identified two different types and sources of weather data. The first type contains information about the actually observed or "past" weather at the local modelling site. The second one is the data about possible "future" weather for the remainder of the growing season, which are important during online predictive calculations. In the previous work (Badenko et al. 2018), we briefly discussed the problem of information support of crop model massive computations for dynamically adjusted forecasting of agro-landscape productivity in macroscale (for all Russian agricultural regions). Below we explain in more details the specificity of the input data for each enumerated factor for regional models.

The first issue is the choice of the set of base (benchmark) points (sites or locations) where the crop model calculations will be performed. For macroscale calculations, it is permissible to choose any conditional geographic base points, even if they do not correspond to agricultural land. The selection of these points must be done more carefully for the projects of regionalized calculations. They should belong to real agricultural fields and sufficiently reflect the diversity of the soil and climatic conditions of the region under consideration. Another desirable aspect is the presence of several locations in this point set, for which one can obtain both historical and current data from actual field observations or measurements. It allows the preliminary verification of the model as well as its online calibration during the computer experiment for the current growing season. So, in the below-described project of spring barley simulation on the scale of the Crimea, we used three such "verification" points. They correspond to the location of the experimental stations of the Research Institute of Agriculture of Crimea.

In detailed calculations on a farm scale, information on the biochemical and hydrophysical soil properties for each reference site can be obtained directly by laboratory processing of the samples from soil inspections. For the projects of regional calculations, it is practically impossible and, moreover, is not necessary. It is more reasonable for the case to operate not with the specific unique soils, but with some conditional characteristic representatives, which are the most typical ones for the current location. The source of such information may be digital soil maps. For example, in our last research (Badenko et al. 2018) we used the soil map and linked database of soil properties from Unified State Register of Russian Soil Resources (USRRSR), which is designed in Dokuchaev's Soil Institute and is published on the official website of Russian Ministry of Agriculture. It can be also used in region-oriented crop simulation but using more detailed soil maps seems to be preferable. For example, in frames of the project described below, we digitized a paper map of the soils of the Crimea and used the resulting digital version in further calculations.

Similar reasons relate to the assigning of agro-technology in the model (planting date, fertilization rates, etc.). It is necessary to set not a specific known variant, but some kind of averaged regulatory technology recommended for application. For the macroscale projects, this benchmark technology depends on the region. For regional calculations, it can be taken the same for all calculation points.

There are several dynamic variables that determine the initial state of the agroecosystem at the start time of crop model run (soil moisture content, labile forms of nitrogen in soil profile, etc.). It has been repeatedly shown that dynamic agroecological models are highly sensitive to the initial conditions (Poluektov et al. 2005). For the case of rather precise simulation in farm scale, these values can and must be determined by direct field measurements at the beginning of the growing season. There is no such ability for the simulation projects on a large spatial scale. A sane solution for the case would be to start the simulation of the current season not from the sowing, but with a sufficient time lead-in. Then the spring (or autumn for winter crops) time interval of modelling the agroecosystem dynamics "without a plant" (that is, taking into account only abiotic processes) will play the role of "spin up". This method is especially convenient for setting the initial soil water content, since it allows simulating not only the total moisture in the root zone, but also its distribution over the soil profile. Sensitivity analysis of the computer crop model AGROTOOL to the date of the start of simulation is shown in Fig. 28.1. It presents the influence of the modelling start date on model yields (a) and soil moisture in 1 m layer at the sowing (b) for spring barley crops in the 2018 season for three reference points (experimental fields of the Research Institute of Agriculture of the Crimea in Belogorsk, Saki and Krasnogvardeysky districts). It is easy to see that the model crop productivity is strongly dependent on the duration of "spin-up" period, and is closely correlated with the accuracy of the estimation of the pre-sowing soil moisture content.

Recently, there is no problem of obtaining historical data on the actually observed weather for any geographical location (at least for the nearest WMO station). There are also a large number of Internet services (www.worldweatheronline.com, www.aerisweather.com) that provide special API for getting information on short-term synoptic forecasts based on the calculations of atmospheric general circulation models. Sustainable predictive ability of such forecasts is 3–5 days.

In turn, predictive variant simulation for the "future time" or for ensemble statistical calculations require the big number of virtual weather scenarios is to be used as a source of weather information. As such scenarios, one can use the so-called "years-analogues" or the samples of stochastic weather generators. To date, the theory of implementation of stochastic weather generators as well as the successful practice of their integration within computer crop models have been developed (Richardson and Wright 1984; Semenov et al. 1998). All these weather generators are single point, that is, they allow obtaining the series of required meteorological characteristics having necessary temporal trends as well as temporal and structural correlation for the specific geographic point. The specificity of model calculations on a regional scale places additional challenge, namely, the need for not only temporal and structural, but also spatial consistency of generated weather series. Indeed, even the virtual weather during the same year for the closely located points in a region cannot be set independently. This is explained by the fact that, in contrast to the planetary or continental scale, on a regional scale it is possible to speak about a "good season" and a "bad season" in plant production meaning. This fact permits to suggest the following criterion for identifying a rather homogeneous region of agricultural production: ***this is such a geographical area where the inter-annual variability of agricultural***

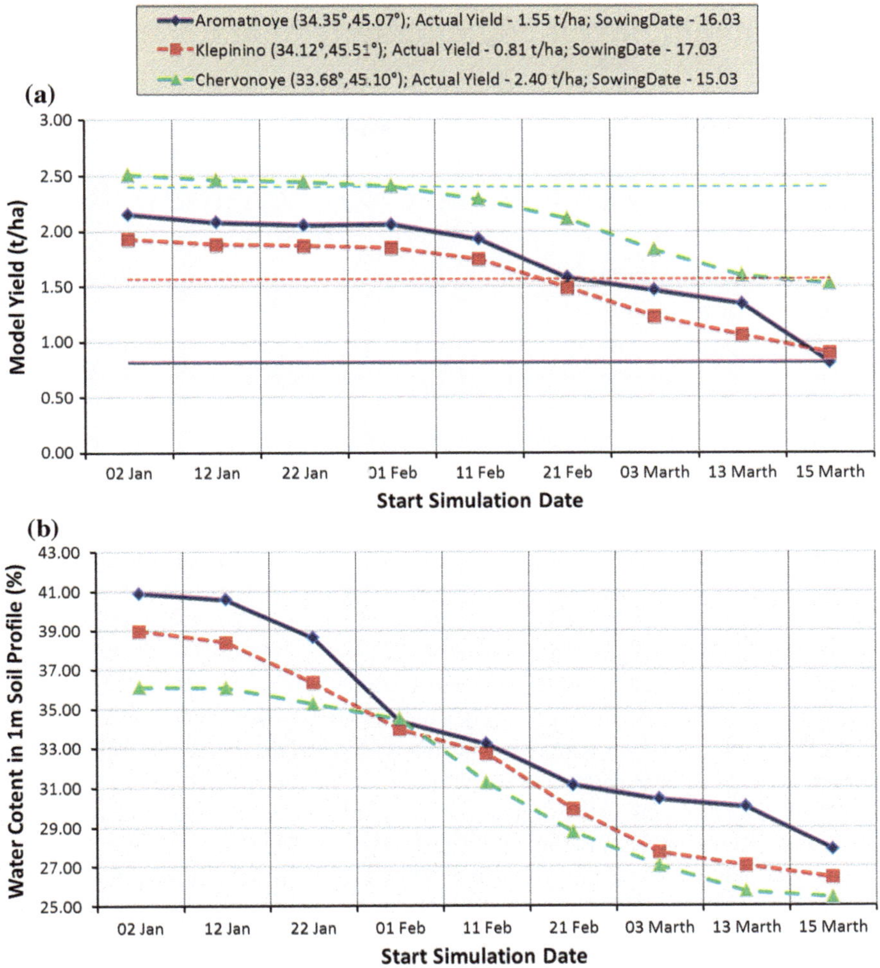

Fig. 28.1 Model yields (**a**) and soil moisture in 1 m layer at the sowing (**b**) as a function of the simulation start day. Thin lines—actual yields (spring barley 2018) for three reference points

productivity significantly exceeds its spatial variability inside the region. The latter concept can be illustrated by the example below. We have processed statistics on the yields of major agricultural crops across the region (Crimea) and across all Russia. Information on crop yields in Crimea districts was provided by researchers from domestic agricultural institute (an example of the initial data is given in Table 28.2). Archival data on the yields in the regions of Russia are taken from the open Internet resources of the state statistical department (Rosstat). In both cases spatial as well as temporal variability is obviously observed. In order to compare their relative significance, one can use the standard statistical analysis. The most proper treatment of this kind of tabbed data is two-factor analysis of variance without replicas. The selected

Table 28.2 Spatial and temporal variability of winter wheat yield (t/ha) for the Republic of Crimea

District	Year						
	1990	2000	2012	2014	2015	2016	2017
Bakhchisaray	4.86	2.24	1.58	2.02	2.19	2.34	2.87
Belogorsk	3.53	1.76	1.62	2.15	1.9	1.87	2.41
Dzhankoysky	3.41	1.75	1.78	2.16	2.53	3	2.88
Kirovsky	4.08	1.81	2.79	2.56	2.77	2.45	2.98
Krasnogvardeysky	3.79	2.25	1.52	2.52	2.65	2.89	3.16
Krasnoperekopsky	3.43	1.9	2.05	2.68	3.93	3.7	2.93
Leninsky	3.07	1.68	1.38	2.27	2.76	3.37	3.84
Nizhnegorsky	4.42	2.61	1.78	2.29	2.95	2.62	2.92
Pervomaysky	3.46	1.77	0.87	2.17	2.7	2.37	3.04
Razdolnensky	3.25	1.79	0.96	2.13	2.48	2.73	3.09
Saki	4	2.12	0.76	2.23	2.66	2.56	2.74
Simferopol	4.51	2.57	1.25	2.5	3.15	3.36	3.64
Sovetsky	3.63	3.16	2.38	2.34	2.76	2.31	2.8
Chernomorsky	3.46	2.11	1.02	2.33	2.03	2.48	2.95

results of the processing are presented in Table 28.3. It is not difficult to notice the fundamental difference. For a relatively small region (in which, nevertheless, several contrasting natural zones are geographically distinguished), the comparative contribution of the inter-annual (temporal) variability to the total dispersion exceeds by an order of magnitude the share of spatial one. While on the scale of a large country, these values are quite comparable with each other.

Above-mentioned reasons entail the substantial modification of the method of stochastic weather modelling as an input of crop models for the purposes of region-oriented simulations. The main improvement is that the generated weather series for base locations should have not only a temporal, but also a spatial correlation. This necessitates the development of a fundamentally new stochastic weather generator based on Monte Carlo simulation of space–time random processes and fields. Currently, our research team is working on the development and parametric identification of this spatial–temporal stochastic generator of regional weather scenarios based on the processing of multi-year observational data at 20 meteorological stations of the WMO network in the Republic of Crimea. The interface of the subject-oriented research software specially developed for this purpose is presented in Fig. 28.2. It shows the results of the calculation and approximation of the empirical variogram (the dependence of the spatial correlation of meteorological characteristics on the distance between the reference meteorological stations). The details of the developed mathematical algorithm for generating spatial random processes and its use in regionalized crop models are going to be presented in the next publications.

Table 28.3 Two-factor analysis of variance of the temporal/spatial yield variability in Crimea and All-Russia scales

Culture	Source of variation	Crimea				Russia			
		SS	df	F	F-thr	SS	df	F	F-thr
Winter wheat	Spatial	558	13	2.50	1.847	31,027	59	22.38	1.350
	Temporal	4,206	6	40.90	2.217	9,983	8	53.11	1.958
Cereals and legumes	Spatial	795	13	3.89	1.847	58,271	74	61.11	1.310
	Temporal	3,602	6	38.18	2.217	5,345	8	51.85	1.954
Barley	Spatial	639	13	2.67	1.847	12,863	15	13.97	1.750
	Temporal	4,768	6	43.16	2.217	1,337	8	2.72	2.016
Potato	Spatial	4,241	13	*1.29*	1.847	563,820	79	16.80	1.299
	Temporal	125,358	6	76.11	2.217	143,400	6	42.19	1.953

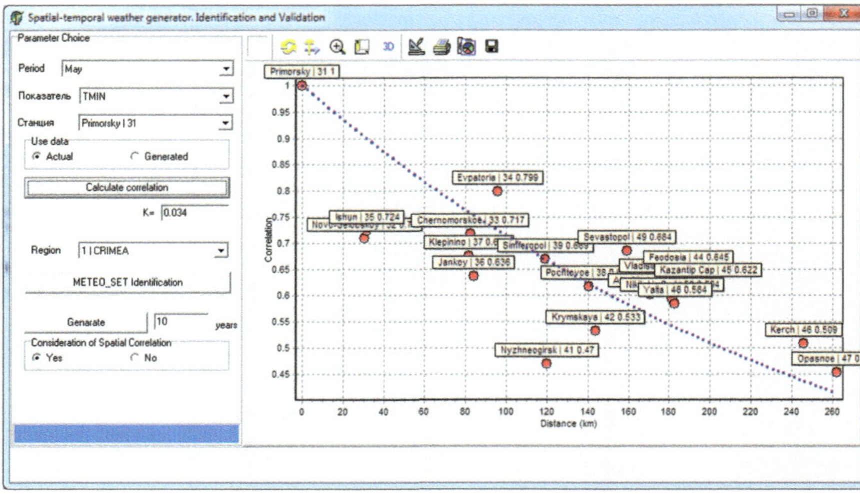

Fig. 28.2 Main window of the research software for identification and analysis of spatial–temporal stochastic weather generator (variogram function approximation mode)

28.2.2 Regionalized Crop Models: Use Cases

Obviously, at the regional level, models are used to solve regional problems. First of all, these problems are connected with the provision of water balance, as well as a balance in the volume and type of agricultural production. The solution to such problems is determined by the integration of crop models and regional GIS. This allows specialists to take into account the interests of agricultural producers in solving the problems of regional planning for different forecast periods in the context of limited resources, primarily soil and water. If we talk about long-term forecasts, the use of simulation models integrated into GIS will make it possible to justify for the region rational crop rotations and the allocation of crops by specific agricultural fields. The use of simulation models allows carrying out an operational forecast and determining the possible development of the situation in the region on the current situation with actually implemented soil tillage and current weather.

Observed global climate change is another major challenge for the sustainable development of agricultural production in the region (Wolfert et al. 2017). This contradiction may be allowed to solve, in particular, by creating fundamentally new varieties of crops, primarily cereals. At the same time, one of the powerful means of significantly accelerating the traditional selection process can be the simulation models of agroecosystems (Lobell et al. 2011). Researches in this direction also are associated with the introduction of the concept of ideotype (Rötter et al. 2015; Kissoudis et al. 2016). Ideotype is a future virtual species of agricultural crop, capable of producing a theoretically possible yield in accordance with bioclimatic potential; in essence, it is a promising goal of breeding. Qualitative assessments of the prospects for an ideal variety as a scientific prediction about a plant and its individual characteristics under specified conditions showed their efficiency, but all the advantages of this approach can be revealed only in combination with the use of simulation models of the agroecosystems. At the same time, genotypic adaptation, which involves the inclusion of new traits, is expected to be one of the most important adaptation strategies for future climate change, and modelling agroecosystems can serve as a basis for assessing the biophysical potential of crops for such adaptation (Ramirez-Villegas et al. 2015). In general, we can talk about two ways of using agroecosystem models in adapting the agricultural sector of regional development to climate change: (1) identifying areas favourable for existing varieties taking into account significant dynamism of climatic changes, and (2) justifying the requirements for traits of new varieties adapted to expected changes in cultivation conditions.

As examples of the first direction, one can analyse results in (Tao et al. 2017; Jeuffroy et al. 2014). The results of (Tao et al. 2017) showed that barley specific genotypes may be promising for the predicted conditions of climate change, and the most favourable zones for specific ideotypes with a combination of several key genetic traits (phenology, leaf growth, photosynthesis, drought resistance, and grain formation) were proposed. In turn, review (Rötter et al. 2015) provides an analysis of papers relating to the second trend. In particular, an analysis of the main limits on the use of simulation modelling of crops to support breeding have been presented, as well as examples of how modelling allows assessment and formation of varieties of crops for future climate conditions. The use of crop models to identify traits necessary for future crop varieties have been analysed in (Semenov and Stratonovitch 2015).

28.2.3 Regionalized Crop Models: Result Processing and Interpretation

So, the simulation results for point simulation models of agroecosystems are always associated with a specific location (agricultural field or management unit). Therefore, the transfer of the results obtained at individual points to the region level requires the use of special algorithms, which are widely represented in the literature. Such algorithms belong to a rather wide class of point-to-area conversion algorithms that are built-in in modern GIS software. Such algorithms can be of two types—non-interpolative methods and interpolative methods. Non-interpolative methods involve the assignment of one or more attributes of a point to a polygon. In the simplest case, a grid of square cells is overlapped with modelling points. The attributes of cells that contain more than one point are determined by aggregating point attributes according to appropriate rule (taking the mean, mode, maximum, etc.). Cells that contain no points are given attributes by aggregating attributes of neighbouring cells (not empty) also according to appropriate rule (Fig. 28.3a). A method that overcomes the problem of polygons either having no points or more than one point is to generate Thiessen (Voronoi) polygons (Fig. 28.3b). Such polygons are generated on a base of

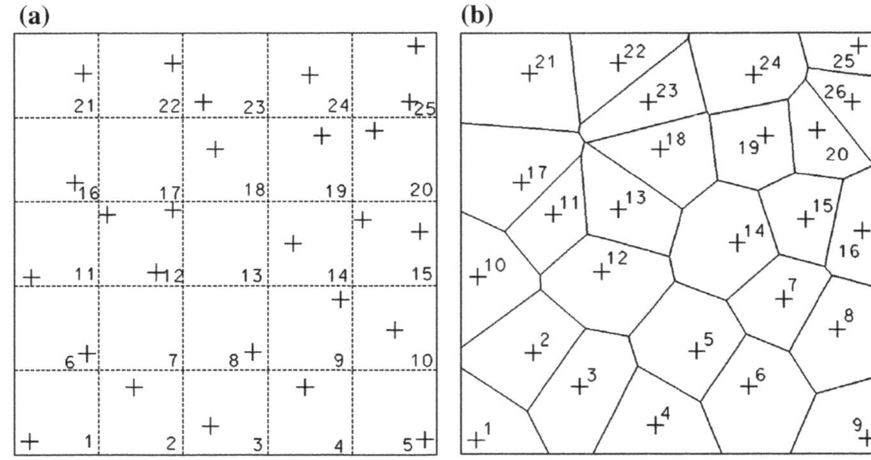

Fig. 28.3 Non-interpolative methods of point-to-area conversion

a single point (base point) and any location within a polygon is closer (in Euclidean distance) to the base point than to any in other polygons.

Interpolative methods are implemented if an attribute of the base point is a spatially continuous “field” variable, mappable as single-valued surface. The interpolation process involves estimating the value of the modelled variable at a succession of point locations, usually on a square lattice (as named “gridding”). The gridded values are then treated as the pixels of a raster image. For calculation of the gridded values (z_g) based on set of known modelled values (z_i), it can be used the method for inverse weighted distance:

$$z_g = \frac{\sum_{i=1}^{n} w_i \cdot z_i}{\sum_{i=1}^{n} w_i}$$

where $w_i = 1/(d_{ig})^2$, or $w_i = 1/d_{ig}$, d_{ig}—is the distance from a point with known modelled values to a point with gridded value.

28.2.4 Regionalized Crop Models: Software Implementation

We use APEX (**A**utomation of **P**olyvariant **EX**periments) software as the main automation tool of the massive calculations of crop models. It is originally developed for design and performing of multi-factor computer experiments with dynamic crop models and encapsulates two basic functionalities: (1) versatile repository of the descriptors of arbitrary crop models, and (2) generic environment for their polyvariant analysis (Medvedev and Topaj 2011). The polyvariant analysis means designing and preparation of multivariate computer case study, performing the model runs in batch mode and applying advanced procedures of statistical treatments for results obtained.

The information model of APEX database can be divided into three principal domains (see Fig. 28.4), there the third one extends the basic APEX structure and functionality for the case of spatially distributed massive computations.

All fixed (invariant) tables in system APEX, which user can't add, modify or remove, have specific prefix reg. Other tables belonging particular models have prefixes corresponding each model. For example, model AGROTOOL-LITE used for this research has table prefix al_, model MONICA has the table prefix mo. Generally, these tables contain data about the registered models: description of specific tables of input data and their columns, and information domains of the model these tables belong. For example, the specific table "Irrigation" of the model AGROTOOL belongs to the information domain "Technology" predefined in APEX. Each predefined factor existing in the model has its root table. As a rule, such tables have only ID and Description columns. All model-specific data is stored in child tables with foreign key to the master table.

Additionally, APEX has a few tables which extend model-specific data and closely relate to the projects of regional simulation. These tables have complex

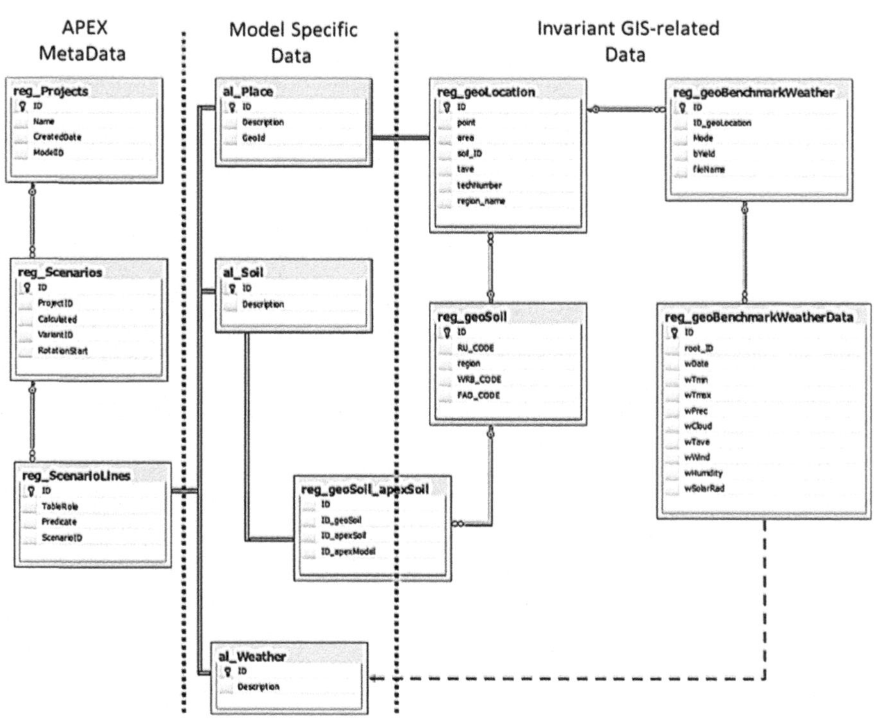

Fig. 28.4 Direct and indirect links between MetaData, Model-specific Data and Invariant GIS-related Data concepts in **APEX** relational DataBase

prefix ***reg_geo***. Based on these tables, semantically rich data analysis can refer to external invariant spatial and weather data to combine it with model-specific data. Interaction is performed by single-direction or cross-links between fixed tables with extended data and model-specific tables created by the user. Most domain-specific algorithms are attached to the predefined information domains "Weather" and "Soil". The corresponding tables are the following:

reg_geoSoil—model-independent table of benchmark soils with spatial linkage. Each its record contains the unique identifier of the soil variety, polygon in geographic coordinates covered by the current soil contour and several codes of alternative classification of this soil variety.

reg_geoSoil_apexSoil—cross-link table between table ***reg_geoSoil*** and model-specific root tables of the predefined factor "Soil". Each model-specific root table of the predefined factor "Soil" can be linked with several records of the table ***reg_geoSoil***, because several polygons with soils can use the same data of the soil horizons, especially when the type codes of the different soil areas are the same.

reg_geoLocation—model-independent table of locations, where spatially distributed computer experiments are performed. It contains the columns describing the latitude and longitude of every reference point and the foreign key to the table ***reg_geoSoil***. Several locations can occur in one soil polygon.

reg_geoBenchmarkWeather—model-independent table of the daily weather data in the invariant format.

This information scheme gives the solid ground for automation the procedure of creation the "spatial" projects in APEX infrastructure. For example, user can start from the selection of several locations under consideration and then automatically form all project records, having the proper internal correspondence between site, soil and reference weather scenarios in terms of used crop model. The screenshot of APEX main window demonstrating the typical structure of the regional project is presented in Fig. 28.5.

Thereby, each scenario of the APEX computational project is associated with specific gradations of predefined factors. This enables multiple approaches for presentation and interpretation of the results of a computer experiment. The most natural way to represent the influence of the "terrain" factor in regionalized projects is to build the thematic digital maps in an arbitrary third-party GIS-system. In turn, the study of predictors having non-geographical nature (soil, weather, technology, etc.) can be done by the tools of factorial and variance statistical analysis built into the APEX system (see Fig. 28.5). Examples of both approaches are set out below.

28.3 Results

Testing of the proposed methods has been carried out for the Crimean Peninsula (hereinafter we will call it Crimea). Figure 28.6a shows an overview map of the Crimea and 55 agricultural enterprises, which are the points at which the modelling was carried out (reference points). Three base calibration locations are marked with arrows. Figure 28.6b shows the soils of the Crimea and 13 meteorological stations

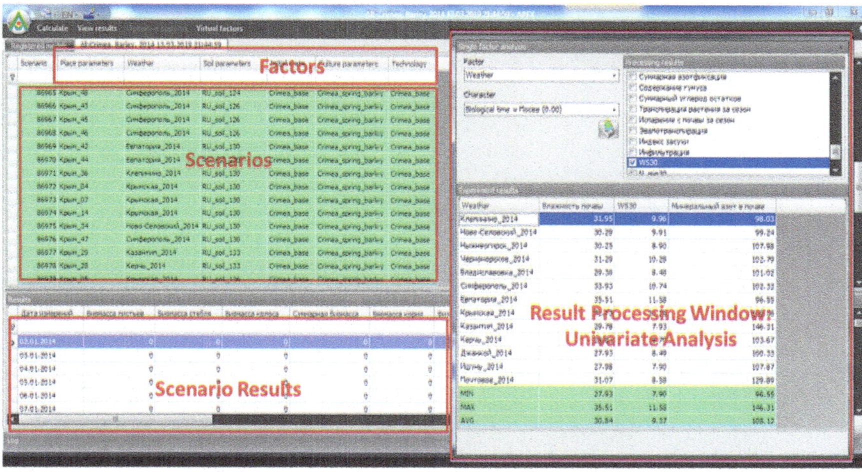

Fig. 28.5 Structure of the regional project and the principles of result processing in terms of various influencing factors in APEX user interface

that were used to simulate the actual weather for the nearest reference points. The map legend in Fig. 28.6b shows 18 soil varieties and the number of soil contours of each variety in brackets, but only 13 of them were used in the calculations, since some soil contours did not contain reference points. Description of corresponding soil varieties is presented in Table 28.4.

The calculations were carried out in the APEX environment for two growing seasons of 2012 and 2014. Spring barley has been selected as a simulated culture. In Fig. 28.7, the barley yield simulation results have been presented. Spatial distribution of barley yield and dates of its ripening (in brackets—the number of modelling points with the same date of barley ripening) have been also presented in Fig. 28.7.

Figure 28.7 shows, that in 2012 barley yield is lower than in 2014, but spatial variation in 2014 is higher than in 2012. For presentation of barley yield, the Thiessen (Voronoi) polygons (see Fig. 28.3b) have been applied.

Spatial distribution of all other model parameters, including not only the final (at the moment of harvest), but also the intermediate ones, can be presented in GIS-based thematic mapping tools. Such thematic maps are an effective media for analysis of process and events related to crops development in the region during the growing season.

For example, the spatial distribution of soil moisture in the region on 15 May and 01 June during 2012 and 2014 growing seasons is presented in Fig. 28.8.

It can be recognized, that soil moisture level in 2014 is more suitable for barley development than in 2012.

An example of post-processing of a computer experiment for "non-cartographic" factors with the help of statistical analysis tools built into the APEX functionality is given in Table 28.5. It presents the selected results of univariate analysis of

(a)

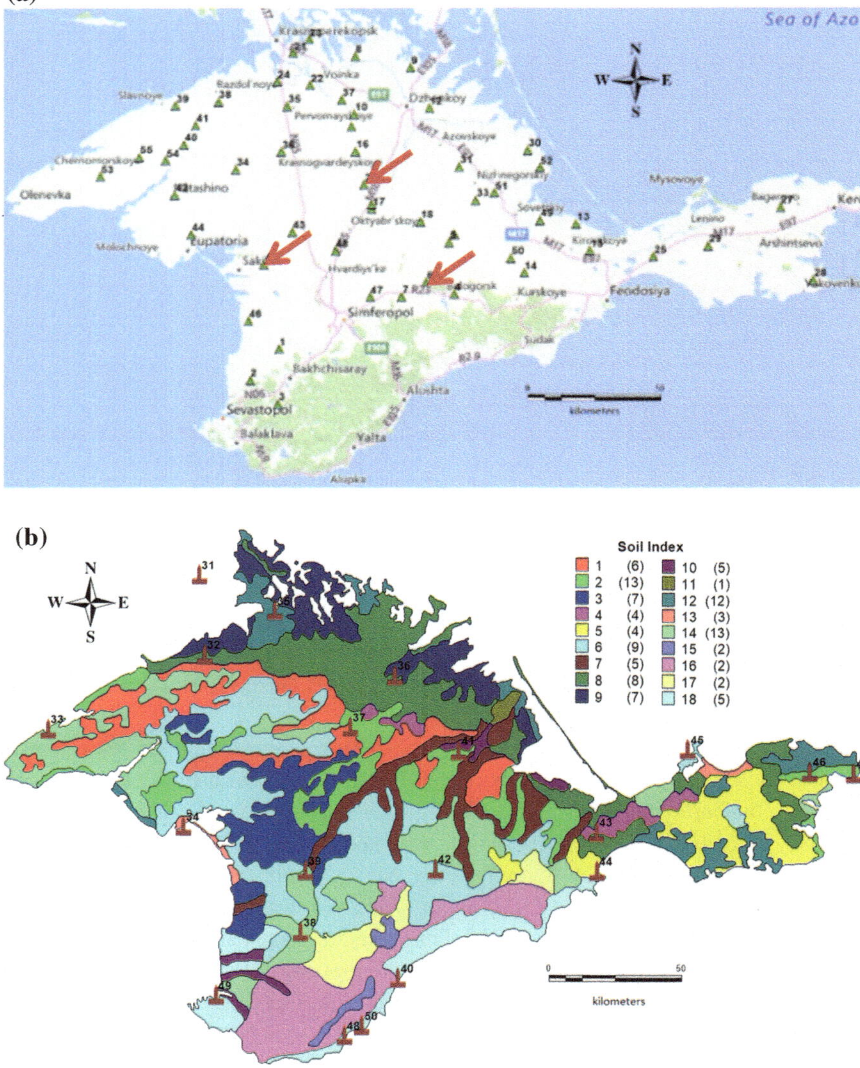

Fig. 28.6 **a**—overview map of the Crimea and 55 reference points; **b**—soils of the Crimea and meteorological stations (symbol)

the above-mentioned studies for the factors "soil" and "weather". The most meaningful conclusion here is that the diversity of soil conditions in the region under consideration has a greater influence on the variation in model yields than the spatial heterogeneity of weather conditions.

Table 28.4 List of soil varieties

1	Chernozems southern	10	Meadows
2	Chernozems southern and ordinary mycelial-calcareous	11	Meadows solonetzic and solonchakous
3	Chernozems ordinary glossic	12	Solonetzes
4	Chernozems solonetzic	13	Mountain primitive
5	Chernozems compact	14	Sod-calcareouses
6	Chernozems residual-calcareous	15	Mountain-meadow chernozem-likes
7	Meadow-chernozemics	16	Mountain forest chernozemic
8	Dark chestnuts	17	Brown forest weakly unsaturated
9	Meadow-chestnuts solonetzic	18	Mountain debrital-organogenuos

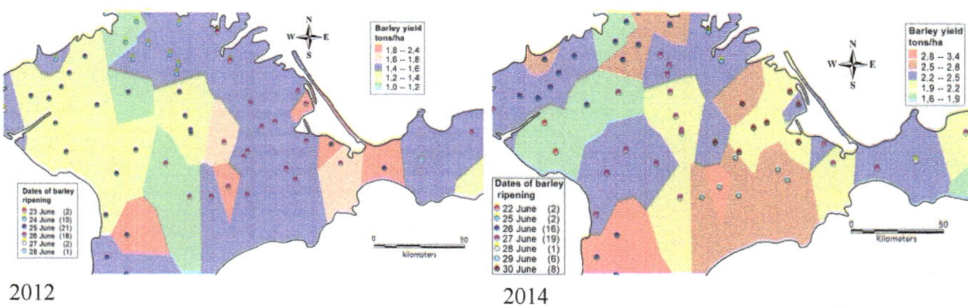

2012

2014

Fig. 28.7 Spatial distribution of barley yield and dates of its ripening (in brackets—the number of modelling points with the same date of barley ripening)

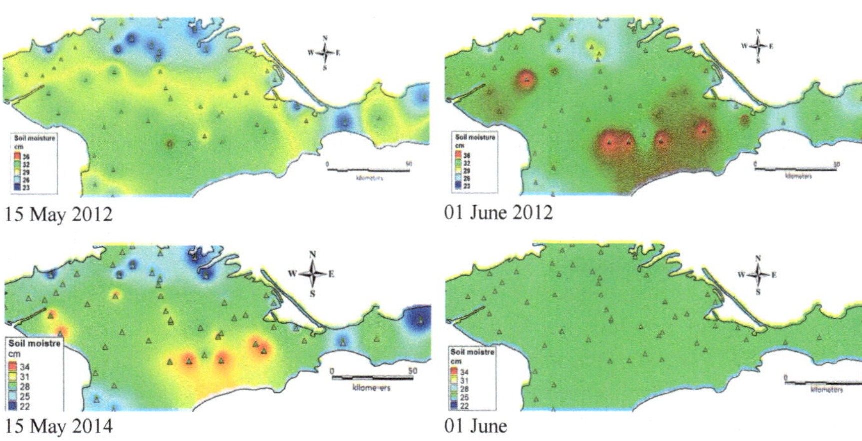

15 May 2012

01 June 2012

15 May 2014

01 June

Fig. 28.8 Spatial distribution of soil moisture in region at 15 May and 01 June during 2012 and 2014 growing seasons

Table 28.5 One-factor analysis of soil—weather impact into barley productivity for Crimea regional project of crop model calculations

Factor: soil variety	Average yield (t/ha)		Factor: weather station	Average yield (t/ha)	
	2012	2014		2012	2014
RU_soil_83	2.23	2.34	Vladislavovka	1.80	2.26
RU_soil_120	1.40	2.20	Jankoy	1.42	2.38
RU_soil_124	1.41	2.36	Evpatoria	1.39	1.72
RU_soil_126	1.23	2.11	Ishun	1.36	2.35
RU_soil_130	1.31	1.94	Kazantip	1.81	2.49
RU_soil_133	1.38	2.19	Kerch	1.36	1.92
RU_soil_136	1.56	2.46	Klepinino	1.35	2.21
RU_soil_143	1.55	2.61	Krymskaya	1.61	2.76
RU_soil_156	1.59	2.80	Novo-Selovskiy	1.37	2.29
RU_soil_178	1.53	1.75	Nyzhneogirsk	1.61	2.15
RU_soil_181	1.22	2.99	Pochtovoe	1.75	2.82
			Simferopol	1.19	2.28
			Chernomorskoe	1.54	2.33
MIN	1.22	1.75	MIN	1.19	1.72
MAX	2.23	2.99	MAX	1.81	2.82
MEAN	1.49	2.34	MEAN	1.50	2.30
St. DEV	**0.28**	**0.36**	St. DEV	**0.20**	**0.29**

28.4 Conclusions

Traditionally, the simplest calculation methods (empirical regression models) are used for the predictive estimates of the productivity of agroecosystems over rather large areas (region-wide). The accuracy of the results obtained with their help is relatively low, and it is impossible to use them for the tasks of supporting operational–technological solutions. The current level of development of computing and information technologies makes it possible to use for this purpose more accurate and detailed dynamic crop models. At the same time, the tasks of choosing the optimal agro-technical plan (by "playing" with the model for explore the effects of various options) and, perhaps, improving the model in online mode based on the results of actual measurements during the growing season (by assimilating of remote sensing data of crops) must also be solved. The volume of corresponding model runs (calculation variants) is determined by simultaneously varying several influencing factors:

1. Geographical factor: a set of reference points (locations) of the calculation, each of which is characterized by its geographical coordinates, soil properties and

varietal characteristics of the studied culture, adapted for cultivation in the region in question.

2. Meteorological factor: in the tasks of online forecast of expected productivity of agroecosystems, possible weather dynamics for the remaining period of the vegetation season is modelled using an "options for possible trajectories" consisting of a representative number of synthetic weather scenarios. For their formation, a stochastic weather generator can be used, which maintains both temporal and spatial correlations between meteorological variables.
3. Technological factor: to form a strategy of proactive management of the production process of crops, it is necessary to analyse the effects of a variety of input effects options (date and quality/quantity of technological operations) in order to select the best one in the context of the selected statistical optimization criterion.
4. Model factor. To obtain trustworthy results, it is desirable to use not one, but several alternative models of the agroecosystems production process with the integration of their results based on the analysis of ensemble calculations.

Simultaneous variation of several factors listed above according to the set of gradations leads to the fact that the resulting dimension of a full factorial computational experiment (the total number of variants of a single model run) makes it impossible to perform all the corresponding calculations in a sequential mode on a standard personal computing technique in a reasonable time. To implement these calculations, it will be necessary to attract modern technologies of distributed parallel computing and supercomputer hardware.

Acknowledgements The reported study was funded by RFBR according to the research project 19-016-00148.

References

Badenko VL, Topaj AG, Medvedev SA, Zakharova ET (2018) An operatively-refined forecast of wheat productivity in agricultural zones throughout Russia on the basis of the simulation model of productivity. AgroecoInfo 3(33):68. http://agroecoinfo.narod.ru/journal/STATYI/2018/3/st_345.doc

Badenko VL, Garmanov VV, Ivanov DA, Savchenko AN, Topaj AG (2015) Prospects of application of dynamic models of agricultural ecosystems in the problems of midterm and long_term planning of agricultural production and land management. Russ Agric Sci 41(2–3):197–200

Badenko V, Badenko G, Topaj A, Medvedev S, Zakharova E, Terleev V (2017a) Comparative simulation of various agricultural land use practices for analysis of impacts on environments. Environments 4(4):92

Badenko VL, Topaj AG, Yakushev VV, Mirschel W, Nendel C (2017b) Crop models as research and interpretative tools. Sel'skokhozyaistvennaya Biol 52(3):437–445 (in Russian)

Benoît M, Rizzo D, Marraccini E, Moonen AC, Gall M, Lardon S, Bonari E (2012) Landscape agronomy: a new field for addressing agricultural landscape dynamics. Landsc Ecol 27(10):1385–1394

Ceglar A, Van der Wijngaart R, De Wit A, Lecerf R, Boogaard H, Seguini L, Baruth B (2019) Improving WOFOST model to simulate winter wheat phenology in Europe: Evaluation and effects on yield. Agric Syst 168:168–180
Chopin P, Blazy JM, Doré T (2015) A new method to assess farming system evolution at the landscape scale. Agron Sustain Dev 35(1):325–337
Dunaeva EA (2018) SWAT agrohydrological model application for crop productivity simulation. In: Collection: Konyaevskyechtenya collected articles of VI international science-practical conference, pp 245–248
Dury J, Schaller N, Garcia F, Reynaud A, Bergez JE (2012) Models to support cropping plan and crop rotation decisions: a review. Agron Sustain Dev 32(2):567–580
Ewert F, van Ittersum MK, Heckelei T, Therond O, Bezlepkina I, Andersen E (2011) Scale changes and model linking methods for integrated assessment of agri-environmental systems. Agr Ecosyst Environ 142:6–17
Fritz S, See L, Bayas JCL, Waldner F, Jacques D, Becker-Reshef I, Rembold F (2019) A comparison of global agricultural monitoring systems and current gaps. Agric Syst 168:258–272
Inwood SE, Dale VH (2019) State of apps targeting management for sustainability of agricultural landscapes: a review. Agron Sustain Dev 39(1):8
Inwood SE, López-Ridaura S, Kline KL, Gérard B, Monsalue AG, Govaerts B, Dale VH (2018) Assessing sustainability in agricultural landscapes: a review of approaches. Environ Rev 26(3):299–315
Jeuffroy M-H, Casadebaig P, Debaeke P, Loyce C, Meynard J-M (2014) Agronomic model uses to predict cultivar performance in various environments and cropping systems: a review. Agron Sustain Dev 34(1):121–137
Jones JW, Antle JM, Basso B, Boote KJ, Conant RT, Foster I, Keating BA (2017) Toward a new generation of agricultural system data, models, and knowledge products: state of agricultural systems science. Agric Syst 155:269–288
Kissoudis Ch, van de Wiel C, Visser RGF, van der Linden G (2016) Future-proof crops: challenges and strategies for climate resilience improvement. Curr Opin Plant Biol 30:47–56
Köstner B, Wenkel KO, Berg M, Bernhofer C, Gömann H, Weigel HJ (2014) Integrating regional climatology, ecology, and agronomy for impact analysis and climate change adaptation of German agriculture: an introduction to the LandCaRe2020 project. Eur J Agron 52:1–10
Lecerf R, Ceglar A, López-Lozano R, van der Velde M, Baruth B (2019) Assessing the information in crop model and meteorological indicators to forecast crop yield over Europe. Agric Syst 168:191–202
Lobell DB, Schlenker W, Costa-Roberts J (2011) Climate trends and global crop production since 1980. Science 333:616–620
Medvedev S, Topaj A (2011) Crop simulation model registrator and polyvariant analysis. IFIP Adv Inf Commun Technol 359:295–301
Medvedev S, Topaj A, Badenko V, Terleev V (2015) Medium-term analysis of agroecosystem sustainability under different land use practices by means of dynamic crop simulation. IFIP Adv Inf Commun Technol 448:252–261
Mirschel W, Schultz F, Wenkel K-O, Wieland R, Poluektov RA (2004) Crop growth modelling on different spatial scales—a wide spectrum of approaches. Arch Agron Soil Sci 50(3):329–343
Najafi E, Devineni N, Khanbilvardi RM, Kogan F (2018) Understanding the Changes in global crop yields through changes in climate and technology. Earth's Futur 6(3):410–427
Nendel C, Berg M, Kersebaum KC, Mirschel W, Specka X, Wegehenkel M, Wieland R (2011) The MONICA model: testing predictability for crop growth, soil moisture and nitrogen dynamics. Ecol Model 222(9):1614–1625
Nendel C, Wieland R, Mirschel W, Specka X, Guddat C, Kersebaum KC (2013) Simulating regional winter wheat yields using input data of different spatial resolution. Field Crop Res 145:67–77
Poluektov RA, Smolyar EI, Terleev VV, Topaj AG (2005) Models of production process of agricultural plants. SPb: SPbGUEds, 396 p (in Russian)

Popovich VF, Molyar SA (2018) Comparative analysis of AGROTOOL and WOFOST models adaptability for conditions of steppe Crimea. In: Collection: Konyaevskyechtenya collected articles of VI international science-practical conference, pp 285–287
Ramirez-Villegas J, Watson J, Challinor AJ (2015) Identifying traits for genotypic adaptation using crop models. J Exp Bot 66(12):3451–3462
Richardson CW, Wright DA (1984) WGEN: a model for generating daily weather variables. USDept. Agric., Agricultural Research Service, Publ. ARS-8
Rötter RP, Tao F, Höhn JG, Palosuo T (2015) Use of crop simulation modelling to aid ideotype design of future cereal cultivars. J Exp Bot 66(12):3463–3476
Rounsevell MDA, Pedroli B, Erb KH, Gramberger M, Busck AG, Haberl H, Kristensen S, Kuemmerle T, Lavorel S, Lindner M, Lotze-Campen H, Metzger MJ, Murray-Rust D, Popp A, Pérez-Sobab M, Reenberg A, Vadineanu A, Verburg PH, Wolfslehner B (2012) Challenges for land system science. Land Use Policy 29(4):899–910
Semenov MA, Stratonovitch P (2015) Adapting wheat ideotypes for climate change: accounting for uncertainties in CMIP5 climate projections. Climate Res 65:123–139
Semenov MA, Brooks RJ, Barrow EM, Richardson CW (1998) Comparison of WGEN and LARS-WG stochastic weather generators for diverse climates. Clim Res 10:95–107
Tao F, Rötter RP, Palosuo T, Schulman AH (2017) Designing future barley ideotypes using a crop model ensemble. Eur J Agron 82:144–162
Van der Velde M, Biavetti I, El-Aydam M, Niemeyer S, Santini F, van den Berg M (2019) Use and relevance of European Union crop monitoring and yield forecasts. Agric Syst 168:224–230
Van Bussel LGJ, Ewert F, Leffelaar PA (2011) Effects of data aggregation on simulations of crop phenology. Agr Ecosyst Environ 142(2):75–84
Wenkel KO, Berg M, Mirschel W, Wieland R, Nendel C, Köstner B (2013) LandCaRe DSS–an interactive decision support system for climate change impact assessment and the analysis of potential agricultural land use adaptation strategies. J Environ Manag 127:168–183
Wolfert S, Ge L, Verdouw C, Bogaardt MJ (2017) Big data in smart farming—a review. Agric Syst 153:69–80

Chapter 29
Landscape Phenology Modelling and Decision Support in Serbia

Branislava Lalic, Milena Marcic, Ana Firanj Sremac, Josef Eitzinger, Ivan Koci, Tara Petric, Mirjana Ljubojevic and Bosko Jezerkic

Abstract An operationally efficient DSS should be designed to operate on different time and spatial scales and to meet the needs of producers and policymakers. Introduction of novel scientific techniques and weather forecast in plant and harmful organism phenology modelling is an important prerequisite for clear and publishable recommendations. The presented examples of monthly and seasonal forecast application in phenology dynamics and its use in PIS as a DSS rely on strong scientific background—models calibrated and validated using biological observations and meteorological measurements. It serves as a reminder of how important it is to stick to the basics: observe events, measure relevant variables and then apply all available tools and techniques to produce high-quality information.

B. Lalic (✉) · A. F. Sremac · M. Ljubojevic
Faculty of Agriculture, University of Novi Sad, Dositej Obradovic Sq. 8, 21000 Novi Sad, Serbia
e-mail: branislava.lalic@polj.edu.rs

A. F. Sremac
e-mail: ana.sremac@polj.edu.rs

M. Ljubojevic
e-mail: mirjana.ljubojevic@polj.edu.rs

M. Marcic · I. Koci · B. Jezerkic
Forecasting and Reporting Service for Plant Protection of AP Vojvodina (PIS), Novi Sad, Serbia
e-mail: milena.marcic@pisvojvodina.com

I. Koci
e-mail: ivan.koci@pisvojvodina.com

B. Jezerkic
e-mail: bosko.jezerkic@pisvojvodina.com

J. Eitzinger
Institute of Meteorology, University of Natural Resources and Life Sciences Vienna, Peter-Jordan Str 82, 1190 Vienna, Austria
e-mail: josef.eitzinger@boku.ac.at

T. Petric
Faculty of Agriculture, University of Novi Sad, Novi Sad, Serbia
e-mail: tara.petric@gmail.com

W. Mirschel et al. (eds.), *Landscape Modelling and Decision Support*, Innovations in Landscape Research,
https://doi.org/10.1007/978-3-030-37421-1_29

Keywords Phenology modelling · Decision support system · Weather forecast · Harmful organism · Agrometeorological modelling

29.1 Introduction

Decision support systems (DSS) for agrometeorological and agricultural applications are considered a valuable contribution to sustainable food production globally. Over the past several decades, due to technological developments in the fields of measurement techniques, automated data collection systems, data transfer and analysis, as well as information transfer to stakeholders and farmers, various types of DSS have been developed and have become operational. Through the digitalization trend in agriculture, a further boost in applications is expected over the next several decades; however, many challenges exist in tailoring specific information to stakeholder needs in specific decision support applications.

A specific focus of forecasting for agrometeorological applications should include the impacts of weather and climate extremes on agricultural production. This focus needs to cover short-term forecasting and should also address seasonal forecasting (weeks to months) and long-term climate scenario assessments. An example of a short-term operational extreme weather warning in Europe is "Meteoalarm", which also provides information for agricultural purposes (e.g. warnings can be used in daily operations for planning and the protection of crops, orchards and vineyards against hazards). Nevertheless, for the purposes of agriculture management, forecasts of more than 5 days can still be very useful for certain applications, despite the increasing uncertainty related to the longer lead times. Clearly, there is a lack of specific tailored warning indicators that could be of great use to farmers.

A recent activity to address this gap is the "Agricultural Risk Information System" (ARIS) that was developed in Austria (https://combirisk.boku.ac.at/), which is an operationally designed monitoring system at high spatial resolution (1 km grid) and a daily time step that covers a range of locally calibrated agrometeorological general and crop-specific risk and damage risk indicators, including forecasts. Research on the application of ARIS indicators to tailor information to precision and smart farming applications and seasonal forecasts is still ongoing. It is especially necessary to implement a bottom-up approach to identify stakeholder needs to derive tailored decision support (e.g. to obtain application-specific information in a useable form with appropriate accuracy at the right time).

Agriculture is a sector that is significantly affected by climate change (CC). The increased frequency and severity of extreme weather events associated with climate change have challenged agricultural production throughout the world every year (IPCC 2014a). Serbia is not an exception (Ruml et al. 2017), even though the CC signal is not as strong and severe as in other regions of the world (IPCC 2014b). Following the climate records published by the Republic Hydrometeorological Service of Serbia (RHMSS), in the case of a typical Pannonian lowland weather station (Novi Sad, Rimski Sancevi) during the last climatological period (1981–2010) with respect to a common reference (1961–1990), an annual temperature increase from 10.9 to 11.4 °C (average maximum and minimum air temperature increases of 0.5 °C and 0.6 °C, respectively) was observed, while annual precipitation increased from 576 to 647.3 mm. Temperature and precipitation increases affected both the winter (temperature: 0.57–1.10 °C; precipitation: 116.7–119.3 mm) and summer (temperature: 20.43–21.20 °C; precipitation: 199–213.2 mm) seasons with July as the warmest month (past: 21.1 °C; present: 21.9 °C) and June as the month with the highest monthly amount of precipitation (past: 82.5 mm; present: 91.4 mm).

These statistics indicate a slight increase in temperature (0.5 °C), which is far below the global values (1.5 °C with respect to the pre-industrial era according to the IPCC (2018)), and a slight increase in precipitation, which is assumed to be an encouraging signal in a changing climate. However, a more detailed inspection of the latest climatology indicates an increased number of tropical days (8 days) from May to August and an increased number of days with precipitation greater than 10 mm (3.4 days) from May to November. This finding is simply an example of the "real face" of CC in Serbia, where the most significant expression is the high variability in precipitation (up to 79% of the average) and the deviation of temperature from normal (up to 2.7 °C in the case of maximum temperature) even during a relatively short time span (Table 29.1).

Therefore, the significant drought damage (for 1981–2010 climatology in 1992, 1993, 1998, 2000, 2003, 2007 and 2012 as driest in the twenty-first century) and flooding in 2014 are not surprising. The registered events and measured trends of temperature and precipitation are in accordance with the expected effects of CC in the northern Serbia (Vojvodina) region (Lalic et al. 2012; Mihailovic et al. 2014). This result implies that the monitoring of micrometeorological conditions, plant conditions (particularly drought or over-moistening damages) and the appearance of harmful organisms (particularly timing and intensity of the already present organisms

Table 29.1 Maximum air temperature (T_{max}), minimum air temperature (T_{min}) and precipitation (H) for March–August from 2006–2014; air temperature deviation (Δ) and relative deviation of precipitation (δ) with respect to the 1981–2010 climatology at the Rimski Sancevi (Serbia) station

Year	T_{max} (°C)	T_{min} (°C)	H (mm)	ΔT_{max} (°C)	ΔT_{min} (°C)	δH (%)	1981–2010
2006	22.0	11.1	466.1	−0.5	0.5	27	T_{max} = 22.5 °C T_{min} = 10.6 °C H = 367.9 mm
2007	24.9	11.6	365.5	2.4	1.0	−1	
2008	23.7	11.6	232.1	1.2	1.0	−37	
2009	23.9	11.6	293.0	1.4	1.0	−20	
2010	22.3	12.0	658.7	−0.3	1.4	79	
2011	23.2	11.4	200.3	0.7	0.8	−46	
2012	25.2	11.6	217.8	2.7	1.0	−41	
2013	23.2	11.2	413.4	0.6	0.6	12	
2014	23.1	11.7	560.8	0.6	1.1	52	

and the appearance of new ones) on a daily basis is very useful for addressing CCs and their effects on agricultural production in real time.

An additional burden of the CC struggle is that Serbia is a developing country with agricultural production accounting for 6% of the national GDP in 2017 (Statistical Office of the Republic of Serbia). Low subsidies and low prices of agricultural products in comparison with neighbouring countries left a very limited window of opportunities for farmers in Serbia to make a profit. Under these circumstances, the timely and effective use of all resources, particularly the available data and knowledge, to ensure stable yields and safe food production is an absolute priority and the main goal of all experts and researchers working in and/or for agriculture.

Per definition, a decision support system (DSS) is established as an IT platform that includes all observed meteorological, biological and soil data, empirical parameters, models and algorithms and connects all input elements in one functional system, which allows agile communication between a large amount of data and human experience, creating additional value for producers and policymakers. Previous studies in developed countries have shown that DSSs tested and applied on a farm scale but allowing upscale results are the best way to offer useful and timely information to producers and policymakers. This chapter introduces the Forecasting and Reporting Service for Plant Protection of Republic of Serbia (PIS) as a decision support system and demonstrates the enhancement to the capacities of both an operational DSS (PIS) and the research group AgMnet$^+$ (PFNS) as a result of the joint work focused on solving the everyday problems of producers using advanced scientific methods and tools and permanent onsite biological and meteorological observations. The phenological models that are currently in use are either modules of crop models already calibrated for local conditions (Lalic et al. 2012) or algorithms designed for a particular fruit based on accumulated heat units (Anderson et al. 1986). As an example of landscape phenology modelling in the framework of DSS, winter wheat growing stages and apple flowering results for the northern Serbia (Vojvodina) region are presented.

The methods used to forecast landscape and harmful organism phenology are still in their testing phase. Encouraging results are obtained when short-range (Lalic et al. 2017a; Firanj Sremac et al. 2018) and seasonal (Lalic et al. 2017b, 2018) weather forecasts are used. The demonstration of the monthly seasonal weather forecast efficacy in predicting winter wheat growing stages and seasonal weather forecasts in predicting both winter wheat growth stages and apple flowering is performed for the Novi Sad (Rimski Sancevi) location.

29.2 Materials and Methods

29.2.1 Forecasting and Reporting Service for Plant Protection (PIS) as a DSS: Philosophy and Performance

Agricultural production is the most important economic activity pursued in the territory of the Autonomous Province of Vojvodina. Recognizing the needs and dispositions to apply plant protection in accordance with accepted standards, in 2010, the Provincial Secretariat for Agriculture, Water Management and Forestry established the Forecasting and Reporting Service for Plant Protection of AP Vojvodina (PIS). In 2012, thanks to the financial and organizational support from the Ministry of Agriculture, Forestry and Water Management, PIS extended its activities to the entire territory of the Republic of Serbia, becoming a national DSS.

In its work, PIS relies on the standards implemented by the European Plant Protection Organization (EPPO) (EPPO 2019). During the implementation of integrated plant protection measures and sustainable pesticide applications, PIS adheres to the basic guidelines outlined in DIRECTIVE 2009/128/EC of the European Parliament and the Council (Directive 2009).

PIS conducts daily monitoring of the production of the most important agricultural crops and the occurrence of harmful organisms. The scope of work also includes the monitoring of meteorological conditions in which the production takes place and the development of harmful organisms. The monitoring is conducted using various tools, and the collected data are used to prepare recommendations for the application of measures necessary to control plant diseases and pests and protect agricultural plants. An objective of PIS is to improve the efficiency of communication with agricultural producers to provide them more time to act upon the received recommendations. The benefits that the service provides include systematic monitoring and implementation of uniform methodology across the entire country and the application of chemical protection measures in accordance with the principles of good plant protection practice.

The diagram presented in Fig. 29.1 shows all relevant elements that comprise the decision-making model in PIS, with the intention to show the correlations between causes and effects. The diagram has been made after the Ishikawa cause-and-effect diagram (Ishikawa 1982), in which effects are typically presented as fish heads while the main causes that lead to the realization of the effects are arranged along the spinal column. These lateral parts can be presented at various levels of detail, i.e. as smaller bones. Factors are grouped into categories. The PIS diagram follows the general rule in that all causes that contribute to the achievement of the desired effects are labelled with arrows that point towards the head, while the causes that impede the realization of the desired effects are labelled with arrows that point towards the tail.

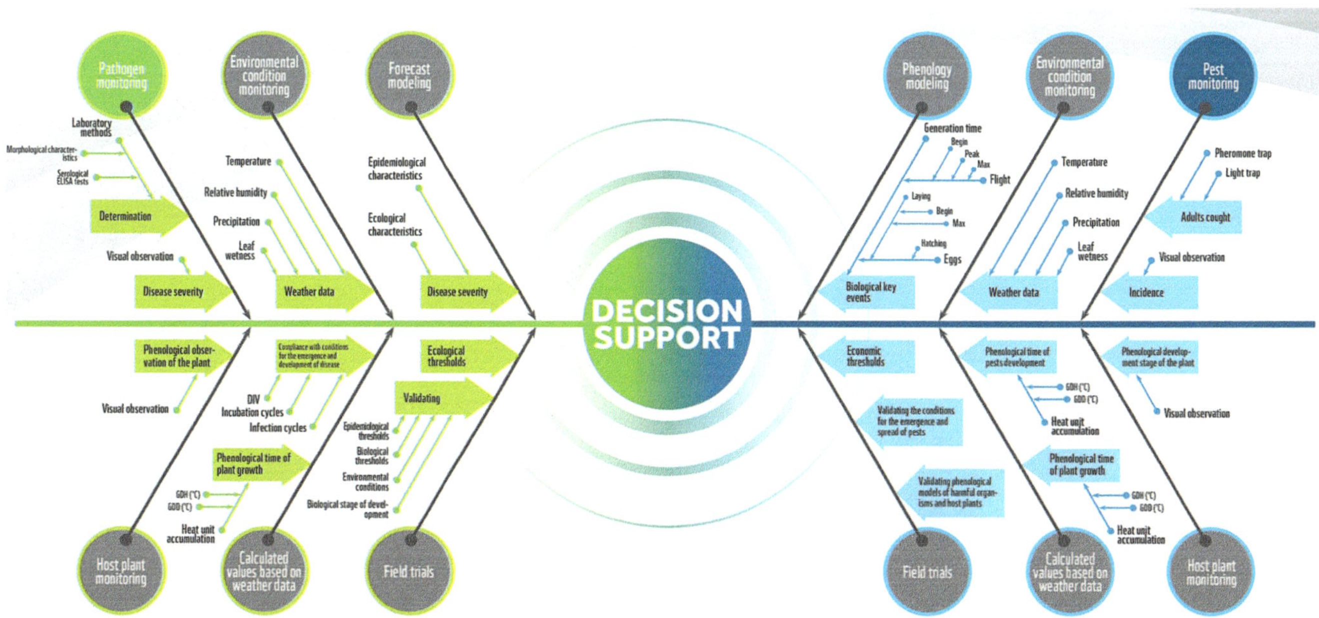

Fig. 29.1 Conceptual design of PIS as a decision support system

To present all relevant elements of the daily activities of PIS, two diagrams were merged into one. Thus, the joint diagram simultaneously shows the categories that belong to pest monitoring on one side and pathogen monitoring on the other (two fish bones arranged head-to-head). Thus, for example, the diagram on the left shows the main components of the categories/processes whose successful completion increases the capacity of PIS to assist farmers in making decisions that concern pest management. These components include pest and host plant monitoring, monitoring of environmental conditions and processing of the obtained data, trials, and the development of phenological models. Similarly, the diagram on the right shows the components/processes that are part of pathogen monitoring and that increase the capacity of PIS to assist farmers in making decisions that concern plant protection against disease-causing agents.

In general, the diagram shows the causes that are positively correlated with the desired effects. For example, this means that in the "environment monitoring" category, the measurements of air temperature are shown as a cause that is positively associated with the effects, while a lack of data on air temperature would be a cause for an effect in the opposite direction. For the sake of clarity and simplicity, the causes with negative impacts on the effects are not shown in the diagram.

29.2.2 Data

Regional PIS offices have been established in local agricultural stations, employing plant protection experts who are in charge of monitoring activities. Twenty-nine regional offices have been put into force across the country: 12 in Vojvodina and 17 in central Serbia. The activities of the regional offices are coordinated by the Provincial PIS Centre and the Republic PIS Centre.

The basic territorial and biological monitoring unit is the "monitored location" and is defined by region, location (site toponym) and observation spot (adjusted sets consisting of a host plant and a harmful organism). During the first 3 years of work, the links between a region, location and observation spot were targeted and marked. The links were then assigned to economic and biological thresholds that were used to select the locations for inclusion in the monitoring plan. Based on these data, a set of representative monitoring sites was established. Modifications to the monitoring plan, which are necessary due to the occurrence of new harmful organisms, changes in the characteristics of harmful organisms, the early start of the vegetation period, the premature end of the vegetation period, etc., are made on an annual basis.

At the monitored locations, regional offices collect observations on a daily basis using tools for biological and meteorological observations and measurements. The collected data are communicated by means of the PIS information system.

29.2.2.1 Meteorological and Biological Measurements and Observations

Automatic weather stations (AWSs) located in crop fields and plantations are used to monitor the local weather conditions. The most important meteorological elements (air temperature, soil temperature, relative air humidity, precipitation, leaf wetness) are monitored by 166 AWSs installed across Serbia (Fig. 29.2).

Biological monitoring tools include pheromone traps, light traps, visual examination of pathogens, pests and host plants, spore catchers, trials and laboratory analyses. Pheromone and light traps are checked on a daily basis. The visual inspection of pathogens, pests and host plants is the most important tool used by PIS. The inspections are carried out in accordance with a uniform methodology at regular intervals, following the specifications developed for the individual parameters under observation. The monitoring covers over 150 harmful organisms in 36 different crops and plant species. Observations are regularly performed at more than 2,000 locations across the territory of Vojvodina (Fig. 29.2).

Field trials serve to determine harmfulness thresholds and biological characteristics of observed harmful organisms. Morphological, serological and molecular identification of harmful organisms is carried out in the PIS laboratory.

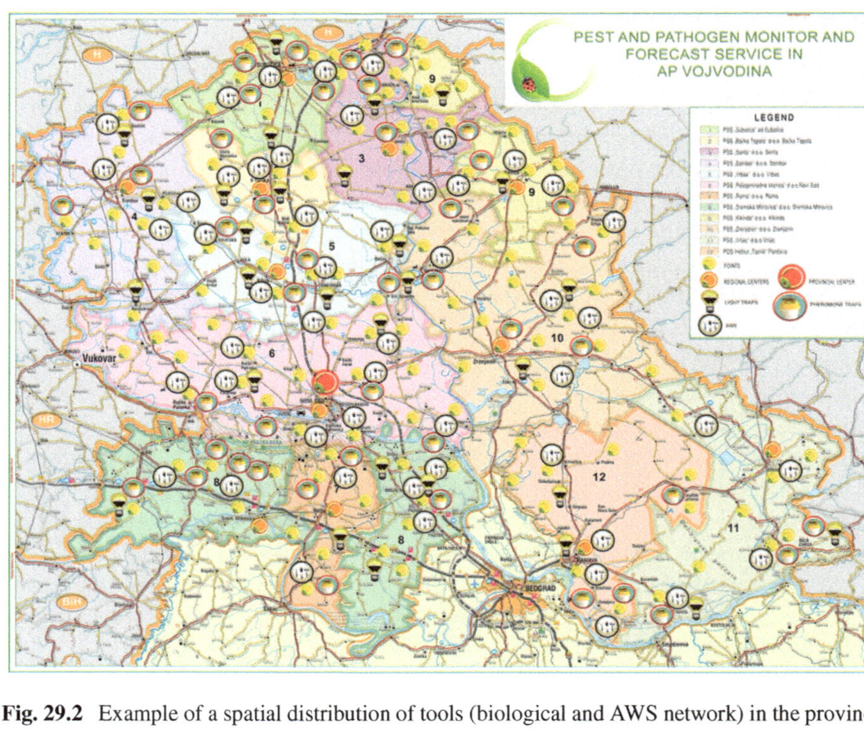

Fig. 29.2 Example of a spatial distribution of tools (biological and AWS network) in the province of Vojvodina in the monitoring conducted by the PIS

For the purposes of data storage, editing, processing and presentation, a PIS information system has been developed. The data collected during field observations and meteorological readings are kept in separate databases.

At the representative monitoring sites, high levels of tool aggregation have been achieved, and all available tools are effectively used. Most reliable information is obtained at monitoring sites where traps for harmful organisms and AWSs are installed, where visual inspections of host plants and harmful organisms are carried out on a regular basis and where all chemical protection measures are implemented according to the model recommended by PIS. The tool labels and tool aggregation schemes are shown in Figs. 29.3 and 29.4, respectively.

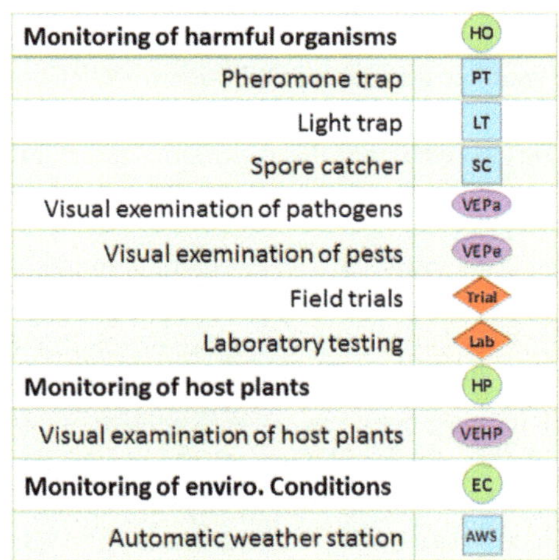

Fig. 29.3 Tool labels

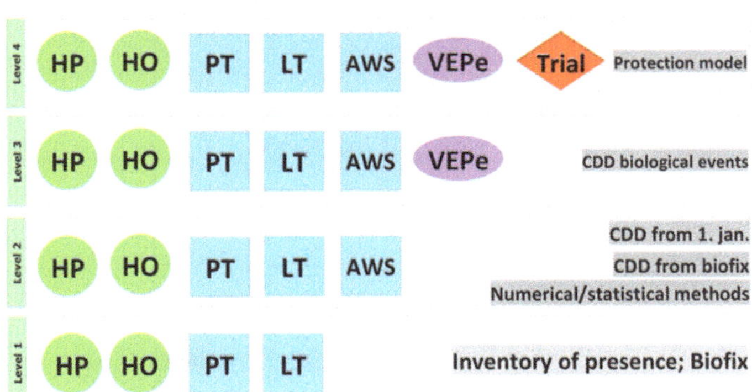

Fig. 29.4 Tool aggregation levels

29.2.2.2 Monthly (MWF) and Seasonal (SWF) Weather Forecasts

The MWF used is the product of the European Centre for Medium-Range Weather Forecast (ECMWF). The product is generated in the form of an ensemble forecast with 51 members. Fifty of the ensemble members are created with perturbations of initial conditions (ENS) and one control run created without any perturbations of the initial conditions (CR). The model resolution is 0.4° × 0.4° latitude/longitude grid. The MWF or medium-range weather forecast is mostly an initial atmospheric value problem. The monthly time range is short enough that the atmosphere still has a memory of its initial condition and long enough that the ocean variability could have an impact on the atmospheric circulation. Therefore, the ECMWF monthly forecasts are produced by a coupled ocean–atmosphere model. To reduce the grid scale, downscaling from 0.4° to 0.2° was performed using MWF as boundary conditions for the Weather Research and Forecast (WRF) modelling system with the non-hydrostatic mesoscale model (NMM) core 3.3 version model run. For the purpose of phenology forecasting, the maximum and minimum air temperatures from the MWF 51 ensemble were selected for the period from 1 March to 30 June 2005 (Lalic et al. 2017a).

The SWF is based on the long predictability of the atmospheric boundary conditions (ocean/land/ice), which have an impact on the atmospheric circulation. Seasonal forecasting is also the problem of initial conditions, particularly of the ocean state. The SWF cannot predict the daily variation in meteorological elements at specific locations, but it can provide a range of possible values that are presented in the form of ensembles and one control run. The system for the ensemble SWF is based on the system of hydrodynamic equations used in MWF forecasting, and ensembles are created with similar perturbations as in MWF. The seasonal forecast system of ECMWF begins with 10 ensemble members (EMs) in 2006 for 6 months and progresses to 50 EMs in 2014 for a 7-month forecast and a CR created without any perturbations of the initial conditions. The resolution of the SWF ensemble forecast data was 0.5° × 0.5° latitude/longitude grid. For the purpose of phenology forecasting, data from 1 March to 31 August are extracted as the 24-h averaged values from the four geographically nearest numerical points (Lalic et al. 2017b, 2018).

29.2.3 *Models*

29.2.3.1 Modelling Winter Wheat Phenology Dynamics

SIRIUS is a dynamical crop growth simulation model that calculates biomass production from intercepted photosynthetically active radiation (PAR) and grain growth based on simple partitioning rules (Semenov and Porter 1995; Jamieson et al. 1995). The model has been previously calibrated for the Vojvodina region (Lalic et al. 2009) and used in seasonal forecasts of winter wheat development and yield (Lalic et al. 2017a) and climate studies (Lalic et al. 2012). The phenological development of wheat in SIRIUS is described through the six stages described in detail in Semenov and Porter (1995).

PIS_PHEN, which was developed in the PIS monitoring framework, is the dynamic concept for the following phenological stages of winter wheat designed and conducted by PIS (Lalic et al. 2017a). This concept is based on the continuous observation of plant growing stages according to the BBCH scale (BBCH 2001). Accumulated GDD are calculated with base air temperatures of 0 °C (from emergence) and 5 °C (from the beginning of tillering). Regression analysis is typically based on the correlation between the BBCH growing stage observations from emergence (9) to ripening (80) and accumulated GDD before the stage is noticed.

29.2.3.2 Modelling Apple Phenology Dynamics

Flowering models of fruit trees are typically based on "chilling" requirements given in the form of chilling hour units (CU) and "forcing" elements represented by growing degree hours (GDH). Both CU and GDH should reach a certain plant-specific level to reach the flowering date. The chilling and forcing periods can overlap or be defined in a sequence. It is assumed that chilling and forcing requirements are fulfilled sequentially, with heat being effective only after sufficient chill has accumulated (Cannell and Smith 1983; Cesaraccio et al. 2004; Fuchigami and Nee 1987; Rea and Eccel 2006). The beginning of chilling accumulation is commonly set according to the local conditions.

For the calculation of forcing period duration, the GDH model developed by Anderson et al. (1986) and Luedeling et al. (2009) was selected. This model performs heat accumulation when the hourly air temperature (T_i) ranges between a base temperature (T_b) and a critical temperature (T_c) with the maximum heat accumulation at an optimum temperature (T_u)

$$
\mathrm{GDH}_{(i)} = \begin{cases} F\left(\dfrac{T_u - T_b}{2}\right)\left[1 + \cos\left(\pi + \pi\left(\dfrac{T_i - T_b}{T_u - T_b}\right)\right)\right], & T_b \le T_i < T_u \\ F(T_u - T_b)\left[1 + \cos\left(\dfrac{\pi}{2} + \dfrac{\pi}{2}\left(\dfrac{T_i - T_u}{T_c - T_u}\right)\right)\right], & T_u \le T_i < T_c \\ 0, & T_i > T_c \text{ or } T_i < T_b \end{cases} \tag{1}
$$

where F is a plant stress factor that is commonly set to 1 if no particular stress exists; T_b, T_u and T_c were set according to the characteristics of the apple varieties used.

29.2.3.3 Modelling Harmful Organism Phenology Dynamics

Different biometeorological indices are calculated to predict phenology dynamics and assess feedback between plant and harmful organism development. The accumulated degree-days, developmental degree-days, daily infection values and infection risks and the duration of a particular phenophase of various moth species are calculated on a regular basis. Reliable values of biometeorological indices are obtained only when the monitoring site reaches tool aggregation level 3 or higher (Fig. 29.4).

Degree-days accumulated from January 1. Using the values of air temperature measured in crop stands, accumulated degree-days are calculated starting from January 1 for all pests for which temperature thresholds (minimum and maximum temperatures) are known.

Degree-days accumulated from biofix. For all harmful organisms monitored by pheromone traps for which degree-days accumulated from January 1 are counted, the degree-days accumulated from biofix are also calculated. When biofix (a biologically significant event) occurs, the number of degree-days accumulated from 1 January is construed as the numerical value of heat units until the day of biofix occurrence. From the day of biofix occurrence, accumulated degree-days are calculated for all incipient biologically significant events (e.g. egg-laying, hatching and first flight) as the basic unit for expressing the phenological time.

Growing degree-days. In the case of wheat, barley, corn, sunflower and soybean, growing degree-days are calculated from the date of sowing and/or date of emergence.

Daily infection rates and infection risks. In the case of sugar beet leaf spot (*Cercospora beticola*), the daily infection rates and infection risks are calculated on the basis of the prediction model developed by Shane and Teng (1983).

Percentage of phenological time. For codling moth, peach moth, plum fruit moth and grapevine moth, the percentages are calculated for phenological times from the beginning of flight of adults, beginning of egg-laying, and beginning of egg hatching. Namely, the values of accumulated degrees-days for generation time, the start of egg-laying and the beginning of egg hatching are used to calculate the percentages (from 0 to 100%) from the beginning of the flight of adults, beginning of egg-laying and beginning of egg hatching by generation.

29.2.4 Simulations and Analysis

The phenological model outputs obtained using measured meteorological data are compared with the observed times of phenological phases to assess the model capacity to reproduce plant growth dynamics. Validation of the model is performed using an error analysis based on the methods employed by Pielke (1984) and Mahfouf (1990). Following these methods, we have (i) calculated the RMSE for the dates of the growing stages and (ii) compared the standard deviations of the observed (σ_o) and calculated (σ_c) growing stages. According to *Pielke* (1984), the simulation is more realistic if (a) the root-mean-square error (RMSE) is less than the standard deviation of the observed values, σ_o, and (b) the standard deviation of the calculated values, σ_c, is close to the standard deviation of the observed values, σ_o.

The efficacies of MWF and SWF in predicting the growing stages of different plant species were tested using the methodology described in Lalic et al. (2017b). This methodology is based on the calculation of the RMSE and SPREAD from the ensemble of model estimates calculated using observed and forecasted weather data. Because each ensemble member is equally probable, the RMSE, as a measure of forecast accuracy, was calculated for particular growing stages, i.e. BBCH values using MWF and SWF data. The SPREAD of phenological model outputs (obtained using ensemble members as input weather data) indicates the uncertainty of the obtained ensemble of growing stage estimates. The phenology forecast obtained using ensemble weather forecasting is considered realistic when the RMSE and SPREAD values are similar.

29.3 Results

29.3.1 Winter Wheat Phenology Modelling and Forecasting

Monthly weather forecast. The validation of the PIS_PHEN algorithm is presented in this study using phenological observations of four winter wheat varieties in the Novi Sad region (Serbia) for the 2014–2016 period. The winter wheat varieties chosen in the study are Zvezdana, Apache, Simonida and NS_40S, which have the longest time series of observed phenological stages. Validation of the model is performed using an error analysis, and the results are presented in Table 29.2. For all varieties obtained, the results indicate that the standard deviation of the simulated values differs up to 2 days from the observed values, while the RMSE is much smaller (2–5 times) than σ_o. Therefore, winter wheat phenology simulations performed with the PIS_PHEN algorithm can be considered sufficient.

To test the efficacy of MWF in forecasting winter wheat phenology dynamics, the input data for the PIS_PHEN are created using observed (OBS) and MWF ensemble members (ENS) of air temperatures. The maximum and minimum air temperatures from the MWF 51 ensemble member were extracted for 1 March to 30 June 2005 for

Table 29.2 Standard deviation of observed (σ_o) and calculated (σ_c) time of growing stages and RMSE for Zvezdana, Apache, Simonida and NS_40S winter wheat varieties in the Novi Sad (Serbia) region calculated using the PIS_PHEN algorithm

Variety	σ_o (day)	σ_c (day)	RMSE (day)
Zvezdana	21.4	22.5	9.3
Apache	18.3	17.6	3.2
Simonida	22.1	21.8	4.5
NS_40S	24.0	22.0	5.6

the Rimski Sancevi weather station in the Novi Sad region. The efficacy of MWF in predicting growing stages is tested by calculating the ensemble RMSE and SPREAD for the day of appearance of the stages from 20 to 99 using OBS and ENS data sets (Lalic et al. 2017a).

The performance of the phenological model and MWF efficacy are presented in Figs. 29.5 and 29.6 for the two temperature thresholds GDD_0 (0 °C) and GDD_5 (5 °C). The impacts of low temperatures during the winter and spring months and photosynthetic radiation limit on crop development are not taken into account. This condition can explain the more uniform values for the RMSE and SPREAD of BBCH_D calculated using GDD_5 in comparison to GDD_0. The difference between the RMSE and SPREAD is less pronounced after BBCH phase 40 for GDD_0, i.e. BBCH phase 60 for GDD_5 for all varieties, indicating better model performance.

Seasonal weather forecast. The SIRIUS model has already been calibrated and validated for different winter wheat varieties in Serbian agroecological conditions (see, for example, Lalic et al. 2012). The verification statistics presented in Table 29.3 confirm the high model performance in simulating winter wheat phenology dynamics.

To test the efficacy of SWF in forecasting the phenology dynamics of winter wheat, the SIRIUS model was run using the OBS and SWF (CR and ENS) data sets as the input weather data to obtain an ensemble of estimates for the phenological development of winter wheat. The maximum and minimum air temperature, solar

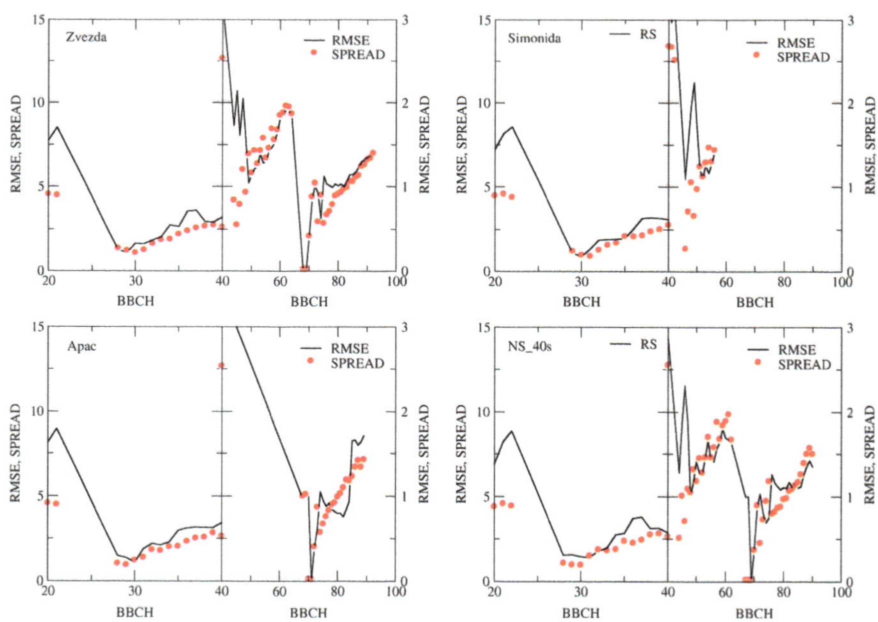

Fig. 29.5 RMSE (line) and SPREAD (circle) of BBCH calculated using OBS and ENS data sets for a 0 °C lower temperature threshold for Rimski Sancevi (Serbia)

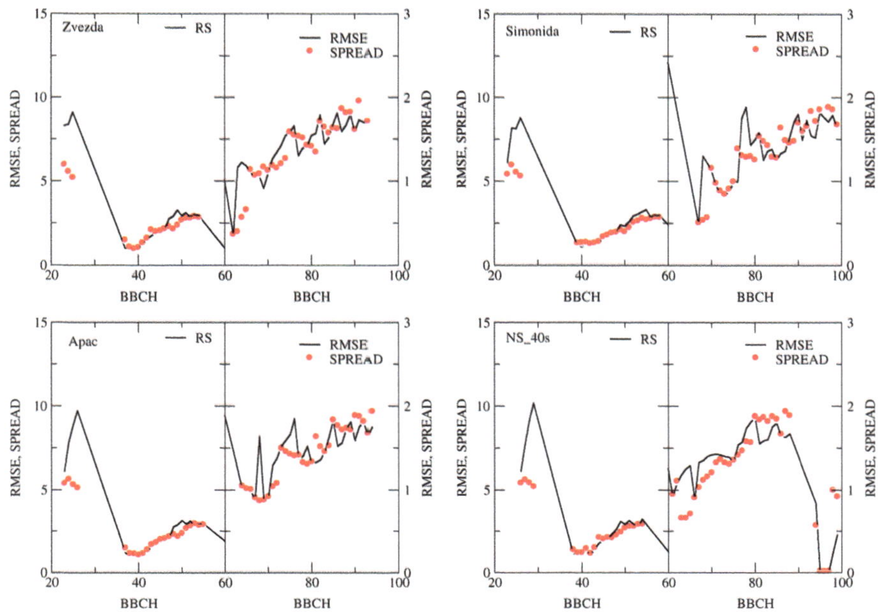

Fig. 29.6 RMSE (line) and SPREAD (circle) of BBCH calculated using OBS and ENS data sets for a 5 °C lower temperature threshold for Rimski Sancevi (Serbia)

Table 29.3 Standard deviation of observed (σ_o) and calculated (σ_c) time of growing stages and RMSE for winter wheat variety Anastasia for Rimski Sancevi (Serbia) location calculated for 2001–2005 seasons using the SIRIUS crop model

σ_o (day)	σ_c (day)	RMSE (day)
71.2	71.8	11.1

radiation, precipitation, air humidity and wind speed were extracted from the 51 SWF ensemble members from March 1 to August 31 (2006–2014) for the Rimski Sancevi weather station in the Novi Sad region. Crop model output ensemble analysis is performed according to the methodology presented in Lalic et al. (2017b). The differences between the anthesis day (AntD) and maturity day (MatD) simulated using ENS (presented in the form of ensemble average EA), and those in the CR were negligible. The notably small difference is the result of summed air temperatures when small underestimates or overestimates of forecasted values exclude each other (Fig. 29.7). The RMSE and SPREAD reflected the impact of the differences obtained for temperature and precipitation (Lalic et al. 2017a). Whenever the maximum air temperature RMSE was large, the RMSE and SPREAD AntD and MatD showed

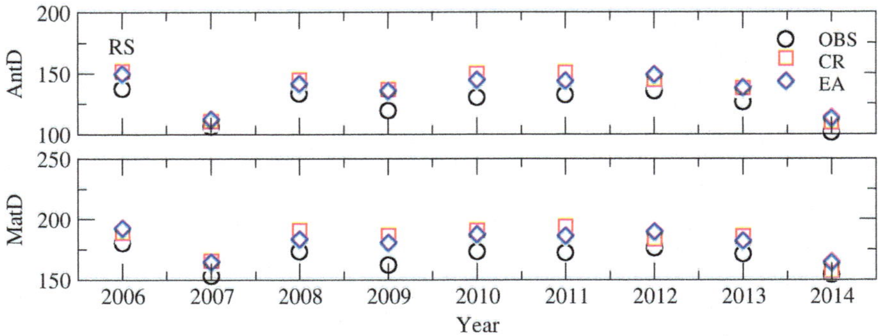

Fig. 29.7 AntD (DoY) and MatD (DoY) calculated using OBS, CR and ENS data sets (EA presented) for the Rimski Sancevi (Serbia) location

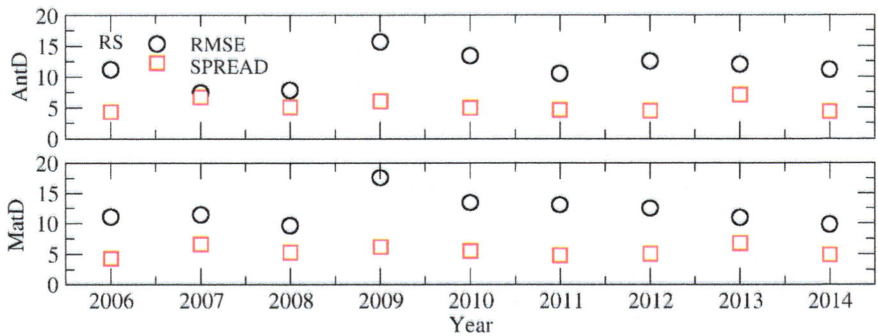

Fig. 29.8 RMSE (circle) and SPREAD (square) in days for AntD (DoY) and MatD (DoY) calculated using the OBS and ENS model outputs for the Rimski Sancevi (Serbia) location

more significant differences. This effect is visible at both locations in the 2006–2007 and 2009–2012 crop-growing periods (Figs. 29.7 and 29.8).

29.3.2 Apple Flowering Modelling and Forecasting

Seasonal weather forecast. The HU model for apple flowering is calibrated and validated by the AgMnet$^+$ (PFNS) group for the agroecological conditions of the Vojvodina region and for two apple varieties, Idared (ID) and Golden delicious (GD). The inventory of data used for model calibration and validation is presented in Table 29.4. It is important to note that phenological observations are performed on a weekly basis. Therefore, the probability that a particular phenological stage occurs on the identified date is equal for a particular day and all days during the previous week. The mean HU values obtained for the period between dormancy and flowering are 1559 °C (ID) and

Table 29.4 Varieties and locations of phenological data series used for validation of the model

Variety	Location	Data series (from–to)
ID	Novi Sad/Cenej	15.11.2013–01.06.2018
ID	Ruma/Irig-Kudos	15.11.2016–01.06.2018
ID	Sremska Mitrovica/Mandjelos	15.11.2016–01.06.2017
ID	Zrenjanin/Sutjeska	15.11.2015–01.06.2018
GD	Novi Sad/Cenej	15.11.2013–01.06.2018
GD	Ruma/Novi Slankamen	15.11.2015–01.06.2016, 15.11.2016–01.06.2017, 15.11.2017–01.06.2018
GD	Kikinda/Kikinda	15.11.2016–01.06.2018
GD	Zrenjanin/Sutjeska	15.11.2015–01.06.2018

1600 °C (GD), with standard deviations of 270 °C and 370 °C, respectively, for all locations and seasons. The parameters used in the study are as follows: $T_b = 7.2$ °C, $T_u = 12$ °C, $T_c = 36$ °C. Stress factor F was taken into account if air temperature dropped below critical for apple flowering (−0.5 °C). In the case of ID, the best result was obtained for $F = 0.6$, while for GD, $F = 0.7$, which is in accordance with the robustness of GD The verification statistics presented in Table 29.5 are calculated while considering the uncertainty of the phenological observations. The RMSE was calculated by assuming that each day of observation during the week could be the date of flowering. For all cases, the average RMSE is 3.2 days for ID and 4 days for GD, while the lowest RMSE is obtained for the sixth day before official recording, and this day is considered the date of flowering in further calculations. The presented verification statistics indicate good performance of the HU model in calculating the date of flowering in the Vojvodina region.

To test the efficacy of SWF in forecasting apple flowering, the HU model was run using OBS and SWF (ENS) data sets as the input weather data to obtain an ensemble of estimates for the date of flowering. The forecasted date of flowering was compared with the observations from the Novi Sad region. The Cenej location is selected as the nearest to the SWF grid point. The SWF verification statistics are presented in Table 29.6. High RMSE obtained with respect to time of flowering calculated using SWF in 2014 can be a result of stress which is not taken into account. Namely, in our HU calculations, we considered only stress which is the result of temperatures below −0.5 °C. It is important but not only stress which affects flowering. In future improvement of flowering model more attention will be devoted to this issue.

Table 29.5 Standard deviation of observed (σ_o) and calculated (σ_c) time of flowering and RMSE for ID and GD apple varieties in the Vojvodina region

Variety	σ_o (day)	σ_c (day)	RMSE (day)
Idared (ID)	7.9	7.1	1.0
Golden delicious (GD)	6.0	6.5	2.9

Table 29.6 RMSE and SPRD for the date of flowering for ID and GD apple varieties in 2014 at Cenej location (Vojvodina, Serbia)

Variety	RMSE (day)	SPRD (day)
Idared (AJ)	10.86	4.14
Golden delicious (GD)	10.67	4.05

29.3.3 Landscape and Harmful Organism Phenology Modelling as a Part of the DSS System

Collected field and meteorological data, phenological model outputs and weather forecasts are essential for the operation of PIS and the design of the most complex output information prepared by the system and communicated to end users—recommendations. Recommendations are part of the protection models that have been defined for the most important agricultural crops grown in the country. Protection models are based on the principle of biological validation of pesticide application, observation of anti-resistance strategies in pesticide application, and adherence to recommendations regarding the maximum number of pesticide treatments and application intervals. Each recommended measure is checked in the field before it is implemented.

Each recommendation includes the following information: (i) stage of development of the host plant; (ii) harmful organism present in the crop or plantation whose presence is expected or whose occurrence is intended to be prevented; (iii) stage of development of the harmful organism; (iv) abundance of the harmful organism; iv) forecast of further development of the harmful organism; (v) harmfulness threshold for the harmful organism, if known; (vi) measure to be undertaken (in the case of protective chemical application, instructions are provided as to when the treatment is to be carried out, which pesticide should be used, and which dose or concentration should be applied); (vii) recent photographs taken on-site, which serve to identify the harmful organism and its development phase (Appendix 1).

Data on the phenological development of plants, occurrence and behaviour of harmful organisms and instructions for the use of control measures are presented to end users through the PIS web portal (www.pisvojvodina.com and www.pissrbija.com), and in the form of SMS messages, TV broadcasts at the national frequency (from March to September each working day), lectures and seminars.

29.4 Discussion and Conclusions

From the stakeholder perspective, the economic benefit of any service provided dominates the decision regarding whether the service will be used or not. From a societal point of view, however, the values may include, in addition to socio-economic benefits, benefits to the environment (e.g. for biodiversity, ecosystem functions, and mitigation effects). Positive effects on ecosystem services, however, will also enhance sustainable agricultural production, which is a long-term aim of any farmer or producer in the agricultural sector.

The first challenge in the context of estimating the "value" of any decision support system or service is that economic evaluation methods differ by the level of criteria applied (e.g. spatial and time scale, type of stakeholder, qualitative vs. quantitative evaluation). There are no guidelines available for the economic evaluation of agrometeorological services and products, which should be better addressed in future research activities.

For example, in a past survey (Sivakumar 2006), significant economic benefits of agrometeorological services were reported, especially in the field of plant protection warning services. The main tasks identified for saving costs in future plant protection are weather forecasts with improved performance and prediction models with finer spatial resolution, practical use of prediction systems, and intense pest and disease control in combination with a powerful organization of stakeholders.

As a result of scientifically based and practically evaluated permanent monitoring, plant and harmful organism phenology modelling and site-specific analysis, PIS was able to produce and manage 1776 recommendations in only 2018 (Table 29.7). Return of investment, as we already elaborated, is difficult to assess in cases when many outcomes should be taken into account. To avoid the use of financial and budgetary information, it will be noted that the annual budget of PIS for the Vojvodina region is 22% of the costs of one sugar beet spraying in Vojvodina on approximately 60 000 ha. This means that by saving one spraying, four years of PIS Vojvodina work can be financed or that work can easily be significantly improved.

The main pillars of PIS as DSS in plant protection are knowledge of the host plant, harmful organisms and meteorological conditions as driving forces of all crucial processes. Plant phenology modelling opens a window of opportunity to assess the growing phase of host plants, even in locations that are not part of permanent phenological monitoring networks. The use of weather forecasts instead of observed

Table 29.7 Number of public communications by PIS for the 2013–2018 period

Year	Public communications			
	Recommendations	Short reports	SMS messages	TV releases
2018	1,776	563	127	122
2017	1,072	483	84	122
2016	1,228	555	84	129
2015	1,189	444	14	130
2014	1,723	487	18	144
2013	1,105	287	11	57

weather as PIS input data raises the entire platform to a higher level and allows the assessment of harmful organism appearance and intensity of attack in the future.

Acknowledgements The authors would like to acknowledge the support received from the Ministry of Education and Science of the Republic of Serbia within the framework of integrated and interdisciplinary research for 2011–2017 for the work on the "Studying climate change and its influence on the environment: impacts, adaptation and mitigation" project (43007). All activities of PIS were financed by the Provincial Secretariat for Agriculture, Water Management and Forestry trough the "Project of establishing Forecasting and Reporting Service for Plant Protection of AP Vojvodina".

Appendix: Example of Recommendation

Each recommendation is based on the original (biological and meteorological) and derived (biometeorological) monitoring data and the data from the protection model. Here is an example of the recommendation for the protection of apples for the Novi Sad region.

In the region of Novi Sad, one of the representative locations for monitoring the production of apples is designated as Novi Sad/Neštin/apple. This location has been equipped with a complete range of tools such as pheromone trap, automatic meteorological station, spore catcher, while visual inspections of plants host, pathogens and pests are performed on a regular basis. The agricultural producer incumbent at this monitoring location applies all measures recommended by PIS, i.e. the apple protection model is applied.

The verbatim recommendation is presented in Table 29.8, where it can be seen that the document was published on 7 June 2018, that it concerned the protection of apples in the Novi Sad region, and that the subject matter was four harmful organisms: apple scab (*Venturia inaequalis*), powdery mildew (*Podosphaera leucotricha*), codling moth (*Carpocapsa pomonella*) and green apple aphid (*Aphis pomi*).

The observations from monitoring tools and the calculated values presented in Table 29.9 relate to the monitoring conducted on 7 June 2018 at the monitoring site Novi Sad/Nestin/apple. These data were the elements used for the preparation of the recommendation.

Biometeorological data obtained on 7 June 2018 for the monitoring site Novi Sad/Nestin/apple, which were basic elements for the preparation of the recommendation, are shown in Table 29.10. In addition to the data that the Service calculates automatically, results of other analyses can also be included in the recommendation. All these data are supplied in order to help growers make correct decisions about the application of chemical protection measures. Explanations of meteorological data registered by AMS can be especially useful when assessing conditions that could lead to a pathogenic infection. In the case of this particular recommendation, the Service analysed the obtained AMS data for the length of foliage exposure to high humidity and the average daily temperatures during this and the previous two days in order

Table 29.8 The recommendation for apple protection issued on 7 June 2018 for the region of Novi Sad

Issued: 7 June 2018, 15:43		
Region: Novi Sad		
Title: Apple protection	Harmful organisms: Apple scab (*Venturia inaequalis*); powdery mildew (*Podosphaera leucotricha*); codling moth (*Carpocapsa pomonella*); green apple aphid (*Aphis pomi*)	
In apple plantations on the territory monitored by PIS Office Novi Sad, apple fruits have reached about half of their final size (BBCH 75)		

Idared BBCH 75	*C. pomonella* egg	*A. pomi* colony
The percentage of empty pseudothecia of the causative agent of the apple scab (*Venturia inaequalis*) was about 90%, indicating that protection measures against this pathogen should be continued. Unstable weather and precipitation forecasted for the weekend will create favourable conditions for infection, so growers are recommended to use preventive fungicides based on captan (Captan 50 WP, Capi, Merpan 50 WP). The recommended dosage rate is 3 kg/ha If symptoms of powdery mildew (*Podosphaera leucotricha*) are observed in the plantation, application of Luna Experience (tebuconazole + fluopyram), at a concentration of 0.075%, is recommended The visual inspection of leaves and fruits indicated the presence of eggs of the codling moth (*Carpocapsa pomonella*). The eggs were in various stages of embryonic development. During the weekend, second-generation larvae of this pest will begin to hatch Also, colonies of the green apple aphid (*Aphis pomi*) were observed on top growth Growers are advised to check their plantations for the presence of these two pests and, if necessary, to carry out protection measures using the combination of insecticides Coragen 20 SC (chlorantraniliprole) 0.02% + Pyrinex 25 CS (chlorpyrifos) 0.25%. Treatment should be performed in evening hours		

to determine whether the climatic conditions were favourable for the development of the apple scab (*Venturia inaequalis*). This piece of information was important for the selection of correct fungicide, whether the one with preventive or the one with curative action.

Based on the above information, the prepared recommendation should offer an answer on the following questions: Why is a particular measure recommended? When the measure should be implemented? Which pesticide to apply and what dose? A short summary is presented in Table 29.11.

Table 29.9 Monitoring tools and the calculated values for 7 June 2018 for the monitoring location Novi Sad/Nestin/apple (NL = number of leaves; NF = Number of fruits; NP = Number of pseudothecia)

Tool	Parameter observed	Observations	Harmful organism
Visual inspection of host plant	Stage of apple development (BBCH pome fruits)	75	–
Visual inspection of pathogen	Number of checked pseudothecia	52	Apple scab (*Venturia inaequalis*)
	NP without formed ascospores	0	
	NP with up to 25% ascospores formed	0	
	NP with 26–50% ascospores formed	0	
	NP with 51–75% ascospores formed	0	
	NP with over 75% ascospores formed	52	
	Calculated percentage of pseudothecia maturity (%)	100	
Visual inspection of pathogen	Number of checked pseudothecia	52	Apple scab (*Venturia inaequalis*)
	Number of empty pseudothecia	48	
	Percentage of empty pseudothecia (%)	92.5	
Spore catcher (SC)	Spores caught by SC after the last rain (+/−)		Apple scab (*Venturia inaequalis*)
	Readings	+	
Visual inspection of pathogen	Number of checked leaves	100	Powdery mildew (*Podosphaera leucotricha*)
	Number of infected leaves	9	
	Percentage of infected leaves (%)	9	
Visual inspection of pathogen	Number of checked leaves	100	Powdery mildew (*Podosphaera leucotricha*)
	Number of symptom-free leaves	91	
	NL with 1–10% infected leaf area	4	

(continued)

Table 29.9 (continued)

Tool	Parameter observed	Observations	Harmful organism
	NL with 11–25% infected leaf area	2	
	NL with 26–50% infected leaf area	2	
	NL with 51–100% infected leaf area	1	
	Calculated index of leaf area infection (%)	4.5	
Visual inspection of pest	Number of checked fruits	100	Codling moth (*Carpocapsa pomonella*)
	NF in category 0 (0 eggs per fruit)	98	
	NF in category 1 (1 egg per fruit)	2	
	NF in category 2 (2 eggs per fruit)	0	
	NF in category 3 (3 eggs per fruit)	0	
	NF in category 4 (more than 3 eggs per fruit)	0	
	Calculated attack index (%)	0.5	
Visual inspection of pest	Number of checked leaves	100	Codling moth (*Carpocapsa pomonella*)
	NL in category 0 (0 eggs per leaf)	99	
	NL in category 1 (1 egg per leaf)	1	
	NL in category 2 (2 eggs per leaf)	0	
	NL in category 3 (3 eggs per leaf)	0	
	NL in category 4 (more than 3 eggs per leaf)	0	
	Calculated attack index (%)	0.25	
Visual inspection of pest	Number of checked leaves	100	Green apple aphid (*Aphis pomi*)

(continued)

Table 29.9 (continued)

Tool	Parameter observed	Observations	Harmful organism
	NL in category 0 (0 aphid imagoes and larvae per leaf)	96	
	NL in category 1 (1–5 aphid imagoes and larvae per leaf)	0	
	NL in category 2 (6–20 aphid imagoes and larvae per leaf)	4	
	NL in category 3 (21–50 aphid imagoes and larvae per leaf)	0	
	NL in category 4 (over 50 aphid imagoes and larvae per leaf)	0	
	Calculated attack index (%)	2.5	
AWS	Mean daily temperature (°C)	23.55	–
AWS	Maximum daily temperature (°C)	30.26	–
AWS	Minimum daily temperature (°C)	16.96	–
AWS	Relative air humidity (%)	86.85	–
AWS	Precipitation (mm)	0.2	–
AWS	Leaf wetness duration (min)	276	–

Table 29.10 Values of biometeorological data for 7 June 2018 for the monitoring location Novi Sad/Nestin/apple (*PPT = Percentage of phenological time)

Biometeorological data	Calculated value	Harmful organism
Accumulated degree-days from biofix	528.25	Codling moth (*Carpocapsa pomonella*)
PPT for the flight of first-generation adults	100	Codling moth (*Carpocapsa pomonella*)
PPT for first-generation ovipositing	100	Codling moth (*Carpocapsa pomonella*)
PPT for first-generation egg hatching	90.92	Codling moth (*Carpocapsa pomonella*)
PPT for the flight of second-generation adults	13.9	Codling moth (*Carpocapsa pomonella*)
PPT for second-generation ovipositing	8.44	Codling moth (*Carpocapsa pomonella*)

Table 29.11 Measures recommended on 7 June 2018

Recommended measure	Question	Harmful organism
Number of empty pseudothecia of the apple scab (*Venturia inaequalis*) was about 90%, indicating the need for a continued control of the pathogen	Why?	Apple scab (*Venturia inaequalis*)
Unstable weather and precipitation forecasted for the weekend will create favourable conditions for infection	When?	
Growers are recommended to use preventive fungicides based on captan (Captan 50 WP, Capi, Merpan 50 WP). The recommended dosage rate is 3 kg/ha	Which pesticide?	
If symptoms of powdery mildew (*Podosphaera leucotricha*) are registered in a plantation, we recommend the application of the preparation Luna Experience (tebuconazole + fluopyram) in the concentration of 0.075%	Why? When? Which pesticide?	Powdery mildew (*Podosphaera leucotricha*)
Visual inspection of leaves and fruits indicated the presence of eggs of the codling moth (*Carpocapsa pomonella*). The eggs were at various stages of embryonic development. Also, colonies of the green apple aphid (*Aphis pomi*) were registered on top growth	Why?	Codling moth (*Carpocapsa pomonella*) green apple aphid (*Aphis pomi*)
During weekend, second-generation larvae will begin to hatch	When?	
Growers are advised to check their plantations for harmful organisms. If presence of pests is confirmed, it is necessary to undertake protection measures by applying the following combination of insecticides: Coragen 20 SC (chlorantraniliprole) 0.02% + Pyrinex 25 CS (chlorpyrifos) 0.25%. Treatment should be performed in evening hours	Which pesticide?	

References

Anderson JL, Richardson EA, Kesner CD (1986) Validation of chill unit and flower bud phenology models for 'Montmorency' sour cherry. Acta Horticiculariae 184:71–75
BBCH Monograph (2001) Growth stages of mono-and dicotyledonous plants, 2nd ed. In: Meier U (ed). Quedlinburg: Federal Biological Research Centre for Agriculture and Forestry, 158
Cannell MGR, Smith RI (1983) Thermal time, chill days and prediction of budburst in Picea sitchensis. J Appl Ecol 20:951–963
Cesaraccio C, Spano D, Snyder RL, Duce P (2004) Chilling and forcing model to predict bud-burst of crop and forest species. Agric For Meteorol 126:1–13
Directive (2009) Directive 2009/128/EC of the European Parliament and of the Council of 21 October 2009 establishing a framework for Community action to achieve the sustainable use of pesticides (Text with EEA relevance), OJ L 309:71–86
EPPO (2019) EPPO global database. https://gd.eppo.int
Firanj Sremac A, Lalić B, Marčić M, Lj Dekić (2018) Toward a weather-based forecasting system for fire blight and downy mildew. Atmosphere 9(12):484
Fuchigami LH, Nee C (1987) Degree growth stage model and restbreaking mechanisms in temperate woody perennials. HortScience 22:836–845
IPCC (2014a) Climate change 2014 - impacts, adaptation, and vulnerability. Part A: global and sectoral aspects. Contribution of working group II to the fifth assessment report of the intergovernmental panel on climate. Cambridge University Press, Cambridge, United Kingdom and New York, NY, USA, 1132 pp
IPCC (2014b) Climate change 2014 - impacts, adaptation, and vulnerability. Part B: regional aspects. Contribution of working group II to the fifth assessment report of the intergovernmental panel on climate change. Cambridge University Press, Cambridge, United Kingdom and New York, NY, USA, 688 pp
IPCC (2018) Summary for policymakers - global warming of 1.5 °C. An IPCC special report on the impacts of global warming of 1.5 °C above pre-industrial levels and related global greenhouse gas emission pathways, in the context of strengthening the global response to the threat of climate change, sustainable development, and efforts to eradicate. World Meteorological Organization, Geneva, Switzerland, 32 pp
Ishikawa K (1982) Guide to quality control, second revised english edition. Asian Productivity Organization, Tokyo
Jamieson PD, Brooking IR, Porter JR, Wilson DR (1995) Prediction of leaf appearance in wheat: a question of temperature. Field Crop Res 41:35–44
Lalic B, Mihailovic DT, Malesevic M (2009) Introduction of crop modelling tools into Serbian crop production: calibration and validation of models. In: Utset Suastegui A (eds) Support water-management decision-making under climate change conditions. Nova Science Publishers, Inc., New York. ISBN: 978-1-60692-033-6
Lalic B, Eitzinger J, Mihailovic DT, Thaler S, Jancic M (2012) Climate change impacts on winter wheat yield change – which climatic parameters are crucial in Pannonian lowland? J Agric Sci 151(6):757–774
Lalic B, Jankovic D, Dekic Lj, Eitzinger J, Firanj Sremac A (2017a) Testing efficacy of monthly forecast application in agrometeorology: winter wheat phenology dynamic. In: International symposium on earth observation for one belt and one road, 16–17 May 2016. The Institute of Remote Sensing and Digital Earth (RADI), Chinese Academy of Science, Beijing, China
Lalic B, Firanj Sremac A, Dekić Lj, Eitzinger J, Perišić D (2017b) Seasonal forecasting of green water components and crop yields of winter wheat in Serbia and Austria. J Agric Sci 156:645–657
Lalic B, Firanj Sremac A, Eitzinger J, Stričević R, Thaler S, Maksimović I, Daničić M, Perišić D, Lj Dekić (2018) Seasonal forecasting of green water components and crop yield of summer crops in Serbia and Austria. J Agric Sci 156:658–672

Luedeling E, Zhang M, Girvetz EH (2009) Climate changes lead to declining winter chill for fruit and nut trees in California during 1950–2009. PLOSOne 4: e6166. https://doi.org/10.1371/journal.pone.0006166
Mahfouf JF (1990) A numerical simulation of the surface water budget during HAPEX-MOBILHY. Bound-Layer Meteorol 53(3):201–222
Mihailović DT, Lalić B, Drešković N, Mimić G, Djurdjević V, Jančić M (2014) Climate change effects on crop yields in Serbia and related shifts of Köppen climate zones under the SRES-A1B and SRES-A2. Int J Climatol 35(11):3320–3334
Pielke PA (1984) Mesoscale meteorological modeling. Academic Press, New York, NY
Rea R, Eccel E (2006) Phenological models for blooming of apple in a mountainous region. Int J Biometeorol 51:1–16
Ruml M, Gregorić E, Vujadinović M, Radovanović S, Matović G, Vuković A, Počuča V, Dj Stojičić (2017) Observed changes of temperature extremes in Serbia over the period 1961−2010. Atmos Res 183:26–41
Semenov MA, Porter JR (1995) Climatic variability and the modeling of crop yields. Agric For Meteorol 73(3–4):265–283
Shane WW, Teng PS (1983) Cercospora Beticola Infection Prediction Model. Sugarbeet Res Ext Rep 14:174–179
Sivakumar MVK (2006) Dissemination and communication of agrometeorological information – global perspectives. Meteorol. Appl. 13(S1):21–30

Printed by Printforce, the Netherlands